ECONOMIC BOTANY

For the Students of B.Sc., M.Sc. and Competitive Examinations

B P PANDEY

MSc, PhD, FPSI

Former Head

Department of Botany

J.V. College, Baraut, Uttar Pradesh

S Chand And Company Limited

(ISO 9001 Certified Company)

S Chand And Company Limited
(ISO 9001 Certified Company)
Head Office: D-92, Sector–2, Noida – 201301, U.P. (India), Ph. 91-120-4682700
Registered Office: A-27, 2nd Floor, Mohan Co-operative Industrial Estate, New Delhi – 110 044, Phone: 011-49731800
www.**schandpublishing.com**; e-mail: **info@schandpublishing.com**

Marketing Offices:

Chennai : Ph: 23632120; chennai@schandpublishing.com
Guwahati : Ph: 2738811, 2735640; guwahati@schandpublishing.com
Hyderabad : Ph: 40186018; hyderabad@schandpublishing.com
Jalandhar : Ph: 4645630; jalandhar@schandpublishing.com
Kolkata : Ph: 23357458, 23353914; kolkata@schandpublishing.com
Lucknow : Ph: 4003633; lucknow@schandpublishing.com
Mumbai : Ph: 25000297; mumbai@schandpublishing.com
Patna : Ph: 2260011; patna@schandpublishing.com

First Edition 1978
Subsequent Editions and Reprints 1980, 84, 88, 90, 93, 95, 97, 2000, 2001, 2003, 2005, 2007, 2009, 2010, 2011
Revised Edition 2012
Reprints 2014 (Twice), 2015, 2016 (Twice), 2017, 2018, 2019, 2020 (Twice), 2021, 2022 (Twice), 2023, 2024

Reprint 2025

ISBN: 978-81-219-0341-7 **Product Code:** H6EBT68BOTN10ENAH12O

PRINTED IN INDIA

By Vikas Publishing House Private Limited, Plot 20/4, Site-IV, Industrial Area Sahibabad, Ghaziabad – 201 010 and Published by S Chand And Company Limited, A-27, 2nd Floor, Mohan Co-operative Industrial Estate, New Delhi – 110 044.

PREFACE TO THE EIGHTH EDITION

The text of the eighth edition of 'Economic Botany' has been revised and enlarged. A few very important new topics, such as: **Natural Resources and Vegetation Types of India; Biodiversity and Biogeographical Regions of India; Global Environmental Change, Green House gases and their Effects; Microbes in Human Welfare; Domestication of Plants; Germplasm Collection and Conservation; New Crops; Under-utilized plants of India; Forests of India; Forest Products of India and Forest Research Centres of India,** have been added that make the text more useful and interesting to the students of Economic Botany and all other readers alike. Many other plants of economic value have been added in this edition. However the general plan of the book remains unchanged. The omissions and the printing mistakes have been removed, and the utility of the book enhances manifold.

My sincere thanks are due to Dr. S. Kumar, Dr. J. Mohan, Dr. Rajeshwari Sharma and Dr. Ashok Kumar of J.V. College, Baraut; Dr. S.S. Sharma and Dr. Shashi, H.P. University, Shimla; Prof. P.N. Purekar, Nagpur University; Dr. V.V. Joseph, Chennai, and Dr. S.R. Sharma, I.A.R.I. New Delhi, for their constant support, in several ways in the completion of this work. Dr. Anjaneylu, C.T.R.I., Rajamundry deserves special mention for providing the photographs of tobacco and coconut plants. Sanjeeva Pandey, I.F.S., Rahul Pandey, I.F.S. and Shanti Priya, I.F.S. deserve special mention for their valuable suggestions. I also express my deep sense of gratitude and indebtedness to my friends, botanists and educationists who have accorded their full co-operation and well wishes during the preparation of this edition.

I am thankful to the Management Team and the Editorial Department of S.Chand & Company Ltd., for all help and support in the publication of this edition of Economic Botany.

The healthy suggestions and criticisms from all corners are invited for the improvement of subsequent editions.

'Neelkanth Kuteer'
Subhash Nagar
BARAUT - 250 611.
August 25, 2011

Dr. B.P. Pandey
Former Reader and Head,
Department of Botany
J.V. College, Baraut, 250 611

PREFACE TO THE FIRST EDITION

The economic botany deals with the application of botanical knowledge to the well-being of mankind. The primary necessities of man are threefold—*food*, *clothing* and *shelter*. The most essential need of man is food. This food primarily comes from plants in the form of cereals, millets, pulses, vegetables and fruits. For clothing again plants are indispensable. The plants that yield fibres are second only to food plants. To meet an ever-increasing demand for food and clothing, an application of the knowledge of botany has been considered to be of great importance for better utilization of plant-products. From the time immemorial, shelter from the whether inclemencies and as a protection against natural enemy, has been felt too much. In this respect, the forest products have been of service to mankind from the very beginning of his history. The most familiar and most important of these products is wood, which is used in all types of construction work. With the advancement of civilization, the need of man is also ever on the increase. To meet his requirements he has tried to tap plants as sources for his comforts and varied uses by exploiting his scientific knowledge and has been successful in this direction to a great extent. Other useful plant-products include—drugs, tannins, dyes, paper, sugar, starch, gum, resin, rubber, vegetable fats, fatty oils, essential oils, tea, coffee, cocoa, tobacco, spices, condiments, cork, etc. Utility of bacteria, yeasts, other edible and antibiotics producing fungi, algae and lichens cannot be denied. It becomes quiet evident that a knowledge of botany and its proper application led to well-being of humanity in several ways.

The present text is the outcome of several years of experience in presenting a comprehensive account dealing with economic plants. An attempt has been made to include the most important plants of India and other parts of the world. The text has been written in simple way and profusely illustrated with photographs and line diagrams. The photographs and other figures are provided with descriptive legends.

The book is primarily an elementary text for degree and postgraduate students and not a text for researchers, and therefore, the bibliography has been kept at minimum. A glossary of medical terms has been given at the end of the text. A detailed index is also worked out and given in the last which makes the consultation of the subject easier.

I am deeply indebted to those institutions and sources who have made me available several materials in the form of brochures, bulletins, leaflets, periodicals, journals, annual reports, advertising literature and photographs. In this connection my thanks are due to Indian Council of Agricultural Research, New Delhi, Council of Scientific and Industrial Research, New Delhi, Ministry of Food and Agriculture, New Delhi, Ministry of Information and Broadcasting, New Delhi, Tea Board of India, Calcutta, Coffee Board of India, Bangalore, Cardamom Board, India, Cochin, Tobacco Board of India, Guntur, Rubber Board of India, Kottyam, Forest Research Institute, Dehradun, Indian Agriculture Research Institute, New Delhi, Coir Board, India, Cochin, Jute Technological Research Laboratories, Cotton Textiles Report Promotion Council, India, National Botanic Gardens, Lucknow and Sugarcane Breeding Research Institute, Coimbatore, India.

In conclusion I wish to express my deep sense of gratitude to those individuals who helped me directly or indirectly during the compilation of this text. Especially I am indebted to Dr. N.P. Saxena, D.N. College Meerut, who allowed the diversion from research to this work. My sincerest thanks are also due to Dr. Anand Swarup Bhatnagar, Principal, I.P. College, Bulandshahar, for the encouragement and appreciation. I also wish to express my appreciation to Professor Jitendra Mohan, who suggested several things of interest from time to time. I am really grateful to Sri T.N. Goel, my publisher, who managed to bring out the book in this shape. In the last but not the least I wish to express highest appreciation to my son Sanjeeva who provided me the photographs from AGRI EXPO 77 and other sources by his excellent photography.

The suggestions and healthy criticisms for further improvement of the book will be most welcome and thankfully acknowledged.

Department of Botany

J.V. College, Baraut 250611

B.P. PANDEY

1

Introduction

The economic botany deals with application of botanical knowledge to the well-being of mankind. The primary necessities of man are threefold—*food*, *clothing* and *shelter*. The most essential need of man is food. The food primarily comes from plants in the forms of cereals (*e.g.*, rice, wheat, maize, oat, barley and rye), millets (*e.g.*, sorghum, pearl millet, finger millet, etc.), pulses, vegetables and fruits. For clothing again plants are indispensable. The plants that yield fibres are second only to food plants. From the times immemorial the man needed some form of clothing, and this need was fulfilled only by fibre-yielding plants. In this respect cotton occupies an all important position, supplemented by jute and some other fibres for coarse clothing. To meet an everincreasing demand for food and clothing an application of the knowledge of botany has been considered to be of great importance for better utilization of plant products. From the time immemorial shelter from the weather inclemencies and as a protection against natural enemy has been felt too much. In this respect the forest products have been of service to mankind from the very beginning of his history. The most familiar and most important, of these products is wood, which is used in all types of construction work. Other products like bamboo, reed, cane, thatch grass, etc., are of equal value for house making. With the advancement of civilization the need of man is also ever on the increase. To meet his requirements he has tried to tap plants as sources for his comforts and varied use by exploiting his scientific knowledge and has been successful in this direction to a great extent. Other useful plant products include—wood (for furniture, boat building, bridge building, railway sleepers, fuel, etc.), fibres (for gunny bags, rope, cordage, matting, canvas, carpet, etc.), drugs, tannins, dyes, paper, sugar, starch, gum, resin, rubber, vegetable fats, fatty oils, essential oils, tea, coffee, cocoa, tobacco, spices, condiments, cork, etc. Utility of bacteria, yeasts, other edible and antibiotics producing fungi, algae and lichens cannot be denied. It becomes quite evident that a knowledge of botany and its proper application led to the well-being of humanity in several ways. The more important plants and their products have been grouped in several important categories. In the present text the plants and their products have been grouped as follows:

FOOD PLANTS

Cereals and millets. *Cereals*: The cereals are the most important sources of plant food for man. They constitute the most important group in the food plants of India. The cereals are the members of family Gramineae. There are six true cereals rice, wheat, maize, barley, oat and rye. They contain a high percentage of carbohydrates, together with a considerable amount of proteins and some fats. Even vitamins are present.

Millets: They are also known as small grains. They are considered to have been cultivated in India from prehistoric times. Some of the commonly grown millets in India are sorghum, pearl millet and finger millet.

Legumes and nuts. *Legumes*: The legumes or pulses are next in importance to cereals as sources of food. They belong to the family Leguminosae. They contain more protein material than any other vegetable products. Carbohydrates and fats are also present in the legumes. The pulses form an important item in India where the majority of the population consists of vegetarians. The important Indian pulses are gram, black gram, green gram, pea, pigeon pea, lentil, etc.

Nuts: A nut is a one-celled, one-seeded dry fruit with a hard pericarp. Only a few nuts of commerce fulfil this description, *e.g.*, chestnut. The others included among nuts may be peanuts, almonds, coconuts, cashewnuts, walnuts, pistachio nuts, etc. These "so called" nuts make a valuable food material. The food value is due to a high protein and fat content.

Vegetables. The term vegetable is usually applied to edible plants which store up reserve food in roots, stems, leaves and fruits and which are eaten cooked, or raw as salad. The vegetables rank next to cereals as sources of carbohydrate food. The nutritive value of vegetables is tremendous, because of the presence of indispensable mineral salts and vitamins.

Fruits. Morphologically a fruit is the seed-bearing portion of the plant, and consists of the ripened ovary and its contents. Simple fruits are derived from a single ovary, and compound fruits from more than one. The aggregate fruits are formed from numerous carpels of the same flower, while composite fruits develop from ovaries of different flowers. In economic botany only those

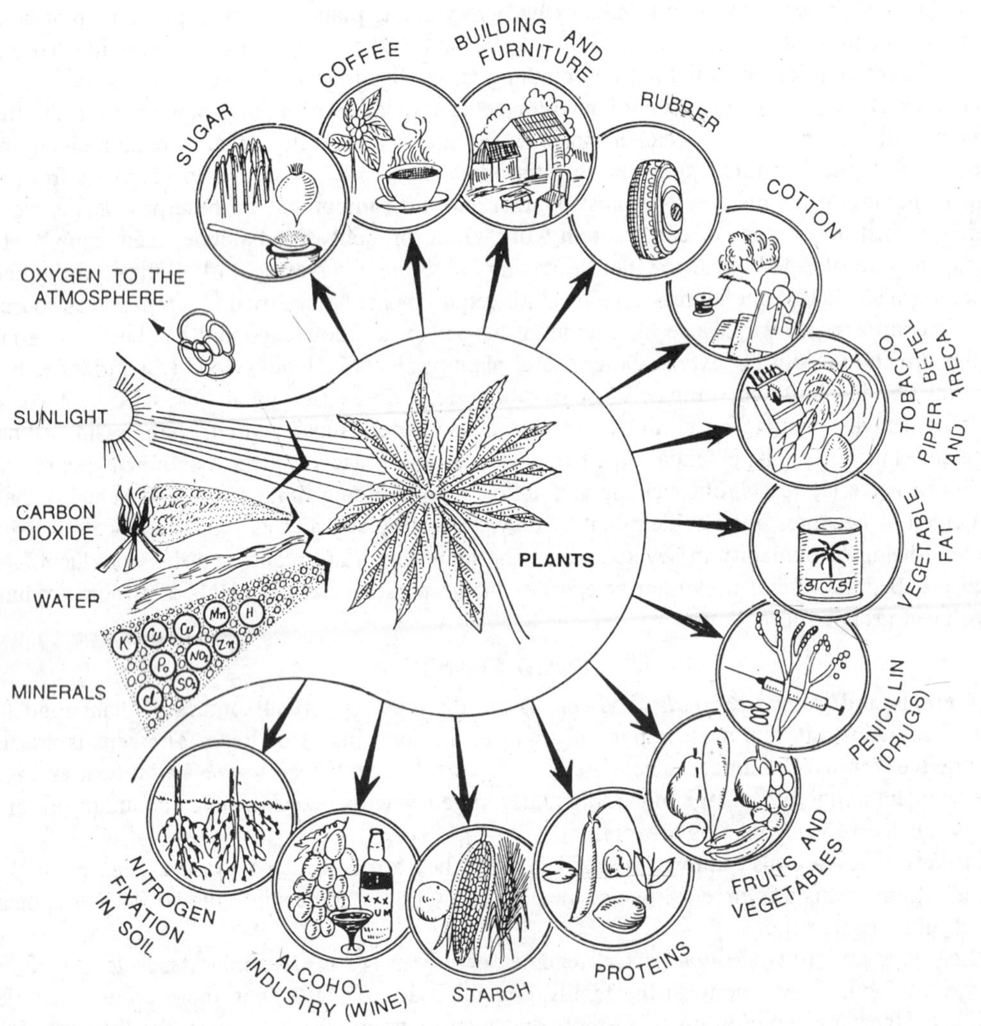

Fig. 1.1. Plants and human welfare.

fruits are considered which are usually eaten without cooking. For convenience the fruits have been divided into two groups *tropical fruits* (*e.g.*, mango, citrus fruits, litchi, banana, guava, sugar apple, fig, papaya, pine-apple, etc.) and *temperate fruits* (*e.g.*, apple, pear, plum, peach, strawberries, grape, etc.).

PLANTS AND PLANT PRODUCTS OF INDUSTRIAL VALUE

Fibres and fibre-yielding plants. The fibre-yielding plants rank second only to food plants in their usefulness to human kind. The utilization of fibres is directly related to the advancement of civilization. As civilization advanced, the use of vegetable fibres increased greatly. They are of enormous value in our daily life. About two thousand species of the plants yield various fibres all over the world. The chief fibres of commercial importance have been classified in six groups—textile fibres (*e.g.,* cotton, flax, etc.), brush fibres, filling fibres, rough weaving fibres, natural fabrics and paper-making fibres. The most important fibre-yielding plants belong to the families—Malvaceae, Tiliaceae, Linaceae, Leguminosae, Musaceae, Bombacaceae, Liliaceae, Palmae, Gramineae, Amaryllidaceae, Urticaceae, and Bromeliaceae.

Wood and cork. *Wood:* In modern days wood is the most widely used commodity outside of food and clothing. It is one of the most versatile of the raw materials of industry. Wood is cheaper, lighter, and more easily worked with tools. Moreover the wooden structure can be readily altered or rebuilt. Wood is very strong for its weight and unique in strength, elasticity and toughness. It is a bad conductor of heat and electricity and does not rust. The wood is also used in the form of thin sheets or *veneers*.

Cork: The cork consists of the outer bark of the tree. It is used for many purposes. In some cases the natural cork is utilized whereas in others composition cork, made of coarse pieces treated with adhesives and molded. The most important cork made articles are stoppers, hats, floats, mats, tiles, metal tops for sealing bottles, gaskets, inner soles for shoes, etc.

Tannins and dyes. *Tannins:* Tannins make a heterogeneous group of complex compounds of widespread occurrence in plants. They are organic compounds chiefly glucosidal in nature, which have an acid reaction and are very astringent. It is a colourless solid, dissolves in water to solutions of astringent tastes. They are of commercial importance because of their property of forming an insoluble colloidal compound (leather) with the hides of animals. They also react with the salts of iron to form dark-blue or greenish-black compounds, the basis of common inks.

Dyes: The dyes are coloured compounds capable of being fixed to fabrics and which do not wash out with soap and water or fade on exposure to light. Therefore, a coloured organic substance is not necessarily a dye. Natural dyes and stains, obtained from the roots, leaves, bark, fruit or wood of plants, have been used from ancient times.

Rubber and its products. Rubber is obtained from the latex, of various plants of tropical and sub-tropical regions. Most of the rubber plants belong to the angiospermic families—Euphorbiaceae, Moraceae and Apocynaceae. The most important rubber plant is *Hevea brasiliensis* of family Euphorbiaceae. Latex occurs in latex tubes or latex vessels of the plants. Latex is a varying mixture of water, hydrocarbons, resins, oils, proteins, acids, salts, sugar, etc., the substance used as the source of rubber. The Hevea or Para rubber tree is the source of 95% to 98% of the rubber production of the world. The most important rubber products are tyres and tubes for automobiles and cycles accounting for about 75% of the total rubber consumption, about 6% is used for footwear and about 4% for wire and cable insulations.

Fatty oils and vegetable fats. The fatty oils are liquid at ordinary temperatures and usually contain oleic acid. Chemically they consist of glycerin in combination with a fatty acid.

The fats are solid at ordinary temperatures and contain stearic or palmitic acid. When a fat is boiled with an alkali, it decomposes and the fatty acid unites with alkali to form soap.

Usually the fatty oils are stored up in seeds of the plants belonging to many angiospermic families. Sometimes, to a less extent they are also stored in fruits, stems and other plant organs. Several fatty oils are edible and used as cooking media. These edible oils contain both solid and liquid fats.

Organic acids. They are widely distributed in plants, particularly in fruits and vegetables. The organic acids may be found as salts of calcium, potassium or sodium and in combination with alcohols. These acids play a part in metabolism and growth, and therefore they are supposed to be of great food value for man.

Essential oils. The essential oils, also known as volatile oils, evaporate in contact with the air and possess a pleasant fragrance. Chemically the essential oils are very complex. They are found in many species of plants of various families. All aromatic plants contain essential oils. Generally the oils are secreted in oil glands.

Sugars and starches. *Sugars:* The glucose manufactured by the green plant in photosynthesis, is almost universally present in plant cells. This basic material of metabolism, the glucose, has the formula $C_6H_{12}O_6$. Fruit sugar or fructose is another product of photosynthesis which has the same formula. The more complex sugars are built up from these simple sugars. The most important complex sugar is sucrose or cane sugar, which has the formula $C_{12}H_{22}O_{11}$. This sugar is accumulated in abundance in sugarcane and sugar beets. Today, the sugar industry is the second largest in India, next only to textiles.

Starch: Starch occurs in all green plants. It is a complex carbohydrate with the formula $(C_6H_{10}O_5)\ n$. Like the sugars, they are derived also from glucose and constitute the first visible product of photosynthesis. This is the commonest type of reserve food in green plants and plays the most important role in their metabolism. Commercial sources of starch are wheat, barley, maize, potatoes and arrowroot. In these plants starch occurs in the form of starch grains which vary in shape and size.

Cellulose. It is the highest type of carbohydrate and usually found in the cell walls. It is of little food value.

Reserve cellulose. They resemble cellulose physically, but differ in their chemical properties. The reserve celluloses include pectins, gums, mucilages and hemicelluloses. Sometimes these compounds serve as reserve food. The hemicelluloses may gradually convert into pectins and thereafter in gums.

Hemicellulose. They make the extra layers of the cell walls. They are generally found in the seeds of tropical plants (*e.g.,* in date palm). They are of less food value. However, they are utilized in the industries.

Pectins. They are also known as fruit jellies. They occur in most plant cells, mainly in fruits and vegetables. Guava jelly is famous for its tastefulness. Pectins are readily soluble in water and can be used as food. They increase the water-holding capacity of the cells. Pectins solidify after their extraction from the fruits and jellies are prepared from them because of this property.

Pulp and paper. The paper is a cellulose product. Cellulose is the most complex carbohydrate and is universally present in the cell walls of plants. One of the important uses of cellulose is in the manufacture of paper. Paper can be made from any natural fibrous material.

Gums and resins. *Gums:* The true gums are formed as the result of disintegration of internal tissues, for the most part from the decomposition of cellulose. The gums are insoluble in alcohol but soluble in water, readily swell up in it, and form a viscous mass. They are colloidal in nature. They contain a large amount of sugar and are closely allied to the pectins. The gums are formed in various kinds of plants. *Acacia senegal* yields the best gum-arabic of commerce. The gums exude naturally from the stems, or in response to wounding. The commercial gums are the dried exudations of the plants.

Resins: They represent oxidation products of various essential oils and are very complex and varied in their chemical composition. They are mostly found in the stems of the trees, and occur in abundance in special canals or ducts. They are yellowish solids, insoluble in water but soluble in alcohol, turpentine and spirit. It normally oozes out through the bark and hardens on exposure to the air.

Mucilages. They are gum-like and widely found in the plants. When water is added to them they do not dissolve, but form a slimy mass. They are secreted in hairs, sacs and mucilage canals. They are variously used. They are chiefly used in medicine.

Proteins. The proteins are derived from the carbohydrates through the formation of amino acids. The proteins contain high nitrogen content. Sulphur and phosphorus are also present. A typical protein known as gliadin, is found in wheat; its empirical formula is $C_{736}H_{1161}N_{184}O_{208}S_3$. The proteins are generally found in seeds in the form of aleurone grains. They make an important food for both plants and animals. The proteins are essential part of man's diet. They are known as muscle and nerve builders. The pulses are the good source of proteins.

MEDICINAL PLANTS AND DRUGS

Medicinal plants. The branch of medical science, which deals with the drug plants, is known as *pharmacognosy*. This branch of medical science is concerned with the history, commerce, collection, selection, identification and preservation of crude drugs and raw materials. Most of the drugs are obtained from wild plants growing in all parts of the world, and especially in tropical regions. The medicinal value of drug plants is due to the presence in the plant tissues of some chemical substances that produce a definite physiological action on the human body. The most important of these substances are alkaloids. Some of these chemicals are powerful poisons and therefore the drugs should be prepared and prescribed only by expert physicians.

Fumitories and masticatories. From the time immemorial the human beings have smoked or chewed various substances for pleasure. For example, the bulk of tobacco is used for smoking in the form of cigarette, *bidi*, cigar, cheroot and *chuttas* and in pipe and *hookah*. Large quantities of tobacco are also consumed for chewing.

On the other hand betel leaves and arecanuts are widely used as a masticatory. The basic preparation for chewing purposes consists of betel leaf smeared with hydrated lime, and catechu to which scrapings of arecanut are added; flavourings such as coconut shavings, clove, cardamom, fennel, powdered liquorice, nutmeg, and also tobacco are used according to one's taste. Chewing of betel leaves with various adjuncts is an ancient practice in India and other countries of East Asia.

Alkaloids. They are vegetable bases containing nitrogen, and they are supposed to be decomposition products of proteins. They are secreted in special cells and occur in several plants of different families. They have a marked physiological effect on animals and, therefore, they are of much value in medicine and drugs. They include some of the most powerful plant poisons and narcotics. Caffeine and theobromine, are usually classed as alkaloids.

Glucosides. They are similar to alkaloids, but they are derived from carbohydrates and not from proteins. They are commonly used in the manufacture of medicine and drugs for man.

FOOD ADJUNCTS

Spices and condiments. The spices cannot be grouped as foods, for they contain less nutritive value. They stimulate the appetite and increase the secretion and flow of gastric juices. They give a good flavour and aroma to food, and add greatly to the pleasure of eating. For this reason they are commonly known as *"food adjuncts"*. The aromatic value of the spices is due to the presence of the essential oils. Sometimes the term *"spice"* is restricted to hard parts of plants, which are generally used in a pulverized state. *Condiments* are spices or other flavouring substances which possess a sharp taste, and are commonly added to food after it has been cooked.

Vitamins. The vitamins are absolutely essential for well-being of both plants and animals. They are formed by plants. They are necessary for normal metabolism, growth, development and reproduction. Vitamins are also essential for the prevention of several human diseases. Fruits, vegetables and seeds of various plants are good sources of vitamins. Seaweeds contain nearly all the known vitamins. The yeasts also contain several important vitamins.

Non-alcoholic beverages. The beverages containing caffeine are used all over the world for their stimulating and refreshing qualities. Caffeine is an alkaloid, which has definite medicinal values and acts as a diuretic and nerve stimulant. The most important non-alcoholic beverages are—tea, coffee and cocoa. Caffeine is harmful in large quantities, it is present in these beverages in very small amounts, not exceeding two per cent. The other beverages which do not contain alcohol are commonly known as *soft drinks*. They contain high sugar content and make a good source of energy. The fruit juices are the simplest soft drinks.

Alcoholic beverages. From earliest time the alcoholic drinks were used to celebrate the occasions and for the man's own pleasure. The alcoholic beverages are the result of the natural process of fermentation. Today the consumption of alcoholic beverages is world wide. Alcohol is a poison and when taken in excess, it is proved very harmful and injurious to health and life. The important alcoholic beverages are—wine, beer, whisky, brandy, rum, gin, etc.

LOWER PLANTS IN ECONOMIC BOTANY

Fungi. Some fungi have been esteemed as articles of food from early times. Edible fungi have delicate flavours and are often valued delicacies. The important edible fungi are—*Agaricus* spp., *Cantharellus* spp., *Volvaria* spp., *Lycoperdon* spp., *Morchella* spp., *Tuber* spp. and many others.

The fungi, like yeasts are responsible for fermentation. The baking industry depends solely on the enormous budding of the yeast cells.

Penicillium is best known to the non-botanist because it is the source from which the antibiotic penicillin is extracted. The famous drug, ergotine is obtained from *Claviceps purpurea,* the causal organism of plant disease, ergot of rye.

Lichens. As food lichens have been used from the earliest times as animal feed and also as food for human consumption in times of scarcity. Lichens were once employed for dyeing wool and silk and colouring materials derived from them were highly valued before the advent of synthetic dyes. Lichens were formerly used as sources of fermentable sugars for the production of ethyl alcohol. Several lichens possess medicinal properties.

Algae. From the earliest times several sea weeds have been used as direct source of food to human beings. High mineral content up to 5% of the wet material, in which all the mineral elements important in human and animal physiology are found, makes sea weeds a unique supplement for a well balanced diet. The sea weeds (marine algae) are the richest source of vitamins. The vitamins A, B and E are found in abundance in sea weeds. Several important foodstuffs prepared by algae are—suimono, mitsu, dulse, seatron, laver, kompu, carrageen, etc.

2

Cereals and Millets

CEREALS

The cereals are the most important source of plant food for man. They constitute the most important group in the food plants of India. The cereals are the members of family Gramineae, and possess the characteristic fruit, the caryopsis. In this fruit the wall of the seed becomes fused with the ovary wall to form the husk. The term 'grain' is applied to this type of fruit. There are six true cereals—rice, wheat, maize, barley, oats and rye. Of which rice, wheat and maize, are most important, and they have played an important part in the development of civilization. Sometimes the millets and sorghums are erroneously referred to as cereals. Cereals contain a high percentage of carbohydrates, together with a considerable amount of proteins and some fats. Even vitamins are present.

Rice

Oryza sativa Linn; family—Gramineae (Poaceae); Eng.—rice; Hindi—*chaval, dhan;* Sanskrit—*dhanya, vrihi, nivara, syali;* Bengali—*chal;* Marathi—*tandula, dhan, bhat;* Gujarati—*choka, dangar;* Telugu—*vadlu, varidhanyamu, biyyamu;* Tamil—*nellu, arisi;* Kannada—*akki, bhatta, nelu;* Malayalam—*nellu ari.*

History. The earliest mention of rice cultivation is connected with China. According to Stanislas Julien, a ceremony was established in that country about 2800 B.C., by the Emperor Chin-nung, in which the sowing of five kinds of grain is the chief observance. The reigning Emperor himself has to sow the rice, but he may delegate the sowing of the other four kinds to the princes of his family. However De Candolle, affirms that rice cultivation in India, though subsequent to that of China, in point of date of first record, has been a valued crop since classic periods. One of its Sanskrit names *Dhanya* means the supporter or nourisher of mankind. By the Hindus it is regarded as the emblem of wealth or fortune.

According to Dutt's *Materia Medica of the Hindus,* there are three principal classes of rice—*Sali,* or that reaped in the cold season, *Vrihi,* or that ripens in the rainy season, and *Shashtika,* or that grown in the hot weather in low lands. It may be said that rice enters extensively into the ordinances of Hinduism. Several kinds of rice are used as votive offerings at many religious ceremonies. The Sanskrit word *syala* denotes the custom of the bride's brother scattering fried grains at the marriage ceremony. In the *Ordinances of Manu* the word *food* is given as synonymous with *rice*. Wild rice in Sanskrit works is spoken of as *nivara,* and various Sanskrit names are given to the cultivated rices, such as *dhanya, vrihi, syali, jiva-sadhana, tanonu* and *shashtika.* Most of these names are traceable to roots which denote life, existence and subsistence. As a possible historic fact, which would have a bearing on the cultivation of rice in Upper India, it may be said the word *vrihi,* is met with in the *Atharva Veda.*

The most general vernacular name for wild rice is, however, *uri* and the plant occurs mainly in Tamil Nadu, Orissa, Bengal and Kerala. The area of wild rice, corresponding to that in which *uri* and other such names are used leads to the concept that in India rice cultivation may have spread from there all over the rest of the peninsula and ultimately across the Himalayas.

The chief wild habitat of the plant is from south India. The cultivation of rice appears to have spread from there eastward to China, perhaps 3,000 years before the Christian era, and at perhaps a slightly more recent date, westward and northward, throughout India to Persia and Arabia, and ultimately to Egypt and Europe.

The Chinese were the first to cultivate rice (approximately 4,000 B.C.). In classical Chinese the words for agriculture and rice culture are synonymous. It is believed that the cultivation of rice in India dates back to ancient periods even earlier than 3,000 B.C. (Parthasarathy, 1954). The cultivation of rice was introduced into India before the time of Greeks, and then reached Syria and Northern Africa. The rice was first cultivated in Europe in 1468 in Italy. In America this was first grown in 1694 in South Carolina.

Rice is predominantly a crop of Asia, 95% of the world's area being concentrated in south-eastern Asia extending from West Pakistan to Japan. India has the largest area, 79 million acres in the world under rice cultivation, and production is about 28 million tons per year. India and Indo-China are supposed to be the centres of origin of rice (Ramiah, Ghose and Vachhani, 1952).

Fig. 2.1. *Oryza sativa*—paddy (*dhan*)—Sabarmati, an improved variety, developed by I.C.A.R., New Delhi.

The traditional use of rice in the religious ceremonies of the Hindus, associated with birth, marriage and death is a testimony to its great antiquity as also to the intimate place it has in the life of the people. The crop is cultivated in almost all the states of India, extending from the delta region of the south to the higher altitudes of 3,000 to 5,000 feet above sea level in the Kashmir valley in the north. It is mostly cultivated in the valley of the Ganga and its delta and the low-lying coastal areas in southern and north-eastern India.

Of all the principal crops grown in the country rice has the largest area under it. Andhra Pradesh, Assam, Bihar, Maharashtra, Madhya Pradesh, Tamil Nadu, Orissa, Uttar Pradesh and West Bengal are the main producing states; together they account for more than 90% of the total area under the crop.

Cultivation. The rice grows best on damp soils underlaid with a semi-impervious subsoil in places where it can be flooded. A considerable interest has been aroused in India during last two decades in the Japanese method of rice cultivation. This combines improved cultural practices with proper manuring and plant protection measures. The increased acre yields obtained by this method have naturally led to its popularity. The results of breeding for high yield have also been very satisfactory, and many improved varieties have been evolved which give an enhanced yield. It has also been realized that the newer intensive methods of cultivation and manuring demand varieties which are capable of giving higher yields. With this aim in view, there has been started an intensive hybridization programme between the *japonica* and *indica* varieties of rice. The *japonica* varieties are high yielding, while the *indica* varieties have a better tillering capacity and are resistant to disease and adverse climatic conditions.

The fields are ploughed and the rice is transplanted from seed beds when 9 or 10 in. high. The young plants are covered with water, and the water is kept in circulation. As soon as ripening starts, the water is drawn off and the field is allowed to dry out. Rice is harvested and the stalks are stacked up to dry.

Fig. 2.2. Rice *(Oryza sativa)* variety 'IET 4106' takes about 102 days for maturity with a yield potential of 40–45 q/ha.

Investigations have been carried out to determine the role of blue-green algae in rice nutrition and their effect on the fertility of soil. The results indicate that the fixation of nitrogen is increased by the activity of blue-green algae.

It has been established that incorporation of green matter in paddy soil improves yield. It provides the cheapest way of manuring rice fields wherever facilities for raising green-manure crops are available. *Sesbanea aculeata (dhaincha)* has been found to be the most suitable crop for rice areas as it can stand the soil and water conditions prevailing in these areas. In Kashmir valley, green manuring of rice crop with *Lens esculentus* has been found to be very effective. In U.P. *Crotalaria juncea* is becoming popular.

Milling. The grains of rice are removed by threshing. The grains are to be husked when they are to be used, and are then pounded in a mortar *(okhli)* with a wooden mallet *(musal)* and winnowed. The resulting grain is very nutritious for it contains protein and fat as well as starch.

In the commercial preparation the impurities are removed and the paddy is passed between millstones to break up the husk. This chaff is moved by blowers. Now the grain is pounded in huge mortars and a portion of the bran layer and embryo is removed. The waste is called rice bran. The white rice is then scoured by friction and polished. In polishing the embryo which contains proteins, vitamins and oils is knocked out due to lack of strong link between the embryo and the endosperm. During this process the outer, more nutritive parts of the grain are removed.

Uses. The chief use of rice is as food, and more people use it than any other cereal. The rice is generally eaten with pulses (legumes) or some other food rich in proteins. A diet of rice and soyabeans makes the food of millions. The rice straw is used for making straw boards, paper and mats. Rice bran oil is used for making soaps and cosmetics. Rice starch is much used in European

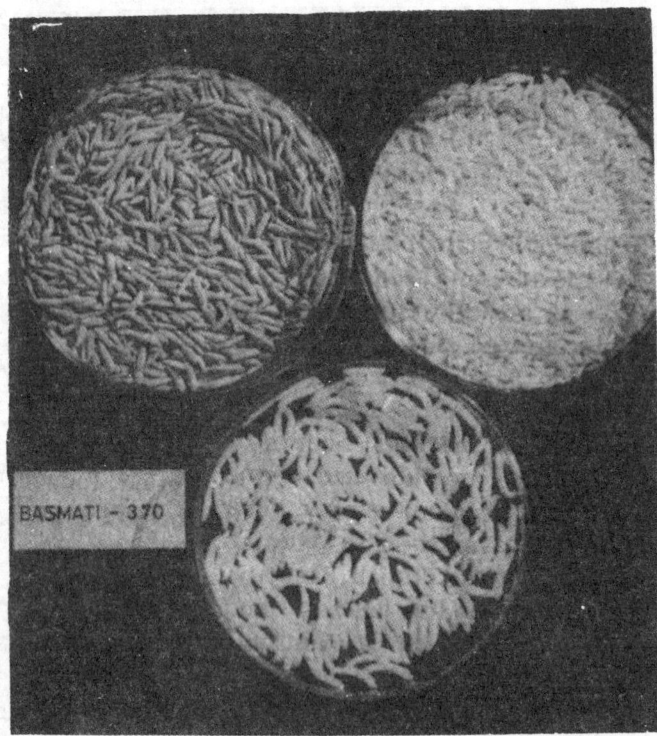

Fig. 2.3. *Oryza sativa* (Paddy—**Dhan**). The rough rice (paddy), polished rice and fragrant cooked rice of 'Basmati 370'.

countries. In several tropical countries intoxicating beverages are prepared from rice. Important beverage of Japan, *Sake* is prepared by fermenting rice. In India also in some parts of Orissa and Andhra Pradesh intoxicating beverages are prepared from rice.

This starch-rich food is eaten after boiling. It may also be converted into parched rice, beaten rice or rice flakes, and puffed rice; mixed with black gram (*urd*) powder, used for fermented preparations like *idli* and *dosa*. Broken grains obtained during milling, used as human and cattle food, for making alcoholic beverages, and as a source of starch and rice flour. Paddy husk is used as fuel; the bran is also used as cattle feed. The fatty oil obtained from bran is used for edible purposes.

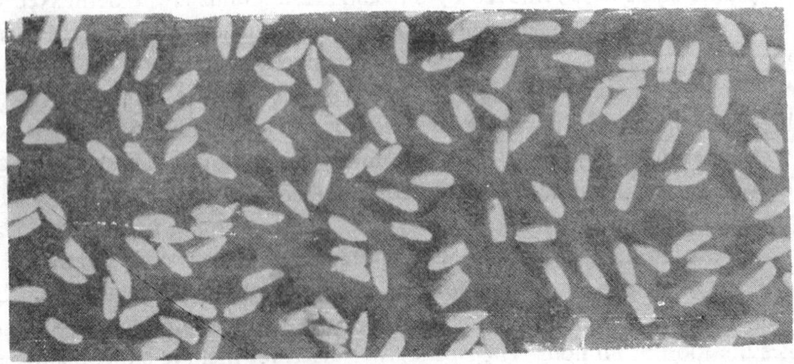

Fig. 2.4. Unhusked grains of rice.

Wheat

Triticum aestivum Linn.; family—Gramineae (Poaceae); Hindi—*Gehu* ; Sanskrit—*Godhuma*; Bengali—*Giun, gom, gam*; Marathi—*Gahum, gahung* ; Gujarati —*Ghavum, gawn, govum*; Telgu—*Goodhumalu*; Tamil—*Godumai, godumbayarisi*; Kannada—*Godhi*; Malayalam—*Gendum, kotanpam, godamba*.

History. According to M.A. de Candolle, the cultivation of wheat is prehistoric in Old World. Very ancient Egyptian monuments show the establishment of this cultivation. However, when Egyptians or Greeks speak of its origin, they attribute it to mythical personages—Isis, Ceres, Triptolemus, etc. A small-grained wheat has been found at the earliest lake dwellings of Western Switzerland, the inhabitants of which were at least contemporary with the Trojan war, and perhaps earlier. The same form of wheat, *i.e., Triticum vulgare antiquorum* Heer, was found by Unger in a brick of the pyramid of Dashur in Egypt, to which he assigns a date of 3359 B.C. Another small-grained wheat, *T. vulgare compactum muticum* Heer, was less common in Switzerland in the earliest stone age, while a third intermediate form was cultivated in Hungary at the same period. However, De Candolle believes that the wheat cultivation in the temperate parts of Europe, Asia and Africa is probably older than the most ancient known languages. The Chinese certainly grew it in 2700 B.C., and considered it a gift direct from heaven. In the annual ceremony of sowing five kinds of seeds, instituted by the Emperor Shen-nung or Chin-nung, wheat was one of the species employed.

De Candolle has written in his famous work the *Origin of Cultivated Plants* that the original home of wheat in very early prehistoric times was in Mesopotamia. According to Berosus, the earliest of all Western Historians, and a Chaldean priest who wrote some twenty-three centuries ago, it occurred wild there. According to Strabo, who was born 50 B.C., a grain very similar to wheat grew wild upon the banks of the Indus.

According to De Candolle there is strong evidence in favour of India being the home of some of the forms of wheat as can be shown for any other part of the globe. India possesses perhaps as comprehensive a series of time honoured forms of wheat as can be shown for any other country. India has its hard wheats and soft wheats ; its starch wheats and speltwheats ; its bearded and beardless wheats. It has been established that most of these have been grown for countless ages on the same fields as they are to be found at the present day. It has also been proved that wheat cultivation in India is as ancient as in Europe or any other part of the world. The wheat grains discovered as a result of Indus valley excavations at Mohenjo-Daro indicate that north-western India was one of the ancestral lands of this cereal (Pal, B.P., 1954). The carbonised grains resemble those of *Triticum sphaerococcum* Percival, an endemic species which is still found in a few places in India.

According to Vavilov, the wheats have had a multiple origin, the soft wheats have come from the mountains of Afghanistan and the southwestern Himalayas ; the durum wheats from Abyssinia, Algeria and Greece; and einkorn from Asia Minor. The archaeological evidences establish that the wheat has been cultivated for at least 6000 years.

However, it must be kept in mind, while discussing the question of the origin of this ancient, cultivated cereal, that it is very nearly impossible, whatever the original home of wheat may have been, to determine with accuracy the character of the first parent from which it was derived.

India accounts for about 3.5 per cent of the global wheat production. The area under wheat constitutes roughly 14.0 per cent of the total area under cereals, and 10.0 per cent of the total area under food grains in the country. The crop is grown mainly in the northern and central parts. It is not much significance in the south.

It is cultivated as a food crop mainly in Uttar Pradesh, Punjab, Madhya Pradesh, Maharashtra, Bihar and Rajasthan.

Wheat is the second staple food crop of India and occupies about 29 million acres of land. It is consumed mainly by the people in the north. The wheat grains discovered as a result of Indus Valley excavations at Mohenjo-Daro indicate that north-western India was one of the ancestral lands of this cereal (Pal, B.P., 1954). The carbonised grains resemble those of *Triticum sphaerococcum* Percival an endemic species which is still found in a few places in India. The wheat growing area in India can be divided into three zones—the Gangetic alluvium, the Indus valley, and the black soil tract of Peninsular India. Wheat is also cultivated in Gujarat. A special feature of Indian wheat is the relatively short season in which it completes its life cycle. While the crop stands in the field for 9 to 10 months in some western countries, in India it is ready for harvest in four to six months after sowing. However, in the hilly tracts of Northern India the growing season for wheat is about nine months but the area under the crop in this region is very small.

Of the 18 recognized species under the genus *Triticum,* five are cultivated in India (Mirchandani and Khan, 1953). Of these *Triticum aestivum* Linn., commonly known as bread wheat, occupies the largest area and is mainly grown in the northern regions. Next in importance is the macaroni wheat, *T. durum* Desf., which is the predominant species in Central and Western India. The area

Fig. 2.5. *Triticum aestivum* : Wheat **(Gehu)** 'Moti'. A trigenic variety of wheat

under *T. dicoccum* Schrank is grown in South India. *T. turgidum* Linn., the rivet wheat, and *T. sphaerococcum* Percival, the dwarf wheat are even less important.

Wheat was the first crop to attract the attention of Indian breeders. The main object was to evolve high-yielding varieties with good grain quality. A notable achievement of the last few years has been the evolution of a new series of improved wheats. Their special characters include adaptability to certain regions, high yield, good grain quality, tolerance to rusts, and resistance to loose smut.

Cultivation, harvesting and threshing. In northern India sowing of wheat is done in October-November. Wheat may be sown broadcast, either by hand or by sowing machines. Germination begins immediately and the first leaves appear within a fortnight. Wheat is properly manured and irrigated. The crop is harvested by cutting the plants with a sickle close to the ground, in March–April. Threshing is the next process, and this involves the separation of the grain from the spike. Threshing is generally done under the feet of bullocks or by threshing machines. After threshing the wheat is winnowed and sifted. In the Punjab, recently the *combines* are introduced for the purpose. The combines reap, clean, thresh, winnow and sift the grains ; wheat must be stored in firmly built structures, and it must be well ventilated.

Milling. In India in most of the motor driven flour mills there remains a fixed lower stone upon which a movable upper one revolves. The grains are dropped in the openings in the upper stone and gradually worked out between the stones which possess grinding surfaces cut in radiating lines. In this process whole grain is used.

The roller process of milling is an advanced and perfect process. The first step in this process comprises cleaning and scouring. After making the grains clean they are thoroughly washed and scoured. The next step is tempering. In this process a little water is added, which toughens the bran and prevents its breaking up, so that it will flake out all in one piece. In the last the conditioned and tempered wheat is submitted to breaking, grinding and rolling. The grain is first cracked or crushed gradually through a series of four to six pairs of chilled iron break rolls. The surface of the break rolls is made rough by sharp lengthwise foldings. The break flour is separated out by sieves, while the main portion goes to the second break. This process is repeated until five sets of rollers have been utilised. All bran is removed during this process and the purified material is passed to smooth rollers for final granulation. The final product is the flour.

Uses. There are three main kinds of flour—*suji, maida,* and *ata* which are used for various purposes. The flour is used chiefly for making 'bread' and 'chapatis'. The flour is also used for making biscuits, cakes, pastry and similar articles. Wheat flakes are used as breakfast food.

Wheat is also used in the manufacture of beer and other alcoholic beverages. Wheat straw is used for seating chairs, stuffing mattresses, etc. It makes a good food for livestock. Wheat straw is also used as fodder.

This makes a staple food in most parts of the world. Properties of gluten in the grains are such that it produces bread-stuffs generally superior to those from any other cereal grains. By products of wheat milling, such as bran, germ and middlings constitute valuable feed for stock, readily eaten; supplementary feeds are provided to supply protein and minerals in which the straw is deficient. The straw is used as bedding for cattle; it is also used for padding, as in mattresses, for packing fragile goods, for thatching and many other purposes. It may be used also for production of furfuryl alcohol. Straw-pulp is utilized for the manufacture of paper, straw-board, and building - board.

Non-feed industrial uses of wheat include the manufacture of starch, industrial alcohol, malted wheat, and core-binder flour; only small quantities of wheat are used for starch and gluten manufacture. Grain is regarded as a stand-by for alcohol production. Low-grade flours are utilized in the preparation of pastes for wall papering and ply-wood adhesives, and in iron foundries as a core binder.

Wheat products include peeled wheat ; *Bulgur,* a parboiled wheat product; *Wurld wheat,* similar to *Bulgur,* but of lighter colour ; *Instant* or *agglomerated* flour; *Farina* or *semolina*; *Wheat flakes; Shredded wheat* ; *Puffed wheat* ; *Grape-nuts,* prepared from toasted slices of malted bread ; *Gluten,* used in special breads; and *wheat germ,* rich in vitamin E.

Maize

Zea mays Linn., family—Gramineae (Poaceae) ; Eng.--Maize, corn, Indian corn; Hindi—*Makai, Makka, Bhutta* ; Bengali—*Janar, bhutta, jonar*; Marathi—*Maka, makai, buta*; Gujarati—*Makkari, makkai*; Telugu—*Mokka-janna, makka jonnalu*; Tamil—*Makka-cholam*; Kannada—*Mekkejola, musukojola, goinjol*; Malayalam—*Cholam*; Oriya—*Maka, buta*; Assam—*Gomdhan, makoi*; Manipur—*Chujak, nahom.*

History. It is now universally admitted that Maize or Indian corn is a native of America. This species probably originated in a wild state in the tropical South America. In Andes its cultivation goes back to prehistoric time. It has been established by evidence that the plant must have been grown for many centuries prior to the period of the Inca civilization. The maize was grown by the Indians in new Mexico as early as 2000 B.C. De Candolle on this subject writes—"No one denies that maize was unknown in Europe at the time of Roman Empire but it has been said that it was brought from the east in the Middle ages". The various names which it bears in Europe, Egypt and Asia only show that in each country it was supposed to come from some not very distant region. Thus it is Turkish-wheat, Indian Corn, Roman Corn, Sicilian Corn, Barbary and Guinea Corn, etc. The Turks call it Egyptian Corn and the Egyptians speak of it as Syrian grain. Its most general name in India may be rendered Mecca Corn *(Makkai)*. However, there is no authentic

Sanskrit name for the plant, nor is the grain in any way associated with the Hindus. Crawfurd also writes, — "Maize is, beyond all question, a native of America, and before the discovery of the New World was wholly unknown to the old." Royle says that the Portuguese very probably introduced the richest products of America into India, such as maize, capsicum, guava, custard-apple and pine-apple. Abul Fazl writes in *Ain-i-Akbari* about *kewarah*, where he says that its leaves are like those of "maize". Rox-burgh wrote about the beginning of last (nineteenth) century, that Indian-Corn was "cultivated in various parts of India in gardens, and only as a delicacy; but not any where on the continent of India, so far as I can learn, as an extensive crop." Buchanan-Hamilton (1819) wrote about Kangra—"The poor people live much on Maize." Graham (1839) wrote about Western India that maize was commonly cultivated. Dalzell and Gibson (1861) wrote that maize was extensively grown in the early part of the rains, especially near large towns. It is thus very probably that in the Upper India maize was much more extensively grown at the beginning of the last

Fig. 2.6. 'Protina Opaque 2 Composite', a protein-rich variety of maize (*Zea mays*) (developed by Govind Ballabh Pant University of Agriculture and Technology Pantnagar).

(nineteenth) century. It is a field crop upon which at least the bulk of the aboriginal tribes of the hilly tracts of India are very largely dependent for subsistence.

According to De Candolle, the Maize was brought to China after the discovery of America (1492). The Portuguese came to Java in 1496, that is to say four years after the discovery of America and to China in 1516, Magellan's voyage from South America to Philippines took place in 1520. It is thought, that the seeds of maize may have been taken to China by navigators from America or from Europe. Maize had reached Europe a short time before the dates mentioned for India and China. Dodoens (1583), Camerarius (1588) and Matthiole, adopted the name *Mays,* which they knew to be American.

Conclusively this can be said that Maize was first introduced into Europe by Columbus, and into Asia by the earlier Portuguese explorers. Maize had acclimatized wherever the climate permitted, and has now spread all over the world.

Maize also known as 'Indian corn' or simply as 'corn', is a food crop of considerable importance in many parts of the world, especially in the U.S.A., which produces nearly 57 per cent or the total maize grown in the world as compared to about 1.4 per cent produced in India. The other important maize producing countries are—China (4.4 per cent), the Russia (3.3 per cent), Manchuria (2.9 per cent), Yugoslavia (2.4 per cent) and Mexico (2.3 per cent).

A tall annual cultivated grass ; it has been introduced in this country from tropical America about the beginning of the seventeenth century. The area under this crop measures 9.6 million acres with an annual total production of three million tons. Maize is mainly grown as 'kharif crop'. Although it is grown throughout India, the chief concentration is in the north-eastern parts. The crop is of special importance in the hilly and submontane regions where it makes the staple food of people. It is grown as a food crop mainly in Uttar Pradesh, Punjab, Madhya Pradesh, Bihar, Andhra Pradesh, Jammu and Kashmir.

Cultivation and harvesting. Maize is a summer annual. It thrives best in fertile, well irrigated, medium, heavy loamy soil. It is also commonly grown in the coarse gravel soils of hilly tracts. In India the crop is generally sown in June–July and harvested in September–October. The maize

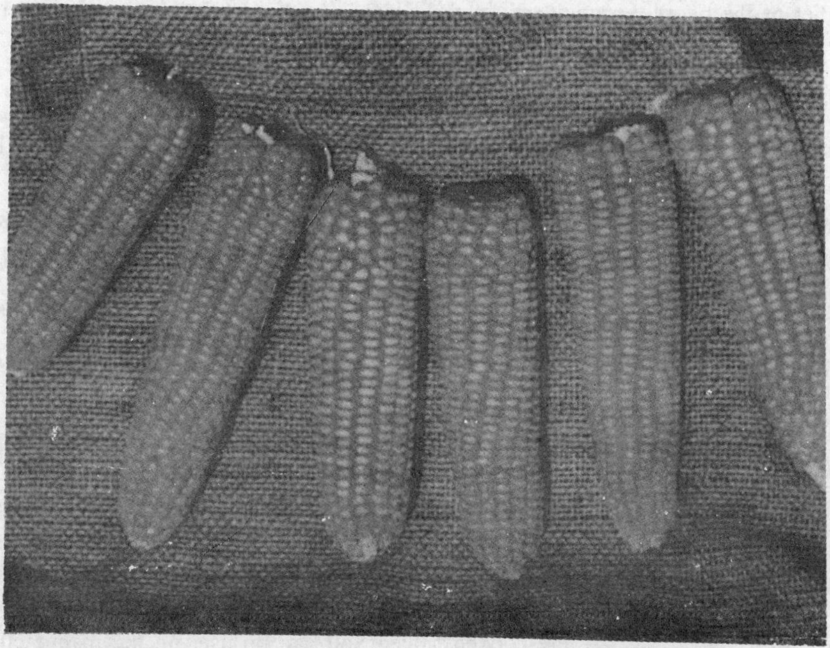

Fig. 2.7. *Zea mays* (corn-**makka**). Unhusked cobs.

stems are cut close to the ground with the help of corn knife or sickle. The stalks are stacked to allow the grain to ripe further. After a month of this curing process, the ears (cobs) are husked by hand or by machine. Maize must be stored in well ventilated bins so that the excess of moisture is evaporated.

The table shows that maize contains all the essential nutrients of food when compared to Wheat and other foodgrains.

Uses. The chief use is as a food for man and livestock. The grain is very nutritious, with a high percentage of carbohydrates, fats and proteins. Not only is the grain valuable as a stock feed, but the plant as a whole is an important fodder crop. The immature cobs are largely eaten after roasting. The grains are also used in making corn starch and industrial alcohol. The glucose is also manufactured from the grain. The corn oil is prepared which is used for soap making, lubrication and as salad oil. Corn flakes make a good breakfast food. The fibres in the stalks are utilized for making paper and yarn. Zein, the protein which occurs in maize grain, is utilized for making artificial fibres with good tensile strength and wool-like qualities.

Cereals	Proteins (Gms)	Fat (Gms)	Calories	Calcium Mgs	Iron Mgs	Vit 'A' I.U.	Vit 'B' Rib-oflavin Mgs	Vit 'C' Mgs
Maize (Tender)	4.7	0.9	125	9	1.1	54	0.17	6
Maize (Dry)	11.1	3.6	342	10	2.0	1502	0.10	-
Ragi	7.3	1.3	328	344	17.4	70	0.10	-
Rice (milled)	6.8	0.5	345	10	3.1	-	0.03	-
Jowar	10.4	1.9	349	25	5.8	79	0.28	-
Wheat (whole)	11.8	1.5	46	41	4.9	108	0.12	-

It is of special importance in the hilly and sub-montane regions of the country where it forms the staple diet of the people, particularly in the winter months. In the northern parts, it is also extensively grown as a fodder.

	Maize Products	**Uses**
1.	**Maize grits**	Can be used directly for maufacture of breakfast foods like corn flakes. This can also be used in the manufacture of starch.
2.	**Maize rice**	Can be used in replacement of rice or in conjunction with rice. This can also be used in the manufacture of starch.
3.	**Poultry grits**	Can be used in replacement of rice and other cereals in breweries and distilleries.
4.	**Brewer's grits**	Can be used in replacement of rice and other cereals in breweries and distilleries.
5.	**Maize flakes**	Can be used to prepare various types of dishes in place of beaten rice. This can also be used in the preparation of soup and in breweries and distilleries in replacement of rice and other cereals.
6. 7.	**Maize soji** **Pre-cooked Maize soji**	Can be used in the preparation of sweet and savoury dishes like Kesaribhath, Kharabhath, Idli, Dosa, Vermicelli, etc. Pre-cooked Maize Soji can be used in the manufacture of Balahar and snack foods.

8.	**Maize maida**	Can be used in the preparation of Bread, Biscuit, Vermicelli, Nippottu, Obbattu, Bajji, Pakoda, Kodubale, etc. This can also be used by mixing with Wheat, Maida and Bengal gram flour for preparing the above said dishes. Also used in gum manufacturing industry.
9. 10.	**Maize flour** **Pre-cooked Maize flour**	Can be used in the preparation of Ragi balls by mixing with ragi flour. Can also be used in the preparation of sweet and savoury dishes like Pakoda, Bajji, Kodubale, Chapaties, etc. As raw materials in the manufacture of binders (Gum) for being used in the foundry and paper industry. Pre-cooked flour can be used in the preparation of Soup, Curry, Sambar, Majjigehuli (khadi) and in the manufacture of cereal and other foods for babies.
11.	**Maize oil**	Used as edible oil. This oil has got certain advantages like easy blood circulation, reducing body fat, etc. This can also be used in soap manufacturing industry.
12. 13. 14.	**Maize cake** **Animal meal** **Bran**	Can be used in Cattle and Poultry Feed industry.

Other uses. However, due to low gluten content, it does not form dough with elastic properties, but dough with good elastic properties can be made from flour made from dehulled grain. Maize starch is extensively used as a sizing material in the textiles and paper industries. In the food industry it is used in the preparation of pies, puddings, salad dressings and confections. Maize starch is used for the production of dextrose and corn syrup; also employed as a diluent for pharmaceutical preparations, dusting material to prevent articles like surgeons gloves, from sticking together, ingredient of oil-well drilling muds, and as a depressant in ore-floatation process. In cosmetics, maize starch form an ingredient of various forms of toilet powders. Dried germs yield a semi-drying oil, known as maize oil or corn oil used as salad or cooking oil; it may also be used with linseed oil for paints. Zein, recovered from maize gluten used as a binder for cork particles in forming composition cork. A textile fibre has been produced from zein under the trade name *Vicara;* it combines the virtues of cotton and wool. Cobs are rich in pentosans and used for furfural production. They may also be used for making building-boards which are water and fire-resistant. Maize silk (styles) is astringent, diuretic and chloretic. Some of the maize products include dextrose, maize starch, syrup, corn flakes and popcorns.

Barley

Hordeum vulgare Linn.; Family—Gramineae (Poaceae); Eng.—Barley; Hindi—*Jau*; Sanskrit—*Yava*; Bengali—*Jab, jau*; Marathi—*Java*; Gujarati—*Jau, jav, ymvah;* Telugu—*Barlibiyam, yavaka;* Tamil—*Barliyarisi;* Kannada—*Jave godhi;* Bihar—*Jowakhar*; Punjab—*Jaon.*

History. Barley is amongst the most ancient of cultivated plants. Proof exist in abundance, however, that barley in one or more of its forms was cultivated in the remotest times. According to Bretschneider it was included in the list of five cereals sown by the Emperor Shen-nung of China who reigned about 2700 B.C. Theophrastus was acquainted with several sorts of barley. It is frequently mentioned in the Bible, and must have been an important article of food in the time of Solomon (1015 B.C.). According to De Candolle the variety *hexastichum* which was known to Theophrastus, "has also been found in the earliest Egyptian monuments; in the remains of the lake dwellings of Switzerland (stone age), of Italy and Savoy (bronze age). A six rowed barley is also represented on the medals of South Italy, six centuries before Christ. It has been established that the variety most cultivated in antiquity, was *Hordeum hexastichum* Linn. *H. distichum* Linn, was also grown and employed as a food-grain in prehistoric times, while the cultivation of *H. vulgare* seems to date from more modern times.

The only variety found wild, in any part of the globe, is *H. distichum* Linn., which is probably indigenous to Western Temperate Asia. Since *H. distichum* is the only variety which exists in wild condition, it would seem that the four and six-rowed barleys are either cultivated varieties of the two-rowed form or that they owe their origin to a wild ancestor of their own type which has since become extinct.

Little is known regarding the introduction or the origin of barley's cultivation in India, but it is well known and valued by other Eastern peoples, it is probable that it was also grown at least in the Northern India in very remote times. This has been confirmed by the intimate connection of the grain with several of the rites and beliefs of the Hindu religion. It is further supported by the antiquity of the Sanskrit name, *Yava,* which in the earliest times was probably applied as a general term to any grain-yielding plant.

Barley reached the Western Hemisphere in the sixteenth or seventeenth century.

A herb, cultivated as a food crop. It is of great antiquity as a cultivated cereal and was used even before wheat. It is thought to be the oldest of all cultivated plants. It was known to all the ancient civilizations of the old world. It is an important rabi cereal of northern India. It is either grown pure or as a mixed crop. It is specially important in the hilly tracts. The area under barley cultivation in India is 8.4 million acres and the annual production 2.7 million tons. The cultivation and harvesting of barley is similar to that of wheat. It is mainly cultivated in Uttar Pradesh, Punjab, Rajasthan and Madhya Pradesh, Bihar and West Bengal.

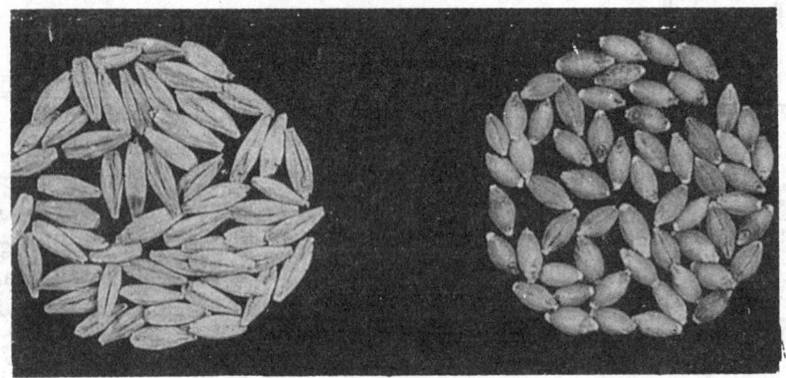

Fig. 2.8. *Hordeum vulgare* (Barley—**Jau**). A comparative view of huskless (left) and hulled (right) barley grains.

Barley is cultivated in the *rabi* season. Sowing takes place between October and December, and harvesting between February and April. It is grown mostly as a dry crop in retentive soils. It is raised either by itself or mixed with a variety of crops, principally leguminous crops like peas and lentils, and with wheat or mustard or linseed.

Uses. The grain is chiefly utilized in the form of flour for making 'chapatis'. Barley is also used for making breakfast food and children's food. The six-rowed kinds of barley have a higher protein content, and are used for food purposes, both for man and animals. The barley is used as a source of malt to be used for making alcohol, whisky, beer and other alcoholic beverages. For making malt the two-rowed types of barley are preferred.

The grains of barley mainly used in the form *sattu*, a cooling drink; also mixed with wheat and gram flour for making *chapatis*. Barley grain is easily assimilable, used in the dietary of invalids and convalescents. Barely water used as a diuretic and demulcent drink. Powdered grain is much employed in the form of a gruel in cases of painful and atonic dyspepsia. Used as a source of malt in manufacture of beer. Straw used as a fodder of livestock, and as stable bedding.

Oat

Avena sativa Linn.; Family—Gramineae (Poaceae); Eng.—Oat; Hindi—*Javi.*

History. The origin of oat is unknown, but it is thought to have been originally a native of northern Europe. It probably had a multiple origin, some coming from Abyssinia, others from the Mediterranean area; and still another from China. The oats were grown by lake dwellers, though not by the early Mediterranean nations. It is thought, that the two wild varieties, *A. fatua* or *A. brevis*, have given rise to the cultivated forms (*i.e., A. sativa.*)

An annual grass. They vary in height from 2 to 5 feet. The inflorescence is one-sided or spreading panicle which may be either erect or drooping. *Avena sativa* L. is commonly cultivated in the temperate climate of northern Europe (Chatterjee and Randhawa, 1952). This is grown to a limited extent in North-West Himalayas. The Indian oat belongs to the species *Avena sterilis* var. *culta*. A hybrid has been raised between Indian oat and *A. sativa*.

Uses. Oats are consumed as human food and fodder for cattle. They have a high fat, protein and mineral content. It is used chiefly in making cakes, biscuits and breakfast food. Oatmeal is prepared by grinding the grains rather coarsely between the stones.

It is a source of very nutritious oatmeal; used as rolled oats, porridge, breakfast foods, and in cakes and biscuits. Oats are particularly rich in fat and protein, but protein does not occur in form of gluten. Plant is used as fodder. Oat hulls are a source of furfural, a valuable industrial solvent.

Rye

Secale cereale; Family—Gramineae (Poaceae); Eng.—Rye.

History. According to De Candolle, the original area of rye *Secale cereale* Linn. was in the region comprised between the Austrian Alps and the north of the Caspian Sea. However, this was not found in India to any extent either wild or cultivated. *Secale montanum,* a wild species of Afghanistan and Turkistan, has been thought by Aitchison to be the wild ancestor. Some authorities consider *S. anatolicum* of Asia Minor to be the progenitor. There are no traces of this cereal among the ruins of Egypt or the Lake Dwellings, however, this was known to Romans and Greeks.

It is of more recent origin than the other cereals. It is thought to be native of the Black and Caspian Seas region of Central Eurasia. It is related to both barley and wheat and resembles the barley in habit. Its grains look like that of wheat. The tillers are slender reaching a height up to six feet. The leaves are somewhat bluish in colour. A large number of spikelets are found in a head. The spikelets are produced singly at the joints of the axis and each one contains two fertile flowers.

Uses. In European countries it is used chiefly for bread, as the grains contain gluten. The bread of rye is also known as black bread because of its dark colour.

The straw of rye is used for making paper and hats. The grain is also used for stock feed. 90 per cent of the world's crop is produced and consumed in Europe. Two-thirds of the crop is fed to livestock and the rest is used for making flour whisky and alcohol.

Other uses. Rye is also grown for production of medicinal Ergot *(Clavieps purpurea* Tul.*)*, or as green manure. Due to lack of gluten in the grain, the dough lacks elasticity. Rye biscuits are popular in some countries. It is used also for making *chapaties,* porridge, and alcoholic products. It makes a good cover crop. Straw is used for thatching and as packing material ; also employed to some extent in paper manufacture.

MILLETS

The millets are considered to have been cultivated in India from prehistoric times. Their importance as an article of human food can be realized from the fact that about 30 million acres in India fall under millets. Millets are generally grown as mixed crops in regions of low rainfall, the other crop grown with them being usually one of the legumes. Most of the millets grown in our country are of short duration, taking three to four months from sowing to harvesting. Some of the commonly grown millets in India are sorghum, pearl millet and finger millet.

Fig. 2.9. *Sorghum vulgare* (Sorghum–**Jowar**). A sorghum hybrid with a high harvest index.

Sorghum

Sorghum vulgare (L.) Moench ; Family Gramineae (Poaceae)—Eng.—Sorghum ; Hindi—*Jowar*; Telugu and Oriya—*Jonna*; Tamil and Malayalam—*Cholam;* Kannada—*Jola*; Marathi—*Jowari*.

History. The sorghums were among the first of the wild species to be domesticated by man. It was grown in Egypt prior to 2200 B.C., and since then has continued as an important crop in that country. Sorghum was cultivated in China and India at an early date. From their native home in Africa and Asia they have been dispersed to all warm countries, in temperate regions as well as in the tropics. It is thought that the cultivated form of sorghum *(S. vulgare* Pers.*)* has been derived from the abundant wild species *S. halepensis*. It is also thought that some of the cultivated sorghums had been developed in India. Its Persian name *juar-i-hindi* shows that it reached Persia, at least, from India. An ancient cultivation may be inferred from the very extensive series of forms, recognized by distinct names and adapted to the climates and soils of the particular tracts of India in which they occur. It is assumed that the Sanskrit people first learned of this grain in India. The religious associations of the grain, the observances of cultivation, and the multiplicity of forms of the crop, all point to an antiquity as great as can be shown for most other articles of the phases of life. Conclusively, it can be said that many of the forms of this millet are beyond doubt natives of India.

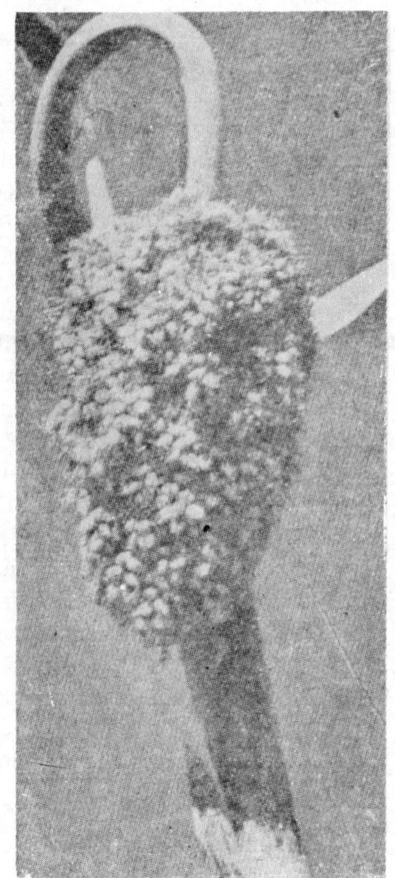

Fig. 2.10. *Sorghum vulgare (Jowar)*—Ear (*bhutta*) of an improved variety.

India is one of the leading countries in the world for the production of sorghum. The area under this millet is 42.6 million acres, and the production of the grain is estimated at about 7.4 million tons. It is chiefly cultivated in Uttar Pradesh, Punjab, Madhya Pradesh, Andhra Pradesh, Maharashtra and Rajasthan.

In India, *jowar* is mainly a crop of plains, and is grown as both a *kharif* and *rabi* crop. The kharif crop is sown

Fig. 2.11. *Pennisetum typhoides* (Pearl millet—**Bajra**). Hybrid bajra.

between May and July, and harvested between October and December. The *rabi* crop is sown between September and November, and harvested between January and March.

The plants are tall coarse annuals, growing to a height of 3 to 15 feet. The inflorescence is a dense head or panicle. The grains are small, round and pinkish white in colour. They can easily be grown in less irrigated and arid regions.

Uses. The grain is eaten by breaking it and cooking it in the same way as rice or by grinding it into flour and preparing unleavened bread from it. The stem and leaves are used as cattle fodder especially in north India.

Other uses. Several sorghum varieties with juicy and palatable stalks have been developed and grown exclusively for fodder purposes; the plant is cut after flowering or seeding and fed to cattle green or after drying, or ensiling, Green stalks are chopped before feeding. Grain may also be used as feedstuff. Leaves and stalks of most varieties contain a cyanogenetic glucoside, dhurrin, which decreases with maturity of the plant. In addition to grain and fodder, Sorghum Oil, obtained from the germ fraction separated in the wet-milling process, is used after refining in salads or for general cooking. A hydrogenated product has also been produced from the oil.

Pearl Millet

Pennisetum glaucum (Burm.) Stapf. & Hubbard; Family—Gramineae (Poaceae); Eng.—Pearl millet; Bulrush millet, Spiked millet, Hindi—*Bajra*; Bengali—*Bajra, lahra*; Marathi and Gujarati—*Bajri*; Telgu—*Sajja, gant* ; Tamil—*Kambu*; Kannada—*Sajje.*

This is the second important millet of India. The area under the crop is 28.2 million acres, and the annual production of grain is about three million tons. It is chiefly cultivated in Uttar Pradesh, Punjab, Andhra Pradesh, Tamil Nadu, Rajasthan, Maharashtra and Gujarat.

It is suited to regions of low rainfall, and can be grown even in tracts which receive only 17 to 20 inches of rainfall. It is seldom grown in areas where the rainfall is heavier than about 35 inches annually. Outside Asia, the grain is largely cultivated in Africa, where it is a common article of food among both the Arabs and the natives.

Fig. 2.12. *Pennisetum typhoides* (Pearl millet—**Bajra**). Ears.

In India the crop is grown mainly in *kharif* season. Sowing takes place between May and September and harvesting between September and February.

The plants are tall annuals, growing to a height of 6 feet to 15 feet. The inflorescence is a dense spikelike head or panicle 6 to 15 inches long and 1 inch or so in diameter. In India it is grown as a rainy season crop.

Uses. It furnishes an important food for the poor people and labour class and is particularly valuable in cold weather because of its heating qualities. The flour made from the grain is very nutritious and is used for making bread. The crop has an enormous yield of forage.

Other uses. Grains are used for porridge, or eaten after parching (*akohi, bhunja, lahi* or *phula*. Green ears also roasted and eaten. It is suitable also for making malt. It is sometimes grown for green fodder. The straw is used for thatching and as fuel.

Finger Millet

Eleusine coracana (Linn) Gaertn; Family—Gramineae (Poaceae); Eng.—Finger millet; African millet, Ragi; Hindi—*Mandua;* Sanskrit—*Rajika*; Bengali—*Marua*; Marathi—*Nagli, nachoni*; Gujarati—*Bavto, nagli* ; Telgu —*Ragulu*; Tamil—*Ragi*, *kelvaregu*; Kannada—*Ragi*; Malayalam—*Muttari.*

History. According to De Candolle, *ragi* (*Eleusine corocana* Gaertn.) is native of India. In Egypt the ancient monuments bear no trace of its cultivation in early times, and Greek-Roman authors, who knew the country, do not speak of it. It is mentioned by Sanskrit writers under the name of *Rajika* or *Ragi*; the word *coracana* comes from *kurakkan*, its Ceylon name. Its nearest allied species in the wild state is *E. aegyptiaca* which bears very close resemblance to the cultivated *E. coracana.*

This is the third important millet of India. The area under the crop is 5.9 million acres and the annual production of grain is about 1.9 million tons. About 75% of the area under the crop lies in South India. It is also grown to a limited extent, in the hilly tracts of northern India. It is cultivated as a food crop in Andhra Pradesh, Tamil Nadu, Karnataka, Orissa, Bihar, Uttar Pradesh and Maharashtra.

It has many valuable features. It is one of the hardiest crops suited for dry farming. It can grow under conditions of very low rainfall, and can withstand very severe drought. It can be grown as a dry crop as well as under irrigation. Unlike other crops, both the plants and grain remain free from pests and diseases. *Ragi* grain can be stored for many years even up to fifty years without damage, if kept away from moisture.

In India, *ragi* is sown in the *kharif* season. Sowing takes place between May and August, and harvesting between September and January. In Andhra Pradesh and Tamil Nadu, sowing and harvesting continue almost throughout the year.

This tall grass, has tufted stems, each with four to six spikes. The important variations observed are in ear-head shape and plant pigmentation. The seed coat is generally brown in colour ; a few types with white grains also occur.

Uses. Its grain can be made into cakes, porridge and sweetmeats. A beer is brewed from the grains by the hill-tribes. The flour is used for puddings. It is a grain of great nutritive value, and is considered more sustaining for people doing hard physical work than any other grain. The straw is considered a valuable food for the milk animals.

The grains are tonic and astringent, useful in biliousness; specially recommended for diabetics as a wholesome food.

Fig. 2.13. *Ragi* has many valuable features, which mark it off sharply from other foodgrains. It is a grain of great nutritive value and is considered more sustaining for people doing hard physical work than any other grain. The straw is considered a valuable food for the cattle.

Italian Millet

Setaria italica (L) Beauv.; Family—Gramineae (Poaceae); Eng.—Italian millet, foxtail millet, Hungarian millet; Hindi—*Kangni, Kangu, Kakun*; Sanskrit—*Chinaka, kangu, kanguni, pitatandula, priyangu*; Bengali-*Kangu, kora*; Marathi—*Kangu*; Gujarati—*Kang, karang*; Telgu—*Korralu*; Tamil—*Tenai*; Kannada—*Navane, ksongu*; Malayalam—*Tena, thina*; Oriya—*Tangun, kangu*; Assam—*Kaon;* Punjab—*Kangui, kusht*; Kashmir—*Shali, pingi*; Santal and Mundari-*Erba*; Andamans—*Tanahal.*

It is native of Eastern Asia. It must have been domesticated in the Orient ages ago, for it was one of the five sacred Chinese plants as early as 2700 B.C. At the present time it is extensively grown in India, Japan, china, the East Indies and other parts of Asia. In India, it is chiefly cultivated in Andhra Pradesh and Tamil Nadu.

Uses. They are used for human food and fodder. When used for food, the grains are boiled or parched. It is used as diuretic and astringent and also used externally for rheumatism.

The grains are consumed in the form of cakes or porridge. Grain is said to enhance intoxication of beer. Plant is used as a sedative to the gravid uterus. Grain is a popular medicine for alleviating pains after parturition. Straw is a good fodder; also makes good hay and used for thatching.

Bristly Foxtail Millet

Setaria verticillata (L.) Beauv.; Family—Gramineae (Poaceae); Eng.—Bristly foxtail millet; Hindi—*Laptuna;* Bengali—*Dorayra*; Telgu—*Chick-lenta*; Punjab—*Chir chira, bar chitta, kulta*; Santal—*Bir kauni.*

Commonly cultivated in Tamil Nadu and Andhra Pradesh. The grains are used as food.

Young grass grazed by cattle; on maturity the bristles become rigid and such spikes may be used to cover granaries as a protection against rats. The grains are consumed as food by the desert tribes.

Proso Millet

Panicum miliaceum Linn.; Family—Gramineae (Poaceae); Eng—Common millet, proso millet, hog millet; Hindi—*Chin, morha, anu*; Bengali—*Cheena*; Marathi—*Varo, vari*; Gujarati—*Chino, vari*; Telgu—*Varagalu, variga*; Tamil—*Panivaragu*; Kannada—*Varagu.*

Cultivated mainly in Uttar Pradesh. The grains are used as food and the straw is used as fodder.

Green plants make excellent fodder for cattle and horses, also used as hay. Grains contain 10–18 per cent of protein. Starch is the chief carbohydrate in the grains and it is similar to corn starch in appearance and other properties; suitable as a sizing agent in the textile industry. The husked grain eaten whole, boiled and cooked like rice; sometimes made into flour for *chapaties*; also parched into *marha* or *mard*. It is used for porridge; fermented into a kind of beverage.

Little Millet

Panicum miliare Lamk.; Family—Gramineae (Poaceae); Eng.—Little millet; Hindi—*Sawa, kutki*.

Cultivated in Utter Pradesh and Madhya Pradesh for its edible grains and fodder.

PSEUDO CEREALS

In various parts of the world the seeds of the other plants are used like cereals and millets as source of human food. Such plants are not grasses. In India buckwheat is grown on the large scale in the Himalayan tracts.

Buckwheat

Fagopyrum esculentum; Syn. *Polygonum fagopyrum*; Family—Polygonaceae; Hindi—*kutu*; Punjab—*Daran, obul, phaphar*; Assam—*Doron*; Darjeeling—*Titaphapur*.

It is native of Central Asia, but now extensively cultivated in the hills of Northern India. It is normally a plant of cool, moist; temperate regions and thrives best in a sandy well-drained soil. The plant is small branching annual. The inflorescence is a raceme bearing small white or pinkish flowers. The seeds are hulled and ground and the starchy flour is used for porridge, soups and *puris*. In India, the flour is commonly used by Hindus on festive occasions to break their fast. Seeds are also fed to livestock and poultry.

Hulls yield a dye. A promising source of rutin which reduces increased capillary fragility. Leaves and young shoots are boiled and eaten as spinach.

Gorgan Nut, Fox Nut

Euryale ferox; Family—Nymphaeaceae; Hindi—*Makhana*; Bengali—*Makhana*; Telugu—*Mellunipadmamu*; Oriya—*Kuntapadmu*; Punjab—*Jewar*.

This is a prickly aquatic herb, found commonly in Kashmir, Bihar, Assam, Manipur, Tripura, Bengal and Uttar Pradesh. The seeds are roasted and eaten. They are added in sweetmeats and confectionery. It makes a light food with milk.

The flour of the seeds is used as a substitute for arrowroot. It is nutritious and easily digestible, recommended for invalids. The seeds are considered tonic and deobstruent.

Duckwheat, Tartary or Indian Buckwheat

Fagopyrum tataricum; Hindi—*kaspat*; family—Polygonaceae; Punjab—*Brapu, chim, ugal*.

Uses. Better source of rutin than *F. esculentum*. Duckwheat is used in the same manner as common buckwheat. The leaves are used as a pot-herb in summer when other greens are not available.

3

Legumes and Nuts

LEGUMES

The legumes or pulses belong to the family Leguminosae. The legumes are next in importance to cereals as sources of human food. They contain more protein materials than any other vegetable product. The pulses form an important item in India where the majority of the population consists of vegetarians. Carbohydrates and fats are also present in the legumes. The proteins occur as aleurone grains in the same cells with the starch grains. The high protein content is related with the presence, on the roots of legumes with nodules containing nitrogen fixing bacteria. The pulses are also important from the point of view of animal nutrition, to which they contribute by their seeds, hulls and the green parts. The legumes have been cultivated and used for food for centuries all over the world. The pulses figure prominently in crop rotations and in the mixed cropping commonly practised in Indian agriculture. About one-seventh of cultivated area in India is under pulses (Mehta, T.R. 1954). Of this about 37% is occupied by gram and 13% by the pigeon pea. The other important Indian pulses are—black gram (*urd*), green gram (*moong*), lentil (*masur*), pea *(matar)*, dew gram (*moth*), etc.

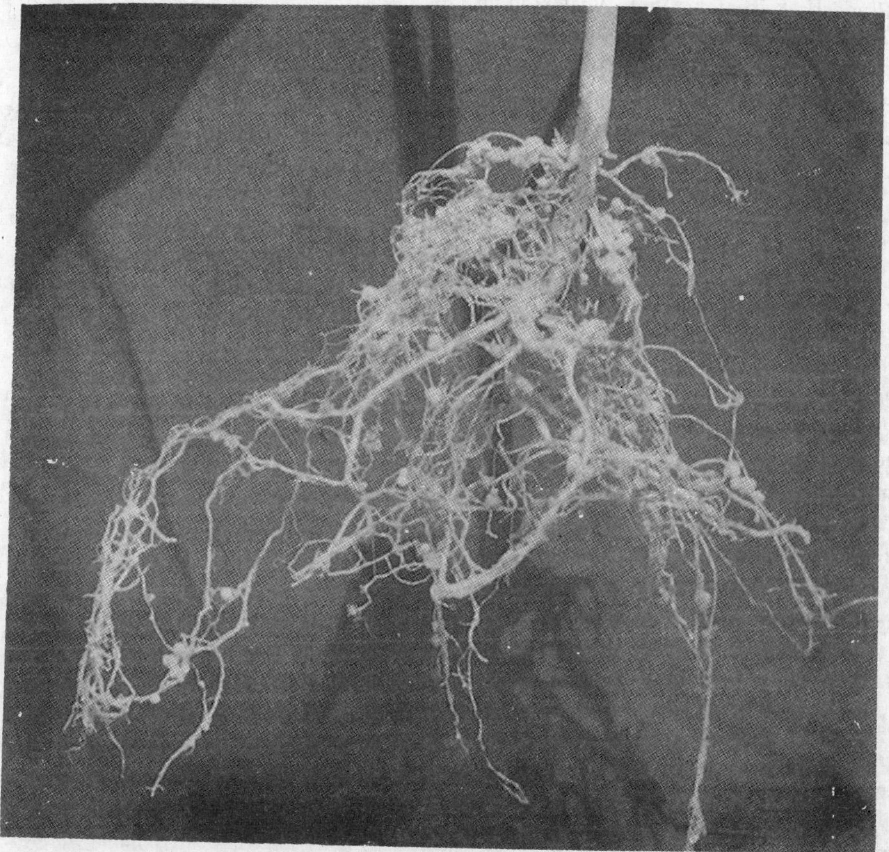

Fig. 3.1. *Rhizobium* nodules of a leguminous crop (French bean).

Gram

Cicer arietinum Linn.; Family—Leguminosae (Papilionaceae/Fabaceae); Eng.—gram; Bengal gram, chick pea; Hindi—*chana*; Sanskrit-*Chanaka*; Bengali—*Chola, but*; Gujarati—*Chana, chania*; Marathi—*Harbara*; Tamil—*Kadalai*; Telugu—*Sanagalu*; Kannada—*Kadale*.

History. In ancient times the gram (*Cicer arietinum* Linn.) was known to the Greeks, under the name *Erebinthos*, and to the Romans as *Cicer*; and the existence of other widely different names show that it was early known and perhaps indigenous to the south-east of Europe. It is thought that the gram has been cultivated in Egypt from the very earliest times of the Christian era. It is most likely that it was introduced into Egypt from Greece or Italy. Its introduction into India is of more early date, for there is a Sanskrit name (gram—*chanaka*) and several other names in modern Indian languages. It is thought that the Western Aryans carried it into India.

This is the most important pulse in India. It is native of South Europe, now commoly grown in Uttar Pradesh, Punjab, Maharashtra, Rajasthan, Bihar and Madhya Pradesh. These together account for more than 90 per cent of the total area under it. The main producing areas are the upper basins of the Ganga and the adjacent tracts of Central India. States of lesser importance are Andhra Pradesh, Karnataka and West Bengal. There are two broad groups of gram—brown and white. The brown or *desi* type is most widely grown. The white or *kabuli* gram is characterized by larger seeds.

In India, gram is sown as *rabi* crop at the end of the rainy season. Sowing takes place between September and November, and harvesting between February and April.

Uses. The gram is consumed in several ways. *Dal* is prepared by splitting the whole grain into two and removing the husks (the seed coat). This is one form in which gram is used in India. The flour of *dal* is known as *baisan*. It is used in the preparation of unleavened bread and sweets. The whole grain is eaten raw, roasted, parched or boiled. Gram is also used as cattle feed. It makes a nutritious feed for horses.

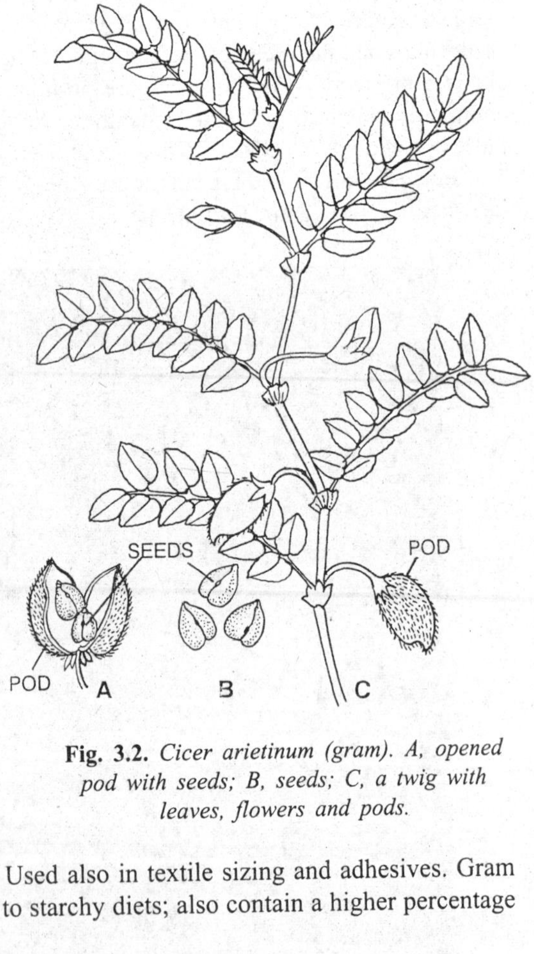

Fig. 3.2. *Cicer arietinum (gram). A, opened pod with seeds; B, seeds; C, a twig with leaves, flowers and pods.*

Germinated gram is used as a prophylactic against deficiency diseases, scurvy in particular. Used also in textile sizing and adhesives. Gram is nutritive pulse and is used as a protein adjunct to starchy diets; also contain a higher percentage of oil, 4–5 per cent, than other pulses.

Pigeon Pea

Cajanus cajan (Linn.) Millsp; Family—Leguminosae (Papilionaceae); Eng.—Cajan pea, pigeon pea, red gram, Congo pea; Hindi—*Arhar*; Sanskrit—*Adhaki, tuvari, tuvarika*; Bengali and Marathi—*Arhar, tur, tuver*; Telugu—*Kandulu*; Kannada—*Togare*; Malayalam—*Thuvara*; Tamil—*Thovaray*.

History. According to De Candolle (*Origin of Cultivated Plants*) pigeon pea (*Cajanus cajan*) is more probably a native of tropical Africa, and introduced perhaps 3,000 years ago into India.

This was first domesticated in Asia or Africa, and is now widely cultivated in the tropics and subtropics. It is particularly grown in the East Indies, India and West Indies. In India, it is chiefly grown in Madhya Pradesh, Bihar, Andhra Pradesh, Maharashtra, Uttar Pradesh, and Karnataka.

Grown mostly as a *'Kharif'* crop and used in form of *'dal'*. This is the second important pulse crop of India.

It is grown as a dry crop mixed with millets like *jowar*, *bajra* and *ragi*. In North India, however, the crop is irrigated over considerable areas.

It is sown in the *kharif* season from May to July, and is harvested in about six to eight months' time (*i.e.,* from December to March). The area covered by it is about six million acres and annual production is about 2 million tons.

There are two main varieties grown in India—(i) *Cajanus cajan* var. *bicolor* DC. (*arhar*) and (ii) *Cajanus cajan* var. *flavus* DC. (*tur*). *Arhar* comprises of late-maturing, large, bushy plants, bearing purple streaked, yellow flowers and dark coloured pods, each having four or five seeds. *Tur* on the other hand comprises early maturing and smaller plants having yellow flowers and

Fig. 3.3. *Cajanus cajan* (Pigeon pea—**Arhar**). A high yielding variety.

plain pods, each containing two or three seeds. *Arhar* is commonly cultivated in north-eastern India and *tur* in the peninsular India. The important types of India are—N.P. 15, N.P. 16, N.P. 64, N.P. 80 and type 17 of U.P.

Uses. Both the immature and ripe seeds are used for human food as a good source of protein. The leaves and twigs are used for fodder. The pericarp and husk, separated in threshing are used as cattle feed. The enzyme urease, obtained from it, is required for estimation of urea in blood, urine, etc. Livestock and poultry are very much fond of it. It is chiefly consumed in south Indian homes.

This contains *two* globulins: *cajanin* and *concajanin*. The leaves are used in Madagascar for rearing silkworms. Green pods are used as a vegetable. Husk makes useful fodder. Green leaves are used as green manure.

Soybean

Glycine max (Linn.) Merr.; Family—Leguminosae (Papilionaceae); Eng.—Soybean; Hindi—*Bhat, Ramkurthi*; Bengali—*Garjkalai*; Assam—*Patnijokra*; Khasi Hills—*Rymbai-kutung*.

History. The soybean *(Glycine max)* is one of the oldest crops grown by man. It was cultivated in China centuries before the first written records in 2838 B.C. It is a native of south-eastern Asia.

It is native of South-East Asia. Since many years, soybean is being grown in India. Yet, only a few farmers grow it. The area under this crop is small.

The soybean can be grown in drier areas of the country where the rainfall is 35 inches or less. It can be grown at elevations up to 6,000 feet above sea level, especially in Assam, Orissa, West Bengal, Manipur, the Khasi and Naga hills and the Kumaon hills of Uttar Pradesh. Several varieties have been found suitable for growing in this country. *Palmetto, Monetta, Clemson, Creole* and *Charlee* are some of the good ones. *Type 33,* an early maturing variety is recommended for the Himalayan districts of Uttar Pradesh.

Uses. Soybean is a very good food. The bean has a high oil content, and the kernel is rich in proteins. It also contains important minerals; more of calcium and phosphorus than any cereal, more than even peas, and beans. It is rich in iron, potassium and magnesium and in vitamins.

Fig. 3.4. *Glycine max* : **Soybean**—Unripe fruits.

Soybean oil can be used for cooking. Oil-cake is used as cattle feed.

Soybean flour or cooked bean is very good for diabetic patients.

Green pods and mature soybean grains are a richer food than peas and beans. *Chapatis* can be prepared with one part of soybean flour mixed with one of wheat flour. Tasty biscuits and cakes can also be prepared from the flour.

Vegetable milk and curd is also prepared from soybean. There is a plant for making soybean milk in New Delhi. Both milk and curd have a good taste and are nutritious.

Soybean oil finds many uses in industry. It is used for paints and varnishes and for soaps, and also for food products. The meal is used in making plastics and a good ghee.

It can be fed to poultry as protein food. It gives a good hay. Cattle fed on this hay give more milk.

Soybean, being a legume, makes the soil more fertile when it is grown.

The beans are consumed as a vegetable, in salads, and canned. The flour is also used for making ice-creams and chocolate bars. Among the useful products may be included Soya-Sprouts, Soya-Milk and Soyabean oil from the seeds, extensively used in paints, varnish and enamel industries; refined oil is used as a salad or cooking oil. Soyabean Lecithin (total phosphatides), used as a wetting and stabilizing agent in foods, cosmetics, pharmaceuticals, leather goods, paint and plastics industries, and soaps and detergents. Soybean meal is used as a fertilizer.

Soybean meal and soybean protein are used in adhesives, water paints, leather finishes, textile sizes, insulating and wall board coating, insecticidal sprays, and fire-fighting compounds.

Soybean meal is also used in the manufacture of plywood glue. A synthetic fibre comparable to commercial casein fibre and suitable for blends with rayon or cotton has been produced from the protein. Residue left after extraction of protein is used in the manufacture of phenolic type plastics. It is also grown as a forage plant.

Fig. 3.5. *Glycine max* (Soybean). The beans are rich in protein contents.

Garden Pea

Pisum sativum Linn. var. ***arvense*** Poir.; Family—Leguminosae (Papilionaceae); Eng.—Garden pea; Hindi—*Mattar*; Sanskrit—*Satila*; Bengali—*Matar*; Marathi and Gujarati—*Watani*; Telugu—*Patanlu*; Tamil and Malayalam-*Pattani*; Kannada—*Batani, batagadle*.

History. The common pea *(Pisum sativum)* is a native of South Europe. It is thought that it has been cultivated even before the beginning of the Christian era. Peas were known to Greeks and Romans. However, they were cultivated in Europe for the first time in the middle of the seventeenth century. The peas were brought to America by the earliest colonist. According to De Candolle, the Aryans introduced it into Europe, but it perhaps existed in Northern India before the arrival of Eastern Aryans. It is probably a cultivated race or variety of *Pisum arvense*.

It is native of South Europe but now cultivated chiefly in Punjab, Uttar Pradesh, Delhi and Himachal Pradesh. It is annual, glaucous, tendril-bearing, climbing or trailing herb with white or coloured flowers and pendulous pods.

Uses. The garden pea seeds are eaten green or are used for canning. The seed is also used for human consumption in the form of pea meal or split peas. The peas are used as pulses, and they are good source of proteins. The plants are used for forage and green manuring.

The peas are also available as canned, frozen and dehydrated peas. Peas have high content of proteins, *i.e.*, upto 28 per cent or more, and also supply adequate quantities of vitamins and minerals, and potassium and phosphorus. Pea oil when given parenterally showed possibility of

Fig. 3.6. *Pisum sativum* (Peas—**Matar**). Peas L-116 'Hans'.

preventing pregnancy, the active principle being *m*-xylohydroquinone. Trials on women with capsules containing 300–350 mg of *m*-xylohydroquinone, twice a month, for variable periods showed 50–60 per cent reduction in pregnancy rate. The drug is non-toxic and has no side effects. In trials on men, the drug caused 50 per cent reduction in the number of spermatozoa.

Black Gram

Vigna mungo (Linn.) Hepper, syn. *Phaseolus mungo* Roxb.; Family—Leguminosae (Papilionaceae); Eng.—Black gram; Hindi—*Urd*; Sanskrit—*Masha*; Bengali—*Mash kalai*; Gujarati—*Adad, arad*; Marathi—*Udid, maga*; Telugu—*Minimulu*; Tamil—*Ulundu*; Kannada—*Uddu*; Malayalam—*Ughunnu*.

A herb, cultivated as a pulse crop in Uttar Pradesh, Madhya Pradesh, the Punjab and Bengal.

Uses. The small oval black seeds are highly nutritious. It is good source of protein for vegetarians. The seeds are used in the form of *dal* (splitted seeds). It is largely used by south Indians to make *vada idili*, *dosa,* etc. Black gram is the main ingredient of these items. The plants are used as cattle fodder.

This pulse is a good source of phosphorus. It is the chief constituent of wafer-biscuits (papds). Also cooked as a vegetable. Fried and salted seeds are eaten as a snack. Pulse is used in rheumatism and nervous and hepatics diseases. Also used in dropsy cephalagia as a diuretic. Root is narcotic, and used for aching bones.

Green Gram

Vigna radiata (Linn.) Wilczek, syn. *Phaseolus aureus* Roxb.; Family—Leguminosae (Papilionaceae). Eng.—Green gram; golden gram; Hindi—*Moong*.

Sanskrit—*Mudga*; Bengali—*Mung*; Marathi—*Mug*; Gujarati—*Mag*; Telugu—*Uthulu, patchapessalu*; Tamil—*Pasi-payaru*; Kannada—*Hesaru*; Malayalam—*Cherupayaru*.

History. Green gram *(P. aureus* Roxb.) is a native of India, and has been cultivated for at least 3,000 years.

This is grown as a pulse crop in Madhya Pradesh, Uttar Pradesh, Bihar, Rajasthan and Bengal. It is one of the very ancient legumes of India and is still an important crop.

Fig. 3.7. *Vigna radiata* (Linn.) Wilczek. (Green gram—**Moong**)—A high yielding variety.

Uses. The small oval seeds are highly nutritious, and the green pods are also eaten. The seeds are eaten as a *'dal'*. It makes a good source of protein. The plants are used as cattle feed.

It ranks high among the pulse crops in India. Tender pods are consumed as vegetable. Mung is parched and made into a porridge. Often fried and used as a snack. Sprouted seeds are eaten and sometimes seedlings are candied. Decoction of seeds is given in beri-beri as a diuretic. Seeds are also used for vertigo. Mungo extract is said to have curative properties in polyneuritis gallinarum. The plants are also used for hay.

Lentil

Lens culinaris Medic. Family—Leguminosae (Papilionaceae); Eng.—Lentil; Hindi—*Masur*; Bengali, Marathi and Gujarati—*Masur, masser, masuri*; Telugu—*Misurpappu, chirisanagalu*; Kannada—*Massur, chanangi*; Punjab—*Masur, malka, musri*; Assam—*Masurmoha*.

A twining herb. It is native of South West Asia, now cultivated mainly in Andhra Pradesh, Maharashtra, Madhya Pradesh, Uttar Pradesh, Rajasthan and Bengal. It is one of the most ancient of food plants. The lentil is frequently referred to in the Bible. The herb is slender, tufted, much-branched annual with tendrils. The pods are short and broad, with small lens-shaped seeds. It thrives best in black clayey soils.

Uses. The seeds are used as *'dal'*. The plants are also used for fodder.

The seeds are also used in soups. The young pods are eaten as vegetable. They are valued for high protein content. They may be employed as a source of commercial starch for use in textile and calico printing industries. Husk is used as feed for livestock.

Moth Bean

Vigna aconitifolia (Jacq.) Marechal syn. *Phaseolus aconitifolius* Jacq.; Family—Leguminosae (Papilionaceae); Eng.—Moth bean, dew bean; Hindi—*Moth*; Sanskrit—*Makushthaka*; Bengali—*Kheri*; Marathi—*Matki, mati*; Gujarati—*Mut, math*; Telugu—*Kuncumape-salu*; Tamil—*Tulukkapayir*; Kannada—*Madike*; Punjab—*Moth, bhioni*; Santal—*Birmung, moch*; Mundari—*Mugirambara*.

A herb; native of India. It is chiefly grown in Uttar Pradesh, Andhra Pradesh, Maharashtra, Punjab, Bihar and Bengal.

Uses. The seeds are used as *'dal'*. The famous *'dal-moth'* of Agra is prepared from its seeds. The seeds are a good source of protein. The plants are used as fodder.

The tender pods are consumed as a vegetable. The seeds are boiled or parched and mixed with condiments to eat. The seeds may be ground into flour and mixed with flours of other grains for making unleavened bread. This pulse is given as diet to cases of flatulence and fever. Plants also make an excellent hay.

Lima Bean

Phaseolus lunatus Linn.; Family—Leguminosae (Papilionaceae); Eng.—Lima bean, sieva bean; Hindi—*Lobiya*; Tamil—*Khasi kollu*; Punjab—*Lobia*; Marathi—*Daful*.

A herb. It is native of tropical America, now grown in Northern India.

Uses. The seeds are used as pulse. The young pods are used as vegetable.

The seeds are eaten fresh or after drying, also fried in oil. Dried beans are boiled or baked for table use. In USA, the beans are canned and also frozen. The beans are rich in minerals and proteins. Sprouted seeds are consumed boiled, used as a vegetable, or in stews. Also grown as a cover crop and green manure.

Kidney Bean

Phaseolus vulgaris Linn.; Family–Leguminosae (Papilionaceae/Fabaceae); Eng.—Kidney bean, french bean, garden bean, faraz bean, dwarf bean, haricot bean; Hindi—*Vilayati sem, bakla, lobiya, frash bean, rajmah* (seed).

Marathi—*Shravanghevda*; Gujarati—*Phanasi*; Telugu—*Barigalu*; Kannada—*Tingalavari*; Punjab—*Babri*.

They are natives of America; now cultivated throughout India. A fertile soil, rich in lime is necessary for a good yield.

They are low, erect or twining annuals with small white or coloured flowers, trifoliate leaves and slender pods.

Uses. The young pods, the unripe seeds and the dried ripe seeds are all used for human consumption. They are the good source of protein. The plants are used for fodder.

The beans are eaten fresh as a vegetable in the unshelled condition, or as a pulse after shelling. Discoloured pods are fed to live stock. Bean straw is also used as an animal feed. Dried beans *(Rajmah)* is used in the similar way as cowpea and gram. It is nutritious, comparing closely with meat as a source of protein. Green pod shells are used as a diuretic especially in kidney and heart troubles.

Cowpea

Vigna unguiculata (Linn.) Walp. syn. *V. sinensis* (Linn.) Savi ex Hassk; Family—Leguminosae (Papilionaceae/Fabaceae); Eng.—Cowpea; Hindi—*Lobia, riantshi, rawan, santa*; Sanskrit—*Nishapava, dirghabija, chavala*; Bengali—*Barbati, ramhikolai*; Gujarati—*Chorap, chola*; Marathi—

Chaoli; Telugu—*Bobberlu, alusendi, duntupesalu*; Tamil—*karamani*; Kannada—*Tadagunn, kursan-pyro, alasandi*; Malayalam—*Kottapayuru, vellapayaru*; Assam—*Urohi, mahor-pat, urhi-mah.*

It is cultivated throughout India. It is a bushy or trailing annual with cylindrical, pendant pods. It is a very old crop, probably native to Central Africa.

Uses. The young pods and seeds are used as vegetable. The dry seeds are used as pulse. The seeds are also fed to cattle and poultry.

The seeds are known as cowpeas. There are three types of cowpeas; common cowpea; catjang or Hindu cowpea; and Asparagus or yardlong beans. The plants are also valued as a cover or green-manure crop. Ripe seeds are boiled and eaten. The seeds are pounded into flour, used for sweetmeats, cakes, and porridge. Sprouted seeds are eaten. Leaves and young shoots of some varieties are eaten as a pot-herb, and also used in salads. Starch obtained from the seeds may be used for textile-sizing and as a thickening agent in calico-printing. Cowpea is considered antibilious and prescribed in hepatic troubles and jaundice. Seeds contain stigmasterol and may be used in place of soyabean in the preparation of steroidal hormones.

Broad Bean

Vicia faba—Linn.; Eng. Broad bean, Windsor bean; Hindi—*Bakla*; Family—Leguminosae (Papilionaceae/Fabaceae).

Kannada—*Kadu huralikayee*; Delhi—*Bakla sem*; Punjab—*Chastang, kabli bakla, mattz-rewari, raj-rawan*; Kashmir—*Katun*; N.W. Himalayas—*Chastang raiun*; Ladakh—*Nakshan*; Kumaon—*Bakla*; Mundari—*Hende matar.*

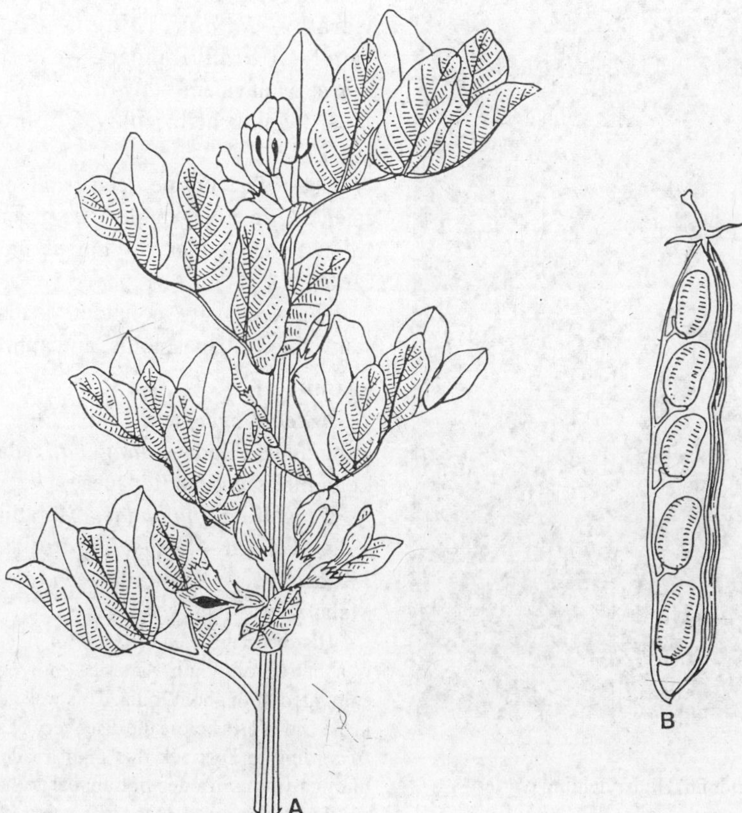

Fig. 3.8. *Vicia faba* (broad bean). A, twig with leaves and flowers; B, pod with edible seeds.

The plant is a strong erect annual, 2 to 4 feet in height, with flat pods and large seeds.

Uses. The seeds and young pods are eaten as vegetable. The broad bean was the only edible bean known in Europe before the time of Columbus. It is still an important crop in England. The plant is used as a fodder.

Ingestion of fresh uncooked or partially cooked beans may result in a condition called *Favism,* characterized by hemolytic anaemia, hemoglobinuria, jaundice, etc. Plant contains high amount of carotene in the stooling stage and also at the beginning of flowering. Decoction of leafy shoots is used as a diuretic. The seeds are aphrodisiac.

Jack Bean

Canavalia gladiata (Jacq.) DC.; Eng. Jack bean, Sword bean; Hindi—*Bara sem, makhan sem*; Family—Leguminosae (Papilionaceae).

Bengali—*Makhan shim*; Marathi—*Abai*; Tamil—*Segapu thambattai*; Telugu—*Yerra tamma;* Kannada—*Shembi avare, tumbekai.*

The plants are bushy annuals with long sword-shaped pods. Cultivated throughout India.

Uses. The pods are used as vegetable. The whole plant is used as green forage. This is also used as a green manure and a cover crop.

Fig. 3.9. Groundnut. High yielding variety.

Lablab

Dolichos lablab Linn; Eng. Lablab, Hyacinth bean; Hindi—*Sem*; Family—Leguminosae (Papilionaceae).

Bengali—*Shim*; Gujarati—*Val*; Marathi—*Pavta*; Telugu—*Chikkudu*; Tamil—*Avarai*; Kannada—*Chapparada avare*; Malayalam—*Avara.*

A twining herb; cultivated throughout India.

Uses. The unripe seeds and pods are eaten as green vegetable. The green plants are used as fodder for horses and other cattle.

The seeds are considered febrifuge, stomachic, antispasmodic, and aphrodisiac.

Peanuts

Arachis hypogaea Linn,: Eng. Peanut, groundnut, Hindi—*Mungphali;* Family—Leguminosae (Papilionaceae).

Bengali—*Chini badam*; Marathi—*Bhui mug*; Telugu—*Verusenagalu*; Tamil—*Verkadalai*; Kannada—*Nela-gadale*; Malayalam—*Nelakadala.*

History. The peanut (*Arachis hypogea* Linn.) is a native of South America, now generally cultivated throughout India. This was not known in the Old World before the discovery of America. According to Dymock this plant reached India through China. It does not appear to have been cultivated for more than 150 years. This was brought to western India from Africa.

It is native of Brazil. It is widely grown in South India, Maharashtra and Uttar Pradesh. North

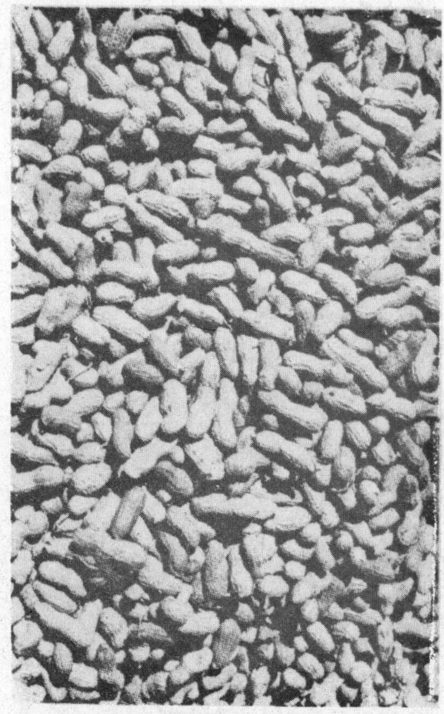

Fig. 3.10. Roasted pea nuts for sale.

Gujarat is famous for peanut cultivation. The plant is a bushy or creeping annual with the peculiar habit of ripening its fruit underground. A sandy soil is best for its cultivation. The soil must be friable so that the ripening fruit can be buried, and it must be well fertilized.

Uses. The peanuts (seeds) are used for roasting or salting and for the preparation of peanut butter. Peanuts are a very nutritious food. One lb. of peanuts yields 2700 cal. The filtered refined oil is used for cooking and in making margarine. Peanut oil is an important food oil. The oilcake is used as fodder. The protein in peanuts is used in the manufacture of ardil, a synthetic fibre. The vegetable *ghee* is made from the peanut oil after hydrogenation.

The kernels are also used in various foods and confectionery. They are ground and made into peanut butter. Peanut flour is prepared by grinding the finest grades of peanut cake; it is used for supplementing the white flour. Cake is used as feed for cattle and other farm animals; also used as manure. Cake has high nutritive value. Seedcoats are mixed with groundnut husk and the product is called groundnut bran. Some commercial products are groundnut milk, peanut ice-cream and peanut massage oil for infantile paralysis. Hulls are used as filler for fertilizers, or ground into meal for insulation blocks, floor sweeping compounds, bedding the stables, etc. Peanut oil also finds some use as a lubricant, and blends with mineral oil have been developed.

NUTS

A nut is a one-celled, one-seeded dry fruit with a hard pericarp. Only a few nuts of commerce fulfil this description, *e.g.,* chestnut. The others included among nuts may be peanuts (legumes), almonds (drupe), coconuts (drupe), cashewnuts, walnuts, pistachio nuts, etc. All these 'so-called nuts' will be considered here. Nuts make a valuable food material. Because of their low water content they are a concentrated food. The food value of nuts is due to a high protein and fat content. The nuts also contain starch and sugar. They are also rich in mineral elements. The nuts may be eaten raw or cooked, or in the form of nut butters or pastes. Nuts are marketed as unshelled or shelled. Generally the nuts are kept in three categories (i) nuts with a high protein content, (ii) nuts with a high carbohydrate content and (iii) nuts with a high fat content.

NUTS WITH HIGH PROTEIN CONTENT

Almonds

Prunus amygdalus Batsch; Eng. Almond; Hindi—*Badam*; Family—Rosaceae.

Bengali, Marathi and Punjabi—*Badam*; Tamil—*Vadamkottai*; Telugu—*Badam vittulu*; Kannada—*Badami*; Malayalam—*Vatam-kotta*.

The almond is a native of the eastern Mediterranean region and is grown throughout Southern Europe, California and South Africa. In India, it is cultivated chiefly in the Punjab and Kashmir.

The almonds are obtained from a small tree. The almond fruit is, an inedible drupe, with a tough fibrous rind surrounding the stone (shell) and seed (nut). There are many varieties, some with paper shells (*kagzi badam*) and some with hard shells.

Uses. The seeds are delicious when eaten green. Usually they are roasted and salted or made into a paste to be used in making bread or cake. Almond extract is also prepared. The seeds are demulcent, stimulant and nervine tonic. Seeds of almond *(badam)* are also used in sweetmeat and the paste of it mixed with is delicious and tasteful.

Almond kernels are extensively used in confectionery and for almond milk. Almond oil extracted from sweet as well as bitter almonds is extensively used in confectionery and pharmaceutical and cosmetic preparations. Almond flour and almond butter are free from starch and used in foods for diabetics, and in cosmetics. Almond kernels are also used as lithontriptic and diuretic, and their poultice is useful for irritable sores and skin eruptions. Kernels form a part of the diet also for patients with peptic ulcers. Unripe fruits are used in astringent application for gums and mouth. Chief protein in kernels is globulin. Almond shells are used for fur cleaning and metal polishing. Burnt shells are powdered for cleaning teeth. Gum, exuded from trunk and branches is employed in place of tragacanth. Oil cake of bitter almonds yields an essential oil called bitter almond oil.

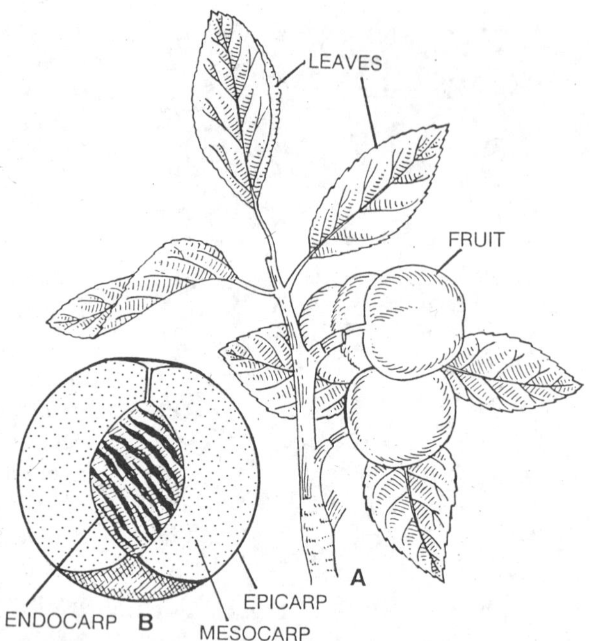

Fig. 3.11. *Prunus amygdalus* (almond). A, twig with fruits; B, sectional view (longitudinal section) with seed. The ternel of seed is edible.

Green Almond

Pistacia vera Linn., Family—Anacardiaceae; Eng.—Green almond; Hindi—*Pista*.

A small tree; native of West Asia, has been cultivated in the mediterranean countries for nearly 4,000 years. It is chiefly grown in Iran, Afghanistan, the Southern United States and California. It is also cultivated in North India. It is wildly grown in the forest of West Orissa. The fruit is a drupe. The seed contains two large green cotyledons with a reddish covering.

Uses. They are highly prized for their green colour and resinous flavour. They are much used in mixed nuts and as a flavouring material in ice-creams, candy and sweetmeats.

The seeds yield a low-melting fatty oil used to a small extent in confectionery as spice oil and in medicine. Galls on the leaves (Bokhara galls) are used for dyeing and tanning.

Beech Nuts

Fagus sylvatica; Eng. European beech; Family—Fagaceae.

A tall tree, found in Kullu and the Nilgiris. The nuts are small, traingular and very sweet.

Uses. The nuts are eaten and used for the edible oil of beech.

Raw nuts are poisonous due to the presence of a saponin, the toxin could however, be removed by baking and cooking. The fatty oil obtained from the nuts is used for the culinary purposes and soap making, also as an illuminant. Press cake from decorticated nuts is used as feed for cattle pigs, and poultry.

Beechwood is used in Europe and America for furniture, tools, planes, keys and cogs in machinery, shoe lasts, toys, brush backs, and saddle trees; used also in paper and table fibre industries. Yields beechwood Creosote, employed externally as an analgesic and antiseptic. Beech nuts are edible and sweet.

NUTS WITH HIGH CARBOHYDRATE CONTENT

Chest Nuts

Castanea sativa Mill.; Family—Fagaceae; Eng.—European chestnut.

A tall tree; extensively cultivated in Southern Europe. In India it is grown in Darjeeling, the Khasia hills, Punjab, Himachal Pradesh and Kashmir. They are commonly grown on dry hillsides.

Uses. The nuts are eaten raw or are roasted, boiled, or used for stuffing or flour. The nuts are often cooked like potatoes.

The leaves are tonic and astringent, useful in paroxysmal coughs and other irritable condition of the respiratory organs. Wood is durable, used for constructional work, furniture and cask staves. Wood pulp of some foreign species is suitable for paper and rayon manufacture. It yields, 8–13 per cent tannin.

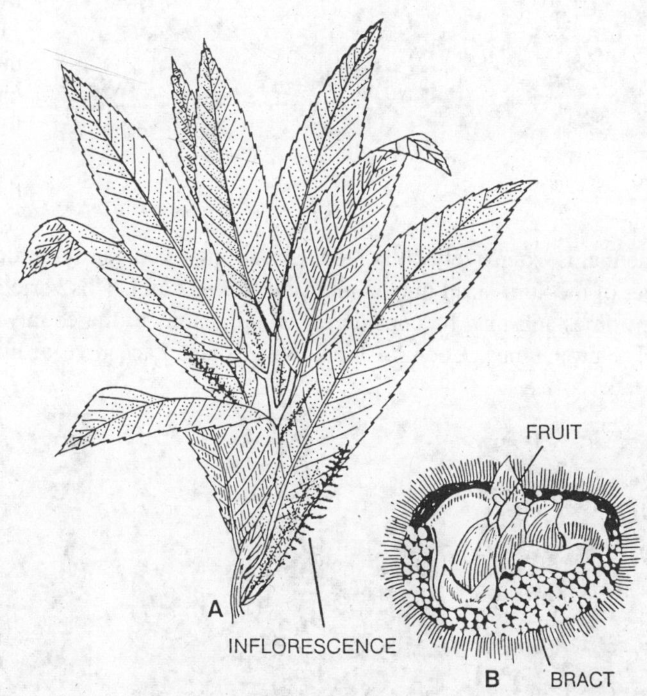

Fig. 3.12. *Castanea sativa* (chest nut). A, twig with inflorescence; B, ripe fruit of chest nut, showing three nuts inside.

NUTS WITH HIGH FAT CONTENT

Coconut

Cocos nucifera Linn.; Family—Palmae/Palmacea/Arecaceae; Eng.—Coconut; Hindi—*Nariyal*; Sanskrit—*Narikela*; Bengali—*Dab, narikel*; Marathi—*Narel, naral*; Telugu—*Kobbarichettu, narikelamu, tenkaya*; Tamil—*Tennaimaram, tenaki*; Kannada—*Tengu*; Malayalam—*Thenna, thenga, marikelam*.

History. De Candolle argues in favour of an American as well as those of an Asiatic origin for coconut tree *(Cocos nucifera* Linn.*)* and concludes that it most probably belongs to the Indian Archipelago. According to him its introduction into India, Sri Lanka and China does not date further back than three thousand years, but the transport by sea to the coasts of America and Africa took place perhaps in a more remote epoch.

A tall palm; cultivated chiefly in Kerala, Tamil Nadu and Karnataka. The coconut has been called a *Kalpavriksha* because it provides a large range of useful products, among which ripe coconut, copra, coconut oil and coir are important. The value of this tree has been recognized

from very ancient times. Indeed, the region from South Kanara to Cape Comorin is known as *'Kerala'*, meaning the land of the coconut. In Malayalam or Sanskrit *"Keralam"* means land of coconut. The Kerala state in Malayalam is known as *"Keralam"*. In these parts, the coconut plays a vital part in the life and economy of the people. The coconut is mainly grown for nuts which it yields almost at monthly interval all through its life of about eighty years.

Varieties and cultivation. Of the varieties of coconuts grown in India, the West Coast variety, is the one that is extensively cultivated. It is a long-lived, hardy, multipurpose palm, yielding a large number of medium-sized nuts in almost every leaf axil. This variety has been cultivated from very ancient times and may be considered as indigenous to the county. Among other-varieties are the Dwarf, New Guinea, Cochin-China, Java, Siam, Laccadive ordinary and Laccadive small. The

Fig. 3.14. Germinating coconut.
I, II and III stages

Fig. 3.15. Germinating coconut. IV and V stages.

Fig. 3.16. Germinating coconut. Sixth stage

New Guinea variety yields about 65 nuts a year, the Cochin-China about 90, the Java about 95, the Laccadive ordinary about 127 and the Laccadive small 160.

The primary requirements for the successful cultivation of coconut are a tropical climate, abundant rainfall and a well-drained soil. Seashores, river banks and hill-slopes afford ideal situations. In India, the principal areas of cultivation are Kerala, Karnataka, Tamil Nadu, Andhra Pradesh, Maharashtra, West Bengal and Orissa; 85 per cent of the total output comes from Kerala State.

Uses. The tender nuts are in demand for their liquid which provides a very refreshing and delicious drink. It is said to have laxative and diuretic properties. The soft kernel inside the tender nut is also scraped and eaten. The mature nuts are used mainly to make copra (dried meat or kernel) and for edible purposes as fresh kernel in food preparations.

The copra is crushed to obtain coconut oil which is in great demand for edible

Fig. 3.17. *Cocos nucifera.* VHC-1 coconut hybrid.

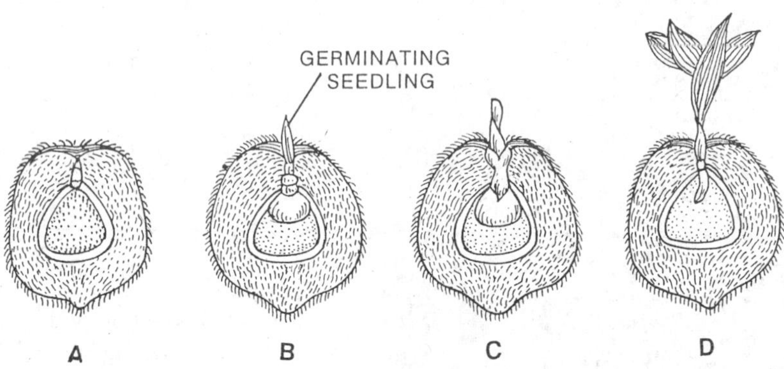

Fig. 3.18. *Cocos nucifera* (coconut). A–D, successive stages (line diagram) showing germination of coconut (seedling stage).

purposes. It is used as an adjunct for milk-fat, as a filler and thickener for milk and as an ingredient of artificial milk for infants. The cake left after the extraction of oil is largely used to feed cattle and poultry. It is occasionally used as food by poor people. The coconut cake helps to increase the milk content when fed to cows and the butter made from such milk has a firmer texture and better flavour than when other cakes are used.

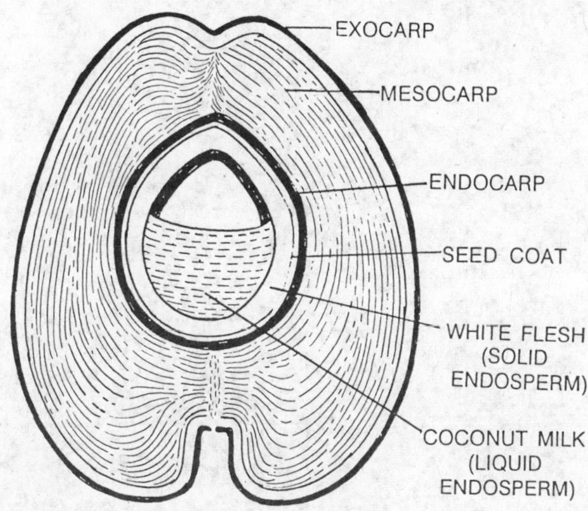

Fig. 3.19. *Cocos nucifera* (coconut). L.S. of fruit with seed.

The coconut inflorescences (spadices) are tapped to extract the juice which is converted into sweet or fermented toddy. Vinegar, jaggery, sugar or arrack (alcohol) are also made from the juice.

Shell of coconut is used for handicrafts; mesocarp (fibre) is used for coir manufacturing. The Kerala State is second to Philippines in the production of coir materials. The dry woven leaves are used for thatching the house and for temporary sheds and pandals in Southern States especially in Kerala.

Cashewnut

Anacardium occidentale Linn.; Eng. Cashew; Hindi—*Kaju* of family Anacardiaceae.

Sanskrit—*Kajutaka*; Bengali—*Hijli-badam*; Marathi—*Kaju*; Telugu—*Jidi-mamidi, muntha-mamidi*; Tamil—*Mindiri*; Kannada—*Geru bija*; Malayalam—*Andi paruppu.*

History. The cashewnut (*Anacardium occidentale*) is native of West Indies. It has been originally introduced into India from South America. It is probable that the Portuguese on its introduction to the west coast of India called it *kaju* as a rendering of the Brazilian name *acajau.*

The cashew tree is an evergreen, of spreading, somewhat straggling habit, low branched and reaching a height of 35 to 40 feet. The wood of the tree exudes a yellow gum. The most conspicuous product of the tree is the 'false fruit', the swollen peduncle or hypocarp commonly known as the 'cashew apple'. This apple is actually the receptacle for the true fruit, the cashewnut, which is lightly attached to its apex and which attains full size prior to the enlargement of the receptacle. The shell of the nut closely resembles a miniature boxing-glove. The shell contains a caustic oil which acts as a powerful vesicant on skin and prevents the retailing of unshelled cashewnuts. Entirely free from the caustic element is the curved, or somewhat kidney-shaped, white kernel which remains protected by a thin, easily removed, testa, reddish brown on the outside and crimson within. The kernel is the cashewnut of the commerce and is considered one of the most flavourful of nuts.

The cashew industry is almost completely an export oriented one. The export earnings of the industry are enormous. Apart from earning valuable foreign exchange it also provides employment to about 2 lakh workers mostly women. The cashew industry thus plays a vitral role in India's economy particularly of the Southern States where the industry is concentrated.

Distribution. The cashew tree is native to north-eastern Brazil. It has been thoroughly dispersed throughout the tropical lowlands of northern South America, Central America, Mexico and the West Indies. In the sixteenth century, Portuguese traders introduced the tree into India,

Fig. 3.20. Cashewnut. Cashew apple and nuts.

where it was planted along the shore to bind to soil and soon spread as an escape throughout India, Ceylon, Malaya and the Andaman Islands.

The horticulturists of India are now strongly urging the selection of superior, high-yielding types and their propagation by inarching, or approach grafting and air-layering.

Commercial cultivation. The cashew ranks second only to the almond among the nine tree nuts that are of importance in world trade. India, the world's chief source of cashews, produces, nearly 80,000 tons of the nuts annually. In Kerala alone there are more than 200 cashewnut processing factories. Seventy-five per cent of Indian production is exported and 25% consumed domestically. The ranking cashew states of India at present are Kerala, Andhra, Tamil Nadu, Karnataka and Maharashtra. A government programme is aimed at establishing the cashew industry in every state, each striving to meet a high quota with the aid of subsidies and cooperative organizations. Cashew plantations are intended to cover 50,000 acres on the west coast and 60,000 on the east coast.

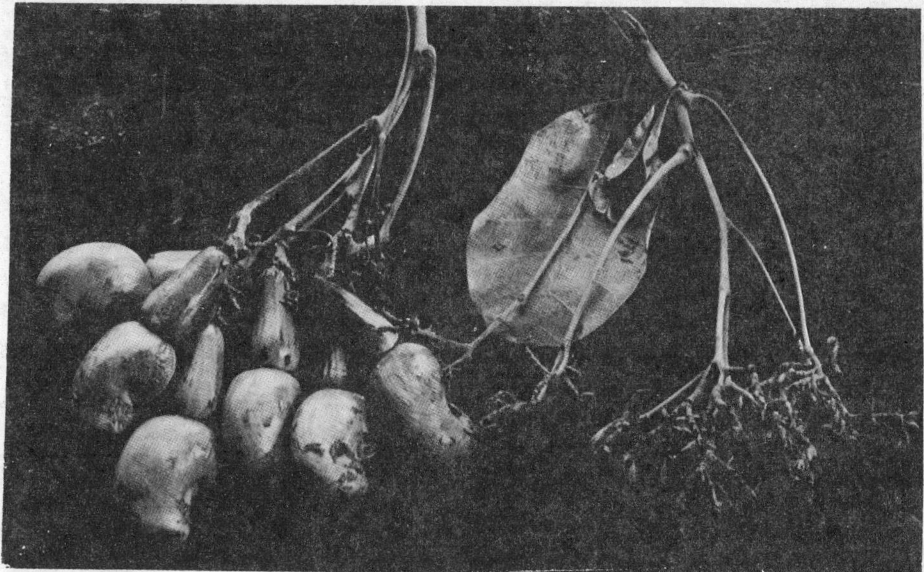

Fig. 3.21. Cashewnut. Inflorescence, and nuts with thalamus.

Yield and harvesting. Cashew trees flower 2–2½ months before the maturing of the fruit crop. In India, the cashew crop is harvested in April, with a few trees bearing a light crop in October and November. The yield varies from 10 lbs. to 200 lbs. of nuts per tree. An average tree is said to bear 7 lbs. of nuts when 3 years old, increasing to 70 lbs. in the 15th year.

Usually the cashew apples with nuts attached fall from the trees when ripe and are gathered from the ground. The nuts are then removed by the hand and the 'apples' discarded. After collection, the nuts are allowed to dry for two days in the sun to reduce moisture content from 6 to 7%. They are then stored in large sacks for use throughout the year. Prior to processing, they are passed through rotary cleaners to remove foreign particles and then are dipped in a water tank, and stored in moist heaps for 12 hours to make the shells brittle.

Processing. About 5% of the cashewnuts produced in India, especially in South Kanara, Tanjore and South Arcot district of Tamil Nadu State, are simply dried and stored after harvest and then dried in the sun for 2 or 3 days and shelled without roasting. This results in kernels

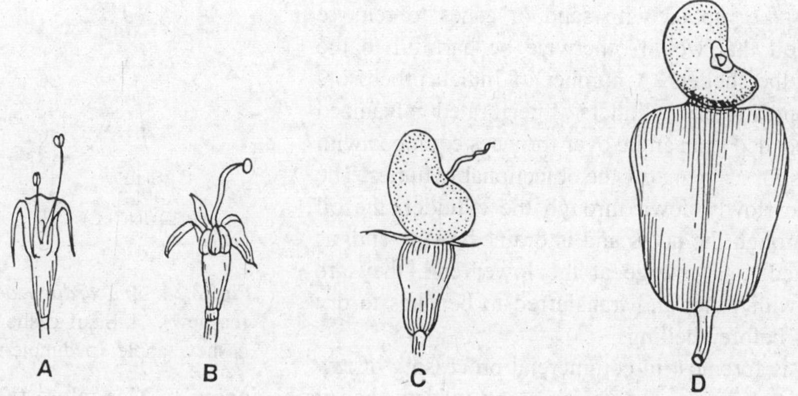

Fig. 3.22. *Anacardium occidentale* (cashew). A–D, successive stages—development of cashew.

Fig. 3.23. *Anacardium occidentale.* Cashewnut **(Kaju)**—The nut makes a valuable food whereas the cashew apple is famous for a unique liquor obtained from the exclusive distillation of cashew apple juice.

of inferior quality which can only be used locally. Generally, the caustic oil in the shell is dispelled by heat in some form before the nuts are opened. The nuts are roasted in shallow pans over open fire. However, all open fire roasting is intensely disagreeable due to spurting oil and the fumes which are highly irritating to the eyes, nose, throat and skin. After roasting, the nuts must be rubbed with sand or ashes to remove residual oil that would otherwise be harmful to the hands of the shellers. A number of Indian processors have adopted rotary cylinders of perforated galvanized iron suspended at an angle over furnaces equipped with chimneys for carrying off the objectionable fumes. The nuts slide slowly down through the cylinder, the oil passes through the holes and is drained away, and, as the roasted nuts emerge at the lower end, they are sprayed with water and transferred to benches to dry and cool before shelling.

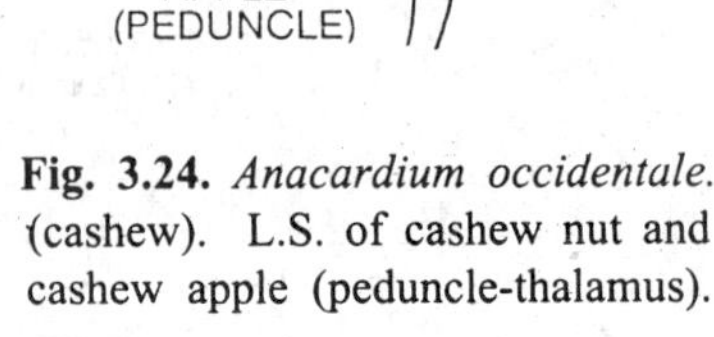

Fig. 3.24. *Anacardium occidentale.* (cashew). L.S. of cashew nut and cashew apple (peduncle-thalamus).

India is foremost in commercial processing of raw cashewnuts. Cashew processing as an industry had its beginning in Mangalore (Karnataka) but soon it moved to Quilon in Kerala, where a majority of processing units were located until recently.

Cashewnuts - 210 count wholes

Cashewnuts - 500 count wholes

Cashewnuts - Splits

Fig. 3.25. Grading of cashewnuts. Cashewnut is the wonder nut that is at once a food and a snack the world over.

In recent years, a large number of new processing units have been established in other parts of the country; particularly in Kanyakumari district of Tamil Nadu. It is estimated that there are near-about 500 processing units in India employing about 2,00,000 workers.

Shelling. Shelling of the nuts is almost entirely a hand operation. In India, it is usually done on stones embedded in the ground and by squatting women using wooden mallets, and there is considerable breakage of kernels. It is customary to dip the hands frequently in lime ash, linseed oil or castor oil to protect them from any trace of shell oil remaining.

Roasting of raw kernels. When raw kernels are obtained by the above-described methods, they may be roasted in olive oil or peanut oil in 5 lb. lots placed in metal trays over a coal furnace. The trays are taken out 3 or 4 times at 5-minute intervals, shaken and replaced until the roasting is half completed. The kernels are then mixed in a bucket with a small amount of liquid paraffin, replaced on the tray and given another 5-minute roasting, sprinkled with a solution of salt, gum acacia and water, and finally reheated to remove excess moisture.

Husking, grading and packing. After roasting, the kernels must be dehusked, that is, testa is removed by hand. The cashewnuts are graded into 25 clearly differentiated grades according to size, colour and other characteristics—Wholes, Splits, Brokens, Butts, etc. For export, the kernels are packed in 25 lb. tins lined with oiled paper, or 112 lb capacity wooden boxes lined with tin. Distributors, after processing and seasoning of kernels to suit consumer tastes, package for retailing in small cans, cellophane bags.

CASHEWNUT GRADES

There are 25 grades of cashewnuts. They are as follows:

1. 180 Count Wholes;
2. 210 Count Wholes;
3. 240 Count Wholes;
4. 280 Count Wholes;
5. 320 Count Wholes;
6. 400 Count Wholes;
7. 450 Count Wholes;
8. 500 Count Wholes;
9. Scorched Wholes (S.W.)
10. Scorched Wholes Seconds (S.W.S).
11. Dessert Wholes (D.W.)
12. Butts (B)
13. Splits (S)
14. Large White Pieces (L.W.P.)
15. Small White Pieces (S.W.P.)
16. Baby Bits (B.B)
17. Scorched Butts (S.B.)
18. Scorched Splits (S.S.)
19. Scorched Pieces (S.P.)
20. Scorched Small Pieces (S.S.P.)
21. Scorched Pieces Seconds (S.P.S.)
22. Dessert Pieces (D.P.)
23. Dessert Small Pieces (D.S.P.)
24. Dessert Butts (D.B.)
25. Dessert Splits (D.S.)

At present the best grade of cashews is 180 Count Wholes.

Quality control. Quality control is of utmost importance in the manufacture and export of cashew kernels. A compulsory quality control scheme for cashew kernels was introduced by the Cashew Export Promotion Council first in 1963. This was subsequently taken over by the Export Inspection Council of India set up by the Government. Under the scheme cashew kernels for export are subjected to stringent quality tests and only approved consignments are allowed to be exported.

Exports. India is the world's biggest exporter of cashew kernels. Commercial exports of cashew kernels from India started some 75 years ago. From a small quantity less than 50 m.t. in 1920, exports rose at a steady pace in subsequent years. Presently the export averages between 50,000 and 60,000 tonnes per annum earning over Rs. 100 crores. India exported cashew kernels worth Rs. 110 crores in 1977 and worth Rs. 147.61 crores in 1978. Indian cashews are exported to over 50 countries the world over. U.S.A. and Russia are the two major buyers of Indian cashews. The other important markets are Japan, Canada, Australia, Netherland, Hong Kong, U.K., Switzerland, German Fed. Rep., Kuwait, Singapore, France, etc.

Uses. Cashew kernels, sugared or salted, are consumed primarily as tablenuts and some are used in bakery goods and confectionery. Cashew contains proteins, vitamins, calcium, iron, etc.

Anacardic oil derived from cashew kernels is light-yellow and similar to almond oil. It is used for hardening the chocolate and it is given medicinally as an antidote for irritant poisons.

Cashewnut shell liquid (CNSL), a by-product of cashew industry, is extracted during the roasting of raw cashews in the course of processing. This liquid is used in the manufacture of brakes, linings, paints, foundry, electrical insulators, resins, etc. Annual export of CNSL is about 5000 tonnes.

Cashew apple, which is not a fruit, but thickened peduncle is used for jams and for preparation of wines and beverages.

Cashew shells yield a drying oil used as water proofing agent, preservative in painting of boats and fishing-nets, and light wood work. Oil has high heat resistance and is an excellent lubricant in magnets armatures in aeroplanes; also used in varnishes, inks, termite proofing of timber, and insulation coatings.

Tree produces a gum, which is used in varnishes and as protection for books, and wood work against insects. Peelings obtained during the preparation of kernels may be utilized as a wholesome poultry feed. Milky sap of the bark turns black on exposure to air and is used as an indelible ink for marking linen. Wood is used for packing cases and boat-building.

Pecan

Carya illinoensis (Wang.) Koch.; Family—Juglandaceae; Eng.—Pecan.

A tree; native of the South-eastern United States and Mexico. In India, it is grown in Uttar Pradesh, Himachal Pradesh and Punjab.

Uses. Pecans have a higher fat content than any other vegetable product, over 70 per cent. They are used as dessert nuts and in candy, ice creams, cakes, etc. By-products include Pecan oil and a tannin obtained from the shells.

English Walnut

Juglans regia Linn; Eng. walnut, Persian walnut; Hindi—*Akhrot*; Family Juglandaceae.

A large tree; native of Iran. In India, it is grown in Kashmir, Himachal Pradesh, the Khasia hills and in the hills of Punjab and Uttar Pradesh. The beautiful trees are usually planted in rows. The outer limbs produce perfect nuts. The edible kernels are easily freed from the pericarps. The characteristic furrowed kernels are the cotyledons of the seeds, no endosperm being present.

Uses. The kernels are eaten raw. Walnuts yield an excellent oil for table use. They retain their flavour when cooked and have a food value four times greater than meat. They are used in the preparation of candy and ice creams. The oil cake is a good feed for livestock.

The young fruits are pickled, a rich source of ascorbic acid. Mature kernels yield a fatty oil, walnut oil, used for edible purposes, small quantities for artist's oil colours, printing inks, varnishes, and soap-making. Green hulls are used for dyeing and tanning. The leaves are astringent, tonic and anthelmintic. They yield an essential oil.

Pine Nut

Pinus gerardiana Wall.; Family—Pinaceae; Eng.—Pine nut, Edible pine; Hindi—*Chilgoza*.

A tree; in India, it is found in North-Western Himalayas. The cones are roasted so that the scales fall apart and the seeds are separated.

Uses. The seeds are edible. The nut has a rich delicious flavour. They are tasty and nutritious. They are either eaten by natives or by rich people in winter season. The seeds contain a fatty oil upto 50 per cent.

Turkish Hazlenut

Corvlus colurna Linn.; Family—Betulaceae; Eng.—Turkish hazlenut; Hindi—*Bhutia badam*, *urni*; Punjab—*Thangi, urni*; Kashmir—*Virin*; Kumaon—*Kapasi*, *bhotia badam*.

A tree, native of Europe. In India, it is found commonly in Kashmir and Kumaon. The nuts (seeds) are edible.

Queensland Nut

Macadamia ternifolia F. Muell.; Eng. Queensland nut; Family—Proteaceae.

A tree, the native of Australia. The nuts (seeds) are edible. The kernels possess a sweet flavour and are rich in oil.

The kernels are deliciously sweet and pleasantly crisp resembling Hazel nuts and almonds in flavour and consistency; eaten raw or roasted; also used in confectionery. It makes a good source of vitamin B_1. Kernels yield an oil resembling Olive oil; may be used as salad oil and also for high grade soaps and medicinal purposes.

Wood is suitable for turning and for cabinet-work and veneers.

4

Vegetables

The term vegetable is usually applied to edible plants which store up reserve food in roots, stems, leaves and fruits and which are eaten cooked, or raw as salad. The vegetables rank next to cereals as sources of carbohydrate food. The nutritive value of vegetables is tremendous, because of the presence of indispensable mineral salts and vitamins. India grows a large variety of vegetables belonging to the tropical, sub-tropical and temperate groups. However their present availability is only 1.3 ounces per adult per day. On the other hand a balanced diet requires ten ounces of vegetables per adult per day. A good number of the vegetables in India are introductions from foreign countries. For convenience the vegetables may be classified in three categories (*i*) underground vegetables, (*ii*) herbage vegetables, and (*iii*) fruit vegetables.

Fig. 4.1. Vegetables—Eat green and fresh vegetables. Each person requires 230–300 gm of vegetables per day out of which 85 gm consist of root vegetables and 110 gm leafy vegetables.

UNDERGROUND VEGETABLES

In these vegetables the food is stored in underground parts. The storage organs may be true roots or modified stems, such as rootstock, tubers, corms and bulbs. From earliest time roots and tubers have furnished food for man. The underground vegetables may be classified in two groups—(*i*) roots and (*ii*) underground stems.

ROOT VEGETABLES

Sweet Potato

Ipomoea batatas (Linn.) Poir.; Eng. Sweet potato; Hindi—*Shakarkand, Mitha alu;* Family—Convolvulaceae.

Bengali—*Lal alu, ranga alu*; Marathi—*Ratalu*; Gujarati—*Kanangi, sakaria*; Telugu—*Chelagada*; Tamil—*Sakkareivelleiki-langu;* Malayalam—*Chakarakilangu*; Kannada—*Genasu*; Oriya—*Kanda*; Punjab—*Shakarkandi*; Santal—*Sakarkenda.*

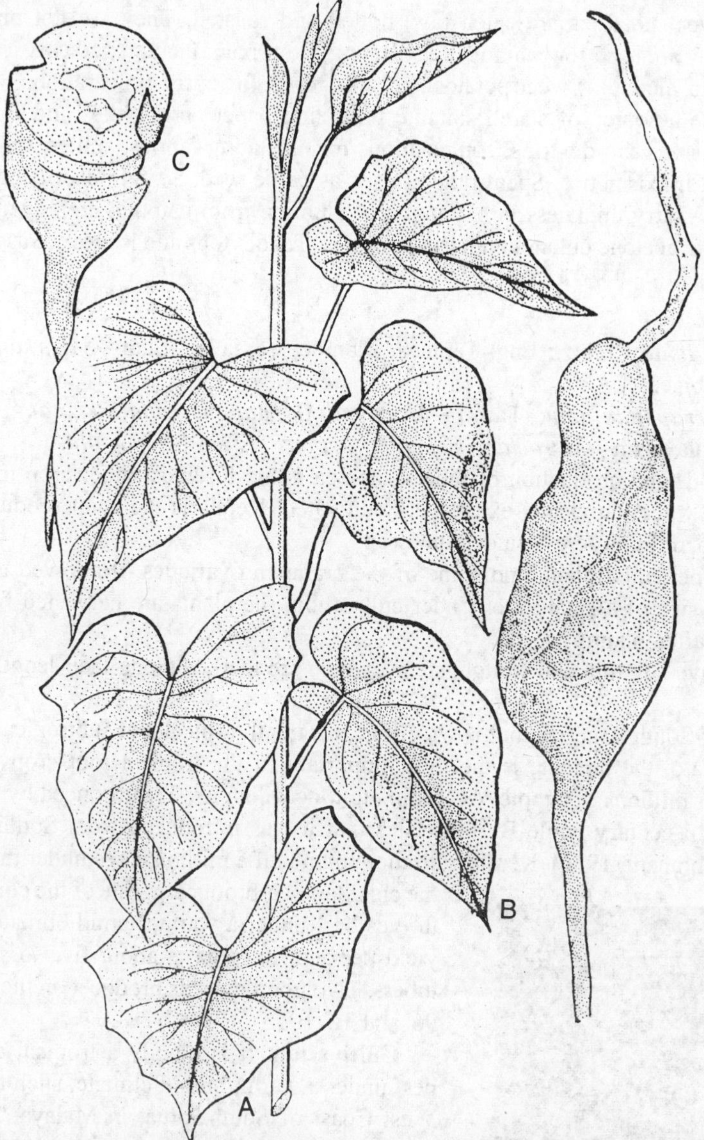

Fig. 4.2. *Ipomoea batatas* **(Shakarkand)**—The sweet potatoes are eaten raw, boiled and roasted. A, a branch; B tuberous edible root; C, T. S. of tuberous root.

The sweet potato is a native of tropical America. Now it is widespread in all tropics and some parts of the temperate zone, and found abundantly in the South Seas, China, Japan, Indonesia and India. It is cultivated throughout India.

The plant is a twining, trailing perennial vine with adventitious roots that end in swollen tubers (tuberous roots). The sweet potato requires a sandy soil and a warm moist climate. The plants are grown as annuals and propagated vegetatively by using vine cuttings.

People in various parts of India seem to have a traditional preference for one (white) or the other (red) colour of the skin of the tuber (tuberous root). In the Punjab white sweet potato is hardly grown, while in Karnataka it is the white varieties which are preferred. In other parts of India, both red and white varieties are popular. 'Pusa Suffaid' and 'Pusa Sunehri' are supposed to be best varieties in yield, quality and nutritive value.

Uses. The sweet potatoes are eaten raw, boiled and roasted. They are not only a common vegetable, but they are used for canning, dehydrating, flour manufacture, and as a source of starch, glucose syrup and alcohol. Sweet potatoes are also used for curry preparation.

The tubers are a source of starch suitable for sizing paper and textiles and for laundry use; also used in adhesives and dextrins, compositions for insulating fabrics, and coating formulations for dry cells, and in cosmetics. Spent pulp is used as cattle feed. Sweet potato flour is employed as a coagulant in slurry thickness in extraction of alumina from bauxite; employed for industrial alcohol, lactic acid, acetone butanol, vinegar and yeast. Tender tops and leaves are used as vegetable or salad.

Tapioca

Manihot esculenta Crantz.; Eng. Tapioca, Manioc, Cassava; Hindi—*Sakarkanda, Maravuli*; Family—Euphorbiaceae.

Telugu—*Karrapendalamu*; Tamil—*Maravalli kizhangu, ezhali kizhangu*; Kannada—*Maragenasu*; Malayalam—*Marachini kizhangu, kappa*; Assam—*Simal alu.*

In acreage and total production of tubers, tapioca is the foremost tuber crop in India. Kerala is the only place where tapioca is extensively cultivated. Kerala is the chief producing area with more than half a million acres under the crop.

Tapioca is a perennial plant, and some of the branching varieties, if allowed to grow indefinitely, attain the size of a small tree. Under cultivation, the plants are harvested from the eighth to tenth month after planting.

The plants have tall, thin and straight stems, which are marked along their length by numerous leaf scars.

It is native of South America, but is widely grown in all tropical and subtropical regions. Next to the sweet potato, the tapioca, is the most important of the tropical root crops and furnishes the basic food for millions of people. According to MacMillan it was introduced by the Portuguese in the seventeenth century while Burkill states that it was introduced from South America into India in 1840 (Abraham, 1956). Kerala with more than half a million acres under the crop remains the chief tapioca-producing state of the country. The crop thrives best under a warm, humid climate. The average yield per acre in India is about five to six tons of raw tubers. Two good hybrids produced in the country are H. 96 and H. 105.

Fig. 4.3. *Manihot esculenta:* Tapioca **(Sakarkanda)**—The tubers (tuberous roots) are exploited commercially to obtain starch, sago, semolina and flour.

Cultivation. It is mainly a tropical crop. It thrives best under a warm, humid climate, such as exists on the West Coast of South India, in Malaya, Java and other places. The crop does not stand much cold, and even a light frost kills all the leaves. The crop can be grown up to an elevation of 3,000 feet, in the tropical belt.

The crop is usually harvested from the eighth to tenth month after planting by pulling out the plants with hands. The tubers are then severed from the stem with a sharp knife, collected in baskets and marketed.

Preservation. As the fresh tubers do not keep in good condition for more than a week, they are usually preserved as dry chips. The tubers are cut into thin round slices, and dried in the sun after parboiling (by plunging the slices in boiling water for about 10 minutes).

Uses. Tapioca forms the chief item of food for a considerable proportion of the population on the west coast of South India. It is commonly used as fresh tubers, boiled and prepared in the same way as potatoes. The tapioca tubers are exploited commercially to obtain starch, sago, semolina and flour.

Tapioca can be considered as the second staple food of the Kerala State. Deep fried chips are very delicious and tasty. It is prepared like potato chips. Out of tapioca dried flour a delicious preparation known as "puttu" is made and used as a common breakfast in Kerala and some parts of Tamil Nadu. Dried flour is also used for making "pappads," wafers, glucose and starch.

The starch made from tubers is employed in sizing and finishing of textiles, laundering, paper-making, and in adhesives, and cosmetics, puddings, biscuits and confectionery. Glue made from the starch is used for preparation of syrup, acetone, etc. Plant is also a source of alcoholic beverages.

Yams

Dioscorea alata Linn.; Eng. White yam, Greater yam; Hindi—*Khamalu, Chuprialu*; ***D. bulbifera*** Linn; Eng. Potato yam: Hindi—*Gaithi, ratalu;* ***D. deltoidea*** Wall,; Hindi—*Kirta; D. esculenta* (Lour.) Burkill; Eng. Lesser yam; Hindi—*Susnialu;* ***D. pentaphylla*** Linn.; Hindi—*Kanta alu*; Family—Dioscoreaceae.

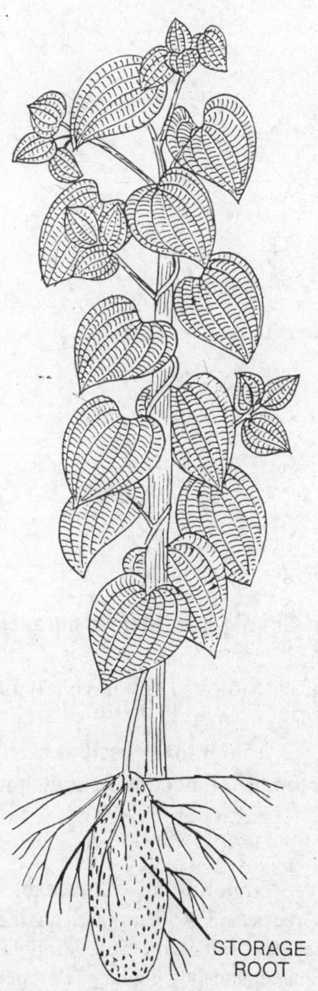

Fig. 4.4. *Dioscorea alata* (Yam). A plant with edible storage root.

Fig. 4.5. *Dioscorea esculenta*. The tuberous roots are starchy and free from dioscorine. They have a sweetish taste; in flavour and mealiness they closely resemble potatoes.

The true yams belong to the genus *Dioscorea*. The most commonly cultivated species is *Dioscorea alata* Linn. They are climbing shrubs. Usually cultivated through out India. They possess large storage roots. They require deep soil, but are quite drought resistant.

Uses. The tuberous roots are used as vegetable. They make the cheap food of millions of people all over the world. They are also used to feed livestock.

Salsify

Tragopogon porrifolium Linn.; Eng. Salsify Vegetable—oyster; Family—Compositae (Asteraceae).

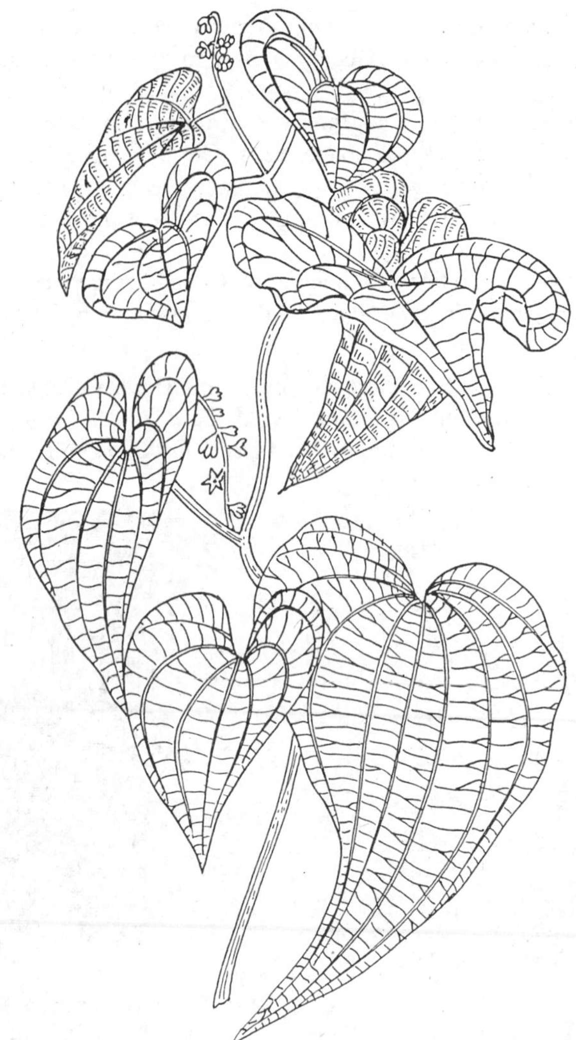

Fig. 4.6. *Dioscorea alata* (Yam). A flowering twig with large cordate leaves.

A herb. It is native of South Europe. In India, it is found in Shimla and Maharashtra. It is a hardy biennial with a large fleshy taproot.

Uses. The roots are eaten as vegetable. The roots are cooked or eaten raw. They have a typical flavour of oysters.

The shoots of cultivated plants are also edible. Tender shoots, 12–15 cm long, used as salsify. The roots are antibiotics, deobstruent, diuretic, pectoral, and expectorant. They contain aoutchouc, used for making chewing gum.

Parsnip

Pastinaca sativa Linn.; Eng. Parsnip; Hindi—*Gujur;* Family—Umbelliferae (Apiaceae).

History. The parsnip (*Pastinaca sativa* Linn.) was used by the Greeks and Romans and has been cultivated in Europe ever since. It is a native of Europe. It reached the West Indies in 1564, Virginia in 1609, and was grown by the American Indians in the beginning of the eighteenth century. Europeans introduced it into India in the beginning of nineteenth century.

A herb. It is native of Europe but also grown in India.

Uses. The roots are eaten as vegetable. They contain a considerable sugar content and even some fat. They are consumed by man and livestock. They are also used for making wine.

Fleshy edible roots are known as parsnips. They are consumed as a vegetable and salad, sometimes employed in soups and wines. Herb is diuretic and carminative. Fruit is used for stomach trouble and stone in the bladder. All parts contain an essential oil; mericarps yield a fatty oil.

Beet Root

Beta vulgaris; Hindi—*Chukandar*; Eng. Beet root; Family—Chenopodiaceae.

Variety. Detroit dark Red: Crops in 80–100 days. Leaves small dark green, tinged with maroon, roots smooth, round, deep red. Flesh very dark blood red with light red gaining, fine grained.

Uses. Generally used in vegetable salad. Refreshing juice is extracted. Sugar is manufactured from juice in European countries.

Roots are eaten boiled or in salads, Var. *rapa* Dum., the Sugar Beet, is a white-rooted biennial, used for sugar manufacture (sugar content 20 per cent). Byproducts of beet sugar industry are tops, the pulp or slices, the filter-cake and molasses. Tops and pulp are used as stock feed and filter-cake as manure. Molasses are used for alcohol manufacture and as a sweetner or animal feeds. Leaves of some varieties are eaten as spinach and leaf-stalks as asparagus.

Carrot

Daucus carota Linn. var. sativa DC; Eng. Carrot; Hindi—*Gajar*; Family—Umbelliferae (Apiaceae).

Sanskrit—*Shikha-mula*; Bengali, Punjabi, Gujarati and Hindi—*Gajar*; Marathi—*Gazara*; Telugu—*Gajjaragedda, pitakanda*; Tamil—*Gajarakkilangu*; Kannada—*Gajjari*.

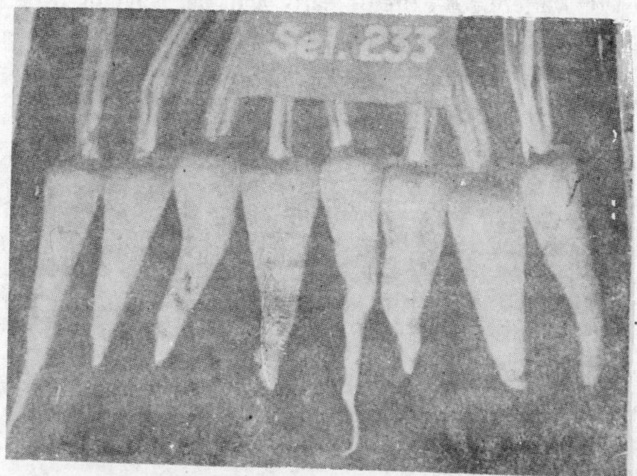

Fig. 4.7. *Daucus carota* **(Gajar)**—A popular, vitaminized root vegetable. An improved variety.

An annual or biennial herb. The pinnately compound leaves are very characteristic. The numerous varieties of carrots differ in size, shape, colour and quality. A deep sandy loam gives the best results. Most of the food is stored in the outer cortical portions of the taproot. They are chiefly grown in the Punjab, Uttar Pradesh and Madhya Pradesh.

Variety: 1. *Pusa kesar*—An improved type of asiatic red carrot with reduced foliage. Roots with self-coloured narrow

Fig. 4.8. *Raphanus sativus.* Radish **(Muli)**—The roots, young leaves and the fruits are used as vegetable.

centrai core. Tolerates higher temperature than Nantes. Suitable for early autumn season. (N.S.C. India).

Variety: 2. *Nantes*—Half-long slim, well shaped, orange, cylindrical, stump roots; good flavour, fine grained, tender, sweet, self-coloured core. Flesh orange scarlet. Suitable for cooler month. (N.S.C. India).

Uses. Carrots are eaten raw or cooked. They are also used or flavouring soups and stews. The yellow colouring matter extracted from the roots is used for colouring butter. It is good cattle feed.

Tender roots are pickled; rich source of carotene. The seeds are aromatic, stimulant, diuretic and carminative, useful in dropsy and kidney troubles, yield an essential oil, *i.e.*, Carrot Seed Oil.

Radish

Raphanus sativus Linn. ; Eng. Radish; Hindi—*Muli*; Family—Cruciferae (Brassicaceae).

Sanskrit—*Mulaka*; Bengali, Marathi, Gujarati and Hindi—*Muli, mula, mura, muri*; Telugu, Tamil, Kannada and Malayalam—*Mullangi*.

History. According to Bentham the radishes (*R. sativus* Linn.) may possibly have come from the British wild plant *R. raphanistrum* Linn., and De Candolle is of the same opinion. *R. raphanistrum* is an European plant and not found in Asia, and is not likely, therefore, to be the species that has furnished the inhabitants of India, China, and Japan with the radishes which they have cultivated for centuries. However, it is thought that the European radish may have arisen from *R. raphanistrum,* but the Indian form, from an Indian and Chinese indigenous wild plant, now apparently lost.

They are annual or biennial plants with a fleshy tap root. They are grown all over the world, and are highly esteemed for their pungent flavour. Many varieties are cultivated differing in size, shape and colour of the roots. In India, they are chiefly cultivated in Uttar Pradesh, Punjab, Maharashtra and Baroda.

Variety: 1. *Pusa Chetki*—A hot season variety for extreme early and late sowings. Roots medium long, pure white stumpy and soft in texture. Ready for harvesting in 40–45 days.

Variety: 2. *Pusa Rashmi*—Roots white with green tinge on top, 30–45 cms. in length. Suitable for early sowing in cooler months. Tolerates slightly higher temperature. Good size roots ready in about 55–60 days.

Variety: 3. *Japanese White*—Roots 22 to 25 cm long, 5 cm in diameter. Cylindrical, skin pure white and smooth; flesh snow white crisp, solid and mild in flavour. Ready in 55–63 days. Suitable for cooler months.

Variety: 4. *Pusa Himani*—Roots semi-stumpy, greenish at the top, white crisp and sweet flavoured with mild pungency. Top short. Suitable for spring sowing in plains and spring and summer sowing in the hills. Matures in 55 days.

Variety: 5. *Rapid Red White Tipped*—Crops in 25–30 days (extra-early) roots fine, smooth, small, round bright red with white tip; flesh pure white crisp and snappy. (N.S.C. India).

Uses. The roots, young leaves and the fruits are used as vegetable. Usually they are eaten raw, but may be cooked like other vegetables.

The roots are used as diuretic in urinary troubles, piles and gastrodynia. Juice of fresh leaves is diuretic and laxative. The seeds are expectorant, diuretic and carminative. Seeds yield a non-drying fatty oil suitable for soap-making; also for edible purposes and as an illuminant. Hydrogenated oil is used in Japan in the manufacture of crayon. Seed cake is rich in proteins and appears to be suitable for use as manure and after removal of isothiocyanates as a feed stuff.

Turnip

Brassica rapa Linn.; Eng. Turnip; Hindi—*Shalgam*; Family Cruciferae. (Brassicaceae).

Bengali—*Shalgam*; Marathi and Oriya—*Salgum*; Telugu and Kannada—*Turnip*; Malayalam—*Seemamullangi*; Assam—*Salgom*; Punjab—*Ganglu, shalgam, thipper.*

Annual or biennial herbs. They are mainly grown in Punjab and Uttar Pradesh.

Uses. The roots are eaten as vegetable. They are eaten raw or cooked.

The roots are a popular vegetable, and the leaves are used as fodder. Turnips are also used in soups. Turnip greens are a good source of calcium and vitamin C.

The leaves are stomachic. Turnips are diuretic and aperient, used in haemorrhages after parturition. They are also used for treatment of exacerbations, tumours, and carcinoma. Seeds yield a fatty oil, used in dengue fever; also rubbed on chest in bronchitis. An embrocation is prepared by mixing the oil and camphor which is useful in muscular rheumatism and stiff neck. Oil is also used as lubricant and for soap-making.

STEM VEGETABLES

Potato

Solanum tuberosum Linn.; Eng. Potato; Hindi—*Alu;* Family—Solanaceae.

Bengali and Hindi—*Alu*; Marathi and Gujarati—*Batata*; Telugu—*Bangaladumpa, uralagadda*; Tamil—*Uralakilangu, wallaraikilangu*; Kannada—*Batate, alu-gidde*; Malayalam—*Urulan kizhangu*; Punjab, Kumaon, Bihar, Orissa and Assam—*Alu.*

History. It has been firmly established that at the time of the discovery of America, the cultivation of the potato (*S. tuberosum* Linn.) was practised in the temperate regions extending from Chile to New Grenada. In Europe the potato was first introduced, at some period between 1535 and 1585, into Spain when its cultivation spread into Portugal, Italy, France, Belgium, and then into Germany. The introduction into Ireland in 1585 or 1586, was independent but after the era of its first cultivation in Spain.

The date of the introduction of the potato into India is uncertain. According to Roxburgh, it has been cultivated in India before the beginning of the eighteenth century (*i.e.*, middle or end of seventeenth century). The probability is that the cultivation of potato was introduced into India from Spain, sometimes in the seventeenth century.

Fig. 4.9. Potato crop in hills.

This is one of the most important food plants of the world. It is native of South America now grown, chiefly in Uttar Pradesh, Assam, Punjab, West Bengal and Bihar.

Potato is also a cash crop of great importance. With proper care, it can be stored for several months and transported without much damage from the producing centres to the consuming area. The crop is mainly suited to colder latitudes and temperate zones. India accounts for only 0.7 per cent of the world production. The important potato producing countries are the Russia (32.0 percent), Poland (14.6 per cent), West Germany (10.5 per cent), France (5.7 per cent), the U.S.A. (4.6 per cent), United Kingdom (3.9 per cent),

Czechoslovakia (3.0 per cent) and China (0.8 per cent).

Fig. 4.10. Potato, germinating tubers. Healthy sprouts emerge from healthy tubers.

The potato was introduced into India in the seventeenth century. The tubers are consumed in India as a vegetable rather than as a staple food. The per capita annual consumption of potatoes is estimated at nine lbs., while in some of the Western countries the figure is as high as 440 lbs. (Pushkernath, 1955). With an area of 702,000 acres the annual production is estimated at 1,674,000 tons (Maheshwari, 1959). About 90% of the crop is grown in the north Indian plains in winter and the rest in hills in the summer. In the Nilgiris, the crop is cultivated all the year round. In the plains, two successive crops are sometimes raised on the same land. The commonly grown varieties are 'Phulwa', 'Darjeeling Red Round', 'Up-to-date', and 'Gola'. Several newly evolved high-yielding hybrids, such as 208, 209, 2236 and No. 45, have been found to be very suitable for cultivation in north Indian plains.

The plant is a spreading annual. It has pinnately compound leaves, fibrous roots, and numerous rhizomes which are swollen at the tip to form the tubers.

They can be cultivated in many soils and many climates. The best environment is a cool moist climate, and a rich light soil. They are usually propagated vegetatively by means of tubers, or

Fig. 4.11. Potato. A heap of healthy potatoes.

parts of tubers, the so-called 'seed potatoes'. The essential parts of the seed potatoes used for propagation are the eyes. The eyes are really groups of buds situated in the axils of aborted leaves. There is usually a central bud in each eye, surrounded by smaller lateral ones. The eyes are more numerous toward the apex of the tuber. Pieces of the tubers are cut at right angles to the main axis. The larger the piece, the more vigorous is the vegetative growth, and there is greater yield. In any case, at least one eye must be present. The tubers require a rest period of several weeks for maturation. Potatoes contain about 78% water, 18% carbohydrates, 2% proteins, 9.1% fat and 1% potash.

Uses. The tubers are used as vegetable. In European countries they make a universal table food. Small tubers are utilized for the production of starch and industrial alcohol. The potatoes are also fed to livestock.

Potato chips made from fresh potatoes are famous in the South while *"katri* and *patri"* are quite common in Maharashtra and Gujarat.

Katri and patri. The fresh potatoes are made into potato chips and shavings after boiling the potatoes. These are dried in the sun. These dried chips are deep fried as and when required.

Surplus and cull potatoes are used as feed for livestock and also as raw material for manufacture of starch, ethyl alcohol, and some other industrial products. Potato is among richest foods in potassium. Biological value of potato protein is 68 and protein efficiency ratio 1.9 and it decreases on storage, more so at room temperature than at 4°C. Cooking improves digestibility of potato protein. Tuberin has a high nutritive value, being superior to wheat protein. Potatoes contain phosphorylase, amylase and several other enzymes including proteolytic enzymes. Potato flesh, even at its natural high moisture content, is superior to 72 per cent extraction of wheat flour as a source of vitamin B_1, riboflavin and nicotinic acid. Concentration of thiamine is lower and that of riboflavin and folic acid is higher at the periphery than in the interior of tuber. Potato leaf haulms are useful source of a-cellulose. Potatoes are used as antiscorbutic, aperient, diuretic, and galactagogue. Extract of leaves is used as antispasmodic in cough. Tuber is ground and made into a paste for application to burns with much benefit.

Fig. 4.12. Potato. Healthy tubers of 'Kufri Alankar' variety.

Fig. 4.13. Potato. Healthy tubers of Kufri Chandramukhi variety.

Potato products include potato chips, frozen French fries, potato granules, potato flakes, potato flour, spaghetti-like potato, canned potatoes, potato starch and fermentation products.

Potatoes are used for the preparation of nutrient media, *e.g.*, potato dextrose agar media (PDA), for culture of microorganism and for potato broth, used in microbiological work.

Variety: *Kufri Alankar*—Flourishes well in the plains of Northern India. Plants are tall, erect and vigorous. Flowers white in colour. Tubers are oval, white with medium deep to fleet eyes. Flesh is white with floury texture on cooking. It takes 80–90 days to mature. In the plains, the average yield is about 250–300 quintals per hectare. Early variety.

Fig. 4.14. Potato. 'Kufri Sheetman'. It is best suited for the plains of Punjab, Western Rajasthan, Haryana and Tarai region of Uttar Pradesh.

Variety : *Kufri Jyoti*—It is most suitable for hills and plains of Northern India. The plants are tall, erect, vigorous and medium compact and the flowers are white and abundant. The tubers are oval, white with fleet eyes. Flesh is dull white. It takes about 100 days in the plains and 120 days in hills to mature. Average yield 150 quintals per hectare. Early variety.

Variety : *Kufri Sheetman*—It is best suited for the plains of Punjab, western Rajasthan, Haryana and Tarai region of Uttar Pradesh. The plants are tall, vigorous, erect, open and the flowers have blue purple colour with white tips. The tubers are round oval slightly flattened with fleet eyes. The flesh is dull white and texture floury on cooking. It takes about 100–110 days to mature in the plains and 140–150 days in hills. It yields 230 quintals per hectare on an average. Medium variety.

Jerusalem Artichoke

Helianthus tuberosus Linn.; Eng. Jerusalem artichoke; Hindi—*Hathichuk.*; Family—Compositae Asteraceae. Bengali—*Brahmokha*; Punjab—*Hathipich*.

History. The Jerusalem artichoke (*H. tuberosus* Linn.) is a native of North America and has been cultivated there for centuries. The English name Jerusalem artichoke has been derived from the Italian word for sunflower, *girosole*. It was first introduced into Europe at the Farnese Garden at Rome about 400 years ago, and rapidly spread over Europe. It has been introduced into India by Europeans about two centuries back.

This is native plant of North America, now grown in Bengal, Assam, Maharashtra, Uttar Pradesh and Andhra Pradesh. It is a perennial sunflower 6 to 12 feet in height. The tubers somewhat resemble potatoes, but with larger eyes.

Uses. The tubers are cooked, pickled or eaten raw. The carbohydrate food is in the form of insulin, which makes a good food for diabetics. It is also used as a source of levulose and industrial alcohol. The plants are also used as a forage crop.

The tubers are also made into chips or ground into flour. Green tops and tubers are used as feed for stock, also ensiled; has aroused much interest as a commercial source of levulose used as a sweetening agent by diabetics. Stalks are used as raw material for paper manufacture.

Taro

Colocasia esculenta (Linn.) Schott; Eng. Taro, Arum; Hindi—*Arvi, Ghuiyan, Kachalu*; Family—Araceae.

Sanskrit—*Kachu*; Bengali—*Kachu*; Marathi—*Alu*; Tamil—*Seppankizhangu*; Telugu—*Chamadumpa, chemagadda*; Kannada—*Kachchi, shamagadde*; Malayalam—*Shembu*; Oriya—*Saru*.

This is a herb. It is native of South-East Asia but cultivated throughout India. It has huge peltate leaves. The tops of the corms are used for vegetative propagation. They require a wet rich soil and a long season in which to mature.

Uses. The corms are edible. They are roasted or boiled to destroy the acrid calcium oxalate crystals in the raw tubers. The leaves are also used as vegetable.

Young leaves are eaten like spinach, also consumed when bleached. Juice of the petioles is used as an astringent and styptic. Taro mucilage may be used as a size for impermeable paper. Tubers may be used for production of industrial alcohol. Taro flour can be used for soups and gruels, gravies, and puddings. Taro-lactin and Taro-malt, prepared from flour, are good foods for infants and invalids.

Onion

Allium cepa Linn.; Eng. Onion; Hindi—*Piyaz*; Family—Amaryllidaceae.

Sanskrit—*Palandu*; Bengali—*Pyanj*, Marathi—*Kanda*; Gujarati—*Dungari*; Telugu—*Nirulli, ulligaddalu*; Tamil—*Vengayam*; Kannanda—*Nirulli, irulli*; Malayalam—*Chuvannaulli*.

History. The onion *(A. cepa)* is very old, its use going back over 4000 years. It is probably a native of South Asia or Mediterranean region. In Egypt it was worshipped before the Christian era. It has long been used in China and India.

The onion is the food plant in which the food is stored in a bulb. It is the native of Southern Asia or the Mediterranean region. They are cultivated over large areas in temperate and even tropical climates. They thrive best in cool moist regions with a sandy soil.

Variety: 1. *Pusa Red*—medium size, round red bulbs, less pungent than moist red varieties. Keeps very well in storage. Characteristically free from bolting tendency. Short to intermediate day length type.

Variety: 2. *Early Grano*—Bulbs large, yellow with mild pungency. Storage life is rather short. (N.S.C. India).

Uses. The bulbs are used as food and as flavouring substance.

Bulbs as well as fresh herb yield an essential oil. Onions are considered stimulant, diuretic and expectorant, used against flatulence and dysentery. Roasted onions are applied as poultice.

Fig. 4.16. Garlic bulbs.

Garlic

Allium sativum Linn.; Eng. Garlic; Hindi—*Lasan, Lahsun*; Family—Amaryllidaceae.

Sanskrit—*Arishtha, lashuna*; Gujarati and Hindi—*Lasan*; Bengali and Marathi-*Lasun*; Telugu and Malayalam—*Velluli, tellagadda*; Tamil—*Vellaipunda*; Kannada—*Bellulli*; Punjab—*Thom*.

This is a perennial plant with narrow flat leaves and several small bulbs, known as cloves *(puti),* enclosed in a white skin.

Uses. The bulbs are used as a condiment and flavouring substance. Garlic powder is extensively used as a condiment and also serves as carminative and gastric stimulant. They also possess antiseptic and bactericidal properties.

Preparations of garlic are used in pulmonary phthisis, gangrene of the lung, and whooping cough. Laryngeal tuberculosis, lupus, and duodenal ulcers are treated by garlic juices. The garlic cloves are also used for flatulence, colic and atonic dyspepsia. Juice is applied in skin troubles and used as ear drops.

East Indian Lotus

***Nelumbo nucifera*:** Hindi—*Kamal*; Family—Nymphaeaceae. Commonly found in Northern India.

Sanskrit—*Ambuja, padma, pankaja, kamala*; Bengali—*Padma*; Marathi—*Kamal*; Gujarati—*Suriyakamal*; Telugu—*Kalung, erra-tamara*; Tamil—*Ambal, thamarai*; Kannada—*Kamala, tavaregadde*; Malayalam—*Thamara, senthamara*; Oriya—*Padam*; Kashmir—*Pamposh*; Punjab—*Kanwal, pamposh*; Mundari—*Salukid ba, upal ba, Kombol ba*; Assam—*Podum*; Khasi Hills—*Soh-lapudong*.

A common aquatic herb with large, pink flowers. The rhizomes, seeds and young leaves eaten as vegetable.

Farinaceous rhizomes *(kamal-kakdi, bhen)* are used as a vegetable. Fruiting torus *(kamal-gatta, kaul chapani)* contains round or oblong carpels which are eaten after removing the outer covering and bitter embryo. Carpels are sweet and eaten raw, roasted, boiled, candied, or ground into flour; considered more nutritive than cereals. Young leaves, petioles and flowers are also eaten as vegetables. Rhizomes yield a kind of nutritious arrowroot, given to children in diarrhoea and dysentery. Carpels are demulcent and nutritive.

Leaf stalks yield a fibre. Petioles, pedicels and embryos contain an alkaloid, nelumbine, which acts as a cardiac poison.

White Lotus

Indian red water lily—***Nymphaea nouchali***: Hindi—*Kamal Kakri*; Family—Nymphaeaceae..

Bengali—*Shaluk, rakto-kambal, nal*; Marathi—*Lalakamal, raktakamal*; Gujarati—*Kanval, nilophal*; Telugu—*Allitamara, tella-kalava*; Tamil—*Allikamarai, vellambal*; Kannada—*Nyadale huvu*; Malayalam—*Periambal, neerambal;* Oriya—*Dhabalakain, rangkain*, Punjab—*Chota kanwal*; Mundari—*Pundi salukid*; Assam—*Mokuva*.

Commonly found in Northern India. This is a floating aquatic herb.

Uses. Starchy rhizomes are eaten raw or boiled, sometimes baked. Flowering stalks and unripe fruits are used as a vegetable; also used in salads and stews. Rhizome demulcent used in dysentery and dyspepsia. Flowers are cardiotonic, and used for the preparation of *Ghillad* and *Gulkand*. Seeds are employed in cutaneous diseases.

Indian Blue Water Lily

Nymphaea stellata Willd.; Family—Nymphaeaceae.

Sanskrit—*Nilopala*; Hindi—*Nilpadma, nilkamal*; Bengali—*Nilshapla, nilpadma*; Marathi—*Krishnakamal, poyani*; Gujarati—*Nilkamal*; Telugu—*Nallakalava, nitikulava*; Tamil—*Karu neythal, nilotpalam*; Malayalam-*Sitambel*; Orya-*Subdikain*; Punjab-*Bambher, nilpadma*; Delhi-*Chotakamal*.

Uses. Rhizomes, tender leaves, and peduncles are used as vegetable. Powdered rhizomes are given in dyspepsia, diarrhoea, and piles; infusion emollient and diuretic, and given in diseases of urinary tract. Seeds are stomachic and restorative. Decoction of flowers is narcotic.

Kaseru

Cyperus esculenta; Hindi—*Kaseru*; Family—Cyperaceae. Commonly found on the muddy banks of the ponds. The rhizomes and tuberous roots are eaten raw and after cooking.

Sanskrit—*Kaseruka*; Hindi and Punjabi—*Chichoda, Kaseru, dila.*

Uses. The tubers are edible; accredited with stimulant and aphrodisiac properties.

HERBAGE VEGETABLES

They have the nutrient materials stored in parts of the plant found above ground. Any part of the shoot of the plant may be utilized for storage. The leaves are used in spinach, cabbage and lettuce; condensed inflorescence in cauliflower; condensed stems in *ganth-gobhi*; leafstalks in celery, etc. These vegetables contain more proteins but smaller amount of carbohydrates. They also contain a sufficient amount of mineral salts and vitamins, which make them an essential part of human diet.

Cabbage

Brassica oleracea Linn. var. ***capitata*** Linn.; Eng. Cabbage; Hindi—*Bandgobhi*; Family—Cruciferae (Brassicaceae).

Bengali—*Bandhakapi, kopi;* Marathi—*Kobi*; Gujarati—*Kobij, kobia*; Telugu—*Aalugobi, cabbage*; Tamil—*Muttai kose*; Kannada—*Kelekou, kogu gadde*; Malayalam—*Muttakose*; Oriya and Assam—*Bandha kobi;* Mundari—*Arakubi, kubiara*; Punjab—*Bandgobi.*

History. Cabbage *(B. oleracea)* is one of the most ancient herbage vegetables. These are supposed to have been derived by cultivating the European wild colewort or wild cabbage. The cabbages have been produced from this plant by selection or mutation. Its cultivation is very old about 2500 B.C. The cabbages were known to Greeks and Romans. The ancient Germans, Saxons and Celts were the first to grow cabbage in Northern Europe and it became an important herbage vegetable at a very early date in Scotland and Ireland.

In this plant the stem is so short that the great mass of thick overlapping leaves tend to form a head. The older leaves surround the younger smaller, more tender leaves, and the cabbage resembles a huge bud. Cabbage contains 91% water, with some sugar and starch, considerable protein and valuable salts.

Fig. 4.17. *Cabbage.* The cabbage thrives in a relatively cool moist climate. It is grown mainly as a winter crop in the plains of India. The main crop is grown in Northern India. The common varieties are—Golden Acre, Pride of India, Copenhagen Market, Mammoth Rock Red, Express, Pusa Drumhead, Jersey Wakefield and Chieftain.

Variety: 1. *Golden Acre*—Early; plants small and compact with few outer leaves. Short stemmed with small cup shaped leaves. Heads very uniform, solid and round; interior clear white of excellent quality. Suitable for cooler months.

Variety: 2. *Pride of India*—Early; heads round, solid and medium large sized weighing about 1 to 1.5 kg. very good for marketing.

Variety: 3. *Large Drum Head*—Late variety 110–120 days short stemmed small frame, few outer leaves, heads uniform, large flattened drum shaped and solid. (N.S.C. India).

Uses. It is eaten raw or cooked. It is also used for feeding livestock and poultry.

It is eaten either as salad or cooked; red cabbage is preferred for pickling. Steaming is preferred to boiling. Cabbage is included in the diet of patients, particulary the ones suffering from fistula and liver troubles. Red cultivars are a fairly good source of vitamins A, B and C. Sinigrin is present in the cabbage. A large number of cabbage varieties are cultivated.

Yellow Sarson

Brassica campestris Linn. var. sarson Prain; Eng. Yellow sarson; Hindi—*Sarson*; Family—Cruciferae. (Brassicaceae).

An annual herb. It is grown mainly in Uttar Pradesh, Punjab, Bihar and Assam.

Uses. The tender leaves and shoots *(sarson ka saag)* are used as vegetable.

White Mustard

Brassica hirta Moench; Eng. White mustard; Hindi—*Safed rai*; Family—Cruciferae (Brassicaceae).

Uses. The young leaves and tender shoots are used as vegetable.

Leaf Mustard

Brassica juncea (Linn.) Czern. & Coss var. ***cuneifolia*** Roxb.; Eng. Leaf mustard; Hindi—*Rai*; Family—Cruciferae. (Brassicaceae).

Commonly grown in terai areas of Nainital, North Bengal and Assam.

Uses. The young leaves and tender shoots are used as vegetable.

Fig. 4.18. Cabbage. 'Pusa Mukta', a high yielding variety, yields about 49 tonnes per hectare.

Kale

Brassica oleracea Linn. var. ***acephala*** DC.; Eng. Kale, Borecole; Hindi—*Karam-sag*; Family—Cruciferae (Brassicaceae).

A herb. Cultivated in Assam, Kashmir, Maharashtra and Baroda.

Uses. The young leaves and tender shoots are eaten as vegetable.

Shoots and broad leaves are a popular vegetable in Kashmir; also used in soups and as greens. Kale can be converted into silage. A type of kale known as Hungarian Gap-kale is grown for fodder in some parts of Uttar Pradesh. When fed to sheep, it helps in better growth of wool.

Brussels Sprouts

Brassica oleracea Linn. var. ***gemmifera*** Zenk; Eng. Brussels sprouts, Bud-bearing cabbage; Hindi—*Buttam gobhi*; Tamil—*Marakkhose*; Family—Cruciferae (Brassicaceae).

A herb. Cultivated chiefly in Maharashtra.

Uses. The buds, young shoots and leaves are used as vegetable.

The sprouts are cooked like cabbage and eaten; also used for fodder. It is fairly rich in vitamin A and C and contain appreciable quantities of riboflavin, niacin, calcium and iron. Sprouts are used with wine in diseases of liver and cancer.

Cauliflower

Brassica oleracea Linn. var. *botrytis* Linn.; Eng. Cauliflower; Hindi—*Phulgobhi*; Family—Cruciferae/Brassicaceae.

Bengali—*Fulkapi*; Marathi—*Phulkobi, fulvar*; Gujarati—*Fulkobi*; Telugu—*Pookobi, cauliflower*; Tamil—*Gospoovu*; Kannada—*Kosuguddae*; Malayalam—*Cauliflower*; Oriya—*Fulakobi*; Assam—*Poolkobi*; Punjab—*Phulgobi.*

Fig. 4.19. ***Brassica oleracea*** var. ***botrytis***: Cauliflower **(Phulgobhi)** —
The compact inflorescence is eaten as vegetable. Pusa Hybrid 2, an improved variety.

In this plant a short erect stem is produced with an undeveloped inflorescence. The whole inflorescence forms a large head of abortive flowers on thick hypertrophied branches. Cultivated all over Northern India.

Variety: 1. *Pusa Deepali*—Early season asiatic variety, suitable for sowing in hot season. Head ready in 100–120 days, plants medium tall with erect waxy green short leaves; self branching type, curd medium size, compact and white.

Variety: 2. *Improved Japanese*—Mid season variety, crops in 90–95 days. Curds large size, compact and white.

Variety: 3. *Pusa Snowball*—Late variety, produces medium sized solid curds of attractive snow white colour. Suitable for cooler months.

Variety: 4. *Pusa Snowball-1*—About 10 days late in maturity than Pusa Snowball. Leaves straight, upright, inner leaves tightly covering the curd. Curd very compact of medium size with good staying quality.

Uses. The condensed inflorescence is eaten as vegetable.

Curds are not only used as a vegetable but also used in soups and pickles. Leaf stalks are also cooked as vegetable. A good number of broccolies are canned and stored in frozen condition. Leaves and stems of cauliflowers, obtained as wastes, are utilized as livestock feed.

Fig. 4.20. *Knolkhol*. The fleshy edible portion is an enlargement of the stem, which develops entirely above ground.

Kohlrabi

Brassica oleracea Linn. var. *gongylodes* Linn.; Eng. Kohlrabi, knol-khol; Hindi—*Ganthgobhi*; Family—Cruciferae (Brassicaceae).

Bengali—*Oldkapi*; Marathi—*Nawal kol*; Gujarati—Nolvol; Telugu—*Gaddagobi, noolkhol*; Tamil—*Noolkhol*; Oriya-*Olkobi, ganthikobi*; Assam—*Olkobi*; Punjab—*Gobi.*

Knol khol

Brassica oleracea var. ***caulorapa.***

Variety: *White Vienna*—Early, crops in 70 days, leaves and stem medium dwarf bulb globular, light green and smooth, flesh creamy white and tender.

A herb with swollen stem. It is cultivated in Maharashtra, Uttar Pradesh Baroda and the Punjab.

Uses. The short swollen stem is eaten as vegetable.

Edible tuberous portion of the aerial stem is spherical and turnip-like; used for human consumption and as a stockfeed. Small heads may be cooked like cabbage or served as salad. Leaves are also eaten, they are rich in calcium.

Garden Asparagus

Asparagus officinalis Linn; Eng. Garden Asparagus; Hindi—*Halyun, Seetmuli. Merchuba*; Bengali—*Hillua*; Family—Liliaceae.

It is a native of temperate Europe and Western Asia. Also grown in India. The plant has perennial roots, which send up each year an erect branching stem several feet in height. Instead of true leaves, modified branches, the cladodes occur. It thrives best in fertile well-drained soil in moist temperate regions with abundance of sunshine.

Uses. The delicate shoots are eaten fresh. For the best flavour asparagus should be cooked within 12 hours of picking. There is 94% water in it, but it still contains more protein.

New succulent shoots which come up every year constitute the Asparagus; large quantity is canned. Young stems are eaten green or bleached after boiling, also used in soups. It contains asparagine which is diuretic and used in cardiac dropsy and chronic gout.

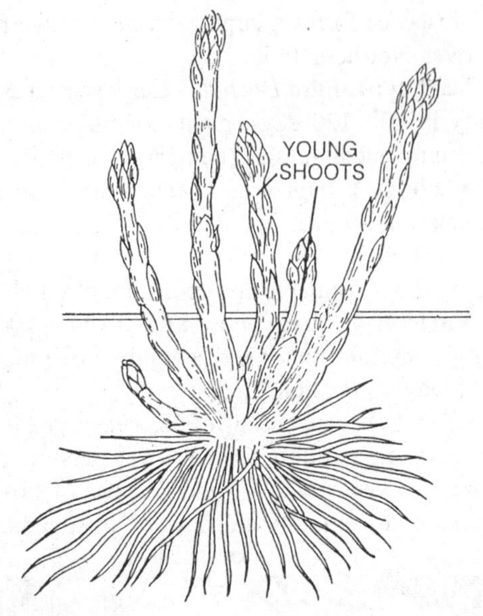

Fig. 4.21. *Asparagus* (garden asparagus). The young shoots are eaten as vegetable.

Spinach

Spinacia oleracea Linn; Eng. Spinach; Hindi—Palak; Family—Chenopodiaceae.

Bengali—*Palang, pinnis*; Marathi and Gujarati—*Palak*; Telugu—*Dumpabachhali*; Tamil—*Vasayleykiray*; Kannada—*Spinachsoppu, Spinaksoppu*; Oriya—*Palaksag, mithapalanga*; Punjab—*Palak*; *isfanak*; Assam—*Palangsag*.

It is native of South-Western Asia. This herb is commonly grown throughout India especially in cool regions where there is abundant supply of water.

Uses. The leaves are eaten raw or cooked as a vegetable. They are rich in mineral salts and proteins.

The leaves are nutritious. They are also employed in soups and salads. They are diuretic; also employed as a source of chlorophyll; lipids present in the leaves possess antibacterial action. Fruits are demulcent and diuretic, employed in fevers and inflammation of bowels.

Spinach Beet

Beta vulgaris; Hindi—*Palak*.

History. The various beets in cultivation are referred to as single species *Beta vulgaris. Chard* is the oldest type of beet. It was known as early as 300 B.C. Common beets have been known since the beginning of the Christian era.

Variety: *All Green*—Uniformly green tender leaves; heavy yielder. Very early, ready for first cutting in 40 days. Suitable for sowing round the year.

Uses. Used as herbage vegetable. Rich in minerals and vitamin.

Lettuce

Lactuca sativa Linn; Eng. Lettuce; Hindi—*Salad*; Family—Compositae (Asteraceae).

Bengali—*Kahu*; Telugu—*Kavu*; Tamil—*Salattu*.

A herb; native of South Europe and Western Asia but cultivated throughout India.

Variety: 1. *Chinese Yellow*— Non-heading, loose leafed and early, leaves light green crisp and tender.

Variety: 2. *Great Lakes*— Heading type; head large, firm, well folded and flattened at poles; leaves dark green, serrated edge, crinkled, brittle in texture and crisp. (N.S.C. India)

Uses. The leaves are eaten as vegetable and salad.

In nutritive value it is classed with cauliflower, celery and asparagus, chiefly valued for its mineral and vitamin content. It yields lactucarium used as hypnotic in bronchitis and asthma.

Garden Celery

Apium graveolens Linn. var. dulce DC.; Eng. Garden celery; Hindi—*Ajmud*, *Karas*, *Salari*; Family—Umbelliferae (Apiaceae).

Bengali—*Randhuni*; South India—*Ajmod* or *ajmoda.*

It is native of Europe, now cultivated in the north-western Himalayas and in the hills of Uttar Pradesh, Punjab and South India. It is a biennial forming a fleshy root and clump compound leaves with long leafstalks. These leafstalks constitute the celery of commerce. The stalks are large and succulent. Celery requires a rich sandy loam and lot of water.

Uses. This is used as a vegetable. The leaves are used in salad and soups. Cremocarps (fruits) are used as spice; they are stimulant, carminative sedative, and nervine tonic, decoction used in rheumatism. Fruits yield an essential oil, used as a spasmodic and nervine stimulant.

Endive

Cichorium endivia Linn; Eng. Endive; Hindi—*Kasni*; Family—Compositae (Asteraceae).

A herb; native of India. Cultivated mainly in Northern India. The plant is an annual or biennial.

Uses. The young basal leaves, which often have curled margins, are eaten as a salad. The leaves are also used as vegetable.

Chicory

Cichorium intybus Linn.; Eng. Chicory; Hindi—*Kasni*; Family—Compositae (Asteraceae).

A herb. It is native of Europe but commonly found in Punjab and parts of Andhra Pradesh.

Uses. It is used as a salad plant for greens.

Chicory is also cultivated for its roots in Gujarat for its commercial value. The roots are dried and powdered and mixed with coffee powder known as chicory blend coffee.

Fenugreek

Trigonella foenum-graecum Linn; Eng. Fenugreek; Hindi—*Methi*; Family—Leguminosae (Papilionaceae).

Sanskrit—*Methika*, *chandrika*, *asumodhagam*; Bengali—*Methi*, *methi—shak*, *methuka*; Marathi—*Methi*; Gujarati—*Methi*, *methini*; Telugu—*Mentikoora* (herb), *mentulu* (seeds); Tamil—*Vendayam*; Kannada—*Menthya*, *mentesoppu*, *mente*; Malayalam—*Uluva*, *venthiam*.

A herb. It is native of South Europe now grown mainly in Northern India.

Variety: *Pusa Early Bunching*—Shoots upright, vigorous; height 40 to 50 cm Heavy yielder. Suitable for bunching.

Uses. The leaves are used as vegetable. They are commonly consumed in winter season.

Methi Kasuri

Trigonella corniculata; Family–Papilionaceae.

Sanskrit—*Malya*; Hindi—*Kasuri methi*, *kasturi methi*, *marwari methi*, *champa methi*, Bengali—*Pirang*; Rajasthan—*Chirawa*; Assam—*Piring sak.*

Variety: *Pusa kasuri*— A late flowering variety, rosette type; yields several cuttings. Heavy yielder. Leaves have special fragrance.

Grown as a pot-herb and for flavouring. Herb contains diosgenin.

Pigweed

Chenopodium album Linn.; Eng. Pigweed, Lamles—quarters; Hindi—Bathua; Family—Chenopodiaceae.

Bengali—*Chandan betu*, *bethusag*; Tamil—*Parukukkirai*; Telugu—*Pappukura.*

A common annual herb. Found as a weed in wheat fields.

Uses. The leaves and tender twig are eaten as vegetable in winter season. Used as a pot-herb and accredited with laxative and anthelmintic properties. It yields an essential oil.

Purslane

Portulaca oleracea Linn; Eng. Purslane; Hindi—*Kulfa*; Family—Portulacaceae.

Sanskrit—*Brihalloni, lonika*; Bengali—*Baraloniya*; Marathi—*Kurfah*; Gujarati—*Moti loni*; Telugu—*Ganga-pavilikura*; Tamil—*Karikeerai*; Kannada—*Dooddagooni soppu*; Malayalam—*Kariecheera*; Oriya—*Purunisag*; Assam—*Noniya*; Punjab—*Lonak, kundar.*

A common herb. Grown in N. India.

Uses. The leaves and tender twigs are used as vegetable.

Used as a pot-herb, also consumed as salad and employed in soups. Fleshy stems pickled. May also be used as fodder. Herb is refrigerant, vulnerary, antiscorbutic, aperient and diuretic; the diuretic action being due to the presence of high percentage of potassium salts. Roasted seeds are eaten. Purslane being rich in minerals can be used as green manure.

Amaranths

Amaranthus blitum Linn. Var. ***oleracea*** Hook. f.; Eng. Amaranth; Hindi—*Chaulai*; Family—Amaranthaceae.

Sanskrit—*Bashpaka, marisha*; Bengali—*Sadanatya*; Marathi—*Bhaji*; Gujarati—*Dambho*; Telugu—*Totakura*; Tamil—*Tandukirai*

A herb. Found throughout India.

Fig. 4.22. Amaranth. 'Pusa Kirti' variety.

Uses. The leaves and tender twigs are used as vegetable.

Used as a pot-herb; also seeds roasted and eaten. Herb fairly rich in iron, 18.18 mg/100 g.

Amaranthus tricolor; Eng. Amaranth; Hindi—*Bari chaulai*; Family—Amaranthaceae.

Variety : *Bari chaulai*— Good green tender leaves without spines. Stem green, medium thick and tender, petioles green, inflorescence terminal and medium-sized. Ready for first cutting after 35 days of sowing.

Mountain Ebony

Bauhinia variegata Linn.; Eng. Mountain ebony, Variegated Bauhinia; Hindi—*Kachnar*; Family—Caesalpiniaceae.

Marathi—*Kachnar*; Bengali—*Raktakanchan*; Tamil—*Segpumanchori*; Kannda—*Kanchavala*; Malayalam—*Chuvannamandaram*.

This is a tree found in Punjab, the western Peninsula and Assam. It possesses white flowers having red and yellow stripes.

Uses. The young leaves and flower buds are eaten as vegetable.

Flower buds are pickled. The young pods are also eaten as vegetable. Dried buds are used for diarrhoea, dysentery and piles. Flowers are laxative.

FRUIT VEGETABLES

Tomato

Lycopersicon esculentum Mill.; Eng.—*Tomato*; Hindi—*Tamatar*; Family—Solanaceae.

Bengali—*Tamatar*, *Vilayithi baingan*; Marathi—*Velbangi*; Gujarati—*Vilayithi vengan*; Tamil—*Takkali*.

History. The tomato (*Lycopersicon esculentum*) is native of Peru-Ecuador area, from which it spread north ward to Mexico, where it was first cultivated. Thereafter the plant was carried to Southern Europe by Spanish explorers. It was eaten for a long time in Southern Europe before it was used by the people of Northern Europe and the United States. The plant was first domesticated in India in the last century.

The cultivation of tomato on a commercial scale in India began towards the close of the last century. It has now become a popular vegetable and is cultivated extensively, particularly in the vicinity of large towns and cities.

Variety 1: *Pusa Ruby*—Early variety; fruits ripen in 60 days after transplanting; withstands hot and humid climate; fruits flat round in shape, medium sized, uniform deep red; heavy yielder slightly acidic. Suitable for both hot and cold months. Good for fresh market and for making ketchup.

Variety 2: *Pusa Early Dwarf*—Early variety; fruits ripen in 60 days after transplanting; withstands hot and humid climate; dwarf in growth; fruits medium sized, roundish, uniform red coloured and heavy yielder; slightly earlier and fruits slightly bigger and smoother than *Pusa Ruby*. Suitable for both hot and cooler months.

Variety 3: *Sioux*—Early maturing with indeterminate habit of growth; fruit medium to large size whitish green when unripe and attractive red when ripe. Fruits round and smooth.

Variety 4: *Marglobe*—Late maturing with indeterminate habit of growth. Unripe fruits having dark green stem end. When ripe, fruits uniformly deep scarlet in colour, round, juicy with less seeds.

Variety 5: *Roma*—Plants dwarf with luxuriant foliage cover and excellent bearing, producing pear-shaped fruit in clusters; fruits ripe in 95–100 days.

Variety 6: *Punjab Chhuhara*—Plant height medium. Fruits pear-shaped 50–60 gms. in weight. Red in colour, 2–3 locules, pericarp thick. Flowers in 90 days and maturity up to 130 days. Fruits well protected under foliage. Sets fruits under high temperature conditions.

Uses. Fresh ripe fruits are refreshing and appetizing and are consumed raw in salads or after

Fig. 4.23. *Lycopersicon esculentum* (tomato). An improved variety in fruiting.

cooking. Unripe fruits are cooked and eaten. Large quantities of fruits are canned. Tomatoes are consumed also in the form of juice, paste, ketchup, sauce, soup and powder.

Tomato Products—Canned tomatoes: Ripe tomatoes of medium size, regular shape, uniform red colour, solid meat and of good flavour are selected for canning. They are washed scalded in boiling water or steam for 2-3 minutes, peeled and canned.

Tomato juice: Ripe tomatoes with bright red colour and high acidity are used for the expression of juice. Tomato juice is highly esteemed as an appetizing and nourishing beverage. It is sometimes seasoned to produce cocktail known as Tomato juice Cocktail. Tomato juice powder is prepared by dehydrating the juice.

Tomato ketchup: Tomato ketchup is prepared from ripe tomatoes of deep red colour by cooking the pulp in kettles with spices (chopped onion and garlic, cloves, cardamom, black pepper cumin, mace, and cinnamon). Cooking is continued till the desired consistency is obtained. Vinegar, salt, sugar and sometimes pectin are added.

Tomato soup: For preparing tomato soup, the fruit pulp is partly neutralized by adding sodium bicarbonate solution and then concentrated in a pan; spices, arrowroot and butter or cream are added. When the desired consistency is obtained, salt and sugar are added and the mass boiled for another two minutes.

Brinjal

Solanum melongena Linn.; Eng. Brinjal, Egg plant; Hindi—Baigun.

Sanskrit—*Vatigana, bhanṭaki*; Bengali—*Begun, kuli begun*; Marathi—*Vangi*; Gujarati—*Vengni*; Telugu—*Chirivanga, niruvanga*; Tamil—*Kathirikai*; Kannda—*Badanekayi*; Malayalam—*Vazhuthana*; Oriya—*Baigun*; Kashmir—*Vangun*, Punjab—*Baingan*; Assam—*Jati bengani.*

Fig. 4.24 Brinjal. Azad Hybrid, a round variety.

Fig. 4.25. Brinjal. An improved oval variety.

It is native of India. The plant is an erect branching herb or small shrub, sometimes several feet in height. The fruit is large, ovoid, whitish or purple berry. The plant is cultivated as an annual. Several taxonomic varieties are recognized. Usually *long* and *round* types are found.

Variety 1: *Pusa purple long*—Extra early; plant semierect; fruits glossy, purple, 25-30 cm long, smooth and quite tender. Suitable for all areas. Very high yielder.

Variety 2: *Pusa purple cluster*—Plants tall, erect, quite compact and sturdy having purplish leaves and stems; fruits available in 75 days after transplanting. Fruits 10–12 cm long; deep purple in colour and borne in clusters of 6–9. A heavy yielder and fruiting period can be extended for longer period if regular pickings are made. Suitable for all areas.

Variety 3: *Pusa kranti*—Plants erect, prolific bearer. Fruits uniform thick, oblong in shape, 15–20 cm long, dark purple in colour with shining green calyx.

Uses. The fruits are given as vegetable. They are usually cut into slices and fried or broiled.

Besides being consumed as a vegetable, brinjals are also pickled; sliced fruits are dried and stored. Brinjals are rich in iodine contents, they are given in liver complaints; they stimulate interhepatic metabolism of cholesterol. Aqueous extract of fruits inhibit choline esterase activity of human plasma. Percentage of vitamin B_2 is higher than many other vegetables. The roots are antiasthmatic and general stimulant; juice employed for otitis; pounded and applied to ulcers in the nose. The leaves are sialagogue, and used in bronchitis asthma and dysuria. Seeds yield a fatty oil.'

Jack Fruit

Artocarpus heterophyllus Lamk; Eng. Jack fruit, Jack tree; Hindi—*Kathal*; Family—Moraceae.

Sanskrit and Telugu—*Panasa*; Bengali—*Kanthal*; Marathi—*Phanas*; Tamil—*Pilapalam*; Kannada—*Halasu*; Malayalam—*Chakka*.

This is a tree reaching a height of 60 to 70 feet. It has entire leaves and huge fruits, 1 to 2 feet long and weighing from 20 to 40 lbs., which are borne on the trunk.

Jack fruits are sometimes more than two feet long. In Kerala the fruits of 3 feet are very common. The fruits are also borne on branches. It is wildly grown throughout Kerala. In Orissa it is chiefly found in Western district of Koraput and in some coastal districts.

It is supposed to be native of Western Ghats of India. They are grown chiefly in Uttar Pradesh, Bengal, Bihar, Andhra Pradesh and Orissa.

Uses. The fruits and seeds are used as vegetable. The carbohydrate content is especially high in the fruit.

In Kerala various preparations are made out of it. Delicious and tasty deep fried chips of jack fruit are available in the market. The fruits are also eaten ripe. Brahmins of Western Coast prepare "papads" and sweet bread out of ripe fruit.

There are *two* varieties : *Kapa* with sweet and fleshy pericarp; and *Barka*, inferior with mucilaginous sour pericarp. Unripe fruits are used as vegetable and pickled; ripe fruits are eaten fresh or preserved in syrup. Seeds are rich in starch and eaten after roasting or boiling.

Wood is used for general carpentry work; wood yields a yellow dye.

Fig. 4.26. *Artocarpus heterophyllus*. Jack fruit (Kathal)—The fruits and seeds are used as vegetable and the bark is used for tanning. The heartwood yields a yellow dye.

Bread Fruit

Artocarpus communis Forst; Eng. Bread fruit; Hindi—*Barhal*; Family—Moraceae.

It is a native of Malaya and Pacific Islands but grown in Maharashtra, South India and along the West and East Coast. The tree reaches from 40 to 60 feet in height, with deeply incised leaves. The melon-sized prickly fruits, are brownish yellow when ripe with a fibrous yellow pulp. They are often borne in small clusters.

Uses. The fruit is eaten cooked. It is either baked, boiled, roasted, fried or ground up and used for bread. The fruits are used as vegetable. The ripe fruits are useless.

Avocado

Persea americana Mill.; Eng. Avocado, Alligator pear; Family—Lauraceae.

A small tree; native of South America, now cultivated in Bangalore, the Pulney hills, the Nandi hills and the Nilgiris. The fruit is a one-seeded berry. The pulp surrounding the large seed has a butter-like consistency and up to 30% fat. The vitamin content is also high.

Uses. The fruits are used as salad or vegetable.

Fruits are eaten as salad flavoured with pepper, salt and sugar, also used in ice-cream. Fruit is rich in fat with appreciable amount of protein, minerals and vitamins.

Avocado oil is used in cosmetics and pharmaceuticals and as a high grade salad oil. Cake is used as manure. The leaves contain an essential oil. Roots yield an antibiotic used as a food preservative.

Trapa

Trapa bispinosa; Hindi—*Singhara*; family Trapaceae.

Sanskrit—*Shringata, sringataka, trikonaphalam, jalaphala*; Hindi, Marathi, Tamil, Gujarati and Kannada—*Singhara*; Bengali—*Paniphal*; Telugu—*Kubyakama*; Malayalam—*Karimpolam.*

A floating aquatic herb commonly cultivated in the ponds of Northern India. The fruit is a drupe. Each fruit is provided with two stiff spines. The cotyledons are fleshy, white and rich in starch. They are eaten raw or boiled.

Fresh tender kernels are sweet, delicious and farinaceous, also nutritious and a good source of minerals. Eaten fresh, or after boiling or roasting. Meal, prepared by grinding the kernels, is used as a substitute for cereal flour. Fruits are also canned.

Lady's Finger

Abelmoschus esculentus (Linn.) Moench; Eng. Okra, Lady's finger; Hindi—*Bhindi*; Family—Malvaceae.

Bengali—*Dheras*; Gujarati—*Binda*; Marathi—*Bhendi*; Telugu—*Venda bendi*, Tamil—*Vendai*; kannada—*Bhende*; Malayalam—*Bendai, venda.*

It is a native of tropical Africa, now cultivated throughout India. The plant is a stout annual. The young pods are mucilaginous.

Variety. *Pusa Sawani*— Pods long, 5 ribbed, dark green, and spineless. Prolific bearer; suitable for all seasons.

Uses. The fruits are used as vegetable. It can also be dried and canned. The stalks are sometimes used for making fibres.

Tender pods are also used for thickening soups and gravies because of the mucilage content. Flowers are eaten in soups. Tender leaves are boiled and eaten like spinach. Ripe seeds are roasted for use as coffee substitute; also used in curries and chutneys. Seeds are rich in protein (18–26%); they are powdered and mixed with maize flour. A vegetable gum, called Okra Gum is obtained from the plants, and used as a combination flavouring and bodying agent in vegetable soups and gravies. Seeds yield a fatty edible oil.

A mucilaginous preparation from the fruits has found application as a plasma replacement or blood volume expander. Immature capsules are emollient, demulcent and diuretic; Seeds are stimulant, cordial and antispasmodic. The leaves yield an essential oil. Seed cake is rich in proteins.

Capsicum (Sweet Pepper)

Capsicum annuum var. *grossum.*

Variety: *California wonder*— Crops in 75 days, medium height, erect and sturdy. Fruits square, smooth, heavy, thick, deep glossy and green. Flesh very thick, sweet and fine flavoured, bright scarlet when mature.

Uses. Used as green vegetable.

CUCURBITS

Red Pumpkin

Cucurbita maxima Duch; Eng. Red pumpkin, Winter squash; Hindi—*Sitaphal*; Family—Cucurbitaceae.

Tamil—*Parangikayi*; Telugu—*Gummadi*; Kannda—*Kumbalakai*; Malayalam—*Mathan*; Mumbai—*Lal bhopli, lal dudiya.*

Annual vine with large yellow flowers and fruits that rest on the ground. Commonly cultivated in Northern India.

Uses. The large fruits are used as vegetable. They are also used in making ketchups.

All parts of plant are edible; tender shoots and leaves are eaten as salad; flowers and fruits are cooked as a vegetable. Seeds are used as a taeniacide, tonic and diuretic. Fruit pulp is used as poultice on boils and burns.

Summer Squash

Cucurbita pepo Linn; Eng. Pumpkin summer squash; Hindi—*Kumra*; *Safed Kaddu.*

Large vine. Commonly cultivated in Northern India.

Uses. The large fruits are used as vegetable and for making ketchups. The famous sweetmeat *petha* is prepared from these fruits.

Bottle Gourd

Lagenaria siceraria (Mol.) Standl. Syn. *L. vulgaris* Seringe; Eng. Bottle gourd; Calabash cucumber; Hindi—*Lauki, ghia kaddu*; Family—Cucurbitaceae.

Sanskrit—*Alabu*; Bengali—*Lau*; Marathi—*Bhopala, dudhya*; Gujarati—*Dudhi, tumada*; Telugu—*Sorrakaya*; Tamil—*Shorakkai*; Kannada—*Halagumbala*; Assam—*Lau, bogalau*; Punjab—*Ghiya*

A large vine with white flowers. Cultivated throughout India.

Variety: *Pusa Summer Prolific Long*— Prolific fruiting; fruits 40-50 cm long, 20-25 cm in girth when young, yellowish green in colour. Suitable for both wet and dry (hot) seasons.

Uses. The fruit is eaten as fresh vegetable.

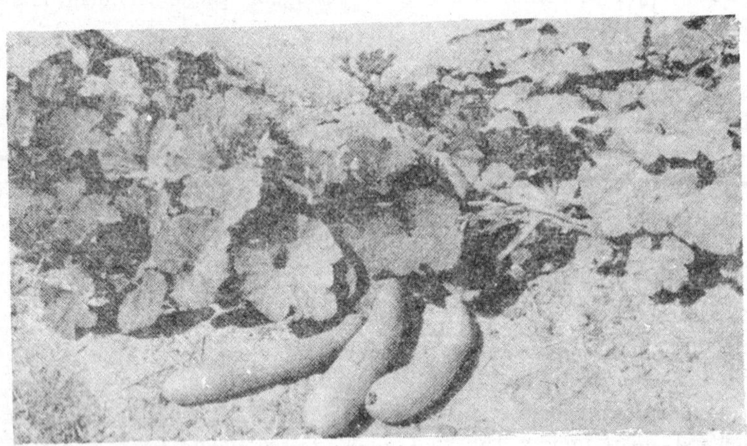

Fig. 4.27. *Lagenarai siceraria.* Bottle gourd (*Lauki*)—*A* climbing or trailing herb. The fruits are used as vegetable.

This vegetable is grown all the year round; the bottle gourds being gathered while tender for use as vegetable. Fruit is a good source of vitamin B and fair source of vitamin C. Seeds are edible, also yield an edible fixed oil. Shell is used as water bottle and bowl. Fruit pulp is cooling diuretic, and antibilious. Seeds are used in dropsy and as anthelmintic. Seed oil is used for headache.

Vegetable Sponge

Luffa acutangula (Linn.) Roxb.; Eng. Vegetable Sponge; Hindi—*Kali torai, Nasdar torai, Jhinga*; Family—Cucurbitaceae.

Sanskrit—*Jhongaka, koshataki*; Bengali—*Jhinga, sataputi*; Marathi—*Shirola*; Gujarati—*Ghisoda*; Telugu—*Birakaya*; Tamil—*Pirkankai*; Kannada—*Hirekayi*; Malayalam—*Pichenga.*

A trailing herb; cultivated throughout India.

Variety: *Pusa Nasdar*— Fruits ridged, light green, medium in size. 15–20 fruits per plant.

Uses. The fruits are used as vegetable. The fibrous material obtained from the dried fruits is used as a substitute for bath-sponges.

Leaves are used as a poultice, in hemorrhoids, leprosy, and splemitis, juice is used in conjuctivitis; decoction for uraemia and amenorrhoea; ripe seeds are emetic and purgative. Seeds yield a fatty oil. Cake is rich in nitrogen and phosphorus and may be used as manure.

Luffa cylindrica (Linn.) M. Roem.; Eng. Vegetable sponge; Hindi—*Ghia torai.*

Sanskrit—*Rajakoshataki, dirgha patolika*; Bengali—*Dhundal*; Marathi—*Ghosali*; Gujarati—*Turia*; Telugu—*Guthibira*; Tamil—*Mozhaku pirkankai*; Kannada—*Tuppahirekai*; Malayalam—*Kattupeechal.*

A trailing herb; cultivated throughout India.

Variety: *Pusa Chikni*—Prolific bearer, fruits smooth, tender and attractive green.

Uses. The fruits are used as vegetable. The fibrous material obtained from the dried fruits is used as a substitute for bathsponges.

The tender fruits are considered diuretic and lactagogue; ripe fruit is used after burning and pulverizing as a carminative and anthelmintic; juice is purgative. Mature seeds are bitter, emetic and cathartic. Seeds yield a fatty oil. Fibro-vascular net-work of ripe fruits affords a sort of sponge widely used for scrubbing and cleaning purposes combined with plaster and varnished over, they make sound-proof and heat-proof wall boarding. They have also been tried as a raw material for paper-pulp.

Round Gourd

Citrullus vulgaris Schrad. Var. ***fistulosus*** Duth. And Full.; Eng. Round gourd; Hindi—*Tinda*; Family—Cucurbitaceae.

A small trailing herb with yellow flowers and round fruits. Cultivated mainly in Uttar Pradesh, Punjab, Rajasthan and Bihar.

Variety 1: *Tinda arka*—Fruits round, light green in colour and medium in size, 15–20 fruits per plant.

Variety 2: *Tinda* S-48—Crop in 70–80 days. Vigorous plant growth, fruit medium sized, weighing on an average 50 gm round flattened and light green flesh white, 8–10 fruits per vine.

Uses. The fruits are used as vegetable.

Snake Gourd

Trichosanthes anguina Linn; Eng. Snake gourd; Hindi—*Chichenda*; Family—Cucurbitaceae.

Sanskrit—*Chachinda*; Bengali—*Chichinga*; Marathi—*Padwal*; Gujarati—*Padavali*; Telugu—*Lingapotla, Potlakaaya* (fruit); Tamil—*Pudal*; Kannada—*Padavalakayi*; Malayalam—*Patavalanga*; Oriya—*Chhachhindara*; Punjab—*Galartori*; Mumbai—*Chikonda*; Mundari—*Lilkaetha.*

A trailing herb, cultivated throughout India.

Uses. The fruits are cooked as vegetable.

They improve appetite and are beneficial in biliousness. Roots and seeds are used in diarrhoea and as a vermifuge.

Patol

Trichosanthes dioica Roxb; Eng. Patol; Hindi—Parwal;
Family—Cucurbitaceae.

Sanskrit—*Putulika*; Hindi and Punjabi—*Parwal*; Bengali—*Potol*; Gujarati and Oriya—*Patal*; Telugu—*Kommupotla*; Tamil—*Kombu—pudalai*; . Kannada—*Kaadupadavala*; Malayalam—*Patolam.*

A trailing herb, cultivated mainly in Northern India, Assam and Bengal. It is also cultivated widely in Orissa and Gujarat.

Uses. The fruits are eaten as vegetable. The leaves are also eaten.

The fruits are also used in confectionery, and pickled. They are particularly useful for convalescents, as they are laxative and easily digestible.

Fruits show some prospects in the control of some cancer-like conditions. Seeds contain a fatty oil.

Bitter Gourd

Momordica charantia Linn; Eng. Bitter gourd; Hindi—*Karela*; Family Cucurbitaceae.

Sanskrit—*Sushavi*, Bengali—*Karela*; Marathi—*Karle*; Tamil—*Pakal*, *pavakka*; Kannada—*Hagal*; Malayalam—*Kaippa*, *Kaippavalli.*

A trailing herb. Cultivated throughout India.

Variety 1: *Coimbatore Long*—Vines prolific, spreading; fruits long, tender white in colour. Very heavy bearing, suitable as a rainy season crop.

Variety 2: *Pusa Do Mausami*—Vigorous vines, fruits medium thick, green 7–8 cm long duration 35 days, suitable for dry as well as rainy season.

Uses. The fruits are eaten as vegetable. They are famous for their bitter taste.

The young fruits may be sliced and preserved as vegetable. They are also pickled and used as a flavouring. Seeds from ripe fruits are used as a condiment. Tender shoots and leaves are also used as vegetable. The leaves are a source of calcium, carotene, riboflavin and ascorbic acid. Seeds yield an edible fatty oil. Fruits are tonic, stomachic, carminative and cooling; used in rheumatism, gout and diseases of liver and spleen, also for diabetes. Juice of leaves is given in bilious affections. Root is used for hemorrhoids.

Momordica cochinchinensis Spreng; Hindi—*Bhat Karela, Kakrol,* Cultivated throughout the country.

Sanskrit—*Karkataka*; Bengali—*Go-kakara*; Telugu—*Adavi kakara*; Marathi—*kakana;* Gujarati—*Karapata*.

Uses. The fruits are used as vegetable.

Besides tender fruits, the young leafy shoots are also cooked and eaten.

Seeds yield a fatty oil used as an illuminant, also employed in paints and varnishes. Seeds are used as aperient and in the treatment of ulcers. Fruits and leaves are used in external applications for lumbago, fracture and ulceration.

Momordica dioica Roxb. ex Willd; Eng. Small bitter gourd; Hindi—*Jangli karela*, *musela*, A common climbing or spreading shrub.

Sanskrit—*Vahisi*; Bengali—*Ban-karela*; Marathi—*Kartoli*; Telugu—*Agakara*; Tamil—*Tholoopavai*; Kannada—*Karlikai*; Assam—*Bhatkarela*; Punjab—*Kakaura*; *kirara*, *dhar karela*.

Uses. The fruits are used as vegetable.

Kernels yield a semi-drying oil used in paint and varnish industry. Roots are used in bleeding oils, bowel affections and urinary complaints.

Cucumber

Cucumis sativus Linn; Eng. Cucumber; Hindi—*Khira*; Family—Cucurbitaceae.

Sanskrit—*Sukasa*; Bengali, Hindi and Marathi—*Khira*; Telugu—*Dosakaya*; Tamil—*Vellarikkai*, *kakrikai*.

It is indigenous to Southern India. It has been cultivated for 4,000 years. It is a trailing vine with yellow flowers.

Variety 1: *Poinsette*—Fruit dark green and ends blunt. Prolific bearer starts fruiting in 60 days.

Variety 2: *Japanese Long Green*—Extra early, crops in 45 days, fruits dark green, 25–35 cm in length. Flesh light green and crisp.

Uses. The fruits are eaten raw, pickled or cooked.

Seeds are diuretic, tonic, refrigerant. Fruits are much used in salads. Odorous principle of cucumber is extractable with alcohol used in blending certain bouquet perfumes.

Muskmelon

Cucumis melo var. ***reticulatus.***

Variety 1: *Hara Madhu*—Vigorous plant growth, large sized fruits weighting on an average 1 kg. almost round in shape, light yellow in colour with distinct green ribs when ripe. Late in maturity. Flesh green, crisp, very sweet. Non-slipping type. Maturity 100–105 days.

Variety 2: *Punjab Sunehri* (*Pb. Golden*)—Vine prolific, foliage dark green. Female flowers complete. Fruit globular round 700–800 gm each. Rind brown in colour with intense netting, very hard, stands transportation. Slip when fruit mature. Flesh orange with green tinge towards rind, very sweet, cavity small; fruit maturity 90 days.

Cucumis melo Linn. var. ***utilissimus*** Duth. Full; Hindi—*Kakri.* Commonly cultivated in Punjab and Uttar Pradesh.

Uses. The fruits are eaten raw, pickled or cooked as vegetable.

Kovai Fruit

Coccinia cordifolia (L.) Cogn.; Syn. *C. indica* Wight and Arn.; Eng. Kovai fruit; Hindi—*Kanduri, kudroom*; Family—Cucurbitaceae.

A twining or spreading shrub, cultivated in Assam, Bihar, Orissa, Bengal, Maharashtra, Andhra Pradesh and Tamil Nadu.

Uses. The fruits are used as vegetable.

Legumes

The seeds and pods of several legumes are used as greens and vegetables. The important leguminous vegetables are—***Dolichos lablab*** L.; Eng. Lablab, Hyacinth bean; Hindi—*Sem.* ***Pisum sativum*** Linn. var„ ***aravense***; Eng. Pea; Hindi—*Matar.* ***Phaseolus vulgaris*** Linn; Eng. Kidney bean, Faraz bean; Hindi—*Vilaiti sem.* ***Cyamopsis tetragonoloba*** (Linn) Taub; Cluster bean; Hindi—*Guar. Vicia faba* Linn.; Eng. Broad bean; Hindi—*Bakla.* ***Vigna sinensis*** (Linn.) Savi ex Hassk; Eng. Cowpea; Hindi—*Lobia.* ***Cicer arietinum*** Linn; Eng. Gram; Hindi—*Chana* of family Leguminosae.

Peas

Pisum sativum; Hindi—*Matar* (Papilionaceae).

Variety 1: *Bonneville*—Mild season variety; plants medium tall; crop ready in 85 days; double podded; pods 8 cm long, sweet and well filled. Heavy yielder. Seeds wrinkled.

Variety 2: *Jawahar*—Plant medium with light green pod. Pod length 7–8 cm. with 6–7 seeds. First pod is produced at ninth node. Maturity 80–85 days. Seed wrinkled.

Variety 3: *Arkel*—Wrinkled seed very early variety, pods sweet, well filled. Ready for first picking in 60–65 days.

Cowpea

Vigna sinensis; Hindi—*Lobia.*

Variety 1: *Pusa do Fasli*—Bushy in habit, produces yellowish green tender pods which are more than 15 cm long, heavy yielder. Day neutral suitable for all seasons. Ready for picking in 50 days.

Variety 2:. *Pusa Barsati*—Very early variety, ready for picking in 45 days, 2 to 3 flushes of pods; pods 25-27 cm long, heavy yielder. Suitable for growing under long day conditions only.

Variety 3: *Pusa phalguni*—Bush type. Pods dark green, 12 cm long, ready for picking in 60 days, heavy yielder; grown only under short day conditions.

Guar

Cyamopsis tetragonoloba

Sanskrit—*Bakuchi, dridhabija, goraksha phalimi, gorani*; Hindi—*Guar*; Marathi—*Bavachi*; Tamil—*Kothaveray*; Telugu—*Gorchikudu*; Kannada—*Gorikayi*; Punjab—*Kulti, guar, kuwara.*

Variety: *Pusa Navbahar*— Non-branching, single stemmed, plant about one metre tall. Pods in branches, bearing profus, pod length 10–12 cm soft, light green, very meaty. Variety suitable for all seasons.

The pods of *Guar* are used as vegetable. They are commonly used in Northern India. The pods are generally found in the rainy season. They are commonly eaten with *chapatis* of maize (*makka*). The *makka* and *guar* make the food of poor people whereas the delicacy of rich.

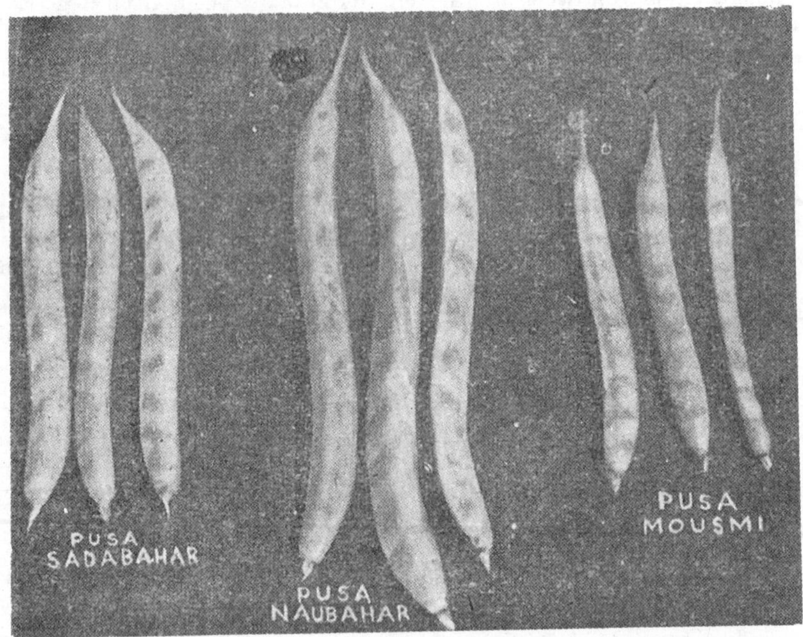

Fig. 4.28. Three improved varieties of cluster bean (*Guar*).

"*Guar*" is also not uncommon in the markets of South India. Recently it has been established by scientists that "*Guar*" is a very good vegetable for 'diabetics'.

The plants and pods are commonly utilized for fodder.

Seedflour, a source of 'guar gum' is used in food preparation, cosmetics, and paper and textile industries. It has high viscosity at low concentrations. Seeds are highly valued as cattle feed.

Bean Dolichos

Dolichos lab-lab; Hindi—*Sem.*

Tamil—*Seeme kattirikkai*; Kannada—*Seembadane.*

Variety : *Pusa Early Prolific*—Early field type; pods long and thin; bearing in bunches heavy yielder.

Bean French

Phaseolus vulgaris.

Variety 1: *Contender*— Bush variety; ready for picking in 50 days; pods round, green, very long, curved at the tip, thick, meaty and stringless; very prolific bearer; suitable for most areas except very hot and dry regions.

Variety 2: *Pusa Parvati*— Prolific, bush type, pods attractive, light green in colour, round and meaty. Early variety, pods available 5 days earlier than contender.

DRYING AND STORAGE OF VEGETABLES AND FRUITS

A person requires 200–300 gm of vegetables per day. Out of which 85 gm consist of root vegetables and 110 gm leafy vegetables. But we consume only a small amount of the necessary vegetables, and one of the reason for this is non-availability.

In season, we have a bounty of delicious fruits and vegetables. And once the season is past, we start feeling the pinch for them. Very often a good deal of the fruits and vegetables is wasted. Probably, the idea of keeping away a part of the bounty for the rainy day has not struck many of us so far. Indeed all the surplus vegetables and fruits can be carried over provided they are preserved in time.

The method is simply the drying of fresh and mature fruits and vegetables. Done correctly, their taste will be greatly changed, not their value.

Drying of vegetable has advantage too. When we dry the vegetables, they are reduced in bulk even up to 90 per cent. Ten kg of fresh cabbage may be reduced to one kg of dried cabbage., and this makes the storage of these very easy. A little space will be needed to store them. Cheap containers like alkathene bags or tins with airtight lids will do for storing.

We can dry fruits and vegetables by any of the three means: *sun drying*, *oven drying* or *mechanical drying.*

Whatever the means we employ there are a few basic rules to be followed. The fruits or vegetables we want to store should be clean so also the equipment we use and the drying place.

For preserving, one should select always the best quality of fruits or vegetables.

One should ensure that the fruits are fully developed, ripe but firm and in ready-to-eat stage. Vegetables should be mature and tender and juicy.

If the vegetables and fruits are obtained from the kitchen garden, one should pluck them early in the morning. This helps keep their freshness and flavour longer. If they are to be purchased from the market, go marketing early in the morning and try to dry all fruits and vegetables thus collected within six hours.

A few vegetables and fruits are dried whole but many of them have to be peeled, sliced or cured. Use a stainless steel knife for the purpose. This prevents discolouration.

Most of the vegetables have to be sliced 1/8 to 1/4 inch thin. Thin slices dry fast and this would mean more flavour retained. Before slicing, wash the vegetables and fruits thoroughly. You can clean them by wiping with a wet cloth also. Peel or pit fruits as needed.

To check the fruits and vegetable slices getting discoloured, they are to be dipped in a weak salt solution (3 tablespoonfuls in one litre of water).

Most of the vegetables need steaming or blanching to retain their flavour, colour and nutritive quality. Sulphuring is used for fruits to achieve the same results.

Blanching is done this way. After washing, peeling and slicing vegetables, put them in boiling water for some time varying from two to eight minutes, while in water, vegetable pieces should be stirred to keep them from sticking to the pan or to each, other. The water in the pan should be twice the quantity of vegetables.

Blanching can also be done by steaming the vegetables by suspending them in live steam over vigorously boiling water. There are two methods of steaming. For the first method, a deep container with a tight-fitting lid is required. In case a proper lid is not available, cover the container with a board which will cover the container completely so that the steam does not escape.

Put a rack at the bottom of the container. The rack should be 4 cm high. A wooden rack can be used or it can also be made from bamboo. It should be such that it permits water to boil freely from it.

Place the vegetable in a container on the rack. The container can be a colander, wire basket or reed basket. If this is not available, place another pot over the rack inside the bigger container. Pour water to the level of about 3 cm in the bigger container and bring it to boil. Place the basket with vegetables on the rack. Count time as soon as the steam is formed. Each vegetable has a different optimum steaming time, as is given in the chart.

Instead of putting the vegetables in a basket, they can also be put in a loosely woven clean cloth. Tie the vegetables loosely and suspend them in the container for the required length of time.

After blanching, shift the vegetables on trays for drying. A tray with holes in the bottom ensures quick and thorough drying. A wooden frame, at least four inches off the ground with wire netting at the bottom and mosquito curtain as the cover will be the best.

Now comes the sulphuring of fruits. This is done to avoid discolouration during the drying process. It also protects the fruits and vegetables from insects during storage. Sulphur absorbed by the fruits is almost completely driven off when the product is cooked.

Take a small quantity of about 2.5 gm of sulphur for every 10 kg of fruits. Keep the fruits on wooden trays. You will, need a wooden or a card board box to cover the trays for sulphuring. It should be deep enough to cover the stacked trays plus the raised platform over which the trays are kept.

First raise a platform by stacking bricks, blocks or stones on which trays can be stacked. It should be 15 to 20 cm high.

VEGETABLE DRYING CHART

Vegetables	*Preparation*	*Blanching time*	*Condition when dry*
Beans-green	Snip, cut into 1" lengths	5–8 min.	Brittle greenish black
Beans-lima	Shell, wash	3 min.	Hard, wrinkled
Cabbage	Cut in 1/8' shreds	3–4 min.	Brittle
Carrots	Cut in 1/8" shreds	3–4 min.	Brittle
Cauliflower	Break into 1" flowerettes	4–6 min.	Tough, brittle
Brinjals	Slice 1/4" thick	1 min.	Tough
Bitter gourd	Cut into 1/4" long pieces	6–7 min.	Tough
Lady's fingers	Slice 1/4" thick	3 min.	Brittle
Potatoes	Peel, cut into slices 1/4" thick	2–3 min.	Brittle
Peas	Shell	3–5	Hard, wrinkled
Pumpkin	Peel, seed, slice 1/8" thick	3–6 min.	Tough, brittle
Spinach or other greens	Cut off roots, wash trim	2 min.	Crisp, brittle
Sweet potatoes	Peel, wash, slice in 1/2" strips	6–8 min.	Hard, brittle
Turnips	Slice 1/4" thick	1–2 min.	Tough, brittle

Place the fruits on trays and place in a stack over the platform. The fruits should be kept only in one layer. Stack the trays one on top of the other leaving a space of about 3 cm between two trays. This can be done by putting strips of wood between the trays to separate them.

Place sulphur on a small piece of paper, roll loosely and twist ends so that the end can be lighted. Put this in a small metal or crockery container. Set it by the side of the stack of trays and light the paper containing sulphur.

Quickly cover the stack and the sulphur container with the box and keep it thus for the length of the time the fruit has to be sulphured. Different fruits and vegetables need different periods of sulphuring. For this, chart is being given here.

After blanching and sulphuring, the vegetables and fruits are now ready for drying. The cheapest and time-honoured method of drying is sun-drying. Spread the fruits and vegetables evenly on trays in thin layer in the morning so that most of the moisture is removed during the first day itself. Turn then every two hours on the first day and twice thereafter Just before sun-set, bring in the trays and turn the fruits and vegetables. The time for complete drying will take between two to five days, depending upon the fruits and vegetables.

SULPHURING OF FRUITS

Fruit	*Preparation*	*Time*	*Condition when dry*
Apples	Peel, core, slice or cut in eights, dip in light brine solution	1—1½ hrs.	Springy
Apricots	Cut in halves, remove pits	1—1½ hrs.	Leathery
Berries	Dry whole	—	Springy
Cherries	Remove the pits	½ hr.	Leathery
Pears	Cut in halves, core	1½ hrs.	Leathery
Mango (raw)	Peels, remove the stone, cut in pieces	1½ hrs.	Leathery
Bananas (raw)	Peel, cut into halves lengthwise or slice 1/4" thick	½ hr.	Brittle

One can dry by other methods also such as oven drying and mechanical drying.

Oven drying is done by spreading fruits and vegetables on trays or plates in a single layer and kept in a hot oven. The heat is to be kept at 140° to 150°F. Trays should be kept for 5 minutes in the oven and taken out in a draft (under a fan) and kept for cooling for 15 minutes. Again the tray is kept in the oven. This is repeated till the food is dried enough. Oven drying is the best for quick drying but needs constant care.

Mechanical drying needs special machinery which is out of reach for an ordinary housewife. But in case the whole community is interested the machinery can be procured through the help of government or private agencies.

After drying, keep the fruits and vegetables in moisture proof and clean container. One can use alkathene bags, tin cans with tight fitting lids or glass jars. See that the container is completely dry. Fill it as full as possible. Small containers are better than the large ones because once you open the tin the contents will be used in a much shorter time so there is less likelihood of spoilage. Check foods often to see that these are all right. It will pay well to keep the dehydrated food in sun once in a while.

This all is being done to preserve the fruits and vegetables when they are surplus in season.

Before using preserved food, soak them in as much water as they can absorb, not more, not less. You will now see them spring back to their freshness. Green and leafy vegetables need not be soaked as they soak sufficient water when being cooked.

As most of the vegetables are pre-cooked or blanched before drying, they need less water for cooking. Cook the food in the same water in which it has been soaked.

Dried fruits need less sugar than fresh fruits. When sugar is used, it should be added in the end, otherwise fruit will not absorb water. A pinch of salt helps in brining out the natural sweetness of the fruit.

Dehydrated fruits and vegetables are next only to fresh ones yet they are the best buy when fresh vegetables are scarce in the market.

SOME OTHER IMPORTANT VEGETABLES

Chayote

Sechium edule Sw.; Eng. Chayote, Christophine; Hindi—*Launku*; Family—Cucurbitaceae.

A trailing or climbing vine, which produces a gourdlike fruit much prized a vegetable. The applelike fruits are technically pepos, berries with a spongy pulp and a hard firm rind. The plant is a perennial with large tuberous roots. The plants grow vigorously and have a prolific yield. The tubers and fruits are used as vegetable, whereas the foliage can be used for greens or forage.

The plants are mainly cultivated in the hilly regions of Northern India.

Unripe fruits are used as vegetable and also consumed in salads. Roots are a source of starch, used as a substitute of arrowroot starch. Seeds are cooked in butter and eaten. Tender shoots are also eaten as vegetable. Fruits, vines and tubers are used as fodder. Woody stems yield a fibre.

Rhubarb

Rheum rhabarbarum; Eng. Bucharian rhubarb; Family—Polygonaceae. A herb. The leaves are consumed as vegetable. *R. rhaponticum* L.; Eng. Garden rhubarb. A large herb cultivated in Tamil Nadu and West Bengal in India. The thick leaf-stalks are used as vegetable.

Pleasantly acidic petioles are used in sauces and pies, added to preserve for tartness and mixed with fruits for flavour. Stewed stalks are eaten. Juice extracted from leaf stalks is used for making wines and beverages. Leaves are used as feed for stock.

5
Fruits

India grows a large variety of fruits belonging to the *tropical* and *temperate* groups. However, their present availability is only 1.5 ounces per adult per day, whereas a balanced diet requires 3 ounces of fruits per adult per day. A good number of the fruits grown in India are introduced from foreign countries.

From the very ancient times the man has had close contact with the fruits in some or the other way. It is interesting to note that a large number of the fruits grown today had their origin in the same part of Asia that was the earliest home of man. The man began to cultivate the fruits in good old days, choosing the fruits of best taste and largest yield, as well as those which were easiest to grow. Later the civilization progressed and the nutritive value and other qualities of the fruits were studied, our present-day fruits were thus developed gradually. Many varieties were known to the Indians, Greeks, Romans and Egyptians and fruit growing was an important part of their civilization. The mango is the most ancient among the tropical fruits indigenous to India. Few other tropical fruits have the historic background that the mango has and few others are so intimately connected with Indian folk-lore.

Morphologically a fruit is the seed-bearing portion of the plant, and consists of the ripened ovary and its contents. Simple fruits are derived from a single ovary, and compound fruits from more than one. The aggregate fruits are formed from numerous carpels of the same flower, while composite fruits develop from the ovaries of different flowers of an inflorescence. In the following paragraphs only those fruits will be considered which are usually eaten without cooking. For convenience the fruits have been divided into two groups—1. Tropical fruits in India and 2. Temperate fruits in India.

TROPICAL FRUITS IN INDIA

Mango. ***Mangifera indica*** Linn.; Family—*Anacardiaceae*; Eng.—mango; Verna. *Am*. Sanskrit—*Amra chuta* ; Bengali—*Am*; Hindi—*Am, amb*; Marathi—*Amba*; Gujarati—*Amri*; Telugu—*Mamridi, mavi*; Tamil—*Manga, mau*; Kannada—*Mavu*; Malayalam—*Amram, cutam, mavu*.

History. The mango *(Mangifera indica)* is one of the oldest tropical fruits and has been cultivated for nearly 6000 years. From its indigenous home in India, somewhere at the base of the Himalayas, the cultivation must have spread at an early age over the Indian Peninsula. According to Rumphius, it has been introduced into certain islands of the Asiatic Archipelago within the memory of living men, while in others it has existed from a very remote epoch. It is closely connected with Sanskrit mythology and is mentioned in many of the old tales and folk lore of the Hindus. Abul Fazl, in the *Ain-i-Akbari* about 400 years back, describes a large number of cultivated races and states that "mangoes are to be found everywhere in India, especially in Bengal, Gujarat, Malwa, Khandesh and the Deccan."

The mango is the most popular and the choicest fruit of India and occupies a prominent place among the best fruits of the world.

It is by far the most important fruit crop of the country, occupying about 60% of the total area under fruits. The total area under mango is estimated at 2 million acres. The largest acreage is in Uttar Pradesh ; Bihar ranks second ; other important mango growing areas are distributed in West Bengal, Andhra, Tamil Nadu and West Coast.

Varieties. Mangoes are grouped under two broad categories—seedling types and horticultural clones, propagated by budding or grafting. A large number of mango types estimated at over 1,000 are grown in various parts of India, each having its own peculiar taste, flavour and consistency of pulp.

Fig. 5.1. Mango. Langra variety, unripe fruits.

Cultivation. Mangoes thrive in all parts of India where temperatures as high as 115—120^{0}F prevail during the summer. The mango thrives in a wide variety of soils. It grows in rich clayey loams, as well as in poor sandy and gravelly soil, provided it is fairly deep and well drained. Some of the best mango groves are found in the Indo-Gangetic plains and also on the banks of rivers in Peninsular India, where the soil is much alluvial loam of great depth.

Uses. The mango fruit is one of the most highly prized dessert fruits of the tropics. It has a rich, luscious, aromatic flavour and a delicious taste in which sweetness and acidity are delightfully blended. Young and unripe fruits are usually acidic and used in pickles, chutney, *amchur* and culinary preparations. Ripe fruits are preserved by canning or used in the manufacture of juice and squash, jams and jellies, preserve (*murabba*) and *ampapur*. Sucrose, glucose and fructose are the principal carbohydrates present in ripe mango; maltose is also present. Unripe fully developed mangoes of pickling varieties contain citric, malic, oxalic, succinic and two unidentified acids. As the fruit ripens the acidity gradually decreases with a steep fall at ripe stage. Ripe fruits constitute a rich source of vitamin A; some varieties contain fairly good amounts of vitamin C also. The fruit is a rich source of potassium.

Fig. 5.2. Mango tree bearing fruits.

Ripe mango fruit is considered invigorating, refreshing and fattening. The juice, along with aromatics, is recommended as restorative tonic. It contains vitamins A and C and is useful in heat apoplexy.

Citrus Fruits. They are supposed to be natives of Eastern and Southern Asia. Some of the citrus fruits have been cultivated for over 3,000 years. The plants are thorny aromatic shrubs or small trees. The coriaceous evergreen leaves are glandular dotted and unifoliate—compound with a joint between leaf blade and petiole. The white or purplish flowers are solitary, but produced in abundance, and often very fragrant. The fruit is a modified berry known as a hesperidium. The citrus plants are cultivated in such places where there is sufficient amount of moisture. The fruits ripen throughout the year.

The citrus fruits contain considerable amounts of the essential vitamin C, the antiscorbutic vitamin as well as fruit acids. They are also used in the preparation of juices, squashes and other canned products.

Several citrus fruits have been described here:

Lime

Citrus aurantiifolia (Christm.) Swing., Family—Rutaceae; Eng.—Lime; Hindi—*Kaghzi-nimbu*; Bengali—*Kaghzinimbu, patinebu*; Gujarati—*Khatalimbu*; Tamil—*Elumichai*; Telugu—*Nimma*; Kannada—*Limbe, nimbe*; Malayalam—*Erumichinarakam*.

This is a large shrub or small tree. It is cultivated throughout India, but chiefly in Andhra Pradesh, Maharashtra, Karnataka, Assam, Bihar, Uttar Pradesh, Punjab, West Bengal, Madhya Pradesh and Rajasthan.

Uses. The fruits make a good source of vitamin C. They are used raw and pickled. The oil distilled from the peel is used in confectionery, pharmaceuticals and toilet preparations.

Fruits are extensively used for culinary purposes; for flavouring jams, jellies, marmalades, and alcoholic drinks, and as a garnish. Considered appetizer, stomachic, and antiscorbutic.

Citrus reticulata Blanco; Family—Rutaceae; Eng.—Mandarin; Hindi—*Santara*; Bengali—*Kamala*.

Small tree. Cultivated in the Khasi hills, Darjeeling, Garhwal, Dehradun, Sikkim, Tripura, Himachal Pradesh, Punjab, Nilgiris, Palni and Yercand hills in Tamil Nadu and Maharashtra.

Uses. The fruits are delicious. They are rich in vitamin C. The peel is used for marmalades. The essential oil is distilled mainly in Nagpur and Coorg and used in confectionery, pharmaceuticals and toilet preparations.

It is most valued commercial orange. Fruit is used mainly as dessert and in the production of orange juice. Petitgrain oil is obtained from leaves and twigs. Peels yield Mandarin Oil.

Sweet Orange

Citrus sinensis (Linn.) Osbeck; Family—Rutaceae; Eng.—Sweet orange; Hindi—*Musambi, malta*,; Bengali, Gujarati, Marathi and Hindi—*Musambi, narangi*; Kannada—*Kittile, sathgudi*; Tamil—*Sathagudi, chini*; Telugu—*Battavinaringa, naranji, satghudi, sini*.

It is grown in the Deccan, Maharashtra, Punjab, Rajasthan and Tamil Nadu.

Uses. The fruits are edible. They are good source of vitamin C. The peel is the source of orange oil.

Fruit is sweet and juicy, nutritious, highly esteemed dessert fruit. Peels are the source of an essential oil called Orange Oil. Flowers also yield an essential oil, Neroli Oil. Leaves and young shoots are another source of Petitgrain Oil. Fruit juice is useful in bilious affections.

Sour Orange

Citrus aurantium Linn.; Eng.—Sour orange; Hindi—*Khatta*; Tamil—*Narangam, narattai*; Telugu—*Mallikanarangi*; Kannada—*Heralay*; Malayalam—*Karna*.

Fig. 5.3. *Citrus.* They are a rich source of vitamin C. The citrus fruits also contain vitamin P, which keeps the small blood vessels in our bodies in a healthy condition and helps in the assimilation of vitamin C.

A small tree. Found scattered along with other citrus.

Uses. The fruits are edible. The leaves are the source of an essential oil, which is used in confectionery, cosmetics and perfumery.

Used in the preparation of confections, marmalades, liqueurs, and other drinks. Rich source of provitamin A and B_1. Lime oil obtained from the fresh rinds is the source of Bergamot Oil extracted from the peels. Leaves and tender twigs yield Petitgrain oil.

Shaddock

Citrus maxima (Burm.) Merrill; Eng. Shaddock ; Hindi—*Chakotra* ; Benglai—*Mahanibu, sadaphal*; Gujarati—*Obakotru*; Marathi—*Pains, papnasa*; Malayalam—*Pamparamasam*; Kannada—*Chakotre, sakkota*; Tamil—*Pambalimasu* ; Telugu—*Pampalamasam*; Family—Rutaceae.

Small tree. Grown in Tamil Nadu, Karnataka, Maharashtra, Punjab and Uttar Pradesh.

Uses. The fruits are edible. They are good source of vitamin C (ascorbic acid).

Fruits esteemed for dessert; made into jams and marmalades; considered nutritive and refrigerant. Leaves are used in epilepsy, chorea, and convulsive coughs.

Citron

Citrus medica Linn.; Family—Rutaceae; Eng.—Citron; Hindi—*Baranimbu, madhkunkar*; Bengali—*Bara nimbu, begpura*; Gujarati—*Turanj, bijoru*; Marathi—*Mahalunga, mavalung*; Malayalam—*Gilam, rusakam*; Kannada—*Madala, mahaphala*; Tamil—*Kadaranarathai;* Telugu—*Lungamu.*

Small tree. Found in the Khasia hills of Assam, submountainous Himalayan ranges in Garhwal, Kumaon in Uttar Pradesh and Punjab, Wyanad areas of South, Pachmari (Madhya Pradesh), Sikkim and the Western Ghats.

Uses. The fruits are edible. They are enriched with vitamin C.

Fruits are used mainly for pickling, also candied. Peel is made into marmalades and other preserves. Preserved rind is used in dysentery. Citron oil is obtained from fresh rinds.

Sweet Lime

Citrus limettioides Tanaka; Family—Rutaceae; Eng.—Sweet lime; Hindi—*Mithanimbu*; Bengali, Gujarati and Hindi—*Mithanebu, mithalimbu*; Tamil—*Kolumichangai,*; Telugu—*Gajanimma*; Kannada—*Gajanimbe, imbe.*

Small tree. Cultivated in Northern India and Madhya Pradesh.

Uses. The fruits are edible and good source of vitamin C. It is also used as a root stock for sweet oranges and mandarins. The leaves yield an essential oil known as 'petitgrain oil', which is used in confectionery and cosmetics as a flavouring substance.

This is valued as a refrigerant in fevers and jaundice. Eaten either fresh or after cooking, also used in preserves.

Grapefruit

Citrus paradisi Macf.; Family—Rutaceae; Eng.—Grapefruit.

Small tree. Commonly cultivated in Punjab, Uttar Pradesh, Madhya Pradesh and Bengal.

Uses. The fruits are edible. Rich in vitamin C.

Also a fair source of vitamin B_1. Used mainly as a breakfast fruit. Rinds yield grapefruit oil, employed in perfumery and as a flavouring. Grapefruit juice is recommended for building up resistance to common colds. Dry and fortified wines, brandies and cordials are prepared from the fruits.

Lemon

Citrus limon (Linn.) Burm. f.; Family—Rutaceae; Eng.—Lemon; Hindi—*Nimbu*; Bengali—*Baranebu, goranebu* ; Gujarati—*Motulimbu*; Marathi—*Idalimbu, thoralimbu*; Kannada—*Bijapura, bijori*; Tamil—*Periya yelumichai*; Telugu—*Bijapuram*.

It is native of South-Eastern Asia. It had reached India at an early date for there is a Sanskrit word for it. The lemon is a small tree, 10 to 20 feet in height, with short spines and large white and purple flowers. The small, light yellow, oval fruits end in a blunt point. The fruit is picked green, as it deteriorates if allowed to ripen on the tree. It is ripened in storage. Lemons contain 0.5% sugar and 5% citric acid.

Uses. The fruits are eaten raw and pickled. The juice is used for lemonade and other beverages and as a flavouring substance. Bleaching agent and stain remover. The rind is the source of oil of lemon. Lemon oil is used in perfumery and toilet soaps. It is also utilized in confectionery.

Used for culinary purposes and in the preparation of beverages. Citric acid, pectin, and lemon oil are obtained as byproducts. Oil of lemon is also used as carminative and for flavouring liqueurs. Lemon juice is very useful for scurvy. Pickled fruit is useful in hypertrophy of spleen.

Kumquat

Fortunella japonica (Thunb) Swing; Family—Rutaceae; Eng.—Kumquat.

A shrub or small tree. It is native of East Asia, now grown mainly in Northern India. The fruits are globose.

Uses. The fruits are edible.

The fruits are used also for marmalades, jellies, chutneys, pickles, and as candied fruit, or in the preparation of drinks. They yield a volatile oil.

Litchi

Litchi chinensis (Gaertn.) Sonn. of Family Sapindaceae, Eng.—Litchi ; Verna. *Lichi*.

Litchi is reported to have been introduced into India from China towards the end of the eighteenth century. It is now cultivated in a number of countries, outside China. India and South Africa are the largest producers outside China. The total area under litchi in India is estimated at 23,950 acres, of which 23,600 acres are in North Bihar, mostly in Muzaffarpur and Darbhanga districts. The rest of the area under litchi lies in the sub-montane districts of Saharanpur, Dehradun and Muzaffarnagar in Uttar Pradesh.

Cultivation. For successful cultivation the following requirements are considered essential : (1) humid atmosphere, (2) freedom from injurious frosts, (3) abundance of soil moisture and (4) deep loamy soil. It does not fruit well in tropical regions, except at elevations high enough to be cold. It thrives best when planted in deep rich soil near banks of irrigating canals. The soils of litchi growing areas in Bihar and Uttar Pradesh are rich in lime.

Harvesting. Fruits are harvested from the middle of May to the end of June or July in most areas of Northern India. The maturity of fruit is indicated by colour, flatness of tubercles, and brittleness of epicarp. A tree at its prime may give 4,000–5,000 fruits per annum. Individual trees are reported to have given as many as 15,000 fruits (1,000 lb.), while in China yield as high as 1,500 lb. per tree has been reported.

Uses. Litchi fruits are usually consumed fresh. They remain in fit condition for 3–5 days. The litchi fruit consists of peel, aril and seed. The aril which can be readily separated from seed, is soft and juicy with a delicious flavour and is generally eaten fresh. Litchi arils can be preserved by canning with syrup.

Fig. 5.4. ***Banana plantation.*** Banana is one of the oldest fruits of the world. Its names "Adam's fig" and "Apple of Paradise" suggest its antiquity. It can give up to 50 tons of fruit per acre within a year or 15 months. The ripe fruit is a good source of vitamin A and fair source of vitamins C, B_1 and B_2. It is also rich in minerals like Mg, Na, K and P and is a fair source of Ca and Fe.

Banana

Musa paradisiaca Linn.; Family—Musaceae; Eng.—Banana; Hindi—*Kela*; Sanskrit—*Kadli, rambha*; Telugu—*Arati, anati*; Tamil—*Vazhai*; Kannada—*Bale*; Malayalam—*Vazha*;.

It is native of India and Malaya. It is grown in the places where the climate is warm, humid and rainy. Kerala, Tamil Nadu, Andhra Pradesh, Karnataka, Gujarat, West Bengal, Bihar, Assam Jalgaon district (Maharashtra) and the coastal areas are ideal for growing bananas.

It is very ancient plant, possibly the world's oldest cultivated crop. It was important in Assyria in 1100 B.C., and it was well known to all other early civilizations.

There are many varieties of bananas. The most popular ones are—*Poovan* of Tamil Nadu which is also known as *Lalvelchi* in Maharashtra, *Champa* in Bengal and *Karpura Chakkerakeli* in Andhra Pradesh. *Basrai Dwarf* also known as *Dwarf*, in certain places, *Vamankeli* in South India. *Kabuli* in Bengal and Orissa, *Bhusavali* in Maharashtra and Madhya Pradesh. *Harichhal*, also known as *Bombay green* in certain areas. *"Nendran"* and *"Kadali"* varieties of Kerala are also important.

Fig. 5.5. Banana—Fruit of *Chakkerakeli*.

Fig. 5.6. Banana—Fruits of *Basrai Dwarf.*

The banana is one of the tallest of the herbs. The tree-like stem is composed of the sheathing spiral leaf bases. At the top of the 10 to 30 feet stem there is produced a crown of large oval deep-green leaves. The leaves are up to 12 feet in length and 2 feet in width, with a prominent midrib. Each plant produces a single inflorescence. The flower stalk develops from the rootstock and pushes its way through the pseudostem emerging in the center of the crown. The drooping inflorescences develop into the bunches of bananas. Bunches consist of from six to fifteen clusters (hands). Each hand contains 10 to 20 individual bananas (fingers). The fruit of the cultivated banana is a seedless berry. As soon as the tree bears, it dies or is cut down, and suckers develop from the rhizome, which give rise to new plants. Some of these suckers have pointed, narrow and long sword-like leaves and are hence called 'water suckers'. The sword suckers are selected for planting. The roots of the suckers are pruned before planting. November to February and June to July are the best months for planting.

Uses. The fruits are edible. They have a high content of carbohydrates with some fats and proteins. Their food value is three times that of wheat. Green bananas may be cooked and eaten as vegetable. Banana powder can be used as baby food and in the manufacture of chocolate and biscuits.

The Kerala banana chips famous all over India, are prepared from raw "Nendran" varieties. Ripe "Nendran" when cooked in steam and dried under the Sun, can be kept months together like dried date palm fruits and it forms a very good breakfast.

Edible bananas of hybrid origin valued for their seedless fruits. Unripe fruits are eaten as vegetable. Fruit-pulp is dried and made into flour; used also for jams and jellies, sugar coated chips and several Indian confections. It makes a fair source of minerals and vitamins particularly of B group. Peels are used as cattle feed. Inflorescence before opening is used as a vegetable. Core of pseudostem is eaten after cooking; starch in pseudostem is used for finishing of textiles. Banana fruit is laxative and used in intestinal disorders, uraemia, nephritis, hypertention and other vascular diseases. It is a very nutritious fruit.

Guava

Psidium guajava Linn.; Family—Myrtaceae; Eng.—Guava; Sanskrit—*Mansala*; Hindi—*Amrud, safed safari* ; Bengali—*Goachhi, peyara, piyara*; Marathi—*Jamba, tupkel*; Gujarati—*Jamrud, jamrukh, peru*; Telugu—*Ettajama, goyya, tellajama;* Tamil—*Koyya*; Kannada—*Sebe hannu, jama phala*; Malayalam—*Pera, kioyya.*

It is native of Central America. It was probably introduced into this country by the Portuguese. In India, it is cultivated mainly in Uttar Pradesh, Punjab, Bihar, Maharashtra and Andhra Pradesh.

Fig. 5.7. Guava fruits. It is the fourth most important fruit of India in area and production. It is a very rich and cheap source of vitamin C and contains a fair amount of Calcium. It makes an excellent jelly and does not lose the vitamin C in the preserved forms.

The plant is a small tree with large white flowers. The fruits are round or oval and possess a variously coloured flesh.

Uses. The fruits are edible. It is aromatic, sweet, juicy and highly flavoured. It contains acid, sugar and pectin. It is one of the richest sources of vitamins A, B and C and of ascorbic acid. It is commonly used for making jellies, jams and pastes.

Fruits are also canned, preserved, spiced or made into jam, butter, marmalade, pies, ketchups and chutneys. Seeds yield a fatty oil. Leaves contain an essential oil used as a flavouring. Leaves are used as an astringent for bowel troubles. Decoction of bark is given in diarrhoea. Fruits are tonic, cooling, and laxative, useful in colic and bleeding gums.

Sugar Apple

Annona squamosa Linn.; Eng. Sugar apple; sweet sop; Hindi—*Sharifa*; Family Annonaceae.

Sanskrit—*Sitaphal*; Bengali—*Ata*;

It is medium-sized tree. It is native of South America and the West Indies. In India it is cultivated in Assam, Bengal, Uttar Pradesh and Maharashtra. The yellowish-green tuberculate fruit is about 2 or 3 inches in diameter. It has a white custard-like pulp. The fruit is a fleshy syncarp, formed by the fusion of numerous ripened ovaries and the receptacle.

Fig. 5.8. *Annona squamosa* (*Sharifa–Sugar apple*). Fruits, an improved variety.

Uses. The fruits are edible and quite popular. The pulp of fruit is eaten.

The edible pulp is juicy white or cream-yellow delicately flavoured, sweet flesh. They can be made into drinks and fermented liquor. Seeds yield a fatty oil.

Anjeer

Ficus carica Linn.; Eng. Fig; Hindi—*Anjeer*; Sanskrit—*Anjira*; Bengali, Hindi, Gujarati and Marathi—*Anjir*; Telugu—*Anjuru, manjimedi*; Tamil—*Simaiyatti, tenatti*; Kannada—*Anjura*; Malayalam—*Simayatti*; Family—Moraceae.

It is small tree. It has been cultivated since earliest times. It is native of Southern Arabia. It is frequently mentioned in the Bible. Theophrastus was familiar with many varieties of fig. He has given a detailed account of fig cultivation in his 'Enquiry into Plants'.

Today figs are cultivated in subtropical regions. The fruit is a syconus, a fleshy hollow receptacle with a narrow aperture at the tip. The true fruits (achenes) are borne on short stalks on the inside of the syconus.

In India, the figs are cultivated chiefly in Uttar Pradesh, Rajasthan, Punjab, Andhra Pradesh and Maharashtra.

Fig. 5.9. Fig. (*Anjeer*). A large number of cultivated forms are known in which the fruits vary in shape, size, colour of skin, colour and flavour of flesh, and period of ripening. Some of the forms grown in India are—Black Ischia, Brown Turkey, Turkish White, Kabul and Marseilles.

Uses. The fruits are edible. The figs are used fresh or are dried. The fresh fig latex is used in making cheese and junkets.

The fruits are delicious with high nutritive value. They are considered laxative, emollient and diuretic and makes a source of fig syrup; also used in the preparation of fig wine and brandy. Latex is used as an anthelmintic.

Jujube

Ziziphus mauritiana Lamk.; Syn. Z. *jujuba* Lam. non Mill.; Family—Rhamnaceae; Eng.—Jujube, Chinese date; Hindi—*Ber, Pemdi Ber*; Sanskrit—*Ajapriya, badara, karkandhu, kuvala, madhuraphala*; Bengali—*Kool, ber, boroi;* Marathi—*Bor, bera;* Gujarati—*Bor, bordi;* Telugu—*Reegu, gangareegu*; Tamil—*Elandai, elladu*; Kannada—*Yalachi, elanji*; Malayalam—*Elentha*; Oriya—*Barkoli, bodori.*

It is native of China and has been cultivated in that country for at least 4000 years. In India, it is cultivated in Punjab, Uttar Pradesh, Bihar and Rajasthan. It can be grown in all tropical and sub-tropical countries. This is a large shrub or small spiny tree with a small dark-brown fleshy drupe, which has a white, crisp, rich flesh.

Uses. The fruits are edible. The fruit is used fresh, dried or preserved. The bark is used for tanning and the leaves are used as fodder.

Large-sized fruits, which have just began to turn yellow are chosen for candying. Fruits of wild trees are considered cooling, anodyne, and tonic. They enter into the preparation of *Joshanda,* a medicine used in chest troubles. Kernels are sedative, used as a soporific and to stop vomiting. Seeds are given in diarrhoea. *Badari* is mentioned in old literature in the list of contraceptives. Seeds yield a fatty oil. Leaves eaten with catechu as an astringent.

Ziziphus nummularia (Burm. f.) Wight and Arn.; Syn. Z. *rotundifolia* Lamk.; Hindi—*Jharber*; Sanskrit—*Bhukamtaka, sukhshamaphala*; Marathi—*Gangar, janglar, junglaber*; Gujarati—*Chanyabor, adbaubordi*; Telugu—*Neelareegu*; Tamil—*Korgodi*; Punjab—*Bal, birar, malaber*; Rajasthan—*Bordi, pala*; Delhi—*Badber, jharberi*; Mumbai—*Chanibor*; Family—Rhamnaceae.

It is found in North-Western India and Andhra Pradesh. This is a small spiny undershrub.

Uses. The fruits are edible. The bark is used for tanning.

Tender parts are grazed by animals, but later when plant becomes hard and thorny only sheep and goat eat it. Leaves are used in scabies and other cutaneous diseases. Dried leaves are burnt and smoke inhaled for coughs and colds. Fruits are cooling and astringent, and used in bilious affections.

Fig. 5.10. *Ber.* The Chinese date, Chinese fig, Ber or Bor is an important minor fruit of India. It is the most hardy fruit—tree cultivated all over India and is often called the poor man's fruit. Most of the trees grown isolated are of poor quality. However, many plantations of budded tress have now come up all over India which fetch higher prices. They are rich in vitamin C and sugar contents.

Fig. 5.11. Loquat. Many varieties of loquat are cultivated in India. The colour of their fruit and flesh varies from pale yellow to orange. The taste also varies considerably. The varieties are-Golden yellow, Improved Golden Yellow, Thames Pride, Large Pale Yellow and Large Agra. California Advance and Tanaka are late varieties.

Loquat

Eribotrya japonica (Thunb.) Lindl.; Family—Rosaceae; Eng.—Loquat; Hindi—Lokat; Tamil—*Ilakotta* ; Kannada—*Lokkote*.

It is native of China, now cultivated chiefly in Uttar Pradesh and Punjab. It is grown in most tropical and sub-tropical countries. It is a small evergreen tree with broad leaves and white flowers. The fruits are pyriform berries. They are small, round, downy and yellow-orange in colour.

Uses. The fruits are edible. The fruit is used fresh, and is made into jellies, jams and sauces. Fruits are also used for making pies and preserves. It is considered to be a sedative. Flowers are used as an expectorant.

Papaya

Carica papaya Linn. ; Eng.—Papaya ; Hindi—*Papita*; Bengali—*Pappaiya, papeya*; Marathi—*Papaya*; Gujarati—*Papayai*; Tamil—*Pappali, pappayi*; Telugu—*Boppayi*; Kannada—*Parangi-mara*; Family—Caricaceae.

It is native of the West Indies and Central America. In India, it is cultivated chiefly in Uttar Pradesh, Punjab, Rajasthan, Gujarat, Maharashtra, Tamil Nadu and the Nilgiris. The papaya tree is up to 25 feet in height. It is dioecious. The stem is succulent, with a crown of large, deeply seven-lobed leaves and yellow flowers. The fruits are fleshy berries, resembling melons in size.

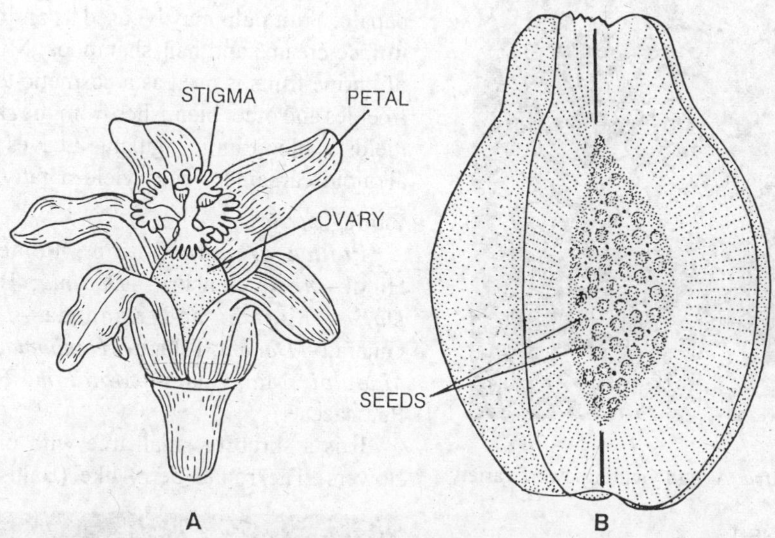

Fig. 5.12. *Carica papaya* (papaya). A, female flower; B, sectional view of fruit with seeds.

Fig. 5.13. *Papaya.* Punjab sweet

Fig. 5.14. *Papaya*. A high yielding dwarf variety

Uses. It is an excellent breakfast fruit. The orange coloured flesh possesses sweet musky taste. It is also used for salads, sherbets and confections. Unripe fruits are cooked as vegetable and pickled. The latex, obtained from the fruits, is used in preparing chewing-gums and for tenderizing meat. It is also used in medicine as an anthelmintic. The latex of papaya is known as papain which is used medicinally.

The ripe fruits are source of vitamins and papain. Fruit pulp may be used as an ingredient in face creams and hair shampoos. Milky juice of unripe fruits is used as a cosmetic to remove freckles and other blemishes from the skin. Plant yields a blood anticoagulant. Leaves yield an alkaloid carpaine. Seeds yield a fatty oil.

Pomegranate

Punica granatum Linn.; Pomegranate; Hindi—*Anar*; Sanskrit—*Dadima*; Bengali—*Dalim*; Marathi and Kannada—*Dalimba*; Gujarati—*Dadam*; Telugu—*Danimma*; Tamil—*Madulai*; Malayalam—*Matalam*; Family—Punicaceae.

It is a shrub or small tree with orange-red flowers. The round berry-like (Balusta fruits) brownish-yellow or reddish fruits are 2 to 4 inch in diameter and are crowned with the thick persistent calyx. The fruit has a hard rind and juicy seeds. The juicy testa is edible part.

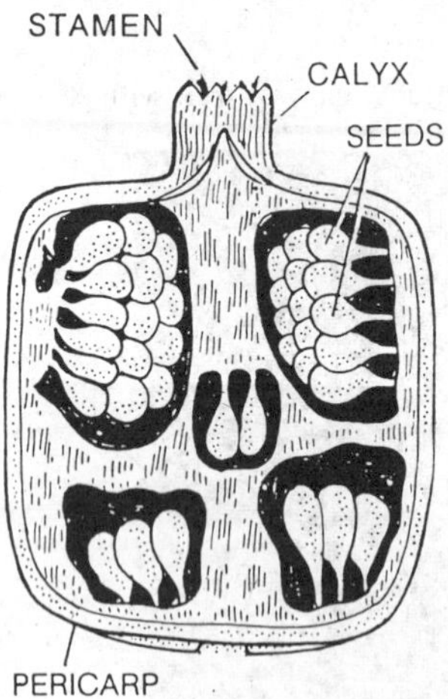

Fig. 5.15. *Pomegranate*. Sectional view of fruit.

It is native of Iran but now cultivated throughout India. It has been cultivated for

Fig. 5.16. Pomegranate (*Anar*). A refreshing fruit.

centuries and early spread to the Mediterranean region and South Asia. It was grown in the Hanging Garden of Babylon.

Uses. This is a very refreshing fruit. The bark and fruit shells are used for tanning. The peel of the fruit is used medicinally in dysentery and diarrhoea.

Fleshy testa is edible. Among the numerous types grown, *Bedana* and *Kandhari* are considered the best. Seeds of wild trees are sour and dried ones constitute *Anardana*, used as condiment. Fruit is a good source of sugars, vitamin C, and a fair source of iron, but poor in calcium. Seed juice is a favourite drink, and may also be used for making wine.

Bark is used to expel tapeworms, *iso-pelletierine* is the most potent among the active principles; given as decoction. Rind is used as an astringent in diarrhoea and dysentery. Flower-buds are used in bronchitis.

Fruit rind is rich in tannin and used as a tanning material, also yields a dye. Flowers yield a red dye.

Sapodilla

Achras zapota Linn. ; Eng. Sapodilla; Hindi—*Chiku*; Family—Sapotaceae.

It is a large tree of 75 feet in height. It has white flowers and a rough brown fruit 2 to 3 inch in diameter.

It is native of tropical America but now commonly cultivated in Bengal, Tamil Nadu and Maharashtra. South Gujarat is famous for its cultivation.

Uses. The yellowish brown flesh is translucent and very sweet. It is rich in carbohydrates and vitamins. The milky juice is known as chicle and used in the manufacture of chewing gum.

The edible sweet fruits are rich in fine flavour. The bark contains latex, 20–25 per cent of which consists of gutta-percha like substance, *i.e.,* chicle gum, used as a base for chewing gum. Other uses of chicle are in dental surgery, as a substitute for gutta-percha, and for making transmission belts. Seeds yield a fat.

The bark also contains tannin (11.8 per cent) used by fisherman for colouring sails and fishing tackles.

Pineapple

Ananas comosus (L.) Merr.; Eng.—Pineapple; Hindi—*Ananas*; Family—Bromeliaceae.

It is native of South America. The pineapple is one of the most important commercial fruit plants of the world. Pineapple is grown in most tropical and subtropical regions of the world, the most important producing countries being Hawaii, Brazil, Mexico, Formosa, Malaya, Australia, South Africa, the Philippines, India, Ceylon and Cuba.

Fig. 5.17. Pineapple Fruit.

Fig. 5.18. Pineapples planted in rows.

Amongst these, maximum development of pineapple cultivation has been made in Hawaii. In India this fruit was probably introduced in 1548 probably by Portuguese. The plant was carried by the Spanish and Portuguese people to the old world and spread all over tropical Asia, Africa, the East Indies and Polynesia.

The area under pineapple in India was estimated at 7,830 ha in 1962. Assam, with 2,620 ha is leading in pineapple growing. Kerala and Tripura are two other important states with 2,150 and

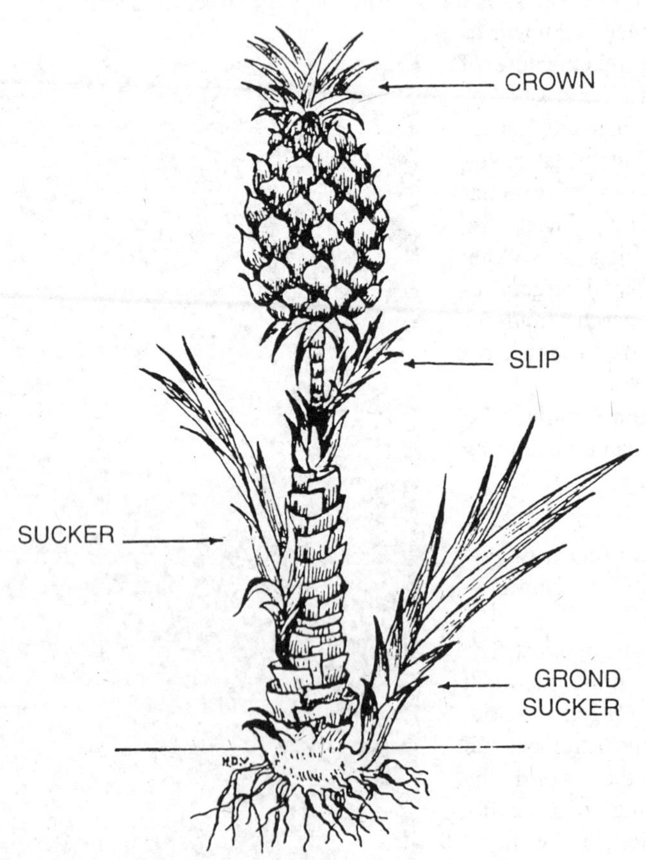

Fig. 5.19. Pineapple. A plant with leaves removed.

1,010 has respectively under this fruit. The rest of the area is distributed among West Bengal. Tamil Nadu, Andhra Pradesh and Karnataka.

The plant is a biennial, with a short stem and rosette of stiff leaves, 3 feet in length, with spiny ends and prickly margins. The flowers are found to be arranged in compact heads and are crowned by a tuft of leaves. The large fruits are syncarps. These composite fruits are formed from the whole inflorescence; the individual ripened ovaries remain embedded in a fleshy mass formed from the bracts, sepals and axis of the inflorescence.

For growing good pineapples in plenty, three things are important–*heavy manuring, enough water* and a *good shade.*

Soil. Pineapple grows best in lighter soils. An acid soil is ideal for this crop. In Assam, Tripura and other places; good crops are grown on lateritic soils.

Climate. Pineapple is grown at an altitude below 1,500 to 2,000 feet, where the prevailing temperature is comparatively higher. It produces fruits which are bigger in size, are more sweet and keep longer than the ones grown at higher altitudes. They thrive well in partial shade.

Variety. *Kew, Giant Kew* and *Queen* are some good varieties to grow. The fruits of *Kew* and *Giant Kew* are big and weigh from 5 to 7 pounds each. These are sweet and have good flavour. The fruit has a mottled yellow colour, and the flesh a light yellow. Both the varieties are ideal for canning.

The *Queen*, a small spiny plant bears rather small sized fruits, weighing two to three pounds each. The fruit is juicy and has sweet flesh, the sweetest of all varieties, and is golden yellow in colour when ripe.

Cultivation. Mother plants develop a large number of suckers and slips. Detach these after harvesting is over. Before planting, the leaves must be removed from the base of the suckers.

The best time for planting pineapple is the early rainy season. Where the rainfall is 80 inches or more, it should be planted in the later part of the monsoon.

Harvesting. The pineapples should be picked when fully ripe to get quality fruits. Fruits picked under-ripe are not as sweet as fully riped ones and also do not have the flavour and the right colour. When the eyes of the fruit become yellow or deep orange in colour and the rind of the fruit less angular, the fruit is ready for picking. The fruit of the first crop becomes ready for harvest in about 18 to 22 months of planting the sucker and the second or ratoon crop, a year later.

Yield. A well-cared garden for plantation gives 6 to 10 tons of fruit per acre. The yield goes down to three tons as the plantation becomes older. For higher yield pineapples should be planted after five to six years.

Uses. The fruits are edible. In addition to the content of sugar and fruit acids, a valuable digestive ferment, *bromelin*, is present. The slices of pineapples and pineapple juice are canned for export and domestic consumption. In the fresh form the fruit is a good source of vitamins A and B and is very rich in vitamin C. It is highly valued as a fresh fruit.

In countries like Hawaii, Australia, etc., where the canning industry is well developed, as much as 40% or more of the total small quantities are consumed by the industry.

The dried waste after extraction of juice, known as pineapple bran, is a valuable stock feed. Alcohol, calcium citrate, citric acid and vinegar are other products for the manufacture of which pineapple juice is utilized, although in small quantities. The leaves are the source of a strong fibre, which is made into fabrics.

Pineapple waste is used for making vinegar. Juice from ripe fruits is diuretic and antiscorbutic (vitamin *C* 63 mg/100 g.); that from unripe fruits, purgative and abortifacient.

Leaves yield a fibre, Pina, a delicate fabric of Philippines, is made from the fibre. Waste material, after extraction is used for paper making.

Date

Phoenix dactylifera Linn.; Eng.—Date palm; Hindi—*Pindkhajur*; Bengali, Marathi, Gujarati and Hindi—*Khajur, pindkhajur*; Telugu—*Ita, Kharjuramu*; Tamil—*Perichchankay, karchuram*; Kannada—*Kharjura*; Malayalam—*Ittappazham, tenitta*; Oriya—*Khorjjuri*; Family—Palmae ; (Palmaceae/Arecaceae).

Fig. 5.20. The dried dates are known as *'chhuhara* It is used as a tonic and laxative with milk.

This is a tall tree reaching 70 to 100 feet in height. It has a crown of stiff, pinnate leaves. The numerous flowers of the inflorescence are surrounded by spathe. Plant is dioecious. The fruit is a one-seeded berry with fleshy pericarp.

It is a plant of hot sunny climate with low humidity. For the successful cultivation of dates, propagation must be confined to the planting of suckers collected from mother trees.

It is one of the oldest of crops, going back at least 5,000 years. It was of great importance in Babylonia, and had reached Egypt long before the Christian era.

It is a native of India or Arabia. Even so, the current commercial production of its valuable fruit is practically negligible in India. At present it is cultivated mainly in Uttar Pradesh, Gujarat, Rajasthan, Punjab and Madhya Pradesh. India thus depends for its requirements of date largely on imports.

The date palm bears its first crop in its fourth year. In its adult life, the tree yields nearly 100 lb. of fruit. The average bearing life of a palm is about fifty years.

Harvesting. The dates are generally harvested before they are fully mature to avoid damage by rain. The harvested produce is processed or cured to make them fit for consumption. This is done by dipping the fruit in one per cent *lye* solution and then placing it in a wooden box lined with tin sheets into which holes are cut for circulation of air.

Uses. The date fruit is very nutritious, containing in its dried condition nearly 75 to 80 per cent sugars, mainly glucose and fructose. It is a good source of iron and potassium, besides being rich in nicotinic acid. The dates are used as a table fruit and in jams, pastes, cooking and alcoholic beverages. The leaves are the source of a fibre which is used for making ropes, baskets and cordage. The leaf stalks are used for making walking sticks.

The dates are also used in bakery and made into preserves. Brandy of good quality is prepared from dates. Sap is sweet, nutritive and laxative; used for preparation of jaggery and sugar. The dates are demulcent, expectorant and also used in respiratory diseases and fever.

Seeds when ground or softened by soaking in water are used for feeding camels, goats and horses and have been used as a poultry feed. Date seeds can serve as raw material for preparation of oxalic acid. Terminal leaf-bud is consumed as vegetable.

Purple Grandilla

Passiflora edulis Sims; Eng.—Purple grandilla; Hindi—*Jhumkalata*; Family—Passifloraceae.

It is a shrubby vine. The flowers are white, with a white and purple crown. The deep-purple fruit is about 2 in. in length.

It is native of Brazil. In India it is cultivated in Uttar Pradesh, Punjab and Tamil Nadu.

Uses. The fruits are edible. It is used as a table fruit and in sherbets, candy and beverages.

It bears two types of fruits, yellow and purple, however, the yellow ones have inferior flavour. Juice is highly acidic and is preserved for use in blends with less acidic fruit juices and in the preparation of squashes, cordials, syrups, carbonated beverages, jellies, etc. It is also used for flavouring candy, ice-creams, cake fillings, etc. Passion fruit juice concentrates and powders are also prepared. Juice has been successfully spin-pasteurized.

Peels can be used for recovery of pectin, as stock feed and as manure. Seeds yield a semi-drying oil, suitable for use in paints and varnishes. Seed cake is used as animal feed and manure.

Mangosteen

Garcinia mangostana Linn., Eng. Mangosteen; Hindi—*Mangustan*; Family—Guttiferae.

It is native of Malaya. In India it is cultivated mainly in the Nilgiris, Maharashtra as along the Eastern Coast of Tamil Nadu. It is also cultivated for commercial purpose in Quilon district of Kerala.

The tree is small. The fruit is a dark-purple berry, 2 or 3 in. in length, with adherent sepals at the base. The pulp is cream coloured with crimson veins and exudes a yellow juice emitting an excellent flavour.

Uses. It is a best flavoured fruit. The flesh is quite delicate and melts in the mouth like ice cream.

The arils are eaten as dessert or made into preserves and squashes. The rind is used for jellies, contains tannin 7–14 per cent, also used in diarrhoea and dysentery and cutaneous affections.

Seeds yield a fatty oil.

Japanese Persimmon

Diospyros kaki Linn. f.; Eng. Japanese Persimmon, Kaki; Hindi—*Halwa tendu*; Family—Ebenaceae.

It is native of China. In India it is grown in Kulu, Himachal Pradesh, Kumaon, the Nilgiris and West Bengal.

It is a large tree, with orange-red fruits 3 in. in diameter. These are berries, with an enlarged calyx at the base.

Uses. The fruits are eaten fresh or dried.

Carambola

Averrhoa carambola Linn.; Family—Oxalidaceae; Eng.—Carambola; Hindi—*Kamrakh*.

It is native of South-eastern Asia. It is cultivated throughout India.

Uses. The fruits are quite acid. They are eaten raw or cooked with sugar. Chutney is prepared from it.

The fruits are used in compotes, jellies, jams, preserves and pickles; also sliced and used in salads. For culinary preparation they are made into stews, curries, puddings and tarts. Sweeter fruits are eaten as dessert.

Averrhoa bilimbi which is also having the same use of *A. carambola*, is largely found in Kerala. Pickles are prepared out of it.

Karanda

Carissa carandas Linn.; Family—Apocynaceae; Eng.—Karanda; Hindi—*Karaunda*.

It is native of South Africa. A spiny shrub, found throughout India.

Uses. The small pinkish white fruits are eaten raw or cooked or used for jellies and preserves. The fruits are quite acidic and used for preparing chutney and pickles.

Elephant Apple

Feronia limonia (Linn.) Swingle; Syn. *F. elephantum* Correa; Eng.—Elephant apple; Hindi—*Kaith*; Family—Rutaceae.

A tree, found all over India.

Uses. The acid fruits are edible. They are used for making chutney and pickles. The gum obtained from the trunk and branches is used like gum-arabic.

Tamarind

Tamarindus indica Linn.; Eng.—Tamarind; Hindi—*Imli*; Family—Caesalpiniaceae.

It is native of tropical Africa or Southern Asia. It is a large tree with a dense crown. The fruits are brown pods, 3 to 8 inches in length.

Uses. The pulp contains 12% tartaric acid, as well as 30% sugar, and so has a sour taste. Tartaric acid and its salts are extensively used in various foods, chemical and pharmaceutical industries.

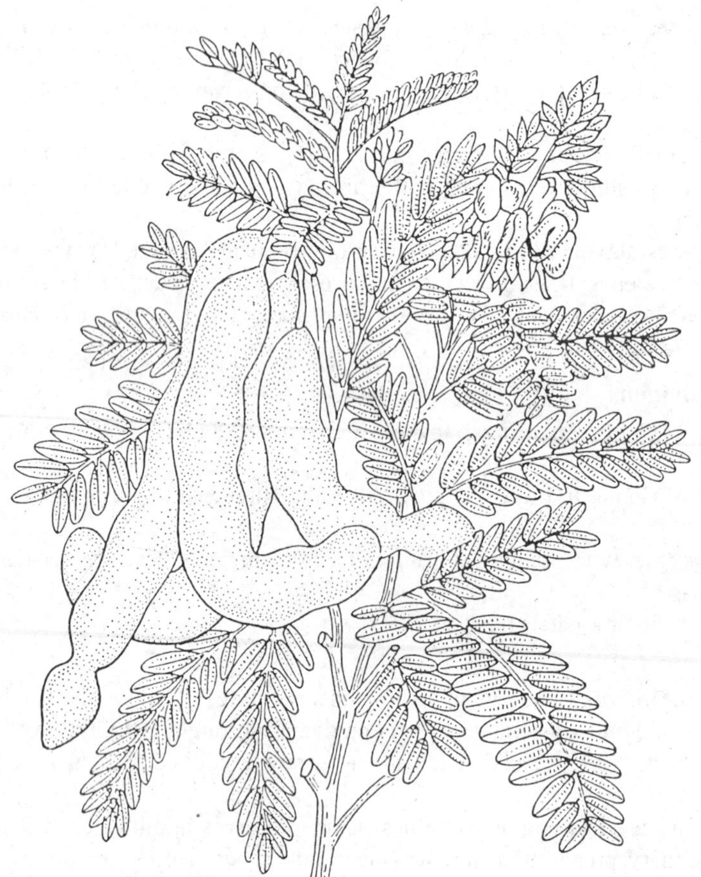

Fig. 5.21. *Tamarindus indica* (tamarind). A twig with fruits.

The seeds are the source of jellose, which is used as a sizing material in the cotton and jute industries. The polyose obtained from the seeds is good substitute for fruits. The bark and leaves are used for tanning.

Jambolana

Syzygium cumini (Linn.) Skeels; Family—Myrtaceae; Eng.—Java plum, Jambolana; Hindi—*Jamun*; Bengali—*Jam, kalajam*; Gujarati—*Jambu, jamli*; Marathi—*Jaman, jambul* ; Telugu—*Neereedu*; Tamil—*Neredam, naval, sambal*; Kannada—*Nerale*; Malayalam—*Naval, perinnaral;* Oriya—*Jamo*; Punjab—*Jammu* ; Nepal—*Kalajam*; Lepcha—*Phoberkung*; Andamans—*Thabye, jamun*.

A large tree. Cultivated throughout India.

Uses. The fruits are edible. They are rich in Iodine. The vinegar is prepared from fruit juice. The seeds are used as fodder.

A spirituous liquor as well as a wine is prepared from the ripe fruits. The fruits are also used for making preserves, jams, squashes and jellies.

Khirni

Manikara hexandra (Roxb.) Dub.; Syn. *Mimusops hexandra* Roxb.; Hindi—*Khirni*; Bengali—*Khirkhejur*; Marathi—*Ranjana, raini*; Gujarati—*Rayan, khirni*; Telugu—*Manjipala*; Tamil—*Palla*; Kannada—*Bakula*; Malayalam—*Pala*; Oriya—*Khirakuli*; Family—Sapotaceae.

A tree. It is cultivated in Delhi, Madhya Pradesh, Uttar Pradesh, Gujarat, Maharashtra and Andhra Pradesh.

Uses. The fruits are edible. They are rich in sugar and proteins.

Phalsa

Grewia asiatica Linn.; Eng.—Phalsa; Hindi *Phalsa*, Bengali—*Phalsa, shukri*; Gujarati—*Phalsa*; Marathi—*Phalsi*; Telugu—*Jana, nallajana*; Tamil—*Palisa*; Kannada—*Tadasala*; Oriya—*Pharasakoli*; Family—Tiliaceae.

This is a large shrub or small tree. It is cultivated throughout India.

Uses. The fruits are edible. They are used for making cold drinks.

The fruits are also used in the preparation of beverages, and pickles; astringent, cooling and stomachic.

Wood is used for golf shafts, shoulder poles bows, spear handles, and shingles. Bark yields a fibre used for ropes. Santhals use the bark in rheumatism.

Bael

Aegle marmelos (Linn.) Corr.; Eng.—Bael; Hindi—*Bel*; Sanskrit—*Bilva*; Bengali, Hindi and Marathi—*Bel*; Gujarati—*Bil*; Telugu—*Maredu*; Tamil and Malayalam—*Vilvam*; Kannada—*Bilpatre*; Family—Rutaceae.

A tree. Cultivated all over India.

Uses. The ripe fruits are edible. The cold drinks and squashes are prepared from the fruits. The unripe fruits are eaten after roasting. The fruits are used as astringent, stomachic and also in the treatment of diarrhoea and dysentery. The mucilaginous substance of the fruits is used as gum and is also used as varnish.

The pulp of ripe fruits is aromatic and cooling and used in the form of sherbet. Marmelosin is the active constituent; it acts as a laxative and diuretic, in strong doses a cardiac depressant.

Dried fruits, freed from pulp, are used as pill-boxes. Stem yields a gum. Leaves contain an essential oil.

Fig. 5.22. Water melon fruits.

Water Melon

Citrullus vulgaris Schrad. ex Eckl. & Zeyh.; Eng. —Water melon; Hindi—*Tarbooz*; Family—Cucurbitaceae.

It is native of tropical Africa. Cultivated throughout India. It is an annual extensive vine. The fruits are large in size and may weigh 50 lbs. or more. The red pulp is very sweet and juicy, with white or black seeds. The plant required a fertile sandy soil with abundant sunshine. The fruits are picked when fully ripe.

Uses. The red pulp of the fruit is eaten. The white solid flesh of the fruit is used for making jams, jellies and preserves.

The seeds are used as food. They are considered cooling, tonic, diuretic; yield a fatty oil. Fruit juice forms a cooling and refreshing beverage; also considered diuretic.

Melon

Cucumis melo Linn.; Eng.—Melon; Hindi—*Kharbuza*; Sanskrit—*Kharbuja, madhupaka*; Punjabi, Gujarati, Marathi and Hindi—*Kharbuja*; Bengali—*Kharmuj*; Tamil—*Mulampazham*; Telugu—*Kharbujadosa, putzakova*; Family—Cucurbitaceae.

It is native of Southern Asia. Several kinds of melons are cultivated in North Western India. The fruit has a soft rind and netted markings on the surface.

They require a fertile soil, high temperature and plenty of moisture and sunlight. They are nearly ripened on the vines, as this increases the sweetness and flavour.

Uses. The fruits are edible. They are rich in sugar and proteins.

The melons are eaten as dessert. The seeds are edible and contain a fixed oil. The seeds are diuretic, refrigerant and nutritious. The pulp is useful in chronic eczema.

Snap Melon

Cucumis melo Linn. var. ***momordica*** Duth. & Full; Family—Cucurbitaceae; Eng. Snap melon; Hindi—*Phunt, Kachra*.

A vine with yellow flowers; monoecious. The fruits are large elongated and cylindrical. The rind of ripe fruit cracks at several places.

Uses. The fruits are edible. They are rich in sugar, carbohydrates and proteins.

TEMPERATE FRUITS IN INDIA

The temperate fruits are popularly known in India as '*hill fruits*' or '*cold region fruits*'. They are commonly grown in several states at elevation of 1,214 metres and above in the hills. Their cultivation is confined at present to Jammu and Kashmir, Punjab, Himachal Pradesh, Assam and Uttar Pradesh. They are also grown to a small extent, at suitable elevations in the south as in the Nilgiris of Tamil Nadu State.

The temperate fruits have a low to medium heat requirement and can resist an intensive winter cold. They shed their leaves with the approach of winter and remain in a leafless or dormant condition throughout the winter months. To enable this class of fruit trees to start their normal growth in spring, severe winter temperatures are essential. When grown late in conditions of mild winters, these fruits will start their growth in spring, thus interfering with their normal flowering and fruit-setting.

The '*hill fruits*' are broadly divided into two categories, based on the climatic conditions under which they do best. In the first category may be grouped the apple, the pear, the plum, cherry, etc., which do best under severe winter conditions usually experienced at elevations above 1,517 metres in the Himalayan hills. Depending upon this aspect, they can be successfully grown up to elevations of 2,426 metres or even higher.

In the second category may be grouped the peach, the Japanese plum, the apricot, the grapes, the persimmon, etc., which thrive in comparatively warmer situations of elevations of 1,062 to 1,517 metres.

Some varieties and species of temperate fruits are indigenous, but the cultivated varieties grown in the country have largely been imported from time to time from foreign countries.

Apple

Malus pumila Mill.; Eng.—Apple; Verna.. *Seb*; Sanskrit—*Seba*; Bengali and Hindu—*Seb, sev*; Kannada—*Sebu, sevu*; Punjabi—*Seo*; Ladakh—*Kushu;* Family—Rosaceae.

The apple has been in cultivation from time immemorial. It is regarded as the native of the mountainous regions between Western Himalayas on one side and Caucasus and Asia Minor on the other. Apple seeds have been found in the remains of the Lake Dwellers. Twenty-two kinds of apple were known to the Romans.

Fig. 5.23. An apple tree bearing fruits.

Apple occupies the most important position among the fruits of temperate regions and is widely cultivated in many parts of the world. In India apple is a commercial crop in the hilly areas of Kashmir, Kulu and Kumaon. The more important areas of cultivation are Srinagar, Kulgam and Uttarmachipura tehsils in Kashmir and Udhampur and Kishtwar in Jammu; Almora. Nainital and Garhwal districts in Uttar Pradesh; Kulu valley and Simla hills in Himachal Pradesh; and Bangalore and Nilgiri hills in South India.

About 6,500 horticultural forms of *M. pumila,* which can be distinguished on the basis of fruit and tree characters, are reported to be under cultivation in various parts of the world. Fruits vary in size, colour, flavour and quality; trees are adapted to different climatic and soil conditions and show difference in periods of maturity and season of bearing.

Cultivation. The apple plant is essentially suited to regions which have a low winter temperature, attended by snowfall. It requires considerable chilling to check the rest period and induce the opening of buds in spring. If the winters are not sufficiently cold, blossoming as well as fruiting are insufficient and uneven. Apple thrives best in well-drained medium loam, but it has been successfully grown on a variety of soils ranging from the deep fertile loams of Kashmir to the light loams of Kulu valleys and the brown or reddish brown sandy loams of Kumaon. The apple is usually propagated by budding or grafting suitable types on selected root stocks.

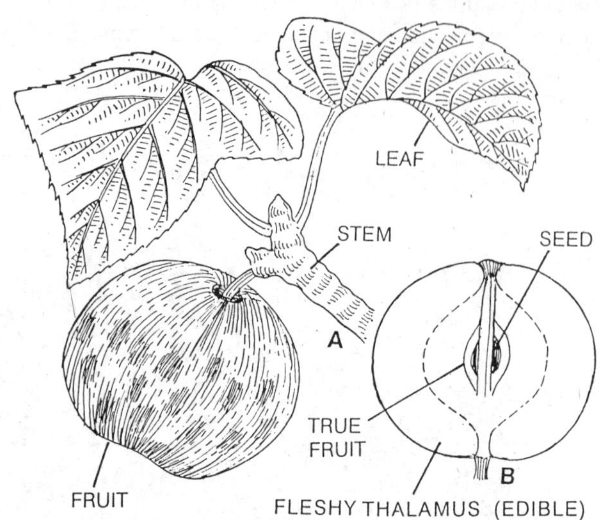

Fig. 5.24. Apple—A, twig with fruit; B, fruit with edible thalamus.

Uses. Apples are valued mainly as dessert fruits. Fruits may be preserved for later use after slicing and drying; they are also canned and jams and jellies are made from them. The juice extracted from the fruits is used fresh or after fermentation into cider wine and vinegar; apple brandy is obtained by distilling cider.

Apples are rich in pectin and are useful in diarrhoea. Apple juice, syrup and vinegar reduce curd tension of milk used in infant feeding. Apple *murabba*, a preserve popular in India, is regarded as a stimulant for the heart; it is reported to relieve physical heaviness and mental strain.

The vitamins, salts and organic acids are concentrated particularly in and just below the skin and the fruit should be eaten unpeeled. Apple is considered a good source of potassium. The edible portion of fresh apples contains Ca, Mg, K, Na, P, Cl, S and Fe. The mineral constituents of the apple are considered valuable for human nutrition.

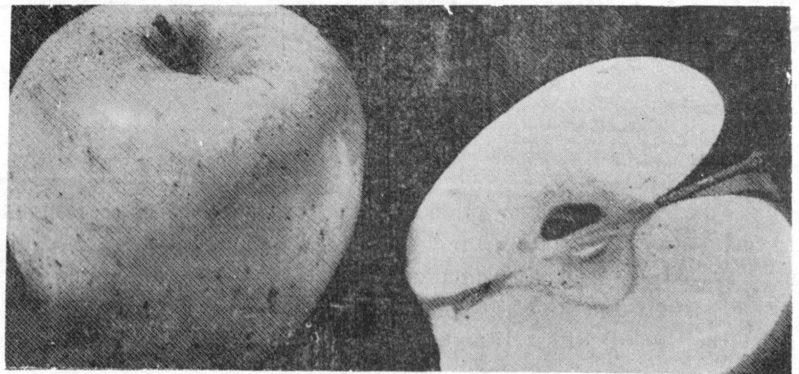

Fig. 5.25. Apple. *Golden* Delicious.

Apple products. Dehydrated apples, apple powder and dried apples in the form of rings, chops, and sauces are available in commerce. Dried apples find use in bakeries.

Canned apples are usually available in large packings. Several varieties are favoured for the production of apple juice.

Apple cider is obtained by the alcoholic fermentation of apple juice. It contains about 5% alcohol.

Apple wines containing not more than 14% alcohol are produced by the fermentation of a mixture of apple juice and sugar; those with higher alcohol content are prepared by adding apple brandy.

Peeled fruits freed from core and used for the preparation of jam. Apple *murabba* (preserve) is prepared from firm fruits which are peeled, cored and cut into sections.

Apple butter is a smooth semi-solid material prepared by cooking the fruits with water until soft, followed by screening.

Pear

Pyrus communis Linn., Eng.—Pear; Hindi—*Nakh*, Family—Rosaceae.

This is a small tree with white flowers. The fruit is pyriform and possesses a persistent calyx. The fruit is sweeter and more juicy than the apple. The flesh of fruit contains numerous stone cells. The pears are grown in heavy soils with humus and good drainage. They are propagated from seed and by grafting.

It is native of Eurasia. In India it is cultivated in Kashmir, Kulu, Kumaon and Himachal Pradesh.

Some of the early varieties, particularly *Bartlett*, are now in very great demand by the canning industry. *Baghu Gosha* is under large-scale cultivation in Kashmir valley. *Nashpati* is another variety that is successfully grown in the plains of Punjab and elsewhere.

Propagation. The method of propagation is Shield Budding, which is done in the months of June and July. The stock is raised either from the seed of commercial varieties or from the wild pear *shegal* (*Pyrus pashia*).

Planting. Maidens or one-year old trees should be used for planting. The best time of planting is autumn.

Picking and ripening of fruit. *Bartlett* variety is picked when the fruits are still green and hard. As a rule, picked for the market without storing, as the time taken in reaching the consuming centres will be sufficient to ripen the fruit properly. Pears ripen at a higher temperature than apples. They can be stored indoors in a cool dry place. They are kept singly on shelves or trays.

Uses. The fruits are edible. They are eaten as such or canned. The fruits are rich in sugars, salts and vitamins.

Plum

Prunus domestica Linn. subsp. ***insititia*** (Linn.) Schneid.; Eng. Plum; Hindi—*Alucha, alubukhara*; Family—Rosaceae.

Fig. 5.26. Plum (*Alucha*). Fruits.

It is a native of Eurasia, and still grows wild in that region. It has been cultivated for more than 2000 years and was known to the Lake Dwellers and the Greeks and Romans. It is the best known and most widespread of all the plums. It is a tree, 30 to 40 feet in height, with variously coloured fruits.

In India it is cultivated in Kashmir, the hills of Punjab, Himachal Pradesh, Kulu and Kumaon. A large number of plum varieties have been tried out in the Himalayan hills. The varieties comprised both the European and Japanese types. Some of the selected ones that were found most profitable for growing are listed here:

Grand Duke. Fruit very large-sized, oval, halves unequal, apex flattened. Skin dark purple. Flesh greenish yellow. Prolific. Good for cooking and jam. Ripens at the end of August.

Fig. 5.27. *Plum 'Fla.* 1-2', bright crimson-coloured fruits.

Victoria. Fruit large, oval, Skin dark red. Prolific. Good for canning, cooking and as dessert.

Early Transparent gage. Fruit rather small and roundish, delicious. Regular and fairly prolific. Ripens from the middle of July onwards.

Santa Rosa. Fruit large, dark purple with whitish bloom. Flesh golden yellow, very juicy, sweet. Prolific and a regular bearer. Picked in July

Wickson. Fruit large, heart-shaped, symmetrical. Colour cherry red. Flesh amber, very juicy, sweet, aromatic. Regular and a good bearer. Picked in August.

Prunus salicina Lindl.; Eng.—Japanese plum.

It is native of China, now grown in the Nilgiris.

Bright Red Burbank. A Japanese variety. Fruit almost globular, halves equal, apex roundish. Skin dark red. Flesh very sweet and of agreeable flavour. Very prolific and a regular bearer. Ripens in July-August.

Uses. Plums are used for fresh fruit, cooking, canning and jams.

Sweet Cherry

Prunus avium Linn.; Eng.—Sweet cherry; Hindi—*Gilas*; Family—Rosaceae.

It is native of Eurasia; now cultivated in Kashmir, Kulu, Kumaon and Himachal Pradesh.

The sweet cherries are well-adapted to the climatic conditions in these places at elevations above 1,500 metres.

This is a tall long-lived tree with yellow or greenish fruit.

Some of the selected varieties of sweet cherries are: *Early Rivers, Governor Wood, Bedford Prolific, Late Black Bigarreau,* etc.

Early Rivers. Fruit large sized, skin black. Flesh deliciously flavoured. Small stone. The finest and the best variety. Early and prolific, ripens in the beginning of May, flowering mid-season.

Governor Wood. Fruit large-sized. Skin light yellow with bright red shade. Flesh tender, juicy, aromatic. Regular and prolific cropper. Early variety. Ripens just after *Early Rivers*. Flowering late.

Bedford Prolific. Skin black. Flesh of good taste and flavour. Ripens by the middle of May. Early flowering.

Late Black Bigarreau. Flesh well-flavoured and of good quality. Very prolific and regular cropper. Ripens in early July. Flowering early.

Propagation. Sweet cherry is grafted on seedlings of wildcherry stock (*Prunus cerasoides* D. Don). Grafting is done in January by the whip and tongue method and the grafts become ready for transplanting in two years.

Cultivation. Cherry orchards are best put under grass. The grass is kept in check by sheep that supply organic manure to the trees.

Picking of fruit. The fruit is picked by hand when dry so that it may travel well to the consuming market. It should be packed in specially-made baskets, and packed in securely to avoid damage in transit.

Sour Cherry

Prunus cerasus Linn.; Eng.—Sour Cherry; Hindi—*Alubalu*; Family—Rosaceae.

This is a small tree with red fruits. It is native of Eurasia, now cultivated in Punjab, Kashmir, Himachal Pradesh and Kumaon.

Uses. Cherries are used as table fruits and in canning. Cherry brandy is distilled from cherry juice. The juice is also used for syrup, cherry cidar and jelly.

Peach

Prunus persica (Linn.) Stokes; Family—Rosaceae; Eng.—Peach; Hindi—*Aru*.

The tree is native of China and has been cultivated there for thousands of years. The peach reached the Mediterranean region very early in history and the Romans knew at least six kinds. In India it is cultivated in Kashmir, the hills of Punjab, Himachal Pradesh, western Uttar Pradesh, Kulu and Kumaon. It is cultivated in most temperate countries, particularly in Southern Europe, the United States, South Africa, Japan and Australia.

The peach is a small tree. It is short-lived and susceptible to frost injury and low temperature. The round fruits have a velvety skin and a compressed, pitted or furrowed stone. It thrives best in sandy soil.

Uses. The fruits are edible. They are the most popular fruit for canning. The oil obtained from the seeds, is used for cooking and as an illuminant.

Fig. 5.28. *Prunus persica* (peach). Fruits.

Apricot

Prunus armeniaca Linn.; Family—Rosaceae; Eng.—Apricot; Hindi—*Khubani, Zardalu.*

It is native of Asia. In India it is cultivated in Kashmir, Himachal Pradesh, Kulu and Kumaon. It is a small tree 20 to 30 feet in height with pink flowers. The fruit is velvety when young and has a yellowish orange flesh. The stone is smooth and flat. The following are some of the varieties that are recommended for growing.

Shipley Early. Fruit medium-sized, roundish. Skin creamy-white with red flesh on the sunny side. Flesh meshy, juicy and of average quality. Ripens in the middle of May. Earliest of all varieties.

New Castle. Fruit roundish, medium sized, highly coloured, orange yellow with a nearly spherical pit. Flesh juicy and sweet. Regular in production. Picked in early June. A popular dessert variety.

Uses. They are used as a table fruit in the regions where they are grown. They are also canned and candied. The oil obtained from the seed, is used for cooking, burning and for the hair.

Strawberries

The strawberry grows wild in the Himalayas, but the fruit of wild strawberry is very inferior in quality. Its cultivated varieties grown in the plains and the hills were all imported from abroad. Strawberry-growing gives the quickest returns in the shortest period of time. The classification of the strawberries may be as follows:

Fragaria chiloensis (Linn.) Duchesne; Eng.—Garden strawberry; Family—Rosaceae. A herb cultivated in Kumaon, Kulu and Himachal Pradesh.

Fragaria indica Andr. Syn. *Duchesnea indica* Focke; Eng.—Indian strawberry; Hindi—*Kiphaliya*; Family—Rosaceae. A perennial herb, commonly found in the temperate Himalayas, Punjab, Assam, the Khasia hills and Nilgiris.

Fragaria nilgerrensis. Schlecht, ex F. Gay ; Eng.—Houtbois strawberry, Alpine strawberry; Family—Rosaceae. Found in the Khasia hills and the Nilgiris.

Fragaria vesca Linn. ; Eng.—Perpetual strawberry; Family—Rosaceae. It is native of Europe but now cultivated in Kumaon, Kulu, Himachal Pradesh, Assam and Maharashtra.

Fig. 5.29. *Fragaria chiloensis* (Strawberries). Ripe strawberries are bright red in colour with a soft melting flesh of sweetish flavour. They are esteemed as dessert and consumed not so much for their food value, as for their flavour. They may be made into preserves, jams, jellies and syrups; they may also be canned. Strawberry wine is also prepared.

The strawberry is a low perennial herb with a very short thick stem and trifoliate leaves. It produces numerous runners, which root at the tip and are used to propagate the plant. The fruit is an aggregate fruit consisting of a number of small dry achenes embedded on the surface of a large fleshy receptacle.

Propagation. For propagation, maiden plants that have not borne any crop should be planted on soils of high fertility. They may be allowed to set many runners as possible, but not allowed to bear any fruit. Given the best attention, a single plant usually produces 12 to 18 runners.

Planting. In the hills, transplantation is done in March-April, but in the plains, the months of January-February may be utilized for this purpose.

Picking of fruit. When the fruit is fully ripe, pick it when dry by pinching off the stalks.

Uses. Strawberries are a dessert fruit primarily, but are also canned and used in jams and preserves and as a flavouring material.

Raspberries

They are small shrubs with prickles. The aggregate fruit separates from the receptacle when ripe, thus leaving a cavity on one side. The raspberries found in India are as follows:

Rubus elipticus Smith; Eng.—Himalayan yellow raspberry. Hindi—*Lal anchu, hisalu*; Family—Rosaceae. A shrub, found in the Western Himalayas, South India, Western Ghats and the Khasia hills.

Rubus lasiocarpus Sm.; Eng.—Ceylon raspberry; Hindi—*Kala hisalu, kala anchu*; Family—Rosaceae. It is found in the Western Himalayas, Sikkim, the Khasia hills, the Western Ghats, Karnataka and the Pulney hills.

Rubus rosaefolius Sm.; Eng.—Mauritius raspberry; Hindi—*Yeshul*. Cultivated in Kumaon, the Khasia hills and Manipur.

Uses. It is used fresh or cooked, and is utilized for jams, jellies, vinegar and as a flavouring material.

Blackberry

It is erect, creeping and prickly shrub. The erect plants die down to the ground every few years and are renewed from the rootstocks. The velvety black fruits are aggregate fruits of small drupelets. When picked the fruit does not separate from the fleshy receptacle.

Rubus fruticosus Linn. Eng.—European black-berry; Hindi—*Vilaiti anchu*; Family—Rosaceae. It is cultivated in Kashmir, Assam and Coonoor.

Uses. The fruits are edible. They are used fresh and for jams, preserves and canning.

Gooseberry

Ribes grossularia Linn.; Eng.—European gooseberry; Hindi—*Amlanch, Baikunti*; Family—Saxifragaceae. This shrub is native of Eurasia, now found in the Western Himalayas from Kumaon to Kashmir. The round fruits may be red, yellow, green or white.

Uses. The fruits are edible. They are used for jams, jellies and sauces.

Currants

Ribes nigrum Linn.; Eng.—European black currant; Hindi—*Nabar*, Family—Saxifragaceae. It is native of Eurasia, now found in Kashmir.

Ribes rubrum Linn.; Eng.—Red currant; Hindi—*Dak*; Family—Saxifragaceae. Found in the Western Himalayas from Kumaon to Kashmir.

Uses. The fruits are edible. They are too acid, and are much used for jams, jellies, sauces and wine.

Mulberry

Morus alba Linn.; Eng.—White mulberry; Hindi—*Tut*; Family—Moraceae. It is a tree. It is a native of China but now cultivated in Punjab, Uttar Pradesh, Kashmir and the North-western Himalayas.

Sanskrit—*Tula*; Hindi, Bengali and Marathi—*Tut*; Gujarati—*Shetur*; Telugu—*Reshme chattu*; Tamil—*Kambli*; Kannada—*Hipnerle*; Oriya—*Tuto*; Kashmir—*Tut*; Punjab—*Tut, tutri*; Kumaon—*Siahtut*.

Morus australis Poir.; Eng. Common mulberry; Hindi—*Tut*. Commonly found in the sub-Himalayan tracts.

The fruit of mulberry is a multiple accessory fruit derived from catkin inflorescence. The actual fruits little achenes or nutlets, are surrounded by fleshy sepals and grouped together with the flesh axis to form *'sorosis'*.

Uses. The fruits are edible. The leaves are the source of food of silkworms.

The fruits are refrigerant, used also for sore throat, dyspepsia and melanchalia. The leaves are rich in calcium and vitamin C, and eaten as vegetable. The fruits are also used in the form of juices, stews and tarts, and fermented to yield spirituous liqueurs.

Grapes

Vitis vinifera Linn; Eng.—Wine grape; Hindi—*Angur;* Sanskrit—*Draksha;* Hindi—*Angur, dakh*; Bengali—*Angurphal, drakhyaluta*; Gujarati—*Darakh, draksha*; Marathi—*Draksha*; Telugu—*Draksha*; Tamil—*Kodimundiri, gostanidraksha*; Kannada—*Angura, draksha*; Malayalam—*Mundiri, gostani*; Oriya—*Drakya, onguro, gostoni*; Raisins—*Kismis, kishmish, drakashai, sougi, manakka*; Family—Vitaceae.

Grape grows wild in many temperate regions of Europe, Asia, Africa and America. The cultivated grapes of the present day have been derived from European and American species. The wine grape (*Vitis vinifera*) is one of the oldest of cultivated plants. It is thought to be native of Caspian sea region of Western Asia. They are frequently mentioned in the Bible. Thc grapes have

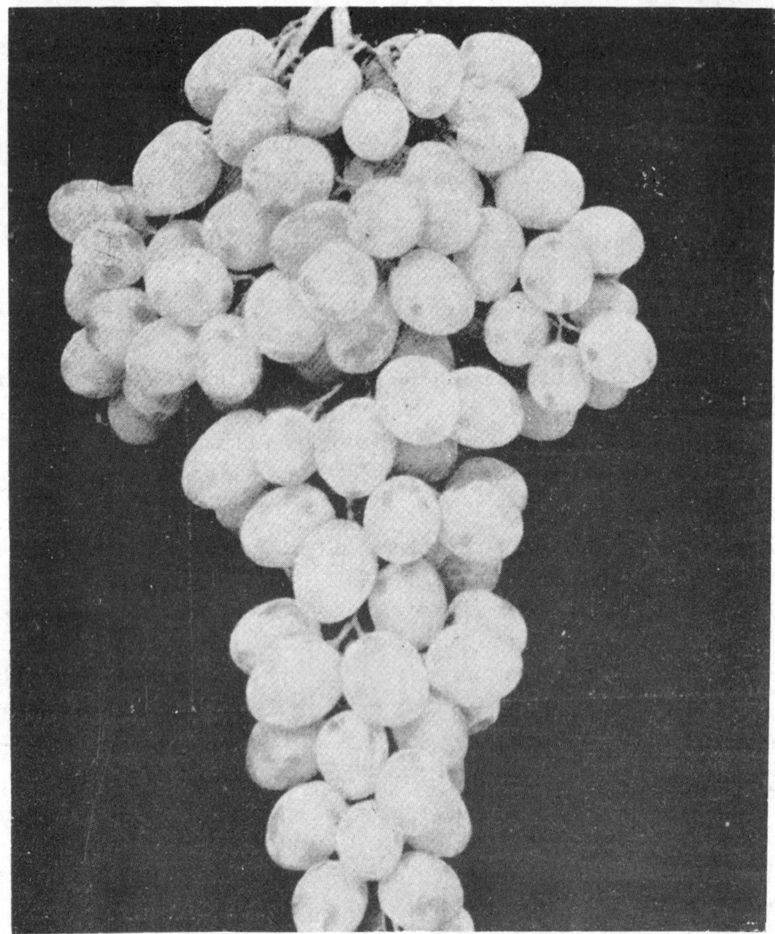

Fig. 5.30. Grapes—*Anab-e-Shahi,* a white oval, sweet grape of excellent eating quality.

been grown in Egypt for 6,000 years. They spread all over Europe with the Roman civilization, and now are cultivated in all temperate parts of the world.

It is a woody, climbing, tendril bearing vine with large palmatifid leaves. The flowers are small and sweet-smelling. The fruits are found in large clusters.

Grape cultivation in India. In India a comprehensive trial for testing the adaptability of grape-wine varieties was conducted at Lyallpur (now in Pakistan) where 112 varieties imported from several countries were planted in 1928. The grape seems to possess a remarkable adaptability to varying climatic conditions. In India, grape cultivation extends from the temperate climate in the Chini tract in Himachal Pradesh to the tropical condition in Hyderabad. The cultivation in the southwest is largely concentrated in Mumbai region.

There is much scope for grape cultivation in India. Already, the grape plantations in Andhra Pradesh, Karnataka and Tamil Nadu States have proved a great success. For instance, *Anab-e-Shahi* and *Pachadrakshai* have given as high a yield as 16,000 kg–32,000 kg. Approximately, *Bhokri* 9,900 kg, and *Bangalore Blue* about 5,400 kg per acre annually. *Dakh* is also a good yielder.

The principal states growing grapes are Maharashtra, Karnataka, Tamil Nadu and Andhra Pradesh. The chief varieties grown in Maharashtra and Andhra Pradesh are *Anab-e-Shahi, Bhokri, Gulabi, Kali Sahebi, Kandhari, Pandhari Sahebi*, and *Phakdi*. In Karnataka State *Bangalore Blue* is widely grown, whereas in Tamil Nadu State *Pachadrakshai* is largely cultivated. Among other varieties considered promising for cultivation are *Black Hamburg, Black Muskat, Black Prince, Gros Colman* (Pusa), *Kishmish,* and *Pusa seedless*.

Anab-e-Shahi. Widely cultivated in Andhra Pradesh and other parts of Deccan is a white, oval, thin-skinned, sweet grape of excellent eating quality. The bunch is medium sized and moderately compact.

Bangalore-Blue. Principally grown in Karnataka State. The fruits are bluish black, round and plump. The bunch is medium-sized and compact, the skin thick and tough, and the pulp greenish, mucilaginous and firm. The taste is sweet and slightly acidic.

Bhokri. This variety is cultivated chiefly in Maharashtra (Nasik District). It bears medium to large compact bunches. The berries are greenish yellow and spherical and have seeds. The taste is slightly acidic.

Black Muscat. This is a popular variety in Delhi. The bunches are small to medium and well filled. The berries, which are deep purple with bluish bloom and small in size, are sweet and have a pleasant aroma.

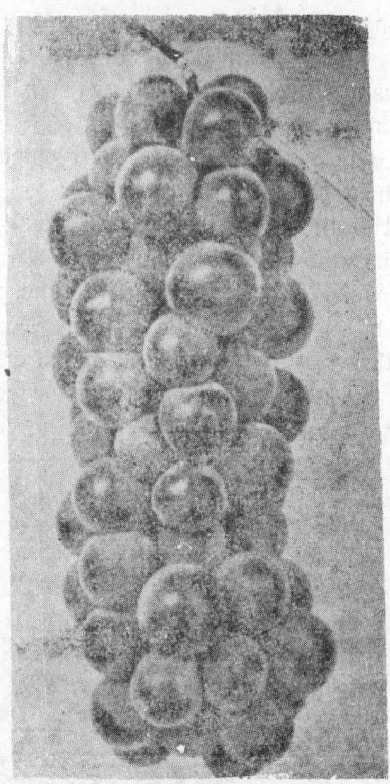

Fig. 5.31. Grapes—*Bangalore Blue*, a Bluish black, round variety.

Fig. 5.32. Grapes. *Pandhari Sahebi*, large white grape of superior quality.

Kandhari. This is a purple, oval grape with medium-sized, compact and shouldered bunches. The skin is thick and the pulp sweet and firm.

Pandhari Sahebi. This is a large, white grape of superior quality. The bunch is medium-sized and compact and the pulp very sweet and firm.

Phakdi. This is an oval, whitish green grape, having an extremely thick skin and soft, sweet pulp. The bunch is large, loose, heavily shouldered on the sides and pendulous in the middle. Being an extremely delicate variety, it does not keep long.

Pusa Seedless. This is seedless variety of very good quality. The berries are medium, oval and greenish yellow. The taste and flavour are excellent.

Cultivation. Grapes can be propagated by the rooting of hard wood cuttings. The cuttings are made from one-year old wood (canes) at the time of pruning when the vines are dormant. The cuttings should be tied in bundles and stored in moist sand for about a month for callusing. The callused cuttings start well in the nursery. One year old vines are fit for planting out in the field.

Grape vines should be transplanted in the months of January and February when they are in dormant condition.

The grape is a fairly hardy plant. The vines should, however, be irrigated regularly during summer months to maintain moisture in the soil during the growing period. The vineyard should be regularly irrigated from the time growth starts in spring till the crop is harvested.

During the early year of the development of the trunk, all vines have to be properly supported.

Harvesting, picking and packing. Grapes should not be harvested till they are fully ripe, as they do not improve either in sugar content or in flavour after picking. They should not be left even too long on the vine, for they will not carry well. The right stage of picking of grapes has to be determined according to the variety. The colour of the skin, the colour of the peduncle and the test of the berries are some important indices of the maturity of grapes. The stem of the bunch at maturity begins to shrivel and often so much that it can be bent. The coloured varieties develop their characteristic colour, whereas the white varieties become more nearly pale yellow as they ripen.

Ripe grape bunches should be clipped from the vine with the help of knife or scissors. Picking should be done in fair weather when vines and fruits are not wet either with dew or rain, otherwise they are likely to spoil soon. After picking, the unripe, bruised or diseased berries should be removed and the bunches packed in baskets or wooden boxes. The bottom of the package should be padded with dry grass. The bunches should be packed closely and the lid fixed tightly. Grapes keep in excellent condition when packed in boxes with sawdust.

Uses. Grape is a delicious, refreshing and nourishing fruit. It is classed as a protective food, is easily digestible and has large quantities of sugar, minerals like calcium and iron, and vitamin B.

Grapes good for eating fresh are called "table grapes". Some varieties are known as 'raisins,' can better be dried and preserved, while others are used in making juices and wines. Black grapes are also dried and preserved like raisins. They are available in plenty in South India.

Fig. 5.33. Grapes. *Pusa seedless,* a high quality grape, well known for its taste and flavour.

Fig. 5.34. Grapes. *Phakdi,* an oval, whitish green grape of extremely delicate variety.

Out of the total production of grapes 80 per cent is crushed for wine, and seven per cent dried into raisins; some quantity is consumed as fresh grapes or in the form of juice. Byproducts of wine-making are used for grape-seed oil, tannins, and potassium hydrogen tartarate. Selected fruits are used for canning. Fresh grapes are laxative, stomachic, diuretic, demulcent, and cooling. Popular tonic, *Drakshasava,* is made from grape juice. Similarly grapes form a constituent of *Chayvanaprasa*.

The leaves are astringent, sometimes used in diarrhoea. Sap obtained from young branches is used in skin affections.

Quince

Cydonia oblonga Mill.; Eng.—Quince; Hindi—*Bihi*; Family—Rosaceae. A small tree. It is native of West Asia, now grown in Punjab, Kashmir and the Nilgiris. The fruit is round or pear-

Fig. 5.35. Grapes. *Kandhari,* a high quality medium cropper.

shaped and quite large. The leaves are densely tomentose beneath, and even the fruit is wooly when young. The golden-yellow flesh is hard and rather unpalatable. The fruit is rarely eaten raw, but is used for jelly and marmalade, often combined with apples and pears. Also used as a rootstock for pear.

Fig. 5.36. Fruit preservation industry. The industry is becoming popular in India day by day.

FRUIT PRESERVATION INDUSTRY

There is a network of fruit preservation factories all over the country. They have a large range of products such as fruit juice, fruits in syrup, squashes and syrups, jams and jellies, pickles, tomato products, etc. These products are supposed to be nearest to nature. The common products are:

Fruit juices. Apple, mango, pineapple, guava, tomato.

Fruit drinks. Apple, apricot, mixed fruit, strawberry, mango and pineapple.

Canned fruits. Mango, pineapple, peach, pears.

Squashes. Mrange, lemon, mango, bel, lichi, rhododendron.

Syrups. Pineapple, khus, rose, sandal.

Jams and *Jellies.* Peach jam, mango jam, pineapple jam, apricot jam, mixed fruit jam, apple jam and strawberry jam; guava jelly.

Murabbas. Apple, anwla, bel.

Pickles. Mango pickle, lemon pickle, mixed pickle etc.

HIMALAYAN TEMPERATE REGION

Name of the fruit	*Varieties successfully grown*	*Places where grown*
Apple Bot. name—*Malus pumila* Fam.—Rosaceae Type—Pome Edible part—Fleshy thalamus.	Red Delicious, Golden Delicious, Starking (Royal) Delicious, Cox's Orange Pippin, Richard and Red June. Granny Smith, Baldwin.	Himachal Pradesh
	Ambri Kashmiri, White Dotted Red, Blood Red, Baldwin, Red Delicious.	Kashmir Valley
	Blenheim Orange Pippin, Delicious, Early Shanbury, James Grieve, Jonathan, Rome Beauty, Winter Banana, Golden Pippin, King of Pippins, Rymer.	Kumaon hills (U.P.)
	Red Delicious, Cox's Orange Pippin, Baldwin, Golden Delicious, Granny Smith, Royal Delicious, Richard, Red June.	Kulu Valley (H.P.)
	Rome Beauty, Glengyle Red.	Nilgiris and the South.
	William's Favourite, James Grieve, Prince Albert, Lady Sudley, Charles Ross.	Assam
Pear Bot. name—*Pyrus communis* Fam. Rosaceae Type—Pome Edible part—Fleshy thalamus.	Williams Bon Chritien (Bartlett), Clapp's Favourite, Thomsons, Easter Beurre, Winter Nelis, Conference, Dr. Jules Guyot, Marie Louise d'uccle, Baghu Gosha (Citron Des Carines), Emile d'Heyst, Jargonellee, Louise, Bonne of Jersey.	Elevations above 5,000 feet.
Cherries	Early Rivers, Governor Wood, Bigarreau de Schrecken Elton, Monstrueuse de Mezel, Bigarreau Napolean, Late Black Bigarreau, Early Amber.	Elevations above 5,000 feet.
Peach Bot. name.—*Prunnus Persica* Fam.—Rosaceae Type—Drupe Edible part—Mesocarp	Alexander, Early Rivers, Duke of York, Noblesse, Late Devonian, Elberts, J. H. Hale, Triumph, Alton, Prince of Wales, Sharbati.	Elevations between 3,500 and 5,000 feet.
Plum	Grand Duke, Bright Red Burbank, Early Transparent Gage, Victoria, Santa Rosa, Beauty, Kelsey, Satsuma, Hall, Plum Red.	Elevations between 3,500 and 5,000 feet.
Apricot	Shipley Early, Kaisha, New Castle, St. Ambroise Royal, Safeda.	Elevations between 3,500 and 5,000 feet.

Name of the fruit	*Varieties successfully grown*	*Places where grown*
Strawberry	Trafalgar Laxton's, Latest, Royal Sovereign, Early Cambridge, Katrain Sweet, Pusa Early Dwarf, Premier, Gem.	Elevations between 3,500 and 5,000 feet.
Persimmon	Fuyu, Hachiya, Hyakume	Elevations between 3,500 and 5,000 feet.
Grape Bot. name—*Vitis vinifera* Fam.—Vitaceae Type.— Berry Edible part—Complete berry	Thompson's Seedless, Sultana, Kishmish white.	

NORTHERN DRY REGION

Name of the fruit	*Varieties*	*Places of Cultivation*
Mango Bot. name—*Mangifera indica* Fam.—Anacardiaceae Type— Drupe Edible part—Mesocarp	(i) Dushehri, Langra, Samarbehist Chousa, Safeda Lucknow, Malihabad, Bombay Green, Bombay Yellow, Fajri Zafrani, Gopal Bhog, Fajri Rataul, Sehroli, Malda Handel, Yakutia (Hushnara) Baramasia (all the year round).	Grown extensively in the cenral and western districts of Uttar Pradesh especially in Kanpur, Lucknow, Meerut, Saharanpur, Bareilly and Hardoi districts.
	(ii) Dushehri, Langra, Malda, Sindhuria, seedling mangoes and Sehroli.	Mainly grown in Hoshiarpur, Ambala, Gurdaspur, Karnal districts of Punjab and Delhi state.
Citrus fruits Bot. Name—*Citrus spp.* Fam.—Rutaceae Type—Berry, hesperidium. Edible part—Juicy hairs of endocarp.	1. MANDARINS (Santara) (i) Desi, Nagpuri, Emperor and Laddu.	Desi seedlings mostly in Kumaon, submontane tracts and Saharanpur and Dehra Dun districts in Uttar Pradesh.
	(ii) Desi, Coorg, Narrtjee, Lahore Local.	Desi mostly in the hills near Pathankot and Himachal Pradesh, others mostly in Punjab.
	(iii) Nagpur.	In north-western districts of Madhya Pradesh.
	2. TIGHT SKINNED ORANGE (Malta)	
	(i) Mosambi, Blood Red, Jaffa, Navel, Vanile, Common, Early Blood, Joppa, Friefly Navel, Majorca, Tardiff, Mediterranean Sweet.	Mostly grown in Saharanpur and Meerut districts of Uttar Pradesh.
	(ii) Blood Red, Pineapple, Hamlin, Jaffa, Common, Valencia Late.	All over Punjab.

Name of the fruit	*Varieties*	*Places of cultivation*
	3. ACID AND SWEET LIMES	
	Kagzi Nimbu and Lime.	Saharanpur and Meerut districts of Uttar Pradesh and to certain extent in submontane tracts of Himalayas viz., Kumaon Hills. Grown also in Himachal Pradesh and Punjab.
	4. LEMONS	
	Baramasia, Eureka, Villa-Franca, Ponderosa, European, Hill Lemon.	Mostly in parts of Punjab and Uttar Pradesh.
	5. GRAPE FRUIT	Mostly in Saharanapur and Meerut districts of Uttar Pradesh.,
	(i) Saharanpur Local, Marsh Seedless, Triumph, Duncan.	
	(ii) Marsh Seedless, Foster, Duncan, Poona Budded.	Punjab.
	6. PUMMELO	
	Red Flesh and Gills Jeolikote.	Western districts of Uttar Pradesh and Punjab.
Date Bot. name. (Arecaceal) *Phoenix sylvestris* Fam.—Palmae Type—Berry Edible part—epicarp, and mesocarp	Hillawi, Khudrawi, Shamran and Zaidi.	Western districts of Punjab and North-Western Rajasthan.
Fig	Black Ischia, White Fig and Poona Fig, Brown Turkey.	Successful in U.P., Rajasthan, and parts of Punjab.
Guava	Safeda, Hafzi, Lucknow 49, Seedless, Karela, Dholka, Nasik.	Grown mostly in U.P. and to a certain extent in Punjab.
Grandilla (Passion Fruit)	Selections from Palampur and Saharanpur.	Localised in Saharanpur and Chakrata in U.P. and Palampur in the Himachal Pradesh.
Grape	Thompson Seedless, Black Prince, Dakh, Khalili, Foster's Seedling and Bangalore Blue, Bhokri, Pusa Seedless.	Mostly in Punjab and certain parts of U.P.

Name of the fruit	*Varieties*	*Places of cultivation*
Jujube	Local selections of round and large varieties; large fruited Nazook, Bedana and Benarsi.	Punjab, Uttar Pradesh and Rajasthan
Loquat	Golden Yellow, Pale Yellow, Large Agra, Fire Ball, Tanaka, Thames Pride, California Advance.	Mostly grown in Western parts of U.P. like Saharanpur, Meerut and Dehra Dun ditricts and Gurdaspur district in Punjab.
Litchi	Early Bedana, Late Bedana, Desi variety, Seedless Dehra Dun, Rose Scented, Late Large Red, Calcutta.	Mostly in Saharanpur and Dehra Dun districts of U.P. and Gurdaspur district in Punjab.
Papaya	Honey Dew, Washington and Saharanpur selections, Barwani.	Common in U.P. and Punjab and parts of Rajasthan and H.P.
Pecan	Mehan, Success, Nellis and Halbert.	Grown in the submontane tracts of U.P., Himachal Pradesh and Punjab.
Phalsa	Local selections.	Common all over.
Pomegranate	Kabuli, Kandhari, Paper Shell, Muscat Red, Bedana.	Found all over.
Persimmon	Fuyu Tanenasni, Hachiya.	Grown in U.P. and the Kulu Valley (H.P.)
Pear	Baghu Gosha, China, Country pear, Kicffer, Bartlett, Nashpati Chunsi.	Grown in western parts of U.P., Himachal Pradesh, Punjab and Delhi.
Peach	Bidwill Early, Early Cream, Early Agra, Elberta, Early Grawford, J. H. Hale, Alexander, Waldo, China Flat Chakli.	Mostly cultivated in western districts of U.P., Lucknow, H.P., Jaipur in Rajasthan and Jammu.
Plum	Early Large Red, Japan Gold, Alubokhara, Large Yellow, Rubio, Alucha Red, Sharp's early, Shalimar, Kelsey, Burbank, Satsuma, Czar, Victoria, Damsons, Ciyan, Titron.	Mostly grown around Chaubatia in U.P., in Punjab and Himachal Pradesh.
Sapota	Round, Oval, and Sharanpur selections.	Cultivated in some districts of U.P. and Jaipur in Rajasthan.
Mango	(i) Bombai, Fazli Malda, Safeda Calcutta, Himsagar.	Grown intensively in Burdwan, Hooghly, Murshidabad, Malda and 24 Parganas districts of West Bengal.
	(ii) Bihar Alphonso, Langra, Gulabkhas, Kishna Bhog, Jardalu, Mithua, Kaithki, Khas-ul-khas, Sukul.	Chiefly grown in Muzaffarpur, Darbhanga and Bhagalpur districts of Bihar.

EASTERN WET REGION

Name of the fruit	*Varieties*	*Places of cultivation*
Citrus.	(iii) Langra, Shadwala, Shahpasand, Sakkarchina, Brindabani, Fajri, Safdar Pasand, Dusheri, Samar Behisht Chowsa, Taimuria, Bombay Yellow, Buda-ka-kelwa.	Mostly in eastern districts of U.P. viz., Allahabad, Bara Banki, Faizabad, Gonda and Sultanpur.
	(iv) Mainly seedling mangoes and some varieties like Langra, Dushehri and Chowsa.	In Bilaspur, Jabalpur and Hoshangabad districts of Madhya Pradesh.
	(v) Mostly seedling mangoes and some varieties from Andhra Pradesh like Neelum, Banganpalli, Bangalore and Dophool.	Orissa
	(vi) Suvarnarekha, Rajumanu, Banganpalli, Jehangi, Pedda Rasam, Chinna Rasam, Cherukurasam, Panakalu, Himayuddin, Firangiludha, Kottapalli, Kobbari, and Neelum.	Intensively cultivated in Godavari, Krishna and Visakhapatnam districts of north-east Andhra Pradesh.
	1. THE SANTARA ORANGE (Mandarin)	
	(i) Khasi Orange or Darjeeling Orange.	Darjeeling district of West Bengal and Assam.
	(ii) Nagpur Orange	Nagpur, Nimar district of M.P.
	2. Mosambi	Ranchi district of Bihar and Banaras district of U.P.
	3. Kagzi Lime	All over to a considerable extent.
Guava	(i) Allahabad Safeda (ii) Hafzi (iii) Smooth Green	Mainly grown in Allahabad, Banaras, Bareilly and Faizabad districts of U.P., Bhagalpur, Muzaffarpur and Champaran districts of Bihar.
Banana	(i) Rasthali, Martaban, Chini Champa, Jahaji Manohar, Dig Joa, Sapri.	Mostly grown in Khasi Jaintia Hills, Cachar, Kamrup and Goalpara districts of Assam and to a lesser extent in Makir Hills, Lushai Hills, Nowgong, Sibsagar, Lakhimpur and Garo Hills.

Name of the fruit	*Varieties*	*Places of cultivation*
Sapota	(ii) Martaban, Singapore, Champa, Dudsagar Kunnan.	Extensively cultivated in Malda, Murshidabad, Burdwan and 24 Parganas districts of West Bengal and also grown in Hooghly, Nadia and Bankura districts.
	(iii) Malbhog, Chini Champa, Champa, Alphan, Sabja, Agheshwar, Dhudsagar.	Grown extensively in Muzaffarpur, Darbhanga, Bhagalpur and Champaran districts of Bihar.
	(iv) Basrai, Chini Champa, Champa, Raikela.	Easten districts of U.P., viz,. Allahabad, Banaraas and parts of Madhya Pradesh.
	(v) Mauritius (Basrai) Martaban, Champa.	Orissa
	(vi) Vamankeli, Ney Poovan, Poovan, Rasthali, Martalean, Monthan, Singapore, Chakkarakeli Nana Nendren.	Extensively grown in East Godavari, Krishna and Visakhapatnam districts of north-east Andhra Pradesh.
	Kirthiparthi, Thagarampudi, Dwarapudi and Palsapota.	Chiefly grown in Vishakhapatnam district of north-east Andhra Pradesh.
Cashewnut	Seedling trees only.	Visakhapatnam District of north-east Andhra Pradesh and Ganjam Pur, Cuttack and Balasore districts of Orissa.
Pineapple	(i) Giant Kew, Queen, Mauritius, Ceylon, Jaldhup.	Assam
	(ii) Giant Kew, Queen, Mauritius.	West Bengal
Jack fruit	(i) Khaja, local ordinary	Grown in eastern districts of U.P.
	(ii) Seedling trees	Grown in Santhal Parganas, Hazaribagh, Ranchi, Manbhum and Singbhum districts of Bihar, Coastal districts of Orissa and north-east districts of Andhra, viz., Vishakhapatnam, Godavari and Krishna.
Pomegranate	(i) Early Bedana, Desi, Late Bedana, Rose Scented, China, Muzaffarpur, Purbi.	Mainly cultivated in Champaran, Muzaffarpur, Darbhanga and Bhagalpur districts of Bihar.

Name of the fruit	Varieties	Places of cultivation
	(ii) China and Desi	West Bengal.
	(iii) Early Bedana, Early Large Red, Late Large Red, Late Bedana, Rose Scented, Calcutta.	Eastern districts of U.P. especially extensively grown in Faizabad district.
Papaya	Washington, South Africa, Ceylon Round, Ranchi and Honey Dew	Grown all over the region.
Ber	(i) Nagpuri, Thornless, Narkale, Ordinary Round	In Bihar especially in Shahabad district.
	(ii) Banarasi	Mostly in Varanasi district of eastern U.P.
Mango	(i) Alphonso, Pairi, Rajapuri, Kaiser.	Extensively grown in Saurashtra, Baroda, and Surat districts of Maharashtra State.
	(ii) Azum-us-Samar, Fakrus-Samar, Langra, Dushehri, and Mulgoa.	Mostly grown in Andhra Pradesh.

SOUTHERN REGION

Name of the fruit	Varieties	Places of cultivation
	(iii) Bangalora, Neelum Banganapalli, Mulgoa, Allumpur Baneshan, Salem Bangalora; Suvarnarekha, Rumani, Fernandin, Jehangir, Himayuddin, Khudadad, Cherukrasam, Panakalu and Rajumanu.	Chiefly grown in Andhra Pradesh, Tamil Nadu and Karnataka.
Citrus	1. Orange : Sathgudi, Mosambi and Batavian. 2. Lime : Mostly seedlings 3. Lemon : Italian, Nepali Oblong, Nepali Round, Eureka and Lisbon.	Mostly grown in Aurangabad and Parbhani districts of Hyderabad, Ahmednagar and East Khandesh in Maharashtra, Guntur, Kurnool, Cuddapah, Chitoor districts in Andhra Pradesh and Madurai district in Tamil Nadu.
Banana	(i) Basrai, Safed Velchi (Ney Poovan), Lal Velchi, (Poovan), Rajeli Monthan, Chakkarakeli.	Mostly grown in Kolaba, Surat, Kaira and East Khandesh districts of Maharashtra Osmanabad, Parbhani, Aurangabad, and Gulbarga in Andhra Pradesh, Salem and Coimbatore districts of Tamil

Name of the fruit	*Varieties*	*Places of cultivation*
	(ii) Yellakhibala, Rasabala	Nadu and Amravati and Akola districts of Maharashtra.
Guava	Dholka, Lucknow, 49, L. 46 Allahabad and Seedless (Arbhavi)	Mostly grown in Karnataka. Mainly cultivated in Aurangabad, Bihar, Parbhani and Nanded districts of Hyderabad; Nasik, Ahmedabad, Poona, Ahmednagar, East Khandesh and Dharwar districts of Maharashtra; Cuddapah and Kurnool districts of Andhra.
Grape	Bhokri, Fakadi, Pandhari Sahebi, Pachadrakashai, Anab-e-Shahi and Bangalore Blue.	Mainly grown in Nasik district of Maharashtra; Aurangabad, Hyderabad districts, and Anantapur district of Andhra; Salem in Tamil Nadu and also in Madurai district.
Papaya	Honey Dew, Washington, Round, Long, Gujarat.	Mainly grown in Poona and Ahmedabad districts of Maharashtra and parts of Tamil Nadu state also.
Fig	Poona Prolific, Light Purple, Penukonda, Coimbatore Fig, Daulatabad, Maisavani Black Ischia, Brown Turkey.	Grown in Anantpur district of Andhra; Poona district in Maharashtra; parts of Hyderabad and Karnataka also.
Pineapple	Mauritius, Desi, Giant Kew, Kew.	Is cultivated to a limited extent everywhere especially in Baroda (Maharashtra state), Madurai in Tamil Nadu and parts of Karnataka.
Custard apple and allied species	*Anona squamosa, A. reticulata* and *A. muricata.*	Grows in a wild condition in Andhra Pradesh, and is cultivated to a limited scale in Madurai district of Tamil Nadu.
Pomegranate	Dholka, Habshi, Uthukuli, and Alandi or Vadki, Paper Shell.	Cultivated on a small scale everywhere.
Jack fruit	Mostly seedling trees.	Grown all over.
Sapota	Round, Oval, Kalapatti, Calcutta.	Grown everywhere on a small scale, it is quite popular in Kolaba and Ratnagiri and Baroda districts of Maharashtra; it is also grown in certain districts of Tamil Nadu.
Cashewnut	Mostly seedling trees.	Is grown in Kolaba and Ratnagiri districts of Maharashtra state and Guntur distrct of Andhra Pradesh.

COASTAL WET REGION

Name of the fruit	*Varieties*	*Places of cultivation*
Mango	Alphonso, Pairi, Kallapady, Mundappa, Bangalore, Mulgoa, Neelum, Olour, Black Alphonso, Khudadad, Bennet Alphonso, Suvarnarekha, Fernandin.	Are grown all over the region.
Bread fruit	Seedless varieties.	Is popular in the West Coast but is also grown to a lesser extent on the East Coast.
Pineapple	Giant Kew, Mauritius, Country or indigenous types, Kew, Mauritius, Queen, Coorg Local.	Are cultivated all over but common on West Coast.
	Seedling trees.	Grown all over uniformly.
Cashewnut	1. MANDARIN : (Loose Jacket)	Mostly in Coorg and Wynad tract in Tamil Nadu.
Citrus	(i) Kamla or Coorg Orange	Localised to Kukul Valley in the Nilgiris.
	(ii) Kukul Orange	In Palni Hills and Kodaikanal areas.
	(iii) Seedlings	
	2. SATHGUDI (tight skinned orange)	Mostly confined to Malnad or hilly tracts of Kodur in Andhra Pradesh and Shimoga district of Karnataka.
	3. LEMONS AND LIMES	
	Italian, Nepali Oblong, Nepali Round, Eureka, Lisbon, seedlings of Kagzi Nimbu.	In Coorg, the Nilgiris, Malnad or hilly tract in Andhra, North and South and South districts of Tamil Nadu.
Mangosteen	Seedling trees.	Confined only to tropical ranges in the Nilgiris in Tamil Nadu State.
Banana	Poovan, Nendran, Mauritius, Rasthali, Safed Velchi, Monthan, Chakkarakeli, Kunnan.	Mostly cultivated in Madras, Kerala, Tanjore and Tiruchirapalli.
Apple	Allsops Early, Irish Peach, Carrington Winterstein and Rome Beauty.	Grown mostly in the Nilgiris and Bangalore in Karnataka.
Pear	Keiffer, Jargaonells, Beure Glifford.	--do--
Papaya	Round, Oblong and Honey Dew.	Common all over this region.

FOOD VALUE OF FRUITS

In temperate regions fruits are considered more as an agreeable addition to the diet than as a staple food. However, in the tropical regions, the fruits may often be the chief, and even the only source of food, as in the case of banana, mango, date, fig, coconut, and breadfruit. The fruits of temperate regions have only a slight nutritive value. The water content is about 80 per cent and the remainder consists of cellulose, and a solution of sugars, starches, pectin and organic acids, flavoured with essential oils and aromatic ethers. Carbohydrates are most abundant. organic acids, chiefly malic, citric and tartaric are present in greater amounts than in any other plant products. Pectin and mineral salts are also present in considerable amounts.

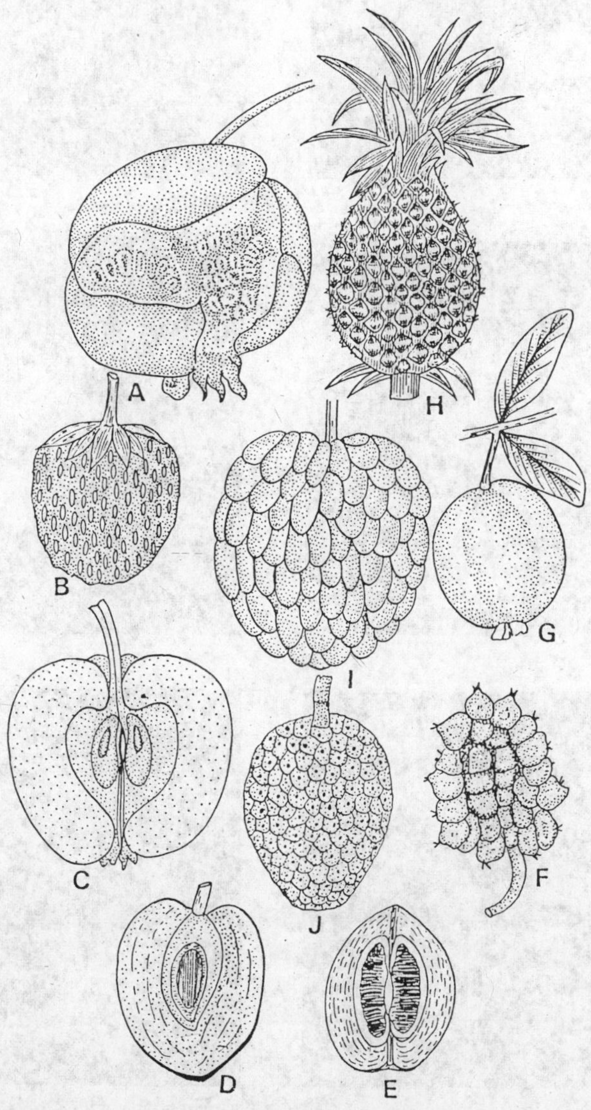

Fig. 5.37. Some edible fruits. A, balusta of pomegranate (*Punica granatum*); B, an etaerio of achenes of strawberry (*Fragaria vesca*); C, pome of apple (*Malus sylvestris*); D, drupe of peach (*Prunus persica*); E, Berry of sapodilla (*Achras zapota*); F, sorosis of mulberry (*Morus alba*); G, berry of guava (*Psidium guava*); H. sorosis of pinc apple (*Ananas comosus*); I, one-seeded nut-litchi (*Litchi chinensis*), edible part aril; J. *Annona squamosa* (custard apple).

SOME HYBRIDS OF TOMATO

Fig. 5.38. Some hybrids of tomato; the production of these hybrids is prolific; in this country the tomato is used both raw and cooked and several products, such as ketchup, sauce, juice, etc., are prepared from tomatoes. Some important hybrids are : A, Pusa Hybrid-1; B, Selection 18-Hisar Lalima; C, Selection 120-I.A.R.I.; D, Rashmi-IAHS (Indo-American Hydrib Seeds); E, Karnataka - IAHS; F, Arjuna of Sungro.

6

Fibres and Fibre Yielding Plants

It goes without saying that the fibre yielding plants rank second only to food plants in their usefulness to humankind. The utilization of fibres is directly related to the advancement of civilization. From the earliest times the man required three things for his living—food, clothing and shelter. Though other things were at hand to fulfil these demands yet they were insufficient and incomplete too without the genuine aid of the plants. As far as the clothing was concerned the skins and hides were not sufficient. He needed somewhat more and that gap was fulfilled by the plant fibre. As there was advancement of civilization the needs of man multiplied several fold and the use of the plant fibres increased greatly. Till now more than two thousand species of plants have been worked out which yield fibres. However, the commercial fibre yielding plants are few in number. The fibres from other plants are used by primitive tribal people of the world.

Economically the fibres may be grouped as follows:

Textile fibres. Fibres utilized for the manufacture of fabrics, netting and cordage, etc.

Brush fibres. Fibres utilized in the manufacture of brushes and brooms.

Rough weaving fibres. Fibres utilized in making baskets, chairs, mats, etc.

Filling fibres. Fibres used for stuffing mattresses, cushions, pillows, etc.

Natural fibres. Natural fibres used as cloth without weaving.

Paper-making fibres. Wood fibres, etc., utilized for paper-making.

As regards their structure, the sclerenchyma cells are grouped into fibres. Fibres are described as long cells with thick walls, correspondingly small cavities, and usually pointed ends. The walls of fibres usually consist of lining as well as cellulose. The fibres may occur singly or in small groups. The fibres may be found in any part of the plant—stems, leaves, roots, fruits and even seeds.

The plants belonging to several families of angiosperms yield fibres. They are chiefly tropical plants. The important families of the fibre yielding plants are—Malvaceae, Linaceae, Tiliaceae, Gramineae, Musaceae, Bombacaceae, Leguminosae, Palmae, Liliaceae, Urticaceae, Amaryllidaceae, Moraceae, Asclepiadaceae, and Bromeliaceae.

TEXTILE FIBRES

These fibres are supposed to be most important because they are connected with the textile industry. This industry is concerned with the manufacture of fabrics, netting and cordage. For making fabrics and netting, the flexible fibres are twisted together into yarn and thereafter woven, spun and knitted. For making cordage the individual fibres are twisted together rather than woven. The textile fibres are always long and possess high tensile strength, together with cohesiveness and pliability. They possess a fine, uniform, lustrous staple. The chief textiles are included in three categories—(i) surface fibres, (ii) soft fibres and (iii) hard fibres.

Surface fibres, *e.g.,* Cottons.

Soft fibres, *e.g.,* Flax, hemp, jute and ramie.

Hard fibres, *e.g.,* Sisal, coconut, pineapple, abaca, etc.

SURFACE FIBRES

Cotton

Gossypium Linn.; Eng.—Cotton; Hindi—*Kapas*; Bengali, Hindi, Gujarati, Punjabi and Marathi—*Kapas, rui, tula;* Kannada—*Hathi*; Telugu—*Patti, karpasamu*; Tamil and Malayalam—*Paruthi, panji*; Oriya—*Karpaso, kopa*; Family—Malvaceae.

History. The history of cotton *(Gossypium* sp.*)* is most interesting, and perhaps no more remarkable example of a sudden development exists in the whole history of Economic Products, than in the case of cotton. Only a few hundred years ago cotton and its products were practically unknown to the civilized nations of the West. However, evidences exist amongst certain of the aboriginal tribes of India which indicate a much more ancient knowledge of the plant. For example, the Khonds grow a cotton-bush in the place selected for a new settlement, and in after years the village plant is tended as sacred and carefully watered. This custom is probably of very ancient origin, and may denote a superstitious regard for the plant, probably derived from the knowledge and appreciation of its valuable properties.

The Sanskrit word translated 'cotton' is perhaps first mentioned in the *Institutes of Manu*, where it is stated that, "the sacrificial thread of a Brahman must be made of cotton, so as to be put on over his head in three strings." The word used in that passage and translated cotton is *Karpasi* from which has been derived, the *Kapas*.

However, it is certain that at the time of *Institutes of Manu* spinning and weaving were not only known, but the art of starching or weighing a textile was also practised, and the admissible amount prescribed. A still earlier notice of the process of starching probably occurs in the *Rigveda*. The picture thus raised of the character of the textile industries of India, some two thousand years ago. *Institutes of Manu* also deal with the regulations concerning washerman. We are thus led to believe that the arts of spinning, weaving and washing were perfectly understood in the East, at a time when the textile industries of the West, through ignorance of cotton, were in a much more backward condition.

Katan is an old Arabic word for flax, but may also have been used for cotton, since it apparently gave origin to the European words such as Cotton *(Eng.)* Cotone *(Ital.)*, Cotn *(Fr.)*, etc. In Greek the word *Karpason* is generally translated cotton. It is significant to mention that, with writers at the beginning of the Christian era, it was customary to speak of cotton and cotton goods by names not in use prior to that date. Herodotus is perhaps the first who seems to refer to the fibre. But perhaps the first unmistakable reference to cotton is by Theophrastus, subsequent to the expedition of Alexander into India. He mentions, that the plant was cultivated, at least in North-Western India, at that early date.

The muslins of Dacca seem also to have been known. They are described as superior to all others, and are said to have been called *gangitiki* by the Greeks, a name indicative of their manufacture on the banks of the Ganges.

The knowledge of the manufacture of cotton appears to have extended into Arabia and Persia about the beginning of the Christian era. From India, there is no doubt, the cultivation of cotton spread into Persia, Arabia and Egypt, from where it probably extended into Central and Western Africa. From Persia the culture migrated into Syria and Asia Minor, also into Turkey in Europe and thereafter into other parts of southern Europe. The earliest account of its cultivation on the European shores, is to be found in the works of Eben el Awan of Seville who lived in the twelfth century.

In the new world, we find that cotton has probably been used from the earliest times. It is connected with some of the most ancient beliefs of the aboriginal peoples of South America. Columbus found cotton in use among the natives of Hispaniola, but only in the most primitive way. Cortez found the manufacture in a much more advanced condition in Mexico.

With the rise of Muhammadanism the knowledge of spinning and weaving cotton probably spread to Europe, consequently the delicate cotton fabrics of the East were first imitated in Italy and Spain during the twelfth or thirteenth century. From these localities the art of cotton manufacture became diffused and extensively established all along the southern shores of the Mediterranean, but it was confined to that area till sixteenth century. From the low countries the industry passed over to England in the seventeenth century. Towards the end of the seventeenth century cotton printing was established in England. Up to the commencement of the nineteenth century America contributed but a very insignificant portion of the cotton was consumed. At the close of the war in 1815, the production of cotton in America received a fresh impetus, and the subsequent progress was rapid and continuous. This great increase in the importance of America, as a cotton-growing country early led the East India Company to commence operation for the purpose of improving the quality and increasing the quantity of Indian cotton for export to England. The imports into Great Britain from India during the years 1800 to 1809 averaged 12,700 bales per annum.

This is the world's greatest industrial crop, the chief fibre plant. It was known to the ancient world long before written records were made. References to it are to be found in the works of the Greek and Roman writers. Cotton has been in the use in India since 1800 B.C., and from 1500 B.C. to A.D. 1500 India was the centre of the industry. The Hindus were the first people to weave cloth. Cotton was introduced into Europe by the Mohammedans. This was first grown in the United States soon after the first settlements were made. The first cotton mill was established in 1787.

All the cultivated cottons fall under four species, two belonging to the Old World and two to the New World. These are *Gossypium arboreum, G. herbaceum; G. hirsutum,* and *G. barbadense.*

G. arboreum. This is most widespread of all the species of Old World cottons, being distributed throughout the rain-fed savannah areas from Africa, through Arabia and India, to China, Japan and E. Indies. Its origin is obscure, but it is obviously Asian, since the area of its greatest variability is found around the Bay of Bengal. This species comprises a large number of varieties and races including many of the cultivated cottons in and around India.

Fig. 6.1. Staple cotton. **'Kanchana'** variety.

The staple is coarse and very short, only 3/8 to 3/4 inch in length, but it is strong.

G. herbaceum. This is also an Old World species. It occurs in Africa, Middle East countries, Central Asia and Western India. It has been grown in India from time immemorial. Commercially the cottons belonging to this species constitute a fairly large percentage of medium staple cotton grown in India. The major part of the cotton crop in Maharashtra state comprises this species. This species comprises a large number of cultivated races.

It is utilized for low-quality fabrics, carpets, and blankets and is especially suitable for blending with wool.

G. hirsutum. This is a New world species. The centre of variability for this species is Central America. It comprises a large number of varieties or races. Only three races—*punctatum*, *marie galante* and *latifoliam*—extend beyond Central America. The last one is of great importance agriculturally, as it comprises the Upland cotton, which has spread over vast areas in America, Asia and Africa. The types of cotton belonging to this race constitute the bulk of long staple cottons grown at present in India. The fibres are white with a considerable range in staple length, from 5/8 to $1^3/_8$ inch.

G. barbadense. This is a New World species. It includes perennial shrubs or small trees, 3–15 feet high, or annual shrub moderately high.

The centre of origin of this group is tropical South America particularly its north-western parts.

Two distinct types of cotton belongs to:

Sea-island Cotton. This is one of the most important selections which yields lint of perhaps the highest quality which the genus *Gossypium* has so far proved capable of yielding. Its fine, strong, light cream-coloured fibres are more regular in the number and uniformity of the twists and have silkier appearance than those of other cottons. Sea-island cotton was brought to the United States from the West Indies in 1785. The finest types yielded staples 2 in. or more in length, surpassing all the others in strength and firmness. Another form of sea-island cotton is grown along the coast in Georgia and Florida and in the West Indies and South America. This has a staple form 1½ to 1¾ in. in length.

Egyptian Cotton. The second important linc of annual types derived from the perennial stock is the Egyptian cotton. Egyptian cotton is derived from a hybrid stock between a perennial *G. barbadense* and the annual sea island cotton. It is ecologically quite distinct from sea island cotton and is well adapted for irrigated conditions in the subtropics. Its practical value lies in its greater earliness. Some of the important commercial types at present in cultivation are Karnak, Menoui, Ashmuni, and Giza. The lint length of these cotton ranges from $1^1/_8$ in. (Ashmuni) to 1½ in (Karnak). Considerable quantities of these Egyptian types are imported every year into India for supplementing indigenous long staple cotton supplies. Besides Egypt, significant quantities of this cotton are produced in Sudan, U.S.A., Peru, French North Africa and Russia.

Because of its staple's length, strength and firmness this cotton is used for thread, underwear, hosiery, tire fabrics and fine dress goods.

Cultivation. Cotton is essentially a tropical crop, but its cultivation is carried on successfully over many parts of the world, far removed from the tropics. The limits of cultivation may be said to be the 40th latitude on both sides of the Equator. It is grown either at sea level or at moderate elevations not exceeding 3,000 feet. Cultivation is confined largely to flat open country and rough hilly tracts, where the minimum temperature does not fall below 70°F. Higher temperatures are very favourable, and the upper limit may go up even to 105°F in the picking season. The crop thrives well in moderate rainfall. Rainfall exceeding 35 inches is supposed to be harmful to the crop. The lower limit for a purely rain-fed crop is 20 inches. On black cotton soils, hardly any rainfall is needed over most of the growing period provided good showers have been received before the crop was sown and a satisfactory start has been made. Cotton is grown both as a dry crop and as an irrigated crop. If the rainfall is distributed over both the monsoons, the extraordinary fertility of the black cotton soil allows a wide variety of crops to be grown, and also taking of two crops in the year—one in the north-east monsoon period, and the other in the south-west monsoon period. On the other hand, if the rainfall is low and is confined to the north-east monsoon period, the only one crop is grown in the year. There is considerable mixed cropping practice with cotton. Pulses such as *arhar*, blackgram and green-gram and other crops such as groundnut and the castor, into the mixture. The 'New World' cottons are, however, grown pure—whether as dry crops or as irrigated crops.

THE COTTON INDUSTRY

Several operations are necessary in order to prepare the raw cotton fibre, as it comes from the field, for use in the textile industry. In brief these operations are as follows—ginning; baling; transporting to the mills; picking, a process in which a machine removes any foreign matter and delivers the cotton in a uniform layer; lapping, an operation whereby three layers are combined into one; carding, combing, and drawing during which the short fibres are extracted and the others are straightened and evenly distributed; and finally twisting the fibres into thread.

Harvest and yield. The cotton crop is usually harvested in three or four pickings, taken at suitable intervals. Picking is carried out by hand, mostly by women, the amount of cotton collected ranging from 20 to 50 lb. per day per person. Cotton should be picked only when the bolls are fully mature, fully open and the floss has puffed up consequent on exposure to sun.

The yield per acre is low in India as compared to yield in other countries.

Cotton marketing. The bulk of cotton produced in India is sold as *Kapas* or unginned cotton. *Kapas* is transported to the local markets or ginneries mainly in carts or, sometimes, on pack animals. The cultivator sells his cotton in the village market. The purchasers are village merchants, or agents of ginneries, spinning mills or exporting firms. In some states, co-operative societies organized by cultivators have taken up the purchase and sale of the cotton.

Ginning. Kapas or seed cotton collected from the field contains both lint and seed. For use in industry, cotton should be cleaned and the lint separated from the seed. A small amount of seed cotton is ginned in villages by the use of *charkha* gin. The bulk of it, however, is ginned in factories by power-driven machinery. The yield and quality of lint depend on the type of cotton and the machinery used for ginning.

Fig. 6.2. Cotton to market.

Baling. Cotton is packed for trade purposes both in loose and compressed bales. Loose packing is adopted for inland transit of ginned cotton to a pressing factory, while compressed packing is adopted for transporting ginned cotton to the market and for storing in the godowns. Each loose bale contains 200 to 300 lb. of cotton. The usual weight of compressed bale is 392 lb. net with density of 40 lb. per cubic foot.

DEVELOPMENT OF FIBRE IN COTTON

A cotton fibre is a delicate tubular prolongation of the peripheral wall of the seed. The fibres are formed by the outer epidermis of the outer integuments. These hairs are single celled and thick walled and attain a length of up to 45 mm These fibres also show characteristic twists and are called *lint fibres*. Intermingled with the lint hairs are some *fuzz hairs*. The maximum length attained by the latter is 10 mm They are thick walled with the narrow lumen and lack the twists.

USES OF COTTON

The bulk of cotton production is consumed in the manufacture of woven goods, alone or in combination with other fibres. The principal types of woven fabrics are—print cloth, yarn fabrics, sheetings, fine cotton goods, napped fabrics, duck, tyre fabrics and towels. Products in the form of yarn and cord include unwoven tyre cord, thread, cordage and twine and crochet yarns. Unspun cotton finds use in mattresses, pads and upholsteries. Cotton constitutes one of the basic raw

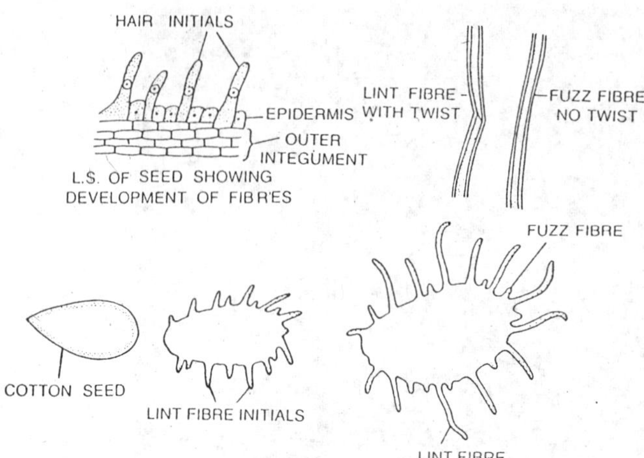

Fig. 6.3. Development of fibre in cotton.

materials for cellulose industries including plastics, rayon and explosives. Sterilized absorbent cotton finds use in medical and surgical practice.

Yarns of varying size and fineness are needed in the production of fabrics. Coarse yarns are spun from short staple cottons and fine one from medium and long staple types. Long and uniform staples are utilized for yarns of high counts required for fine fabrics.

Cotton waste is a by-product of the spinning and weaving mills and consists principally of short fibres rejected by combing and carding machines, floor sweepings, odds and ends from weaving and various scraps. The amount of waste given by cotton is an important factor in its quality evaluation. Cotton waste of good grade is employed in making cotton blankets, sheets, towels and flannelettes. Cylindrical strips from carding machine, which are constituted of fibres of good strength, are used for warps, twines, ropes and nets; they are also useful for wadding, padding for upholstery, bed quilts, etc. Strips from Egyptian cottons are mixed with wool for making mixed woollen goods. Floor sweeping and fibres unfit for spinning are bleached and used for gun-cotton, cellulose and artificial silk. Short remnants and thread waste that can not be respun are used as wiping and polishing material.

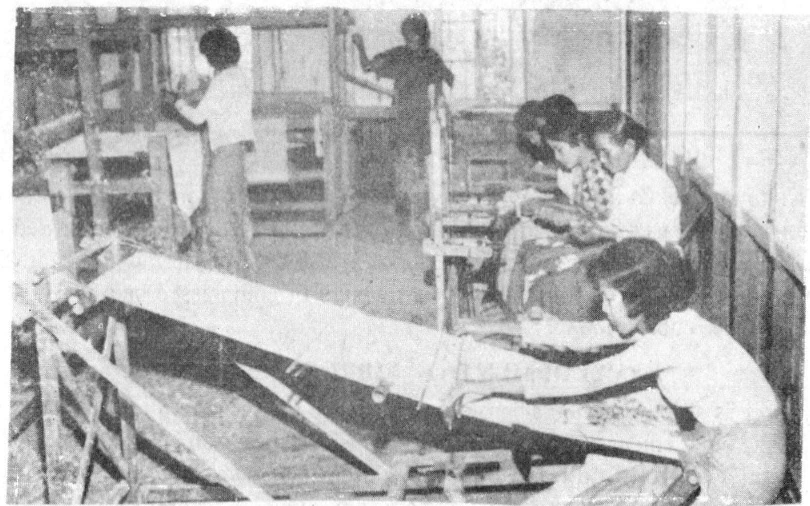

Fig. 6.4. Uses of cotton. Cottage *'kargha'* industry. The Mizo people at handlooms.

The *stalks* of plant contain a fibre that can be used in paper making or for fuel, and the *roots* possess a crude drug. The seeds are of the greatest importance and every portion is utilized. The *hulls* are used for stock feed; as fertilizer; for lining oil wells; as a source of xylose, a sugar that can be converted into alcohol and for many other purposes. The *kernels* yield one of the most important fatty oils, cottonseed oil and an oil cake and meal which are used for fertilizer, stock feed, flour, and as a dyestuff.

THE COTTON TEXTILES EXPORT PROMOTION COUNCIL OF INDIA—REPORT

As an exporter of quality textiles India's reputation dates back to the pre-Christian era. India's textile goods were shipped to distant corners of the world and they drew appreciation of high and mighty. To this day, these fabrics are known for the novelty and creativeness. The textile industry of India in those days was a cottage industry and the cloth was woven on handlooms. The skill and craftsmanship were handed over from generation to generation and this imparted a rare aesthetic quality to product. The glory of the beautiful handwoven cloth suffered a sort of eclipse in the wake of industrial revolution in the middle of the nineteenth century which created a vogue for mass produced textile products manufactured by mechanised mills. The beginning was made in India when a modern cotton mill was set up in 1854 in Bombay. The industry passed through many vicissitudes and pressure of competition from abroad in the process of its development until 1947 when India gained independence. The conditions created by the Second World War and the demand from many countries for their supplies of cotton textiles from India created an outlook for exports among the coteries of cotton mill industrialists. Cotton textiles are India's major source of foreign exchange. In 1974 cotton textiles brought in foreign exchange worth more than rupees 3,000 million.

A great need was felt to systematise the efforts to boost exports of textiles and the Government of India decided to sponsor an organisation towards the end. The Cotton Textile Export Promotion Council was sponsored in 1954 by the Government of India with active cooperation of the industry and the trade and is designed to function as an autonomous non-profit-making promotional body. The objectives of the Council extend to the promotion of exports of cotton piecegoods, yarn made-up articles, readymade garments and hosiery articles.

The Council in association with reputed manufacturers and exporters participates in specialized trade fairs and international exhibitions in a number of countries to highlight the excellence of India's fabrics, new articles of apparels, both woven and knitted and the new range at home textiles.

India's cotton goods have built up a reputation for quality in all markets of the world and gained acceptance in the most sophisticated markets. As a result of the functions carried out by Cotton Textiles Export Promotion Council, India's exports have been helped by qualiative improvement, in realising higher foreign exchange.

COTTON TEXTILE INDUSTRY

Cotton Textile industry is one of the oldest and largest industries of India, which has made rapid strides during the current century. The industry is well organised in our country either in terms of value of annual output or number of labour employed. India has the world's highest acreage under cotton, but in terms of production among fifty-five major countries, it ranks only fourth after the U.S.A., Russia and China. Cotton is grown in our country over an area of about eight million hectares and average production during the last few seasons had been around eighty seven lakh bales. At present the industry provides direct employment to nearly twelve lakh workers. It also provides indirect employment to many millions like the cotton growers, processors, handloom and powerloom weavers who alone are estimated over three million and innumerable cloth dealers and shopkeepers. The industry contributes in ever increasing measures to the Central and State Governments by way of taxes and duties. A large domestic market and plentiful supply of domestic cotton were the two basic factors responsible for the growth and development of cotton textiles into a major key industry. The first factor still holds good. With an annual growth of 2.5 per cent in population the domestic demand for textiles is rising. But the second factor has been weakening and has tended to undermine the strength and stability of the industry. It may be said that the future of this premier industry primarily depends on our ability to increase the cotton production efficiently and economically. In this industry, concentration of management and ownership is mostly in the hands of thc privatc sector.

Location

Though cotton textile industry being the largest industry in India has spread practically over all parts of the country, it is mostly localised in the states of Maharashtra and Gujarat. This is obvious from that these two states claim more than half of the spindlage and about 2/3 of the loomage. At the end of 1983 there were 805 mills in India. Out of this, 525 were spinning units and 280 were composite mills. Tamil Nadu, with 303, had the largest number of textile mills. But out of these only 24 were composite mills and the remaining made yarn. In Maharashtra, there were 108, out of these, 80 were composite mills. There were 89 composite units in Gujarat, out of a total of 114 mills. These three States account for the block of the production of yarn and growth in the organised sector. Gujarat and Maharashtra put together account for two-third of the mill loomage.

Reasons for this location are (i) availability of raw material both in good qualities and quantities; (ii) cheap and developed transport facilities including ports like Bombay and Kandla; (iii) availability of commercial and financial facilities like export and import houses and banking institutions; (iv) suitable climate and (v) enterprising capital. In recent years, the cotton textile industry has also spread to a number of other states like Madhya Pradesh, Bihar, Kerala, Andhra Pradesh, Tamil Nadu and Uttar Pradesh.

Size

On the whole, Indian market favour the organisation of the small or moderate sized mills. Cotton mills in Tamil Nadu, Kerala, Karnataka, West Bengal, Orissa, Assam and Bihar are small in size. Size of the cotton mill industry in these states ranges from ten to fifteen thousand spindles and 200 to 400 looms. In Andhra Pradesh, Uttar Pradesh, Madhya Pradesh and in Maharashtra the units are considered as medium sized. The size of model units in these regions ranges from fifteen to thirty thousand spindles and 400 to 600 looms. Big sized units with a range from 30 to 60 thousand spindles and 750 to 1500 looms are localised at Mumbai.

Progress Under the Five Year Plans

Taken as a whole, during the period 1951 to 1985, there has been a definite increase in the total cloth production. It increased from 4,740 million metres in 1951 to 12,498 million metres in 1985. During the year 1989 the total cloth production was 14,500 million metres. It is also clear that gradually the percentage of mill cloth in the total cloth production has been on the decline, while the percentage of handloom and powerloom cloth to total cloth production has been increasing year after year.

Problems of the Industry

Low yield of raw material. India has the largest area under cotton in the world, but ranks fourth as the producer. This is because yield of cotton per hectare is very low in India compared to those in U.S.A., Russia, Egypt, etc. At present the country is importing about six to eight lakh bales of foreign cotton, resulting in an expenditure estimated at Rs. 80 to 100 crores.

Substitution effect. Rapid strides in the field of science and technology resulting in a variety of synthetic materials are giving a stiff competition for cotton textiles. Now there is a tendency to go for terre-cotton and in fact, the demand for these is increasing. For blending with synthetics, usually extra long staple types are required and for fabrics of special finishes superior quality cotton with high strength is needed.

Rise in costs and low profit. The industry is suffering from cost inflation. Raw cotton accounts for about 40 percent of the cloth of cost and 65 percent of the cost of yarn. Next to the cost of cotton, wages make the most important item accounting for 27 per cent of the total cost of producing cloth. Other costs include dyes, chemicals, etc. (15 percent), power, fuel (6 per cent) and overhead and other expenses (12 per cent). Thus the profit margins have been poor.

Replacement of plant and machinery. If we want to reduce costs and increase the efficiency of the industry, rationalisation by means of replacement of plant and machinery becomes urgent necessity. Because of low profits and insufficient reserves, the industry is struggling to find enough resources for carrying on its day-to-day operations, the vast majority of mills could pay little or no attention to modernisation of their equipment.

The Textile Policy

The textile policy announced by the Government in August 1978 emphasises the need for encouragement of handloom sector. The policy stated that the government would freeze the loomage of mills and powerlooms and the responsibility of meeting the additional requirements arising from the increase in population and also any improvement in the per capita consumption will be entrusted to the handloom sector. This policy is aimed at in view of the fact that this would create additional jobs in the rural areas. A number of handloom development centres have been proposed to be started. As far as the policy on controlled cloth is concerned, it has been stated that mill made controlled cloth would be limited to 400 million sq. metres and would be produced by both the National Textile Corporation and the private sector mills. The policy further laid that, "it is intended that the handloom sector should to the maximum extent possible and in the shortest period of time, meet the requirements of cloth for the weaker sections of the population. The manufacture of controlled cloth in the organised sector would be phased out over a period of time consistent with the growth of production of the required variety of cloth in the handloom sector.

Again, the Government of India announced its textile policy on the sixth June 1985, indicating significant changes in the Government's approach to the growth and development of the textile industry. It aims at free and unrestricted growth of the industry by removing sector-specific and fibre specific rigidities in production and capacity creation, prevention of unregulated growth of power looms, treating mill and powerloom sectors at par, providing protection to handloom sector, restructuring of the industry in terms of stages of manufacturing process and rationalisation of excise duty structure. It also emphasises rehabilitation of viable sick units, modernisation and upgradation of technology, as well as improvement in product quality and reduction in cost of production.

SOFT OR BAST FIBRES

Flax

Linum usitatissimum Linn.; Eng.—Flax; Hindi—*Alsi*; Sanskrit—*Atasi*, Hindi—*Alsi, tisi*; Bengali—*Masina*; Marathi—*Javas*; Gujarati—*Alsi*; Telugu—*Avisi*; Tamil—*Alivirai*; Kannada—*Agasi*; Family—Linaceae.

History. According to Fluckiger and Hanbury, the history of flax (*L. usitatissimum* Linn.), its textile fibre and seed, is intimately connected with that of human civilization. The whole process of converting the plants into fibre, fit for weaving into cloth, is frequently depicted on the wall paintings of the Egyptian tombs. The grave clothes of the old Egyptians were made of flax and the uses of the fibre in Egypt may be traced back, as far as the twenty-third century B.C. The fabrics woven of flax have been discovered, together with fruits and seeds of the plant, in the remains of the ancient pile-dwellings bordering the lakes of Switzerland.

The propagation of flax in Northern Europe was promoted by Charlemagne. It seems to have reached Sweden and Norway before the twelfth century.

According to Oswald Heer, *L. usitatissimum* was a cultivated race derived from *L. augustifolium.* The annual flax (*i.e., L. usitatissimum*) has not been found in a wild state. It is the cultivated form met with in the greater part of India. However, *Linum augustifolium* grows wild, especially on the hills throughout the Mediterranean region.

The word *lin, llin, linu, linon, linum, lein, lan* exists in all the European languages of Aryan origin of the centre and south of Europe, but as it is not common to the Aryan languages of India, the cultivation of flax most probably took its origin with the Western Aryans, and before their arrival in Europe. The name *flachs* or flax of the Teutonic language comes from the old German *flahs*. However, it is not known precisely at what epoch the cultivation of the annual flax *(L. usitatissimum)* in Italy took the place of that of the perennial *L. angustifolium,* but it must have been before the Christian era. Flax has been found in a tomb of ancient Chaldea prior to the existence of Babylon, and its use in this region is lost in the most remote antiquity. Thus the first Egyptians, of white race, may have imported the cultivation of flax or their immediate successors may have received it from Asia before the epoch of the Phoenician colonies in Greece, and before direct communication was established between Greece and Egypt under the fourteenth dynasty. Thus the western Aryans and the Phoenicians may have introduced into Europe a flax more ádvantageous than *L. angustifolium,* during the period from 2500 to 1200 B. C.

According to DeCandolle this may be summed up as follows:

1. *Linum angustifolium*, usually perennial, found wild from the Canary Isles to Palestine and the Caucasus, was cultivated in Switzerland and the north of Italy by people more ancient than the conquerors of the Aryan race. Its cultivation was replaced by that of the annual flax *(L. usitatissimum).*

2. The annual flax (*L. usitatissimum*), cultivated for at least four thousand or five thousand years in Mesopotamia, Assyria and Egypt was wild in the area included between the Persian Gulf, the Caspian sea and the Black sea.

3. This annual flax (*L. usitatissimum*) appears to have been introduced into the north of Europe by the Finns (of Turanian race), afterwards into the rest of Europe by the western Aryans and perhaps here and there by Phoenicians, lastly into India by the eastern Aryans, after their separation from European Aryans.

4. These two principal forms of flax exist in cultivation, and have probably been wild in their modern areas for the last five thousand years at least.

Very little information of a historical character can be given regarding flax in India. It is thought that the Muslims have always given more attention to it than the Hindus. According to most writers certain Sanskrit names, which occur in some of the early works, are assigned to it. If the *Kshauma* garments alluded to in the Ramayana and Mahabharata, be accepted as having been made, as many writers maintain, of *Kshauma, i.e.,* linen—synonyms *Uma*, *Haimavati*, *Atasi* and *Marsina*—then the fibre must have been well known to the Sanskrit-speaking people from very ancient times. In the *Institutes of Manu*, mantles of woven *Kshauma* are those directed to be worn by theological students of the military class. In the Puranas references to the *Kshauma* cloth are frequent. In the *Ain-i-Akbari* a linen cap is described as part of the dress of a Brahmachari. Anyhow, it is quite evident that as in other countries, so in India, the true flax was known and manufactured from very early times, but that within the last two to three hundred years it has entirely lost its ground.

Flax is extracted from the stalks of linum types specially grown for this purpose and harvested when the capsules are immature. Fibre types are generally grown in temperate regions, they require a cool, moist climate, somewhat cloudy during the growing period.

They grow well in moist, well drained soils, rich in organic matter. They mature in 70–100 days.

Cultivation. The linseed crop is generally confined to sea level areas or low elevations and the plains as distinct from hilly tracts. As a seed crop, it can be cultivated even up to a level of 2,500 feet. However, this can also be grown as fibre crop at such elevation. In India, the crop is sown between August and November and harvested between January and April. It is mainly grown in Madhya Pradesh, Uttar Pradesh (Bundelkhand region), Maharashtra, Bihar and Rajasthan which together account for about 90 per cent of the total area sown to this crop. Karnataka and Andhra Pradesh are the other important linseed producing states.

Harvesting. The crop is harvested before the capsules are mature. The stage for harvesting is reached when the lower part of the stalk turns yellow and bottom leaves begin to droop. In most places harvesting is generally done by pulling the plants by roots. In America, Russia, Canada and some other countries mechanical appliances are also applied for pulling the plants.

Retting. Harvested stalks are partially or completely dried and seeds removed by rippling, *i.e.,* drawing the heads through a coarse comb. Deseeded stalks are then subjected to retting; the methods in use: water retting, dew or grass retting, mixed retting and snow retting. Water retting is most common. Stalks are bundled together and immersed in water after weighing. The time required varies from 2 to 3 weeks, depending upon the temperature and softness of water. Fibres are separated from retted stock by scutching and drying.

Characteristics. The colour of raw fibre varies from creamy white to grey. Fibre strands range in length from 15 to 100 cm (usually 37.7 to 62.5 cm) and are made up of bundles of small elements cemented together by a gummy substance. The ultimate fibres have an average length of 25–30 mm and an average diameter of 15-18μ. They are round to polygonal in cross section, more or less cylindrical in shape, with pointed ends, uniformly thick walls and narrow and irregular lumen; the surface is smooth. The fibre is valued for its outstanding strength, fineness and durability. It is stronger and more durable than cotton. It is soft, lustrous and flexible and possesses high water absorbency. It has low elasticity and is stronger when wet than when dry. Flax has good resistance to moisture and mildew. It takes crease-resistant and other special finishes well. The fibre does not react to mordants and dyes as readily as cotton but is less resistant to high temperatures. Fabrics made from it launder well; they feel colder to the touch than the cotton fibres.

Uses. Flax fibre is woven into fine fabrics, such as lawns, cambrics, hollands, crashes and damasks, and also into canvas, drills, ducks and buckrams. It is used for linen stitching, fishing twines, binding twines, ropes, fishing nets, and other purposes where strength is a primary requirement. It is also used for carpet making, house furnishing, sails and wrapping cloth. Other items of linen manufactures include handkerchiefs, table and bed linens, dress interlinings, men's suitings, towels, tents, draperies, upholstery covers, surgical dressings, fire fighting hoses and water holding bags.

Jute

Corchorus Linn.; Eng. Jute; Hindi—*Patsan*; Sanskrit—*Kalasaka*; Hindi—*Narcha*; Bengali—*Chinalitapat, narcha, nalita, titapat*; Assam—*Titamara*; Family—Tiliaceae.

History. J. F. Duthie found *C. capsularis* on the banks of the Gumpti near Judalpur in a wild state. However, Roxburgh reported the non-jute yielding species of *Corchorus* as natives of India. A special enquiry was, in 1873-74 in Madras of wild or cultivated jute found in that Presidency, and *C. olitorius* was reported to have been discovered both wild and cultivated. However, De Candolle was not convinced that the species existed in a truly wild state in the north of Calcutta. It might be found possible to produce forms of *Corchorus* from some of the truly wild species which would closely approximate to *C. capsularis* and *C. olitorius.* There are no Sanskrit names for these plants. There are neither Arabic nor Persian names for the species of *Corchorus*, known to the people of India. Mr. Hem Chánder Kerr reviewed all the reports and early books of travel that refer to fibre, and found that in none of these publications did there occur any mention of the word jute until 1796. In several works *pat* was mentioned as a fibre in India as a form of hemp. By the beginning of the nineteenth century the word *pat* was completely superseded by jute in all commercial correspondence. With the facts of this kind it is believed that jute plant is a comparatively modern development from some wild stock which was unknown to the Sanskrit writers. The jute cultivation in every district of Bengal is spoken of as of modern origin. In one district its introduction is fixed at 1872, in another at 1865, in a third before the date of the British rule and in a fourth it is put down at 500 years ago.

Fig. 6.5. A Jute plant.

The commercial fibre jute is obtained from either one or both of the following species of *Corchorus,* viz., *C. capsularis* Linn., grown in West Bengal and Bangladesh, and *C. olitorius* Linn., raised in the vicinity of Calcutta. Little or no jute is produced in the other states of India.

Cultivation. In India, the crop is sown between March and May, and harvested between July and September, It is grown mainly in West Bengal, Assam and Bihar, which together account for 90 per cent of the total area sown. Orissa, Uttar Pradesh and Tripura are the other states, which contribute the remaining 10 per cent area.

Jute is a bast fibre obtained from the secondary phloem. The bast fibre is obtained from *C. capsularis*, a species with round pods which is grown in lowland areas subject to inundation. Fibre from *C. olitorius*, an upland species with long pods, is but little inferior.

In India, the time for harvesting the crop depends entirely upon the date of sowing; the season commences with the earliest crops about the end of June, and extends to the beginning of October. The crop is harvested within three or four months after planting, while the flowers are still in bloom.

Separation of fibre by retting. The fibre is separated from the stems by a process of retting in pools of stagnant water. The crop is stacked in bundles for two or three days, to give time for the decay of the leaves, which are said to discolour the fibre in the retting process. The period of retting depends upon the nature of the water, the kind of fibre, and condition of the atmosphere. It varies from two to twenty–five days. The operator has therefore to visit the tank daily, and ascertain, by means of his nail, if the fibre has begun to separate from the stem. This period must not be exceeded, otherwise the fibre becomes rotten and almost useless for commercial purposes. The bundles are made to sink in the water by placing on the top of them sods and mud. When the proper stage has been reached, the retting is rapidly completed. The cultivator, standing up to the waist in the foetid water proceeds to remove small portion of the bark from the ends next the roots. Grasping them together, he strips off the whole with a little management from end to end without breaking either stem or fibre. Thereafter the washing is done to remove the impurities.

Fig. 6.6. Jute—Physical properties of fibres. Examination of quality and texture of fibre. Ultrasonic emulsifier with pilot plant.

It is now wrung out so as to remove as much water as possible, and then hung up on lines prepared on the spot, to dry in the sun.

The very long pale-yellow fibres from six to ten feet in length, are quite stiff, as they are considerably lignified; they possess a silky lustre. They are very abundant, but are not particularly strong, and they tend to deteriorate rapidly when exposed to moisture, to which they are quite susceptible. Besides these disadvantages the jute is cheap and easily spun makes it valuable.

Utilization. Jute is used chiefly for rough weaving. The thick cloth made from jute fibre is used for making gunny-bags. Another type of fine cloth prepared from jute fibre is chiefly used as a cloth to sleep on. Another type of coarse cloth is largely used for making the sails of country boats, and also for bags to hold large seeds or fruits.

Jute is extensively used in the manufacture of carpets, curtains, shirtings, and is also mixed with silk or used for imitating silk fabrics. The fibre is also used for making twine and ropes. Short fibres and pieces from the lower ends of the stalks constitute jute butts, which are used to some extent in paper making.

India not only grows most of the jute, but it is the largest manufacturer and exporter of jute products.

JUTE INDUSTRY

In the industrial economy of India, Jute industry occupies an important place both in terms of value of annual foreign exchange and the number of people employed. This industry accounts for Rs. 300 crores by way of capital employed, provides employment to nearly four lakh workers and exports more than 75 per cent of the production. The 75 per cent of the production accounts for nearly 5 per cent of total foreign earnings of all commodities. This industry has an installed capacity of 13.0 lakh tonnes spreading over 44,516 looms and 5.92 lakh spindles. Besides, India is the largest producer of jute goods in the world. Apart from all these, about three million families of farmers derive a substantial part of their income from jute cultivation. In 1885, India's first jute factory was set up at Serampore in Bengal, followed by several others. However, after country's partition, nearly 75 per cent of India's jute growing areas went to East Bengal (Bangladesh) whereas most of the jute mills remained on the Indian territory, which gave a severe blow to the industry. In spite of this the industry made serious effort to increase jute production through both extensive and intensive cultivation. As days passed on Bangladesh became most powerful competitor which has built considerable jute manufacturing capacity over the years by putting up more productive modern machines as against India's old factories.

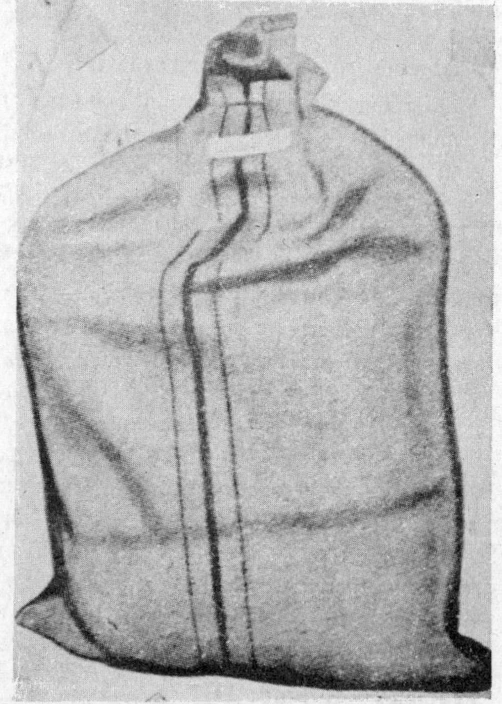

Fig. 6.7. The jute industry. A gunny bag, a jute product.

Location

Jute industry has been located in five states in India, *i.e.,* West Bengal, Andhra Pradesh, Bihar, Uttar Pradesh and Madhya Pradesh. Though the units engaged in the industry are located in all these five states, the main concentration of the industry is only in the State of West Bengal particularly about sixty miles long and two miles broad along both the banks of Hoogly, above and below Calcutta. The reasons for nearly ninety per cent of concentration of this industry around Calcutta are (i) raw jute is cheap and cannot afford much transport cost, as such the mills were forced to establish around Calcutta because of the availability of raw material in abundance in that area; (ii) availability of cheap power especially coal in West Bengal; (iii) transports, port, financial and commercial development of export market in Calcutta. The Central Government as well as some State Governments are encouraging the farming of jute with a view to feed our jute industry and to establish new units in various locations throughout the country. For example, Tamil Nadu proposes to launch a scheme for cultivation of jute with the ultimate aim of starting a jute mill.

Size

Market and managerial integration determine, to a large extent, the optimum size of units in the jute industry. The loomage capacity of a vast number of Indian mills ranges from 1,400 to 5,000 though mills outside West Bengal have less than 500 looms each. The reason for such large size in West Bengal is mainly influenced by location advantage. It is because of availability of adequate supplies of raw jute. Marketing and managerial integration has accelerated the situation and created an atmosphere which made the units big. It is not possible to say that a unit working with a particular loomage capacity is at optimum level because the problem of optimum size for the Indian jute industry becomes more complicated. It is complicated because of scarcity of raw material, the growing adoption of substitutes in other countries and increasing rate of development of technology and modernisation in the competing countries. If we say today that a particular unit is at optimum level it may not hold good tomorrow due to the dynamic conditions arising in day-to-day life of jute industry.

Production, consumption and exports

A production target of 1.5 million tonnes of jute manufacture is projected for 1984–85 including requirements for export (estimated at 0.55 million tonnes). For achieving the export target, efforts need to be made for improving the quality of secondary backing, production of specialised sacking constructions and the greater use of lighter. The demand for jute goods has been estimated at 16.25 lakh tonnes for 1989–90, out of which 13.35 lakh tonnes to be consumed domestically and 2.70 lakh tonnes to be exported. Jute manufacturers occupy a pride of place among the export commodities. Now–a–days, however, the jute industry has been passing through a critical period. Although the exports of jute goods rose from Rs. 135.15 crores in 1966–67, a declining trend set in thereafter. The exports declined to Rs. 190.44 crores in 1970–71. Even though there was an improvement in exports in 1971–72 when they rose to 265.27 crores but consituted only 12.0 per cent of total exports. In subsequent years the exports further declined though there was a slight increase in 1974–75. However, the jute exports, as a percentage of total exports, continued to decline and they constituted only 6.3 per cent of the total exports of Rs. 3941.62 crores in the years 1975–76. During the year 1976–77 and 1977–98, the share of jute goods in total exports was only 3.9 and 4.2, respectively. Again in the year 1978–79, it slumped to 2.8 per cent and rose to 4.6 per cent in 1979–80. From 1980–81 to 1983–84, jute goods exports in value as well as in percentage declined sharply. In 1984–85, the exports improved and could reach 2.45 per cent of the total exports.

Problems of jute industry

Raw materials. Though jute industry in India is well developed and well organised, yet it faces several problems. The fluctuations in the production of raw jute have continued following the partition of the country, when nearly 75 per cent Indian jute growing area went to East Pakistan (now Bangladesh). The fluctuating trend is attributed to the fact that being an agro-based industry the production of jute is subjected to natural vagaries which greatly influence the volume of agricultural output. The production of raw jute which was 75.9 lakh bales in 1967–68 declined sharply to 37.6 lakh bales in 1968–69. Though it was larger at 97.5 lakh bales in 1969–70 it fell short of the raw material requirement of the industry. In 1971–72 the production of raw jute was 68.4 lakh bales. It declined to 59 lakh bales in the year 1972–73. But the year 1973–74 was a year of bumper crop for jute production. The production was 80 lakh bales. But again in the year 1974–75 the production slumped to 55 lakh bales. During the year 1977–78 and 1978–79 it again rose to 75 lakh bales and 82 lakh bales respectively.

Obsolete machinery. The another problem facing the industry is the existence of obsolete machinery. Productivity is very low when compared to the jute mills in Bangladesh, Brazil, Philippines and some African countries who are producing low-priced jute goods. It would be rather difficult to compete with new mills set up in foreign countries on modern lines, if we fail to modernise our plants and equipments, and therefore, it becomes absolute necessary to modernise our machinery, but it needs adequate resources to meet out the financial problem.

Labour unrest. Another serious problem of this industry is the frequent labour unrest in the country. For example, as many as 13.2 million man-days were lost due to labour unrest during January–June 1979, in West Bengal where the jute industry is centralised to a greater extent.

Competition from substitutes. The Indian jute goods are becoming dearer day by day because of increase in the cost of production due to increase in the price of raw material. Many foreign countries have developed the cheap substitutes of the jute goods. In recent years, the synthetic materials and other alternatives such as paper have emerged as major competitors of jute goods in the field of packing. Unless and until we don't reduce our cost of production and offer the jute goods at cheap rate, there will be no bright future for our jute industry.

Cost of production. In the last few decades the cost of production has gone up by about 75 per cent. Among the various components of cost, there has been a steep rise in wages while the increase in the cost of raw material is of lesser magnitude. The cost of other components such as dyes, chemicals and power charges have also increased.

Competition. The main competitor of India in this industry is Bangladesh. However, the industry has witnessed a boom in 1971–72 due to disturbances in Bangladesh. But this boom was for a short duration as the jute industry in Bangladesh has staged a significant recovery in later years. Bangladesh produces raw jute of fine quality and manufactures a variety of jute goods. Bangladesh has been underquoting Indian jute goods in foreign markets. In most of cases the foreign buyers have been placing orders first with Bangladesh; and India gets only residual orders. It is being stated that the jute goods of Bangladesh are cheaper than Indian jute products.

Modernisation. The survival of Industry is mostly dependent on its ability to accelerate the pace of modernisation. The Indian jute industry lags behind Bangladesh both in capital equipment and techniques. In 1956, the Planning Commission had pointed out that the existing machinery had become to a large extent worn out and obsolete. Jute mill modernisation mainly has three problems. Firstly, the industry has no reserves. Secondly, it lacks the capacity to repay and thirdly, because of its inability to make profits, the industry is fast becoming unviable.

Sunn-Hemp

Crotalaria juncea Linn., Eng. Sunn hemp; Hindi—*Sunn, san, sannai;* Sanskrit—*Sana*; Bengali—*Shonpat, shon, ghore sun*; Marathi—*Tag*; Tamil—*Sannappu, sanal*; Telugu—*Janumu*; Kannada—*Sanabu*; Malayalam—*Wuckoo nar*; Family—Papilionaceae.

Sunn or *san* is applied to the true hemp plant, a prefix being given to distinguish the fibre of *Crotalaria juncea*, such as *Phul-sunn, Bhaga-sunn*, and *Badal-sunn*. In Bengal this is known as *Chunpat*. But *patsan* and *mestapat* are generally applied to *Hibiscus cannabinus*. There would seem to be every chance that the earliest writers allude under *Sana* to the fibre of *C. juncea*.

Habitat. The flora of British India gives the habitat of this plant as "Plains from Himalaya to Ceylon, but often planted for its fibre." *C. juncea* is not found in wild state anywhere in India, although it may sometimes exist as an escape from cultivation. *C. juncea* is cultivated more or less in every province of India.

Cultivation. *San* is grown by itself or at time is cultivated in strips or around the margins of fields. It is never cultivated as a mixed crop. Throughout India as a whole it is a *kharif* crop—that is to say, it is sown about the commencement of the rains and cut at the end of September or beginning of October. It is sown on the high sandy lands less suited for the more important crops.

Harvest. Usually the crop is harvested after the flowers have appeared, but in certain localities the plants are left on the field until the fruits have begun to form, and in some instances even until they are ripe. In most cases the plants are pulled up by the roots, in others the stems are cut with a sickle close to the ground.

Separation of the fibre. After removal from the ground, the stems are tied in bundles (20 to 100 in each), but the leaves are left until quite dry, the leaves either fall off naturally as are removed by the stems being beaten. It is a common practice to place the bundles of stems erect in 2 or 3 inches of water for 24 hours, so as to give the thicker and lower ends a longer submersion. But the length of time required for retting depends largely on the temperature of both the atmosphere and water. In August and September two or three days will generally suffice. Since the bark on the butts of the stalk is thicker and more tenacious than that on the upper portion, and requires, therefore, longer exposure to fermentation. Small pools of clean water, well exposed to the sun, seem best suited for steeping in, because heat hastens maceration, consequently preserves the strength of the fibres, while the clean water preserves their colour. Deep water, being cooler, requires more time for the operation.

After the necessary degree of retting has been attained, the cultivator, standing in the water upto his knees, takes a bundle of the stems in his hand, threshes the water with them until the tissue give way and the long clean fibres separate from the central canes.

Washing the fibre is very tedious, and a man rarely works for more than three hours at a time but it is relieved by turns; he will clean 15 kg a day, which represent the fibre obtained from two quintals of stem.

Process of cleaning. When the fibre has been separated and thoroughly washed, it is the usual custom to hang it up over bamboos to be dried and bleached in the sun. When dry it is combed if required for textile purposes or for nets and lines, but if for ordinary use—*e.g.*, ropes and twine—it is merely separated and cleaned by the fingers while hanging over the bamboo.

Uses of sunn hemp. The chief purpose for which this fibre is utilized is the manufacture of a coarse cloth (*tat patti*) or canvas used chiefly for sacking. A large amount of fibre is used up in native of cordage trade. The waste tow and old materials are made into paper.

Sunn stalks are used chiefly as fire wood.

Bimli Jute

Hibiscus cannabinus Linn; Eng. Bimli Jute, Deccan hemp, Kenaf; (Verna. *Patsan*) : Sanskrit—*Nalita*; Hindi—*Ambari, patsan, pitwa;* Bengali—*Mestapat*; Marathi—*Ambadi, ambada*; Gujarati—*Ambari, sheria*; Telugu—*Gogu, gonkura*; Tamil—*Pulichhai, pulimanji, kasini;* Kannada—*Pundi;* Malayalam—*Kanjaru*; Oriya—*Kanuriya*; Bihar—*Kudrum*; Punjab—*Sankokla*; Family—*Malvaceae.*

This is an erect herbaceous annual with straight, slender, glabrous or prickly stem, 8–12 feet high; lower leaves cordate, upper leaves deeply palmately 5–7 lobed; flowers axillary, large 3–4 inches in diameter, yellow with crimson centre; capsules globose, pointed, seeds large, brown, nearly glabrous.

The plant is indigenous to India. It has been under cultivation in India since long and is found to be distributed throughout the country up to an elevation of 3,000 feet in the lower Himalayas.

It is cultivated mainly as a fibre crop in the drier tracts of Deccan, comprising Andhra Pradesh, Karnataka, Maharashtra, Madhya Pradesh and Bihar. In other areas it is cultivated on a limited scale, as a supplementary crop in mixture with others.

Cultivation. It is a tropical crop, thriving best in a humid climate. It thrives best in well-drained, sandy loam soil. It does not grow under water-logged condition. It is generally grown mixed with other crops such as bajra, jowar, rice and cotton and also a border crop in sugarcane and cotton field. It is cultivated mainly as a *Kharif* crop, *i.e.,* sown from May to July and harvested in October to November.

Harvesting. The crop is ready for harvesting 3–5 months after sowing. For fibre purposes, the harvesting is done at the flowering stage before seed setting. The plants are cut close to the ground with a sickle, tied into bundles of 30 to 40 stalks, left on the field for a few days to dry and then steeped in water for retting.

Retting. The method of retting is the same as that adopted for jute or sunn hemp. The leafy tops are cut off the bundles steeped vertically in water for 2–3 days to soak the thick basal parts. They are then steeped in a horizontal position, after weighting with logs, stones, clods of earth, mud, etc. The period of retting varies from 6–10 days depending upon the maturity of the crop at the time of harvesting, the temperature of the water and the types of microorganisms present. The progress of retting requires watching. Under-retting does not facilitate the easy removal of fibre and the fibre obtained is harsh brittle whereas overretting weakens the strength of the fibre. Retted bundles are removed and the bark peeled off from the root upwards. The strips are gently beaten with a stick and rinsed in water to separate the fibre from adhering tissue. The clean fibre is washed, dried in the sun and made into bundles for marketing.

Yield. The yield of dry fibre ranges from 200 to 1,200 lb. per acre, according as the crop is grown mixed or pure. Yields as high as 2,000 to 2,700 lb. or more per acre have been reported from crops in other countries.

Characteristics. The fibre is extracted from the inner part of the cortex, outside the cambium layer. The fibre strands range in length from 5 to 10 feet and are composed of ultimate fibres 1.5–6.0 mm in length and 12–33μ in diameter. The ultimate fibres are cylindrical in shape with thickened walls and blunt or pointed ends, polygonal and rounded in cross section with small or large lumen. The fibre is comparable to that of jute in lustre, but is somewhat coarser, harsher and more brittle and inflexible; it is more resistant to rot. It can be spun alone or in admixture with jute.

Uses. It is widely used for rope and cordage. Considerable quantities are used in making fishing nets and strings for tying rafters. It is also used for coarse canvas, sacks and gunny bags, floor matting, rug backing, chair backing, etc. Fibre of poor quality and cuttings is used in the manufacture of paper.

The fibre is known as mesta or kenaf which competes with jute in lustre, but is coarser, inflexible, harsher, and more brittle; used for the same purposes as jute.

The leaves are used as pot-herb; tender twigs as cattle fodder. Seeds are used as cattle feed. They are considered stomachic and aphrodisiac; yield a fatty oil used for manufacture of soap, linoleum, paints and varnishes, and after refining, for edible purposes.

Roselle

Hibiscus sabdariffa var. ***altissima*** Wester; Eng. Roselle or Rama: Hindi—*Patwa*; Bengali—*Lal-mista, patwa, chukar*; Hindi--*Lal-ambari, patwa*; Marathi—*Lal-ambadi, patwa;* Telugu—*Yerra gogu;* Tamil—*Gogu;* Kannada—*Pundibija;* Malayalam—*Polechi*; Assam—*Chukiar*; Family—Malvaceae.

Roselle has attained prominence as a jute substitute and attempts are being made to extend its cultivation in India. It is particularly successful in Bengal. It is cultivated also in Bihar, Assam, Tamil Nadu and Andhra.

The plant is well adapted to all types of soils. It requires a moist climate, high temperature throughout the growth period, and medium rainfall.

Harvesting and retting. The crop is harvested at the bud stage when both outturn and quality of fibre are good. Stalks are tied into bundles of convenient size, left on the field for 3–4 days, and retted as in the case of *H. cannabinus*. The retted fibre is washed in running water to remove foreign matter, dried in the sun, combed and baled. The average yield of fibre is reported to be 1,000 lb. per acre.

Fibre characteristics. The fibre has a length of about 12 feet, the ultimate fibre being 1.2 to 3.3 mm long and 0.01–0.03 mm broad. It is silky, soft, lustrous, white to pale yellowish brown in colour. It is comparable to jute in strength and durability but is somewhat harsh and coarse. It can be mixed with jute and spun on jute machinery.

Uses. Roselle fibre is used for sacking, cordage, rope fishing nets and generally for all purposes for which jute is used. Bags made of roselle fibre are extensively employed in Jawa for packing sugar.

Hemp

Cannabis sativa Linn.; Eng. Hemp; Hindi—*Ganja, bhang;* Sanskrit—*Bhanga, vijaya*; Hindi, Bengali and Gujarati—*Bhang, ganja, charas, siddhi, jia*; Telugu—*Ganzai*, Family—Cannabinaceae.

This is true hemp. It is native of Central Asia. In India, it is cultivated in Uttar Pradesh, Bengal, Maharashtra, Madhya Pradesh, Tamil Nadu and Orissa. The hemp plant is an under-shrub. It is bushy, branching annual from 5 to 15 feet height. It is a dioecious plant with palmate leaves. It thrives best in a mild humid climate and a rich loamy soil with humus.

The fibre develops in the pericyclic region of stem. It is valuable for its length (3 to 15 feet), strength and durability. The plants are harvested by hand or machine. The fibres are separated by retting, either in dew or in water. They are then broken, scutched and hackled by hand. The best grade of fibre is obtained from the stems of male plants.

Uses. The fibre is used for ropes, twine, carpets, sailcloth, yacht cordage, sacks, bags, etc. The finer grade of hemp can be woven in a coarse cloth. Hemp waste is used for making paper.

Country Mallow

Abutilon indicum (Linn.) Sweet; Eng. Country mallow, Indian Abutilon; Hindi—*Kanghi*; Sanskrit—*Atibala;* Bengali—*Potari*; Telugu—*Tutturabenda*; Tamil—*Paniyarattutti*; Kannada—*Tutti*; Malayalam—*Velluram*; Family—Malvaceae.

A shrub. The stems, on retting yield a fibre.

Uses. The fibre is used for making ropes, twine and cordage.

The fibre is white, lustrous, coarser than jute, with an anti-clockwise drying twist.

The plant is used as a febrifuge, anti-emetic, and anti-inflammatory; also employed in urinary troubles. The bark is astringent and diuretic. Roots are used as a nervine tonic and anti-pyretic; also used in piles. Seeds called *Balbij* are rich in mucilage and used as a laxative and demulcent; contain a fatty oil.

China Jute

Abutilon theophrasti Medic.; Eng. Indian mallow, China jute; Hindi—*Nahani khapat*; Family—Malvaceae.

A herb. It is found in North-West India, Kashmir and Bengal. The stems on retting yield 'China jute' or 'Tientsin jute'.

Uses. The fibre is used for making rugs. It is also used for making ropes and cordage.

Ramie

Boehmeria nivea (Linn). Gaud.; Eng. Ramie; Hindi—*Kankhura*; Bengali—*Kankura;* Assam—*Rhea*; Family—Urticaceae.

A 3 to 6 feet high shrub with cordate leaves. It is native of Asia. In India it is grown in Assam and Bengal. The fibres are obtained from the bast. The fibres are long, strong and durable. The stems on retting yield the fibre. The bark is peeled off, and the outer portions and green tissue are scraped off by hand.

Uses. The fibre is used for sacks, thread, cordage paper and gas mantles.

The fibre is strong, lustrous, excellent and durable, used for fabrics, plushes, and knit material. Ramie is generally used mixed with wool, silk and cotton; its special use being in the manufacture of lustrous, non-erasable fabrics. Gas mantles of good quality are made from it.

Himalaya Nettle

Girardinia heterophylla Decna.; Eng. Himalayan nettle; Hindi—*Bichua*, *Awa, alla, chikri*; Lepcha—*Kazoobi*; Khasi hills—*Taintham, ting thap;* Nepal—*Ullu, sishnu*; Family—Urticaceae.

This is a perennial herb. It is found commonly in Kashmir, Assam and the Khasia hills. The fibre is obtained from the bark of the stem.

Uses. The fibre is used for making rough cloth, rope and twines.

East Indian Screw Tree

Helicteres isora Linn; Eng. East Indian screw tree; Hindi—*Marorphali*; Sanskrit—*Avartani, mriga-shinga*; Hindi and Punjabi—*Marorphali, jonkaphal, bhendu*; Bengali—*Atmora*; Marathi—*Kevan, kevani, varkati*; Gujarati—*Murdasing;* Telugu—*Nuliti, symali;* Tamil—*Valampiri, kaiva;* Kannada—*Yedamuri*; Malayalam—*Isvarmuri;* Oriya-*Murmuria, murimuri*; Family—Sterculiaceae.

It is a shrub or small tree. It is found commonly in North-West India, Madhya Pradesh and Western Peninsula. The fibre is obtained from the stem.

Uses. The fibre is used for making ropes and cordage, and the stalks and twigs for writing and printing paper.

HARD FIBRES

Manila Hemp or Abaca

Musa textilis Nee.; Eng. Manila hemp, Abaca; Tamil and Malayalam—*Naaru vazhai*; Kannada—*Natal bale*; Family—Musaceae.

It is native of the Philippines but cultivated in South India. The plant consists of clump of 12 to 30 sheathing leaf stalks 10 to 20 feet high with a crown of spreading leaves 3 to 6 feet in length. The fibre is obtained from the outer portions of the leafstalks. The mature stalks are cut off at the roots and split open lengthwise. The pulp and fibres are removed. The fibres are washed thoroughly and dried. The fibres are lustrous, whitish and 6 to 12 feet in length.

Uses. The fibre is used for making cordage, especially marine cables. It is also used for twine bagging and wrapping paper.

Fibre is also used for making '*dhotis*' and '*saris*'.

Also used for making a lustrous cloth known as sinamy. The fibres are not injured by salt and fresh water.

Sisal Hemp

Agave sisalana Perri.; Eng. Sisal Agave; Family—Agavaceae.

It is native of Mexico and Central America. In India it is found in Assam, Bihar, Bengal, Tamil Nadu Maharashtra, and Karnataka. The plant is drought resistant. The coarse, stiff, yellowish white fibres are obtained from the leaves. The fibres are cleaned, dried and packed.

Uses. The fibre is used for the manufacture of ropes and twine. Sisal wax is used in the preparation of polishing compositions and in the manufacture of carbon papers.

Manila Maguey

Agave cantala Roxb.; Eng. Manila maguey, Cantala; Family—Agavaceae.

This is a short stemmed woody herb. It is native of Mexico. In India it is found in Maharashtra, Bihar, Tamil Nadu, Uttar Pradesh and the Punjab. The fibre is extracted from the leaves.

Uses. The fibre is used for making ropes, cordage, and twines.

Mauritius Hemp

Furcraea foetida (Linn.) Haw.; Eng. Mauritius hemp, Green Aloe; Family—Amaryllidaceae.

This is a large shrub. It is native of tropical America. In India, it is found in Assam, Bengal, and South India. The fibre is obtained from the leaves. The fibres are exceedingly long, 4 to 7 feet and they are white, soft, very flexible and elastic.

Uses. The fibre is used for making cordage, mats, sacks and soles of shoes.

New Zealand Hemp

Phormium tenax; Eng. New Zealand hemp; Family—Agavaceae.

It is native of the swampy regions of New Zealand, but it is now found throughout the tropics and in temperate regions. The fibres are quite long, 3 to 7 feet in length, and possess a good lustre.

Uses. The fibres are used chiefly for twine, cordage, mattings and towlines.

Indian Bowstring Hemp

Sansevieria roxburghiana Schult. f.; Eng. Indian bowstring hemp; Hindi—*Murhari*; Sanskrit—*Murva, maurvi* (fibre); Hindi—*Marul, murva;* Bengali—*Murba, gorachakra*; Marathi—*Ghonasaphan*; Telugu—*Niyanda, sagal*; Tamil—*Marul*; Kannada—*Manjinamaru*; Oriya—*Murga*; Family—Liliaceae.

A herb. In India it is found along the Coromandel Coast. The fibre is obtained from the succulent leaves.

Uses. The fibre is used for cordage and matting.

Fibre is also used for bowstring and fine cloth.

Ceylon Bowstring Hemp

Sanservieria hyacinthoides (Linn.) Druce; Eng. Ceylon bowstring hemp; Hindi—*Murva*; Family—Liliaceae.

A herb. It is native of Ceylon. In India it is found in the coastal regions from Bengal to Tamil Nadu. The fibre is obtained from the leaves.

Uses. The fibre is used for making mats, cordage and paper.

Coir

Cocos nucifera Linn.; Eng. Coir; Sanskrit—*Narikela*; Hindi—*Nariyal*; Bengali—*Dab, narikel*; Marathi—*Narel, naral*; Telugu—*Kobbarichettu, narikelamu, tenkaya*; Tamil—*Tennaimaram, tenkai*; Kannada—*Tengu*; Malayalam—*Thenna, thenga, narikelam;* Family—Palmaceae/Arecaceae.

A tall palm, cultivated chiefly in Kerala, Tamil Nadu and Karnataka. Coir is the term applied to the short, coarse rough fibres which make up the greater part of the husk of the fruits of coconut palm. The husks of coconuts are soaked in salt water for several months to loosen the fibres. They are then beaten to separate the fibres, which are then washed and dried. The coconut fibres are very light and elastic, and exceedingly resistant to water. The coir extracted from the husks of coconuts, is dried in the sun and spun into coir yarn either by hand or machine.

COIR BOARD, INDIA—A REPORT

The coir industry is an agro-based cottage industry, but coir in its many colourful forms has transcended its own origin and is, today, making life brighter and easier for many modern homes, offices, hospitals, and clinical shops and schools all over the country.

Coir is a paradox in that it comes from the village but is used in the most sophisticated urban setting. The young housewife doing up her new home with bright coir mat and carpet would hardly know that the coir industry is highly labour-intensive, that the coir fibre and yarn that spell the beginning of any coir product are produced exclusively in the cottage sector.

Fig. 6.8. Coir. Coconut husks being pounded for their fibre. Coir making is a traditional industry of Kerala.

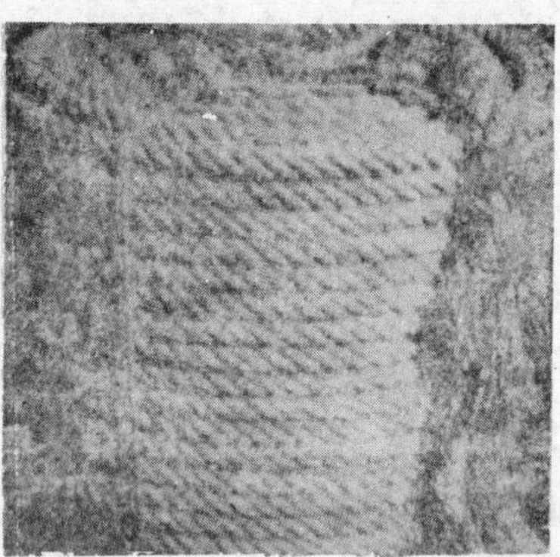

Fig. 6.9. The coir industry—Rope made from coir yarn deserves special mention. Tough, totally impervious to damage by water, and low priced, it is used extensively in transportation, agriculture, shipping and general haulage.

In spite of its simple beginnings—or perhaps because of it, coir products have been welcomed enthusiastically in India and abroad. Colourful doormats, sound-absorbent, hardwearing and high-value matting, bright and tough mourzouks, crush and moth-proof pile carpets, and smart car mats are in extensive use in the United Kingdom, West Germany, France, Belgium, the Netherlands and of course, in our own country. Also the wonder cushioning Rubberised Coir, a comfortable, low-cost blend of coir and latex. It is fascinating how the modest covering of the tropical coconut extends itself to aid modern living at budget prices.

Quality is assured. The Coir Board has instituted quality control measures for coir yarn—as well as mats and matting sold through Coir Board showrooms and Sales Depots and Accredited Dealers. Under the Coir-mark scheme, Coir Board inspectors scrutinize and put their seal of approval. Only then do products leave the main production centres at Alleppey, Sherthalai and Cochin. There are several Coir showrooms all over the country.

Uses. The yarn is used as such or made into ropes, mats, mattings, nets, bags, etc. Coir is also used for bristles for brushes, doormats, coarse textiles, stuffing for rail and bus seats, etc. In certain parts of the country, such as Kerala, a big industry employing thousands of workers has been built up around the manufacture of coir fibre and coir products both on a cottage and large industrial scale.

VARIOUS USES OF COCONUT PALM

Coir. Coir is a natural fibre extracted from the husk of the coconut. It is a wonder fibre, rot proof, sound proof, water damp and pest resistant, a good insulant that stays cool in hot weather and warm in cold weather, tough-hard wearing and economical. It is a rural employment and export-oriented industry. Items like fibre, yarn, mats, mattings, rugs, carpets, rope, rubberised coir, etc., are exported. The total exports from the country are approximately 59,000 tonnes.

The coir industry is concentrated in Kerala and is one of the most important cottage industries in that state. Nearly a million people depend on it for their livelihood.

Coir dust. A waste product in the coir industry. It can be used as manure or soil conditioner. It is an excellent surface soil mulch. In combination with cement, the dust is a thermal insulating material. It can also be used for building slabs, or for hard boards.

Shell powder. It is used as a compound filler for synthetic resin glues and surface smoothing of moulded articles. It is a filler in Bekalite products.

Fig. 6.10. Coconut. A coconut palm bearing coconuts.

Shell charcoal. Formed by burning shells. It has a high absorption capacity. Can be used as a deodorizer and decolourised. It is extensively used by goldsmiths for melting and softening gold and silver.

Leaf blade. Dried leaf blades, cut into pieces and can be used as packing material.

Buttons. Tender nuts are used in Ayurvedic medicine.

Tender husk. Orange dwarf, tender nut husk, and its juice is extensively used in Ayurvedic medicine.

Root tips. Used for medicine. The decoction of the root is a good astringent and used as a mouth wash.

Coconut water. Boiled coconut water, to a sticky consistency is used as medicine on wounds.

Copra. It is the dried meat and is a highly valued commodity for oils and fats. Various methods are used for drying, like sun drying, smoke drying, etc. Copra is of two types—*cup copra and ball copra.*

Coconut oil. It is an essential ingredient in manufacture of soap, margarine and other industries. The oil is extracted by crushing copra. This oil is extensively edible as a substitute for *ghee.*

Coconut cake. It is left after extraction of oil and is used as cattle and poultry feed.

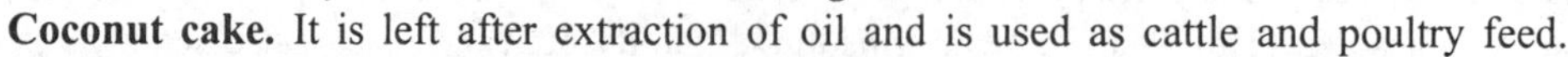

Kombu Chembu. A vessel of coconut shell. This type of vessel is used for marriage ceremonies and pujas in the South. A whole coconut with mango leaves is kept at its mouth.

Snuff grinder. Bowl made of coconut shell, with a stem timber grinder.

Mat. Made of coconut leaf blade used for sleeping on.

Baskets. Made of netted leaf blade, normally used for house-hold requirements.

Hair oil. Pure coconut oil perfumed and used as hair oil is healthy for the hair and scalp.

Shampoo. Made from coconut oil for clearing the hair.

Comb. Made of stem timber, mostly used by villagers.

Kadavur. Slippers made of stem timber, used by sadhus.

Grated dehydrated kernel. It is made by drying kernel of fully matured nuts, which are dried, sliced, washed, sterilized and transferred to the disintegrator which forms a fine wet meat. This is dried and packed. This form of coconut meat is exported.

Coconut milk cream. Cream is an emulsion of coconut oil in water. It is a milk-like liquid obtained from the coconut gratings by squeezing. The pasteurized milk is canned for export.

Coconut milk powder. The skimmed milk is spray dried, to yield a product similar to non-fat dry milk.

Coconut oil container. Made of whole coconut shell, with a covering made of netted coir rope; used in villages for carrying and storing coconut oil.

Vinegar. Coconut water or coconut toddy is fermented, kept in the sun and decanted. After adding acetic acid, the clear solution obtained is vinegar.

Coconut water. When fermented it is used as a rubber coagulator and also as a fermenting agent in household preparations.

Coconut syrup or honey. It is prepared from coconut milk, by adding sugar and boiling it to the required consistency. It is used as an excellent, instant cool-drink when mixed with water, and is a substitute for honey.

Coconut candy. Prepared from coconut toddy. During the process of preparation of jaggery, half boiled liquid separated, and preserved for a long time, forms candy, a sweet crystal.

Coconut toffee. Various kinds of delicacies like coconut toffees and coconut cookies are prepared.

Jaggery. Prepared by evaporating fresh toddy through careful boiling, in open vessels, till it is saturated, while the impurities are skimmed. The solution is poured into moulds, and the jaggery is allowed to harden.

Spongy ball. When nuts are stored for a long period, a spongy ball develops inside the nut. It is eaten raw, and utilized for the preparations of various products like jam and marmalade. It is also sliced and preserved in sugar syrup for use as a dessert, or as a constituent of fruit salad.

Toddy. It is a sugar containing juice, obtained by tapping the unopened spadix of the coconut palm. It is an excellent beverage, and a rich source of sugar and of baker's yeast. When it undergoes fermentation it forms an alcoholic drink.

Fenny. It is the product obtained by the distillation of fermented toddy. The distilled product of toddy, when preserved for a long period, forms a delicious intoxicating drink.

Water carrier. Made from trunk (bole) commonly used in villages for lifting water by '*picota*'.

Window curtain. Made of coloured coir ropes and stem beads.

Market bag. Made of coloured coir ropes.

Hand fan. Made of coconut leaf by netting the leaflefts, mostly used in villages.

Cot. Made of carved stem timber with coir rope nettings.

Bed. Made of rubberised coir.

Broom mat. Broom sticks netted together with coir ropes.

Spice box. Made of trunk 'bole'; common in the South.

Teria. Made of coir rope, used in the kitchen for keeping hot vessels.

Fuel. Every part of the coconut palm can be used as fuel.

Broom. Made from midribs of leaflets.

Seed container. Made of the whole shell, for storing vegetable seeds.

Torch. Made of the dried petiole sheath, covered with split spadix sheath, tied together and lighted at the rim. The fire burns for a longer period and is therefore used by villagers, at night, as torches.

Sanitary brush. Made of coir.

Brush. Used for white-washing; made of coir.

Broom. Made of broomsticks, used for cleaning roads.

Hand brush. Made of coir rope for cleaning cattle.

Fish net. Made of netted coir, tied on the back while shore fishing. It is unaffected by salt water.

Basket. Made of plaited leaves used for bringing fish from the market. The basket made of leaf stalk, is used for carrying articles like rice, etc.

Mouth gag. Made of coir rope, used on cattle while ploughing in the fields.

Koori. Made of broom sticks, used for catching fish in running water from paddy fields.

Net. A large coir net, commonly used for carrying coconuts or dried leaves in the South.

Panoli. Made of leaflets, used against the rain by villagers while working in the paddy fields.

Float. Barren nuts are used as float for fish nets and while learning swimming.

Pineapple

Ananas comosus (Linn.) Merr.; Eng. Pineapple; Hindi—*Ananas*; Family—Bromeliaceae.

This is a biennial herb. It is native of South America. In India, it is cultivated in Tamil Nadu, Andhra Pradesh, Kerala, Karnataka, West Bengal, Tripura, Uttar Pradesh and Maharashtra.

The leaves are the source of a strong fibre. The fibres are shiny white, durable, flexible and water resistant. Usually two-years old leaves are cut and the fibres scraped out by hand. After being dried and combed out, the fibres are tied end to end and can then be woven.

Uses. The fibre is made into fabrics. The ropes, fishing nets and strings are also made from these fibres.

Pina, a delicate fabric of the Philippines is made from the fibre. Waste material, after extraction is used for paper-making.

Patana Oak

Careya arborea Roxb.; Eng. Patana oak; Hindi—*Kalikatbhi, Kumbhi*, Family—Myrtaceae.

This is a large tree. In India, it is found in Bengal, Madhya Pradesh, Maharashtra and Orissa. The fibre is obtained from the bark of the tree.

Uses. The fibre is used for cordage and sacking.

Rajmahal Hemp

Marsdenia tenacissima Wight & Arn.; Eng. Rajmahal hemp; Hindi—*Tongus*, *Jiti*; Family—Asclepiadaceae.

A climbing shrub. It is found in the North Western Himalayas, Madhya Pradesh and Bihar.

Uses. The fibre obtained from the stem is used for bowstrings and netting.

Marsdenia Roylei. Wight.; Hindi—*Murkula Pathor*. This is a climbing shrub found in the Western and Eastern Himalayas, Shimla and Kumaon. The stem fibre is used for strong ropes and fishing nets.

Marsdenia tinctoria R. Br.: Hindi—*Rion*.

Cultivated in Andhra Pradesh, Assam and Khasia hills.

Uses. The fibre obtained from the stem is used for ropes.

BRUSH FIBRES

The fibres which are used in the manufacture of brushes and brooms are known as 'brush fibres'. Such fibres are quite strong, stiff, elastic and flexible. Sometimes whole twigs, stems or roots are utilized as brush fibres, whereas in other cases the fibres are obtained from leaf stalks.

Palmyra Palm

Borassus flabellifer Linn.; Eng. Palmyra palm; Hindi—*Tar*; Family—Palmaceae.

A tall palm, found along the coastal areas of Bengal, Bihar and the Western and Eastern Peninsulas.

Uses. The fibre is used for preparing brushes and brooms and the leaves for making fans, umbrellas, baskets and mats. The fibres are also used to make ropes, twines and paper.

Toddy Palm

Caryota urens Linn; Eng. Toddy palm, Wine palm; Hindi—*Mari*, *Ramgoah*; Family—Palmaceae/Arecaceae.

A tall palm, found in Orissa, North Bengal and Assam. The fibre is obtained from the leaf sheaths.

Uses. The fibre is used for ropes, basket and soft brushes. The fibre is finer, softer and more pliable.

Cocount

Cocos nucifera; Eng. coconut; Hindi—*Nariyal*; Family—Palmae/Arecaceae.

The mid-ribs of the leaves are used for broom sticks; dried and green leaves are used for making baskets. Leaves are also used for thatching.

ROUGH WEAVING FIBRES

These fibres comprise the stems of reeds, rushes, grasses, willows, bamboo, rattan and many other plants. The leaves and roots may also be utilized. These materials are used entire or split. The fibres are woven or twisted together and are used for making hats, chair seats, baskets and other such articles.

Mats and Baskets

Cyperus articulatus Linn.; Family—Cyperaceae. A perennial herb commonly found in Bengal, Tripura, Manipur and Gujarat. The stems are made into mats.

Cyperus corymbosus Rottb.; Eng. Chinese mat grass; Family—Cyperaceae. This grass is found commonly in Bengal, Karnataka and Tamil Nadu. The stems are used for mats.

Cyperus iria Linn.; Hindi—*Burachucha*; Family—Cyperaceae. It is found throughout India. The culms (stems) are used for making mats.

Cyperus malaccensis Lamk.; Family—Cyperaceae. It is found in South India. The stems are used for mats, hats and baskes.

Cyperus tegetum Roxb.; Family—Cyperaceae. The culms are used in the manufacture of mats.

Juncus effusus Linn.; Eng. Rush; Family—Juncaceae. This is a perennial herb, found in the Eastern Himalayas and the Khasia hills. The stems are used for making mats and baskets.

Pandanus tectorius Soland, ex. Parkinson; Eng. Screwpine: Hindi—*Keora*; Family—Pandanaceae.

A large shrub or small tree. It is found along the sea coat of both the Peninsulas, Orissa, Uttar Pradesh and the Andaman islands. The fibres are obtained from the leaves.

Uses. The fibres are used for making cordage, nets, mats and sacks. The leaves are also used in paper making.

The leaves are bleached and cut into pieces. The excellent mats are made out of it and are used by rich and poor alike in Kerala in large scale for sleeping purpose. "Charpai" system is not common in Kerala.

Calamus acanthospathus Griff.; Family—Palmaceae. A common climber. It is found in North-East India. It is used for making tea baskets.

Daemonorops jenkinisianus Mart.; Hindi—*Gola bet*; Family—Palmaceae. A tall palm. It is found in Assam, the Khasia hills and Sikkim. The long soft stems are used in the manufacture of baskets.

Bamboos

Bamboos are also used for window curtains. *Bambusa tulda* is very commonly found in Kerala state. *Bambusa vulgaris* is also wildly found in Kerala forests. (Curtains are made from thin sticks of bamboo).

The bamboo is used in various forms by the people of hill tribes especially in Mizoram and other surrounding hilly states. The bamboo caps of Mizoram are famous for their artistic touch. The Mizo girls are always busy in making bamboo baskets and other artistic articles. Different types of decorative items are made of bamboo such as flower pots, trays, vases, caskets and even ornaments.

Bamboo also finds a place in cultural activities of Mizo people. The most colourful and distinctive dance of the Mizo is called *Cheraw*. Long bamboo staves are used for this dance therefore many people call it a *bamboo dance*. *Cheraw* is a dance of skill and alert minds.

Dendrocalamus hamiltonii Ness & Arn.; Hindi—*Kaghzi bans*; Family—Gramineae. A tall, tufted grass cultivated chiefly in Dehra Dun. It is used in the manufacture of mats, baskets and paper.

Dendrocalamus strictus (Roxb.) Ness; Eng. Solid bamboo; Hindi—*Bans Kaban*; Family—Gramincae. It is found in Northern India, Madhya Pradesh and South India. It is used in the manufacture of baskets, brushes and paper.

Arundinaria falcata Nees; Eng. Himalayan bamboo; Family—Gramineae. It is found in the Western Himalayas and Sikkim. The stems are used for making baskets, arrow and fishing rods.

Arundinaria racemosa Munro; Family—Gramineae. It is found in Bhutan and Western Himalayas. The culms are used for making mats and baskets.

Arundinaria spathiflora Trin.; Family—Gramineae. It is found in Western Himalayas at an elevation of 2,000 to 3,000 metres. The culms are used for making baskets and mats.

Bambusa tulda Roxb.; Eng. Bamboo; Hindi—*Peka, Chau*; Family—Gramineae. It is found in Assam and Bengal. The mature culms are used for making baskets, fans and paper.

Fig. 6.11. Bamboo industry. The Mizo girls are always busy in making bamboo baskets and other artistic articles.

Bambusa vulgaris Schrader; Eng. Feathery bamboo; Hindi—*Basini bans*; Family—Gramineae. It is cultivated in the hotter parts of the country. The split culms made into mats and baskets.

Melocanna bambusoides Trin.; Eng. Tarai bamboo; Hindi—*Tarai bans*; Family—Gramineae. A woody grass, found in the Khasia hills. Its stems are used for making mats, baskets and paper.

Fig. 6.12. *Bamboo cottage industry.* The bamboo caps of Mizoram are famous for their artistic touch.

Pinreed Grass

Erianthus arundinaceus (Retz.) Jesw. ex heyne; Syn. *Saccharum arundinaceum* Retz.; Eng. Pinreed grass; Hindi—*Ramsar, Sarkanda, Sara*; Family—Gramineae.

A perennial grass, cultivated throughout India. A fibre is obtained from the leaf sheaths.

Uses. The fibre is used for ropes, twines and paper. The stems are used for making chairs, stools, tables, baskets and screen.

Munja

Erianthus munja (Roxb.) Jews.; Syn. *Saccharum munja* Roxb.; Hindi—*Munj sentha, sarkanda, munja*; Family—Gramineae.

It is commonly found in the Punjab and Uttar Pradesh. The fibre is obtained from leaves and stem.

Uses. The fibre is used for making baskets, mats and cordage. The leaves are used for thatching, and are also the source of paper.

Plume Grass

Erianthus ravennae (Linn.) P. Beauv.; Eng. Plumegrass. Ravenna grass; Hindi—*Moonj, Sarkara*; Family—Gramineae. The stem fibre is used for making chairs, muddas, chhappars and ropes.

Elephant Grass

Typha angustata Bory & Chaub.; Eng. Elephant grass; Hindi—*Gond-pateri*; Family—Typhaceae. A small shrub, found in North-Western India and Assam. It is used for making mats, ropes and baskets.

FILLING FIBRES

Many plant fibres are used for stuffing cushions, mattresses, pillows, furniture and other such articles. These fibres are also used in the manufacture of stuff for building purposes, as stiffening for plaster, as packing for machine bearings and for protection of fragile articles during transportation. The silk cottons are most important filling fibres.

Redsilk Cotton

Bombax ceiba Linn.; Eng. Redsilk cotton; Hindi—*Semul, simur*; Sanskrit—*Salmali*; Bengali—*Simul, roktosimul*; Marathi—*Saur, simlo*; Gujarati—*Sawar, simalo, shemolo*; Telugu—*Booruga, konda-booruga*; Tamil—*Mullilavu, illavam;* Kannada—*Booruga, mullubooruga*; Malayam—*Mullialvau, pula—maram*; Oriya—*Bouro*; Garo—*Boichu, panchu*; Mundari—*Edelsong*; Family—Bombacaceae.

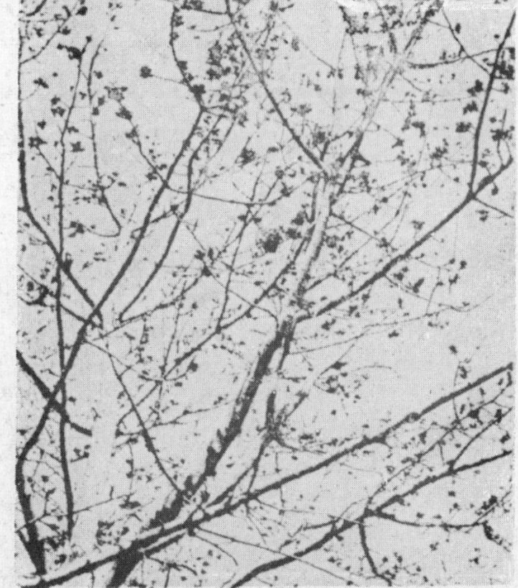

Fig. 6.13. *Bombax ceiba (Semal)*—The red silk cotton or semal, a very large and ornamental tree, supplies a reddish floss known as 'Indian kapok' which has been an important stuffing material in India for centuries.

A tall tree, cultivated in the Punjab, Uttar Pradesh the Eastern and Western Peninsulas. The silky floss is obtained from the inner wall of the fruit.

Uses. The floss is used for stuffing pillows, cushions, mattresses, etc.

Floss is also used for stuffing life-belts, upholstery, and quilts. Also used as an insulating material for refrigerators, sound-proof covers and walls; it is better than cotton-wool for packing fragile materials. Fibre is spun into yarn used for the manufacture of plushcs.

Fig. 6.14. *Ceiba pentandra*. Kapok or silk cotton tree. Note the pods.

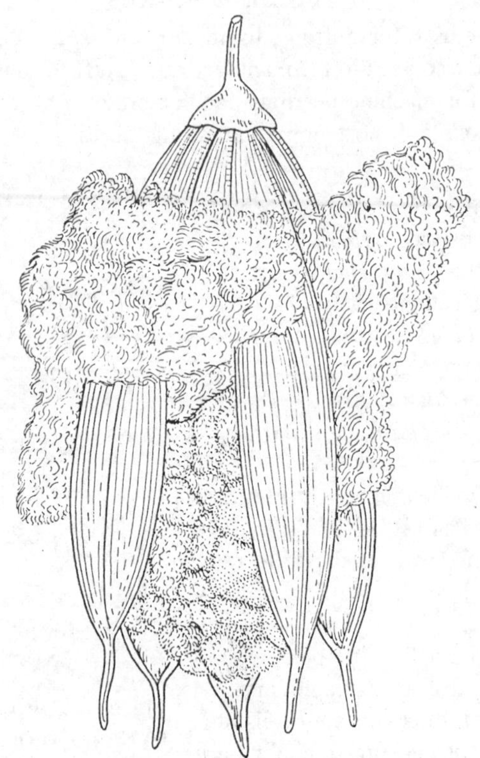

Fig. 6.15. *Ceiba pentandra* (kapok). Splitting mature pod, releasing fibre.

Kapok

Ceiba pentandra (Linn.) Gaertn.; Eng. Kapok, Silk cotton; Hindi—*Safed simal*; Sanskrit—*Sveta salmali*; Bengali—*Schwetsimul;* Marathi—*Salmali, pandhari*; Tamil—*Ilavum;* Telugu—*Tella buraga*; Kannada—*Buraga, biliburaga*; Malayalam—*Ilavu, mullilavu*; Family—Bombacaceae.

A tree, found in South-Western India. It is often found planted around villages and temples. The plants can be grown in avenues and on boundaries of fields. Once established, the trees will thrive without much attention. It has strikingly large pods, and each pod has an average length and girth of 26 and 15 cm respectively. The pods are borne in a profusion of clusters on a well-branched tree. The pods just split when mature and dried, but do not burst open. The floss is not, therefore, carried away by wind.

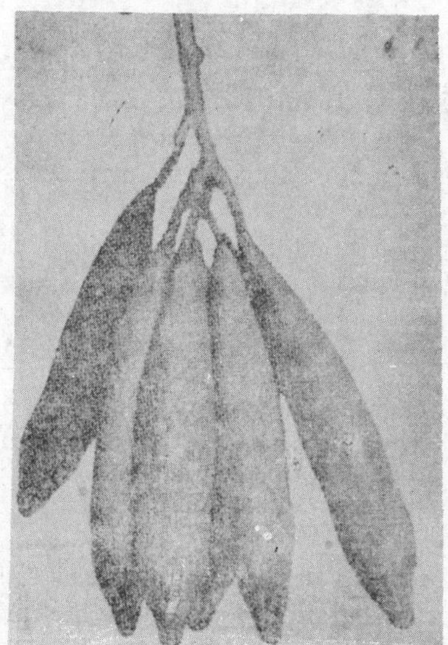

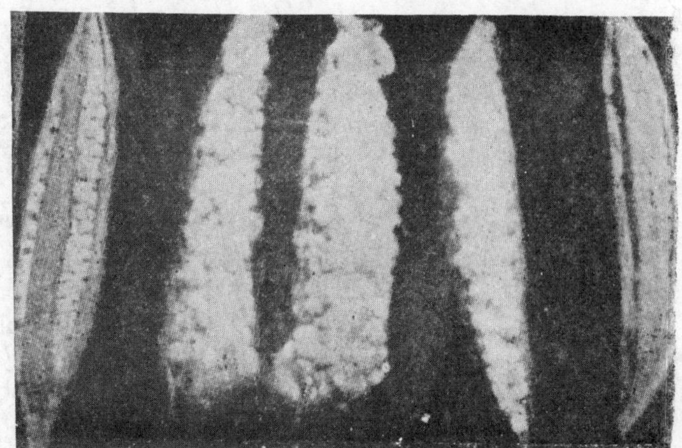

Fig. 6·32. *Ceiba pentandra*—Kapok or silk cotton. The pods split when mature, the floss is not carried away by wind.

Fig. 6.16. *Ceiba pentandra.* Silk cotton, a bunch of pods.

Fig. 6.17. *Ceiba pentandra.* Kapok or silk cotton. The pods just split when mature, the floss is not carried away by the wind.

The tree yields better when well-supplied with moisture.

Uses. The floss, obtained from the fruit, is of very good quality, being very elastic and not liable to 'bunch' when used in upholstery. It makes excellent stuffing material for mattresses, pillows, cushions, etc. Extensively used abroad in the manufacture of life-belts and life-buoys owing to its extreme buoyancy and resistant to water-logging. Also used for insulating refrigerators and sound-proofing rooms. There is everywhere a ready market for such cotton.

The kapok fibre is used in bedding and upholstery industries, and in the production of thermally insulated and sound proof covers and walls.

White Silk Cotton

Cochlospermum religiosum (Linn.) Alston; Eng. White silk cotton; Hindi—*Pilikapas*, *kumbi*; Marathi—*Ganeri, ganglay*; Telugu—*Kondagogu, kongu*; Tamil—*Kongilam, tanakku*; Kannada—*Adviburaga*; Malayalam—*Appakutakka*; Oriya—*Kontopalas*; Family—Cochlospermaceae.

A tree. It is native of India. It is found commonly in Andhra Pradesh, Bihar and Madhya Pradesh. The fibre is obtained from the flowers.

Uses. The fibre is used as stuffing material. It is stuffed in cushions, pillows and mattresses. The tree also yields an edible gum.

The gum is used as a substitute for tragacanth gum obtained from *Astragalus gummifer* Labill; employed in the cigar paste, calico printing, leather dressing, and ice-cream industry.

Cat's Tail

Typha angustifolia Linn.; Eng. Cat's tail, Pith grass; Hindi—*Pater*; Bengali—*Kaw, hogla;* Marathi—*Pankanis;* Gujarati—*Ghabajario, pario*; Telugu—*Jammugaddi*; Tamil—*Sambu*; Kannada—*Aapu, maribala*; Kashmir and Punjab—*Patera, pitz, kai*; Family—Typhaceae.

A large grass. The hairs on the fruits are used for stuffing pillows and cushions. The leaves are used for mats and baskets.

Rhizomes, young shoots, and inflorescence are edible, and eaten in various ways. Flowers are made into a sweet meat. Rhizome is astringent and diuretic. The seeds yield a fatty oil.

Madar

Calotropis gigantea (Linn.) R. Br. ex. Ait.; Eng. Madar; Hindi—*Ak*; Sanskrit—*Arka, mandara*; Bengali—*Akanda;* Marathi—*Rui*; Gujarati—*Akado*; Telugu—*Jilledu*; Tamil—*Arkkam*; Kannada—*Arkagida*; Malayalam—*Erikku*; Family—Asclepiadaceae.

A perennial shrub. It is found commonly in Northern India. The floss obtained from the seeds is used for stuffing purposes.

The bark yields a fibre which is white, silky, strong and durable, used for making fishing nets and lines, bow-strings and twine.

Akund

Calotropis procera (Ait.) R. Br.; Eng. Akund, Swallow wart; Hindi—*Safed ak*; Sanskrit—*Alarka*; Marathi—*Mandara*; Tamil—*Vellerukku*; Family—Asclepiadaceae.

A large herb. It is found in Andhra Pradesh, Punjab and Uttar Pradesh. The floss obtained from the seeds is used as stuffing material. The stem fibre is made into cordage.

NATURAL FABRICS

The basts of certain trees can be separated from the bark in layers or sheets. Such sheets can be utilized by tribe men as a substitute for cloth.

Paper Mulberry

Broussonetia papyrifera (Linn.) L, Heritier in Vent.; Eng. Paper-mulberry; Family—Moraceae.

A small tree. It is found in Northern India. The bark is the source of a strong lustrous fibre, which is used for making writing paper, paper lanterns and umbrellas. The pulp is used for news-print paper.

Wood yields paper-pulp; also used for ply-wood, cheap furniture, toys, shoe-heels, cigar boxes, sports goods and packing cases, and for hard boards.

Sack Tree

Antiaris toxicaria (Pers.) Lesch.; Eng. Upas tree, Sack tree; Hindi—*Jasund, Chandal*; Family—Moraceae.

A tree, found in the Western Ghats and Tamil Nadu.

Uses. The bark fibre is used for cordage and matting.

Cuban Bast

Hibiscus tiliaceus Linn.; Eng. Cuban bast, Mahoe; Hindi—*Chelwa*; Bengali and Hindi—*Bola, chelwa*; Marathi—*Belapata;* Telugu—*Etagogu*; Tamil and Malayalam—*Nirparathi*; Oriya—*Baniah*; Andaman—*Safed chika*; Family—Malvaceae.

A small bushy tree. Commonly found along the coasts of both the Peninsulas and Sunderbans.

Uses. The bark fibre is used for ropes and mats. The inner bark is removed in long ribbon-like strips. These bark strips are used for tying cigars.

The fibre is more resistant to water than sunn hemp and jute. Bark is used for making wrapping paper.

VEGETABLE SPONGES

Vegetable Sponge

Luffa acutangula (Linn.) Roxb.; Eng. Vegetable sponge; Hindi—*Kali torai, Jhinga*; Sanskrit—*Jhongaka, koshataki;* Bengali—*Jhinga, sataputi;* Marathi—*Shirola;* Gujarati—*Ghisoda*; Telugu—*Birakaya*; Tamil—*Pirkankai*; Kannada—*Hirekayi*; Malayalam—*Pichenga*. A trailing herb, cultivated throughout India for its fruits, which are used as vegetable.

Uses. The fibrous material, obtained from the dried fruits, is used as a substitute for bath-sponges.

Vegetable Sponge

Luffa cylindrica (Linn.) M. Roem.; Eng. Vegetable sponge; Hindi—*Ghia torai*; Sanskrit—*Rajakoshataki, dirgha patolika*; Bengali—*Dhundal*; Marathi—*Ghosali*; Gujarati—*Turia*; Telugu—*Guthibira*; Tamil—*Mozhuku pirkankai*; Kannada—*Tuppahirekai*; Malayalam—*Kattupeechal*; Family—Cucurbitaceae.

A trailing herb. Cultivated throughout India for its fruits, which are used as vegetable.

Uses. The fibrous material, obtained from the dried fruits, is used as a substitute for bath sponges.

ARTIFICIAL FIBRES
(Artificial Silk or Rayon—A Cellulose Product)

The term rayon includes all synthetic fibres manufactured from cellulose. The raw material of the rayon industry is high polymer alpha cellulose prepared in a pure form from wood pulp of cotton linters. There are four processes for its manufacture; (1) cellulose nitrate process, (2) celluolse acetate process, (3) cuprammonium process, and (4) viscose process.

In modern times most of the rayon is manufactured by the viscose process and the term rayon is most often applied to viscose yarn. The cellulose needed for the manufacture of rayon is obtained from wood-pulp and cotton-linters. Yarn manufactured from other sources is inferior in quality.

Cellulose nitrate process. Cellulose is nitrated with a diluted mixture of nitric and sulphuric acid when mono and dinitrates are obtained and are known as *pyroxylin* in the solid state. Pyroxylin is dissolved in a mixture of alcohol and ether to get collodion. This is forced through glass capillary tubes into the air when the solvent evaporates leaving behind threads of cellulose nitrate. These are digested with caustic soda or sodium bisulphate solutions when the cellulose nitrate gives cellulose.

Cellulose acetate process. Cellulose is treated with acetic anhydride in the presence of sulphuric acid when cellulose triacetate is obtained. This is decomposed with water to get the diacetate which is net washed, dried and dissolved in acetone mixed with some other solvents. The solution is forced through a spinneret into a warm chamber when the solvent evaporates leaving behind threads of cellulose acetate. The fibre obtained burns with difficulty. The product is termed *celanese silk*.

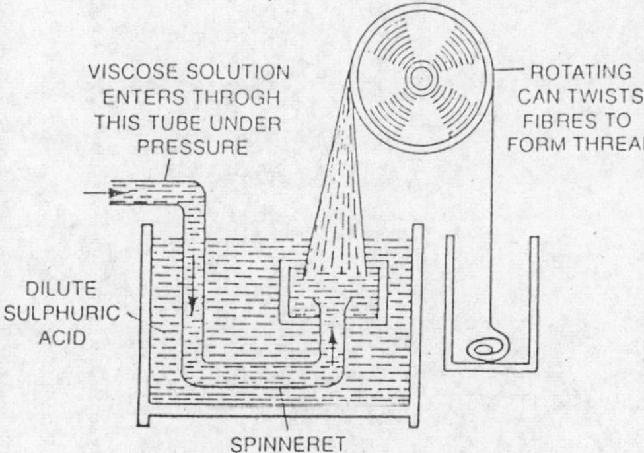

Fig. 6.18. Artificial fibre—Spinneret used in making rayon.

Cuprammonium process. Cellulose is dissolved in ammonical copper hydroxide solution and the solution is forced through a spinneret into a sulphuric acid bath where the cellulose gets precipitated as fine threads. The fibre is known as *cupra silk*.

Viscose process. In the viscose process cellulose is soaked in a 20 per cent caustic soda solution for about three hours. The alkali solution is removed and the product is treated with carbon disulphide. This gives a mixture of cellulose xanthates which is dissolved in sodium hydroxide solution. The alkaline solution of sodium cellulose xanthates has a high viscosity and is, therefore, called *viscose solution*, which is carefully filtered and forced through a spinneret into a bath of dilute sulphuric acid which hardens the gum-like threads into rayon fibres. The silk obtained by the process is termed viscose rayon. This is the cheapest process and commonly used.

Rayon fibres vary in tensile strength and elasticity. They take dyes more readily than do the natural fabrics.

Uses. Rayon has over 500 uses. It may be used alone or in combination with natural silk or fibres of plant origin. The more important products of rayon are: underwear, shirts, hosiery, dress and suit fabrics, neckties, ribbons, and many other household and industrial textiles.

GUNCOTTON

When cellulose is treated with concentrated nitric acid in the presence of sulphuric acid, several types of cellulose nitrate are formed. The higher cellulose nitrates are called guncotton. It is prepared from cotton linters, and during the process the cellulose is completely nitrated. It is used as an ingredient of many high explosives. For example, smokeless powder is made from a mixture of guncotton and the lower nitrates. When properly made, guncotton is one of the safest of the explosives to handle, particularly when it is wet. Pyroxylin is produced by the partial nitration of cellulose. This process is carried on under different conditions from those which result into the formation of guncotton.

7

Wood and Cork

From the time immemorial food, clothing and shelter have been the three great necessities of mankind. The wood is the most familiar and most important forest products. The wood has contributed a lot to the advancement of civilization. Today the wood is the most widely used commodity other than food and clothing. It is one of the most important and versatile of the raw materials of the industry. It has so many advantages over the metals.

Fig. 7.1. *The main building of Forest Research Institute, Dehradun.* The Institute is famous for research work on various forest products. The research is going on at least in fourteen disciplines of plant sciences.

It is cheap, light and may easily be worked with tools. It is very strong for its weight and it is embodiment of strength, toughness and elasticity. It is a bad conductor of heat and electricity. Wood is also used in the form of thin sheets or *veneers*. The wood is definitely superior to metals in several respects.

DIAGNOSTIC FEATURES OF WOODS

On the basis of several important diagnostic features the commercial woods may be identified to some extent. The more important diagnostic features are mentioned here:

Porous and nonporous woods. The presence or absence, and the nature and arrangement of *pores*, serve as a ready means of classifying woods. The coniferous woods do not possess pores, and are known as *nonporous* woods, whereas the angiospermic woods possess numerous pores and are termed as *porous* woods. On the basis of the distribution of pores, the woods may be of two types—*ring porous* and *diffuse porous* woods. In ring porous woods (*e.g.*, ash, elm, oak, etc.) the pores are found to be arranged in concentric circles, the outer and inner portions of which differ with regard to the number and size of the pores. In diffuse porous woods (*e.g.*, beech maple, walnut, etc.) the pores are small and nearly of the same size and are found to be scattered uniformly throughout the wood.

Early wood and late wood. In temperate regions, every year new wood is formed in a limited growing season, with the result definite growth layers develop, which show two distinct areas within each layer. The wood thus formed in the spring is called the *spring wood* or early wood, and that formed in winter is called the *autumn wood* or late wood. There is a sharp contrast between the late autumn wood and the early spring wood, and this makes the successive rings distinct. The growth ring of a single year is called an annual ring and the number of these annual rings gives an indication of the age of tree. Annual rings of successive years may vary greatly in width. Wide rings are formed under favourable conditions of growth of the tree, and narrow ones are formed when conditions are unfavourable.

Sapwood and heartwood. The outer region of the wood which is of lighter colour is known as the *sapwood,* and this alone is used for conduction of water and salt solutions. The cells of this region are alive and physiologically active. In old trees the central region of the secondary wood is filled up with tannin and other substances which make it hard and durable. This region is known as the *heartwood.* It looks black owing to the presence of tannins, oils, gums, resins, etc., in it. The vessels often become plugged with tyloses. The function of the heartwood is no longer conduction of water, but it simply gives mechanical support to the stem. The heartwood usually takes good polish and is used for cabinet work, furniture and other high grade wood-working industries.

Texture, grain and figure. *Texture* refers to the relative size and quality of the various woods, while *grain* refers to their structural arrangement. *Figure* is applied to the design or pattern which appears on the surface of wood.

Rays. The rays are made of parenchyma cells that are oriented at right angles to the main axis of the stem. They vary greatly in width, height and arrangement.

MECHANICAL PROPERTIES OF WOOD

Wood possesses some important mechanical properties which either alone or in combination, determine its usefulness and suitability for various purposes. These properties may differ in different species. The mechanical properties enable the wood to resist various external forces which tend to change its shape and size and produce deformations. The important mechanical properties of wood are mentioned here.

Strength

The strength is restricted to the ability to resist certain definite forces which may be termed—crushing strength, tensile strength, shearing strength and cross-breaking strength. They are as follows:

Crushing strength. It is the resistance offered to forces that tend to crush wood.

Tensile strength. It is the resistance to forces that tend to pull wood apart.

Shearing strength. It is resistance to those forces which tend to make the fibres slide past one another.

Cross-breaking strength. This is the resistance to forces which cause the beams to break, and all the above-mentioned forces are involved.

Stiffness. It is the measure of the ability of wood to resist forces that tend to change its shape.

Toughness. It is referred to the ability of wood to absorb a large amount of energy, and so resist repeated, sudden sharp blows or shock.

Hardness. It is the measure of the power of wood to resist indentations, abrasion and wear.

Cleavability. It is an expression of the ease with which wood can be split.

STRENGTH AND OTHER PROPERTIES OF WOOD

The strength of wood is the most important property in determining the value of any species for structural purposes. It is a very variable property, and is influenced by the density of the wood, the moisture content, the presence of defects and many other factors. Suitability figures for eight different properties have been calculated by taking into consideration the various strength functions of both green and seasoned timber as shown in the table.

Suitability	*Strength and other properties of wood*
1. Weight	Specific gravity based on weight oven dry and the volume of seasoned timber (approximately 12%) moisture content.
2. Strength as a beam	Modulus of rupture and fibre stress as the elastic limit in static bending. Fibre stress at the elastic limit in impact bending.
3. Stiffiness as a beam	Modulus of elasticity in static bending. Modulus of elasticity in impact bending.
4. Suitability as a post	Maximum crushing strength and strength at the elastic limit in compression parallel to grain. Modulus of elasticity in static bending.
5. Shock-resisting ability	Work to maximum load and total work in static bending.
6. Retention of shape	Shrinkage green to oven dry in volume and in the radial and tangential directions.
7. Shear	Shearing strength in the radial and tangential directions.
8. Hardness	Fibre stress at the elastic limit in compression perpendicular to the grain. Radial, tangential and end hardness.

SEASONING OF WOOD

Wood always contains moisture, the amount of which varies from 40 per cent to 100 per cent or more of the dry weight. The water found within the wood is known as *hygroscopic water*. The amount of hygroscopic water constitutes from 20 to 35 per cent of dry weight. This property of wood is known as *hygroscopicity*. The moisture content of wood has an important bearing on its weight, density, and often on its strength. It is only the loss of hygroscopic water which is responsible for the increase in strength, that accompanies seasoning, as the drying out of wood is called. The loss of hygroscopic water causes the shrinkage of wood. This tendency of wood to shrink as it dries is one of the great drawbacks to its use. As the result of uneven shrinkage, warping and other defects may develop, which tend to counteract any increase in strength. However, dried or seasoned wood is for the most part stronger, harder, stiffer and more durable than unseasoned wood. Artificial methods, of wood seasoning are employed. There are two chief types of artificial wood seasoning—1. *air seasoning* and 2. *kiln drying*.

Fig. 7.2. *Wood.* Logs of *Shisham* lying in open.

Air seasoning. Here, the moisture is removed by exposure to air without resorting to artificial heat. It is carried out in open until the wood ceases to lose weight. The ultimate water content varies from 12 to 30 per cent. The main objects of air-seasoning are to reduce weight, the amount of shrinkage and other defects. This is a long-time process.

Kiln drying. Here, heat is applied to wood in an enclosed space. In this process, the moisture is removed more rapidly and more completely. The final moisture content varies from 4 to 12 percent. If green wood is kiln-dried, warping and other defects may be prevented. This process is completed in short duration.

PRINCIPAL WOODS OF INDIA

Babul

Acacia nilotica (Linn.) Del.; Syn. *A. arabica* Willd.; Trade name—Babul; Hindi—*Babul, kikar*; Bengali—*Babul*; Hindi—Punjab and Uttar Pradesh—*Kikar;* Gujarati—*Baval*; Telugu—*Nallatumma*; Tamil—*Karuvelei*; Kannada—*Jaali, gobbli*; Malayalam—*Karivelan*; Family—Mimosaceae.

The sapwood is yellowish-white and usually wide. The heart-wood when freshly cut, is a pinkish or old rose colour, but it darkens on exposure to a dull red or reddish brown. It is often mottled with darker streaks. The wood is dull and without taste or smell, hard, fairly close-textured and with straight or slightly twisted grain.

There are several varieties of *babul.* The two best known are the *telia* and *kauria* varieties, the wood of the former having a better reputation than the latter.

Seasoning. Babul is a wood which can be air-seasoned with fairly good results if care is taken. One year is sufficient to air-season babul boards up to 2 in. in thickness. It can be kiln-seasoned without difficulty.

Strength. Babul is an extremely strong, hard and tough wood. It is nearly twice as hard as teak and has a very good shock resisting ability.

Durability. The sapwood is not durable. The heartwood is durable in most situations but not to the same degree as teak and sal.

Working qualities. An easy wood to convert and to saw when green, but it becomes harder and tougher when seasoned. It works well by hand, and machines and finishes to a good surface. It takes a fair polish but requires careful filling.

Uses. It is popular for parts of carts (body work, spokes, naves, axles, felloes, yokes and shafts), and for agricultural implements such as ploughs, harrows and clod-crushers. It is a useful tool handle wood but not for all types of handle, and it can be used for tent pegs. It can be described as one of the best Indian utility woods where hardness and toughness are required. It is a good turnery wood.

Sources of supply. Babul is usually available in small logs only, but in some districts larger logs are available. It is found throughout the drier regions of North, Central and South India.

Simul

Bombax ceiba L.; Syn. *Salmalia insiginis* (Wall.) Schott & Endl.; Hindi—*Semul*; Sanskrit—*Salmali, rakta-pushpa;* Bengali—*Simul, roktosimul;* Marathi—*Saur, simlo;* Gujarati—*Sawar, simalo;* Telugu—*Booruga, konda-buruga*; Tamil—*Mullialava;* Kannada—*Booruga*; Malayalam—*Pula-maram*; Oriya—*Bour;* Garo—*Panchu*; Mundari—*Edelsong*; Family—Bombacaceae.

A large tree found in the Western Ghats, Assam and the Andaman Islands. The wood is very soft creamy white or pale pink timber, very light in weight and with large open pores.

Seasoning. Quick drying can be achieved by exposing freshly converted timber directly to the drying action of the sun and wind or by kiln-seasoning. If air-seasoning is to be done, the best method is to stack the freshly sawn stock in a vertical position against a building or railing with the planks facing the sun. The boards should be turned from time to time so that both surfaces are exposed, kiln-seasoning is the surest and best method of drying *simul.*

Fig. 7.3. *Bombax ceiba.* A semul tree, the wood is very soft creamy white or pale pink timber, very light in weight, generally used for making matches, packing cases, ply-wood, etc.

Uses. The wood is used for matches, packing cases, penholders, veneers and ply-wood. Wood is most widely used in match industry, especially for match boxes. Suitable for shingles, canoes, toys, scabbards, cooperage, bush handles, tea-chest plywood and picture frames. Also used for cushioning mine-props and for inside portions of opium chests. Floss used for stuffing life belts, mattresses, cushions and pillows, upholstry and quilts. Tree yields a gum called *mocharus*.

Albizia

There are several common commercial *Albizia* woods. The best known is perhaps ***Albizia lebbeck*** (Linn.) Benth.; the others *A. odoratissima, A. procera*, and *A. stipulata*.

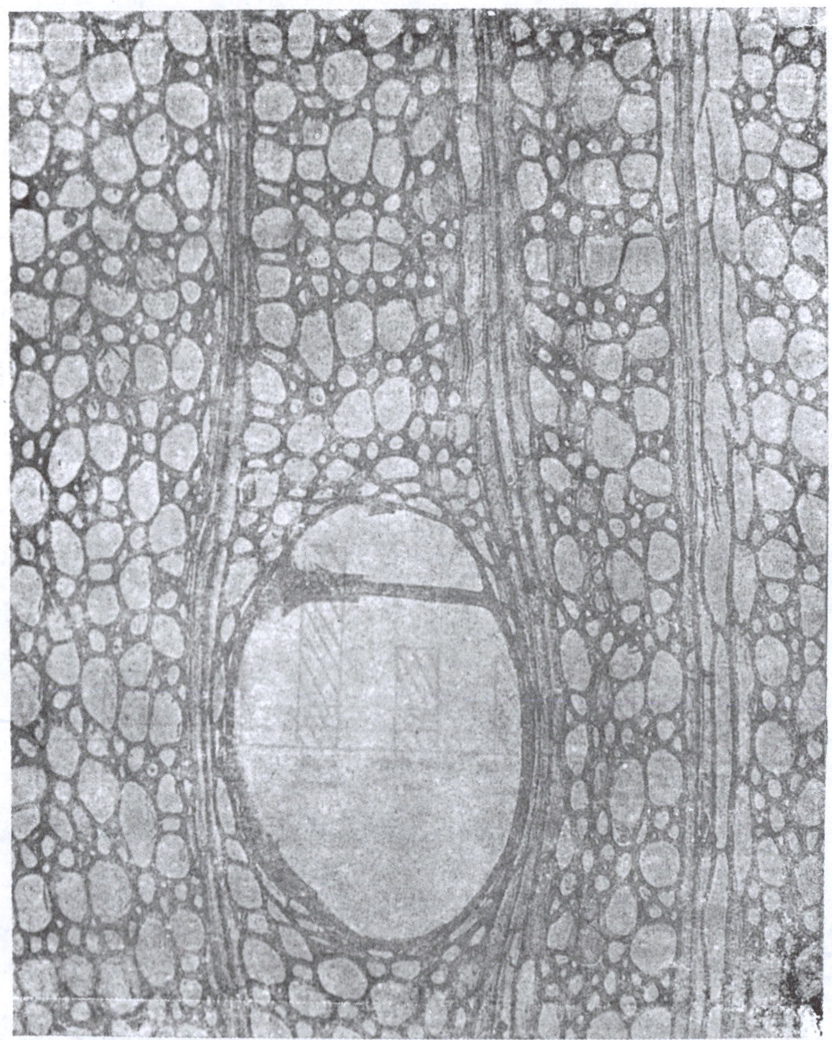

Fig. 7.4. *Bombax ceiba* (**Semul**). Micrograph X110, of wood.

Albizia lebbeck (Linn.) Benth.; Eng. Kokko; Hindi—*Siris*; Sanskrit and Marathi—*Sirisha;* Bengali—*Sirish*; Gujarati—*Pilo sarshio*; Telugu—*Dirasana*; Tamil—*Vagei*; Kannada—*Begemara*; Malayalam—*Vaga*; Family—Mimosaceae.

Albizia odoratissima (Linn. f.) Benth.; Eng. Black siris; Hindi—*Kala siris.*

Albizia procera Benth.; Eng. White siris; Hindi—*Safed siris.*

Description of the woods. All three of the *siris* species have very similar timber. The sapwood is white and often fairly wide. The heartwood is a rich walnut brown, often handsomely streaked with lighter and darker markings and a pleasing golden lustre which shows up well under polish. The wood has very large pores which need to be well filled to obtain a good polish. The texture otherwise is very even, but the grain may be straight or interlocked in broad bands.

Seasoning. The wood can be air-seasoned. Kiln-seasoning presents no difficulties, and it is preferable to kiln-dry these woods if kilns are available. Once seasoned they are very steady.

Strength. *Kokko* is a strong wood, being about the same weight and hardness as teak. Black *siris* is a heavier, harder and stronger wood than teak, while white *siris* is lighter and slightly weaker.

Working qualities. Due to interlocked fibres the *siris* timbers are not always too easy to saw and machine. They can, however, be finished to good surface by hand, but they all require careful and repeated filling on account of the larger pores.

Uses. These woods are worthy of special notice. They are excellent for high class furniture, interior decoration, and panelling. All three of the *siris* woods can be extremely handsome if a proper selection is made, and the golden sheen, which they retain permanently, enhances their value. They can be recommended for all furniture such as desks, tables, chairs, screens, almirahs and chests of drawers.

Sources of supply. The largest supplies of *Kokko* come from the Andamans. It is also found in more limited quantities in Bengal, Assam, Maharashtra, Tamil Nadu, Madhya Pradesh, Punjab and Uttar Pradesh.

Black *siris* is available in Madhya Pradesh, Assam, Maharashtra, Uttar Pradesh, Orissa and Tamil Nadu being fairly common in some districts and rare in others.

White *siris* is available from Uttar Pradesh, Assam, Bengal and Maharashtra.

Yon

Anogeissus acuminata (Roxb. ex. DC.) Wall.; Eng. Yon; Hindi—*Dhaura*; Bengali—*Chakwa*; Oriya—*Pasi*; Telugu—*Pansi*; Tamil—*Nunnera*; Trade—Yon; Family—Combretaceae.

Sapwood greenish or yellowish to light brownish grey. Heartwood chocolate or purplish brown, usually small. The wood is heavy, strong, tough and elastic, with a fairly fine texture and straight or slightly interlocked grain.

Seasoning. The best procedure is to convert logs during or just after the rainy season, and to stack the converted material without delay under cover with protection on all four sides for slow drying.

Strength. It is due to the great strength of this wood, especially in shock resistance, transverse strain and hardness, that they have achieved a well-earned reputation as substitutes for ash and hickory for shafts, helves and tool handles.

Uses. Mainly used for handles of tools.

Axle Wood

Anogeissus latifolia Wall.; Eng. Axle wood, Gum ghatti; Hindi—*Bakla, dhaura*; Marathi—*Dhaura*; Gujarati—*Dhavdo*; Telugu—*Yella maddi*; Tamil—*Vellay naga;* Kannada—*Dinduga*; Malayalam—*Marukinchiram*; Trade—Axle-wood; Family—Combretaceae.

A deciduous tree occurring in the drier parts of India. It is found in the sub-Himalayan tract, Madhya Pradesh, the Western Peninsula and the Nilgiris.

Uses. The wood is used for making wheels and tool handles.

Red Cedar

Toona ciliata Roem.; Eng. Cedrela tree, Red cedar; Hindi—*Toon*; Sanskrit—*Nandivriksha, tunna*; Hindi and Bengali—*Tun, mahanim;* Marathi—*Kuruk*; Tamil—*Santhanavembu, tunumaram*; Telugu—*Nandichettu*; Kannada—*Mandurike*; Malayalam—*Malarveppu*; Assam—*Poma*; Family—Meliaceae.

A tree cultivated in the Punjab, Assam, Bihar, the Western Ghats, Maharashtra, Madhya Pradesh and the Nilgiris.

Description of wood. The wood is a pinkish brick red, when freshly cut, toning to a light brownish red on exposure. The wood is a light weight, and usually straight grained timber. The wood has a distinct cedar like smell.

Seasoning. *Toon* is a fairly easy wood to air-season if care is taken with the stacking. Kiln-seasoning can also be done.

Strength. *Toon* is a timber of only moderate strength qualities. Hill *toon* is strongest and has about 80% the strength of teak.

Working qualities. *Toon* is one of the easiest of Indian woods to saw and work. If properly finished and filled, it takes a very good polish. It makes up into a good type of plywood. It makes high quality congealated wood.

Uses. The wood is used for furniture, tea-chests, shuttles and picking sticks which are used in the textile industry, and cigar boxes.

Walnut

Juglans regia Linn.; Eng. English walnut, Persian walnut; Hindi—*Akhrot*; Family—Juglandaceae. Found in Kashmir, Himachal Pradesh, the Khasia hills and in the hills of Uttar Pradesh.

Indian walnut varies considerably in colour, sometimes being a dull grey, while other wood may be a dark brown with even darker markings. It is relatively light wood for its strength and it works very easily and finishes to a fine surface. Its most important quality, however, is that, when once dried it does not shrink, swell or split. This, combined with its lightness, strength and good working qualities gives it pre-eminence as a wood for rifle parts, gun stocks, high class cabinet making, and delicate carvings.

Seasoning. Walnut wood seasons slowly and springs considerably while drying out, but apart from this, it is a model wood in so far as seasoning is concerned, both in air-drying and kiln-drying. Green conversion, followed by stacking under cover with a good air circulation is the method of seasoning recommended.

Strength. For its weight, walnut is a relatively strong wood. It is about 85% the weight of teak and in shock resistance is about equal to teak.

Working qualities. An extremely easy and pleasant wood to saw and work to a fine finish. It takes an excellent polish and needs very little filling.

Uses. The wood is used for musical instruments and cabinets work. Used extensively in Kashmir and North India for carving. Well known as an excellent veneer and plywood timber. It is also used for gun stocks.

Mango

Mangifera indica Linn.; Eng. Mango; Hindi—*Am;* Family—Anacardiaceae.

A tree, grown extensively in Uttar Pradesh, Punjab, Maharshtra, Andhra Pradesh, West Bengal and Tamil Nadu.

Description of the wood. A grey or greyish-brown wood. No taste or smell. Coarse-textured with occasional interlocking of fibres and curly grain, medium weight, fairly strong wood which can be used for a variety of purposes. A very steady wood and one which retains its shape extremely well.

Seasoning. A fairly easy wood to season. Green conversion, followed by careful stacking in a well ventilated place so as to allow the timber to dry quickly, is the method to adopt. It can be kiln-seasoned without any difficulty.

Strength. Mango is a fairly strong wood. In weight it is very slightly lighter than teak and in shock resistance and shear it is equal or slightly better than teak. In other strength functions it is about 80% the strength of teak. It is very steady wood when seasoned.

Working qualities. It presents no difficulties in so far as sawing and working are concerned. It can be easily finished to a clean surface, and with proper filling takes a good polish.

Uses. It can used to advantage for rotary veneer work and plywood making, and is one of the woods being used for this purpose all over India. Its chief uses are for cheap furniture, planking, floor and ceiling boards, the chests and other box and create work, boat building, agricultural implements, parts of carts and for plywood manufacture. It is a good wood for dry cooperage and shoe heels.

Laurel

Terminalia tomentosa Wight & Arn.; Eng. Laurel; Hindi—*Sain*, *Asan*; Bengali—*Assan*; Marathi—*Ain*; Gujarati—*Sadar*; Telgu—*Tari*; Tamil—*Karramarda*; Kannada—*Sadada*; Oriya—*Sahaju*; Trade—*Laurel*; Family—Combretaceae.

Found throughout India except Punjab and Assam.

Description of the wood. Sapwood creamy white. Heartwood sometimes dull light brown, sometimes reddish brown, sometimes, dark brown with good figuring or blackish streaks. No characteristic smell or taste, fairly straight-grained and of medium to coarse texture. It is a good strong hard and fairly durable wood capable of giving good service in constructional work.

Seasoning. The best procedure for drying the wood is to convert the logs green, preferably in damp or mild weather, and stack the converted material under cover and protect it from too rapid drying. In kiln-seasoning it can be dried with complete success.

Strength. Laurel is 25 to 30 per cent heavier than teak, and in hardness it stands 50 to 60 per cent higher than teak. In shock resistance it is usually well above teak. It is good strong tough wood suitable for most constructional purposes.

Working qualities. A very easy wood to saw and work. It finishes to quite a good surface.

Uses. It has been used for plywood tea chests in Assam. It is also used for ceiling boards and other domestic purposes. Very suitable for cheap utility plywood and also used in match factories for both boxes and splints. It is also used for rice pounders, ship and boat building.

Wood is also used for beams, joints, rafters, door and window frames, and boarding; also used in the construction of carts, toys, furniture, oil mills, rice pounders, engine brake blocks, electric casing and rough carpentry. The wood is suitable for tool handles, agricultural implements, and for plywood manufacture. Timber is also suitable for uses as telegraph and electric poles and yields pulp for manufacture of printing and wrapping paper. Wood is resistant to fire and is used in fire-proof building. Wood also yields good quality charcoal.

Pyinkado

Xylia xylocarpa (Roxb.) Taub.; Eng. Pyinkado; Verna—*Irul*; Sanskrit—*Scimsapa, kanakakuli;* Hindi—*Jambu, suria*; Marathi—*Jamba, suria;* Telugu—*Eravalu*; Tamil—*Irul*; Kannada—*Tirawa*; Malayalam—*Irumul*; Oriya—*Boja, tangini*; Trade—*Irul*; Family—Mimosaceae.

A tree, found in the Western Peninsula, Orissa, Andhra Pradesh and Madhya Pradesh.

Description of the wood. Sapwood pinkish white, narrow heartwood reddish brown to brown, darkening somewhat on exposure. No odour or taste. Irregulary interlocked fibres and medium to fine texture. A heavy and very hard wood of good quality.

Seasoning. To obtain the best results, trees should be felled during or just after the rains, and the logs should be converted green. The converted material should then be carefully stacked under cover and well protected from hot dry winds and sun.

Strength. It is a very hard tough wood. It is nearly twice as hard as teak.

Working qualities. Being a very hard wood, it is not very easy to saw, but it can, with care, be brought to a very smooth surface.

Uses. In South India, *irul* is a well-known wood and one much used for railway sleepers, heavy constructional work, piles, pitprops, railway wagon floor boards, ship building, bridges and general utility work.

Also favourite for boats and canoes and knees, crooks and keels of ships. A good substitute for *Sal* and *Teak* for beams, posts and scantlings in house constructions. Wood lasts well under water and used for wooden bridges. It yields material for paper pulp. A good fuel wood; its charcoal is highly prized by iron smelters.

White Willow

Salix alba Linn.; Eng. White willow; Hindi—*Bis*; Kashmir—*Vivir*; Punjab—*Bis*, *madnu, bhushan*; Family—Salicaceae.

A small tree, found in the North-Western Himalayas and in the valley of Kashmir. The twigs are used for making baskets. Cricket bats are made of the wood.

Used also for house building, match-boxes and splints, shoes, tool-handles, agricultural implements, boats, boxes, combs and toothpicks; suitable for paper pulp and charcoal making.

Mulberry

Morus alba, Linn; Eng. White mulberry; Hindi—*Tut*; Family—Moraceae. A tree, native of China. In India, it is grown in Punjab, Uttar Pradesh, Kashmir and the North-Western Himalayas. It is also cultivated in Karnataka state for silk worm cultivation.

Fig. 7.5. *Cedrus deodara* (**Deodar**). A tree at Forest Division, Chakrata, U.P. ;

Description of the wood. The sapwood is white and sharply delimited from the heartwood, which is a bright yellowish brown when freshly cut. No odour or taste. Straight-grained and of rather open medium coarse texture.

Seasoning. The best results have been obtained by storing the logs for some months before final conversion. The wood can be kiln-seasoned without difficulty. In practice, the most of the mulberry used in Punjab is used in a green state by the sports goods manufacturers. The timber is converted green, steam bent, and the dried out in a bent form tightly held in a clamp.

Strength. It is approximately the same in weight as teak, and in shock resistance, shear and hardness it is considerably higher than teak.

Uses. The primary use is for sports goods, for which purpose it is eminently suitable. It is used mainly for the manufacture of hockey sticks. It is also used for tennis rackets, badminton and squash rackets, presses, cricket stumps, etc.

Also suitable for house-building, agricultural implements, furniture, spokes, poles, shafts, bent parts of carriages and carts and for turnery. Bark is used for paper making; also yields a textile fibre.

Box Tree

Buxus wallichiana Baill.; Eng. Box tree; Hindi—*Chikri, papri*; Family—Buxaceae. A small tree, found in the Western Himalayas and the Punjab. The wood is used for making croquet balls, mallets, carving work, turnery articles, musical instruments and cabinet work.

Jaman

Syzygium cumini (Linn.) Skeels; Eng. Jambolana; Hindi—*Jamun*; Family—Myrtaceae.

A large tree. It yields a reddish-grey wood sometimes with darker markings. It is of medium weight and texture. It has about the same weight as teak but is 20% harder. It is not difficult to saw and work to a clean finish. The grain is regularly wavy, and this usually gives a mottled effect on finished boards.

Uses. It is an average good wood for construction work and house building. It is also used for furniture and cabinet work.

Kadam

Anthocephalus indicus A. Rich; Eng. Kadam; Hindi—*Kadamba;* Sanskrit, Hindi, Bengali, Marathi and Gujarati—*Kadamba;* Telugu—*Kadambamu*; Tamil—*Vellai-cadamba*; Kannada—*Kadawala*; Malayalam—*Attutek*; Trade—Kadam; Family—Rubiaceae.

A moderate sized tree, found in Assam, Bengal and Western Ghats.

The wood is straight-grained even-textured, white or creamy coloured. A non-ornamental timber of good grade. An easy wood to season. The wood is moderately strong and soft. It is a very easy timber to saw and work to a fine surface.

Uses. It is used in Assam and Bengal for ceiling boards and light construction work. It is a cheap boarding and packing case wood. It is used for making match splints in Myanmar.

Also suitable for dugouts, canoes, carving and turnery. Pulp is suitable for manufacture of cheap quality paper.

Deodar

Cedrus deodara Loud.; Eng. Deodar; Hindi—*Deodar*, *Diar*; Family—Pinaceae.

A tall evergreen tree found in the North-Western Himalayas from Kashmir to Garhwal.

Description of the wood. It is of light yellow-brown colour and possesses distinctive odour. It is a medium weight wood which is very sturdy in use and durable. It is usually even-grained and of medium to fine texture, but the presence of large knots is a common feature.

Seasoning. It is an easy wood to air-season. It can also be kiln-seasoned.

Strength. Deodar is the strongest of the Indian conifers. Its weight is 20% less than teak and its strength is also about 20% less.

Working qualities. It is an easy timber to saw and work to a smooth finish.

Uses. The timber is used for construction work and for railway sleepers. It is also suitable for beams, floor boards, ports, window frames, light furniture and shingles.

This is strongest of Indian coniferous woods, also used for doors, furniture, packing cases, masts, spars and also for wooden bridges. Wood yields an oleoresin and a dark coloured oil used for ulcers and skin diseases.

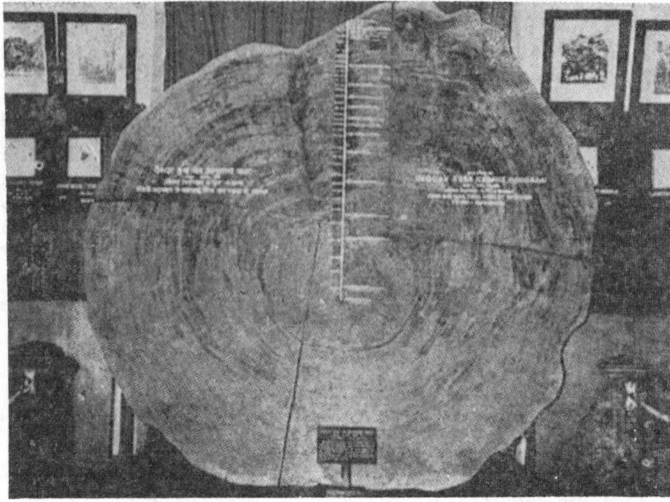

Fig. 7.6. *Cedrus deodara* (**Deodar**). T. S. of wood of trunk of 704 year old tree, displayed in museum at F.R.I., Dehradun.

Himalayan Spruce

Picea smithiana (Wall.) Boiss.; Eng. Himalayan spruce; Hindi—*Bajur*; North West Himalayan region—*Rai, rau, re, kachal, salla, tos*; Jaunsar, Garhwal and Kumaon—*Roi, rhai, ragha, morinda*; Trade—Spruce; Family—Pinaceae.

A tree commonly found in the Western Himalayas. The wood is used for construction work, railway sleepers, cabinet making, packing cases and wood pulp.

One of the most useful timber trees of the Western Himalayas. Wood is used for planking, general fittings and joinery, rough furniture, tea-boxes and packing cases. This is one of the best light boxwood of India. Wood is suitable also for match boxes, battery separators, for fence posts, transmission poles, piles and newsprint manufacture.

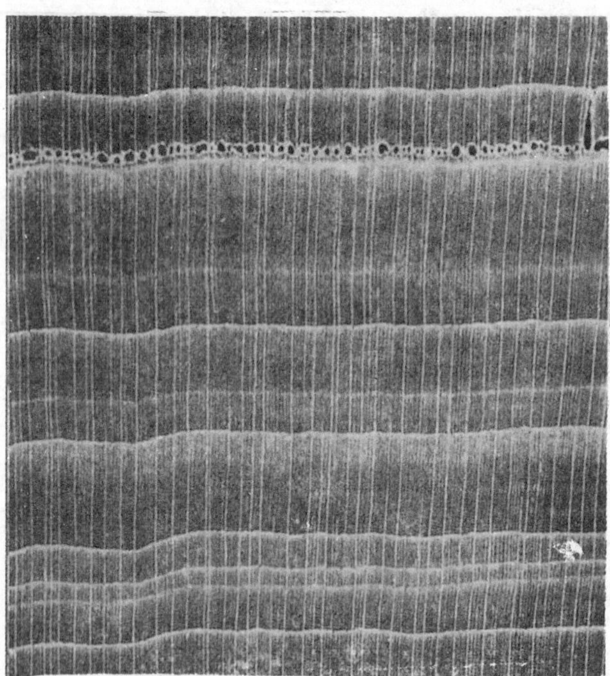

Fig. 7.7. Deodar wood (non-porous), sectional view of micrograph x 10.

Pine Wood

Pinus roxburghii Sar.; Eng. Long-leaved pine; Hindi—*Chir, Salla*; Family—Pinaceae.

An evergreen coniferous tree. It is found in the Western and Eastern Himalayas. The wood is used for construction work, railway sleepers, packing cases, furniture and in match industry.

Pinus merkusii Jungh & *de* Vries. A tree. The wood is very resinous with reddish-brown heartwood and yellow sapwood. It is used for building purposes and general carpentry.

Pinus wallichiana A. B. Jackson; Eng. Indian blue-pine; Hindi—*Kail, Chil.* A coniferous tree. The wood is used for pencils, penholders, splints, match boxes, construction work, railway sleepers, and furniture. Also used for shingles, packing cases, constructional work and house fitments. It is a good wood for pattern making, cores for laminboards, drawing boards, and planetables.

Dalbergia Linn. (Leguminosae).

A genus of trees, shrubs, and woody climbers widely distributed in tropical and sub-tropical regions. About 25 species occur in India; *D. sissoo* and *D. latifolia* are among the most important Indian timber trees.

Indian Rosewood

Dalbergia latifolia Roxb.; Eng. Indian rosewood; Verna. *Kali Shisham*; Sanskrit—*Shishapa;* Bengali—*Sitsal, swetasal;* Marathi—*Shisham, siswa*; Gujarati—*Shissam, kalaruk*; Telugu—*Cittegi,*

Fig. 7.8. *Dalbergia sissoo* (**Shisham**). A tree.

irgudu; Tamil—*Itti, karundorviral;* Kannada—*Bite, todagatti*; Malayalam—*Itti, colavitti;* Oriya—*Sisua*; Trade—Indian Rosewood; Family—Papilionaceae.

A large deciduous or nearly evergreen tree with cylindrical, fairly straight bole and full rounded crown, found in the sub-Himalayan tract of Eastern Uttar Pradesh, Bihar, Orissa and central, western and southern India. It attains its maximum development in the southern region of western ghats, where trees 130 ft. high up to 20 ft. in girth and 70 ft. clear bole are sometimes found. The minimum exploitable size is 6 ft. girth.

The sapwood is narrow and pale yellowish white in colour, often with a purple tinge. The heartwood ranges in colour from golden brown through shades of light rose, purple with darker streaks, to deep purple with rather distant, nearly black lines, darkening with age. It is fragrant, heavy, narrowly interlocked grained and medium coarse-textured. The timber is stronger and much harder than teak.

Indian rosewood ranks among the first wood for furniture and cabinet work. It is a valuable decorative wood suitable for carving and ornamental ply-boards and veneers. It is specially useful for pattern-making, calico printing blocks, mathematical instruments and screws. It is also used for gun carriage wheels, ammunition boxes and army wagons, pulleys, handles, shelves, decorative carriage parts, temple chariots, boat knees, agricultural implements, combs, razor handles and brush backs. Carefully selected and manufactured Indian rosewood ply-boards satisfy aircraft specifications.

Sissoo

Dalbergia sissoo Roxb.; Eng. Sissoo; Verna. *Shisham*; Sanskrit—*Shinshapa, aguru;* Hindi—*Shisham, sissu, sissai;* Bengali—*Shisu*; Gujarati—*Sisam, tanach;* Telugu—*Errasissu, sinsupa*; Tamil—*Sisu itti*; Kannada—*Agaru, biridi*; Malayalam—*Iruvil*; Oriya—*Sisu, simsapa*; Punjab—*Tali, shisham*; Marathi—*Sissu*; Trade—Sissoo, Shisham; Family—Papilionaceae.

A deciduous tree, often with crooked trunk and light crown. Normally the tree attains a height of about 100 ft., a girth up to 8 ft. and a clear bole up to 35 ft. It occurs throughout the sub-Himalayan tract from Ravi to Assam, ascending up to 5,000 ft. It is extensively cultivated in Punjab. Uttar Pradesh, Bengal and Assam. No other timber tree, except teak is cultivated to a greater extent.

The sapwood is white to pale brownish white in colour and the heartwood, golden brown to dark brown with deep brown streaks becoming dull on exposure. It is heavy, narrowly interlocked grained and medium textured. It can be air or kiln-seasoned. However, kiln-seasoning enhances the value of timber as the colour darkness and the difference between lighter and darker bands gets intensified. The wood can be converted into ornamental veneers or commercial plywood of good quality.

Sissoo *(Shisham)*, is a high class furniture and cabinet wood widely used throughout north India. On account of its great strength, elasticity and durability, it is highly valued as constructional and general utility timber and is used for all the purposes for which Indian rosewood is employed in the south. It is esteemed also for railway sleepers, musical instruments, *Charpai* legs, hammer handles, shoe heels, *hookah* tubes and tobacco pipes. Carefully selected and manufactured plywood logs satisfy the specifications prescribed for aircraft, and for this purpose, the wood from trees, growing on canal banks and in plantations is considered to be the best. Sissoo wood is suitable for making laminated skiis. It can be worked into decorative ornamental carvings.

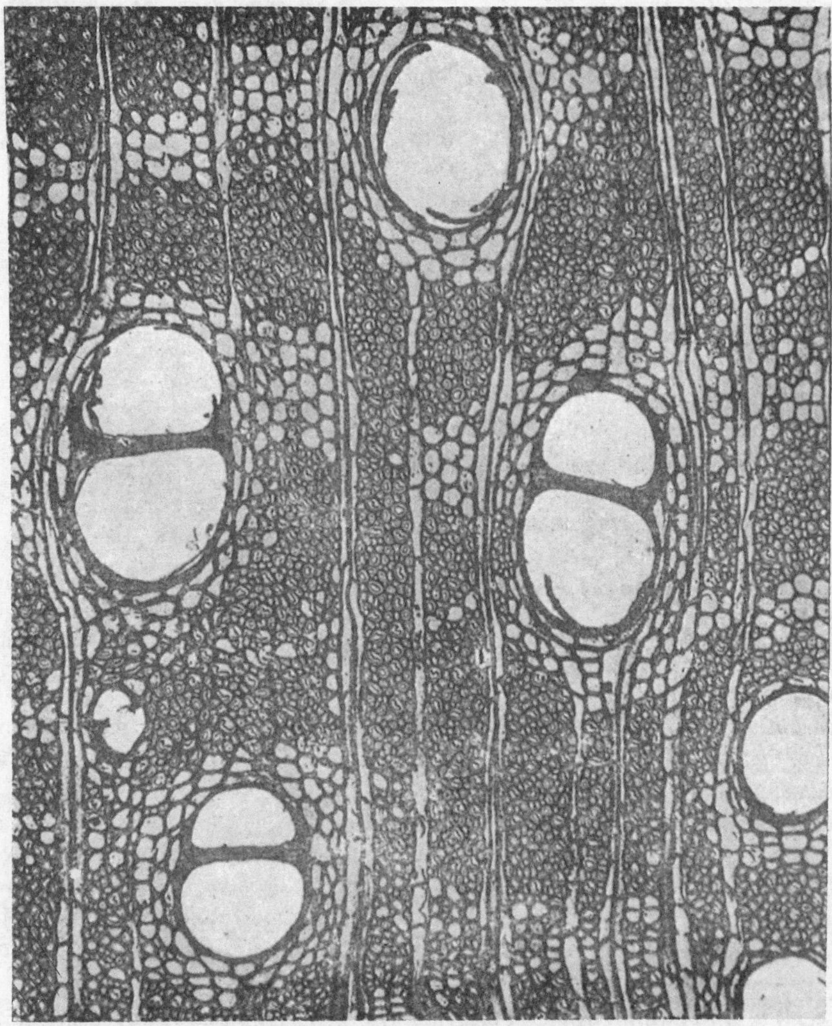

Fig. 7.9. ***Dalbergia sissoo*** **(porous wood) micrograph of wood x 110.**

Dipterocarpus

A genus of large tree with straight cylindrical stems widely distributed from India and Ceylon to Philippines. About 10 species occur in India, mostly in Assam, and in Andaman islands. They are the dominant species in the forests in which they occur. They have relatively small crowns—flat, conical or irregular—made up of a few large branches. Family—Dipterocarpaçeae.

The species of *Dipterocarpus* yield commercial timbers of the non-ornamental structural class, known in the trade under the common name *Gurjan*. The timbers from the different species of *Dipterocarpus* are similar in structure, weight, hardness and colour. They are characterized by numerous longitudinal resin canals. *Gurjan* woods are strong, moderately hard and heavy, fairly straight-grained and medium coarse-textured. They are easy to air season, but liable to high shrinkage, and are durable only after treatment. Though easy to saw and work, they cannot be finished to a very smooth surface.

Gurjan

Dipterocarpus turbinatus Gaertn. f.; Eng. Common Gurjan tree; Verna. *Gurjan*; Bengali—*Teli-gurjan*; Assam—*Gurjan—kuroilsal, kherjong.*

A large tree attaining a height of 120–150 ft. and a girth of 10–15 ft., with a clean cylindrical bole up to 90 feet. It occurs in the ever-green tropical forests of Assam and Andamans.

The wood is dull red or reddish brown in colour with a somewhat coarse texture and resin canals. It possesses an agreeable odour and exudes an oil when heated. The wood seasons slowly and is durable under cover. It is a non-ornamental timber of the first class. The wood is extensively used for internal construction work. It is used for packing cases, tea boxes, panelling and flooring, and carriage and wagon construction. It is a promising sleeper wood after treatment. The other important Indian species are —*D. indicus* Bedd.; *D. macrocarpus* Vesque; *D. tuberculatus* Roxb.; *D. alatus* Roxb.; *D. grandiflorus* Blanco.; *D. costatus* Gaertn.; *D. kerrii* King and *D. bourdilloni* Brandis.

Fig. 7.10. *Dipterocarpus alatus.* It is a lofty evergreen tree of the Andaman forests, attaining a height of 100–150 feet and a girth of 7–15 feet. The wood is quite superior. The oleo-resin obtained from it is a thick fluid when fresh, changing to a semi-plastic mass after long exposure. It is used for making varnishes.

Andaman Redwood

Pterocarpus indicus Willd.; Eng. Andaman redwood; Verna. *Padauk.* It is a large and lofty tree found in the Andaman islands. Sapwood is small. Heartwood is dark-red, close-grained, moderately hard with a slight aromatic scent. It is durable and not attacked by white ants. When thoroughly seasoned it is unaffected by alternate dryness and moisture of the atmosphere. It seasons well, works well and takes a very fine polish. It is used for furniture, carts, guncarriages, and other purposes, and is said to be the most useful wood in the Andamans, and where it grows to an enormous size.

Indian Kino Tree

Pterocarpus marsupium Roxb. Sanskrit—*Pitasara;* Hindi—*Bijasal*; Bengali—*Pitshal*; Marathi—*Asan, bibla*; Gujarati—*Biyo*; Telugu—*Yegi, peddagi;* Tamil—*Vengai;* Kannada—*Honne, bange;* Malayalam—*Venga;* Oriya—*Byasa;* Trade—Bijasal; Family—*Leguminosae.*

The wood is used chiefly for building purposes, such as doors, window frames, rafters, beams and posts. It is also used in railway carriages, wagons, carts, boats, ships, electric poles, pit-props in mines, agricultural implements, drums, tool handles, camp furniture, mathematical instruments, picture frames, combs, and parts of textile looms. It is a very important timber in peninsular India.

Red Sandalwood

Pterocarpus santalinus Linn.; Eng. Red sandalwood; Verna. *Rukhto-chandan*; Sanskrit—*Raktachandana*; Hindi and Bengali—*Raktachandan, lalchandan;* Marathi—*Tambada chandana;* Gujarati—*Ratanjali*; Telugu—*Agarugandhamu, raktagandhamu*; Tamil—*Atti, sivappu chandanam*; Kannada—*Agaru, honne, kempugandha;* Malayalam—*Patrangam*; Oriya—*Raktachandan*; Trade—Red Sanders.

A small tree of South India, chiefly found in Cuddapah, North Arcot, and the southern portion of the Karnul district. It favours a dry, rather rocky soil, and a hot fairly dry climate. Sapwood is white. Heartwood is purplish-black dark orange-red when fresh cut, extremely hard, the shavings giving a blood red-orange colour. It is used for building and for turning, and is said to be much

Fig. 7.11. *Pte ocarpus marsupium.* Indian Kino Tree. The trees in Karnataka State.

prized because it is not subject to the attacks of white ants. It is much used for carvings in Andhra Pradesh especially at Tirupati.

Wood is highly prized for house posts; also used for agricultural implements, poles, shafts and bent rims of carts, picture frames, boxes and other joinery work. In Japan, it is used for a musical instrument called **Shamisen**. Wood is ground and used for dyeing wool, cotton and leather, and staining other woods; *santalin* is the colouring principle.

Diospyros Linn. (Ebenaceae.)

A genus of tree or shrubs, chiefly tropical. About 41 species occur in India mostly in the evergreen forests of South India, Assam and Bengal; only a few are found in north India. All the species yield useful timbers, the best known among them being *D. ebenum*. True ebony is the black heartwood of *D. ebenum* obtained by stripping off the peripheral light coloured wood of felled trees. *D. tomentosa* and *D. melanoxylon* are also ebony yielding trees. The heartwood of most species is, however, only streaked or mottled with black, and of these variegated ebonies the best known are *D. marmorate* and *D. quaesita*.

The heartwoods of ebony trees are extremely hard and tough. They are difficult to season, but are naturally durable and can be finished to a smooth polished surface. They are fancy and ornamental woods and have long been esteemed for carving, turning, marquetry, musical instruments, brush backs, umbrella handles, walking sticks, ornamental cabinets, furniture making, etc.

The sapwoods of ebony tree are of light colour turning reddish brown on exposure. They can be dried with comparative ease, but are not durable under severe conditions. They are strong, tough and shock-resistant, and are valuable commercial woods.

Ebony

Diospyros ebenum Koenig.; Eng. Ebony; Verna. *Abnus*; Hindi—*Ebans, abnus*; Telugu—*Nallavalludu*; Tamil—*Tumbi, karunkali*; Kannada—*Karemara*; Malayalam—Karu, *vauari*; Oriya—*Kendhu*; Trade—Ebony.

It is a moderate to large sized evergreen tree with a dense crown, sometimes attaining a height up to 80 ft. and a girth of 7 ft. The girth of heartwood is not more than 4 ft. It is found in the dry evergreen forests of Deccan, Karnataka, Coimbatore, Malabar, Cochin and Travancore. It prefers well-drained rocky soil and grows well on sandy loam. It is found sparsely scattered in the forests in mixture with other trees.

This is perhaps the best ebony yielding tree and the only one which yields jet black heartwood without streaks or markings, the true ebony of commerce, *D. ebenum* is not very commonly found in India, nor do the trees, where they occur, attain any large size. Only limited supply of ebony is available in the area of its occurrence.

The sapwood is light yellowish grey, often streaked with black. The heartwood is jet black, rarely with a few dark or light brown or golden streaks. It has a metallic lustre when smoothed. It is heavy, straight-grained or somewhat irregular or wavy-grained, fine and even-textured.

Ebony is difficult to season. It is best to convert the green logs to the smallest size permissible and to store them under cover, well protected from hot winds.

The heartwood is resistant to attack by insects and fungi. It is very durable. The value of this wood depends on its intense blackness, working quality and the high finish it takes. It is difficult to work, especially when dry, but finishes to a glossy surface. It requires little hand finishing and takes a high lasting polish. It is one of the most valuable woods in the fancy wood market. It is used for ornamental carving and turnery, and for special purposes in decoration. It is used for veneers, inlaying, musical instruments, sports goods, mathematical instruments, piano keys and caskets.

Diospyros virginianum Linn. is widely used for weaving shuttles and is the standard wood for the heads of golf clubs.

Saffron Teak

Adina cordifolia Hook. f.; Eng. Saffron teak; Verna, *Haldu*, Hindi—*Haldu*; Bengali—*Petpuria, dakom;* Marathi—*Heddi*; Telugu—*Pasupukadamba*; Tamil and Malayalam—*Manjakadamba*; Kannada—*Arsintega, yettega*; Trade—Haldu; Family—Rubiaceae;

This makes an important Indian timber, used for a great variety of purposes; Pearson and Brown (1932) state that it is one of the best turnery wood in India and that selected pieces are attractively

Fig. 7.12. *Adina cordifolia* (**Haldu**). Trees (32 year old) of Saffron Teak.

figured. The wood is commonly used for carvings, construction work, flooring and for railway carriages.

Timber is easy to saw, seasons well and takes good polish; durable under cover, resistant against subterranean termites, but highly susceptible to drywood termites. *Haldu* is accepted as grade I commercial and moisture-proof plywood timber.

Wrapping, writing and printing paper is manufactured from the wood pulp.

Kaim

Mitragyna parviflora; Verna. *Kadam;* Hindi—*Kaim*; Bengali—*Gulikadam*; Marathi—*Kalamb, kuddam;* Gujarati—*Kadamb;* Telugu—*Nir kadambe*; Tamil—*Chinna kadambu*; Kannada—*Kongu, kadaga*; Malayalam—*Vimbu*; Oriya—*Mur, gudikaima;* Trade—Kaim; Family—Rubiaceae. This is a tree found throughout India. The furniture, agriculture implements, and carvings are made from its wood.

Wood is used also for planks and rafters in construction of buildings, and for furniture, agricultural implements, cooperage, utility, brushes and boot lasts; used also for turnery, match–boxes, and splints, calico-printing blocks, and slate frames, and has been recommended for pen-holders, mathematical instruments and shuttles.

Garcinia Linn. (Guttiferae).

A large genus of evergreen tree or shrubs distributed in tropical Asia, Africa and Polynesia. About 30 species occur in India.

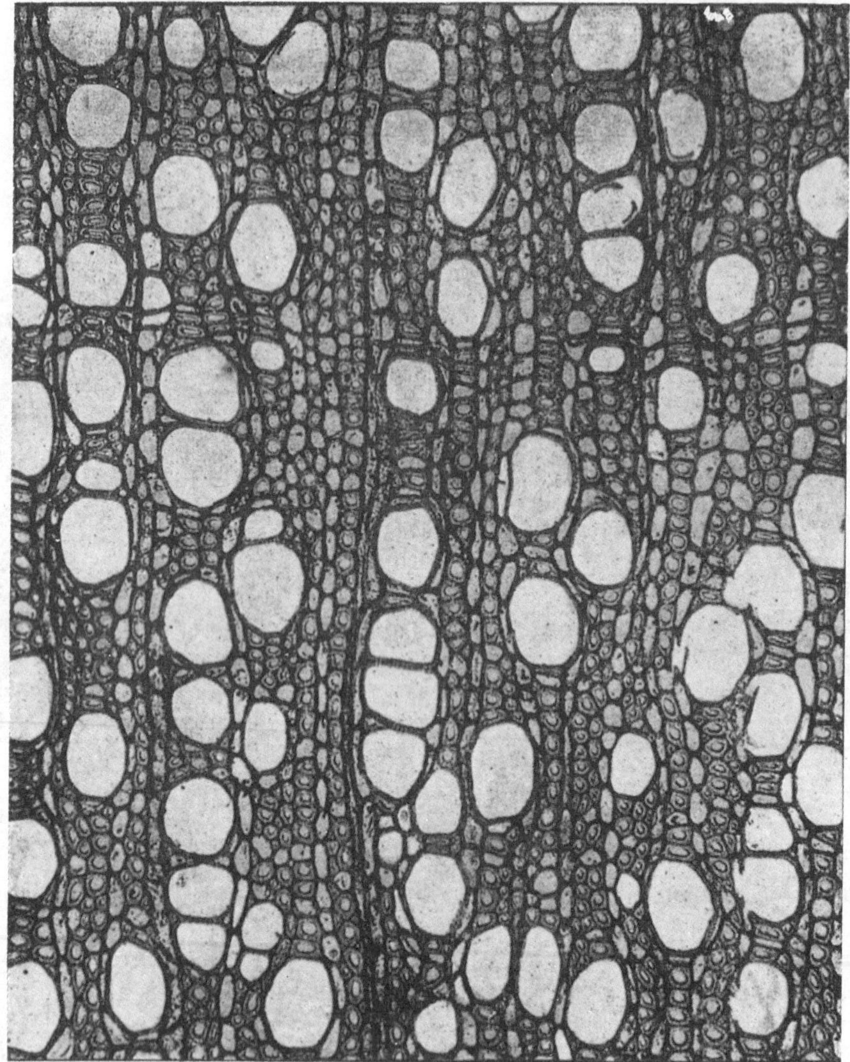

Fig. 7.13. *Adina cordifolia* (**Haldu**). Micrograph x 110, of porous wood.

Mangosteen

Garcinia mangostana Linn.; Verna. *Mangustan*; Eng. Mangosteen. A small or medium sized tree 20–45 ft. high, with deep green leathery leaves, 6–20 in. long; flowers terminal solitary or in pairs, 1–2 inches in diameter with pink fleshy petals. It has been successfully established in S. India. on the lower slopes of Nilgiris, between 1,200 ft. and 3,500 ft. The total area under mangosteen does not exceed 25 acres in S. India. It is a tropical tree adapted to regions of heavy and well-distributed rainfall. It requires a wet but well-drained soil and a humid atmosphere.

The wood is dark brown, heavy rather hard and fairly durable. It is suitable for cabinet work, building purposes, rice pounders and spear handles.

Garcinnia spicata Hook. A medium sized or tall tree, up to 70 ft. high. It is found in the evergreen forests of Western Ghats, at low elevations. The wood is yellowish white to brown, medium-coarse to fine-textured, hard to very hard and heavy. It is a strong timber useful for general construction purposes where strength is the main criterion.

Gardenia Ellis (Rubiaceae)

A genus of shrubs and small trees distributed in tropical and sub-tropical regions. About 6 species are found in India. Several *Gardenia* species yield timbers used as substitutes of boxwood.

Gardenia gummifera Linn. (Verna. *Dikamali*). A small tree, often with a crooked stem (5–6 ft. high, 1 ft. in girth). The wood is yellowish white in colour, somewhat lustrous with smooth feel. It is hard, heavy, straight-grained and fine and even-textured. It is liable to fine end-splitting during seasoning. Tree should be felled soon after monsoon and the material protected from dry winds. End-splitting can be checked by smearing tar or cowdung and leaving the bark on. The timber is fairly durable and takes a good polish. It is suitable for turnery work and is used for combs, tool handles, rulers, penholders and other small articles. It is useful substitute for boxwood.

Boxwood Gardenia

Gardenia latifolia Ait; Verna. *Papra*; Eng. Boxwood Gardenia, Hindi—*Papra; ban pindalu*; Marathi—*Ghogari*; Telugu—*Pedda bikki*; Tamil—*Kumbay*; Kannada—*Kalkombi;* Oriya—*Kota ranga, jantia*.

A small deciduous tree. Stem 12–14 feet high, 2–4 ft. in girth. The wood is creamy to yellowish white in colour with no distinct heartwood, lustrous with smooth feel; hard, strong, heavy, fine, and even-textured. It is easy to saw and work, takes a fine polish and is durable under cover. The timber is used as a substitute for boxwood. It is used for making combs and turnery articles. It is also used for engraving, light furniture, camp beds, tobacco jars, shuttles, mallet heads, toys, mathematical instruments, egg cups, etc.

Gluta Linn. (Anacardiaceae).

A genus of trees distributed in South-east Asia. One species occurs in India.

Gluta travancorica Bedd. A large evergreen timber tree, up to 100 ft. high and 15 ft. in girth, found in dense moist forests in Travancore, up to an altitude of 3,500 ft. The sapwood, which forms a large proportion of the wood, is reddish grey in colour. The heartwood is dark red, beautifully mottled with orange and black streaks, hard, strong, heavy; somewhat interlocked-grained and coarse-textured. The timber seasons well. It is rather difficult to saw, but works to a smooth surface and takes a smooth and lasting polish. The wood is considered to be the one of the finest and most beautiful timbers in India. It is valued for furniture, cabinet work, house fittings, turnery, carving and in-lay work.

Teak

Tectona grandis Linn.; Eng. Teak; Hindi—*Sagaun*; Sanskrit—*Saka*; Bengali—*Segun*; Marathi—*Sag, sagwan*; Gujarati—*Saga*; Telugu—*Adaviteeku, teeku*; Tamil—*Tekkumaram, tekku*; Kannada—*Jadi, sagwani*; Malayalam—*Thekku, tekha;* Oriya—*Singuru*; Assam—*Chingjagu*; Lepcha—*Ripnyok*; Family—Verbenaceae.

A large, deciduous tree, indigenous to both peninsulas of India. In Western India it does not extend far beyond the Mhye. In Central India it attains its northern most point in the Jhansi district, and from that point the line of its northern limit continues in a south-east direction to the Mahanadi river in Orissa. It is found wild in Assam. It is, however, cultivated throughout Bengal, Assam and Sikkim, and in North-West India without difficulty as far as Saharanpur. The forests richest in large timber on the west side of the Peninsula are the Travancore, Anamally, Wynad, South-west Karnataka and North Kanara forests. In the centre of the Peninsula the Godavary forests are most compact and valuable.

Structure and utility of wood. Sapwood white and small; the heartwood when cut green, has a pleasant and strong aromatic fragrance and a beautiful dark golden yellow colour, which on seasoning soon darkens into brown, mottled with darker streaks. The timber retains its fragrance to a great age, the characteristic odour being apparent whenever a fresh cut is made. It is moderately hard, exceedingly durable and strong, does not split, crack, warp, shrink or alter its shape when once seasoned; it works easily, takes a good polish. Teak owes its chief value to its great durability, which is ascribed, probably with justice, to the circumstance that it contains a large quantity of fluid resinous matter which fills up the pores and resists the action of water. (At the Karli caves near Poona the teak-wood-work, two thousand years old, seems perfectly good at the present day).

The many uses of teak are well known. In India it is highly prized for construction, ship building, and for making sleepers and furniture.

Wood is very durable and resistant to fungi. It is used for poles, beams, trusses, columns, roofs, doors, window frames, flooring, planking, panelling, stair cases, and other constructional work. It is one of the best timbers for furniture and cabinet making, wagons and railway carriages. Due to its better shape-retention ability, teak is popular in marine constructions and is a class by itself for boat and ship–building, particularly for decking. On account of its resistance to chemicals, teak articles are used in chemical industries and for making laboratory bench-tops; suitable for casks and vats for shipping corrosive liquids and for storing vegetable oils, fruit syrups, chutneys, etc. Teak is employed for sound-boards of musical instruments, keys, etc., and for different grades of plywood. Wood waste in the form of wood-shavings and sawdust is used for chip-boards, fibre-boards and plastic-boards.

Sal

Shorea robusta; Eng. Sal tree; Hindi—*Sal*; Bengali and Hindi—*Sal, sakhu*; Marathi and Gujarati—*Ral*; Telugu—*Gugal*; Tamil—*Kungiliyam*; Kannada—*Kabba*; Malayalam—*Maramaram*; Oriya—*Sal, sagua*; Punjab and Haryana—*Sal*; Lepcha—*Taksal–Kung*; Assam—*Sal, bolsal*; Trade—Sal; Family—Dipterocarpaceae.

Fig. 7.14. Scientifically managed Sal Forest, at Dehradun.

A large gregarious tree, often covering certain interrupted tracts without the existence of connecting patches. It occurs along the base of Tropical Himalayas from the Sutlej to Assam, in the eastern districts of Central India, and on the Western Bengal Hills.

Structure and utility of wood. The *sal,* one of the most valuable timber trees in India, has a distinct sapwood which is small in amount, whitish and not durable. The heartwood is brown in colour, finely streaked with darker lines, coarse grained, hard, strong and tough, with a remarkably fibrous and cross-grained, structure. The fibres of successive concentric strata do not run parallel, but at oblique angles, to each other, so that when the wood is dressed the fibres appear interlaced. It does not season well, but warps and splits in drying, and even when thoroughly seasoned, absorbs moisture with avidity in wet weather. During the process of seasoning, it dries when first cut, and evaporation goes on afterwards with extreme slowness. *Sal,* when once thoroughly seasoned, stands almost without a rival, as a timber, for strength, elasticity, and durability, which qualities it retains without being sensibly affected, for an immense length of time. Average weight of the seasoned *sal* about 35 lb. per cubic foot (Brandis; Gamble).

The timber is the one most extensively used in Northern India. It is in constant request for piles, beams, planking and railing of bridges, doors, and window posts of houses, for the bodies of carts, and above all, for railway sleepers. In Assam it is favourite wood for boat-building, and in the hills of Northern Bengal where it is found, perhaps, of the largest size available, the trunks are hollowed out into canoes. Owing to the fact that when unseasoned it is not floatable, difficulty is experienced in most *sal* forests in getting the timber out of the forests in log. This is, however, overcome by floating the logs either with the assistance of the boats or with floats of light wood or bamboos (Gamble).

Sal wood ranks with teak and deodar as one of the best sleeper woods in India; also used in form of bellies and poles. After treatment, the poles are suitable for overhead electric, telegraph and telephone lines. As domestic timber it is used for beams, scantlings, rafters and floors; also used for piles, mine work and pit props, bridges, dug-out boats, carriages and wagon building, spokes, fellows and hubs of wheels, agricultural implements, tool-handles, tent pegs, liquid storage vats, and beer and oil casks.

Grewia Linn. (Tiliaceae).

A genus of trees and shrubs distributed in tropical regions of the Old World. About 40 species occur in India. Some species are valued for timber.

Grewia tiliifolia Vahl; Verna. *Dhamni*; Sanskrit—*Dhamni, dhanuvrikha*; Hindi and Bengali—*Dhamni, dhamin*; Marathi—*Daman, damani*; Gujarati—*Dalmon, dhamana*; Telugu—*Charachi, ettatada;* Tamil—*Sadachi*; Kannada—*Thadsal*; Malayalam—*Chadicha*; Oriya—*Dhaman*; Trade—Dhaman.

A moderate-sized to large tree found in the sub-Himalayan tract from Jumna to Assam and in central, western and southern India. The tree attains its best development in the valleys and slopes of the southern hills, attaining a bole length of about 30 ft. and a girth of 7 ft. or more. The sapwoood is white to pale-yellow; heartwood reddish brown with dark streaks, often marked by white spots. It is dull with smooth feel, heavy, strong, elastic, even straight or wavy-grained in the radial plane, and medium-textured. It can be seasoned well. Green conversion, immersing the stock in water, for about six weeks, followed by seasoning under cover gives the best result. Kiln-seasoned stock retains the original brightness for an indefinite period. The wood is durable in the open and under cover. It does not require any antiseptic treatment. It is easy to saw and work and takes a high polish.

This timber is used, wherever available for shafts, poles, frames, panels, masts, tool handles, agricultural implements, bent parts of carts and carriages, spokes, horizontal bars, etc. It is an ornamental timber suitable for furniture. It is used for picker arms in textile mills and for shuttles, rabbet tubes, bobbins, etc. It is used as side props in mine shafts and galleries. It has also been employed for cooperage, golf shafts, billiard cue shafts, and cricket stumps and bails.

Gmelina Linn. (Verbenaceae).

A genus of trees and shrubs. Three species occur in India.

Gmelina arborea Roxb. Verna. *Gambhar*; Sanskrit—*Gambhari*; Hindi—*Gambhar*; Bengali—*Gumbar*; Marathi and Gujarati—*Sewan, shewan*; Telugu—*Gummadi*; Tamil—*Kumadi*; Kannada—*Shivani*; Malayalam—*Kumbil*; Oriya—*Gambari*; Assam—*Gomari*; Punjab—*Kumhar*; Nepal—*Gambari*; Trade—Gumhar.

It is an unarmed tree, 60 ft. high with a clear bole of 20–30 ft. and a girth of 5–7 ft. found scattered in deciduous forests throughout the greater part of India and the Andamans, up to an altitude of 5,000 ft. It prefers most fertile valleys with good drainage. The wood is yellowish to reddish white when first exposed, aging to light yellowish brown, light to moderately heavy, hard, strong, elastic, lustrous with a smooth feel, straight or irregular and interlocked-grained and medium coarse-textured. The heartwood is not distinct. The wood seasons well without cracking, but it is slow to dry. The timber is durable in contact with water. It is easy to saw and works to a smooth finish and takes paint and polish well. It is one of the best and most reliable timbers of India. It is used for furniture, planking, carriages, printing blocks, carving, musical instruments, shafts, axles, boxes, picture frames, stethescopes, artificial limbs, jute bobbins, loom parts, callipers, mine props and a variety of other articles. It is also employed for bridges, well work and ship building. The wood is suitable for tea chest plywood, paper making, match sticks and match boxes.

WOODS FOR VARIOUS USES

Agricultural implements. Agricultural implements is a term of rather wide application but it refers in the main to such appliances as ploughs, harrows, rollers and clod crushers. A strong, hard tough timber is required for this type of work. *Acacia nilotica (Babul)*, *Anogeissus latifolia* (axle wood), *Syzygium cumini (Jaman), Grewia tiliifolia (Dhaman), Shorea robusta (Sal), Xylia xylocarpa (Irul),* etc., are used for agricultural implements of different kinds.

Axe helves and tool handles. Woods for helves and tool handles must be strong and tough and must also possess great shock resisting abilities. *Dalbergia* spp. *(Shisham), Grewia* spp. *(Dhaman), Diospyros* spp. (ebony), etc., are recommended for this purpose.

Boat and ship-building. Timber used in all small boats and large ships is subjected to very great strains, and is often employed under circumstances which tax its durability to the utmost. For this reason any timber used for ship and boat-building should be strong, elastic, durable and free from defects. *Tectona grandis* (Teak—*Sagaun*) is the best ship-building timber in the world, due to its relatively small co-efficient of expansion and contraction and to its durability. The other recommended timbers for this purpose are—*Acacia nilotica (Babul), Dalbergia* spp., *Dipterocarpus* spp., *Shorea* spp., *Xylia* spp., *Grewia* spp., *Morus* spp., *Terminalia* spp., *Bombax* spp., etc.

Boot lasts and shoe heels. There is a considerable demand in India for boot and shoe lasts, and recently demand for shoe heels, and especially ladies shoe heels, has grown enormously. For boot lasts and shoe heels, a tough wood which is not too hard is required. In addition, the wood must be able to stand repeated nailing. The important woods used for the purpose are—*Dalbergia sissoo (Shisham), Gardenia* spp., *Mitragyna parviflora (Kaim), Mangifera indica (Am),* etc.

Cart and carriage building. The various parts of a cart or carriage are subjected to different kinds of stresses and strains, and require different qualities of wood for real efficiency. The important woods recommended for various parts of the carts are—*Dalbergia* spp., *Shorea* spp., *Dipterocarpus* spp., *Syzygium* spp., *Acacia* spp., *Grewia* spp., *Anogeissus* spp., *Terminalia* spp., and several others.

Construction and general joinery work. Constructional woods are those timbers used for superstructures, which include all parts of houses and buildings, bridges and similar structures not actually in contact with water or the earth. For this purpose a timber should be strong and durable. Lightness of wood is sometimes an asset if strength and durability are not sacrificed.

Floor and wall planking should be non-warping and non–shrinking and the wood for interior work and panels should be ornamental. There are three very important woods which stand out above all others as building timbers. These are—*Tectona grandis* (Teak—*Sagaun*), *Shorea robusta (Sal)* and *Cedrus deodara (Deodar)*. The other important timbers used for this purpose are—*Acacia* spp., *Toona ciliata (toon), Dalbergia* spp., *Mangifera indica (Am), Pinus* spp., *Terminalia* spp., *Xylia* spp.

Cooperage. Cooperage or barrel-making consists of two types, *"tight cooperage", i.e.,* barrels used for liquids, and "loose or slack cooperage" used for dry goods like cement. The oaks, *dhaman* are used for tight cooperage for beer and liquor casks. *Bombax ceiba (Simul),* mango spruce, etc., are used for slack cooperage.

Electric poles. The qualities necessary to make a good pole are that it should be straight, *i.e.,* without crooks and bends, that it should not split or crack excessively, and that it should have the required strength for the work it has to do. The woods used in India for use as electrical transmission poles are—*Pinus roxburghii (Chir), Shorea robusta (Sal), Tectona grandis (Teak), Cedrus deodara (Deodar), Terminalia tomentosa* (Laurel), etc.

Engraving and printing blocks. *Dalbergia sissoo (Shisham*) in North India, and *Dalbergia latifolia* (Rosewood) in South India, are the most popular woods for calico-printing blocks. Sissoo *(Shisham)* is considered to be excellent for the purpose. *Toona ciliata* (toon) and *Tectona grandis* (teak) are also sometimes used. *Tamarindus (Imli)* is also said to be good for printing by the Indian method.

Furniture, cabinet making and panelling. For high class furniture, cabinet-making and decorative panel work, there are several very ornamental and excellent woods in India. The chief characteristics required for these uses are non-liability to crack and split, retention of shape, ease of working and good colour figure and grain. The most commonly used and recommended timbers for the purpose are *Albizia* spp., *Toona ciliata* (toona), *Dalbergia* spp., *Juglans regia* (walnut), *Tectona grandis* (teak), *Terminalia tomentosa* (Laurel), etc.

Artocarpus spp., is wildly grown and commonly used for making furniture in South India especially in Kerala it is a valuable wood. It is also preferred in temple works in Kerala.

Match splints and boxes. A good match wood for splint manufacture must be soft, straight-grained, white and cheap. India possesses many woods which are suitable for box making and splints. The woods are—*Bombax ceiba (simul), Anthocephalus indicus* (kadam), *Salix* spp., etc.

Mathematical instruments. The better class of mathematical instruments such as set squares, rulers, etc., are usually made of *Buxus wallichiana* (box wood), *Juglans regia* (walnut) and *Aesculus indica* (horse chestnut), but cheaper instruments intended for school use are often made of *Adina cordifolia* (*Haldu*), *Toona ciliata* (Toon), *Gardenia* spp., etc. For set square and ruler edgings, good quality hardwoods such as rosewood sissoo and ebony are most frequently used.

Musical instruments. In India, many woods are used for making musical instruments as tanpura, sitars, violins, etc., *teak, toon, sissoo, mulberry, haldu* being among the commonest. For *sitars*, teak is used for the long neck and *deodar* or *shisham* for keys. *Toon* is almost universally used for the bodies. For *banjos*, teak is most commonly used. For drums, *mulberry, sissoo, siris* etc, are used. *Canarium euphyllum (Dhup)* is used for guitars.

Packing cases and boxes. Wood for these purposes must be light, easily worked and cheap. They must have good nail-holding power and should preferably be of whitish colour. Tea is now generally packed in plywood chests, which are strong, light and cheap. *Hollock (Terminalia myriocarpa)* and *hollong* are used for making good plywood tea chests. The commonest Indian woods for packing cases and boxes are—*Dalbergia* spp., toon, teak, mango, siris, kadam, *Terminalia chebula, Pinus* spp., ctc., Cigar boxes are used made in South India of *Toona ciliata, Melia* *azedarach*, etc. *Terminalia chebula* is a popular wood in South India for coffee boxes, while poplar (*Populus* spp.) is much used in Kashmir for fruit crates.

Pencils and pen-holders. Amongst the timbers used in India for pencil making may be mentioned Cypress, blue pine or *kail, simul, toon, Salix tetrasperma, Melia* spp., etc. Pen-holders may be made of *Haldu, Gardeni*a spp., *Kaim*, etc., Spruce and fir are used in North India for cheap pen-holders.

Picture framing. No very special qualities are required for picture framing, so long as the wood used is well seasoned and not liable to warp. There are two distinct types of picture framing, one in which the wood is used in its natural state, either polished or waxed, and the other in which the framing is painted or covered with a plaster composition which is moulded to the form required and covered with gilt or other colouring media. For the former type of framing and steady ornamental wood is suitable, the most commonly used being teak, sissoo, rosewood, ebony and *haldu*. For cheaper frames, in plain moulded form or painted, light conifer woods such as fir and spruce are the best. School slate frames can be made from a variety of woods, *e.g. Trewia nudiflora* (*gutel*), *Terameles nudiflora* (maina), *Terminalia chebula, Mangifera indica* (*Am*), spruce and fir, etc.

Railway carriage and wagon building. The qualities required of a wood to be in railway carriage and wagon work are that it should be sufficiently strong and durable. It should be free from bad seasoning defects, and it should be available in sufficient quantities. Teak complies with this specification better than any other timber, and for this reason teak is the main timber used by all the railway wagon and carriage works in India. The other important timbers used for this purpose are—*Shorea robusta, Pterocarpus* spp., *Adina cordifolia, Cedrus deodara, Dalbergia latifolia, Toona ciliata, Acacia nilotica, Dipterocarpus* spp., *Pinus* spp., and several others.

Railway sleepers. The woods most commonly used are—sal, deodar, teak, *pyinkado*, and *bijasal*.

Rifle parts. Walnut (*Juglans regia*) is the chief timber used for the manufacture of rifle work all over the world, as it stands up exceptionally well when worked to a fine finish on high speed cutting and drilling machines. The wood is also very steady and is not prone to excessive shrinking, swelling, warping or splitting, once it is properly seasoned.

Shuttles. The subject of shuttles is a very important one in India. All cotton, jute, wool and other textile mills use wooden shuttles in large quantities. The important wood which may be used for making shuttles are—*Diospyros melanoxylon* (ebony), *Gardenia latifola, Acacia nilotica, Mitragyna parviflora, Dalbergia latifolia, Dalbergia sissoo*, etc.

Sporting requisites. Billiard cue shafts in India are made of *Diospyros melanoxylon* (*tendu*), *Grewia tilaefolia* (*dhaman*), *Polyalthia fragrans*. For the butts of cues ebony, *Hardwickia pinnata* and *Dysoxylum glandulosum* are used.

More than 200 manufacturers in India are involved in the production and export of world-class sports goods. These goods are of such superior quality that over 70 countries import from India. Professional players use them in national, international and even Olympic tournaments. There are several good reasons for the tremendous popularity of Indian sports goods abroad: excellent quality, superb finish, carefully selected raw material, enhanced durability. And to top it all master craftsmanship from India, the land of traditional skills.

Cricket bats are made of *Salix* spp. (willlow). *Populus* spp. has also been used for cheap types of bat.

Indian cricket bats have scored many a century for the world champions. Made from strong Kashmir willows (*Salix* spp.), the bats are designed for force and flexibility.

Golf clubs are made of *Terminalia* spp., *Grewia* spp., *Anogeissus* spp., and *Pyinma*.

Hockey sticks are made of *Morus alba* (mulberry). Hockey sticks from India are known the world over for their balance and strong drive.

The timbers employed for the skis are *Dalbergia sissoo* (*shisham*) and *Anogeissus latifolia* (axle wood).

Stumps and bales are made of *Morus* spp. (mulberry) and *Grewia tiliaefolia* (dhaman).

In India, tennis and badminton rackets are made of mahogany and maple wood. The racket frames are made of *Prunus padus, Dalbergia sissoo, Toona ciliata. Melia azedarach*, etc.

Turnery, carving, combs, toys, etc. Very close-grained woods are acquired for high class turnery and carving. Walnut is much in demand in North India and Kashmir for this purpose. The sandalwood carvings of Mysore are equally well known. Sissoo and ebony are also used high class carving and inlay work in North India.

Dalbergia latifolia is commonly used in Kerala for carvings of idols, animals, birds, etc.

Wood carving is an ancient craft in which the West Bengal excels. The items include are dolls and toys, owl, horse, etc. Recently many piece goods including stands for table lamps are being manufactured. *Shisham* wood is used in abundance.

Erythrina spp., *Gyrocarpus* spp., *Bauhinia malabarica* are used for making toys in Karnataka. *Hymenodictyon excelsum* (*Kuthan*) is the best toy wood in U.P.

Indian combs such as those used by Sikhs in the hair are made of *Buxus* spp. (boxwood), *Adina cordifolia* (*Haldu*), ebony and sandal-wood.

VENEERS

The veneers are thin slices or sheets of wood of uniform thickness. The common thickness is about 1/20 in. to 3/8 in. In making veneers or logs pieces of woods are peeled and boiled and then cut with sharp knife. Usually the veneers are cut by **rotary process** which involves turning a log on a lathe against a stationary knife, with the formation of a continuous sheet of veneer. The other process is known as **slicing process**. Here the logs are quartersawed and therefore show a more attractive figure. The logs are first quartered and thereafter sliced with a stationary knife, resulting in the formation of separate sheets. Freshly cut veneers are usually wet and must be thoroughly dried.

Uses. The veneers are used in the manufacture of boxes, baskets, door panels, cooperage, trunks, mirrors, musical instruments and several other articles.

PLYWOOD

To overcome some of the natural defects in wood is the manufacture of so-called *"built up"* stock or plywood. The plywood is made by gluing together from three to nine thin veneers. The grain of each successive layer is at an angle to the next, so that the strength is redistributed. With the result the finished product becomes very strong and stable and does not warp or twist like ordinary wood. The plywood does not split when the nails and screws are driven close to its edge.

The manufacture of a five-ply plywood panel consists of preparing the face, back, cross band, and core sheets; applying the adhesive; pressing the glued stock into a panel; and drying and finishing the product. The soft woods are generally used for the purpose, as they can be glued more easily. Several resins and gums are used in the manufacture of plywood.

Uses. The plywood is extensively used in the home for doors, walls, flooring, cabinets, shelves, partitions, furniture and interior trim. A large quantity of plywood is utilized in making concrete forms, boats, prefabricated houses, airplanes, railroad cars and the body making of station wagons.

LAMINBOARDS

Laminboards are usually made up with cheap inferior woods for the cores, and better quality decorative good for the faces. Such timber as pine, spruce, fir, *Bombax ceiba, Ailanthus grandis, Tetrameles nudiflora* and *Kydia calycina* are very suitable for core work, while better class timbers such as teak, rosewood, sissoo and toon make excellent face veneers. Face veneers are usually sliced.

Uses. The laminboards are commonly used for making doors and cabinets.

CORK

Most commercial cork is obtained from the cork oak (*Quercus suber*). Cork or "corkwood" consists of the outer bark of the tree. This can be removed without injury to the tree. At the age of about 20 years, when the tree is about 40 cm in circumference, this outer layer, known as virgin cork, is removed by stripping. The operation consists of making vertical and horizontal cuts with knives or saws, and then prying off large pieces of the bark. The underlying exposed phelloderm and cortical cells die, and a new cork cambium layer is formed several millimetres deeper in the cortex. By this cork cambium (phellogen) new cork is formed more rapidly than by the first layer, and after 9 to 10 years the new cork layer has attained sufficient thickness to be commercially valuable and is in turn removed. This cork is of better quality than the nearly useless virgin cork, but is not so good as that obtained at the third and subsequent stripings which take place at intervals of about 9 or 10 years until the tree is 150 years old. Usually the trees live for 100 to 500 years, and have an average yield of 40 to 500 lb. per tree. After stripping the pieces of cork are dried for several days. They are first boiled in large copper vats. This process removes the sap and tannic acid, increases the volume and elasticity and flattens pieces. The rough edges are then trimmed off and the flat pieces are sorted and baled.

Cork is exceedingly buoyant due to its composition of dead watertight cells. It is highly compressible. It is durable, a bad conductor of heat and electricity, and is resistant to passage of moisture and liquids. It absorbs sound and vibrations and also possesses some frictional characteristics.

Uses of Cork

It is used for many purposes and for the manufacture of several products. In some cases the natural cork is used whereas in other cases composition cork, made of finely ground pieces treated with adhesives and molded.

The natural cork is used to make stoppers, hats, helmets, cigarette tips, handles of golf clubs, penholders, fishing rods, floats, life preservers, life jackets, base balls, mats, tiles and several other such products. Cork board is used as an insulating material for cold storage plants, houses and refrigerators.

Composition cork is utilized for the lining of crown caps, the metal tops for sealing bottles; gaskets; toes, counters and innersoles for shoes.

LINOLEUM

A type of floor covering is made from cork or wood flour, linseed oil, resins, such as rosin or kauri gum, pigments and burlap. The oil is boiled and allowed to solidify by dripping on pieces of cloth. The solidified oil is ground up and melted with the resins. This mixture is cooled and hardened, and after 'curing' for several days it is mixed with the cork, which has been ground as fine dust, and with the dry colour pigments. It is then pressed into burlap cloth with hydraulic presses. It is then seasoned in ovens and finished by giving it a protective surface of nitrocellulose lacquer. The tiles, known as linotiles are made from ground cork and linseed oil, but thicker than linoleum.

8

Rubber and its Products

Rubber is obtained from the *latex*, of various plants of tropical or sub-tropical regions. Most of the rubber plants belong to the Moraceae, Euphorbiaceae and Apocynaceae families. Though several species are available as sources of rubber, yet *Hevea brasiliensis* supersedes all. The cultivated *Hevea* trees, the so-called plantation rubber furnish about 98% of the supply. Rubber is the most recent of the major crops of the world. The industry is little more than a century old, and cultivation has been carried on about 80 years or so.

The latex occurs in latex tubes or latex vessels found in the various parts of the plants. Generally the commercial rubber is obtained from the lower part of the stem of tree. Latex is a gummy white liquid full of minute globules. It is mixture of water, hydrocarbons, resins, oils, proteins, acids, salts, sugars and caoutchouc. The physical and chemical properties of rubber have been considered separately in the following paragraph:

Fig. 8.1. Natural rubber, 'nature's most versatile vegetable product', is a unique industrial raw material found in the latex extracted from the tree *Hevea brasiliensis*. It has come to stay as an indispensable ingredient of civilized living.

PROPERTIES OF RUBBER

Physical properties. The physical properties of rubber vary with temperature. Natural rubber is soft and translucent at 20°C ; when chilled to 0–10°C, it becomes hard and opaque ; at the temperature of liquid air, it is brittle and transparent like glass. At temperatures exceeding 25°, rubber loses elasticity and becomes sticky. Rubber melts to a viscous fluid at about 200°C. Rubber is insoluble in water and is unaffected by alkalies or moderately strong acids. It is soluble in benzene, naphtha, carbon disulphide, ether, chloroform and chlorinated hydrocarbons. When rubber is dissolved in a solvent, it first swells to gel-like consistency and then forms a solution. The elastic property of rubber is closely related to its molecular structure. Rubber is amorphous in the unstretched state, but shows crystalline behaviour when kept at 0^0 for a long time or when stretched. Rubber is one of the best insulating and dielectric materials available.

Chemical properties. First grade latex plantation rubber contains : rubber hydrocarbon 92–94%, resin 3%, protein 2% and ash 0.2% in Hevea rubber. Rubber hydrocarbon, also known as caoutchouc, is colourless, odourless, transparent and elastic. It consists of long chains of isoprene units connected end to end through 1 : 4 linkages. Rubber forms compounds with halogens, halogen acids and metallic halides, ozone and oxides of nitrogen. Hydrogenation at elevated temperatures and pressures in the presence of a catalyst gives fully saturated hydrorubber. Rubber is oxidised by nitric acid, potassium permanganate and hydrogen peroxide.

Hevea Aubl. (Euphorbiaceae).

A small genus of trees distributed chiefly in the Amazon region of South America. One species *Hevea brasiliensis* ; the source of Para Rubber, has been introduced in India and cultivated as a plantation crop.

Hevea brasiliensis (H.B.K.) Muell. Arg. (Eng. Para rubber ; Verna. *Rabar*). It is a large tree attaining a height of 60–100 ft. or more and a girth of 8–12 ft. The stem is smooth, straight, generally unbranched up to a considerable height with a much branched leafy canopy. It is considered indigenous to the Amazon valley of Brazil, Venezuela, Peru, Ecuador and Colombia.

The rubber plantation industry in South-East Asia dates back to 1876 ; when 1,900 seedlings raised at Kew, were despatched to Ceylon. Plantations were raised and by 1930 a number of strains had been developed by breeding and selection, some of them superior to the stock originally obtained from Kew.

In India, rubber plantations were first started in Kerala state in 1905. Since then many other areas have been brought under rubber cultivation. At present, Kerala is the most important rubber producing state in India with plantations concentrated particularly in Kottayam and Quilon districts. Kanyakumari district in Tamil Nadu state is also an important rubber-growing area; there are small areas in Coimbatore and Nilgiris in Tamil Nadu state, Coorg in Karnataka state and in Andaman Islands. Rubber is produced in India, in plantations ranging from less than one acre to about 3,000 acres in area. Areas above 50 acres are classified as large estates and those of 50 acres and below, as Holdings.

Hevea brasiliensis thrives in the tropical belt of the equator, in which the climate is warm, humid and equable and the temperature ranges from 74°F to 95°F. It requires a well distributed rainfall of 70 to 100 inches per annum. In South India, the plantations are located mostly in areas with a prolonged dry season and severe South West monsoon. The

Fig. 8.2. The rubber plantations in India registered phenomenal growth since independence. Production increased ten-fold in a span of 30 years and area enhanced four-fold while average yield shot up 2½ times. As a result from an importer of 35,000 tonnes of natural rubber in 1963, India gradually attained self-sufficiency and emerged a net exporter since 1973. Natural rubber, has spread now to over 230,000 hectares producing annually over 150,000 tonnes valued at Rs. 1200 million. Over a million people depend on this industry for their livelihood.

temperature ranges from 60^{0}F to 95^{0}F and there is a large variation between winter and summer temperatures.

Hevea brasiliensis thrives best in deep, well drained loamy soils. The soils in the rubber plantations of South India are red lateritic or clayey loams. The soils are generally deep and well drained. They are fairly rich in nitrogen, but poor in mineral constituents.

Rubber is a long duration tree crop which stays on the ground for several years. The economic life of rubber tree is 30–35 years. Newly opened jungle is considered to be the most suitable for rubber cultivation. The area to be planted is cleared of the jungle by cutting down the undergrowth and felling the trees. The cleared area is then lined and marked and terraces prepared. Provision is made for roads and drainage.

Tapping. Para rubber (*Hevea brasiliensis*) is valued for the latex obtained by tapping, *i.e.*, opening up the latex vessels situated in the bark with a sharp incision on the main trunk. When the vessels are cut, the latex flows out, quickly at first, then slowly, and finally coagulating on the cut surface. The tree is rested after each cut for varying periods according to age, climate and condition of the tree. When the tree is tapped again, the coagulated latex of 'plug' from the old cut surface is removed and a thin strip of bark is cut off. During the first tapping, only a small amount of viscous latex exudes. The flow increases with each successive tapping, the cuts being opened always at an angle to the horizontal. Usually, the cut extends half way around the trunk, but quite often, may completely encircle it.

Yield. The average yield of latex per acre in India from unselected ordinary seedling trees is 300 lb. The yield from budgrafted trees of approved clones and from clonal seedlings of approved parentage is twice, or more. An average yield of 700–800 lb. per acre has been obtained with improved planting materials in India.

Latex. Fresh latex is usually milky white in colour. The latex drawn from a tree which has been rested for sometime is yellow. The latex obtained by tapping is essentially a colloidal suspension of rubber particles in an aqueous serum.

Fresh latex is readily susceptible to enzymic and bacterial action leading to acid formation, and preservatives and bactericides are usually added to latex at the time of collection to inhibit acid development and consequent coagulation.

Raw rubber. Rubber of commerce is obtained from latex of *Hevea brasiliensis* by coagulation; rolling and drying. Two forms of raw rubber are known in trade, namely, *Smoked Sheet Rubber* and *Crepe Rubber*.

Grading. The system adopted in our country for grading raw rubber is based on the type description of the Rubber Manufacturers' Association and endorsed by the Rubber Trade Association of New York. These types are recognised as universal standards by producing and consuming countries alike.

Packing. The raw rubber is packed in bales in accordance with R.M.A. packing specifications.

Marketing. The raw rubber is marketed in the form of dry rubber. About two-thirds of total production is marketed by managing agents after grading. Petty merchants and dealers sell the collection to the nearest depot of a purchasing firm or take them to dealers in Kottayam, Trivandrum, Trichur or other market centres.

Uses. The raw rubber as such has few important applications. Its chief uses are in making insulation tapes, shoe soles, adhesives and erasers.

Rubber is used for the production of a wide variety of products utilized in industries and services, and for domestic purposes. It has been estimated that about 50,000 different products are fabricated from rubber directly or indirectly. Tyres and tubes for automobiles and cycles account for about 75% of the total rubber consumption. About 6% is used for footwear ; about 4% for wire and cable insulations. Among miscellaneous manufactured articles mention may be made of rubberised fabrics, motor mountings for absorbing vibrations and shocks, washers and gaskets, transmission and conveyor belting, garden hose, fire hose and hose for gasoline, pneumatic tools, paints, sports goods, household and hospital supplies, such as sheets, hot water bags, ice bags, surgeon's gloves, bathing caps and prophylactics, toys, such as balls, dolls and balloons, erasers and rubber bands and adhesives. Metals are coated with rubber to protect them from wear and corrosion. Sponge rubber from foamed latex finds uses in cushioning, seating and bedding. Elastic fabrics are made from latex threads. It is also employed in the fabrication of battery boxes, fountain pen barrels, tobacco pipe stems, telephones and combs.

Natural rubber latex, in preserved and concentrated forms, finds many commercial applications. Compounded latex is largely replacing rubber solutions for the manufacture of such articles as balloons, gloves, contraceptive appliances, finger stalls and teats. Over 65% of the latex is used in the manufacture of foam rubber and the rest for dipped goods, fabric coatings, impregnation and moulded goods. Uncured latex is used as an adhesive in footwear manufacture. The bulk of latex goods is made from vulcanised latex.

Reclaimed rubber prepared from discarded rubber products and rubber scrap, is used alone or in admixture with raw rubber in the manufacture of heels, soles, mechanical rubber goods, tyres and adhesives for use in fabric dipping, carpet backing, flooring, etc., and hard rubber used for battery containers, tubes and rods.

OTHER RUBBERS

Besides *Hevea*, several plants have been tried from time to time as sources of rubber as shown in the table :

Rubber plants percentage	*Rubber hydro-carbon*	*Resin*	*Protein*	*Ash*
Hevea brasiliensis	94	3	2	0.2
Castilla elastica	93	6	0.5	0.1
Landolphia thollonii	93	5	1	0.5
L. owariensis	93	6	0.1	1.0
L. kirkii	91	8	0.1	1.0
Taraxacum kok-saghyz	90		2	3.0
Funtumia elastica	88	8	3	0.5
Manihot glaziovii	88	4	6	1.5
Cryptostegia grandiflora	88	9	2	0.3
Ficus elastica	78	20	1	0.5
Parthenium argentatum	72	22	3	2.5

Of these the Russian Dandelion (*Taraxacum kok-saghyz* Rodin) and the Mexican Guayule *Parthenium argentatum*. A. Gray) may specially be mentioned. The Russian Dandelion is grown on C. 2,000,000 acres of land in Russia and rubber of good quality is reported to be obtained from the roots of the plant. The Mexican Guayule is extracted from the whole plant; it is of inferior quality. Other rubber producing plants of local importance are — *Manihot glaziovii* Muell. Arg. (Ceara rubber), *Castilla elastica* Cerv. (Castilla or Panama rubber), *Hancornia speciosa* Gomez (Mangabeira rubber), *Ficus elastica* Roxb. (India or Assam rubber) and *Funtumia elastica* Stapf. (Lagos silk rubber).

Ficus elastica Roxb.; Eng. India rubber, Assam rubber; Hindi—*Bor*; A tree. It is native of India and Malaya. In India, it is found in Assam and the Khasia hills. The latex obtained from the bark of the branches, is used for making rubber; Family—Moraceae.

Panama Rubber

Castilla elastica; Eng. Panama or Castilla rubber, Family—Euphorbiaceae.

A tall native tree of Mexico and Central America. *C. elastica* grows in deep loamy soil on high ground and may reach a height of 150 feet. The trees are tapped when eight to ten years of age, adult trees yielding upto 50 lb. of latex. The latex is coagulated with plant juices, alum, and by boiling or exposure to the air. The crude rubber is shipped in flat cakes.

Ceara Rubber

Manihot glaziovii; Eng. Ceara or Manicoba rubber; Family—Euphorbiaceae. A small tree, native of Brazil. It grows well in dry rocky ground. Ceara rubber is now grown in India, Sri Lanka and many other tropical countries. The trees are tapped when four or five years of age and yield a good grade of rubber. The latex is coagulated by exposure to air or smoke. The crude rubber is shipped in flat cakes.

9

Fatty Oils and Vegetable Fats

The fatty oils are liquid at ordinary temperatures and usually contain oleic acid. They are sometimes called 'fixed oils' because, unlike the essential oils they do not evaporate. Chemically, the fatty oils consist of glycerine in combination with a fatty acid.

However, the fats are solid at ordinary temperatures and contain stearic or palmitic acid. When a fat is boiled with an alkali, it decomposes and the fatty acid unites with alkali to form soap.

Usually the fatty oils are stored up in seeds of the plants belonging to many families. Sometimes, to a less extent they are also stored in fruits, stems and other plant organs. Several fatty oils are edible and used as cooking media. The edible oils contain both solid and liquid fats.

Extraction of fatty oils. The seed coats are removed and the material is being converted into a fine meal. The oils may be removed by solvents or by hydraulic pressure. The hydraulic pressure method is commonly used for extracting the edible oils. The residue or oil cake is rich in protein and makes a good cattle feed. The pressure causes the cell walls to break and the fatty oils escape. These oils are filtered and purified. The refined oil is used for food purpose, and the lower grades of oil are used in industries.

Classification of fatty oils. There are three categories of fatty oils–(1) drying and semi-drying oils, (2) nondrying oils, and (3) vegetable fats. The refined grades of drying and semi-drying oils are used as edible oils, whereas the inferior grades are used in making soaps, varnishes, paints, candles and other such articles. The nondrying oils are edible and can be used for soap and lubricants. The fats are solid or semisolid at ordinary temperatures. They are edible and are also used in the soap and candle industries.

1. DRYING AND SEMI-DRYING OILS

Linseed Oil

Linum usitatissimum Linn.; Eng. Linseed; Verna. *Alsi*; Family—Linaceae.

Linseed produced in India is utilized mainly for the expression of oil. Both bullock-driven *ghanis* and power-driven rotary mills, expellers and hydraulic presses are used for this purpose. *Ghani* oil is reported to possess a sweet taste with pleasant flavour and is preferred for edible purposes. The bulk of commercial linseed oil is produced by the use of expellers. Before crushing, the seed is rolled into meal and heated after moistening in a steam-jacketed trough fitted over the expeller.

Characteristics. Cold-pressed oil has a golden yellow colour, while oil obtained by hot-pressing is somewhat turbid and yellowish brown. Fresh refined oil is pale yellow with a odour and pleasant taste. The oil keeps well if stored in non-metallic containers unexposed to air and light. Commercial oil is usually dark in colour with a disagreeable odour and an acrid taste.

Uses. Linseed oil is primarily used in the manufacture of paints and varnishes. It is by far the most important drying oil. It is used in the manufacture of linoleum and oil-cloth, printing and lithographic inks and soft soap. It is used as a solvent for industrial stains and for seasoning bobbins (in jute textiles) and cricket bats and other sports goods. Raw cold pressed oil is used for edible purposes. In India mainly this oil is consumed in Madhya Pradesh and in parts of Orissa, Bihar, Uttar Pradesh, Punjab, Jammu and Kashmir. In some localities it is employed as an adulterant for mustard, groundnut and coconut oils.

Linseed oil is official in some pharmacopoeias and is recommended mainly for external applications. *Lotio Calcii Hydroxidi Oleosa*, a mixture of equal volumes of linseed oil and calcium hydroxide solution, is a popular application for burns. In veterinary practice, linseed oil is employed as a laxative for horses and cattle.

Niger-Seed Oil

Guizotia abyssinica (L.f.) Cass.; Eng. Nigerseed; Hindi—*Kalatil, Ramtil*; Bengali—*Ramtil, sirguja*; Marathi—*Khurasni, karale*; Gujarati—-*Kalatel, ramtal*; Telugu—*Verrinuvvulu*; Tamil—-*Payellu, uchellu*; Kannada—*Gurella*; Family—Compositae (Asteraceae).

It is a herb. It is native of tropical Africa. In India it is cultivated in Madhya Pradesh, Andhra Pradesh and Orissa. The oil is obtained from the seeds.

Uses. The oil is edible and used in cooking. It is also used as an illuminant and in paint, varnish and soap industries. The oil cake is used as fodder.

Hemp Oil

Cannabis sativa Linn; Eng. Hemp; Hindi—*Ganja, Bhang*; Sanskrit—*Bhang, vijaya*; Hindi, Bengali and Gujarati—*Bhang, ganja, charas, siddhi, jia;* Telugu—*Ganjai*; Family—Cannabinaceae.

It is an undershrub. It is a native of Asia. In India it is cultivated in Uttar Pradesh, Bengal, Maharashtra, Madhya Pradesh, Tamil Nadu and Orissa. The drying oil is obtained from the seeds.

Uses. The oil is used in the soap and paint and varnish industries. It is also used as a lamp oil.

Tung Oil

Aleurites fordii Hemsl; Eng. China wood-oil, Tung—oil; Family—Euphorbiaceae.

A tree. It is native of China. In India it is found in Assam, Bengal, Bihar and Karnataka. The oil is obtained from the seeds. The trees are handsome and are often planted for ornament.

Uses. The oil is widely used in the varnish industry as a substitute for linseed oil. The oils are used for waterproofing wood, paper and fabrics. It is a good preservative and very resistant to weathering, so it is particularly valuable for outside paints.

Kekuna Oil

Aleurites moluccana (Linn). Willd.; Eng. Candle nut, Varnish tree; Hindi—*Jangliakhrot, kekuna*; Family—Euphorbiaceae.

A tree. It is native of Malaya and the Pacific Islands. It grows wild in South India and Assam. The oil is obtained from the seeds.

Uses. The oil is used in the manufacture of paints and varnishes. The bark is used as a tan.

Soyabean Oil

Glycine max. (L). Merr.; Eng. Soyabean; Hindi—*Bhat, bhatwar, bhetmas, ramkurthi*; Bengali—*Garjkalai*; Assam—*Patnijokra*; Khasi Hills—*Rymbai-Kutung*; Family—Papilionaceae.

Fig. 9.1. Soyabean--Seeds, oil and oil cake. The seeds are used as pulse.

The oil, obtained from the seeds, is used in cooking, and is also used in the manufacture of candles, varnishes soap, paints and insecticides. The oil cake is used as fodder.

A herb. It is native of South-East Asia. In India, it is cultivated mainly in Punjab, Bengal, the Khasia hills, Assam, Manipur, Himachal Pradesh, Kashmir and Bihar.

Uses. The oil obtained from the seeds, is used in cooking. It is also used in the manufacture of candles, varnishes, soap, paints, greases linoleum, rubber substitutes, cleaning compounds, insecticides and disinfectants. It is also used in making adhesives, plastics, spreaders, foaming solutions and many other products. Thc oil cake has a good percentage of protein content and is used as cattle feed.

Fig. 9.2. Healthy sunflower crop.

Fig. 9.3. Sunflower (head inflorescence) ready to harvest. Sunflower can yield 15 quintals per hectare in rabi.

Sunflower Oil

Helianthus annuus Linn; Eng. Sunflower; Hindi—*Surajmukhi*; Sanskrit— *Surya-mukhi*; Hindi, Bengali and Gujarati—*Surajmukhi*; Marathi—*Surajamakha*, *suryaphul*; Telugu—*Aditya-bhaktichettu*; Tamil, Kannada and Malayalam—*Suryakanti*; Family—Compositae (Asteraceae).

Sunflower (*Helianthus annuus* L.) is an important oil-yielding crop in temperate countries such as Russia, Bulgaria, Romania, Canada and the U.S.A. where it is raised as a spring crop from April–May to September–October. Sunflower is a rich source of oil. The oil content ranges between 46 and 52 per cent. The oil is of high quality and edible, and has non-cholesterol and anti-cholesterol properties. The sunflower was first domesticated in the central part of the USA.

In India research on sunflower crop (*Helianthus annuus* L) was started during 1972-73. Within ten years high-yielding, open-pollinated varieties like 'Surya' and early-duration variety like 'Col' were developed and released for commercial cultivation in Maharashtra and Tamil Nadu respectively. A sunflower hybrid 'BSH 1' has been released for general cultivation in Karnataka.

In India the crop was cultivated in 0.75 to 0.8 million hectares of land during 1983-84. It was grown extensively in Karnataka, Maharashtra, Tamil Nadu and Andhra Pradesh. The crop has become quite popular among the farmers. It is also grown as a mixed crop with groundnut, *ragi* and *urd*. The oil expellers used for groundnut-oil extraction are being used to extract sunflower oil. Owing to non-cholesterol and anti-cholesterol properties of the oil, its demand is on the increase in market.

Sunflower is an important oilseed crop of today. It can give larger quantity of top quality oil per unit area and per unit time.

Sunflower seeds contain 40 to 50 per cent oil. A crop raised under irrigation yields from 10 to 12 quintals of seed per hectare and 4-6 quintals under rainfall conditions. It can be harvested in about 100 days.

Uses. Sunflower oil is a very good cooking medium. It can be used in place of groundnut oil in the preparation of *Vanaspati*. Nutritionally, it is better than many other cooking media. It can be safely used by heart patients. Sunflower oil does not raise the cholesterol level in the blood. High cholesterol is bad for the heart. Sunflower oil contains proteins, vitamins A, D and E. The oil is easily digested. Being of semidrying and stable type, sunflower oil is used in making paint, varnish and soap.

Fig. 9.4. *Helianthus annuus* (*Sunflower-Surajmukhi*). The seeds are used mainly for the extraction of oil. They are also consumed raw, roasted or salted. In U.S.A., roasted kernels are sold in packets like peanuts; only selected large seeds are husked for this purpose. Shelled seeds are ground and the resulting flour used for bread. Sunflower seeds make a nutritious food for cattle, poultry, hogs and cage birds.

Perilla Oil

Perilla frutescens (Linn). Britt, Eng. Perilla; Hindi—*Bhanjira*, *Bhasinda;* Bengali—*Ban-tulsi*; Assam—*Arim*, *angami*, *kenia*; Kumaon—*Jhutela*; Family— Labiatae (Lamiaceae).

A herb, 3-5 feet high with numerous branches. The oil is obtained from the seeds. It is native of India, China and Japan. It is found in Kashmir, Bhutan and Khasia hills.

Uses. The oil is edible and has been used for food proposes from earliest time. It is also used for making varnishes, oil papers, paper umbrellas, waterproof clothes, artificial leather and printer's ink.

Walnut Oil

Juglans regia Linn.; English Walnut; Hindi—*Akhrot*; Family—Juglandaceae.

A large tree. It is native of Iran. In India it is cultivated in Kashmir, Himachal Pradesh, the Khasia hills and in the hills of Uttar Pradesh. The mature and old kernels yield a drying oil.

Uses. The fresh oil and cold-pressed oil have a pleasant smell and nutty flavour and are edible. Hot-pressed oil is used for white paint, artist's oil paints, printing ink and soap.

Safflower Oil

Carthamus tinctorius Linn.; Eng. Safflower; Hindi—*Kusum*; Sanskrit—*Kusambha*; Bengali—*Kusum*, *Kusumphul*; Gujarati—*Kusumbo*; Marathi—*Kardai, kurdi*; Tamil—*Sendurakam*; Telugu—*Kushumba*; Kannada—*Kusumbe, kusume*; Family—Compositae (Asteraceae).

A shrub. It is native of India. It is cultivated chiefly in Eastern Uttar Pradesh, Madhya Pradesh. Andhra Pradesh, Karnataka and Maharashtra. The oil is extracted from the seeds.

Uses. It is used as an edible oil. Edible products like groundnut oil have been produced from the safflower seed oil as a result of refining, bleaching and hydrogenation studies. The oil is also used for soap, paints, varnishes and as an illuminant.

The oil is also used for linoleum and other similar products; also used for edible purposes. It is applied to sores and rheumatic swellings.

Poppy Oil

Papaver somniferum Linn.; Eng. Opium poppy; Hindi—*Post, afim*; Sanskrit—*Ahifen, chosa, khasa*; Bengali—*Pasto*; Marathi—*Aphu, posta*; Gujarati—*Aphina, khuskhus, posta*; Telugu—*Abhini, kasakasa*; Tamil—*Abini, kasakasa, postaka*; Kannada—*Afim, khasakhasi*, Malayalam—*Afium, kashakhasa*; Family—Papaveraceae.

A herb. It is native of West Asia. In India, it is grown in Uttar Pradesh, Jullundur and Hoshiarpur in the East Punjab, Rajasthan and Madhya Pradesh. The plant is famous for its drug opium. The oil is obtained from the seeds.

Uses. The first cold pressing yields a white edible oil, while a second hot pressing yields a reddish oil which is used for burning, soap and , after bleaching, for oil paints.

Fig. 9.5. *Papaver somniferum* (Opium). Flower and fruit. Poppy seeds contain up to 50% of an edible oil. The oil is odourless and possesses a pleasant almond-like taste. In India, the oil is generally extracted by cold-pressing the seed in small presses in homes. The oil is widely used for culinary purposes. It is free from narcotic properties and is used mixed with olive oil, or as a salad oil. Also used for making soaps, emulsions, ointments and paints.

Cottonseed Oil

Gossypium Linn.; Eng. Cotton; Hindi—*Kapas*; Family—Malvaceae.

This is one of the most important semidrying oils. The seeds are carefully cleaned and freed from impurities and the linters and are subjected to hydraulic pressure or expeller presses. The oil is pumped into tanks where the impurities settle out.

Uses. The pure refined oil is used as a salad and cooking oil and for making *Vanaspati ghee*. The residue is the source of various products that have a wide range of industrial uses. It is used for soap, washing powders, oil cloth, artificial leather, insulating materials, roofing tar, putty, glycerin and nitroglycerin. The cotton seed oil cake is used as cattle feed and fertilizer.

Corn Oil

Zea mays Linn.; Eng. Indian corn, maize; Hindi—*Makai*; Family—Gramineae (Poaceae)

This is a tall, cultivated grass. It is native of South America. In India it is grown as a food crop mainly in Uttar Pradesh, the Punjab, Madhya Pradesh, Bihar, Andhra Pradesh, Jammu and Kashmir.

The embryo of the grain contains about 50 per cent of semidrying oil.

Uses. The corn oil can be used for nearly all the purposes to which any oil is put. It is used as an edible oil. The refined corn oil can be used for cooking, in bakeries and for mixing with other oils. The crude oil is used for the manufacture of rubber substitutes; soaps and cheap paints. The oil cake is used as cattle feed.

Sarson Oil

Brassica campestris Linn.; var. ***sarson*** Prain; Eng. Yellow sarson, Indian colza; Hindi—*Sarson*; Family—Cruciferae (Brassicaceae).

A herb. It is grown as an oil-seed crop mainly in Uttar Pradesh, the Punjab, Bihar and Assam. The oil content is 30 to 45 percent, and the oil is extracted by expression or solvents.

Uses. The oil is used for cooking and burning purposes. The oil-cake is used as cattle feed.

Fig. 9.6. Rapeseed mustard (Indian rape). Crop at maturity.

Indian Rape Oil *Brassica campestris* Linn.; var. *toria* Duth; and Full.; Eng. Indian rape; Hindi—*Toria*; Family—Cruciferae (Brassicaceae).

A herb, It is grown as an oil-seed crop mainly in Uttar Pradesh, West Bengal and the Punjab. The oil extracted by expression or solvents.

Uses. The oil is edible and the oil cake is used as cattle feed and manure.

Sesame Oil (Gingelly Oil)

Sesamum indicum Linn.; Eng. Sesame; Gingelly; Hindi—*Til*; Sanskrit—*Tila*; Hindi and Marathi—*Til*; Gujarati—*Tal*; Telugu—*Nuvvulu*; Tamil and Kannada—*Ellu*; Malayalam—*Karuthellu*; Oriya—*Khasa*, *rasi.*, Punjab—*Til*, *tili*; Family—Pedaliaceae.

It is a herb. It is grown mainly in Uttar Pradesh, Madhya Pradesh, Rajasthan, Andhra Pradesh, Tamil Nadu and Maharashtra. The seeds contain about 50% oil, which is easily extracted by cold pressure.

Uses. The oil is edible and used for cooking purposes. It is also used in the manufacture of soap, cosmetics and hair oils. The oil cake is used in fodder. The seeds are used for making sweetmeats. The oil is also used for anointing the body and as fuel for lamps.

This fatty oil obtained from the seeds is called Sesame oil, Gingelly oil or Til oil. The oil is also used as an ingredient of confectionery and in the manufacture of perfumes, insecticides and pharmaceutical products. The oil cake may also be used as a source material for proteins used in glues and sizes. The utilization of sesame cake, from dehulled seeds, as a source of protein for human consumption is an important development.

Fig. 9.7. *Sesamum*. A high yielding variety.

Laurelwood Oil

Calophyllum insophyllum Linn.; Eng. Indian Laurel, Laurelwood; Hindi—*Surpan*, *Sultun champa*; Sanskrit—*Nagachampa*, *Punnaga*; Bengali and Hindi—*Sultanchampa*; Marathi—*Undi, surangi*; Telugu—*Pouna*; Tamil—*Punnai*, *Pinnay*; Kannada—*Vuma*, *honne*; Malayalam—*Punna*; Oriya—*Poonang*; Trade—Poon; Family—Guttiferae.

A tree. It is found in South India and the Andaman Islands. The oil is obtained from seeds.

Uses. The oil is used as in illuminant and for soap making. It is also used in rheumatism. The oil is known as Domba, Laurel nut, Dillo, Pinnay, and Poonseed oil.

Margosa Oil

Azadirachta indica A. Juss.; Eng.; Margosa tree, Neem tree; Hindi—*Nim*; Sanskrit—*Nimba*; Bengali—*Nim*; Marathi and Gujarati—*Limba*; Telugu, Tamil and Malayalam—*Vepa*; Kannada—*Bevu*; Family—Meliaceae.

A tree. It is native of Myanmar but grown all over India. The oil is extracted from the seeds.

Uses. The oil is used as an antiseptic and for burning purposes. The oil cake is used as a fertilizer and manure.

Seeds yield a non-drying oil used for skin-affections. Neem oil may be mixed with other oils and fats for the manufacture of washing soap; medicated soaps with the odour of Nim Oil are available. *Nimbidin* is the chief bitter principle of the oil. Neem toddy is occasionally obtained as an exudation from the upper part of some trees; used as a tonic. Fresh tender twigs of nim are used to clean teeth particularly in pyorrhoea.

2. NON-DRYING OIL

Castor Oil

Ricinus communis Linn.; Eng. Castorbean, Castor oil plant; Hindi—*Arand*; Marathi—*Erandi*; Bengali—*Bheranda*; Gujarati-*Diveligo*; Tamil-*Amanakku*, *Kottai muthu*, Telugu—*Amudamuchettu*; Kannada—*Haralu*; Malayalam—*Avanakku*; Family—Euphorbiaceae.

A large shrub or small tree. It is cultivated chiefly in Andhra Pradesh, Maharashtra, Mysore and Orissa. The castor-oil is obtained from the seeds. The seeds contain 35 to 55 per cent of a thick colourless or greenish oil which is obtained by expression or solvent extraction.

About 60% of the total area of one million hectares under this crop in India occurs in the Telengana region of Andhra Pradesh. The traditional varieties used to be sown with the onset of rains in June and harvested fully by the next March. This long duration of nearly 270 days became an important factor in linking the destiny of the crop with the behaviour of monsoon. Hence a variety named *Aruna* was developed from an earlier strain HC. 6_1 through irradiation with thermal neutrons. This variety matures in 120 to 140 days and hence became popular with farmers immediately. *Aruna* castor not only helped to stabilise the yield of castor but also facilitated double cropping in single crop lands in years characterised by good winter rainfall. M.S. Swaminathan, 1978.

Uses. The castor oil is used as a purgative. It is an excellent lubricant especially for airplane engines. It is also used for transparent soap, textile soap, typewriter inks, perfume aromatics, varnishes and paints. It is water resistant and so is used for coating fabrics and for protective coverings for airplanes, insulation, food containers, guns, etc. It is used for preserving leather and as an illuminant. The oil cake is used as a fertilizer. The writing and printing papers are made of the wood-pulp. Castor stems are used for strawboards and cheap wrappings.

Seeds yield castor oil, a fatty oil used as a cathartic. Purgative action is due to local irritation of the intestines caused by the ricinoleic acid formed by hydrolysis under the influence of the lipolytic enzymes.

Oil in crude form is put to technical use, much is converted into sulphonated castor oil (Turkey Red Oil), used for dyeing cotton fabrics with alizarine; used also for making textile and transparent soaps, typewriter inks, fly-paper, imitation leathers, and in the manufacture of nitrocellulose-baking finishes, and synthetic nylon.

After dehydration used as a drying oil in paint and varnish industry.

Hydrogenated castor oil may be used as a substitute of carnauba wax (*Copernicia cerifera* Mart.) useful in polishes, candles, insulating materials, plastic moulding compositions, and heat-sealing adhesives, and for the preservation of fruits and vegetables.

Castor cake is almost entirely used as manure in India.

Peanut Oil

Arachis hypogaea Linn.; Eng. Peanut, Groundnut; Hindi—*Mungphali*; Family—Papilionaceae.

The chief producers of peanuts are China, India, the United States and Africa. In India it is grown chiefly in South India and Maharashtra. Saurashtra in Gujarat is famous for peanut cultivation and oil market. It is native of Brazil.

The seeds are shelled, cleaned and crushed, and the oil is expressed by both hydraulic presses and expellers.

Uses. The filtered refined oil is used for cooking and in making margarine. Inferior grades are used for soap making, lubricants and illuminants. The oil-cake is used as cattle feed.

Peanut oil, obtained from the kernels, is predominantly used for culinary purposes; considerable proportion of hydrogenated fats, used as Vanaspati, consists of groundnut oil.

Some new commercial products are groundnut milk, peanut ice-cream, and peanut massage oil for infantile paralysis. Peanut oil also finds some use as a lubricant, and blends with mineral oil have been developed.

Hydrogenation of Oils

Most of the housewives prefer solid fats to liquid oils for cooking purposes. Accordingly millions of kilograms of groundnut oil or cotton seed oil are changed to solid edible fat (*vegetable ghee*) each year by hydrogenation in presence of a suitable catalyst. Hydrogenation converts only a part of the glycerides of unsaturated acids into those of saturated acids. Different steps involved in the actual manufacture of vegetable ghee are as follows:

Removal of free acids. The oil is warmed in a pan and treated with a calculated quantity of sodium hydroxide to neutralize the free fatty acids. The salts formed as a result of neutralization come up in the form of a scum along with some of the suspended matter.

Bleaching. The oil from the first tank is decanted into the second and treated with animal charcoal at 90°C or so. The colouring matter is absorbed by animal charcoal and the oil is filtered.

Deodorising. The bleached oil is treated with superheated steam for deodorising.

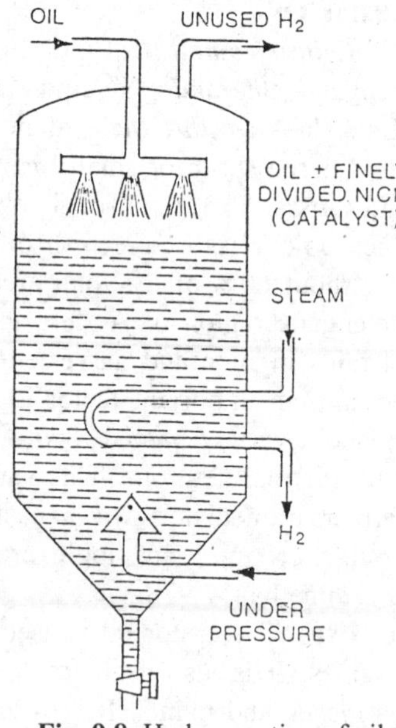

Fig. 9.8. Hydrogenation of oils

Hydrogenation or hardening of oil. The oil purified above is taken in an iron tank surrounded by a heating jacket at 150-200°C. Some finely divided nickel is suspended in the oil and hydrogen gas is passed in under pressure. Finely divided nickel acts as a catalyst in hydrogenation.

The hydrogenation is continued until a fat of the desired consistency is obtained. The hardened oil is taken out and freed from the catalyst by filtration. Some flavouring material somewhat resembling genuine ghee is then added to this before placing it in the market.

Olive Oil

Olea europaea Linn.; Eng. Olive; Hindi—*Zaitun*; Famil—Oleaceae.

A tree. It is native of the Mediterranean region. It is grown in Northern India. The oil is squeezed from the pulp either by hand or mechanically. The finest grades are obtained by the hand method. These oils are golden yellow, clear and odourless.

Uses. It is used chiefly as salad or cooking oil. It is also used in medicine. The inferior grades with greenish tinge are used for soap making and as lubricants. The oil cake is used as cattle feed.

3. VEGETABLE FATS

Coconut Oil

Cocos nucifera Linn.; Eng. Coconut; Hindi—*Nariyal*; Family—Palmaceae/Arecaceae.

A tall palm. It is cultivated chiefly in Kerala, Tamil Nadu and Karanataka. The oil is obtained from the dried meat of coconut. This oil is pale yellow or colourless and is solid below 74°F. The husks of the nuts are removed and the nuts split open and dried. The copra is then easily removed. This is ground up and the oil is expressed. The oil cake is sometimes put through hydraulic presses a second time, and still more oil is removed. The yield is about 65 to 70 per cent.

Uses. Refined coconut oil is edible and is extensively used for food products. Unrefined coconut oil is commonly used for cooking in Kerala. This oil is used for making confectionery and candy bars. Coconut oil has long been used for the best soaps, cosmetics, shaving creams, shampoos and other toilet preparations. It is the only oil used in marine soaps. It is also used as an illuminant. The oil cake is used as fodder and as manure.

Palm Oil

Elaeis guineensis Jacq; Eng. Oilpalm; Family—Palmaceae/Arecaceae.

A tree. It is native of West Africa. In India it is grown in Travancore. The oil is obtained from the nuts. The oilpalm is a very productive tree. It begins to bear at the age of 5 to 6 years, reaches full bearing at 15, and continues until 60 or 70 years of age. Each tree bears 10 bunches of 200 nuts a year. The fibrous pulp of these fruits contains 30 to 70 per cent of fat. The oil is obtained by crude native methods. The oil is yellow-orange or brownish in colour.

Uses. The oil is used in making soap, margarine and as a fuel for diesel motor.

Source of palm oil, obtained from the fleshy pericarp, and palm kernel oil from the seeds; the former is edible and rich source of vitamin A and the latter mostly used for the perparation of margarine, soap and hair oils.

Mahua Oil

Madhuca indica J.F. Gmel. syn. *M. latifolia* Macb.; Hindi—*Mahua, mohwa, mauwa*; Bengali—*Mahua, maul, mahula*; Marathi—*Mahwa, mohwra*; Gujarati—*Mahuda*; Telugu—*Ippa*; Tamil—*Illupei, elupa*; Kannada—*Hippe*; Malayalam—*Poonam, ilupa*; Oriya—*Mahula, moha, modgi*; Family—Sapotaceae.

Fresh mahua oil from properly stored seeds is yellow in colour with a not unpleasant taste. Commercial oils are generally greenish yellow in colour with an offensive odour and disagreeable taste. During the major part of the year, the oil is fluid; in cold weather, the oil solidifies to a buttery consistency.

Uses. Mahua oil is used mainly in the manufacture of laundry soap. Mahua oil is also used for edible and cooking purposes in some rural areas. Crude oil has a deep colour, high acidity and unpleasant odour. Refining and hdyrogenation yield products similar to mutton tallow or cacao butter. Refined oil is used in the manufacture of lubricating greases and fatty alcohols The oil is used also for candles, as a batching oil in jute industry and as a raw material for the production of stearic acid. It has emollient properties

Fig. 9.9. *Madhuca butyracea.* Mahua tree (*Mahua*). A fruiting branch. The seed oil is used for cooking and soap-making. The oil-cake makes good manure.

and is used in skin diseases, rheumatism and headache. It is laxative and considered useful in habitual constipation, piles and haemorrhoids. It is used also as an emetic.

Carapa Oil

Xylocarpus moluccensis (Lamk). Roem.; Eng. Carapa; Hindi—*Pussur*; Family—Meliaceae. A tree. It is found in Andaman islands. The oil is obtained from the seeds.

Uses. It is used for soap and as an illuminant.

Nutmeg Butter

Myristica fragrans Houtt.; Eng. Nutmeg; Hindi—*Jaiphal*; Sanskrit—*Jatiphala*; Hindi, Bengali, Marathi and Gujarati—*Jaiphal* (fruit kernel), *japatri, jotri, jayapatri* (aril); Telugu, Tamil, Kannada and Malayalam—*Jajikai, jadikai* (fruit kernel), *jadipattiri, japatri* (aril); Family—Myristicaceae.

A tree. It is native of the Moluccas Islands. In India it is grown in the Niligiris, Kerala, Karnataka and West Bengal. The seeds contain about 40 per cent of a yellow fat. The seeds are roasted and powdered and the oil is extracted between warm plates.

There are two principal types of nutmeg and mace: East Indian Nutmeg and Mace and West Indian Nutmeg and Mace, the former of better quality, kernels constitute nutmeg; Mace is dried fibrous aril covering the testa.

Both Nutmeg and Mace yield essential oils used for flavouring food products and liqueurs and also cosmetics. The oil is also recommended for inflammation of bladder and urinary passage. Nutmeg butter or fat obtained from nutmegs is used as a mild external stimulant in ointments, hair lotions and plasters, and forms an useful application in cases of rheumatism, sprains, and paralysis, also used in spicy perfumes and in soaps and candles. Leaves and bark also contain essential oils.

Uses. They are used for soap, ointment, perfumes and candles. The oil is used externally for rheumatism.

Pongam Oil

Pongamia pinnata (L). Pierre; Eng. Pongam; Hindi—*Karanj*; Sanskrit, Hindi, Bengali, Marathi and Gujarati—*Karanj*, *karanja*; Telugu—*Gaanuga*, *Pungu*; Tamil—*Ponga*, *Pongam*; Kannada—*Honge*; Malayalam—*Pungu*, *Punnu*; Oriya—*Koranjo*; Punjab and Kumaon—*Sukhchein*, *Karanj*; Assam—*Karchaw*; Family—Papilionaceae.

A tree. It is commonly found near banks of streams in both the Peninusuals, West Bengal and Travancore. The oil is obtained from seeds.

Uses. The oil is used for soap-making as an illuminant, and is also used in the treatment of skin diseases and rheumatism.

Pongam oil is used in tanning industry for dressing leathers; also finds use in the preparation of washing soaps and candles and as a lubricant for heavy lathes, chains, enclosed gears, heavy engines, and bearings of small gas engines. Medicinally the oil is applied in scabies, herpes, leucoderma, and other cutaneous diseases. Internally the oil is used in dyspepsia with sluggish liver. *Karanjin* is active principle. Seed cakes are used as manure.

SOAP SUBSTITUTES

Many plants contain products that can be used as soap substitutes.

Soapnut Tree of South India

Sapindus emarginatus Vahl; Eng. Soapnut; Hindi—*Ritha*; Family—Sapindaceae. A tree cultivated in Northern India. The fruits are used as soap substitute for washing woolen clothes.

Chinese soap berry

Soapnut tree of North India *S. mukorossi* Gaertn.; Eng. Soapnut tree; Hindi-*Ritha*. A tree, grown in North West India, West Bengal and Assam. The fruits are used for washing woollens and silk. They are also used in the preparation of hair tonics.

WAXES

The surface of leaves and fruits often possesses a covering of wax secreted by the plant which protects the excessive loss of water. This wax is similar to fat in its composition. They are harder than fats and possess a higher melting point. They do not become rancid and are less easily hydrolysed. Only a few waxes are of economic importance. They are as follows:

Jojoba Wax

Simmondsia chinensis; Eng. Jojoba wax; Family—Simmondiaceae.

The jojoba bush is an evergreen shrub of the semiarid regions of the south western United States of America and adjacent Mexico. It is unique in having seeds with a fifty per cent liquid wax content.

The jojoba wax is suitable for making waxing compounds, polishes and candles. It is also used as an illuminant, after processing, for several types of food products. The oil cake is used for stock feed. The fruits are edible.

Carnauba Wax

Copernicia cerifera; Eng. Carnauba wax; Family—Palmaceae (Arecaceae).

This is the most important vegatable wax. It occurs as an exudation on the leaves of the wax Palm. The wax palm is the native of Brazil and other parts of the tropical South America. This slender palm is locally known as 'Tree of life', since every part of it is of economic value.

The commercial supply of wax is obtained chiefly from wild trees in northeastern Brazil. The young leaves of the palm tree are carefully selected and gathered before they are fully open. Thereafter they are dried in the sun for several days until the wax is a flourlike dust. This is removed by threshing and is melted down in clay vessels. It is then strained, cooled and made into cakes or broken into small pieces for transportation. Several grades of the wax are formed. The crude product is greenish-gray in colour and is very hard having a higher melting point.

Carnauba wax is generally used in the manufacture of candles, soap, varnishes, paints, carbon paper, batteries, sound films, insulation, ointments and several other useful products.

Another wax, similar to carnauba wax is obtained from the trunk of *Ceroxylon andicola* (Palmae). This tree is known as wax palm of Andes. It is native of tropical South America.

Cauassu Wax

Calathea lutea; Eng. Cauassu wax; Family—Marantaceae.

These tall herbs are native of Tropical America and West Indies. In lower Amazon region, it makes an important source of a commercial wax. The wax is produced on the underside of the large leaves. The leaves are dried in the sun for only 2 to 3 hours and thereafter the tissue-paper thin scales are removed by scraping. New wax-yielding leaves are produced within a year, whereas it takes 8 to 10 years for carnauba leaves to be removed.

Cauassu wax is very similar to carnauba wax and similarly it is used for making candles, soaps, varnishes, paints, ointments, salves and several other products.

Candelilla Wax

Euphorbia antisyphilitica; Eng. Candelilla wax; Family—Euphorbiaceae.

This plant is a low, light bluish-green desert shrub of Texas, Mexico and Northern Central America. It bears tiny deciduous leaves.

The wax exudes from the pores found on the stem. It makes a thin film on the surface of the stem. The amount of wax increases during the winter and therefore, the plants are collected at that time. Wild plants make the main source of the wax. The wax is extracted by solvents or by boiling.

The crude wax is white. However, thc refined product is a light tan colour and possesses a sweetish odour. This wax is softer and contains more resin. It is generally used as an extender wax in mixture with others. However, its quality may be improved chemically and it could be used alone.

Myrtle Wax

Myrica pennsylvanica; Eng. Bayberry; *Myrica cerifera*; Eng. Wax myrtle; Family—Myricaeae.

These plants are the natives of eastern U.S.A. The bayberry and wax myrtle both are covered with a thick layer of wax. This material is actually a fat, rather than a wax. The wax is removed from the fruits by boiling in water. It is used for making candles, soaps, varnishes, paints, etc.

Japanese Wax

Rhus succedanea L.; Eng. Japanese wax tree, wild varnish tree; Hindi—*Kakrasingi*; Family—Anacardiaceae.

A shrub or tree, found in the temperate Himalayas, Kashmir, Sikkim and Khasi hills.

Japan wax is used in manufacture of candles, wax-matches and pencils. It is also employed for leather and furniture polish, waxcrayons, leather and textile finishes, pharmaceutical, soaps, vulcanization of rubber. Also used for pomades and lipsticks.

10

Essential Oils

The essential oils also known as volatile oils evaporate in contact with the air and possess a pleasant fragrance. Chemically the essential oils are very complex. They are found in many different species of plants of various families. All aromatic plants contain essential oils. Generally the oils are secreted in oils glands.

The essential oils may be extracted from the plant tissues by various methods. There are *three* important methods—(1) distillation, (2) expression and (3) by solvents.

1. **By distillation.** In this method the whole or ground material is placed in a still, and steam is introduced. The oil vapourizes and, together with the steam passes into a condenser. After cooling, the oil, collects on the surface of water and is removed and filtered.

2. **By expression**. The pressure is applied in this method. The citrus oils are expressed by this method. The rinds of oranges, lemons and other citrus fruits are pressed in machines to yield the essential oil.

3. **By solvents.** There are two principal methods—(i) enfleurage and (ii) maceration.

(*i*) *Enfleurage*. Here the extraction is carried on in a normal temperature. Glass plates are covered with the cold fat. The flowers are placed on it and allowed to remain for several days. The fat dissolves slowly and absorbs perfume material from the flowers. Jasmine flowers are treated in this way.

(*ii*) *Maceration.* In this process the plant material is digested with hot oil or melted fat and the flower may be broken up to aid the method. Rose flowers are treated in this way. When absorption of the perfumes is completed, the pomades are treated with ethyl alcohol to dissolve out the oil and yield extracts of the flower.

ESSENTIAL OIL YIELDING PLANTS AND ESSENTIAL OILS

Camphor and Camphor Oil

Cinnamomum camphora Linn.; Eng. Camphor; Hindi—*Mushkapur*; Sanskrit—*Karpur*; Telugu—*Karppuram*; Tamil—*Karpurammu*; Family—Lauraceae.

A tree; native of China and Japan. The plant has been grown in India at Dehra Dun, Saharanpur, Calcutta, Nilgiris and Karnataka.

Camphor and camphor oil are obtained from the wood and leaves on steam-distillation. The camphor oil of commerce is the oil from which camphor has been removed. Camphor and camphor oil are separated by filtration and the camphor dissolved in oil is removed by refrigeration and filtration. The oil is distilled in Coonoor.

Uses. They are mainly used in the manufacture of celluloid and the various nitrocellulose compounds. They are widely used in the preparation of medicines. They are also used for making perfumes.

Camphor is extensively employed in external applications as a counter-irritant in muscular strains, inflammations and rheumatic conditions. Camphor is esteemed as an analeptic in cardiac depressions and is used in mycocarditis. It has a calmatic influence in hysteria and nervousness.

Ylang Ylang Oil

Cananga odorata (Lam.) Hook. f. and Thoms.; Eng. Ylang-ylang oil; Tamil—*Karumugai*; Telugu—*Chettu sampangi*; Kannada—*Apurvachampaka*; Family—Annonaceae.

A cultivated tree; a small plantation exists in the Kerala state.

The flowers yield an essential oil on distillation. The first fraction of the distillate is usually the best and is marketed as Ylang-ylang oil.

Uses. It is used for some of the finest perfumery creations in France; cheaper grades are used in soap making.

Cinnamon Oil

Cinnamomum zeylanicum Bereyn.; Eng. Cinnamon; Hindi—*Dalchini*; Sanskrit—*Tamalapatra*; Hindi, Gujarati, Marathi, Telugu and Bengali—*Dalchini*; Tamil—*Ilayangam*; Kannada—*Dalchini, lavangpatti*; Family—Lauraceae.

A tree; native of Ceylon but grown in the Nilgiris; South Kanara, Malabar, Assam and Kumaon.

Cinnamon leaf oil is obtained from the leaves on steam-distillation.

Uses. Cinnamon leaf oil is largely consumed in the flavouring and pharmaceutical industry as well as perfumery. It is used in medicines as a carminative, antiseptic and astringent.

Bark yields an essential oil, Cinnamon Bark Oil, used for flavouring confectionery, liqueurs, pharmaceuticals, and soaps; also used for gastric troubles.

Citrus Oils 1. Lime Oil

Citrus aurantifolia (Christm.) Swingle; Eng. Lime; Hindi—*Kaghzi-nimbu*; Bengali—*Kaghzinimbu, patinebu*; Gujarati—*Khatalimbu*; Tamil—*Elumichai*; Telugu—*Nimma*; Kannada—*Limbe, nimbe*; Malayalam—*Erumichinarakam*; Family—Rutaceae.

A shrub or small tree; cultivated in Tamil Nadu, Maharashtra, West Bengal, Punjab, Madhya Pradesh, Andhra Pradesh, Delhi, Uttar Pradesh and Karnataka.

The oil is obtained from peels on steam-distillation.

Uses. It is mainly used in confectionery, pharmaceuticals and toilet preparations.

2. Petigrain Oil

Citrus aurantium Linn.; Eng. Sour orange; Hindi—*Khatta*; *Citrus limettoides* Tanaka; Eng. Sweet lime; Tamil—*Narangam, Narattai*; Telugu—*Mallikanarangi*; Kannada—*Heralay*; Malayalam—*Karmal*; Hindi—*Mitha nimbu*; Family—Rutaceae. Usually cultivated in Northern India and Madhya Pradesh.

Citrus limettoides. Bengali, Hindi and Gujarati—*Mithanebu, mithalimbu*; Tamil—*Kolumichangai*; Telugu—*Gajanimma*; Kannada—*Gajanimbe.*

Petigrain oil is obtained by the distillation of leaves and young shoots of citrus plants. The yield of oils is higher from young and fresh leaves.

Uses. It has a sweet, sharpish, odour and is invaluable in eau-de-colog..e especially of fancy types. The oil is used as a perfume of skin creams. It is used in cosmetics, confectionery and as a flavouring agent. It is used also for perfuming soaps.

3. Orange Oil

Citrus sinensis (Linn.) Osbeck; Eng. Sweet orange; Hindi—*Musambi*; Bengali, Hindi, Gujarati and Marathi—*Musambi, narangi, kamala nembu*; Kannada and Tamil—*Sathagudi*; Telugu—*Buddasini, satghudi*; Family—Rutaceae.

Usually grown in the Deccan, Maharashtra, the Punjab, Rajasthan and Tamil Nadu.

Orange oil is obtained principally from the peels of fruits by distillation. It is orange yellow liquid with characteristic odour and is most difficult to preserve.

Uses. The orange oil is used in perfumes and flavouring extracts and as a drug.

Peels are the source of an essential oil called Orange Oil. Flowers also yield an essential oil, Neroli Oil. Leaves and young shoots are another source of Petigrain Oil.

4. Mandarin oil

Citrus reticulata Blanco; Eng. Mandarin, Hindi—*Santara*, of family Rutaceae. It is cultivated in the hills of Assam, Darjeeling, Garhwal, Sikkim, Tripura, Himachal Pradesh, Coorg, Punjab, Nilgiris and Maharashtra.

The oil is obtained from the peels by distillation.

Uses. The oil is used in confectionery, toilet products and pharmaceutical preparations.

Cyperus Oil

Cyperus scariosus R. Br.; Eng. Nut grass; Hindi—*Nagarmotha*; Sanskrit—*Nargmusta*; Bengali and Hindi—*Nagarmutha*; Marathi—*Lawala*; Tamil—*Koraikkilangu*; Family—Cyperaceae.

It is found in damp situations in Bengal, Uttar Pradesh, Punjab and Madhya Pradesh.

The oil is obtained from tubers with aromatic odour by distillation. The essential oil is a dark, amber-coloured, viscous liquid with heavy smell.

Uses. The oil is used by some perfumers as a fixative. It is quite useful for this purpose. The oil is also used as a hair tonic.

Eucalyptus Oil

Eucalyptus globulus Labill.; Eng. Blue gum tree, of family Myrtaceae.

It is native of Australia, now cultivated chiefly in the Nilgiris, the Anamalais, the Palni Hills, Simla hills, at Shillong in Assam.

Eucalyptus oil is obtained by the distillation of dried leaves and terminal leaflets. The oil is available in the market in two grades (i) double distilled (pale yellow colour); and (ii) double distilled and extra-refined (colourless).

Uses. The oil is used in the treatment of asthma and bronchitis, and is also used in perfumery. It is largely used as a mosquito and vermin repellent and as an ingredient of germicidial and disinfecting preparations. It is an ingredient of deodorizing and arepticizing composition for use in theatres.

Geranium Oil

Pelargonium graveolens L. Herit.; Eng. Rose Geranium; and other species of Pelargonium, of family Geraniaceae.

It grows wild on the Nilgiris. Cultivated in the Shevaroy hills in Tamil Nadu. The important countries where geranium oil is produced are Algeria, Congo, France and Russia.

The oil is obtained from the leaves and twigs by distillation.

Uses. Geranium Oil makes the basis of several good perfumes. Rhodinol and its esters are prepared from this oil. It is highly prized as an ingredient of soap and toilet perfumes.

Geranium Oil is valued for its pronounced rose-like odour; blends well and largely used as an adulterant of otto of roses, and for flavouring tobacco and pharmaceutical preparations. Terpeneless Geranium Oil makes an excellent base for artificial rose ottos.

Jasmine Oil

Jasminum auriculatum Vahl.; Hindi—*Juhi*; *J. flexile* Vahl., Hindi—*Malti*; ***J. officinale*** Linn. var. *grandiflorum* Linn., Eng. Jasmine, Hindi—*Motiya*; ***J. sambac*** (L.) Ait., Eng. Arabian jasmine, Hindi—*Bela*, *mogra*, of family Oleaceae.

Jasmine oil is obtained from the flowers of *Jasminum* spp., cultivated in Bengal, Orissa, Assam, Maharashtra, Tamil Nadu, Uttar Pradesh, Punjab, Kashmir and Western Himalayas. Kannauj, Jaunpur, Ghazipur and Sikandarpur in Uttar Pradesh are the important centres of production of Jasmine oil.

Indigenous methods for the utilization of jasmine flowers are the modified enfleurage process for preparing perfumed oils and the distillation process for making *attars*. In the former, cleaned and husked *til* (*Sesamum indicum*) seeds are used; fresh jasmine flowers and til seeds are spread in alternate layers on the flour of a cemented pit. Exhausted flowers are replaced by fresh flowers until the seeds are saturated with the perfume. Perfumed oil is obtained from the seeds by expression in *ghanies* and marketed as *Sire ka tel*. For the production of *attars*, the flowers are distilled in earthenware vessels and the vapours absorbed in sandalwood oil.

Uses. The oil is used in the preparation of hair-oils and *attars*. It is used in perfumery.

Khus Oil

Vetiveria zizanioides (L.) Nast; Eng. Vetiver; Hindi—*Khas*; Sanskrit—*Sugandhimula*; Bengali—*Khas-khas*; Gujarati—*Valo*; Marathi—*Vala*; Telugu—*Kurubeeru*; Tamil—*Vettiver*; Kannada—*Vettiveeru*; Malayalam—*Vettiveru*; Family—Gramineae/Poaeceae.

A perennial grass, cultivated chiefly in Rajasthan, Uttar Pradesh, the Punjab and the West Coast.

Khus oil is obtained by the steam-distillation of the roots.

Bharatpur is the biggest centre of Khus oil industry in North India. The production of Khus oil from wild roots in Rajasthan is about 1,500 lb. per annum besides 500 lb. of oil distilled from wild roots in other areas.

Uses. Khus oil is a valuable fixative and is used in perfumery, cosmetics, soaps and for flavouring sherbets.

Fragrant roots yield an essential oil, called Vetiver Oil (Khus Oil). Oil is diaphoretic, stimulant, and refrigerant, used in colic, flatulence, and abstinate vomiting. It affords relief when applied in rheumatism, lumbago, and sprains.

Lavender Oil

Lavandula officinalis Chaix; Eng. Lavender, of family Labiatae.

A small shrub; native of South Europe, now grown in Jammu and Kashmir.

Lavender oil is obtained by steam-distillation of the fresh flowering tops of the plant. The oil is a pale-yellow mobile liquid with a pleasant odour. Its perfumery value is based on its ester content (linalyl acetate).

Uses. They are excellent for use in face and toilet powders and as odoriferant and fixative in soaps and shaving creams. Lavender oil is imported into India from France.

Lavender Oil is used in perfumery production of lavender water, and soap industry; also used medicinally as a carminative and mild stimulant, and for flavouring pharmaceutical preparations.

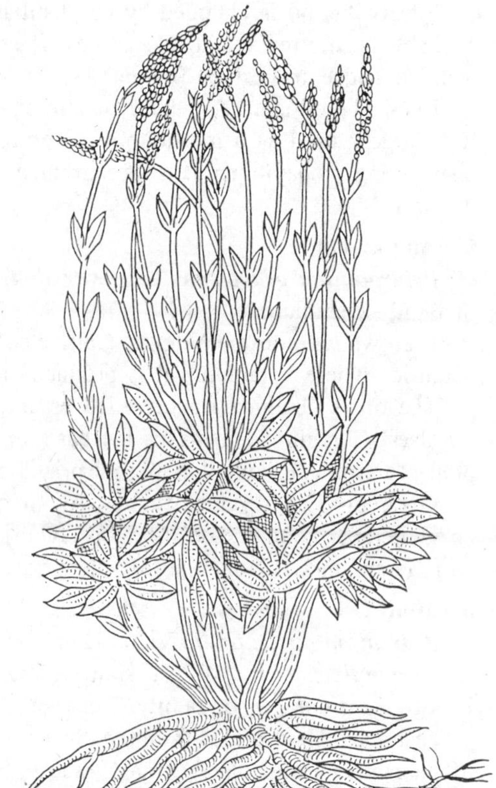

Fig. 10.1. *Lavandula officinalis* (lavander). A small shrub with flowering tops.

Lemongrass Oil

Cymbopogon flexuosus (Steud.) Wats.; Eng. East Indian lemon grass, of family Gramineae.

It is indigenous to India and grown in Madras, Tamil Nadu and Karnataka. It is cultivated chiefly in Travancore and Malabar.

The East Indian or Malabar lemongrass oil of commerce is obtained from the leaves by distillation. The leaves of this plant, yield the normal essential oil containing 75 per cent of aldehydes (chiefly citral). The citral content increases with the age of the plantation though the yield decreases; fresh grass gives more yield than stored grass and the solubility of oil in alcohol decreases on storage.

The current world production of lemongrass oil is estimated to be nearly 1,500 tons. India is at present meeting about 80 per cent of the world demand of lemongrass oil. Almost 97 per cent of the lemongrass oil produced in India is exported. Among the essential oils produced in India, lemongrass oil ranks first in quantity and value.

Uses. This oil makes a basic raw material for the manufacture of many aromatic chemicals which find application in perfumery, soap and cosmetic industries. Ionones required for use in perfumery are prepared from lemongrass oil. The oil is used for the synthesis of vitamin A. The oil is used in many pharmaceutical preparations, such as pain balm and disinfectants.

West Indian Lemongrass Oil

Cymbopogon citratus Staph syn. *Andropogon citratus* DC.; Eng. West Indian Lemongrass; Hindi—*Agin ghas, gandhatrina*; Sanskrit—*Bhustrina*; Bengali—*Gandhabena*; Marathi—*Hirua cha*; Gujarati—*Lili cha*; Telugu—*Nimmagaddi*; Tamil—*Vasanappillu*; Kannada—*Majjigehullu*; Malayalam—*Vasanappulla*; Punjab—*Khawi*; Family—Poaceae.

The West Indian Lemongrass Oil of commerce is obtained from the leaves of the grass by distillation. The oil is used in soaps and as a flavouring. It has powerful fresh lemon scent, and considered a carminative.

Mint Oils

1. ***Mentha arvensis*** Linn.; Eng. Field mint; Hindi—*Pudina*, of family Labiatae.

This is a perennial herb, grown in the Western Himalayas, Kashmir, the Punjab, Kumaon and Garhwal.

The essential oil is obtained from the leaves by distillation process.

Uses. The oil is used in the preparation of certain kinds of cigarettes and pharmaceuticals. Also used medicinally as a carminative, refrigerant and stimulant.

2. ***Mentha arvensis*** DC. var. ***piperascens*** Holms; Eng. Japanese peppermint of family Labiatae.

A perennial herb cultivated in Jammu and Kashmir.

Uses. The leaves yield an essential oil on distillation, which is the chief source of menthol. It is used in the treatment of colds.

3. ***Mentha piperita*** Linn.; Eng. Peppermint; Hindi—*Vilayati pudina* of family Labiatae.

It is an aromatic herb; native of Europe but now cultivated in Kashmir, Maharashtra and Punjab.

The essential oil is obtained from the leaves by distillation process.

Uses. Peppermint oil is one of the most commonly used essential oils. It forms an ingredient of tooth pastes, dental creams, mouth washes, cough drops, chewing gums, tobacco, etc., as also confectionery, liqueurs and medicinal preparations. It is used as carminative, stimulant and for allaying nausea, sickness, and vomiting. It is also used in perfume and soap industries.

4. ***Mentha spicata*** Linn.; Eng. Spearmint; Hindi—*Pahari pudina,* of family Labiatae.

A herb, cultivated in the Punjab, Uttar Pradesh and Maharashtra. The essential oil is obtained by steam distillation.

Uses. The leaves yield an essential oil known as spearmint oil. It is used for flavouring confectionery. It is also used as a substitute of peppermint oil.

Palmarosa and Gingergrass Oils

Cymbopogon martini (Roxb.) Wats. This grass occurs in two varieties known as the '*Motia*' and '*Sofia*'. The essential oil obtained from the '*Motia*' variety is known as Palmarosa oil, and from the '*Sofia*' variety is known as Gingergrass oil; Family—Gramineae.

Cymbopogon martini var. ***motia*** grows abundantly in Madhya Pradesh, Andhra Pradesh, Baroda and some parts of South India. It grows wild in Khandesh, Nasik and Panch Mahal districts of Maharashtra state. It is a perennial plant growing to a height of 6–8 feet. It is propagated from the seeds.

The oil is produced from the stems, leaves and flowers of the grass by distillation in crude direct fired stills.

The most important centres of Palmarosa oil production are Melghat, Betul and Nimar in Madhya Pradesh, Khandesh and Nasik districts in Maharashtra state. Other areas of production are M.P., Surat, U.P., Baroda and A.P. The best grade of palmarosa oil is produced in Melghat, Elichpur and Burhanpur.

Palmarosa oil occupies the third position amongst essential oils exported from India, the first and second being lemongrass and sandalwood oils respectively. The annual production of palmarosa oil in India is estimated to be about 90 tons.

Uses. The oil is extensively used for the adulteration of Attar or Roses and as a base for perfumes and cosmetics. The oil is the principal source of high grade geraniol. The oil is saponified with alcoholic sodium hydroxide and fractionally distilled in vacuum stills, the final distillate being geraniol. A mixture of terpenes, suitable for use in soap blends is obtained in the intermediate stages of distillation.

Cymbopogon martini var. *sofia* grows abundantly in the forests of Melghat (Khandesh Dt.), Yeola (Nasik Dt.), Dohand and Jholad (Panch Mahal Dt.).

Gingergrass oil is obtained by water-distillation of the leaves, stalk and flowers.

Uses. The oil is used in perfume blends for scenting cheaper toilet soaps.

Sandalwood Oil

***Santalum album* Linn.**; Eng. White Sandal wood; Hindi—*Safed Chandan*, of family Santalaceae; Sanskrit—*Chandana*; Bengali—*Chandan, peetchandan, srikhanda*; Marathi—*Chandan*; Gujarati—*Sukhad, sukhet*; Telugu—*Chandanamu, chandanapuchettu, srigandhamu, gandhapu-chekka* (wood); Tamil—*Sandanam*; Kannada—*Srigandha, bavanna*; Malayalam—*Chandanam, chandana-mutti* (wood); Oriya—*Chondono*; Koorg—*Chandana*; Tulu—*Gandha, chandana*; Konkan—*Srigandha*; Punjab—*Chandan*.

A tree; it is a native of the highlands of South India and the Malayan Archipelago.

Sandalwood oil commonly known as East Indian oil of Sandalwood, is one of the most important essential oils produced in India on a factory scale; its annual production being 140 to 154 metric tonnes. Valued as a fixative for perfumes, the oil is distilled from the roots and the heartwood of sandalwood tree.

Fig. 10.2. *Santalum album*. A sandal wood tree (**safed chandan**).

Sandalwood area and cultivation. In India, cultivation of sandalwood is mainly confined to the state of Karnataka. The main sandalwood area spreads over 480 kilometres from Dharwar in the north to the Nilgiris in the south and about 400 kilometres, horizontally from Kurg in the west to Kuppam (Andhra Pradesh) in the east. The evergreen tree thrives on the well-drained loamy soil of the highlands, although the finest wood grows in the comparatively dry regions, particularly on laterite soil. It is also cultivated in large scale in forests of Kerala.

Santalum album is a parasitic plant which previously grew only in the wild state but is now generally propagated by seed. The tree averages a height of 9–12 metres with a girth of 75 to 90 cm. As the tree grows, essential oil develops in the root and the heartwood. The growth is, however, exceedingly slow, and for a tree to develop an 8 cm. core of oil-rich heartwood it may take more than thirty years.

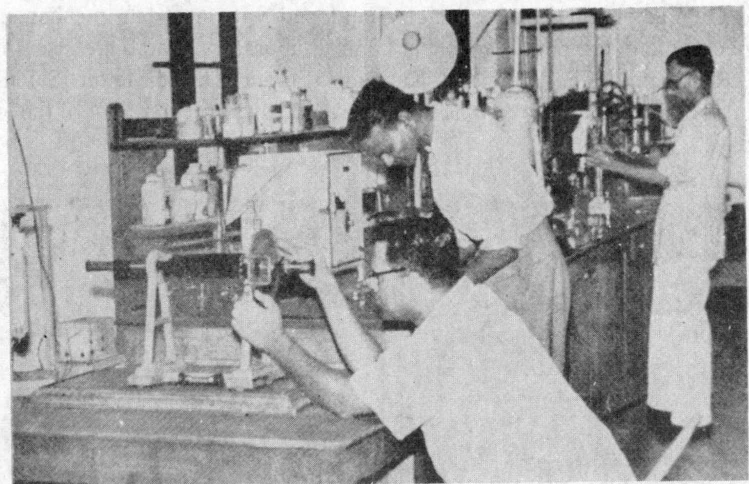

Fig. 10.3. Sandalwood. Laboratory at sandalwood oil factory, Mysore.

Karnataka leads in production. The state of Karnataka is the biggest producer of sandalwood and its oil. Nearly 2,000 to 2,500 metric tonnes of sandalwood or about 65 per cent of India's total production, comes every year from Karnataka forests.

The sandalwood industry in Mysore is a Government monopoly. All sandalwood trees, wherever they may grow, are the property of the state. No person is permitted to cut down a tree, possess, transport sandalwood or distil oil without Government permission. As a rule, dead trees or old trees between 50 and 60 years only are allowed to be cut for distillation. Trees less than 30 years old are never cut.

Next to the state of Karnataka, the most important producer of sandalwood in India is Tamil Nadu. Sandalwood also occurs in Andhra Pradesh, Kerala and Maharashtra.

Distillation of oil. The distillation of oil of sandalwood has been carried on in India from ancient times. Today modern techniques are being employed to distil the oil. The Government

Fig. 10.4. Sandalwood. Sandal oil being separated in the Mysore Sandalwood Oil Factory.

Sandalwood Oil Factory in Mysore is fully equipped with uptodate mechanical and chemical engineering equipment and has an excellent control laboratory. There are also smaller distilleries at Kuppam, Renigunta (A.P.); Bombay; Kannauj, Kanpur (U.P.); Mettur, Salem and Tirupattur (Tamil Nadu). The Karnataka Government factory alone distils nearly 50% of the total oil produced in the country.

Fig. 10.5. Government Sandalwood Oil Factory, Mysore. Sealing phials of sandal oil.

The raw material consists of billets and roots and chips of sandalwood. The wood is first fed into giant chipping machines, which reduce it to a coarse powder. This in turn is fed into disintegrators which further reduce the particle size. The powder is carefully sieved and remixed with coarser powder to form of uniform porous material.

Fig. 10.6. Sandalwood oil. Government Sandalwood Oil Factory, Mysore.
A view of the packing department

The stills are made of copper—rarely of iron—and are provided with 'goose necks' to conduct the vapour, oil and steam into separate tin-lined tubular condenser. The receivers, also tin-lined, are constructed on the principle of old Florentine flasks. The powdered wood is placed in the still on a perforated false bottom, and low pressure steam is let in to pass uniformly over the entire mass of the porous mixture.

Export and quality control. Of the total oil produced in India nearly 33 per cent is used within the country by the manufacturers of indigenous perfumes called *attars*. The rest is exported, particularly to the U.S.A., the U.K. France, Switzerland, Italy, Japan, Australia and Germany.

The Government of India imposed compulsory quality control under AGMARK on oil of sandalwood exported from the country.

Uses. The major demand at present is from the perfumery trade. It is also used medicinally. Kerala sandal soap is well known.

Sandalwood oil forms the base for *attars* (ottos) produced in India. It is also employed as a base for co-distillation of some of the other essential oils of delicate fragrance. Both the wood and oil are diuretic, diaphoretic, refrigerant, and expectorant, finding several applications in household remedies.

Patchouli Oil

Pogostemon perilloides (Linn.) Mansf.; Eng. Patchouli; Hindi—*Pacholi* of family Labiatae/ Lamiaceae; Bengali—*Pachapat*; Marathi—*Patchapan*; Gujarati—*Pachapandi*; Tamil—*Kadirpachai*; Kannada—*Patche tene*; Malayalam—*Pachila*.

A herb, cultivated in Kanara, the Western Ghats and the Nilgiris.

Patchouli oil is obtained from the leaves by distillation.

Uses. The essential oil is used for soaps and perfumes. It makes a good fixative for heavy perfumes.

Also used for scenting tobacco and incenses. It is an excellent masking agent in depilatory creams. It yields an excellent *attar* on blending with sandalwood oil.

Rose Oil

Rosa damascena Mill.; Eng. Damask rose; Hindi—*Gulab* of family Rosaceae.

A prickly shrub, cultivated chiefly in Aligarh, Ghazipur and Kannauj of Uttar Pradesh.

The flowers are picked in the early morning just as they are opening and are distilled as soon as possible. The oil is colourless at first, but gradually turns a yellowish or greenish colour. 20,000 lb. of the flowers are required to make 1 lb. of the essence. The essential oil is known as 'otto of roses' or 'oil of roses'.

Uses. It is used in making first class perfumery. Large quantities of rose water are also made. This consists for the most part of the water left after distillation, which still contains some otto.

Fig. 10.7. A pine tree. Source of turpentine oil.

Turpentine Oil

Pinus roxburghii Roxb.; Eng. Long-leaved pine; Hindi—*Chir* of family Pinaceae.

The principal source of Indian turpentine oil is *Pinus roxburghii* Roxb., an important Himalayan conifer. It grows abundantly in the sub-montane regions of the Himalayas upto 5,000 feet.

Turpentine oil is obtained by the distillation of oleoresin (an exudate obtained by wounding trees) of various conifers. The approximate annual supplies of oleoresin amount to 4–8 lac gallons. The oleoresin contains 14–20 per cent oil. There are three large factories distilling oil at Bareilly, Nahn and Miran Sahib besides a large number of small factories at Hoshiarpur, Someshwar, Bhoali, Rishikesh and Bareilly.

Uses. The oil is used in the manufacture of varnishes, lacquers, disinfectants and paints. It is also used in the manufacture of linoleum, sealing-wax, oil cloth, lubricating compounds and several kinds of inks.

Fig. 10.8. Nutmeg fruits. Wormy nutmegs give a much higher yield of oil than do sound ones. Commercial oil is derived from broken and wormy nutmegs. Oil of nutmeg is a mobile, almost colourless or pale yellow liquid with a characteristic odour.

Clove Oil

Syzygium aromaticum (Linn.) Nerrill and L.M. Perry; Eng. Clove oil; Hindi—*Laung Ka tel*; Family—Myrtaceae.

The clove tree grows abundantly in Zanzibar from where the cloves are supplied to all parts of the world. Clove trees are found in wild state in South India. Also cultivated in Nilgiris and Kerala State.

The oil distilled from the buds is of excellent quality.

Uses. Clove oil is used in soap and other perfumes, in confectionery and in medicine. It is used as stimulant, carminative and in flatulence.

Nutmeg Oil

Myristica fragrans Houtt; Eng. Nutmeg; Hindi—*Jaiphal*; Family—Myristicaceae.

The tree is native of the Moluccas Islands but now grown in the Nilgiris; Kerala, Karnataka and West Bengal.

The oil is obtained from nutmeg (aromatic kernel) and mace (aril) of the fruit on distillation.

Uses. Nutmeg oil is used externally for rheumatism, and is also used in soap and perfume industries. The essential oil, obtained from the leaves, is toxic to weeds, and is used in the preparation of chewing gums, flavouring essences and cosmetics.

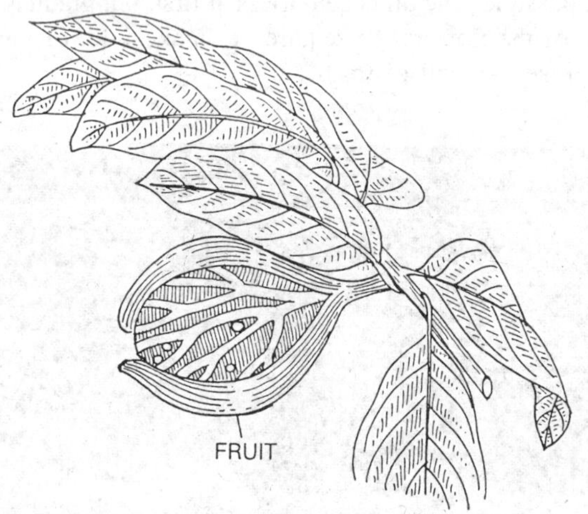

Fig. 10.9. *Myristica fragrans* (Nutmeg). A twig with mutmeg fruit showing mace (aril). Nutmeg oil is used externally for reheumatism.

Champaca Oil

***Michelia champaca*;** Sanskrit—*Champaka*; Hindi and Bengali—*Champa, champaca*; Marathi—*Pivala-*

champa, *sona-champa*; Gujarati—*Champo*, *pito-champo*; Telugu—*Cegagadda*, *veliki*; Tamil—*Manakkarai*; Kannada—*Mullakare*, *gobergally*; Oriya—*Gurbeli*; Assam—*Ketkora*; Family—Magnoliaceae.

Champaca oil makes one of the most famous perfumes of India and other oriental countries. It is obtained from *Michelia champaca*, a large handsome tree of the eastern tropics. The conspicuous yellow flowers are very fragrant and are much worn by the natives. The oil is obtained from the flowers by maceration or extraction. The oil possesses a delicious fragrance.

The flowers are the source of Champaca Oil or Champa Oil. The flowers are employed in preparation of *attars* and perfumed hair oils. The leaves yield an essential oil with odour of Basil (*Ocimum basilicum* L.)

11

Sugar and Starch

SUGARS

Sugarcane

Saccharum officinarum Linn.; Eng. Sugarcane; Hindi—*Ganna, Ikh*; Sanskrit—*Ikshu, khanda, sarkara* ; Bengali—*Paunda, Pundia* ; Telugu—*Cheruku* ; Tamil—*Poovan karumbu* ; Kannada—*Patta patti kabbu* ; Malayalam—*Karimbu*; Family—Gramineae/Poaceae.

History. Charaka wrote the oldest treatise extant on Hindu medicine, at about the beginning of the Christian era. The glossary of Sanskrit names in this work denotes an intimate knowledge with the various forms and preparations of sugar which is considerately more than 2000 years old. The term *Sarkara* appears to have given origin not only to the Arabic, Persian, Greek and Latin classic names, but to the extensive assortment of words in the modern languages of India and Europe which are every nearly the direct equivalents of the English word sugar. However, there are several sources of information that carry the Hindu knowledge in cane-sugar considerably further back than can perhaps be shown for any other country. The sugar-cane plant and the cane-sugar are frequently mentioned in the *Institutes of Manu.* The more generally accepted view is that these *Institutes* were begun two or three centuries before the advent of Christ. It may thus be safely accepted that the reference to a field of sugar-cane, in *Manusmriti*, is more than 2000 years old. It may thus be accepted that sugar-cane was very generally cultivated in India, during the time the *Institutes of Manu* were written. It is thus probable that sugar-cane has been cultivated in India for at least 3,000 years.

The culms of the wild species such as *Saccharum arundinaceum* (of Deccan) and *S. spontaneum* (of Kar-Nicobar) yield a certain amount of saccharine fluid, and it may be possible that one of these wild forms might have given origin to the sugar-cane, the cultivation of 3,000 years having destroyed its original specific characteristics.

In the *Ain-i-Akbari* (Abul Fazl, 1590 A.D.) a description of the methods of cultivation, manufacture of all forms of sugar, and the distillation of spirits from it, is being given in detail.

The Spaniards carried the cultivation and manufacture of sugar-cane to the Canary Islands in the fifteenth century, but prior to that in 1420 the Portuguese had conveyed it from Sicily to Madeira and to St. Thomas Island. In 1506 it was taken from the Canary Islands to San Domingo. The Dutch first established sugar works in Brazil in 1580, and then they carried the art of sugar manufacture to the West Indies in 1655. Sugar was manufactured by the English in Barbadoes in 1643 and in Jamaica in 1664. Sugar-cane was first introduced into the United States of America in Louisiana in 1741.

It is grown chiefly in Uttar Pradesh, Bihar, South India (especially Coimbatore) and the Punjab. The Indo-Gangetic belt previously grew mostly thin canes (*Saccharum barberi* Jeswiet). These are popularly known as Indian canes, have been extensively cultivated in the country for a long time. They are characterized by high fibre and an intermediate sucrose content. In the peninsular belt the thick canes (*Saccharum officinarum* Linn.) are the chief types. They have all the desirable agronomic and milling qualities. They are characterized by a thin rind and are low in fibre and high in sucrose content. The greatest landmark in the sugar industry is the evolution of improved canes in India. Ever since its inception in 1912 the Sugarcane Breeding Institute at Coimbatore has been engaged in the evolution of improved varieties for the sub-tropical regions of India and since 1926 for the tropical regions also. Its programme of hybridization has included various species of *Saccharum* and related genera. The improved Coimbatore canes now occupy 93% of

Fig. 11.1. *Saccharum officinarum.* Sugarcane (**Ganna**). Crop in Uttar Pradesh.

Fig. 11.2. *Sugarcane.* The cultivated sugarcanes belong to *Saccharum officinarum* and *S. sinense* or their interspecific hybrids with wild species of *Saccharum* chiefly *S. spontaneum*. The sugarcane varieties developed in India are identified by the letters Co., which stand for the Coimbatore Sugarcane Breeding Institute, Coimbatore. Sugarcane harvesting in rural area of Meerut district.

Fig. 11.3. Sugarcane (**Ganna**).An improved variety.

the sugarcane area in India. They have increased the yields by at least 50%. Some of the improved types are: Co. 213, Co. 290, Co. 312, Co. 313, Co. 419, Co. 421, Co. 453, Co. 527, Co. 658, Co. 997, and Co. 1148. Of these, Co. 419 yields from 40 to 90 tons of cane per acre and contains 13 to 16% sugar.

The sugar industry in India is the second largest industry in the country, next only in importance to cotton textiles. The chief source of sugar is the sugarcane. This plant is a rapid growing perennial grass, which reaches a height in cultivation of 8 to 12 feet or more and a diameter of about 2 inches. It grows in clumps. The stem is solid with a tough rind and numerous fibrous strands and contains about 10% of juice, the sugar content of which varies greatly.

The commercial sugarcane is not known in a wild state. By 327 B.C. the sugarcane had become an important crop in India. The name "sugar" comes from the Sanskrit "sarkara", meaning gravel, and refers to the crude sugar.

Breeding for effective utilization of sunlight throughout the year. Sugarcane being a C. 4 plant has a very efficient photosynthetic system. Many of the tropical sugarcane varieties take 14 to 18 months from planting to harvest. This long duration also leads to sugar factories remaining idle for considerable periods. Hence, a breeding programme was initiated at the Sugarcane Breeding Institute, Coimbatore for selection of early ripening canes. Some of the cultures give good sugar recovery and yield in about 8 months. The production system with early ripening canes can be made even more efficient with the help of chemical ripeners. Such strains can enable either taking 3 crops (one main and two ratoon) of sugar cane in 2 years or promoting rice-sugarcane (with sugarcane planting in early autumn) and wheat-sugarcane (with sugarcane planting in late spring) rotations. By properly distributing the areas under spring and autumn planting the sugarcane crushing period can be considerably extended. In trials carried out at the Sugarcane Breeding Station, Coimbatore one variety recorded in 1977 about 19% sucrose with 88% purity in 240 days. During 1978 also, trials with short duration varieties gave over 18% sucrose in juice at 7 months, whereas the mid season variety (Co. 6304) used as control recorded only 12.5% sucrose at that time. (M.S. Swaminathan, 1978).

Cultivation. The sugarcane is propagated by cuttings of various sizes made from upper joints of old canes. These cuttings known as *seed*, are placed in trenches and nearly covered with soil. They begin to sprout in two weeks. The crop is harvested from 10 to 20 months after sprouting. The stems are cut close to the ground for the lower end of the cane is richest in sugar. The rhizomes will normally give rise to two or three more crops, known as *ratoons* before another planting is necessary. The canes are brought to the sugar mills by rail road or any available means of transportation.

Fig. 11.4. *Sugarcane.* Sugarcane is cultivated practically in every state in India, but Uttar Pradesh has the largest area under the crop and accounts for about 50 per cent of the total area under sugarcane in the country. Andhra Pradesh, Bihar, Haryana, Punjab, Tamil Nadu and Karnataka are other major cane growing states which account for about 40 per cent of the total cane area in the country. Crop harvest is seen in the picture in Meerut district.

Sucrose or cane sugar. Cane sugar has been known from early days and its crude form was separated from cane juice. India has been the chief producer of cane-sugar from time immemorial. Later on, other countries like Java, Cuba, Philippines, etc., also started producing sugar and they contributed towards the development of better techniques for the production of refined sugar. The main sources of sugar are sugarcane and beet. In cold countries like Russia where beet is in abundance, it is extracted from beet. In tropical countries like India it is mostly obtained from sugarcane. Indian cane contains about 12–13% of sugar.

MANUFACTURE OF CANE-SUGAR

The process of manufacture of cane-sugar as followed in India involves the following steps:

Extraction of the juice. The canes are thoroughly cleaned and cut into small pieces. These cane pieces are dropped over a continuous belt of steel plates called the cane-carrier, moving over roller chains. From the cane-carrier, these are conveyed to a crusher fitted with a set of revolving knives, and cut into very small pieces. These small pieces in the form of a compact blanket are made to pass through two roller crushers and four sets of mills. After major quantity of the juice has been extracted by the crushers and the first two mills, cold or hot water is sprinkled over the *bagasse* whereby residual juice gets diluted and can be easily extracted by further milling. About 90-95% of the juice is usually extracted. The juice extracted is weighed or measured as desired and sent for clarification.

Table--Sugarcane Varieties in Cultivation

Variety	*Parentage*	*Characteristics*	*Areas where cultivated*
Co. 312	Co. 213 x Co. 244	Grown well in all types of soil but lodges in rich loose soil; good ratooner; drought and frost resistant; suitable for Sugar and *Gur*.	Uttar Pradesh, Haryana, Madhya Pradesh and Rajasthan.
Co. 313	Co. 213 x Co. 244	Suited to light loamy soils; requires irrigation or well distributed rainfall; gives *gur* of good quality.	Uttar Pradesh and Assam.
Co. 419	POJ 2878 x Co. 290	Suited to all kinds of soils, but at its best in soils of medium depth and in soils with fairly high pH; good ratooner; good variety from factory point of view; *gur* hard, crystalline.	Andhra Pradesh, Maharashtra, Karnataka, Tamil Nadu, Madhya Pradesh, Orissa, Gujarat and Assam.
Co. 453	Black Cheribon x Co. 285	More suited to heavy soils; moderate tillerer and good ratooner; *gur* not of good quality.	Punjab and Haryana.
Co. 527	Co. 349 x Co. 312	Suited to well-drained rich loam; good tillerer, good for *rab* and sugar: *gur* golden yellow; highly crystalline and hard.	West Bengal, Eastern Uttar Pradesh and Andhra Pradesh.
Co. 785	Co. 443 x Co. 453	Satisfactory under flooded condition; tillering very good.	Kerala.
Co. 1148	Co. 301 x P. 4383	A good ratooner; *gur* outturn high.	Uttar Pradesh, Punjab and Haryana.
Co. 1158	Co. 421 x Co. 419	Mid season maturing cane with good yield of canes and sugar. Canes uniform and erect making harvest easy.	In large-scale cultivation in U.P.
Co. S. 510	Co. 453 x Co. 557	Adapted to sandy and loamy soils; tolerant to drought; tillering heavy, good ratooner; heavy yielder with high sucrose content.	Uttar Pradesh.

Variety	*Parentage*	*Characteristics*	*Areas where cultivated*
B.O. 17	Co. 331 x Co. 326	Suited for loamy soils, tolerant to water logged conditions and *usar* and saline soils; moderate tillering and good ratooning; yields *gur* of good keeping quality.	Uttar Pradesh and Bihar.
Co. L. 9	Co. 312 x Co. 285	Fairly frost resistant; good ratooner; produces *gur* of good quality.	Punjab and Haryana.
Co. L. 29	Co. 312 x Co. 285	Good tillerer and good ratooner; frost resistant; gives *gur* of very good quality.	Punjab and Haryana.

In case sugar is to be extracted from sugar-beet, it is sliced and put in hot water when the sugar present in the beet diffuses out. A number of tanks are used and worked on the *principle of counter currents*—concentrated solution comes in contact with fresh slices and the exhausted slices come in contact with fresh water and lose last traces of sugar in them. The solution obtained is treated like cane-juice.

Clarification of juice. The raw juice is dark opaque liquid containing about 15% sucrose and small quantities of glucose, fructose, vegetable proteins, mineral salts, organic acids, colouring matter, gums and fine particles of *bagasse* suspended in it. Allowed to remain untreated for some

Fig. 11.5. Harvesting of sugarcane, and transport by 'buffalo buggi' to sugar mill, in Meerut district.

time, it begins to ferment and the sucrose present changes to mixture of glucose and fructose. In order to avoid this, the juice is not allowed to stand untreated for a long time.

There are two main processes for clarification of juice:

(i) *Sulphitation.* Adding milk of lime and treating with sulphur dioxide.

(ii) *Carbonation.* Adding milk of lime and treating with carbon dioxide.

Both processes have advantages of their own. In fact combination of both is made use of to get better results.

(a) *Defecation.* The juice is strained to remove suspended matter and treated with 2-3% lime till pH value reaches 7.2 in tanks heated with steam coils.

Heating of the juice helps the coagulation of albuminoids by lime. The vegetable proteins are thus coagulated and the organic acids neutralized. These form a scum on the surface of the juice when the hot juice is transferred to settling tanks called *subsiders*. The scum at the surface and the mud settling at the bottom are mechanically removed by passing the whole through a filter press and the clear solution sent on to the conical tanks for carbonation or sulphitation.

(b) *Carbonation or Sulphitation.* A current of carbon dioxide is passed through the defecated juice, which contains unreacted lime and calcium sucrosate (*carbonation*). This removes the excess of lime as calcium carbonate and decomposes calcium sucrosate.

Very often defecated juice is treated with sulphur dioxide (*sulphitation*) instead of carbon dioxide which serves the same purpose and bleaches the juice in addition, and produces a juice with much lighter colour. It prevents formation of brown mass by oxidation, and brings about coagulation of gums and albuminoids more effectively. During the process of sulphitation the solution is maintained neutral. In some cases '*sulphitation*' follows '*carbonation*'. The juice is filtered again to remove the precipitates.

At present all big sugar factories in India are following double carbonation and double sulphitation process for the manufacture of cane-sugar.

Concentration and crystallisation. The clear juice is concentrated in a multiple effect evaporator. Juice in the first pan is heated by exhaust steam from the engines of the factory. The concentrated juice from the first pan is taken to the second pan and heated there at a lower pressure, by steam from the first evaporator. The concentrated solution from the second pan is taken to the third pan and heated there at a still lower pressure by exhaust steam from the second evaporator.

To this concentrated juice sulphur dioxide is again applied. All through this process a strict control is maintained over the acidity of the solution otherwise there will be losses due to inversion, destruction or even discolouration may be there.

The clear syrupy juice is just boiled in a vacuum pan till formation of sugar crystals begins. The contents of the vacuum pan (*massecuite*) are taken into the crystallizing tank and allowed to cool slowly when the tiny crystals of sugar grow in size.

Separation of crystals. The crystals alongwith the mother liquor (molasses) are whirled in centrifugal machines wherein the molasses is removed. A little molasses adhering to the crystals is removed by spraying cold water and whirling in the centrifuge again. The crystals are given a little blue colour and dried by dropping through a long pipe through which hot air is passing up and bagged.

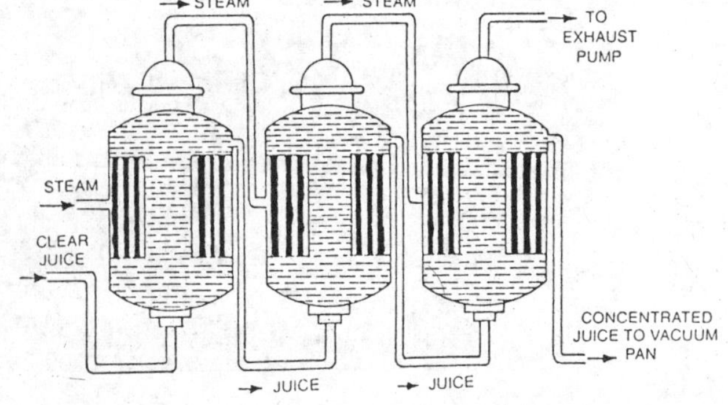

Fig. 11.6. Sugar manufacture. Multiple effect evaporator.

Refining of sugar. Sugar in India is generally put in the market as obtained in the above process. In U.S.A. and other foreign countries it is further refined by dissolving the crystals in hot water and decolorising the solution with animal charcoal or coconut charcoal. It is followed by filtration, concentration under reduced pressure and crystallization as usual.

Recovery of sugar molasses. The molasses still contains sufficient amount of sugar. It is boiled over again to get a fresh crop of crystals. To recover sugar from the final molasses, the latter is treated chemically. It is diluted and treated with hot concentrated solution of strontium hydroxide. Sucrose present forms a precipitate of strontium sucrosate. This is separated, suspended in water

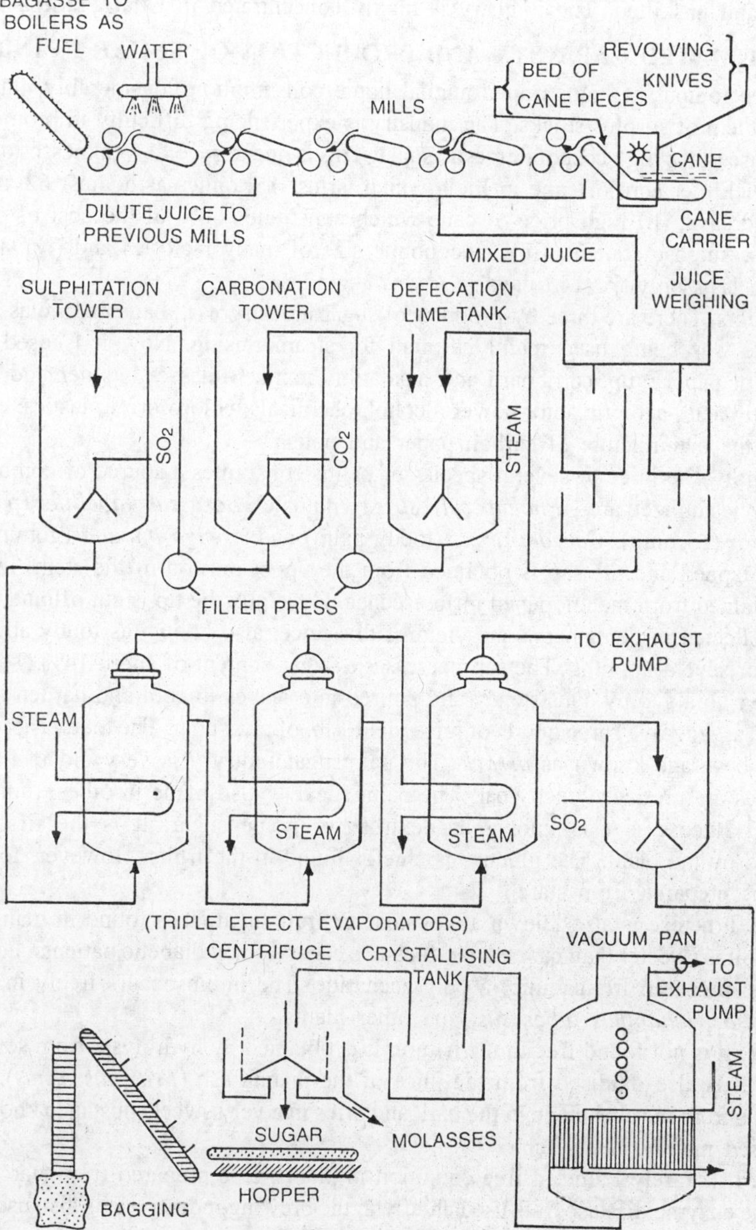

Fig. 11.7. The sugar industry. Flow sheet of the double carbonation and double sulphitation process for the manufacture of cane-sugar.

and treated with carbon dioxide when strontium carbonate separates as a white precipitate and sucrose present in the solution is decolorised, concentrated and crystallized as above.

In India sugar is not extracted chemically from molasses although still about 30 to 35% sugar is there. It is now being sent to distilleries where it is fermented and used in the manufacture of alcohol.

Sugar Industry in India. India is the accredited birth place of sugar cane and sugar. However, the Industry progressed considerably after 1931 when it was granted protection. Today, the industry is the second largest in India, next only to textiles. The industry employs about 3 lakh workers, both skilled and unskilled. The industry is highly concentrated in Uttar Pradesh and Bihar.

ESTIMATED CAPACITY AND PRODUCTION OF SUGAR IN INDIA

The present output has outstripped annual home consumption considerably, thus presenting a serious problem of surplus stocks. The industry is experiencing difficulty in promoting exports as well, because of the high cost of domestic sugar. This is due to several weaknesses of the industry *viz.* (i) low yield per acre of cane in India which is just 15 tonnes as against 62 and 56 tonnes in Hawaii and Java, (ii) high price of cane which constitutes 60% of the cost of production of sugar and low sucrose content, (iii) uneconomic size of many factories and (iv) short crushing season which lasts only for 145 days in a year in India.

By-Products. There are three by-products of sugar industry, *viz.*, bagasse, molasses and press-mud. Bagasse was being used mainly as fuel for steam raising. Now it is used also for the manufacture of paper pulp, cardboard and insulation board. Molasses has demand in the preparation of acetic acid, aconitic acid, power alcohol, chemicals, chloroform, plastics, etc. Pressmud finds use in the manufacture of carbon paper and polish.

Palm Sugar. The juice of several species of palms constitutes a source of commercial sugar. The chief species utilized are—*Phoenix sylvestris* (wild date), *Borassus flabellifer* (palmyra palm), *Cocos nucifera* (coconut), *Caryota urens* (toddy palm) and *Arenga pinnata* (gomuti palm). The date palm is tapped and the sap is obtained from the upper portion of the stem. In other palms the sap is obtained from the unopened inflorescence. Generally the tip is cut off and the sap oozes out and is collected in containers. The yield of this juice, also known as toddy amounts to 3 to 4 qt. a day for several months. The sap possesses a sugar content of about 14%. The sweet juice is boiled down to a syrupy consistency and poured into leaves to cool and harden into the crude sugar, known as jaggery. Three quarts of juice yield 1 lb. of raw sugar. The toddy is often fermented to make the beverage known as *arrack*. The palmsugar industry is very old in India and over one lac tonnes are produced each year. The palm sugar is also made in other tropical countries.

Glucose. Glucose is also known as dextrose or grape sugar. It is the first sugar to be manufactured by the plant. The glucose is chiefly found in the fruits. However, for commercial purposes it is prepared from starch.

Fructose. Fructose is also known as levulose or fruit sugar. It is found in many fruits along with glucose. It is sweeter than cane-sugar. It can be eaten also by diabetic patients. For commercial purposes it is prepared from inulin, a polysaccharide. The inulin occurs freely in the tubers of *Dahlia pinnata*, *Helianthus tuberosus*, and other plants.

Mannose. It is not found free in the nature. It is obtained by hydrolysis from several complex compounds. It is also oxidised from the juice of the manna ash (*Fraxinus ornus*). The juice of this tree oozes from the slits made in the bark and dries in a very sweet substance known as *manna*. Manna is used mainly in medicine.

Maltose. It is rarely found in free condition in plants. It is prepared from starch through the activity of the enzyme diastase. It is used chiefly in the brewing industry. It is also used in medicine as a substitute of glucose. In Japan, the maltose obtained from the rice starch is used as a flavouring material.

Honey or nectar. Many flowers secrete a sweet substance, known as nectar in their special glands, the nectaries. It is composed chiefly of sucrose, fructose and glucose. It is used as food by bees. After partial digestion it is being converted into honey and stored up by bees for future use. During this process the sucrose changes into invert sugar, a mixture of fructose and glucose. Honey is used chiefly in medicine. It has been used by man as a food and sweetening material for countless ages. It makes an excellent food for man, for it is almost pure sugar. Madhuban at Mahabaleshwar in Maharashtra is famous for honey industry.

Uses of sugar. The sugar is used as an article of food and as sweetening agent in sweets and drinks. It is used in food preservation and in the preparation of oxalic acids in the laboratory. It is used also for the manufacture of octa-acetate.

Sugarcane juice is used mainly for three products used in foods and beverages, for sweetening purposes, *viz.*, *Gur* and jaggery, vacuum pan sugar, and open pan sugar or *khandsari*. A good quantity of sugar is also used for chewing. Byproducts obtained during the manufacture of *gur* and sugar are molasses. *Gur* is used for direct consumption by human beings and livestock. Sugar is used for sweetening milk, tea and coffee, and for making sweets.

STARCH

Starch occurs in all green plants. It is a complex, carbohydrate. Commercial sources of starch are wheat, barley, maize, potatoes and arrowroot. In these plants starch occurs in the form of loosely packed granules which vary in shape and size. The underground legumes, nuts and other plant organs may contain appreciable amount of starch.

Manufacture of starch. The process of manufacture of starch from corn or any of the other starchy materials consists in rupturing cell walls when the enclosed starch granules are exposed and washed to remove cellulose.

The grains or other plant parts, from which starch is to be manufactured, is washed with water in a washing machine when any dirt sticking is washed away. Rice if used as raw material, is treated with dilute caustic soda solution before washing to remove gum and gluten (cellulose and protein materials). After washing it is disintegrated mechanically when a paste-like mass called pulp is obtained containing starch granules exposed after disintegration alongwith finely divided fibrous tissue. The mixture is taken to a shifting machine and thoroughly washed with water and then sieved through fine brass-wire sieves when coarser particles of cellulose are left behind on the sieve and are used as fodder. Smaller particles of starch and some fibrous matter pass down as milky aqueous suspension. This is allowed to stand when heavier starch granules settle at the bottom. The lighter particles of cellulose are deposited to a very small extent and that too only in the last layers and are scraped off. The starch pulp so obtained is again treated with water when starch particles settle soon and the supernatant layer is decanted off. It is then centrifuged to free it from the bulk of water present and slowly dried in air or by heating it gently in ovens till it contains only 12–20% moisture when it is ready for the market.

Starch is tasteless, odourless, white amorphous powder insoluble in water. When boiled with water, starch granules swell up and burst to form a colloidal solution called starch paste. Starch granules vary in shape and size with the source.

Uses of starch. The starch and starch products are variously used:

The better grades or corn starch are used for food. The arrowroot starch is very easily digested and so is valuable as a food for children and invalids. The sago starch is used for food purposes.

Starch is extensively used in textile industry to strengthen the fibres and cement the loose ends together, thus making a thread that is smoother and easier to weave. It is also used for giving a finish to cotton fabrics.

Starch is used as a sizing agent in the paper industry.

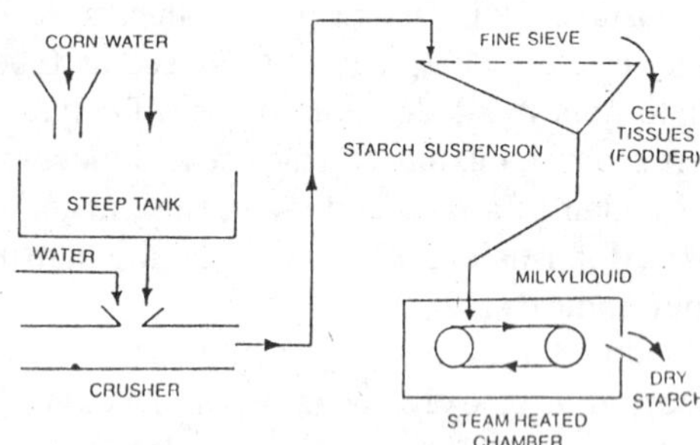

Fig. 11.8. Starch manufacture. Flow sheet for the manufacture of starch.

Large quantities of starch are used for manufacture of glucose. Besides dextrin, industrial alcohol and nitrostarch are manufactured also from starch.

Starch makes the basis of brewing industry as raw material in the manufacture of alcohol. It is used also in manufacture of dextrin and adhesives (starch paste).

It is used in laundry work as a stiffening agent, in medicine, in the preparation of toilet powders, as a binding material for china clay and as a source of many other starch products.

In calico-printing it is used as a thickening agent for colours.

Starch acetate, a derivative of starch, is a transparent gelatin like mass. This is being used mainly for making sweets.

Nitrostarch, another derivative of starch, is used as an explosive for blasting purposes.

Sago starch. It is obtained from the stems of sago palm (*Metroxylon sago*), a tall tree of tropics. It is cultivated chiefly in Malaya and Indonesia. The flowering takes place when the trees are about 15 years of age. Prior to the flowering stage the stems store up a large amount of starch. The trees are cut down and the starch pith is removed. This is ground up, mixed with water and strained through a coarse sieve. The starch is freed from water by the process of sedimentation. After washing and drying it becomes sago flour. The commercial sago is prepared from this by making paste and rubbing it through a sieve in order to bring about granulation. The product is dried in the sun or ovens. It is known as pearl sago. It is used as food. It makes a good food for patients.

The other familiar starches are known as corn starch, potato starch, wheat starch, rice starch, tapioca starch and arrowroot starch. The starch serves as the staple food for animals, and man as well and it has so many industrial applications.

Potato starch. It is obtained from potato tubers. The culls are utilized for making starch. They are washed and reduced to a pulp and the paste is passed through sieves to remove all the fibrous matter. After washing, the solid starch is separated out by sedimentation, centrifuging and then dried. Potato starch is used in the textile industry, and as a source of glucose, dextrose and industrial alcohol.

Rice starch. This is obtained from broken or imperfect grains of rice. The broken pieces of rice are softened by treating with caustic soda, and are then washed, ground and passed through fine sieves. More alkali is added and after some time the starch is sedimented. This is removed, washed and dried. Rice starch is used for laundry work, and for sizing the fabrics.

SUGAR INDUSTRY

The sugar industry is very important to the Indian national economy, because of its multiple contributions in the shape of employment and provision of raw materials to other industries. This industry provides employment to about 35 million cultivators and four lakh skilled and unskilled workers. Further it provides employment to thousands of people in the sugar trade, in the transport of sugar cane and sugar. The byproducts of the industry are used as raw materials in various industries such as alcohol, plastics, synthetics, rubber, fibre board, pharmaceuticals, paper, etc. In recent years, the sugar industry has begun to export sugar, thus earning valuable foreign exchange. Thus the industry ranks second among the major industries of this country, only next to cotton textile industry.

Location

The sugar industry is mostly oriented to sugar cane which forms 60 per cent of the total cost of production, and therefore, this industry is naturally conditioned by the availability of sugar cane and facilities of transportation of raw material to the sugar mill. Proximity to the raw material is essential because the sucrose content of the sugar cane begins to decrease soon after the cane is cut. Since bagasse is used by the factories for generating power and therefore, power is not at all a dominating factor in determining the location of sugar factory. Sugar cane grows both in tropical and sub-tropical regions. In India, Andhra Pradesh, Tamil Nadu, Karnataka, Kerala and Maharashtra come under tropical regions, whereas Uttar Pradesh, Bihar, Punjab, Haryana, West Bengal, Orissa, Rajasthan and Madhya Pradesh come under sub-tropical regions.

There were only 29 factories in India during the year 1931. The number of factories in operation has increased from 29 to 140 in 1950-51, out of which 110 factories were in North India. During the next decade the number of factories increased to 174 out of which 116 factories were in the sub-tropical regions of northern India. In 1965-66 the number of factories was 200 and which has grown to 344 in 1985-86 of which nearly 75 per cent of the factories are located in northern India. The industry is predominantly localised in Uttar Pradesh, particularly in Meerut, Saharanpur, Muzaffarnagar, Bijnor, Bareilly, Moradabad and Rampur districts. Next to Uttar Pradesh the industry is mainly developed in Maharashtra, Bihar and in the eastern coastal districts of Andhra Pradesh. The unique position which Uttar Pradesh enjoys in respect of cane cultivation is due to the advantages conferred by the rich and fertile alluvial soil of the Gangetic plain, the bulk of which contains adequate quantities of lime and potash. The presence of thin varieties of cane admirably suited to the climatic conditions of the region and the existence of cheap and extensive irrigation facilities. On the other hand the concentration of sugar cane crop in compact blocks enables the sugar factories to get fresh supplies of sugar cane direct from the fields.

Now a days the locational factors have influenced the dispersal of sugar industry to the south. The sucrose content in the sugar cane grown in the tropical regions (southern regions) is greater than subtropical regions, and the development of the cane in the south is mainly responsible for bringing about locational changes in the industry.

Problems

Sick units. One of the major problems of the industry is the existence of a large number of 'sick' units. They are unable to run at 'break even' points. The reason attributed to this is use of outdated machines resulting in inefficiency and high cost structure. In northern India, this problem is particularly acute and distressing. To reduce these evils, it becomes essential to rehabilitate and modernise these units and if necessary the old units should be replaced with new ones having bigger capacities. On the other hand if the sickness is due to unfavourable climatic conditions, the units could be relocated in an area where agroclimatic factors are more congenial. Relocation of sugar industry near to the agro-climate factors in the opinion of Indian Sugar Mills Association (ISMA), is only one of the factors influencing the quantity and quality of cane production. The ISMA, therefore, pleaded to give more State assistance for various programmes, such as the improvement of soil conditions, the prevention of water-logging and the increase in the use of agriculture inputs.

Regulatory measures. On the food front, sugar industry has attracted much public attention than any other industry, because it is an agro-industry and an essential item of mass consumption, and therefore, all sections of people, such as legislators, politicians, farmers and the public at large intervene to exert pressure in different directions. Sugar has been put to the most rigid control to which no other industry has been subjected. The factories situated in northern India showed a tremendous progress and expansion in the mid-fifties when the industry was free from controls. The reverse has happened when the control was laid in the north and let the south free from controls. Thus a chain of regulatory controls has emerged. This has adversely affected the overall performance of the factories in almost all the States. Most of the units incurred losses.

Payment of higher prices for cane. The sugar mills have to pay much higher price for the purchase of cane than what statutorily fixed. Sometimes the factories had to pay about fifty per cent more than the statutory cane price. As the cost of cane accounts for about 75 per cent of the total production cost the additional payment over and above what had been statutorily fixed would naturally have an adverse impact on the financial health of the industry.

Beet Sugar

In India, Kinnaur district in Himachal Pradesh has the distinction in the production of quality sugar beet. The unique reason for this is the temperate and dry climate which is ideal for its cultivation. The scientists have proved that sugar produced from sugar beet is of higher quality than that of cane ranges from 10 to 15 per cent while the content varies from 15 to 20 per cent in sugar beet. No additional machinery is necessary in the existing sugar mills for the extraction of sugar from sugar beet except an additional diffuse plant. Sugar beet is grown in winter. It matures in six to seven months. Thus it is a short duration crop when compared with sugar cane which takes as many as 12 to 18 months. It can also be raised as an inner crop with sugar cane. Nutritive food for live stock can be had advantageously from the by-products of sugar beet. Its molasses is also used in the manufacture of citric acid, yeast, antibiotics and others.

12

Pulp and Paper Industry (A Cellulose Product)

Paper plays an increasingly important role in modern civilised society. Traditionally, it has been an indispensable material requirement for the educational and cultural advancement of the people. Writing and printing papers have constituted the principal media of mass communication and expression. Today paper is finding expanding outlets in the industrial sphere. Besides packaging many outstanding developments have taken place in the industrial applications of paper products. Sound recording and producing paper, paper clothing, paper housing units, rust preventors for steel, and quality preserving paper are some conspicuous examples in this field.

Paper is divided into two categories–cultural (printing and writing paper) and industrial (packing and wrapping paper and boards). Cultural paper accounts for 60 per cent of the total demand and industrial 40 per cent. India's consumption of paper (domestic production + imports) rose from 5.75 lakh tonnes in 1965 to 7.73 lakh tonnes in 1970 and further to a little over 9 lakh tonnes lakh in 1977. Paper production in 1976 was 8.80 lakh tonnes and the estimated 1977 production was 9 lakh tonnes against an installed capacity of 11.7 lakh tonnes.

The less developed countries are actively engaged in programmes to upgrade the quality of their human resources through education and cultural development. Writing and printing papers would be increasingly needed to cope with the spread of education and the consequent disappearance of illiteracy in these regions.

The greater part of the industrial papers and paperboards is used in packaging. Rising economic activity stimulates the use of all types of packaging. The industrial papers are and would continue to be much in demand to feed the diversifying distribution outlets of the new and expanding industries.

Three major factors, namely, the degree of literacy, the growth of the industrial sector and the role of mass media, influence the demand for paper and paperboards, and account for its wide variations.

RAW MATERIALS

1. **Wood resources.** The paper and pulp industry draws its raw material requirements mostly from the forest. Wood occupies a position of pre-eminence as a raw material for paper making. In 1963 for example, wood alone provided approximately 77 per cent of the total raw material used in manufacture. In the past, the paper and pulp industry used extensively the long-fibred softwood conifers, which have been most economical and ideal species for paper making.

2. **Non-wood fibres.** Agricultural residues (bamboo, bagasse, straw) form another source of pulpable raw materials. The output of non-wood fibre pulp more than doubled from 1.9 million metric tons in 1950-52 to 4.1 million metric tons in 1961. In spite of this spectacular growth, non-wood fibre pulp accounts for a mere 5 per cent of the total. It is a subsidiary source of supply.

3. **Waste-paper.** The use of waste paper is quite substantial in the major producing areas of the world. It is the principal raw material, particularly in some packaging grades of paper and paperboards. The material furnishes about 25 per cent of the total requirements of the paper and pulp industry in the United States, Europe and Japan. In the developing countries the amount of waste paper available for pulp manufacture is relatively insignificant on account of the limited consumption of paper, and the low recovery rates which vary from 15 per cent in India to 5 per cent in Ceylon and 3 per cent in Pakistan. Large quantities of waste paper are directly re-used as wrapping paper and for various other domestic purposes.

INDIA'S PAPER UNITS

Since the establishment of India's first paper mill at Serampore (West Bengal) in 1832, India's paper has emerged as one of the important core industries. India is almost completely self-reliant in paper. Only some very special varieties are being imported, the quantity being a negligible 15,000 tons per annum. 75 units are engaged in the production of paper. Of these, 60 are small-scale units. The organized sector accounts for over 80 per cent of the total output. Leaders in the industry are: Birlas, Thapars, Somanis and Bangurs. These four groups account for nearly half of the organized sector's production. Most of the new units proposed to be set up both in the organized sector and small-scale sector were actually conceived during the boom period of 1974 and 1975. Mini paper plants are considered to be an effective solution to the various ailments of the paper industry as they involve less capital expenditure and have a short gestation period. They have proved to be technically feasible and economically viable for production capacities in the range of even 15 to 30 tonnes per day. These plants can be erected with fully indigenons know-how and machinery. They are expected to ease the burden on the conventional raw materials as they will be using waste paper, rags, agricultural residues, etc.

SUPPLY OF RAW MATERIALS IN INDIA

1. **The pattern of utilization.** The present pattern of utilization of various pulpable raw materials reflects to a great extent the peculiar endowment of natural resources in India. While wood is the principal fibrous raw material in all the leading paper manufacturing countries, the non-wood fibres constitute the principal source of supply to the Indian paper industry. This has been due to the lack of suitable wood species economically located for the manufacture of paper.

The most important pulpable raw material has so far been a non-wood fibre, *viz.*, bamboo which meets over 60 per cent of the total requirements of raw materials. In fact, bamboo is the mainstay of the paper industry in India. India is the only country where bamboo is used for pulping on such a large scale. Pulpwood is also utilized on a substantial scale. The consumption of bagasse is expanding rapidly. The non-wood fibres are by far the most important raw materials for paper making in India.

2. **Resource availability.** The following resources are available for pulp and paper industry in India.

In India bamboo is the basic raw material and constitutes about 80 per cent of the total supply of raw material. The remaining 20 per cent are met by grass, straw, jute sticks, waste paper, rags, etc. Eucalyptus and other hard woods are also used. The National Commission on Agriculture has estimated the requirements of pulp and bamboo at 5 million and 2.5 million tonnes respectively by 1980. It has suggested an additional investment of Rs. 242 crores by 1980, another Rs. 1000 crores between 1981 and 1990 for clear-felling and plantation of forests for the paper industry alone.

Bamboo. Bamboo is the main long fibre raw material which has been in use since the twenties. It grows abundantly throughout tropical and sub-tropical regions. The net availability, *i.e.*, surplus for industrial purposes has been put around two million tons on the basis of known existing resources. The pulp and paper industry is consuming about half (nearly a million tons) of the available surplus. The indispensability of this raw material can be gauged from the fact that all the available cellulosic raw materials like hardwoods, bagasse, straw, etc., are short fibred materials, the pulp from which will have to be blended with 25% of long-fibred pulp to make paper of good quality and adequate strength.

Grasses. *Sabai* grass grows abundantly in Northern India and possesses qualities closely resembling esparto grass. It is collected from sub-Himalayan tracts, mostly Nepal Terai and Sahibganj (Bihar). The other grasses which are suitable for paper manufacture are--ulla, panni, spear grass and elephant grass. In all, above 300 to 400 thousand tons of various grasses, including sabai, may be annually available in the country out of which between one-fourth and one-third is taken by the pulp and paper industry.

Straw. As an agricultural country, India possesses a huge potential of straw. Out of this enormous quantity in pulp and paper industry takes a nominal quantity of 30 thousand tons.

Bagasse. India, which grows 100 million tons of sugarcane, offers a large potential of bagasse, a by-product of sugarcane processing. The current supplies of bagasse to the paper industry are obtained from large and modern sugar mills which are left with surplus bagasse on account of their improved steam efficiency. However, in relation to the total availability of bagasse, its consumption in the paper industry is still negligible. In 1994-95, the pulp and paper industry procured only 70 thousand tons of bagasse, *i.e.,* 1.5% of the total.

Wood. Wood has been the conventional raw material for paper but in India it has not assumed any major importance due to the limited availability. Species of woods which may be suitable for pulping purposes are widely scattered over a large area. The forests are relatively expansive to harvest, handle and process and hence they have remained untapped. The pulp and paper industry has absorbed approximately 0.23 million cubic metres of wood in 1994 as against the net increment in the supplies of wood fibres assessed at 23.3 million cubic metres. Thus a fraction of total wood supplies (about 10%) is currently going into pulp making.

Waste paper. Waste paper is one of the first sources of the fibrous materials but in India this source is not being tapped as fully as in other countries. The current recovery rate is 15 per cent as against 30 per cent in advanced countries. The sources of waste paper are : (1) printing, converting and packaging establishment, (2) government offices and private business houses, and (3) street sweepings.

The choice of a particular raw material is determined by its economic as well as technical suitability. Various points considered are (*i*) ample supply of the raw materials, (*ii*) availability of raw materials in all seasons of the year, (*iii*) the yield and quality of the fibre it gives, and (*iv*) its cost of collection and transport as well as the cost of converting it into paper.

Fig. 12.1. *Paper Industry.* In India bamboo is the basic raw material and constitutes about 80 per cent of the total supply of raw material.

MANUFACTURE OF PULP

The first step in the manufacture of paper is conversion of the fibrous material into pulp. Two processes used for the purpose are:

1. **Mechanical process.** Mechanical pulp is obtained from relatively soft wood like pine. The short, unbarked logs are subjected to mechanical grinding in presence of water. The pulpy material thus produced contains all the matter (ligno-cellulose) present in the original wood. Mechanical pulp is difficult to bleach and is used for manufacturing inferior paper such as newsprint, wall paper, etc. This paper turns yellow with age and cannot, therefore, be used for printing books, etc.

2. **Chemical process**. In this process the material is cut into chips and digested under pressure ("cooked") with either (*i*) a solution of calcium bisulphite Ca $(HSO_3)_2$ (sulphite process) or (*ii*) a solution of caustic soda (soda process). This dissolved out the lignin

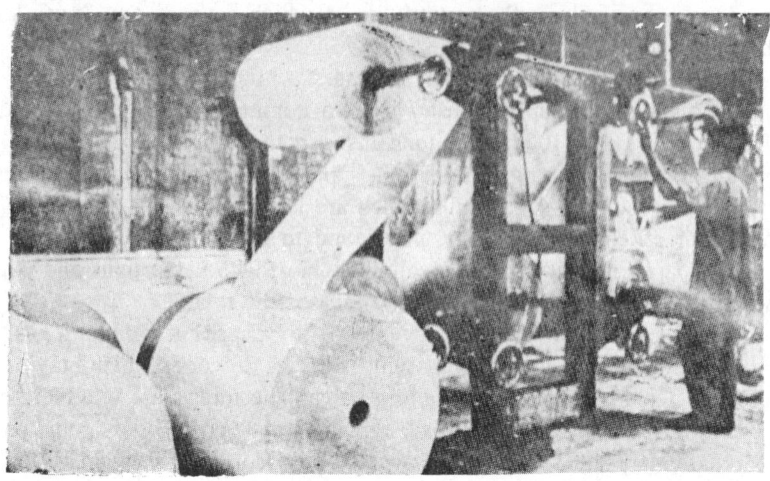

Fig. 12.2. *The paper industry.* Paper plays an important role in modern civilized society. Writing and printing papers have constituted the principal media of mass communication and expression.

which together with cellulose makes up the solid part of its structure. The product obtained is more or less pure cellulose. It is technically known as "half stuff".

The pulpy material so obtained, is washed and screened. It is then bleached by adding chlorine and water or with "bleach liquor". The bleached pulp is beaten with water in a "pulp beater". Beating is the process of reducing the fibres to minute shreds and causing them to absorb and hold more water–an operation often referred to as "hydration".

Filling and sizing. The beater is used as a convenient place to mix filling materials, colouring matter and sizing. Fillers or loading materials are various inorganic substances like clay, calcium carbonate, calcium sulphate, talc, titanium and zinc sulphide. These give the proper body to paper, improve the appearance, make better printing surface, increase the capacity and reduce the costs. Some ultramarine is added to give it the desired shade. Some other colouring materials may be added to give it the desired shade of colour. To prevent spreading of ink as on a blotting paper, the paper requires "sizing". For this purpose some sizing materials like resin, starch, glue or casein are either added to the beater or applied to the sheet of paper during its manufacture.

Manufacture of paper. The thin suspension obtained from the beater is called 'soup' by the paper maker. It is taken to a tank called the 'stock box'. The soup is forced through a thin slit in the bottom of the stock box. It falls on an endless band of fine wire gauge screen in motion. As the screen moves forward, the water present in soup drains through it and the fibres mat together. The process of draining is assisted by vacuum from the suction boxes over which the wire gauge screen travels. The fibrous sheet next passes between heavy press rolls covered with felt. Here the excess of water is removed. The sheet is then passed between drying cylinders which are steam heated. In some cases when sizing materials have not been added in the beater, these are applied here in the drier section or even after the driers. Sizing materials are made into a paste and applied to one or both sides of the paper sheet. The excess material is removed as the sheet passes between squeeze rolls.

Calendering. The surface of the dried sheet is generally rough and irregular. It is passed through a series of very hot and polished rolls. As a result of it the surface becomes compact and smooth and it takes a fine glaze. The process is known as *calendering*. After calendering the sheet goes to the reel where it is wound into large rolls of finished paper. The roll when full is sent to the slitting machine for cutting paper into rolls of proper width and diameter.

Uses. The paper is variously used. Some more important uses are given here:

(*i*) Tissue paper is the lightest weight thin paper used for napkins, toilet paper and light weight wrappings. (*ii*) It is used for making bags and envelopes and as packing paper. (*iii*) It is used for writing purposes. (*iv*) It is used for printing books, journals, magazines and newspapers. (*v*) Thick bulky papers are sometimes used with asbestos and used in construction of buildings. (*vi*) It is used in the manufacture of cardboard finding variety of uses such as making containers, wallboard, bottle caps and for binding purposes.

NAMES OF IMPORTANT PAPER MILLS OF INDIA WITH THEIR LOCATION

Name of the mill	*Location*	*Name of the mill*	*Location*
1. Titaghur Paper Mills Ltd.	West Bengal	9. National Newsprint and Paper Mills Ltd.	M.P.
2. Bengal Paper Mills Ltd.	West Bengal	10. Ballarpur Paper and Straw Board Mills Ltd.	Maharashtra
3. Orient Paper Mills Ltd.	Orissa	11. Sirpur Paper Mills Ltd.	A.P.
4. Straw Products Ltd.	Orissa	12. Andhra Pradesh Paper Mills	A.P.
5. Star Paper Mills Ltd.	U.P.	13. West Coast Paper Mills Ltd.	Karnataka
6. Rohtas Industries Ltd.	Bihar	14. Seshasayee Paper & Board Ltd.	Tamil Nadu
7. Shree Gopal Paper Mills Ltd.	Punjab		
8. Orient Paper Mills Ltd.	M.P.		

Paper Industry in India. The manufacture of machine-made paper in India dates back to 1867 when the first mill was established on the Hooghly. There were 58 paper mills in 1967-70 in West Bengal; 3 in Orissa; 2 in Uttar Pradesh; 2 in Bihar; 4 in Punjab; 5 in Madhya Pradesh, 6 in Gujarat; 14 in Maharashtra; 2 in Andhra Pradesh; 5 in Karnataka; 2 in Kerala and 3 in Madras (Tamil Nadu).

POSITION OF NEWSPRINT IN INDIA

Hopes of India becoming self-sufficient in newsprint are dim in near future. The industry is capital-intensive and the raw materials are available only in few areas–Jammu and Kashmir, Himachal Pradesh, Madhya Pradesh and the North-East. It was proposed to exploit coniferous forests in the Himalayas. But bringing down chipped wood in water slurries appears to be the major obstacle in tapping this resource. Sugarcane bagasse was considered to be a promising source but there are difficulties. Of the approved projects in Kerala, West Bengal, Uttar Pradesh and Karnataka, only Hindustan Paper Corporation's Rs. 100 crore Kerala project with an annual capacity of 80,000 tonnes was materialised. India's present demand for newsprint is around 2.50 lakh tonnes per year. The domestic production from the only unit, Nepa Mills in Madhya Pradesh, is only less than 60,000 tonnes per annum. Hence our dependence on imports will continue for long.

KINDS OF PAPER

Paper serves a multitude of purposes, and hundreds of paper products are in daily use. Different kinds of paper are made from different raw materials. *Fine papers*, include bond, ledger, book, writing and other such papers of similar quality. The best grades of this kind of paper are made from sulphite pulp or rags. The paper of medium grades are made from mixtures of rag and chemical pulp, whereas the paper of poorer grades is made from wood pulp. *Newsprint paper* generally consists of 85 per cent mechanical wood pulp and 15 per cent unbleached sulphite.

Wrapping paper is made from kraft pulp for the most part, while various textile fibres are also used as raw materials. *Blotting paper* is composed of loosely interwoven short fibres, and formerly it was made entirely from cotton rags but nowadays soda pulp is also used for its preparation. *Paper board* is prepared from any low grade materials, often in combination with mechanical or chemical pulp. *Vegetable parchment* is made from the purest alpha cellulose stock. *Bags* are usually made from unbleached kraft. *Carbon paper* is a strong linen or cotton rag paper coated with a coloured wax. *Facial tissue* is made from 100 per cent sulphite pulp. *Cigarette paper* is made from flax. *Wall papers* are manufactured by treating paper with wax, tar, oils, resins or other substances.

PAPER INDUSTRY

The importance of paper and paper products in modern life is tremendous and no other manufactured product possesses such a diversity of use. "Paper pervades all sectors of our activity from books to bullets and from our morning newspaper to nuclear technology. Paper is the basic medium of communication and dissemination of information. Paper serves not only the needs of arts, literature and science but also provides the means of packing and building. It is paper of various descriptions that go into our every day life".

We can safely say that this is 'paper age' and we cannot measure the importance or the utility of the paper and paper goods in terms of money or something else because it is an economic product of necessity. The paper industry has made considerable progress since 1951, when the country embarked upon a programme of planned economic development, and now India ranks as the twentieth paper producer in the world. The industry is providing employment to nearly one lakh persons, with an investment of about Rs. 130 crores. The per capita consumption of paper reflects the degree of economic development and the progress of civilization of a country. When compared to international standards, the per capita consumption of our country is very low. The annual per capita consumption in this country was 2 kg in 1977-78 as against 257 kg in the U.S.A., 134 kg in Canada, 106 kg in the U.K., 73 kg in Japan, 16 kg in the U.S.S.R (now Russia) and 3.4 kg in Sri Lanka.

The paper industry can be divided into three sectors: 1. paper and paper board—mainly engaged in the production of (a) printing and writing paper, (b) wrapping paper, (c) special varieties and (d) paper board, 2. straw boards and other boards, and 3. news print. A new sector, called the security paper has also been added to the existing list.

Location

For quite a long time, the paper industry was predominantly localised in the Hooghly Riverine. Though the Sabai grass was the principal raw material, which had to be carried over a distance of more than 1440 kilometres from places like U.P., Punjab, Nepal and Bihar, yet the industry was predominantly localised in the Hooghly area. This is because of (1) availability of coal in large quantities, and (2) proximity was found more economical to pay freight on two and half tonnes of grass rather than to pay freight on five tonnes of coal for nearly the same distance, and on finished product. Even today the industry is mainly concentrated around Calcutta and Raniganj area as the paper mills have a good market in Calcutta and its suburbs. Most of the paper mills in Calcutta produce stationery and printing paper for which there is a ready market.

Bamboo is another important raw material, and has been developed as a substitute of Sabai grass. This was also responsible for location of paper industry at some places. The paper mills at Brajrajnagar, Dalmianagar, Raniganj, Calcutta, Nepanagar, Ballarpur, Dandeli, *etc.,* have sprung up near the bamboo forests and they consume bamboo as the main raw material for cellulose fibre. About two-thirds of the total produce of bamboo in Bihar is consumed by five mills respectively at Titagarh, Kankinara, and Naihati (suburbs of Calcutta), Raniganj and Dalmianagar. The smaller paper mills in and around Calcutta consume waste paper obtained locally and wood pulp imported from abroad. For these raw materials, these paper mills are suitably located.

At the beginning of the first five year plan (*i.e.,* 1952), there were 17 paper board mills. These increased to 106 in 1979 with installed capacity of 14 lakh tonnes. Of these units, 24 were in Maharashtra, 12 in West Bengal and 16 in Gujarat. These 52 units together accounted for 34 per cent of the total capacity of the industry. There has been a considerable dispersal of productive activity in the paper industry with the increasing use of bamboo pulp and with the development of alternative sources of power particularly hydro-electricity. The use of bagasse as a raw material is also responsible to a certain extent for the dispersal of the paper industry from the concentrated areas. As a consequence of agreement with sugar mills for the requirements of bagasse, few paper mills, such as Dalmianagar have expanded their production and new mills at Belagula and Jitwerpur have been established.

Size

The economic size of a unit in paper industry is determined by the availability of raw materials, proximity and density of markets, availability of cheap power and transport facilities both for assembly of raw materials and distribution of finished products. At the beginning of the first five–year plan, there were only 17 paper and paper board mills with a total annual capacity of about 1.39 lakh tonnes, production in 1951 was 1.34 lakh tonnes. At present there are 106 mills with a total annual capacity of 13.94 lakh tonnes and production is about 11.12 lakh tonnes. In addition there are three pulp mills (two producing rayan grade pulp and the other paper grade pulp) and one newsprint mill with annual capacities of 117,600 tonnes and 75,000 tonnes respectively.

In India, the units of diverse size exist. These are units very well organized and well equipped with a productive capacity of more than 50,000 tonnes and units to small with a productive capacity of 1,000 tonnes. In between these two extremes, the units vary in size, many of them have a productive capacity ranging from six to fifteen thousand tonnes. The principal factors responsible for the wide variations in size of the paper mills are the availability of raw materials like bamboo or grass in required qualities, proximity and density of consuming market, availability of cheap labour, abundant supply of cheap skilled labour and adequate transport facilities for assemblage of raw materials and distribution of finished products.

In view of the large amount of capital required to establish an integrated pulp and paper plant, the Planning Commission has favoured small units during the third plan, so that, they could use the locally available raw materials. Also small units can be located near consuming areas and thus effect an economy in the cost of transportation. Yet small units could not reap the benefits of integrated production of pulp and paper as the size has an important bearing on the efficiency of the industry. In such a case the right policy would be to set up large-scale pulp factories in different regions for the benefits of small size paper mills. However, there has been a sudden rush for setting up mini paper plants with daily capacity ranging from 10 tonnes to 30 tonnes of paper and paper boards.

Problems

Raw materials. The paper industry uses pulp, mainly wood pulp; power, generated mainly from coal; and chemicals like caustic soda, sulphur, chlorine as the main raw materials from its production. The problem of getting adequate supply of cellulosic raw materials on a regular basis at a reasonable cost is the main problem in paper industry in India. Bamboo accounted for 80 per cent of the raw material used in the paper industry of this country. Bamboo resources of the country are almost fully tapped, except for certain areas in Orissa, Madhya Pradesh, Assam, Nagaland, Manipur and Tripura. Consequently a stage has been reached when a departure from traditional methods of pulping and paper making is essential. This implies that fibrous raw material like bagasse, jute sticks and straws would be used in large quantities in future to meet the additional requirements of paper. It is estimated that the total availability of bagasse in our country is eight million tonnes per year, but only a few paper mills are consuming bagasse as raw material which comes to only about one per cent of the total supply of the bagasse. The remaining is burnt in sugar mills as fuel. To release the bagasse to the use of paper mills, incentive must be provided for sugar mills in making available other forms of fuel at a reduced price.

Machinery technology and research. Much has to be done to improve the paper machinery, and technology relating to the designing and development of various equipment required for industry in this country. The industry still depends on imports for most of its modern equipment as indigenous capacity has yet to be developed.

Suggestions. It is suggested in some quarters that the Government should lease suitable areas to the paper industry for plantation of raw materials in the best possible manner. It is said that the industry has always pleaded for long-term leases of existing bamboo forests to paper plants so that careless exploitation by annual contractors can be put to an end and the existing forests can be better harnessed. As prohibitive transport cost inhibit the supply of bamboo from far off forests, it is also suggested that proper care should be taken to conserve and nurse the bamboo plantations near the units.

13

The Medicinal Plants—Pharmacognosy

The branch of medical science which deals with the drug plants is known as *pharmacognosy*. This science is concerned with the history, commerce, collection, selection, identification and preservation of crude drugs and raw materials. The study of the action of the drugs is known as *pharmacology*. There are several thousand drug yielding plants all over the world. Most of the plants are known and utilized by herb doctors and *ayurevdic vaids*. Only a few drug plants are cultivated. Most of the supply of drugs is obtained from wild plants growing in all parts of the world and especially in tropical regions. These drug-yielding wild plants are popularly known as *jari-butis* in India. The drug plants are collected and prepared in crude indigenous way. The medicinal value of drug plants is due to the presence of some chemical substances in the plant tissues which produce a definite physiological action on the human body. The most important chemical substances are alkaloids, carbon compounds, hydrogen, oxygen, nitrogen, glucosides, essential oils, fatty oils, resins, mucilages, tannins, gum, etc. Some of these substances are powerful poisons, and, therefore, the preparation and administering of drugs should be done by competent physicians.

CLASSIFICATION OF DRUGS

The classification of drugs in the present text is based on the morphology of the plant organ from which the drug is obtained. Here the drug plants have been considered and arranged on a morphological basis. The general categories of the drug plants are as follows:

1. **Drugs obtained from roots.**
2. **Drugs obtained from underground stems.**
3. **Drugs obtained from bark.**
4. **Drugs obtained from stem and wood.**
5. **Drugs obtained from leaves.**
6. **Drugs obtained from flowers.**
7. **Drugs obtained from fruits.**
8. **Drugs obtained from seeds.**
9. **Drugs obtained from all parts of plant.**

DRUGS OBTAINED FROM ROOTS

Aconite

Aconitum heterophyllum Wall., of family Ranuncualceae; Eng. Aconite; Hindi—*Atis*; Sanskrit—*Ativisha*.

This herb is commonly found in the subalpine and alpine zones of the Himalayas from the Indus to Kumaon, from 6,000 to 15,000 feet.

Uses. The drug commonly known as aconite is obtained from the roots. The roots are used as astringent, tonic, antiperiodic in diarrhoea, dyspepsia and cough. It contains the alkaloids, atisine, dihydroatisine, heteratisine and hetisine. It is used externally for rheumatism and internally to relieve pain and fever.

Total alkaloidal content is 0.79 per cent. Plant is considered a valuable febrifuge and bitter tonic. Roots are used for hysteria, throat infections, dyspepsia and vomiting, abdominal pain and diabetes.

Celery

Apium graveolens Linn., of family Umbelliferae (Apiaceae); Eng. Cele; Verna. *Ajmod*,;Bengali—*Randhuni*; Hindi—*Shalari*, *ajmud*; South India—*Ajmod* or *ajmoda*.

This is biennial, erect, branching herb. It is found in the hills of Himachal Pradesh and foot of the North-West Himalayas.

Uses. The root is diuretic and alterative. It is given in dropsy and colic. The seeds are carminative, diuretic, stomachic, tonic, aphrodisiac, astringent, cordial, laxative, appetizer, stimulant, anthelmintic and antispasmodic. They are used in asthma, bronchitis and liver and spleen diseases. They are also given in vomiting, urinary discharges, rheumatism, rectal troubles, flatulence, fever with cough, etc.

Indian Belladona

Atropa acuminata Royle ex Lindley, of family Solanaceae; Eng. Indian belladonna; Hindi—*Sag-angur.*

This is tall erect herb, 3-5 feet high. It is found in Kashmir and Himachal Pradesh.

Uses. The roots are used in the manufacture of tinctures and plasters. It is used as an anodyne of rheumatism, neuralgia and local inflammations.

Belladona

Atropa belladonna Linn., of family Solanaceae; Eng. Belladonna; Hindi—*Sag angur.*

This is also a tall erect herb. It is native of Europe, now cultivated in Kashmir.

Uses. The roots are used as sedative, stimulant and antispasmodic. It is used externally to relieve pain and internally to check excessive sweat, cough, etc.

Dried roots and leaves are used as sedative, antispasmodic, and anodyne; also used in ophthalmology to dilate pupils. Leaves are stimulant and narcotic, contains atropine and hyoscyamine, extremely toxic alkaloids.

Pareira Root

Cissampelos pareira Linn., of family Menispermaceae; Eng. Pareira; Hindi—*Harjori*, *Akanadi*; Sanskrit—*Ambashtha*; Bengali—*Akanadi*; Gujarati—*Venivel*; Marathi—*Paharvel*; Tamil—*Appatta*; Telugu—*Adivi banka tige*; Kannada—*Padavali*; Malayalam—*Kathuvalli.*

This is a climbing shrub, found commonly in the Punjab, Uttar Pradesh, Madhya Pradesh and South India.

Uses. The dried roots are a mild tonic, diuretic, stomachic, antilithic, alterative and astringent. The roots are used in diarrhoea, dysentery, cough, bowel complaints, catarrhal disorders and urinary troubles. The roots are used in the form of infusion or decoction. As a blood purifier in skin diseases and syphilis a dilute decoction of the roots is used. A paste of the roots is applied in skin diseases.

Colocynth

Citrullus colocynthis (Linn.), Schrad., of family Cucurbitaceae; Eng. Colocynth; Hindi—*Indrayan*; Sanskrit—*Mahendravaruni*; Urdu—*Indrayan*; Bengali—*Indrayan*, *makal*; Gujarati—*Indrak*; Marathi—*Kaduvrindavana*, *indrayan*; Tamil—*Verithumatti*; Telugu—*Etipuchchha*; Kannada—*Pavamekkekayi*; Malayalam—*Peykommutti.*

This is a trailing herb. It is native of Asia and Africa but found in Northern India, Madhya Pradesh, Gujarat and South India.

Uses. The root is bitter, pungent, cooling, antipyretic, carminative and anthelmintic. It is given in dropsy, tumours, leucoderma, asthma, bronchitis, jaundice, spleen and liver enlargement, urinary troubles, constipation, anaemia, elephantiasis and rheumatism. The fruits are used as purgative.

The fruit pulp, colocynth, a drastic hydragogue cathartic, contains *citrullin*, which was believed to be a glycoside, but is now known to be a mixture of an alkaloid and *citrullol.*

Liquorice

Glycyrrhiza glabra Linn., of family Papilionaceae, Eng. Liquorice, sweetwood; Hindi—*Mulhatti*; Sanskrit—*Madhuka*; Bengali—*Jashtimadhu*; Marathi—*Jestha madha*; Gujarati—*Jethi madha*;

Telugu—*Atimadhuramu*; Tamil—*Atimadhuram*; Kannda—*Atimadhura*; Malayalam—*Iratimadhuram*.

This is a perennial herb; native of the Mediterranean region but grown in Himachal Pradesh, Jammu and Kashmir.

Uses. The root contains glycyrrhizic acid. It is sweet demulcent, emollient, pectoral, alterative; laxative and expectorant; it is given in cough, bronchitis and other chest complaints, catarrhal condition of bowels, urinary passages, asthma, hoarseness of voice, etc. Its infusion is given in sore throat. It is also used in preparation of nauseous drugs.

Liquorice is also chewed with betel leaves. Principal constituents of liquorice to which it owes its characteristic sweet taste is *glycyrrhizin*, 2 to 14 per cent.

Indian Sarsaparilla

Hemidesmus indicus R. Br., of family Asclepiadaceae; Eng. Indian sarsaparilla; Hindi—*Magrabu, anantmul*; Sanskrit—*Anantamula*; Bengali—*Anantamul*; Marathi—*Anantamul*; Gujarati—*Sariva*; Telugu—*Sugandhi-pala*; Tamil—*Nammari*; Kannada—*Karibandha*; Malayalam—*Naruninti*; Oriya—*Onontomulo*.

This is a twining shrub, commonly found in South India.

Uses. The roots are used as substitute for sarsaparilla; they are sweet, demulcent, alterative, blood purifier, diuretic, tonic and diaphoretic. They are given in loss of appetite, dyspesia, fever, skin diseases, syphilis, leucorrhoea, genito-urinary diseases, chronic cough, etc.

Indian Jalap

Operculina turpethum (Linn) Silva-Manso., of family Convolvulaceae; Eng. Indian Jalap; Hindi—*Nisoth*; Sanskrit—*Trivrit*; Bengali—*Dudh kalmi*; Marathi—*Nishottar*; Gujarati—*Nashotar, nahotra*; Telugu—*Tellategada*; Tamil—*Shivadai*; Kannada—*Bangada balli*; Malayalam—*Chivaka*; Oriya—*Dudholomo*.

This is a stout twining herb. Found throughout India.

Uses. The white tuberous roots are cathartic. They are given as a mild hydragogue in chronic constipation, enlargement of the spleen and other disorders where a purgative is required; they are also taken in rheumatic and paralytic diseases. They are effective in dropsy, melancholia, gout, leprosy, rheumatism and paralysis.

Source of a purgative called Turpeth or Indian Jalap is available in two forms: white (*safed Nisoth*) and black (*Krishna Nisoth*). It is almost as active as true Jalap (*Exogonium purga* Benth). White turpeth is preferred, as black turpeth produces drastic purgation. Active principle is a glycosidic resin.

Rosy Leadwort

Plumbago rosea Linn., of family Plumbaginaceae; Eng. Rosy Leadwort; Hindi—*Lal chitra*; Sanskrit—*Chitraka*; Bengali—*Lal-chitra*; Marathi—*Lal-chitrak*; Gujarati—*Lal-chitrak*; Telugu—*Errachitramulam*; Tamil—*Cithiramulam*; Kannada—*Kempacitramulam*; Malayalam—*Chivappukoduveli*; Oriya—*Lal-chita*; Kashmir—*Shitray*; Assam—*Agechhit*.

This is a small shrub; native of India; found throughout India.

Uses. The root is stimulant, diaphoretic, stomachic, sialogogue, abortifacient and vasicant. In large doses it is narcotic and irritant. It must be given in small doses. It is given in dyspepsia, intermittent fevers, piles, diarrhoea, skin diseases, paralysis and rheumatism. The paste of root and some bland oil is applied over rheumatic and paralytic parts of body.

Plumbagin, an orange-yellow pigment, is the active principle; in small doses it has stimulant action on central nervous system, on plain muscles, and on the secretion of sweat, urine and bile.

Chitraka

Plumbago zeylanica; Hindi—*Chitraka*; Family Plumbaginaceae. The plants grow as perennial herbs in the plains of West Bengal and South India. The root has been reported to be a powerful poison. It yields *plumbagin* (chitraka) a crystalline substance. It is when given internally or applied to the *osuteri* causes abortion. Plumbagin has been reported to be a powerful irritant to the smooth muscles and uterus.

Serpentine

Rauvolfia serpentina Benth., of family Apocynaceae; Eng. Serpentine; Hindi—*Sarpagandha*; Sanskrit—*Sarpagandha*; Bengali—*Chandra*; Marathi—*Harkaya*; Telugu—*Paataalagani*; Tamil—*Chivan amelpodi*; Kannda—*Sarpagandhi*; Malayalam—*Chuvannavilpori*; Oriya—*Patalgarur*; Mundari—*Simjenga*; Assam—*Arachontita*; Khasi Hills—*Todong-pait-parao*; Trade—*Rauvolfia*.

Habitat. A large climbing or twining shrub, found in the tropical Himalayas and plains near the foot of the hills from Sirhind and Moradabad to Sikkim. It occurs also in Assam (up to 4000') and in South India Peninsula along the Ghats to Travancore and Kerala. The genus is distributed in the tropics, *i.e.,* Central and South America, Africa, India, Sri Lanka, Burma, China and Japan.

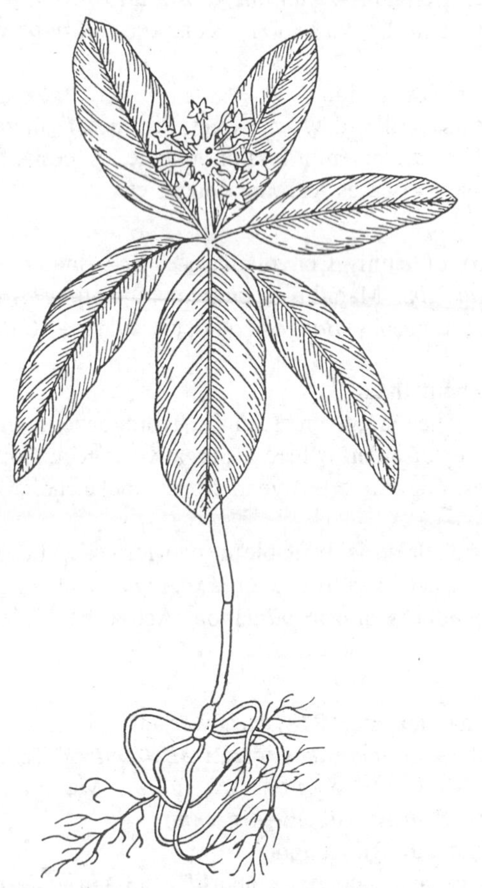

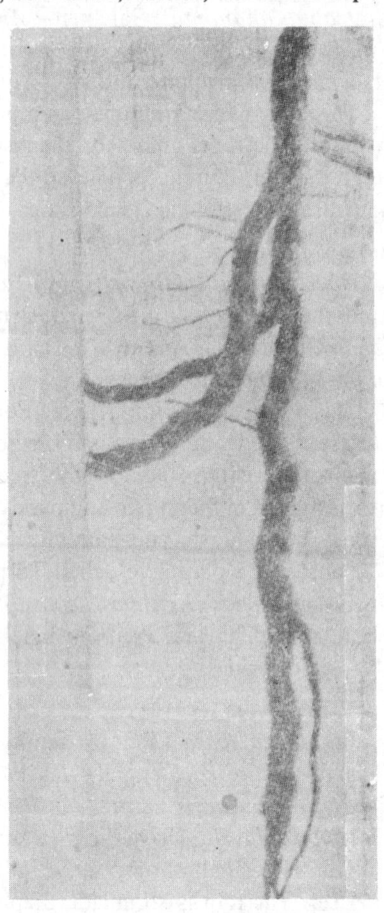

Fig. 13.1. *Rauvolfia serpentina* (*Sarpa-gandha*). The roots are used as medicine. It has been used in indigenous medicine for hundreds of years, has acquired great importance in modern therapeutics following recent scientific research. It is being extensively used in reducing blood pressure in hyperplesis and as a sedative in the treatment of insomnia, hypochondria, mental disorders and certain forms of insanity. It has been recently used to advantage in numerous other diseases like intractable skin disorders such as psoriasis, excessive sweating and itching, gynaecological ailments such as menopausal syndrome; in old-age afflictions such as toxic goitre; and in conditions such as angina pectoris and rapid or irregular heart action. As high blood pressure now is one of the commonest constitutional diseases, the drug is considered a great boon and is assured of an ever increasing demand.

The largest number of species are found in Africa and South America. It is also distributed to Malaysia and Java.

History. The story of *Rauvolfia serpentina* is a most fascinating one. The plant is mentioned in ancient literature including the works of Charaka (1000-800 B.C.) where it is described under its Sanskrit name of Sarpagandha as a useful antidote against snake-bite and insect stings. There are many folk-lores about rauvolfia. One such is that a mongoose would first chew upon its leaves to gain strength before engaging a cobra in combat. According to another, its leaves when freshly ground and applied to the toes, could serve as an antidote for snake poison. A third one has it that the mentally deranged person is relieved of his madness if he eats the pieces of the root. Because of this supposed curative effect in case of insanity the plant has also been known in India as '*Pagal-ki-dava*'. Its generic name was coined towards the end of the seventeenth century by the French Botanist, Plumiers in honour of the well-known sixteenth-century German physician, botanist, traveller and author Leonard Rauwolf of Augsburg.

Medicine. In India and Malaysia the root of this plant has been, from ancient times, much valued as an antidote for the bites of poisonous reptiles and the stings of insects, also as febrifuge, and as a remedy for dysentery and other painful affections of the intestinal canal. In Sanskrit works it is mentioned under the name of *Sarpagandha*.

Rumphius speaks of it under the name of *Radix musteloe*, and says that in his time it was widely used in India and Java as an antidote against every sort of poison. It was administered both internally in the form of a decoction of the root and externally by making a plaster of the roots and fresh leaves and applying them to the soles of the feet. For snake poison, he continues, it was esteemed as specially valuable and the poisonous effects of even cobra's bite were viewed as rendered harmless by the administration of this wonderful root. It is said by him to have been universally employed as an internal remedy against fevers, cholera and dysentery, and the juice of the leaves was instilled into the eyes as a remedy for the removal of opacities of the cornea. He states also that this is the plant to which the mongoose is believed to have recourse when bitten by poisonous snakes. Sir W. Jones gives a similar account of the supposed medicinal virtues of the plant, but expresses a doubt as to whether it really is the so called khneumon plant. Roxburgh states that it is used by the "Telinga physicians, first, in substance, inwardly, as a febrifuge; secondly, in the same manner, after the bite of poisonous animals; and thirdly, it is administered, in substance to promote delivery in tedious cases". Horsfield remarks that the root yields a strong bitter infusion and that its sensible properties indicate considerable activity.

Fig. 13.2. *Rauvolfia serpentina* (*Sarpa-gandha*). Preparation of roots and storage. After digging up the entire plants, the roots are separated from the subaerial parts, thoroughly air-dried and separated from the adhering soil. The fibrous roots being rich in alkaloidal content should not be discarded but collected. The air-dried roots may be broken into pieces about 10 to 15 cm. (4" to 6") in length. The roots may be packed in gunny bags and stored in cool, dry, airy and rat proof godowns pending disposal.

Chemical constituents. Siddiqui and Siddiqui (1931) isolated five crystalline alkaloids. These are divided into the groups: (*i*) *ajmaline group*; —ajmaline, ajmalinine and ajmalicine; and (*ii*) *serpentine group*—serpentine and serpentinine. Today, about 30 alkaloids are known to exist in this plant, the more widely known being reserpine, rescinnamine and yohimbine. Now they can be grouped into three categories: (*i*) deep-yellow coloured quaternary anhydronium bases, (*ii*) the intermediate strong indoline alkaloids and (*iii*) weak basic indole alkaloids; the last mentioned two groups are colourless. Of all the above-mentioned alkaloids, reserpine (serpasil) has gained the greatest prominence in Western countries for its therapeutic action.

Besides the alkaloids, the roots contain oleoresins, a sterol, unsaturated alcohols, oleic acid, fumaric acid, glucose, sucrose, a derivative or oxymethylantheraquinone, a strongly fluorescent substance and mineral salts. Of these, the oleoresin is said to be physiologically active and shows typical hypnotic and sedative action of the drug.

Uses as a drug. Vast developments have been made in the chemistry of its different alkaloids and their role in hypertension, neuropsychiatric and other pharmacodynamics. The two main pharmacological actions of reserpine, *viz.,* lowering the blood pressure and sedative influence, have led to successful trials of the drug in hypertension, neuropsychiatric, gynaecological and geriatrics. With a moderate hypotensive action it controls mainly subjective symptom of high blood pressure when given alone. The drug is also reported to be useful in certain gynaecological conditions, such as menstrual, moliminia, frigidity and women complaining of menopausal syndrome. In the clinical use of reserpine in geriatrics it is reported to be of value in hypertension, nervousness, insomnia, etc.

Some important preparations are as follows:

Rauvolfia serpentina (tablets)—Gluconate Limited Calcutta.

Ralfen (alcoholic preparation) (tablets)—Bengal Chemical and Pharmaceutical Works, Calcutta.

Serpina or S 117—Himalaya Drug Company, Dehradun.

Raudixin (tablets)—Sarabhai Chemicals, Ahmedabad.

Adelphene-CIBA India Limited, Mumbai.

Serpasil—CIBA Pharma Limited, Basle, Switzerland.

Indian Madder

Rubia cordifolia Linn., of family Rubiaceae; Eng. Indian madder; Verna.—*Manjistha*; Sanskrit—*Manjistha*; Hindi—*Manjit*; Bengali—*Manjistha*; Marathi—*Manjeshta*; Telugu—*Taamaravalli, manjestateega*; Tamil—*Manjitti*; Kannda—*Siomalate, manjushtha*; Malayalam—*Manjetti*; Oriya—*Manjistha*; Kashmir—*Dandu*; Punjab—*Manjit*; Assam—*Majathi*; Khasi Hills—*Ryhoi*; Manipur—*Moyum*; Lepcha—*Vhyem*; Bhutan—*Soth*; Mumbai—*Madder*; Nepal—*Manjita.*

This herb is found throughout India in hilly districts.

Uses. The root is astringent, alterative, deobstruent and tonic. It is given as a decoction in jaundice, paralysis, urinary troubles, menstrual disorders, inflammatory condition of the chest, etc. The stem is used in cobra-bite and scorpion sting.

The extract of Indian Madder forms a constituent of the drug *septilin*, used for rhinosinal infections.

Ashwagandha

Withania somnifera Dunal.; Hindi—*Ashwagandha* of family Solanaceae; Sanskrit—*Ashwagandha*; Bengali—*Ashvagandha*; Marathi—*Askandha*; Gujarati—*Ghodakun*; Telugu—*Pulivendram, panneru*; Tamil—*Amukkura*; *aswagandhi*; Kannada—*Panneru, aswagandhi*; Punjab—*Asgand*; Rajasthan—*Chirpotan*; Trade—Aswagandha.

The wild plants are generally erect branching shrubs up to one metre in height. The cultivated plants are morphologically distinct from wild plants. The cultivated plants are "different from the wild ones not only in their therapeutic properties but in all morphological characters like roots, stems, leaves, flowers, pollen grains, mature seeds and the enlarged calyx." (Kaul, K.N., 1957).

Except for a limited collection of roots from *W. somnifera* plants growing wild, the supplies of Ashwagandha roots are obtained exclusively from cultivated sources.

The entire plants are uprooted for the collection of roots. The roots are separated from the overground parts as soon as possible after collection by cutting the stems 1-2 cm. above the crown. The entire root system along with the remains of aerial stems is either cut transversely into smaller pieces or dried in the whole in the sun. The roots undergo further cleaning, trimming and grading before they are despatched to the various drug markets of India.

Uses. This is used as a tonic in geriatrics, being efficacious in relieving hand and limb tremors of elderly people. It is prescribed for all kinds of weaknesses and is supposed to promote strength and vigour, being regarded as an aphrodisiac and rejuvenator. It is used for the treatment of rheumatic pain, inflammation of joints and certain paralytic conditions. Several preparations of this drug are used for all types of nervous disorders and as sedatives in the treatment of insanity and in hypertension. The leaves of the plant are used as a local application for all types of skin lesions, ulcers, boils, and swelling to reduce pus formation, inflammation to promote healing processes.

Roots have long been in use for hiccup, cough, dropsy, rheumatism, and female disorders. The pharmacological activity is ascribed to the presence of several alkaloids. The leaves are used as a febrifuge and applied to lesions, painful swellings and sore eyes. Withaferin A is the most important withanolide to which the curative properties of leaves are attributed.

Asafetida

Ferula asafoetida Linn., of family Umbelliferae (Apiaceae); Eng. Asafetida; Hindi—*Hing*; Sanskrit—*Balhika*, *hingu*; Hindi, Bengali, Marathi, Gujarati and Kannada—*Hing*; Telugu—*Inguva*; Tamil and Malayalam—*Perungayam*; Oriya—*Hengu*; Kashmir—*Yang*; Mumbai—*Hing*.

This is an unpleasant smelling perennial herb, commonly grown in the Punjab and Kashmir.

Uses. The oleo-resin gum that exudes from the root-stock is carminative, antispasmodic, nervine, stimulant, digestive, sedative, expectorant, diuretic, anthelmintic and emmenagogue. It is prescribed in flatulent colic, dyspepsia, asthma, hysteria, convulsions, cholera, chronic constipation, chronic bronchitis, whooping cough, cough, spasmodic disorders of the bowels and angina pectoris. The fried gum is given either as a solution or emulsion or pills, preferably as pills.

DRUGS OBTAINED FROM UNDERGROUND STEM

Greater Galanga

Alpinia galanga Willd., of family Scitamineae (Zingiberaceae); Eng. Greater galanga; Verna. *Kulanjan*; Sanskrit, Hindi and Bengali—*Kulanjan*; Gujarati—*Kulinjan*; Marathi—*Kosht-kulinjan*; Tamil—*Peraratthei*; Telugu—*Pedda-dumparashtram*; Kannada—*Dumbarasme*; Malaylam—*Kolinji*.

This is a herb 6-7 feet high. The leaves are lanceolate, smooth and with white margins. It is found in Eastern Himalayas and South-west India.

Uses. The rhizome is aromatic, stomachic, tonic, carminative and stimulant. It is used in rheumatism, catarrhal affections, especially bronchial catarrh. The rhizome is used as a deodorizer of foul smell from the mouth and other parts of the body.

Dried rhizomes constitute Greater Galangal. Rhizome yields an essential oil, used in perfumery as a source of methyl cinnamate and cineol. The herb is accredited with antitubercular properties. Seeds are used for colic, diarrhoea and vomiting.

Turmeric

Curcuma longa Linn., of family Zingiberaceae; Eng. Turmeric; Hindi—*Haldi*; Sanskrit—*Haridra*; Hindi, Bengali, Marathi and Gujarati—*Haldi*, *Halada*; Tamil—*Manjal*; Telugu—*Pasupa*; Kannada—*Arishina*.

A herb, cultivated commonly in Maharashtra, Bengal, Tamil Nadu, Andhra Pradesh and Orissa for its rhizomes.

Fig. 13.3. The rhizomes/corms of turmeric – a medicinai plant.

Uses. The rhizome is aromatic, stimulant, antiperiodic, tonic, alterative and carminative. It is given in diarrhoea, intermittent fevers, dropsy, jaundice, liver disorders and urinary troubles. The fresh juice of rhizome is used as an anthelmintic. It is commonly used for cleaning foul ulcers. Along with lime or alum it makes a good dressing for sprains, bruises, wounds, etc. Its decotion is an effective eye wash in ophthalmia. The fresh juice is also used as antiparasitic for many skin affections.

Curcuma paper is used as an indicator of chemicals. Rhizomes yield an oil used as a carminative, stomachic and tonic.

Zedoary

Curcuma zedoaria Rosc., of family Zingiberaceae; Eng. Zedoary; Hindi—*Kachura*; Sanskrit—*Sati*; Hindi, Bengali, Marathi, Kannada and Gujarati—*Kachura*; Tamil—*Kichili-kizhanghu*; Telugu—*Kachoram*; Malayalam—*Pula-kizhanna*.

This herb is said to be wild in the Eastern Himalayas and in the moist deciduous forests of the coastal tract of Kanara. It is also cultivated throughout India for its tuberous rhizomes.

Uses. The rhizome is stomachic, cooling, diuretic, aromatic, stimulant, carminative, demulcent and expectorant. It is applied to bruises and pains. It is given in flatulence, dyspepsia and as a corrective of purgatives. Its decoction with long pepper, cinnamon and honey is a remedy for fever, cold, cough and bronchitis.

Mango Ginger

Curcuma amada Roxb., of family Zingiberaceae; Eng. Mango ginger; Hindi—*Amahaldi*; Sanskrit—*Karpura-haridra*; Bengali—*Amada*; Marathi—*Amba haladi*, Telugu—*Mamidiallam*; Tamil—*Mangai inji*.

This herb is grown in Bengal, Andhra Pradesh and Tamil Nadu. Also found wild in parts of Bengal, Konkan and Tamil Nadu.

Uses. The rhizome is carminative, stomachic and cooling. It is applied over contusions and sprains.

Onion

Allium cepa Linn., of family Liliaceae, Eng. Onion; Verna. *Piyaz*. Sanskrit—*Palandu*; Bengali—*Pyanj*; Hindi—*Piyaz*; Marathi—*Kanda*; Gujarati—*Dungari*; Telugu—*Nirulli*; Tamil—*Vengayam*; Kannada—*Nirulli*; Malayalam—*Chuvannaulli*.

This is a bulbous biennial herb with a peculiar smell. It is widely cultivated throughout India.

Uses. The bulb or onion is a stimulant diuretic expectorant. As a diuretic it is eaten raw; its juice is prescribed for piles. Its decoction is given in cough; cooked with vinegar the bulb is given

Fig. 13.4. Onion bulbs and leaves.

Fig. 13.5. Onion bulbs. Onion is a stimulant diuretic expectorant.

in jaundice, splenic enlargement and dyspepsia; taken with salt it is a common remedy for colic and scurvy. It is also used in obstruction of the intestines, prolapse of the anus and as a sedative.

Garlic

Allium sativum Linn., of family Liliaceae; Eng. Garlic; Verna. *Lahsuna*; Sanskrit—*Arishta, lashuna*; Hindi, Gujarati, Bengali and Marathi—*Lasun, lasan*; Telugu and Malayalam—*Velluli, tellagadda*; Tamil—*Vellaipundu*; Kannada—*Bellulli*; Punjab—*Thom.*

This is a strong smelling, glabrous, bulbous-rooted, perennial about a foot in height. The bulbs are short, compressed, breaking up into 10-12 bulbils or cloves, surrounded by a few, dry membranous scales. It is cultivated throughout the country.

Fig. 13.6. Garlic bulbs.

Uses. Garlic is of great medicinal value. It is given in fevers, coughs, flatulence, disorders of the nervous system, pulmonary phthisis, whooping cough and dilated bronchitis. It has diuretic properties. A decoction of garlic made with milk and water is given in small doses in hysteria, flatulence, sciatica, etc. In the form of syrup, it is a valuable remedy for asthma, and disorders of the chest and lungs. It is used as an anthelmintic. Externally it is used as rubefacient, vesicant and disinfectant. Its juice is introduced in the ear for relief of earache. The oil in which garlic has been fried is a useful liniment for rheumatic pains, nervous diseases like scabies and maggot infested wounds.

Ginger

Zingiber officinale Rosc., of family Zingiberaceae; Eng. Ginger; Hindi—*Adrak*; Sanskrit—*Ardraka*; Bengali—*Ada*; Marathi—*Ale*; Telugu—*Allamu, sonthi* (dry); Tamil—*Allam*; Kannada—*Hasisunti*; Malayalam—*Andrakam, inchi.*

This is a perennial herb; the root stock is horizontal, tuberous and aromatic; native of South-East Asia but now cultivated mainly in Kerala, Uttar Pradesh, West Bengal, Maharashtra, Himachal Pradesh and Andhra Pradesh for the rhizomes.

Fig. 13.7. Ginger rhizomes

Uses. The rhizome is used as a stimulant, carminative and flavouring agent. It is given in dyspepsia and flatulent colic. It is administered as an adjunct to many tonic and stimulating remedies. It makes a valuable drug for disorders of the digestive system, rheumatism, piles, pulmonary and catarrhal diseases, dropsy, febrile diseases, neuralgia, etc. The juice of fresh rhizome, with honey is a remedy for coughs and asthma. A paste of ginger is a local stimulant and rubefacient in headache and toothache.

DRUGS OBTAINED FROM BARKS

Mountain Ebony

Bauhinia variegata Linn., of family Caesalpiniaceae; Eng. Mountain ebony; Hindi—*Kachnar*; Marathi—*Kachnar*; Bengali—Raktakanchan; Tamil—*Segapumanchori*; Kannada—*Kanchavala*; Malayalam—*Chuvannamandaram.*

This is a tree with white flowers having red and yellow stripes. Found throughout India.

Uses. The bark is alterative, tonic, blood purifier, anthelmintic and astringent; its decoction is being given in ulcers, syphilis, scrofula, leprosy and other skin diseases. It is also given in diarrhoea and liver complaints.

Indian Oak

Barringtonia acutangula (L) Gaertn., of family Lecythidaceae; Eng. Indian oak; Hindi—*Hijjal*; Bengali—*Hijal*; Marathi—*Piwar*; Telugu—*Kadapa*; Kannada—*Hole kauva.*

This is a small tree with beautiful flowers.

Uses. The bark is astringent; it is used in diarrhoea; as a febrifuge it is given in malaria. The fruit is astringent and useful in nasal catarrh; and the juice of the leaves is a remedy for diarrhoea. A paste of the seeds mixed with the juice of fresh ginger is applied to the chest of children having acute bronchial catarrh.

Cinnamon

Cinnamomum zeylanicum Bl., of family Lauraceae; Eng. Cinnamon; Hindi—*Dalchini.*

This is native of Sri Lanka but grown in the Nilgiris, South Kanara, Malabar, Assam and Kumaon.

Uses. The bark is aromatic, cordial, astringent, carminative and stimulant; it is given in diarrhoea, nausea, gastric irritation, flatulence, vomiting, spasmodic affections of the bowels, toothache, paralysis of the tongue and labour caused by defective uterine contractions. It is administered in the form of a powder, infusion or decoction. A paste made of the bark is applied to the temples to get relief from neuralgic pains and severe headaches. The oil obtained from the stem bark is used as a carminative, antiseptic and astringent.

Fig. 13.8. Cinnamon. The bark quills.

Quinine

Cinchona calisaya Wedd., *C. ledgerina* Moens. ex Trimen; *C. officinalis* Linn; *C. robusta* How., *C. succirubra* Pavon ex Klotzsch. of family Rubiaceae; Eng. Quinine; Hindi—*Kunain.*

The quinine tree is native to the Andes of South America. In India it is found in West Bengal, the Khasia hills, Nilgiris, South India, Sikkim and Madhya Pradesh.

Uses. Quinine is one of the most important drugs known and it is the only adequate cure for malaria. Quinine is obtained from the hard thick bark of above mentioned species of *Cinchona.* The most important constituent of Cinchona bark is quinine, a very bitter, white, granular substance. In addition to its use in the treatment of Malaria, it is a valuable tonic and antiseptic. It is also used in the treatment of other fevers. Some 29 other alkaloids have been isolated from the bark, including cinchonidine, cinchonine, and quinidine, all of which are useful in medicine.

Spanish Cherry

Mimusops elengi Linn., of family Sapotaceae; Eng. Spanish cherry, Indian medlar; Hindi—*Maulsari*; Sanskrit—*Bakula*; Bengali—*Bakul*; Marathi—*Ovalli*; Gujarati—*Barsoli, bolsari*; Telugu—*Pogada*; Tamil—*Vagulam*; Kannada—*Bakula*; Malayalam—*Elengi*; Oriya—*Bokulo baula*; Assam—*Gokul.*

A tree, cultivated in Northern India, the Western Peninsula and South India.

Uses. The bark is astringent and tonic. Its decoction is given in catarrh of the bladder and urethra as an astringent. In fevers it is given as a febrifuge and tonic. The decoction is a useful mouth wash in diseases of the gums and teeth.

Bark and fruits are used in diarrhoea and dysentery. Dried flowers are used as snuff and pounded seeds are used in suppositories for constipation.

Bay Berry

Myrica nagi Thunb., of family Myricaceae; Eng. Bay-berry, Box myrtle; Hindi—*Kaiphal*; Bengali—*Kaiphal*; Marathi—*Kaya phala*; Gujarati—*Kariphal*; Telugu—*Kaidar Yamu*; Tamil—*Marudam*; Kannada—*Kirisńivani*; Malayalam—*Maruta*; Punjab—*Kaiphal, kahela*; Kumaon—*Kaphal*; Assam—*Nagatenga*; Khasi Hills—*Soh-phi*; Nepal—*Kobusi.*

This is a tree. It is native of China and Japan, now found in the sub-tropical Himalayas, the Khasia hills and Assam.

Uses. The bark is aromatic, astringent, stimulant, carminative, tonic and resolvent. It is a good remedy in fevers, catarrh of the mucous membranes, affections of the chest, asthma, diarrhoea, diuresis, typhoid, dysentery, gonorrhoea and bronchitis. The bark is given either as a decoction or powder.

Bark is chewed to relieve tooth-ache; and a lotion prepared from it is used for washing putrid sores.

Lodh Tree

Symplocos racemosa Roxb., of Family Symplocaceae; Eng. Lodh tree; Hindi—*Lodh*; Sanskrit—*Lodhra*; Bengali—*Lodh*; Marathi—*Lodh, lodhra*; Gujarati—*Lodar*; Telugu—*Lodduga*; Tamil—*Velli-lethi*; Kannda—*Pachettu*; Malayalam—*Pachotti*; Oriya—*Ludhu*; Mumbai—*Lodhra, lodh*; Darjeeling—*Khodai*; Assam—*Kavirang*; Khasi Hills—*Lapong-dong*; Lepcha—*Palyok.*

This is a small tree with rough bark found in North-Eastern India, Kumaon, Assam and Bihar.

Uses. The bark, known as totur bark, is cooling, aperient and mildly astringent. The powdered bark with sugar is administered in bowel complaints, discharges due to the relaxed condition of mucous membranes, dropsy, eye diseases, fevers, skin diseases and liver complaints. A decoction of the bark is good mouth wash for bleeding gums.

Indian Red Wood

Soymida febrifuga A. Juss., of family Meliaceae; Eng. Indian red wood; Hindi—*Rohan, Rohini*; Bengali—*Rohina*; Marathi—*Ruhan*; Gujarati—*Rohina*; Telugu—*Sumi*; Tamil—*Shem*; Kannada—*Suani*; Oriya—*Karwi*; Santal—*Ruhen*; South India—*Rohun, Rohunna.*

A tree, found in Western Peninsula, Rajasthan and Bihar.

Uses. The bark is astringent, antiperiodic and febrifuge. It is given in intermittent fevers, dysentery, diarrhoea and malarial fevers.

Asoka Tree

Saraca indica Linn., of family Caesalpiniaceae; Eng. Asoka tree; Hindi—*Ashok*; Sanskrit, Hindi and Bengali—*Asok, ashoka*; Marathi—*Ashoka*; Gujarati—*Ashopalava*; Telugu—*Asoka*; Tamil—*Asogam*; Kannada—*Asokadamara*; Malayalam—*Asokam*; Oriya—*Oshoko*; Khasi Hills—*Dieng-sohkyrkha*.

This is a spreading evergreen tree; native of India; commonly found in Central and East Himalayas, Bengal and West Peninsula.

Uses. The bark is astringent. Its decoction is given in uterine disorders, especially in menorrhagia and leucorrhoea. The decoction is also an efficacious remedy for piles and dysentery. The dried flowers are given in diabetes. The dried flowers of it are fried and used even by common people in Kerala.

Arjun Tree

Terminalia arjuna (Roxb. ex DC.) Wight and Arn., of family Combretaceae; Eng. Arjun tree; Hindi—*Arjuna*.

A tree, commonly found in Madhya Pradesh, Bihar and the Western Peninsula.

Uses. The bark is astringent, febrifuge, cooling, cardiac stimulant, cholagogal, lithontriptic and vulnerary. It is an excellent remedy for heart diseases. The decoction with milk is given every morning on an empty stomach. The decoction is also used for cleaning sores and ulcers.

Indian Barberry

Berberis aristata DC., of family Berberidaceae; Eng. Indian barberry; Hindi—*Rasaut*.

This is a spiny shrub, found in the North-Western Himalayas, the Nilgiris, Kulu and Kumaon.

Uses. An extract, obtained from the root bark, is used in the treatment of ophthalmia. The plant extract (*rasaut*) is alterative, astringent, antiperiodic, deobstruent and diaphoretic. They are as valuable as quinine in malarial fever. The decoction is used as a wash for ulcers, sores, etc. The *rasaut* is used as a purgative for children, blood purifier, tonic and febrifuge. The chief constituent of the plant is berberine, a bitter alkaloid.

Wild Orange Tree

Toddalia asiatica (L.) Lamk., of family Rutaceae; Eng. Wild orange tree; Hindi—*Kanj*.

A shrub or small tree; found in the sub-tropical Himalayas, the Khasia hills, Tamil Nadu and Kumaon.

Uses. The root-bark is the Lopez root of commerce. It is an aromatic tonic, antiperiodic, antipyretic bitter and stimulant. It is given in weak infusion, useful in constitutional debility and in convalescence after febrile and other exhausting diseases.

The plant contains the alkaloids barberine, toddaline and toddalinine.

DRUGS OBTAINED FROM STEMS AND WOODS

Ephedrine

Ephedra gerardiana Wall., of family Gnetaceae or Ephedraceae; Eng. Ephedrine; Hindi—*Khanda*.

A small shrub, found in the drier regions of the temperate and Alpine Himalayas from Kashmir to Sikkim at 7,000 to 16,000 feet elevation.

Uses. The stems are the source of famous drug ephedrine. The liquid extract of stem is used for controlling asthma. The tincture of the plant is cardiac and circulatory stimulant. The decoction of stems and roots is used as a remedy for rheumatism and syphilis in Russia. Juice of berries is used in affections of the respiratory passages. The stems contain the alkaloids—ephedrine and pseudo-ephedrine.

White Sandal Wood

Santalum album Linn., of family Santalaceae; Eng. White sandalwood; Hindi—*Safed chandan.*

A tree, found in the Western Peninsula, Karnataka, Tamil Nadu, and forests of Kerala.

Uses. The wood is cooling, astringent, bitter, sedative, cardiac, tonic and diuretic. It is given in fevers and thirst. The powder of wood is given in gonorrhoea, fevers and bilious disorders. A paste of wood is a cooling dressing in inflammatory eruptive skin diseases. The paste is also applied to the tempie in headache and fever. The essential oil of wood is stimulant, antiseptic demulcent and diuretic. It is used in gonorrhoea, chronic catarrh of the bladder, urethral haemorrhage, chronic bronchitis and intermittent fevers. The oil is applied over scabies and other skin diseases.

Fig. 13.9. *Ephedra*. The stems are the source of famous drug '*ephedrine*'.

Catechu

Acacia catechu Willd., of family Mimosaceae; Eng.. Catechu tree; Verna. *Kattha, Khair.*

Uses. It is a valuable astringent. It is generally used for disorders in which a mild, non-irritating and powerful astringent is required, such as chronic diarrhoea, dysentery, bleeding piles, uterine haemorrhage, leucorrhoea, etc. It is very efficacious for mercurial salivation bleeding or ulcerated or spongy gums, hypertrophy of the tonsils, enlarged uvula, aphthous ulceration of the mouth, etc. It may be used as lozenge made with the addition of gum arabic and sugar. A powder of catechu is used for plugging the cavity of an aching tooth to relieve pain. A tincture of catechu is useful for bed sores and painful mammary glands. A mixture of catechu and myrh, known as *Kathbol*, is used as a tonic, as a galactagogue for nursing mothers. Khersal or catechuic acid, found in small fragments in cavities of the wood, is a useful drug for chest diseases. The juice of the fresh bark is given with asafoetida in haemoptysis.

Long Needled Pine

Pinus roxburghii Sar., of family Pinaceae; Eng. Long leaved-pine; Hindi—*Chir.*

This tree is found in the Western and Eastern Himalayas.

Uses. The oleo-resin popularly known as *gandh-biroza*, is stimulant, stomachic and diuretic. It is given in gonorrhoea, gleet and other disorders of the genito-urinary organs. This is a useful dressing for fowl ulcers. It is a common ingredient of medicinal plasters and ointments.

DRUGS OBTAINED FROM LEAVES

Vasaka

Adhatoda vasica Nees., of family Acanthaceae; Eng. Malabo nut; Verna—*Arusi, rusa.*

This is a perennial shrub. An evergreen shrub about 4-8 feet in height. Leaves elliptic, lanceolate, acuminate, dark green above, pale beneath. Flowers in short dense–axillary peduncled spikes, white with pink or purple stripes. Flowers after rainy season.

It is found throughout the plains of India and in the sub–Himalayan tract up to 4,000 feet. Twigs and flowers of the plants are collected in March-April. The dried drug is sold in the market.

Uses. The leaves contain the alkaloid vascine; they are a powerful expectorant and antispasmodic; they are commonly used in chest diseases, especially phthisis. A decoction of the leaves or their powder is also given, especially for chronic bronchitis and asthma. The leaf juice is given in diarrhoea and dysentery. The leaves are given as a febrifuge in malarial fevers. For relief of asthma the dried leaves are smoked. A poultice of the leaves is applied over fresh wounds, rheumatic joints and inflammatory swellings. A warm decoction of the leaves is used for scabies and other skin diseases.

The flowers and fruits are bitter, aromatic and antispasmodic.

The root is expectorant, antispasmodic, antiseptic, antiperiodic and anthelmintic. It is given in malarial fevers, diseases of the respiratory system, diphtheria and gonorrhoea. The root and the bark have the same medicinal uses as the leaves.

The preparations are—Vansa Khand and Chyavanprash.

Indian Aloe

Aloe barbedensis Mill., of family Liliaceae; Eng. Indian Aloe; Verna. *Gheekunvar, ghrita kumari.*

The plant is about 2-3 feet high. The leaves are fleshly tapering to a blunt point, smooth, pale green, having horny prickles on their margins. The plant is found throughout our country.

Uses. After removing the skin of leaves they are given in fevers, enlargement of the liver, spleen and other glands, skin diseases, gonorrhoea, constipation, menstrual suppression, piles, jaundice and rheumatic affections. In the diseases of liver and spleen the pulp of one leaf is administered with black salt and ginger. The juice of roasted leaf is being given with honey for cough and cold. The pulp of one leaf is being given daily in abdominal tumors, dropsy, carbuncles, piles, sciatica, rheumatism and retention of urine

Fig. 13.10. *Aloe barbedensis*. A common medicinal plant with succulent leaves. The pulp of leaves is given in fever and liver enlargement.

Fig. 13.11. *Crinum defixum*. The juice of leaves is useful in ear and skin diseases.

in fevers. A salad of leaves is eaten in indigestion, constipation and flatulence. The leaf juice is given as a remedy for intestinal worms in children. A poultice of the leaves is applied to tumours, cysts, inflamed parts and scalds.

Poison Bulb

Crinum defixum; Syn. *C. asiaticum*; Eng. Poison bulb; Hindi—*Sukhdarshan*; family—Amaryllidaceae.

This is a large, perennial, bulbous herb; leaves are radical and linear; flowers are found on scape, subtended by involucre of one or more membranous bracts; stamens 6; fruit a capsule with many seeds.

Uses. The seeds are used as purgative and tonic. The warm juice of leaves is useful in ear and skin diseases.

Life Plant

Bryophyllum pinnatum Kurz., of family Crassulaceae; Eng. Life plant; Hindi—*Ghamari*.

This is a succulent perennial plant. Distributed throughout India.

Uses. The leaf juice is given in diarrhoea, dysentery, calculous affections and cholera. The leaves are styptic, astringent and antiseptic; they are slightly toasted before they are applied to wounds, bruises, boils, cuts, ulcers, bites of venomous insects; by its application swelling is prevented, incised wounds heal rapidly and irritation is allayed.

Fever nut

Caesalpinia crista Lam., of family Caesalpiniaceae; Eng. Fever nut; Hindi—*Puti karanj*.

This is a straggling thorny shrub, found to be distributed in Bengal, Maharashtra and Kerala.

Uses. The tender leaves are deobstruent, emmanagogue, astringent, febrifuge in intermittent fevers and anthelmintic. The tender leaves are applied to relieve toothache. Castor oil coated leaves are applied as a dressing over hydrocoele, inflamed piles and other inflammatory swellings. The roots are a febrifuge, anthelmintic and astringent; they are given in leucorrhoea. The seeds are astringent and antipyretic; they are given in fevers, asthma, colic, general debility, etc. The root-bark is used as an antiperiodic in the intermittent fevers.

Fig. 13.12. *Bryophyllum*. A common medicinal plant. The leaves possess antiseptic properties.

Swallow Wart

Calotropis procera R. BR., of family Asclepiadaceae; Eng. Swallow wart; Hindi—*Ak*.

This is perennial undershrub commonly found in Northern India and Andhra Pradesh.

Uses. The leaves are used in dropsy and enlargement of the abdominal viscera. The dry leaves are either smoked or the smoke from the burning leaves is inhaled for the cure of asthma and cough. A decoction of the leaves is used for extracting guinea-worms. The leaf juice is applied to skin diseases. The root bark is a cholagogue diaphoretic, emetic, alterative and diuretic. A paste of the charcoal prepared from the roots, made with some bland oil is applied over skin diseases caused by syphilis, leprosy, etc. The latex is used in leprosy, taenia, dropsy, rheumatism, etc. As an abortifacient it is either taken internally or applied to the mouth of the uterus. The flowers are tonic, stomachic and digestive; they are given in cough, cold, asthma, etc.

Indian Pennywort

Centella asiatica (L.) Urban., of family Umbelliferae (Apiaceae); Eng. Indian pennywort; Hindi—*Brahmi*.

This is a common trailing herb, rooting at the nodes. Commonly found throughout India.

Uses. The weed is alterative, tonic, diuretic, antiphlogistic, blood purifier and local stimulant. It is a remedy for skin diseases, like chronic eczema, chronic ulcers *etc.*, enlargement of glands, chronic rheumatism, chronic nervous diseases, madness, cholera, amenorrhoea and piles. The leaves are a household remedy in early stages of dysentery of children. The powder of leaves is given with milk in small doses in mental weakness and to improve memory. The fresh juice of leaves mixed with milk is given as an alterative in gonorrhoea, jaundice and fevers; this is also useful for children in skin diseases and for improving the blood and nervous system. An ointment made of leaf juice and lanoline is of great value in elephantiasis.

Tasmanian Blue Gum

Eucalyptus globulus Labill., of family Myrtaceae; Eng. Tasmanian blue gum; Hindi—*Karpura maram*.

A tree; native of Australia, now cultivated chiefly in the Nilgiris, the Anamalais, the Pulney hills, Shimla hills and at Shillong in Meghalaya.

Uses. The leaves are the source of an essential oil known as 'eucalyptus oil', which is used in the treatment of asthma and bronchitis. The oil is antiseptic, used in infections of the upper respiratory tract and certain skin diseases; mixed with equal amount of olive oil used as rubefacient for rheumatism; also used in ointments for burns; mosquito repellent. The root is purgative.

Physic Nut

Jatropha curcas Linn., of family Euphorbiaceae; Eng. Physic nut, purging nut; Hindi—*Safed arand*, *Jamalgota*, *Bagbherenda*.

This is a shrub or small tree. It is native of tropical America, now cultivated along the Coromandel Coast and in Travancore.

Uses. The fresh latex is styptic. It is applied to bleeding wounds; it is locally applied to piles, scabies, eczema, ringworm, itch and decayed teeth.

A decoction of leaves is a febrifuge and a mouth-wash for strengthening the gums. A warm poultice of the leaves is a galactagogue; it is applied to the breasts of nursing mothers. The seeds make an acro-narcotic poison. Slightly roasted seeds are used as an active purgative. The oil of seeds is a purgative and alterative. The oil is locally used in skin diseases like eczema, herpes, itch, sores, etc., and bleeding wounds.

Fig. 13.13. A flowering twig of *Tulsi*. The leaves on steam distillation yield a bright yellow volatile oil possessing a pleasant odour. The oil is reported to possess antibacterial and insecticidal properties. It has marked insecticidal activity against mosquitoes. Besides the volatile oil the plant is reported to contain alkaloids, glycosides, saponins and tannins.

Holy Basil

Ocimum sanctum Linn., of family Labiatae; Eng. Holy basil; Hindi—*Tulsi*.

This is an erect softy hairy aromatic herb or undershrub. Found throughout India.

Uses. The leaves are expectorant, stomachic, anticatarrhal, diaphoretic and aromatic; their decoction or infusion is given in malaria, gastric diseases of children and liver disorders; as a prophylactic against malaria fresh leaves are taken with black pepper in the morning. The leaf juice is given in chronic fever, haemorrhage dysentery and dyspepsia; it is also used to check vomiting and as an anthelmintic.

Betel Pepper

Piper betle Linn., of family Piperaceae; Eng. Betel pepper; Hindi—*Pan.*

This is a climbing shrub cultivated for its leaves.

Uses. The leaves are astringent, aromatic, antiseptic, carminative, aphrodisiac, stimulant, expectorant and sialogogue. The leaf juice with honey is given to children in colic, indigestion, diarrhoea and laryngitis. The leaves are chewed to remove foul odour from the mouth. The leaf juice is used as eye drops in ophthalmia, and other painful eye diseases and night blindness. Oil coated leaves are a useful dressing for blistered surfaces and ulcers. The essential oil extracted from the leaves, is given in catarrhal and pulmonary diseases.

Fig. 13.15. *Coriander.* An aromatic herb.

Tylophora

Tylophora asthmatica; Syn. *Tylophora indica*; *Hindi—Jangli Pikavan* or *Antamul*; Family—Asclepiadaceae.

It is a trailing creeper, growing abundantly in Northern India, Bengal, Assam; the Western Ghats, Burma and Sri Lanka. The leaves are thick and deep green, 5 to 12 cm long and 2 to 6 cm wide. In dry state, they are rather, thick and of a pale-yellow colour.

Uses. The chewing and swallowing of leaves by the patient of asthma have shown wonderful results. The leaf was also tried at Vallabhbhai Patel Chest Institute, Delhi on the patients suffering from bronchial asthma, the results were quite satisfactory.

DRUGS OBTAINED FROM FLOWERS

Saffron

Crocus sativus Linn., of family Iridaceae; Eng. Saffron; Hindi—*Kesar.*

This is a herb, cultivated in Kashmir (Pampur), Bhersar and Chaubattia in Uttar Pradesh.

Uses. The dried stigmas and tops of

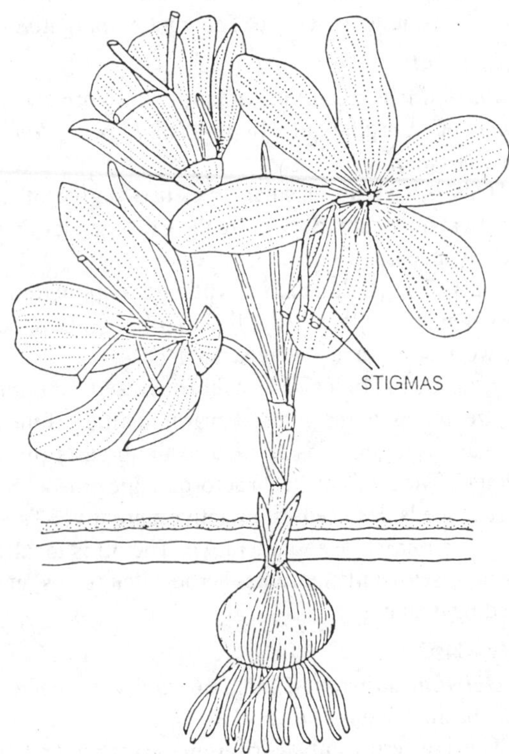

Fig. 13.14. *Crocus sativus* (Saffron). The stigmas and tops of the styles make the saffron of commerce.

the styles make the saffron of commerce. It is of medicinal value. It contains the bitter principle picrocrocin. Saffron is mild stimulant, stomachic, carminative, antispasmodic, nerve sedative, mild narcotic, diuretic and emmenagogue. It is a remedy for promoting menstruation. It is also used in fevers, hysteria, flatulent colic, asthma, leucorrhoea, piles and catarrhal diseases of children.

Iron Wood

Mesua ferrea Linn., of family Guttiferae; Eng. Ironwood; Hindi—*Nagkesar*.

A tree, found in the Eastern Himalayas, Assam, West Bengal, the Western Ghats, Travancore and the Andaman Islands.

Uses. The flowers are astringent and stomachic. The dried flowers are given in vomiting, dysentery, coughs, thirst, irritability of the stomach, excessive perspiration and bleeding pile.

Violet

Viola odorata Linn., of family Violaceae; Eng. Violet; Hindi—*Banafsha*.

A herb; native of Europe but now cultivated in Kashmir.

Uses. The flowers are demulcent astringent, diuretic, emollient, diaphoretic and laxative. They are given in biliousness and lung troubles. The flowers are used for the treatment of coughs, sore throat, kidney diseases, liver disorders and infantile affections.

DRUGS OBTAINED FROM FRUITS

Bael

Aegle marmelos Corr., of family Rutaceae; Eng. Bael tree; Verna. *Bael*, *Siriphal*.

This is a tree.

Uses. The fruit is of great medicinal value. The ripe fruit is aromatic, astringent, cooling and laxative. The unripe or half-ripe fruit is astringent, stomachic, antiscorbutic and digestive. It is best given in sub-acute or chronic cases of diarrhoea and dysentery and in irritation of the elementary canal; it is a useful adjunct in after treatment of bacillary dysentery. The ripe fruit is prescribed in intestinal disorders and certain forms of dyspepsia characterized by alternate constipation and diarrhoea. A sherbet of the ripe fruit is given for chronic constipation and dyspepsia.

Fish Berry

Anamirta cocculus W. & A., of family Menispermaceae; Eng. Fish berry; Verna. *Kakamari*, *Kakaphula*.

This is a large smooth, woody climber. The bark on old stem is spongy and deeply cracked. The fruits are drupes usually in pairs, round or reniform. It is found in Eastern Bengal, Orissa and South India.

Uses. The dried ripe fruit is a powerful narcotic. The juice of the fresh berries is applied to scabies and ulcers. An ointment made of the seeds, after removing the seed coat is a powerful germicide. It is locally used for ringworm and other parasitic skin diseases. It is also used for killing head lice.

Purging Cassia

Cassia fistula Linn., of family Caesalpiniaceae; Eng. Purging Cassia; Hindi—*Amaltas*.

This is an ornamental tree with yellow flowers. Found throughout India.

Uses. The pulp from the pods is of great therapeutic value. It is a mild pleasant and safe purgative, even for children and expectant mothers. Its confection is given in diabetes. The leaves are emollient; their juice makes a useful dressing for ringworm and chilblains. The root is a tonic febrifuge and a strong purgative.

Coriander

Coriandrum sativum Linn., of family Umbelliferae (Apiaceae); Eng. Coriander: Hindi—*Dhaniya*.

This is an aromatic herb; native of the Mediterranean region now cultivated in Madhya Pradesh, Maharashtra, Karnataka, Bihar and throughout India.

Uses. The fruit is aromatic, stimulant, carminative antispasmodis, correctivem diuretic, aphrodisiac, and refrigerant. A decoction of the dried fruits is given in flatulent colic, rehumatism and neuralgia. The volatile oil distilled from the fruits is given in flatulent colic, rheumatism and neuralgia. The watery paste of seeds is used for the cure of ulcers of the mouth and throat.

Fig. 13.15. *Coriander.* An aromatic herb.

Cumin

Cuminum cyminum Linn., of family Umbelliferae (Apiaceae); Eng. Cumin; Hindi—*Zira.*

This is an annula aromatic 1-3 feet high herb. It is native of Mediterranean region but now commonly grown in the Punjab and Uttar Pradesh fro aromatic fruits.

North Gujarat especially Unja is well known for its cultivation and market.

Uses. The fruit is a good source of thymol; it is a stomachic, diuretic, carminative, stimulant, astringent and emmenagouge. It is prescribed in dyspepsia, diarrhoea, and hoarseness of voice. The fruit is given with lime juice to pregnant women for checking nausea; for promoting secretion of milk, cumin seeds are given after childbirth.

Emblic

Emblica officinalis Gaertn., of family Euphorbiaceae; Eng. Emblic; Hindi—*Amla.*

A tree found throughout India.

Uses. The fruit is of great medicinal value. It is one of the richest sources of vitamin C. The fresh fruit is refrigerant, tonic, antiscorbutic, diuretic and laxative. It is used in fevers, vomining, indidestion, habitual constipation and other digestion troubles. The dried fruit is a good astringent, refrigerant, stomachic, antiscorbutic and blood purifier. It is given in diarrhoea, dysentery and haemorrhage. The infusion of seeds is a useful eye-wash in ophthalmic diseases.

Fig. 13.16. *Emblica officinalis* (*Amla*). The fruit is one of the richest sources of vitamin C.

Fennel

Foeniculum vulgare Mill., of family Umbelliferae (Apiaceae, Fennel; Hindi—*Saunf.*

This is an aromatic herb. It is native of Mediterranean region, now cultivated mainly in the Punjab, Assam, Maharashtra and Baroda.

Uses. The fruit is sweet, laxative, aphrodisiac, stomachic, stimulant, aromatic, corrective of flatulence, appetizer, anthelmintic, carminative, lactagogue, diuretic, stimulant and antidotal. It is administered in eye diseases, burning sensation, fever, thirst, wounds, dysentery, diseases of the chest, spleen and kidney, headache, amenorrhoea and bellyache of children; it also promotes female monthly regularity. In dysentery a decoction of fruit is given; in fever a decoction of fruit is given with sugar. The oil extracted from the seed is an anthelmintic against hookworms. The leaves are diuretic, digestive, stimulant and aromatic; they are useful in cough, flatulence, colic and thirst. They are also eaten to improve the eyesight.

Opium

Papaver somniferum Linn., of family Papaveraceae; Eng. Opium; Hindi—*Afim*.

A herb; native of W. Asia, now grown in Uttar Pradesh, in the East Punjab, Rajasthan and Madhya Pradesh.

An inspissated juice obtained by scratching the unripe capsules of *Papaver somniferum* Linn.; and allowing the milky sap, which exudes thereby, to dry spontaneously. The bulk of the medicinal article is obtained from Asia Minor, at the present day, as it has been for nearly the past 2,000 years. Indian Opium is, however, the most important form commercially, being that which is smoked, eaten or drunk, in various preparations by the inhabitants of Asiatic countries, chiefly the Chinese.

In India the opium-poppy is grown here and there throughout the country, but its cultivation is mainly confined to three centres, which afford the following opiums—'Patna Opium' in Bihar, 'Benares Opium' in Uttar Pradesh, and 'Malwa Opium' in Madhya Pradesh and certain parts of Rajasthan.

Fig. 13.17. *Papaver somniferum* (*Afim*)—The latex is obtained by scrapping the capsules. The latex, obtained from the immature fruits; is the source of opium, which is used to induce sleep, relieve pain, and relax spasms.

Medicinal properties of opium. Opium is well known to the medical profession. The medicinal drug of Europe is mainly derived from Asia Minor and is the produce of *P. somniferum* var. *glabrum*.

In the words of Pharmacopaeia of India it may be said that opium in its primary effects medicinally is stimulant, and its secondary action narcotic, anodyne, and antispasmodic. "It operates chiefly on the cerebrospinal system, and, through the nerves arising therefrom, it affects more or less every organ of the body. It tends to diminish every secretion excepting that of the skin, which increases under its use. In overdoses it is a powerful poison." In inflammation, especially of serious membranes, it has been employed extensively either alone, or in combination with calcmel, antimony, and other remedies. As a general rule it is less applicable to inflammatory and other diseases, in which the tendency to death is by coma or by apnoea, then to those which produce death by asthenia. In the latter, as in peritonitis, it holds the first place as a remedial agent. In fevers, especially in the advanced stages, it is of the highest value, either alone or in combination with antimony, camphor, allaying vascular and nervous excitement, and procuring sleep; but as a general rule, it is inadmissible when contraction of the pupil is strongly marked. In painful spasmodic affections, opium in large and repeated doses often affords immediate relief. In various morbid states of the abdominal viscera, *e.g.*, simple or cancerous ulceration of the stomach, chronic gastritis, gastrodynia, nervous and sympathetic vomiting, diarrhoea, dysentery, colica pictonum, strangulated hernia, visceral destructions, it is often given with the best results. In diseases of the genito-urinary system, *e.g.*, cystitis, cistirrhoea, spasmodic stricture of the urethra, also in menorrhagia, dysmenorrohea, irritable states of the uterus, metritis, it is a remedy of the highest value. Tetanus, acute rheumatism, and delirium tremens are amongst the other diseases in which opium has been employed as a sheet anchor. In cholera it has been used but with doubtful results. In cancerous and gangrenous ulceration, opium, by allaying constitutional disturbances, often exercises a favourable influence on the local symptoms. As an external application, opium proves valuable in various rheumatic, neuralgic, and other painful affections; also in ophthalmia and other diseases of the eye".

"*Dose.* From a quarter of a grain to two grains or more according to circumstances, in many painful affections the sole criterion for the regulation of the dose is the amount of relief afforded".

Long Pepper

Piper longum Linn., of family Piperaceae, Eng. Long pepper; Hindi—*Pipar*, *Piplamul.*

This is a slender creeping undershrub. It is native of India and cultivated in the Western Ghats, Karnataka and Tamil Nadu for the fruits.

Uses. The sun-dried unripe fruit (berries) or the spike of the small flowers is known as long pepper. It is a cardiac, stimulant carminative, alterative tonic, laxative, digestive, emollient, stomachic and antiseptic. It is given with honey for indigestion, dyspepsia, cough, bronchitis, asthma, hysteria, cholera, fever, leprosy, piles, gout, etc. To prevent fever after childbirth it is an efficacious remedy. The dried root (*Piplamul*) possesses the same medicinal qualities as the berries but in an inferior degree.

Black and White Pepper

Piper nigrum Linn., of family Piperaceae; Eng. Black pepper; Hindi—*Kali mirch*; *gol mirch.*

This is a stout glabrous creeper; native of the Indo-Malayan region, also cultivated in the Western Ghats, Karnataka, Maharashtra, Assam and Kerala.

Uses. The unripe dried berries are the black pepper of commerce; and the ripe fruit with the outer covering removed is the white pepper of commerce. The berries contain the alkaloid piperine, The fruits are aromatic, stimulant, carminative, digestive, stomachic, nervine tonic, diuretic emmenagogue and antiperiodic. It is given in dyspepsia, flatulence, diarrhoea, cholera, piles, disorders of the urinary sys-

Fig. 13.18. Black pepper. The unripe dried berries.

tem, cough, coma, gonorrhoea and malarial fever. It is given either as an infusion or powder. White pepper is more aromatic and less pungent than black pepper. It is given in dyspepsia and constipation.

Belleric Myrobalan

Terminalia bellirica Roxb., of family Combretaceae; Eng. Belleric myrobalan; Hindi—*Bahera.*

A common tree grown as an avenue.

Uses. The ripe dry fruit known as belleric myrobalan (*bahera*) is astringent, bitter, tonic and laxative. It is given in piles, dropsy, diarrhoea, leprosy, biliousness, dyspepsia and headache. Half ripe fruit is given as purgative and fully ripe as an astringent.

Chebulic Myrobalan

Terminalia chebula Retz., of family Combretaceae; Eng. Chebulic myrobalan; Hindi—*Harara, Haritaki.*

A tree. Commonly found in Northern India. Also found in South India in the Deccan tablelands at 1,000–3,000 feet and up to 6,000 feet in Travancore; higher forests of the Mumbai Ghats, Satpuras, Belgaum and Kanara.

Uses. The fruit is the chebulic myrobalan of commerce. It is an efficacious laxative, astringent, stomachic, tonic and alterative. The fruit pulp is given in chronic diarrhoea, dysentery, flatulence, asthma, urinary disorders, vomiting, enlarged spleen and liver, etc. It is used externally as a local application to chronic ulcers and wounds and as a gargle in stomatitis. Fine powder of fruit is used as a dentrifice and considered useful in various teeth, bleeding and ulcerations of the gums.

Ammi

Trachyspermum ammi (L.) Sprague of family Umbelliferae. Eng. Ammi; Hindi *Ajwain.*

A herb, cultivated throughout India.

Uses. The fruits are carminative, stimulant, antispasmodic and tonic. They are given in colic, flatulence, dyspepsia, indigestion, diarrhoea, cholera, tympanites, hysteria, etc. The fruit yields an essential oil containing thymol.

Solanum

Solanum viarum Dunal; Syn. *Solanum khasianum* var. *Chatterjeeanum*, Sengupta; family Solanaceae.

Solanum viarum, is a wild perennial herb. It grows well in the hilly areas particularly of Khasia and Jaintia hills (Assam) in North East and Nilgiri hills of South India.

Morphology. It is a stout herb, with a woody stem, of about 2 feet in height. Nearly whole body of the plant is covered with yellow hairs (hirsute). The stem bears two types of prickles, hooked and straight. Stem is cymosely branched.

Leaf is 7" to 5" long and is deeply lobed, hirsute and bears only straight prickles.

Flowers are white in colour, and are produced in large number. The berry is about 2.5 cms. in diameter which contains about 350 seeds.

Alkaloid and medicinal uses. The berries of *S. viarum* contain the alkaloid *Solasodine*. It is a glycon which is used by pharmaceutical companies for the preparation of many important drugs. It is a nitrogen analogue of diosyenin and is a good source of *sapogenin*. Sapogenin is used as a base for the preparation of *Cortisone* and allied products. The synthetic substitutes are not known for these drugs, hence the importance of this source.

The gluco-alkaloid percentage of the berries is about 6 per cent which in turn produces one-third or 2 per cent of *Solasodine*. It can be extracted from dry as well as fresh berries.

Cortisone, a steroidal hormone prepared from solasodine, is found to be effective in the treatment of acute stages of rheumatoid arthritis, chronic cases of asthma, leukemia, obesity and many skin diseases.

Induced mutations have yielded favourable results regarding the increases in alkaloid content of the berries. (Dhyansagar and Pingle). *Material received from P.N. Purekar.*

DRUGS OBTAINED FROM SEEDS

Croton

Croton tiglium Linn., of family Euphorbiaceae; Eng. Croton; Hindi—*Jamalgota.*

This is a shrub or small tree. It is native of South East Asia but now cultivated in Assam, Bengal and South India.

Uses. The seeds are the source of croton oil, which is used as a strong purgative. Externally the oil is a stimulant and a powerful irritant and rubefacient. It is used as a liniment and a good stimulant in bronchitis, asthma, paralysis, gout, chronic rheumatism, laryngitis, neuralgia, sciatica and diseases of joints.

Carrot

Daucus carota Linn. var ***sativa*** D.C. of family Umbelliferae (Apiaceae); Eng. Carrot; Hindi—*Gajar.*

An annual or biennial herb, grown chiefly in the Punjab, Uttar Pradesh and Madhya Pradesh.

Uses. The seeds are aromatic, stimulant, carminative, useful in the diseases of the kidney and in dropsy, nervine tonic, aphrodisiac, given in uterine pain.

Lesser Cardamom

Elettaria cardamomum Maton., of family Zingiberaceae; Eng. Lesser cardamom; Hindi—*Chotti ilaichi.*

This is a herb; native of India, cultivated in Kerala, Karnataka, Maharashtra, Assam and Tamil Nadu.

Uses. The seeds are aromatic, stimulant, stomachic, carminative, diuretic, cooling, abortifacient, emmenagogue and antidotal; they are useful in asthma, bronchitis, piles, diseases of the bladder, kidney, liver, uterus, rectum and throat. They are also remedy of scabies, headache, earache and toothache.

Blonde Psyllium

Plantago ovata Forsk., of family Plantaginaceae. Blonde psyllium; Hindi—*Isafgol.*

This is an annual herb. It is cultivated in parts of Rajasthan and Maharashtra.

Uses. The seeds are mildly laxative, demulcent, mucilaginous cooling, emollient, astringent and diuretic. They are used in the treatment of dysentery, constipation and disorders of the digestive system. The mucilaginous matter is found in the seed coat, and therefore, the seed coat or husk is used.

Pongam

Pongamia pinnata (L.) Pierre of family Papilionaceae; Eng. Pongam; Hindi—*Karanja.*

This is a tree, commonly found near banks of streams in both the Peninsulas, West Bengal and Travancore.

Uses. The seeds are used as external application in skin diseases. The seed-oil is used in the treatment of skin diseases and rheumatism. The fresh bark is used internally in bleeding piles. The leaves are applied in the form of poultice to ulcers infested with worms. The juice of roots is used for closing fistulous sores and for cleansing foul ulcers.

Castor Oil Plant

Ricinus communis Linn. of family Euphorbiaceae; Eng. Castor oil plant; Hindi—*Arand.*

This is a small tree, cultivated chiefly in Andhra Pradesh, Maharashtra, Karnataka and Orissa.

Uses. The seeds contain the alkaloid ricinine; the cold drawn oil from the seed is medicinally the most useful part of the plant; it is an effective purgative generally given in acute diarrhoea.

As an enema the oil is given in constipation. The oil is locally applied in conjunctivitis. The leaves are applied to the head to relieve headache and as poultice for boils.

Nux-Vomica

Strychnos nux-vomica Linn., of family Loganiaceae; Eng. Nuxvomica; Hindi—*Kuchla*.

A tree, found in Orissa, Bihar, Andhra Pradesh and Tamil Nadu.

Uses. The dried ripe seeds are the source of a drug nux vomica which is used as a tonic stimulant and in the treatment of paralysis and nervous disorders. The powdered seeds are also given in general debility, chronic rheumatism, dyspepsia, intermittent fevers, diarrhoea, hysteria, general constipation, hydrophobia, impotence, cholera and epilepsy.

Fenugreek

Tirgonella foenumgraecum Linn., of family Papilionaceae; Eng. Fenugreek; Hindi—*Methi*.

A herb; native of South Europe, grown mainly in Northern India.

Uses. The seeds are aromatic, diuretic, nutritive, tonic, lactagogue, astringent, emollient, carminative and aphrodisiac. The seeds are given, boiled or roasted in dyspepsia, diarrhoea, dysentery, colic, flatulence, dropsy, rheumatism, chronic cough, liver and spleen enlargement. The seeds contain the alkaloid trigonelline.

DRUGS OBTAINED FROM ALL PARTS OF PLANTS

Blood Wort

Achillea millefolium Linn., of family Compositae (Asteraceae); Eng. Bloodwort; verna. *Gandana*.

It is an erect, pubescent herb about two feet high. It is usually distributed in Western Himalayas at 6,000–9,000 feet elevation.

Uses. The herb is a bitter aromatic, tonic, astringent, stimulant and diaphoretic. It is recommended for colds, obstructed perspiration, fevers, hysteria, flatulence, heart burn, colic, epilepsy, piles, amenorrhoea, kidney disorders, suppression of haemorrhages, profuse mucous discharges and nervous diseases. Its decoction is a useful mouth-wash, hair-wash and an astringent for bleeding piles, ulcers and sore nipples. A hot infusion of the leaves is given in doses of one to six ounces a bitter tonic. A strong decoction is used as an injection for bleeding piles, vaginal haemorrhages and nasal bleeding.

Kalmegha

Andrographis paniculata Nees; Hindi—*Kalmegha*; family—Acanthaceae.

This is an erect herb up to 3 feet; branches 4-angled or almost winged; leaves 2-3 inches long lanceolate, tapering to the base; flowers small, pink, solitary arranged in lax spreading axillary and terminal racemes or panicles, the whole forming a large paniculate inflorescence 3-4 inches long, tapering at each end; seeds many, subquadrate, rugose, glabrous; flowering time September to December.

The plant is found throughout India, sometimes cultivated. It is found here and there in the forests.

The whole plant is used medicinally.

Uses. The herb is used for bronchitis, dyspepsia, dysentery, influenza, etc. The decoction of plant and the powder of seeds is also used in fevers.

Absinthe

Artemisia absinthum Linn. of family Compositae (Asteraceae); Eng. Absinthe; Verna; *Vilayati Afsanthin*.

This is an aromatic, bitter, herbaceous hairy perennial herb. It is found in Kashmir from 5,000 to 7,000 feet elevation.

Uses. The whole herb is used medicinally but the leaves are preferred. The fresh plant is more efficacious than the dry plant. It is a tonic, stomachic and anthelmintic. Small doses of its infusion are given for dispersing the yellow bile of jaundice from the skin. It is also given in the disease of the digestive system, nocturnal pollution, anaemia, wasting diseases, etc. The poultice of its leaves is applied for fomentation to gouty or rheumatic joints. It is a useful antiseptic and detergent in skin diseases. An infusion of the leaves is given as an enema for killing worms in the rectum.

Neem Tree

Azadirachta indica A. Juss., (Syn. *Melia azadirachta* Linn)., of family Meliaceae; Eng. Margosa tree, Neem tree; Hindi—*Nim.*

It is native of Myanmar but grown all over India.

Uses. The leaves are carminative, expectorant, anthelmintic, antidotal, diuretic and insecticidal. The fresh juice of the leaves is given for intestinal worms; with honey the juice is prescribed for jaundice and skin diseases. An infusion of the fresh leaves is a bitter vegetable tonic and alterative, especially in chronic malarial fevers, because of its action on the liver. Externally the leaves are applied over skin diseases as a discutient, stimulant and antiseptic; they are specially used for boils, chronic ulcers, eruptions of small pox, syphilitic sores, glandular swellings, wounds, etc. The bark is also a bitter tonic, astringent, alterative, anthelmintic, antispasmodic and stimulant. It is used in the same way as the leaves. The root bark has the same properties as the bark of the stem. The *nim* gum is a stimulant and demulcent tonic. The fermented sap that exudes from the trunk is a refrigerant, nutrient, stomachic and alterative tonic.

The fruit is recommended for urinary diseases, piles, leprosy, intestinal worms, etc. The seeds are emetic, laxative and anthelmintic. Seed oil is applied as an antiseptic dressing in leprosy and chronic skin diseases.

Spiny Bamboo

Bambusa bambos Voss., of family Gramineae (Poaceae); Eng. Spiny bamboo; Hindi—*Bans.*

This is a tall woody grass, found throughout India, particularly along river valleys and in moist situations.

Uses. The leaf buds are administered for thread-worms. The leaf juice is given with aromatics in vomiting of blood. A decoction of the leaves is used to induce lochia after child-birth.

The young shoots contain hydrocyanic and benzoic acids; they are stomachic and stimulant; they are used in lung diseases. A decoction of the joints of bamboo stem is a useful emmenagogue; it is also used as an abortifacient.

Banslochan, a silicious crystalline secretion found in the culms of the female plants, is a febrifuge, expectorant, tonic, aphrodisiac, demulcent and pectoral. It is given in hectic fever, phthisis, asthma, paralytic complaints, etc.

Apple of Peru

Datura metel Linn., of family Solanaceae, Eng. Apple of Peru; Hindi—*Dhatura.*

This is a herbaceous plant covered with fine, minute hairs. Commonly found throughout India.

Fig. 13.19. *Datura.* A twig with leaves, flower and fruits. All parts are intoxicant, narcotic, aphrodisiac, toxic, antispasmodic and anodyne.

Uses. All parts of the plant are strongly intoxicant, narcotic, aphrodisiac, toxic, antispasmodic and anodynous. The young leaves and seeds contain the drugs hyoscine, hyoscyamine and atropine. The dried leaves and twigs of the plants are smoked as an antispasmodic in asthma, whooping cough, bronchitis, etc. The juice of the fruits is a useful dressing for the scalp to check dandruff and falling of the hair. The seeds are astringent, antispasmodic, narcotic, anodynous, intoxicating, aphrodisiac, bitter, carminative and stomachic. The paste of seeds is used for decaying teeth, piles, fistula, tumors and parasitic skin diseases.

Achyranthes

Achyranthes aspera Linn., Hindi—*Chirchita*, *Latzira*, *Apamarg*; family Amarantaceae.

This is a common roadside weed with spike inflorescence. Flowering throughout year. It is of common occurrence throughout India.

Uses. The plants are used medicinally for several diseases such as piles, colic, boils, etc. It is pungent, purgative, diuretic and astringent. Roots are used for pyrrhoea. Also used in cough and fevers.

Pedalium

Pedalium murex Linn., of family Pedaliaceae; Hindi—*Bada gokru.*

This is a low, thick stemmed succulent annual herb; found in Gujarat, Konkan and Andhra Pradesh.

Uses. Fresh plants infused in water or milk become mucilaginous. This infusion is diuretic and demulcent; it is given in disorders of urinary system, such as ardour urine, dysuria, gonorrhoea, spermatorrhoea, impotence, nocturnal seminal emission, calculous affections and dropsy. If the fresh plant is not available the dried fruits are used for the diseases for which the mucilaginous infusion of the leaves is recommended.

Black Nightshade

Solanum nigrum Linn., of family Solanaceae; Eng. Black Nightshade; Hindi—*Makoy*.

This herb is found throughout India.

Uses. The herb is a cardiac tonic, alterative, diuretic, sedative, diaphoretic, cathartic, anodyne, expectorant and hydragogue. It is used as a decoction in dropsy; enlargement of the liver and jaundice. The syrup of herb is given as an expectorant. In fevers it is given as a cooling drink. The leaf juice is given in inflammation of the kidneys and bladder and in gonorrhoea, dropsy, heart diseases, piles and enlargement of spleen. Hot leaves are applied over the swollen

Fig. 13.20. *Solanum nigrum*. (Black Nightshade). A common medicinal plant. It makes the medicine for liver enlargement and jaundice.

and painful testicles. The berries are alterative, and tonic. They are given in fevers, diarrhoea and heart diseases.

Indian Solanum

Solanum xanthocarpum Schrad & Wendl., of family Solanaceae; Eng. Indian Solanum; Hindi—*Kateli.*

A common prickly herb found throughout India.

Uses. The herb is expectorant, bitter, stomachic, astringent, diuretic, alterative and anthelmintic. Its decoction is given in fevers, coughs, asthma, dropsy, flatulence, gonorrhoea, pain in the chest and heart disease.

Chiretta

Swertia chirata Buch-Ham., of family Gentianaceae; Eng. Chiretta; Hindi—*Charayatah*

This is a shrub commonly found in temperate Himalayas from Kashmir to Bhutan and the Khasia hills.

Uses. The shrub contains two bitter principles ophelic acid and chiratin. It is a bitter tonic, stomachic, febrifuge, appetizer, anthelmintic, alterative, laxative, antidiarrhoeic, and antiperiodic. It makes a good drug for skin diseases, intermittent fevers, intestinal worms, bronchial asthma and regulating the bowels. It is given as an infusion or tincture.

Heart-leaved Moonseed

Tinospora cordifolia (Willd.) Miers of family Menispermaceae; Eng. Heart-leaved moonseed; Hindi—*Gulancha.*

A common climbing shrub (liane). Found throughout tropical India and the Andamans.

Uses. The fresh plant is antiperiodic, alterative, tonic, hepatic, stimulant and diuretic; its watery extract is very effective in fevers. The plant is commonly used in rheumatism, urinary diseases, dyspepsia, general debility, syphilis, skin diseases, piles, bronchitis, impotence, gonorrhoea and jaundice. The fecula prepared from the roots and stem, known as *sat giloe* is a valued drug for intermittent fevers, chronic diarrhoea, chronic dysentery, jaundice and rheumatism.

Lochnera

Lochnera rosea (Linn.) Reichb.,—*Catharanthus roseus* G. Don. Syn. *Vinca rosea* Hindi—*Sadabahar*; family Apocynaceae.

An erect annual or perennial herb; native of Madagascar but now commonly grown in India as an ornamental plant. Leaves opposite, oval, obovate or oblong, glossy; flowers usually 2-3, in cymose axillary clusters; fruit a cylindrical follicle, manyseeded. Flowers white or pink.

Alkaloids. All parts of the plant, particularly the root bark contain alkaloids; these include three alkaloids of the Rauvolfia group; *viz.*, ajmalicine, serpentine and reserpine. The alkaloids of this plant possess hypotensive, sedative and tranquillizing properties. They also cause relaxation of plain muscles and depression of the central nervous system.

Fig. 13.21. *Catharanthus roseus* (**Sadabahar**). Commonly grown in gardens. It is propagated by seeds or cuttings. It blooms almost throughout the year. The alkaloids possess hypotensive, sedative and tranquillizing properties.

Uses. The plant has been used as a folk remedy for diabetes in various parts of South Africa and also in India and Sri Lanka. It is reported to be toxic to cattle. The juice of the leaves is used as an application for wasp stings. An infusion of the leaves is given in the treatment of menorrhagia. The root is considered tonic and stomachic. The alkaloids possess hypotensive, sedative and tranquillizing properties. The alkaloids inhibit the growth of *Vibrio cholerae* and *Micrococcus pyogenes* var. *aureus*. Vindoline and the other alkaloids from leaves are active against *M. pyogenes* var. *aureus* and var. *albus*, *Streptococcus haemolyticus, Corynebacterium diphtheriae* and a few other bacteria; leaf extracts form a useful anti-bacterial agent for the treatment of staphylococcal and streptococcal infections.

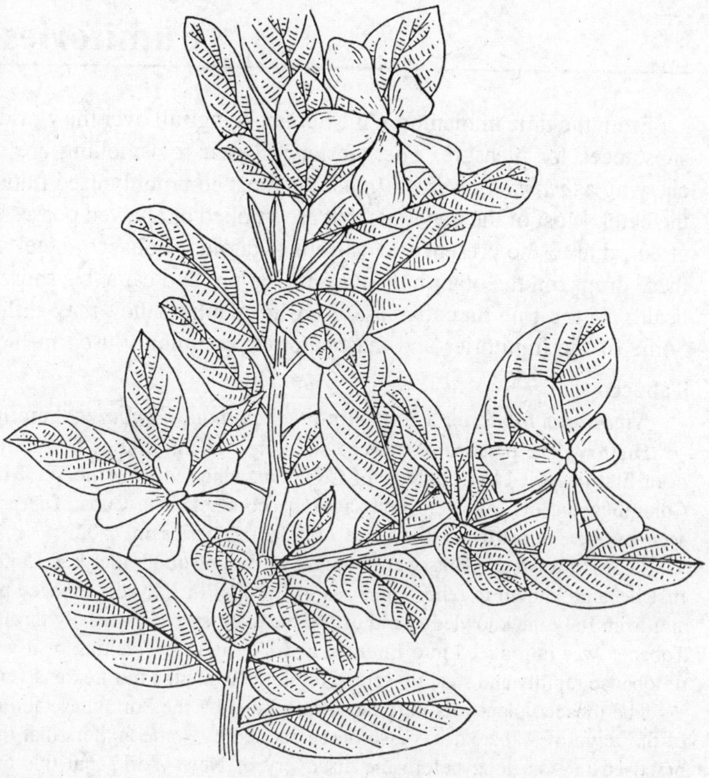

Fig. 13.22. *Catharanthus roseus* (a line diagram). A twig with leaves and flowers. The plant is useful as a medicine in cancer and diabetes.

Wild Liquorice

Abrus precatorius Linn. ; Eng. Wild liquorice; Hindi—*Gungchi*, *rati*, Family—Papilionaceae. A beautiful climber, found all along the Himalayas ascending to 3000 feet, and throughout the plains of India.

Uses. The root is emetic and is used for gonorrhoea. The powdered root with ginger is used for coughs, including whooping cough. A cold infusion of the root is used in leucorrhoea.

The leaves are antiphlogistic in action. The leaves are steeped in warm mustard oil, and applied over the seat of pain in rheumatism.

The seeds are used internally in affections of the nervous system and externally in skin diseases, ulcers, affections of the hair. The seeds are pounded and made up with mercury, sulphur, *nim* seeds, hemp leaves and cotton seeds. The seeds disturb the uterine functions and prevent conception. The boiled seeds possess powerful aphrodisiac properties.

Indian Abutilon

Abutilon indicum (L.) Sweet; Eng. Country mallow, Indian Abutilon; Hindi—*Kanghi*; Family—Malvaceae. A small shrub, found throughout the hotter parts of India.

Uses. The leaves yield a mucilaginous extract which is used as a demulcent. A decoction of the leaves is used in gonorrhoea and chronic bronchitis. The leaves are cooked and eaten in cases of bleeding piles.

An infusion is prepared from roots, which is given in fevers as a cooling remedy. It is also useful in the treatment of leprosy.

The bark is astringent and used as a diuretic.

The seeds are laxative and demulcent, and are used in the treatment of coughs. The seeds are aphrodisiac.

14

Fumitories and Masticatories

From the time immemorial the human beings all over the world have smoked or chewed various substances for pleasure. The substances used for smoking are designated as *fumitories*, and or chewing as *masticatories*. In India, the most commonly used fumitory is tobacco, and masticatory, the betel. Most of the materials that are smoked or chewed possess a distinct stimulating or narcotic effect, due to the presence of various alkaloids. Tobacco, betel and cola are the least harmful of these drugs, on the other hand cocaine, opium and cannabis show the deleterious effects to human health. If the true narcotics are used in quantity, they may inflict the death blow to the addict. Some of the fumitories and masticatories, commonly used in India and abroad are enlisted here:

Tobacco

Nicotiana tabacum Linn; Hindi—*Tamaku*, *Tambaku*; Family—Solanaceae.

History. The practice of tobacco-smoking was made known to Europeans, about the year 1492, having been first observed by followers of Columbus when visiting the West Indian Islands. It is also thought that Columbus and his followers first saw tobacco smoked in Cuba. The practice of chewing tobacco was first seen by the Spaniards on the coast of South America in 1502.

The first tobacco plants were brought to Europe about the year 1560. It was from Portugal that the plant first became known to other countries in Europe. In Italy the tobacco plant was called 'Erlea Santa Croce' and from Italy the knowledge and use of the plant spread gradually throughout Northern and Eastern Europe. Tobacco was introduced into England in 1586, and the smoking of it was fashionable. Nevertheless the use of tobacco rapidly and steadily spread until it became the most extensively used article of luxury in the world.

The tobacco plant was introduced into India by the Portuguese about the year 1605 during the later part of the reign of Akbar. It has also been asserted by some authors that the habit of smoking must have been practised in Asia long before the discovery of New World, but this theory has no proof in support of it. At the beginning of the seventeenth century the influence of the Portuguese in the East was at its highest, and was through their means that the use of tobacco was first made known in Persia, Arabia, India and China. The plant is said to have reached Java in 1801. It has been established from the *Maasir-i-Rapimi* that tobacco came from Europe to the Dakhin, and from the Dakhin to Upper India during the region of Akbar (1556-1605).

The tobacco is thought to be native of northwestern Argentina and adjacent Bolivia. It is believed to have been in cultivation in pre-Columbian times in West Indies, Mexico, Central America and northern parts of South America.

The important tobacco cultivating areas in India lie in Andhra Pradesh, Gujarat, Maharashtra, Karnataka, Tamil Nadu, Uttar Pradesh, Bihar and West Bengal, which together account for 91.0% of the area and 93.0% of the production.

Cultivation. Tobacco is reported to have been introduced into India by the Portuguese sometimes in the beginning of the seventeenth century. It was first grown for commercial purposes in Gujarat and Maharashtra, and cultivation spread to other parts of the country a little later. At present India occupies the third position among the tobacco producing countries in the world and ranks fifth in export trade. From the point of revenue and trade it is of considerable importance to the economy of the country.

Climate. Tobacco thrives well under tropical, subtropical and temperate climates. In India, it is grown under a wide range of conditions from the coastal areas up to an altitude of 900m. In general the crop is raised in South India from October to March when the temperatures are moderate and in the eastern and western parts of the country from September to January; in the Punjab it is grown as an early summer crop.

Fig. 14.1. *Tobacco*. Nursery bed, 45 days old. C.T.R.I. Rajahmundry.

Soil. The best soils are those which are open, well drained and well aerated. Cigarette tobacco is grown in India as a dry crop on heavy black soils of Andhra Pradesh. The soils have high moisture retaining capacity. Cultivation on the lighter soils of Karnataka yields tobacco of better quality.

Propagation. Tobacco is propagated by seed. For several years after the introduction of Virginia cigarette tobacco into India, there existed a belief that to produce crops of good quality, seed should be produced from its original home every other year. At present Central Tobacco Research Institute, Rajahmundry and the Indian Leaf Tobacco Development Company are meeting the entire needs of pure Virginia tobacco seed in the country.

Tobacco seeds are reported to retain their viability for 20 years or more when stored in dry condition at ordinary temperature. Trials in India, however, have shown that they retain viability satisfactorily only upto about three years. Seeds of good quality give about 90% germination.

Nursery practices. Tobacco seedlings are raised in nursery beds. The seed beds must be well manured. Depending on the growing season and type of tobacco, the sowing time in most of the regions extends from the first week of July to the third week of the September; in Punjab, sowing is done in November-December, while in Uttar Pradesh, it is done from the middle of February to the middle March. Seedlings are ready for transplantation in 7-9 weeks after sowing in the case of *Nicotiana tabacum* and in 5-6 weeks in the case of the *N. rustica*. In Punjab where the nursery is raised in the cold season, the seedlings are transplanted after 9-12 weeks.

Transplanting. Transplanting is done by hand for all tobacco in India. Planting is generally done on a rainy day or after irrigating the field. In Uttar Pradesh and Punjab, planting is done in dry soil and the field is immediately irrigated. The number of plants per hectare varies from 12,500 for cigarette tobacco and broad leaf chewing tobacco, to 100,000 for hookah tobacco in Punjab.

The crop is generally intercultured 3-4 times at intervals of about a fortnight. All such operations are completed between 8 and 12 weeks after transplanting as later cultivation is likely to damage the leaves of grown up plants.

Topping and desuckering. The plants are topped when they are 90-100 cm. high or 5-6 weeks old. Following topping, axillary buds grown out from leaf axils as vigorous shoots or suckers and these are removed regularly as they appear or their growth is suppressed by applying coconut oil at the top five or six leaf axils soon after topping. As a consequence of topping there is diversion of nutrients from flower heads to leaves. The time and height of topping have important effects on the development of leaves.

Fig. 14.2. Tobacco—Flue-cured virginia tobacco.

Harvesting and yield. In the case of cigarette tobacco, only leaves which show as light yellowish tinge are harvested. In bidi types, the crop is considered ready for harvest when the majority of leaves have mottled appearance. The cigar and cheroot crop is for harvest when the leaves pucker, become yellowish green in colour and brittle; in hookah types, thick broad yellowish brown flecks appear at this stage on the leaves.

Cigarette tobacco is harvested by priming; only those leaves that are ripe for harvest are removed from the plant at a time and the entire harvest is completed in 5 to 6 primings at intervals of about a week. This method is followed in the case of cigar wrapper tobacco also and in some areas for bidi and hookah tobaccos also. In all other cases, harvesting is done by cutting the plant close to the ground and allowing it to lie in the field overnight for wilting.

The average yield per hectare of flue cured tobacco in Andhra Pradesh is about 790 kg, while in Karnataka it is 1125 kg. A well managed crop of broad leaf cigar type in Tamil Nadu may yield 1335-1680 kg./ha. The yield of *Natu* tobacco ranges from 1790 to 2250 kg./ha. in West Godavary. Among hookah tobaccos the yield of Tabacum types varies from 890 to 1110 kg. of leaf and Rustica types from 1335 to 1610 kg./ha., inclusive of stems.

Fig. 14.3. Tobacco. Burley tobacco crop.

Curing. The harvested leaves are cured before marketing. Curing consists essentially in drying the leaves gradually under conditions which permit certain changes in chemical composition essential for development of the desired quality. Four methods of curing are recognized—(1) flue curing, (2) sun curing, (3) air curing and (4) fire curing.

Fig. 14.4. Tobacco crop ready for harvest.

Flue curing. A large part of the tobacco used in cigarette manufacture is flue cured in barns specially designed for the purpose. The furnace of the barn is fired by coal or wood. A system of flue pipes starts from the furnace and is distributed along the sides of the barn and led out through the wall as a chimney. An improved arrangement for ventilation consists in providing numerous slits in the walls at floor level all round at intervals of 10 cm. The barn is furnished with system of tiers in five levels on which bamboo sticks strung with tobacco leaves are loaded.

The leaves are harvested in the early morning and immediately taken to a cool thatched shed near the barn, where they are strung on to bamboo sticks each of which takes about a hundred leaves. A barn, 5m. x 5 m in size, can take about 650–750 sticks with a total of about 1500–2000 kg. of green leaf.

Flue curing consists essentially in yellowing the leaf at moderate temperature and high relative humidity and then drying the leaf web and stem by increasing the temperature and lowering the humidity in such a way that there is no discolouration. The procedure to be followed and time intervals given for the three principal phases of curing namely, *yellowing*, *fixing the yellow colour* and *drying*, vary with the raw maturity of the raw leaf and prevailing weather conditions.

After curing, the barn is allowed to cool down and the ventilators opened when the leaves absorb moisture from the air and become soft. They are then bundled into hands and bulked in small covered heaps for a few days. The slight tinge of green present in some leaves disappears during this process.

Sun curing. Many types of tobacco are sun cured in India. Entire plants are harvested and left to dry for some days in the field (ground curing). In some areas harvested leaves are made into heaps and periodically opened out to expose the leaves to dew and remade into heaps. Sometimes the leaves are strung on poles or racks and exposed to the sun to dry (rack curing). A combination of ground curing and rack curing is sometimes followed as in the case of cigar and chewing tobacco of Tamil Nadu. *Natu* tobacco of Andhra Pradesh is rack cured. Hookah and chewing tobaccos of Bihar, Uttar Pradesh and West Bengal are cured by wilting the leaves in the field, followed by alternate heaping and drying till the leaves become dark brown. In Punjab, hookah tobacco is cured for about a week in pits.

Fig. 14.5. Mature tobacco plant.

Air curing. *Lanka* tobacco of Andhra Pradesh and wrapper tobacco of West Bengal are air cured. The leaves of *Lanka* tobacco are strung on to ropes in a shed for 8 to 12 weeks and curing is completed in pits. The green leaf of wrapper tobacco is graded according to size and strung on to sticks which are then transferred to barns in which the relative humidity is maintained at 70-80%. The leaves turn yellow and then brown and the curing is completed in 5-6 weeks.

Fire curing. Certain types of chewing tobaccos are cured by smoking. Only limited quantities are cured by this method in India. A peculiarity of this method is the treatment given to smoked leaves with salt water from sea or with a solution of jaggery to impart a distinct taste.

Uses. The bulk of tobacco produced in India is used for smoking in the form of cigarette, bidi, cigar, cheroot and chuttas and in pipe and hookah. Large quantities are also consumed for chewing and for snuff. With the exception of cigarettes, which are manufactured in factories all others are produced in India on a cottage industry basis.

Tobacco powder and extracts of tobacco have been long used as agricultural insecticide. Tobacco waste comprising dust, midribs and stems are utilized for the extraction of the active principle nicotine which, usually in the form of sulphate, is widely used as an insecticide. Nicotine is used also in the production of synthetic acid and nicotinamide.

The seeds of tobacco plant are free of the toxic alkaloid, nicotine, and are used feeding livestock. The seeds yield a semidrying oil which after refining is suitable for edible purposes and for use in paints and varnishes. The seed cake is rich in protein and finds use as a feeding stuff for farm stock. It also serves as a nitrogenous manure.

Tobacco absolute has been prepared from French tobacco by extracting the leaves with volatile solvents. It is commonly employed as a colourless resinoid and is invaluable for imparting an attractive nuance in modern fashionable perfumes.

Fig. 14.6. AGMARK sealing of tobacco meant for export.

The percentage of nicotine, and the proportion which it comprises the volatile base fraction, have important influence on taste quality of smoking products. In flue cured tobacco, a distinctly high nicotine content ($<3\%$) is usually objectionable but a low concentration ($>1.5\%$) also will be unsatisfactory to the average cigarette smoker.

Physiological effects. Tobacco acts as a local irritant. Used as snuff, it excites violent sneezing and a copious secretion of mucus; on chewing, it irritates the mucous membrane of the mouth and increases the flow of saliva. In large doses or in persons used to it, tobacco produces severe ausea, sometimes vomiting accompanied with profuse respiration, and great muscular weakness. The pharmacological activity of tobacco is due almost entirely to its content nicotine which is powerful and rapidly acting poison. Toxic doses of nicotine produce extreme nausea, vomiting, evacuating of bowel and bladder, muscular tremors and convulsions.

Nicotiana rustica Linn. Eng. Tobacco; Hindi—*Vilayati* or *Calcuttia tamakhu*; Family—Solanaceae.

The species is native of South America. It requires a cool climate and its cultivation is confined mainly to northern and north eastern regions of India, *viz.*, Punjab, Uttar Pradesh, Bihar, West Bengal and Assam. It is also cultivated in Gujarat in winter. It accounts for about 15% of the total area under tobacco.

Uses. Rustica types are generally high in nicotine content and are used for hookah, and snuff; they are not suitable for cigarettes, bidis or cigars.

THE INDIAN TOBACCO (TOBACCO BOARD, INDIA—A REPORT)

The World of Indian Tobacco. The story of Indian tobacco dates back to the seventeenth century. India is said to have been introduced to the charms of tobacco by the Portuguese traders. Since then, Indian tobacco farming and industry have withstood the upheavals of time and grown steadily. In fact, today India has emerged as one of the world's most sought-after producers of fine tobaccos.

In world tobacco statistics India has some impressive figures to show. With a production of 360 million kg. India is third largest producer of tobacco in the world, next to U.S.A. and China. In flue cured virginia tobacco production, India ranks fourth—after USA, Brazil and Canada. With a share of 6.3% of the world market, India is third largest exporter of tobacco next only to USA and Brazil. India accounts for 8.5% of the total area under tobacco cultivation and 7% of the total tobacco output in the world.

All of which points to the significant contribution tobacco makes to the national economy: Foreign exchange earnings of rupees 95 crores annually. Excise revenue of rupees 450 crores per year to the national exchequer. Annual returns of rupees 200 crores to the tobacco growing population. Employment opportunities to 7.5 lakh tobacco growers and to another 2 million people in allied activities.

The Many Faces of Indian Tobacco. Climatically and by natural endowments, India is well placed to produce a wide varieties of tobacco. Major among them are the commercially cultivated species *Nicotiana tabacum* and *Nicotiana rustica*.

For cigarettes, India produces the flue-cured virginia, white burley, Natu and Turkish tobaccos. For pipe-smoking the lighter sun-cured country Natu has been found eminently suitable. The Lanka tobacco is widely used for cheroots, and is acclaimed as one of the finest in the world. For a heady hubble-bubble smoke, you can depend on the strength of Moti Hari. But, if your taste, runs to the mild, low-nicotine type, you have the Jutti. For bidis, the Nipani has become very popular. And from the Tanjore District of Tamil Nadu comes an excellent variety of chew tobacco.

Indian Tobacco. *Among the World's Finest.* Indian flue-cured virginia tobacco is well-known for its bright colour, mildness, low nicotine content, negligible pesticide residue, agreeable flavour and excellent burning properties. What's more, the fact that Indian tobacco can be blended well with other varieties has made it very popular in the international market.

India has the resource endowment, the ability and the know-how to produce various types of quality tobaccos, and has shown remarkable flexibility to adapt production to suit varying foreign tastes. The EEC team which visited the country reports that India, with her large variety of agro-ecological conditions, should be able to meet the requirements abroad—both low-nicotine tobaccos and the stronger varieties. For example, the low-nicotine tobacco grown in the southern light soils is one of the highly prized tobaccos in Europe.

Research. *The Genesis of Growth.* Worldwide, there is a definite trend towards greater insistence of quality tobaccos. The research effort and production technology for Indian tobacco farming have been geared to meet this shift in emphasis. As the EEC report observes, a systematic and well organised research and development programme is well under way to grow quality tobaccos for the international buyers. Under the all-India co-ordinated Research Project, several research centres have been established. The centres conduct co-ordinated trials on exportable varieties of tobacco in different agroclimate regions to assess their adaptability and to tackle the endemic problems of specific areas growing different types of tobaccos.

Fig. 14.7. Tobacco. Shipment of tobacco.

Institutional back-up. The Indian Council for Agricultural Research (ICAR) is making determined efforts to improve the quality of Indian tobacco. Besides several regional stations and subsidiary research laboratories, a Central Tobacco Research Institute has been established at Rajahmundry, Andhra Pradesh. Soil surveys and soil testing operations are conducted on a continuing and systematic basis.

A fully developed extension service operates under State Agriculture Department to convey the fruits of research to the farmers.

Indian Tobacco. *At Home and Abroad.* Although four-fifths of Indian tobacco production is consumed at home, India continues to be one of the foremost suppliers of quality tobacco to the world. In 1976, India exported 80 million kg. of tobacco valued at rupees 93 crores. Flue-cured virginia tobacco accounts for 90% of India's total tobacco exports. During the same year India also exported tobacco products like bidi, cigarette and hookah tobacco valued at Rs. 5 crores. Indian tobacco is exported to nearly 50 countries. UK continues to be our largest buyer, followed by Russia Other important buyers are Japan, Bangladesh and Nepal in Asia; Belgium, the Netherlands and Italy in Western Europe; Iraq, Saudi Arabia and Aden in the Middle East and Somalia, Ivory Coast, Senegal and Sierra Leone in Africa.

Quality Assurance. India goes to elaborate pains to ensure that the foreign customers get the best value for their money. A system of stringent quality control has been introduced. The Sea Customs Act and the Agricultural Produce Act, 1937, make it obligatory to grade export consignments of unmanufactured tobacco strictly in accordance with the standards specified in the Tobacco Grading and Marking rules. Quality control measures are enforced and pre-shipment inspection organised by specialist personnel trained under Agricultural Marketing Advisor to the Government of India. Before shipment, every package of export tobacco is inspected, labelled and sealed with the AGMARK stamps.

Tobacco Board. *Committed to the Cause of Indian Tobacco.* Tobacco Board was born of realisation of the vast potentials of Indian tobacco and its significant contribution to the national economy. The board which became operative from January 1976, provides an institutional framework in the form of a single apex organisation to develop tobacco industry by co-ordinating and streamlining production, marketing and exports. Within less than two years of its operations, the board has made encouraging progress through a series of developmental measures: Regulation of production to balance supply and demand. Assurance of fair and remunerative prices to growers. Streamlining of the internal marketing system. Improvement of the quality of produce standardisation of tobacco grading and introduction of plant position grading at farm level. Expansion of marketing opportunities in India and abroad.

Today more than ever before, the Indian tobacco grower is alive to the need to maintain quality standards. Added to this are the massive research and development effort and the determined bid to rationalise the marketing system and production strategies. All of which means that India is poised for a dramatic breakthrough in tobacco farming and exports.

Fig. 14.8. An arecanut garden.

Arecanut

Areca catechu Linn; Eng. Areca nut, Betel nut palm; Hindi—*Supari*; Family—Palmaceae/ Arecaceae.

A slender stemmed palm. It is native of Malaya. The main arecanut-growing areas are concentrated in the South-Western and the North-Eastern regions of the country. Among the principal States that grow arecanuts are Kerala, Karnataka, Assam, West Maharashtra and Tamil Nadu.

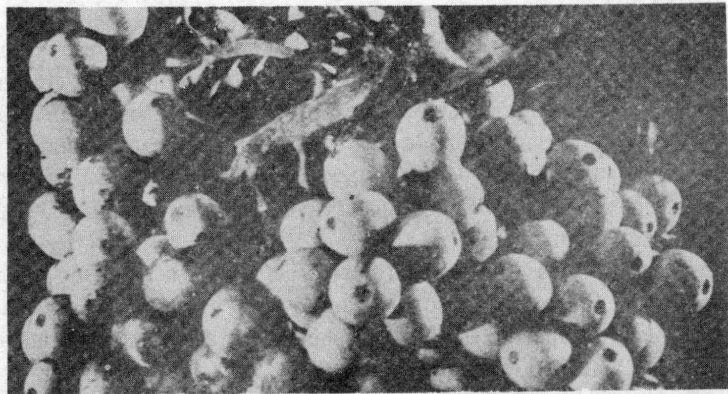

Fig. 14.9. A bunch of Arecanuts.

Fig. 14.10. *Areca catechu*. Arecanut (*Supari*)—Split arecanuts.

Imports and Exports. The production of arecanut in India is about 1 million tons, about two-thirds of her annual requirements. The deficit in her requirements is met by imports, chiefly by Singapore, Malaya and Ceylon. It is, therefore, very important that India should increase her indigenous production in order to achieve sufficiency and avoid the annual drain on foreign exchange.

Cultivation. Arecanut can be classed as "small farmer's cash crop", as it yields direct cash when it is most needed. For successful cultivation, it requires cool moist conditions, and it suffers badly by exposure to sun and wind. The seedlings are raised by selected nuts from palms which are 30 to 40 years of age, heavy yielding and free from diseases. One to two years old seedlings are planted following difficult spacings, generally 12' x 12' with one in the centre. While the cultivated species appear to be the same, there are large, medium and small-sized nuts. Distinct trade types are produced after processing them according to the different stages of maturity at which the nuts are harvested.

Marketing. India was a great connecting link in trade between Western and Eastern Asia for centuries, and arecanut constituted one of the important commodities which was bartered with the other articles. The South-eastern region of India for centuries has been supplying assorted trade types to the entire country. The sundried arecanuts are in great demand in all other States. Efforts are being made to open co-operative societies in the States of Kerala and Assam The existing societies in North Kanara, South Kanara, and Shimoga region of Karnataka State have already established their reputation.

Uses. Arecanut is chiefly used as a masticatory. The custom of chewing *tambula* is not only current throughout India but outside also in countries like southern China, Siam, Indo-China, Malaya and the Eastern Archipelago. In fact, the chewing of arecanut is a popular habit enjoyed by nearly one-tenth of human population.

The tender nuts are used for preparation of boiled and coloured nuts, extensively used in the southern region. The ripe nuts after sundrying for about 40-60 days and thereafter husking yield *chali supari*, most widely used in Northern India. In Kerala and Assam, mature nuts are preserved as such and consumed in raw form.

Betel

Piper betle Linn; Eng. Betel; Hindi—*Pan*; Family—Piperaceae.

A perennial dioecious creeper probably native of Malaysia, cultivated in India for its leaves used for chewing. Stems semiwoody, climbing by short adventitious roots; leaves 5-20 cm. long broadly ovate, slightly cordate and often unequal at the base, shortly acuminate, acute, entire with often an undulate margin, glabrous, yellowish or bright green, shining on both sides; petiole stout 2.0 to 2.5 cm. long, pendulous; fruits rarely produced, often sunk in the fleshy spike; forming nodule-like structures.

The cultivated betel in India is usually the male plant selected from certain races and consequently does not fruit. Several types are grown in different parts of India showing variation in size, shape and colour of leaves, their taste and aroma.

Cultivation. It is mostly confined to small holdings, distributed in almost all States in India, except in the dry north-western parts. Approximately 26,000 hectares are reported to be under this crop, Karnataka, West Bengal and Tamil Nadu each having nearly 4,000 hectares. The other important states are Maharashtra, Kerala, Andhra Pradesh, Madhya Pradesh, Bihar, Uttar Pradesh and Assam.

Climate. The betel vine thrives best under tropical forest conditions having a cool shade, considerable humidity and a good supply of soil moisture.

Soil. The best soil for betel cultivation is clayey loam which is friable, rich in organic matter, and which has good drainage.

Propagation. Betel vine is propagated only vegetatively by cuttings taken from healthy vines at least two years old. These are obtained from vines of the previous year's growth, trimmed into sections of 30-45 cm. length. Each cutting contains 3-5 nodes and planted in such a manner that 2 nodes are buried in the soil and one or more nodes above ground and pointing towards the standards, on to which they will eventually be trailed.

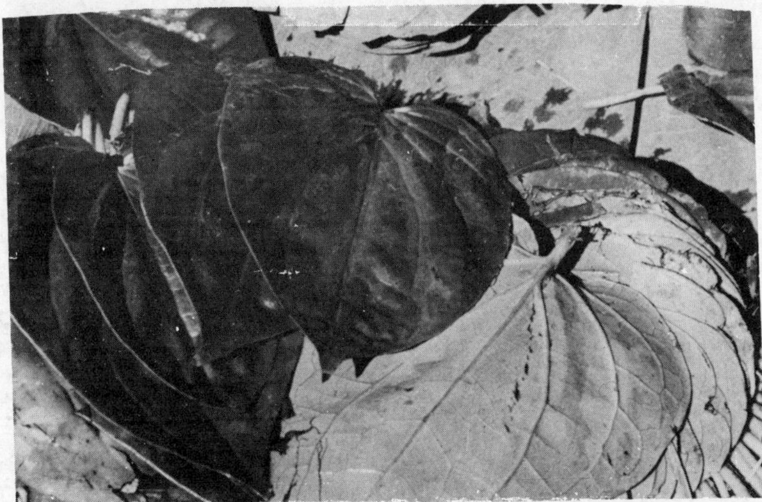

Fig. 14.11. *Piper betle* (*Pan*). The betel for chewing purposes consists of betel leaf smeared with hydrated lime and catechu to which scrapings of arecanut are added; flavourings such as coconut shavings, clove, cardamom, fennel, powdered liquorice, nutmeg and also tobacco are used according to one's taste.

Bleaching. While fresh leaves are generally used for chewing in some areas leaves subjected to a process of bleaching are preferred. Bleached betel leaves normally fetch a higher price as they are believed to possess some improved qualities. Bleaching of leaves on a commercial scale is carried out in parts of North India, particularly at Varanasi, Gorakhpur, Ramtek, and Pune and its neighbourhood. The bleaching process consists in moistening the leaves and allowing them to stand in a warm ventilated place in the absence of sunlight. During bleaching the leaves turn gradually yellowish white. The whole process normally takes 10-15 days in summer and 15-20 days in cold seasons.

Grading and marketing. The picked leaves are washed, cleaned, counted and sorted into different grades, according to size, colour, texture and maturity on which their chewing quality depends. In Uttar Pradesh, the commercial grades recognized, mainly on basis of place of origin are—*Desi* (local), *Maghai* (Bihar), *Bangla* (Bengal), *Jagannathi* (Orissa) and *Kapoori* (Tamil Nadu).

Betel leaves are consumed or marketed as early as possible after harvesting. Small quantities of betel leaves are exported to some countries outside India.

Utilization. Betel leaves have a strong pungent aromatic flavour and are widely used as a masticatory. Generally, mature or overmature leaves which have ceased growing but not yet become brittle are used for chewing. The basic preparation for chewing purposes of betel leaf smeared with hydrated lime and catechu to which scrapings of arecanut are added; flavourings such as coconut shavings, clove, cardamom, fennel, powdered liquorice, nutmeg and also tobacco are used according to one's taste. In some places prepared pan is covered with silver or gold leaf. A beverage called *pan-supari* nectar, has been developed by the Central Food Technological Research Institute, Mysore, and is said to be a good source of calcium.

Chewing of betel leaves with various adjuncts is an ancient practice in India and other countries of East Asia. As a masticatory, it is credited with many properties, it is aromatic, digestive, stimulant and carminative.

Studies on the physiological effects of chewing betel leaves have shown that the initial effects of chewing of betel with arecanut and other adjuncts are the exitation of the salivary glands and the irritation of the mucous membrane of the mouth. The red colouration produced is due to a

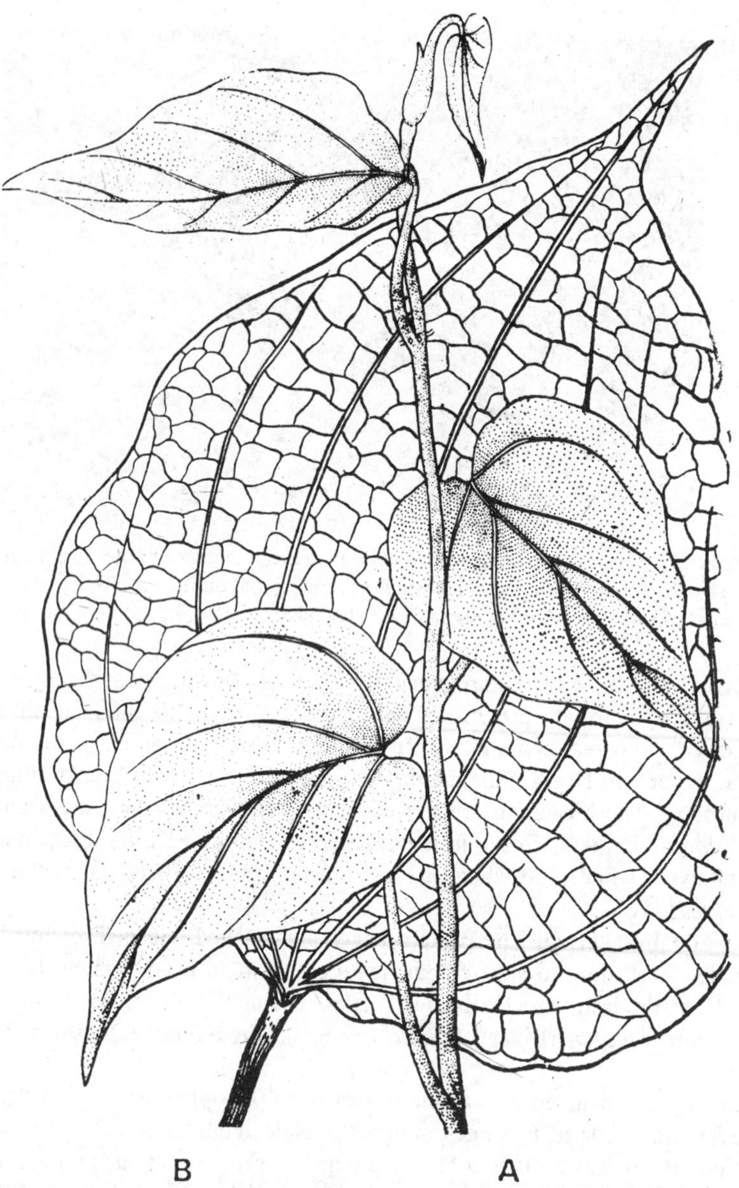

Fig. 14.12. *Piper betle* (*Pan*). A climbing shrub cultivated for its leaves which are used as masticatory. A, twig; B, a leaf.

pigment in the arecanut, which manifests itself under the action of alkali in lime and catechu. A mild degree of stimulation is produced, resulting in sensation of warmth and well-being, besides impairing a pleasant odour.

Betel chewing is considered as a good and cheap source of dietary calcium. The leaves contain good amount of B vitamins (nicotinic acid), ascorbic acid and carotene. The important factor determining the aromatic value of the leaf is the amount and particularly the nature of the essential oil present. Betel leaves from different regions vary in smell and taste.

Catechu

Acacia catechu (Linn. f.) Willd.; Eng. Catechu, Black cutch; Hindi—*Khair*, *Katha*, *Khadira*; Family—Mimosaceae.

A tree found in the Punjab, Madhya Pradesh, Uttar Pradesh, Andhra Pradesh, and Bihar. The katha industry is mostly located in Bareilly, Gwalior and Bombay.

Uses. The *Katha*, obtained from the heartwood, is used as a masticatory. The paste of katha is generally applied on betel leaf. This makes a popular masticatory in India. It is also used as dyeing and preserving agent.

Cola

Cola nitida, Eng. Cola, kola nuts; Family—Palmaceae/Arecaceae.

A tree; with a straight trunk reaching a height of 50 to 65 feet. It grows wild in tropical West Africa. It is cultivated in Brazil, India, Jamaica, Sudan and other parts of Asia. The fruit consists of starch shaped follicles which contain eight hard, plano-convex, fleshy seeds with reddish colour and aroma of roses. The seeds are known as 'cola' or 'kola nuts' of commerce. These so-called nuts are usually chewed directly or after making powder of them. Although bitter at first, the nuts leave a sweet taste in the mouth. The use of this masticatory gives slight stimulation and temporary increases in vigour and vitality. It lessens hunger and fatigue. It contains 2 percent caffeine, an essential oil and a glucoside, Kolanin. The chewing of colanuts does not produce any deleterious results.

Opium

Papaver somniferum Linn; Eng. opium poppy; Hindi—*Afim*; Family—Papaveraceae.

An erect, rarely branched, usually glaucous annual, 60-120 cm. high. Leaves ovate-oblong, or linear oblong, amplexicaul, lobed, dentate or serrate; flowers large, usually bluish white, purple or variegated, capsules large 2.5 cm. diameter, globose, stalked; seeds white or black, reniform.

At present, opium poppy is cultivated mainly in India, Turkey and Russia It is also grown to a small extent in Yugoslavia, Bulgaria, Afghanistan, Pakistan and Japan.

The cultivation of opium poppy in India is controlled by the Government and is now confined to Uttar Pradesh, Madhya Pradesh and Rajasthan. Negligible quantities are also grown in Jammu and Kashmir.

There are numerous varieties of *Papaver somniferum* of which two are under cultivation for the production of opium. They are var. *album* D. C., cultivated in India and var. *glabrum* D.C., cultivated in Asia Minor, Egypt and parts of Iran. There are several forms of var. *album* in cultivation in India: the form with white flowers and white seeds is grown in Uttar Pradesh; the form with red or purple flowers is grown in Madhya Pradesh and Rajasthan (Malwa).

Cultivation. The cultivation of opium poppy in India is entrusted to cultivators under licences issued by the district opium officers of different areas. The licensed cultivators undertake, on behalf of the Government, to sow the poppy, lance the capsules, collect the latex and deliver it at the weighing centres at a price fixed by the Government.

Climate and soil. Opium poppy grows in varying climates but it cannot endure extreme cold. In cold climate, opium yield is greatly diminished. Frost is sometimes very destructive to the crop. It is grown in almost all kinds of soil but it prefers a sandy loam.

Culture. It is cultivated as a rabi crop. Land is prepared in September by repeated ploughing and brought to a fine tilth. Large amounts of fertilizers and manures are added to the field.

Harvesting. Opium is harvested by incising the capsules at a particular phase of plant growth. The period of collection of opium extends from the end of January to April. The capsules are incised with a special type of knife with 3 or 4 small blades designed to ensure uniformity in the depth of incisions. Usually, each capsule is lanced 3 or 4 times until no more latex exudes.

Lancing is done after midday and the exuding latex is allowed to remain on the capsule overnight; during which time it coagulates and becomes somewhat thicker and slightly darker in colour. The latex varies in colour from milky white to bright pink. In the morning, the coagulated latex (raw opium) is collected from the surface of capsules with a blunt-edged small iron scoop.

Storage. Raw opium begins to arrive at the Government Opium Alkaloid Works, Ghazipur, about the second or third week of April in bags or jars. They are stocked in separate vats according to the morphine strength (M.S.) after chemical analysis.

Processed opium. At the Government Opium and Alkaloid Works, Ghazipur, three kinds of opium are processed, *viz.,* Excise Opium, Export Opium and Indian Medical Opium.

Chemical composition. Fresh opium is a brownish, somewhat plastic solid, becoming tough and occasionally brittle on keeping and has a characteristic fruity odour. The total alkaloid content varying from 5-25%. A large number of alkaloids have been isolated from opium, of which at present 25 are known. Morphine, codeine, thebaine, narcotine, narceine and papaverine are the chief opium alkaloids, and of these morphine is the most abundant and by far the most important. Morphine exists in combination with meconic and sulphuric acids in the form of salts readily soluble in water.

Production and trade. India has been producing opium for many centuries, and is at present the largest source of supply of raw opium to the world. Large quantities of opium used to be produced in India during the later half of the last century, the bulk of which was exported to China. Opium is exported for scientific and medicinal purposes chiefly to U.K., U.S.A. and France. The trade and prices of opium and opium alkaloids are entirely controlled by the Government.

OPIUM ADDICTION

The habit of taking opium or morphine has been prevalent for a long time in India, but is now fast dwindling. Opium and its derivatives are consumed in three ways—(1) orally, (2) by smoking and (3) by injection, the first being the commonest.

Orally. The habit of eating opium is acquired as it produces a sense of euphoria and in the belief that it has a remarkable power as an aphrodisiac. Ordinarily crude opium is used, but, sometimes its decoction (*Kasoomba*) is drunk. Once the habit is formed, it is difficult to give it up. The addict has to take more of the drug to combat the feelings of lethargy and mental depression. Its abuse for a prolonged period leads to derangement of appetite and digestion, disturbance of sleep, vomiting, emaciation, impotence, neurasthemic condition, mental perversion of morality, dementia, mania and premature old age.

By smoking. Opium is smoked in India, in the form of *Madak*, *Chandu* or opium dross. *Madak* is prepared by impregnating coarsely powdered leaves of *babul* (*Acacia arabica*) or guava (*Psidium guajava* Linn) in concentrated aqueous extract of opium and rolling the impregnated mass into small balls. *Chandu* is a dark brown concentrated aqueous extract of opium; its morphine content is 8%. The ash from smoking *chandu* is known as opium dross. Excessive opium smoking leads to loss of appetite, a leaden pallor of the skin and extreme leanness of the body, resulting in the addict losing all inclination for exertion.

Fig. 14.13. *Erythroxylon coca* (Cocaine). The leaves are used as masticatory. They possess an alkaloid 'cocaine' in them.

By injection. The addiction of opium or morphine through hypodermic injections is prevalent in Europe and the U.S.A and is now spreading in India also.

Cocaine

Erythroxylon coca; Eng. Cocaine; Hindi—*Cokeen*; Family—Erythroxylaceae.

A shrub, native of Peru and Bolivia. The plant is extensively cultivated in South America, where the leaves are used as a masticatory. The leaves mature in about four years and can then be picked three or four times a year. They possess a bitter aromatic taste due to the presence of the alkaloid cocaine. About 100 lbs. of leaves yield one lb. of the drug. The leaves are generally chewed with lime. It is also cultivated to some extent in Java and India.

People using cocaine can resist physical and mental fatigue and work for long periods without taking food or drink. One adult consumes 25 to 50 grams daily. The narcotic acts directly on the central nervous system and the consumer forgets his hunger or other pains.

Hemp

Cannabis sativa Linn.; Eng. Hemp; Hindi—*Ganja*, *Bhang*; Family—Cannabinaceae.

It is an undershrub. It is native of Asia. In India, it is grown in Uttar Pradesh, Bengal, Maharasthra, Madhya Pradesh, Tamil Nadu and Orissa.

Uses. The use of hemp as a narcotic stimulant is very old, extending back to 3,000 B.C., first in China and thereafter in India. In India, hemp is cultivated as a drug plant. It is much used in India, Mongolia, Southern Asia and tropical Africa.

The female flowering tops and the leaves are the source of a resin (*charas* or *hashish*), which is used as a narcotic or stimulant. *Charas* is smoked. It is the most powerful form of drug. *Bhang* is obtained by drying the young leaves and flowering shoots of either of 1 or both, the male and female plants. Bhang is generally taken with cold drink (*thandai*) in India. It is also smoked. In America this type of hemp is known as *marijuana*. *Ganja* is the dried flowering tops of the cultivated female plants, which become coated with a resinous exudation from glandular hairs during the course of inflorescence. It is also used for smoking. It has a high resin content.

This hemp often produces serious results for the consumer. It produces a stupefying and hypnotic effect, accompanied by hallucinations, agreeable and often erotic dreams, and a general state of ectasy.

Fig. 14.14. *Cannabis sativa.* The female flowering tops and the leaves are a source of resin—*charas*; *bhang* is obtained by drying the young leaves and flowering shoots of male and female plants; *ganja* is the dried flowering tops of the cultivated female plants.

Henbane

Hyoscyamus niger Linn.; Henbane; Eng. Hindi—*Khurasani-ajvayan*; Family—Solanaceae.

An erect, viscidly hairy, foetid annual or biennial, up to 5 feet high, occurring in western Himalayas from Kashmir to Kumaon, at altitudes of 5,000 to 12,000 feet. Attempts have been made to cultivate the plant in Kashmir, Punjab, Uttar Pradesh, Ajmer, Maharashtra and Nilgiris and plants rich in alkaloids have been successfully raised at Yarikash in Kashmir, where systematic cultivation has been undertaken.

Dried leaves and flowering tops constitute the drug henbane. The leaves and tops are picked at the time of flowering. The leaves are dried in the sun for 3-4 days or in the shade for 7-10 days before packing. The yield of dried leaves is of 14 lb. per 100 lb. of green material.

Henbane is valued principally for the alkaloids present in most parts of the plants, particularly in the leaves and flowering tops. Besides the alkaloids, it contains volatile bases, a bitter glycoside hyoscypierin, choline, mucilage and albumin; it is rich in potassium salts. Leaves collected from wild or cultivated plants in Kashmir altitudes above 5,000 feet contain 0.05 to 0.092% total alkaloids. The alkaloid content of the leaves increases with maturity and reaches the maximum at the time of flowering, after which it decreases.

The principal alkaloids present in the various parts of henbane are hyoscyamine ($C_{17}H_{23}O_3N$) and hyoscine ($C_{17}H_{21}O_4N$).

Uses. The drug is used as a sedative, narcotic and also in the treatment of asthma and whooping cough. It has long been used as a poison.

Hyoscyamus muticus Linn.: It is smoked in Northern Africa and India for its intoxicating effect.

Stramonium

Datura stramonium Linn.; Eng. Thorn apple, Stramonium; Hindi—*Dhatura*; Family—Solanaceae.

A common weed, found all over the country. It is native of Asia. The leaves and seeds are narcotic and sometimes used for criminal poisoning. It is also utilized in medicine in the treatment of asthma.

The genus *Datura* has been extensively used in all the continents except Australia for its narcotic and hypnotic properties. The drug stramonium, was known as narcotic as early as 37 A.D. Even today this is one of the favourite sources of "knockout drops" in the tropical regions.

Kavakava

Piper methysticum; Family—Piperaceae.

This species is indigenous to Fiji and other Pacific Islands. It is a bushy shrub 6-8 feet tall, with rounded or cordatae leaves. The thick, knotty, greyish green roots are the important parts. These are dug up and the bark is removed. After the roots are thoroughly cleaned, they are cut into small pieces. These pieces are chewed until they are fine and fibrous and are then placed in a bowl with water and allowed to ferment. After straining, kava is a greyish-brown liquid and is very refreshing.

The beverage kavakava acts as a sedative, a soporific, and a hypnotic, bringing about pleasant dreams and sensations. Excessive use produces skin diseases and weakens the eyesight.

15

Spices and Condiments

The spices and condiments have played a prominent part in all the civilizations of antiquity, in ancient India and China. The majority of spices originated in the Asiatic tropics and were among the first objects of commerce between the East and the West. The quest for condiments and spices was once a powerful force in world history. The search for a sea route to the spice Islands of the East led to the voyages of Columbus, Magellan and many others.

Arabs brought the products of southern India and the spice Islands by caravan to Arabia, and then to Europe. Later other countries took over the spice trade. In the sixteenth century the Portuguese monopolized the trade for two hundred years. Later the British Empire owned most of the spice trade of the world.

The spices cannot be grouped as foods, for they contain less nutritive value. However, they give a good flavour and aroma to food, and add greatly to the pleasure of eating. They stimulate the appetite and increase the secretion and flow of gastric juices. For this reason they are commonly known as "*food adjuncts*". The aromatic value of the spices is due to the presence of the essential oils.

In general, all aromatic vegetable products that are used for flavouring foods and drinks are known as spices. Sometimes the term "*spice*" is restricted to hard or hardened parts of plants, which are generally used in a pulverized state. *Condiments* are spices or other flavouring substances which possess a sharp taste, and are commonly added to food after it has been cooked.

The majority of spices are obtained from the tropics, chiefly from Asia. In India, the major spices are pepper, chillies, cardamom, ginger and turmeric. Besides several others are also quite common, such as: coriander, cinnamon, clove, nutmeg, etc. The pepper plant (*Piper nigrum* L.) is indigenous to the forests of Kerala and has been cultivated in India from time immemorial. Large-scale cultivation of the crop is confined to the Western Ghat of Peninsular India. Chillies (*Capsicum* spp.) are an important condiment crop in India. They are indigenous to South America, and are said to have been introduced into India by the Portuguese. The principal cardamom (*Elettaria cardamomum*) producing countries are India, Ceylon and Indo-China. The big cardamom (*Amomum* spp.) are grown in Sikkim, Nepal, Cambodia, Java and some other tropical countries. Ginger of commerce is a dry product prepared from the rhizomes of the plant *Zinjiber officinale*. The most important producers of commercial dry ginger are India, Jamaica and Sierra Leone. Turmeric (*Curcuma longa* Linn.), is used as an important condiment almost in every Indian house, and has been cultivated in India from very ancient times. Coriander (*Coriandrum sativum* Linn.) is an important condiment used widely in all parts of India, Russia, Central Europe, Asia Minor and Morocco. The cinnamon of commerce is the bark of tree *Cinnamomum zeylanicum* Breyn., which is indigenous to Sri Lanka, Southern India, Myanmar and the Malayan Peninsula. The clove of commerce is the dried unopened flower bud of the clove tree, *Syzyzium aromaticum*. A native of Moluccas, it has spread to a number of countries, *viz.,* Zanzibar, the Spice Islands, the West Indies, the Mascarene Islands, Sumatra, Java, Sri Lanka etc. The cultivation of clove in India is limited to South India. The nutmeg of commerce is the de-arilled dried seed of *Myristica fragrans* Houtt. A native of eastern island of the Moluccas, it is also indigenous to Banda island, Amboyna, Gilolo and Western New Guinea. It was introduced in India towards the end of the eighteenth century. Its cultivation is confined to the south, in the Nilgiris, the Tenkasi Hills and the coastal areas of Kerala. The important minor spices grown in India are ajowan, aniseed, caraway, celery, coriander, cumin, dill seed, fennel, fenugreek, garlic, onion, saffron and vanilla, etc. Several other spice yielding plants of India and abroad have been discussed in the present text.

India is one of the largest suppliers of spices and condiments. India alone contributes 25-30 per cent of the total world trade.

Most of the important spices are grown in India. The classification of *spices* and *condiments* in the present text is based on the morphology of the plant organ which is used for the purpose. The general categories of the plants yielding *spices* and *condiments* are as follows:

SPICES AND CONDIMENTS OBTAINED FROM ROOTS

Angelic

Angelica archangelica Linn.; Eng. Angelica; Family—Umbelliferae (Apiaceae).

It is a stout perennial herb with large pinnately compound leaves. The small, greenish-white flowers are found to be arranged in terminal compound umbels. It is native of Syria. In India, it is grown in Kashmir. All parts of the plant emit pleasing odour.

Uses. The roots as well fruit are dried and used for flavouring food stuffs and beverages. The most important application of angelica roots and seed is in French-type alcoholic liquors and especially in gin absenthe, anesette, etc. In fact, next to juniper berries, angelica root is the main flavouring ingredient of gin. The fresh stems and leaf stalks serve as a garnish and are also used for confectioneries. In Europe, angelica root and stem are candied and eaten as a conserve.

Horse - radish

Armoracia lapathifolia Gilib.; Syn. *Roripa armoracia* (Linn.) Hitch.; *Cochlearia armoracia* Linn.; Eng.—Horse - radish; Family—Cruciferae (Brassicaceae).

It is a tall hardy herb with glossy green toothed leaves and masses of small white flowers. The roots are large, fleshy and cylindrical. It is native of Europe. In India, it is grown in Northern India and the hill stations of Southern India.

Uses. The roots are scraped or grated, and used as condiment, either fresh or preserved in vinegar. The pungent taste is due to a glucoside, sinigrin. It makes a valuable condiment. It aids digestion and prevents scurvy. The roots are used also for flavouring food products. It is credited with digestive and anti-scorbutic properties because of its high vitamin C content. It is highly prized as a condiment specially with oysters and cold meats. When mixed with ketchup, the grated root imparts a refreshing taste to sea-foods, especially shrimp, cocktail and oysters.

Asafetida

Ferula Linn.; English—Asafetida; Hindi—*Hing*; Kashmiri—*Anjudan*; Gujarati—*Hing*; Mumbai—*Hing*; Tamil—*Kyam, perungayam*; Sanskrit—*Hingu*; Family—Umbelliferae (Apiaceae).

This is a genus of perennial herbs commonly distributed from the Mediterranean region to central Asia. Some of the species are important as sources of oleogum - resins, used as condiment and in medicine. Three species are found in India. The Main bulk of asafetida and other ferula gum - resins are being imported in India from Afghanistan.

Asafetida is the dried latex obtained mainly from living rootstocks or taproots of several species of *Ferula*, such as : *F. foetida* Regel, *F. alliacea* Boiss., *F. rubricaulis* Boiss., *F. assafoetida* Linn. and *F. narthex* Boiss., found in Central Asia. *F. narthex* and *F. assafoetida* are found in Himachal Pradesh and Kashmir.

The plants from which the asafetida of commerce is obtained bear massive carrot-shaped roots (5-6 inches in diameter at the crown when they are 4-5 year old). In March-April when the plants are about to flower, the upper part of the root is laid bare and the stem cut off close to the crown. The exposed surface is covered by a dome-shaped structure made of twigs and earth. A milky juice (latex) exudes from the cut surface. After some days the exudate is scraped off and a fresh slice of the root cut when more latex exudes. The collection of the resin and the slicing of the root are repeated until exudation stops (about three months after the first cut). Sometimes the resin is being collected from successive incisions made at the junction of the stem and the tap root. With three incisions, some plants yield 1 kg. or more of gum-resin.

Asafetida (asafoetida) occurs in commerce in three forms—(i) *tears*, (ii) *mass* and (iii) *paste*. The *tears* make the purest form of the resin. They are rounded or flattened, 5-30 mm. in diameter and greyish or dull yellow in colour. *Mass asafetida* is the common commercial form. It consists of tears united into a more or less uniform mass usually mixed with fragments of root, earth, etc. The *paste* also contains foreign matter.

Asafetida is adulterated with gum arabic, other gum-resins, gypsum, chalk, wheat or barley flour, etc.

Asafetida contains 40-64% resin, 25% gum, 10-17% volatile oil and 1.5 to 10% ash. The resin portion chiefly contains asaresinotannol, free or combined with ferulic acid.

Uses. Asafetida is bitter and acrid in taste and emits a strong and peculiar odour. In India, it is used for flavouring curries, sauces and pickles. It is also used in Ayurvedic medicines. It stimulates the intestinal and respiratory tracts and the nervous system. It is useful in asthma, whooping cough and chronic bronchitis. It is used as an enema for intestinal flatulence. It is also used in epileptic and hysterical affections and in cholera. It is also used in veterinary medicine.

Lovage

Levisticum officinale Koth., English—Lovage; Family—Umbelliferae (Apiaceae).

This is a perennial plant, introduced from Europe. It is grown as a garden plant. It thrives best in well-drained soil. It is propagated either by seeds or roots. The roots are dug out in October of second or third year after setting the plants. The freshly dug roots are washed, cut into slices and dried.

Uses. Generally the roots are used for flavouring foods. Sometimes the seeds are used for flavouring confectionery. The leaves and stems are also eaten as salad. The dried roots; on steam distillation yield volatile oil known as '*lovage oil*'. The oil is used for flavouring foods.

Decalepis

Decalepis hamiltonii Wight and Arn.; Tamil—*Mahali kizhangu*; Kannada—*Makali beru*; Family—Asclepiadaceae. It is distributed in the Deccan Peninsula and common in the forest areas of Western Ghats.

This is a climbing shrub with jointed stem and orbicular or elliptic obovate leaves. The roots possess a strong aromatic odour and is marked by fleshy and cylindrical (1-6 cm. diam.). The root has a sweet sarsaparilla - like taste accompanied by a tingling sensation on the tounge. The root contains 92% fleshy matter and 8% woody core. The volatile principle responsible for the aroma and taste of the root is 4-O-methyl-resorcylaldehyde, which is present in a concentration of 0.8% in the air-dried material. It is isolated by steam distillation. The other constituents of the root are - inositol (0.40%), saponins, tannins, a crystalline resin acid, an amorphous acid, a ketonic substance, sterols and resinols.

Uses. The aromatic roots are used as spice and condiment. It is considered to be an appetiser and blood purifier. The root is pickled with limes or as such.

SPICES AND CONDIMENTS OBTAINED FROM UNDERGROUND STEMS

Sweet Flag

Acorus calamus Linn.; English—Sweet flag; Hindi—*Safed bach*; Bengali—*Bach*; Gujarati—*Vekhand*; Sanskrit—*Shadgrantha*; Tamil—*Vashambu*; Telugu—*Vasa*; Malayalam—*Vashanpa*; Kannada—*Baje*; Family—Araceae.

It is a semi-aquatic perennial with indefinitely branched rhizomes. It is native of Europe and North America. In India, it is cultivated in Kashmir, Manipur and Karnataka in damp marshy places, 3,000 to 6,000 fcct above sea level.

The rhizome is indefinitely branched, creeping in mud, with stout joints and large-leaf scars, cylindrical or somewhat compressed, about 3/4 inch in diameter, smooth, pinkish or pale green:

the leaf scars are brown, white and spongy within. It gives off below numerous straight rootlets. Leaves few, distichously alternate forming erect tufts at the ends of the rhizome, tapering into long acute points, entire, smooth; scapes arising from the outer leaves. All parts, but especially the rhizome, aromatic. An essential oil is obtained from the rhizome by distillation. The oil is pale or dark-yellow in colour. It possesses a strong penetrating odour and an aromatic, bitter, burning, camphoraceous flavour.

Uses. The rhizome is used as a condiment for flavouring food stuffs. The aromatic rhizome or root-stock is emetic, stomachic and carminative.

Allium spp. A genus of bulbous, herbaceous plants of family Liliaceae. Bulb tunicated; spathe many-flowered; umbels crowded; flower regular, 6 - merous; segments, or only slightly united below; stamens 6, anthers oblong, attached by the middle and on the back; ovary superior, sessile, 3-celled; stigma 3-fid; ovules mostly 2 in each cell.

Shallot

Allium ascalonicum Linn.; English—Shallot; Hindi—*Gandana*; Bengali—*Gundhun*; Family—Liliaceae.

It is native of Israel. This has been cultivated from the ancient times by all the nations of Asia and Middle East. It is regarded as much milder than garlic. The flowers are greenish-white or purplish white; the bulbs are white. It is mainly found in Tamil Nadu.

Cultivation. Cloves or small bulbs should be planted in October about 6 inches apart, and that by the beginning of the hot season the crop will be ready for use.

Uses. The bulbs (rhizomes) are used as condiment. The bulbs separate into cloves, like those of garlic, and are used for culinary purposes, being of milder flavour than onion. They also make excellent pickle.

Fig. 15.1. Onion. A hybrid VL Piaz 3. It is less-prone to bolting during storage.

Onion

Allium cepa Linn.; English—Onion; Hindi—*Piyaz*; Tamil—*Vella-vengayam, irulli*; Telugu—*Vulli-gaddalu*; Kannada—*Vengayam, nirulli*; Malayalam—*Bawang*. They are cultivated all over India. Onion seed will not keep for certain more than one year. The selected bulbs be planted, and seed obtained from these. If planted in the cold season, they will produce seed in the beginning of the hot season, and if carefully preserved, after being well ripened and dry, the seed obtained in this way will be found to yield a good crop in the following cold season, from October to February.

Uses. The bulbs (rhizomes) are used as food and condiment.

Top onion or Egyptian onion

Allium cepa Linn. var. ***viviparum*** Metz.; English—Top onion; Family—Liliaceae.

It is introduced at Coimbatore and at Coonoor in South India. The bulbs (rhizomes) are used as condiment.

Himalayan onion

Allium leptophyllum Wall.; English—Himalayan Onion; Family—Liliaceae.

The bulbs have a stronger pungency than ordinary onions. The leaves make a good condiment. Found in Himalayan region.

Leek

Allium porrum Linn.; English—Leek; Hindi—*Kirath*, *Vilayati lasan*, *Gandina*; Bengali—*Paru*; Family—Liliaceae.

This succulent plant has been known from time immemorial. It is thought to be a native of Switzerland. It is also probable, like the onion, it came from the Middle East. It was cultivated by Egyptians in the time of Pharoah.

They are best propagated in India by sowing the seed broadcast on a small bed immediately the rains stop. When the seedlings are about 6 inches high they should be carefully transplanted. They should then be planted in rows six inches apart. They require plenty of water and should be earthed up once or twice.

Uses. The bulbs are used as food and condiment.

Garlic

Allium sativum Linn.; English—Garlic; Hindi—*Lasan*, *lahsun*; Bengali—*Lashan*; Sanskrit—*Maha-ushadha*; Tamil—*Vallaipundu*; Telugu—*Vellulli fella-gadda*; Kannada—*Belluli*; Family—Liliaceae. Cultivated all over the country. It is propagated by planting out the cloves singly, in October, in drills about 6-7 inches apart and 2-3 inches deep. The crop is taken up in the hot weather, and after being dried in the sun the bulbs are stored for further use.

Fig. 15.2. Garlic bulbs.

Numerous bulbs remain enclosed in a common membranous covering. Stem simple, about 2 feet high; scape is smooth and shining, solid terminated by a membranous pointed spathe enclosing a mass of flowers and solid bulbils prolonged into leafy points; leaves long, flat, acute, sheathing the lower half of the stem; flowers small, white.

Uses. The bulbs (rhizomes) are used as a condiment and flavouring substance. Garlic powder is used as a condiment. It is a carminative, diuretic, stomachic, alterative, emmenagogue, and tonic.

Chives

Allium schoenoprasum Linn.; English—Chives; Hindi—*Piazi*; Family—Liliaceae. A native of North Europe. It is a cultivated pot-herb allied to garlic, with purple flowers. In India, it is found in the Western Himalayas from Kashmir to Kumaon and also cultivated on the hill stations in South India. It is propagated by the division of the root in the month of October.

Uses. The leaves and bulbs are used as condiment.

Chinese chive

Allium uliginosum Don; English—Chinese chive; Hindi—*Banga gandina*; Family—Liliaceae. In India it is grown in West Bengal.

Uses. It is used as a condiment and flavouring substance for flavouring curries.

Greater galangal

Alpinia galanga (Linn.) Willd.; Syn. *Languas galanga* (Linn.) Stuntz.; English—Greater galangal; Hindi and Bengali—*Kulinjan, kulanjan, bara-kulanjan*; Gujarati—*Kolinjan*; Mumbai—*Malabari-pan-ki-jar*; Marathi—*Kosht-kulanjan*; Tamil—*Pera-rattai*; Telugu—*Pedda-dumpa-rash-trakam*; Malayalam—*Peraratta*; Kannada—*Dumpa-rasmi*; Myanmar—*Podagoji*; Sanskrit—*Dumparastma, kulinjana*; Persian—*Khusrave-durue-kalan*; Family—Zingiberaceae.

A perennial herb; native of Indonesia and Malaya, now cultivated in Bengal and South India.

A perennial with broad, lanceolate, sessile, sheathing leaves, having a short, rounded, ciliate ligule, 12-24 inches long and 4-6 inches broad; *stem* when in flower, 6 feet high, the lower half ensheathed by the smooth leaf-sheaths. *Panicle* terminal, erect, oblong composed of numerous, dichotomous branches, each supporting 2-3-6 pale-greenish white, faintly fragrant flowers. *Calyx* scarcely the length of the corolla-tube. *Labellum* oblong, stalked, arching towards the stamens, lip bifid. *Capsule* the size of a cherry, deep orange-red; *seeds* often only one in each cell.

Uses. The reddish brown rhizomes are used as a condiment and the source of an essential oil. It gives a pungent taste, like a mixture of pepper and ginger. The seeds are also used as spice.

Light Galangal

Alpinia speciosa (Wendl.) K. Schum; Syn. *A. nutans* Rosc.; Eng.—Light galangal, Shell-flower; Bengali—*Punag champa*; Persian—*Kasta-zerambet*; Burmese—*Pa-ga-gyis*; Family - Zingiberaceae.

A native of the eastern Archipelago; found on the Coromandel Coast; Cultivated in gardens.

Leaves lanceolate, short petioled, smooth, *Racemes* compound, by the lower pedicels being two or three-flowered, drooping. *Labellum* broad three-lobed, the lateral lobes incurved into a tube, the exterior one bifid and curled. *Capsule* round; *seeds* few.

Uses. The rhizome is often used as a substitute for greater galangal and even as a substitute for ginger. It is much larger than the greater galangal and not so pungent.

Lesser Galangal

Alpinia officinarum Hance.; English—Lesser galangal; Bengali, Hindi and Mumbai—*Kulinjan, Kolijana, chandapushpi, chhota-pan-ki-jar*; Tamil—*Shitta rattai*, Telugu—*Sanna-elumparash trakam*; Arabic—*Khulanjan*; Persian—*Khusro-daru*; Family—Zingiberaceae.

It is native of Southern China. It is a perennial herb with a raceme of showy flowers and ornamental foliage.

Uses. The rhizomes possess an aromatic, spicy odour and a pungent taste, like a mixture of pepper and ginger. It is used to some extent in cooking, in medicine, and for flavouring liqueurs and bitters. It is also used to impart a pungent flavour to vinegar.

Mango Ginger

Curcuma amada Roxb.; English—Mango ginger; Hindi—*Am haldi*; Bengali—*Amada*; Marathi—*Amba-haladar*; Telugu—*Mamidi allam*; Sanskrit—*Karpura-haridra*; Family—Zingiberaceae.

A herb; grown in Bengal, Andhra Pradesh and Tamil Nadu; flowering during the rainy season.

Uses. The rhizomes are used as spice and condiment. The fresh rhizomes possess the smell of the green mango hence the name *am-haldi*. The rhizome is used as a carminative and promotes digestion.

Turmeric

Curcuma domestica Valet; Syn. *C. longa* Auct. non Linn.; English—Turmeric; Hindi—*Haldi*; Bengali—*Halud*; Punjabi—*Haldar, halja*; Tamil—*Manjal*; Telugu—*Pasupu*; Malayalam—*Mannal, marinalu*; Kannada—*Arishina*; Marathi—*Halede*; Gujarati—*Halada*; Sanskrit—*Haridra, nisa*; Persian—*Zard-chobah, dar-zard*; Family—Zingiberaceae.

Turmeric has been cultivated in India from very ancient times. It prefers sandy and clayey loams for its cultivation. The crop cannot stand water-logging or alkalinity in the soils. The largest supplies of turmeric are obtained from Guntur district of Andhra Pradesh. Orissa is the next important growing area for turmeric where production is concentrated in the districts of Ganjam, Phulbani and Koraput. In Maharashtra, the main centres of turmeric production are in Gujarat, Thane and Khandesh districts. Tiruchirapally, Salem and Coimbatore districts of Tamil Nadu are also important turmeric growing areas. Other important states for this crop are - Uttar Pradesh, Madhya Pradesh, Karnataka, West Bengal, Rajasthan and the Punjab. The total average under turmeric in India has been estimated variously from 60,000 to 100,000 acres, and the production is nearly 100,000 tonnes of rhizomes per annum. A large part is consumed within the country, but a portion is exported to the U.K., Pakistan, Sri Lanka and U.S.A.

Fig. 15.3. Turmeric. Plants with leaves and rhizomes.

Cultivation. The crop is propagated vegetatively by means of *corms* (swollen under ground stems). The preparatory cultivation requires deep ploughing several times (six time or so). It needs a fine tilth and heavy application of manure. The field is prepared into narrow beds with facilities for irrigation and drainage. The corms are planted from April to July in these beds, 6-9 inches apart in the furrows and 16 inches apart between the furrows. After planting the beds are levelled and covered with a mulch of leaves. The corms develop aerial shoots above the ground in a month or so. One or two weedings are given, and the beds are earthed up. By November, the leafy growth is complete and the corms beneath the soil begin to thicken and to develop uniformly deep colour. In the month of February the leaves turn yellow and dry up, which is an indication of maturity of the underground corms. Harvesting is done in March which continues till the end of April.

A great care has to be taken in harvesting operations. The whole clump consisting of both primary and secondary branches known as "*fingers*", and the main thickened portion, the "*bulb*", are to be lifted up without injuring the corms. The leaves and roots are then cut off, and the bulb and fingers separated from each other. Very small quantities are sold in this raw form. In India, the bulb of the crop is marketed as dry cured turmeric (*haldi*).

Curing. The raw produce of turmeric has to be cured properly before the commercial dry turmeric can be obtained. Curing consists mainly of three phases - 1. boiling, 2. drying and 3. polishing. After harvesting, the raw green turmeric is heaped up covered with turmeric leaves, and kept in this condition for some time. The entire produce is then transferred to an earthen pot or larger iron pan which is filled up with water, the water level is kept 2-3 inches above the level of the turmeric. After covering the bulk with dry turmeric leaves, it is boiled over fire. As soon as the rhizomes become soft to touch, they are removed from the vessel, thinly spread out and dried in the sun. After 5-7 days of drying, the produce becomes fit for sale and storage. When it is quite dry, it is cleaned to remove roots and other parts, and then rubbed well between hands. The dry turmeric is also polished by special appliances. Now the produce is sorted out into *bulbs*, *fingers* and *splits*, and graded into large and small sizes according to need.

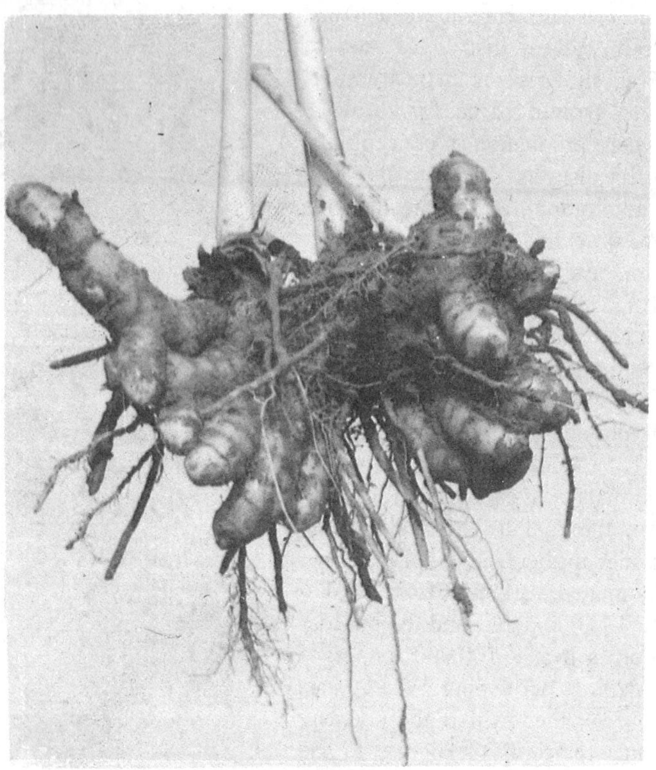

Fig. 15.4. Turmeric. Roots and rhizomes.

The plant of turmeric is a robust perennial with a short stem and tufted leaves. The pale-yellow flowers are found in dense spikes, topped by a tuft of pinkish bracts. The rhizomes, which yield the colourful condiment, are short and thick with blunt tubers. They are cleaned, washed and dried in the sun, it is very aromatic, with a musky odour and yellow colour. It has a pungent bitter taste.

Uses. It is used as a condiment. It is used to flavour and colour pickles, and foodstuffs. It is one of the principal ingredients of curry.

Zedoary

Curcuma zedoaria Rosc.; Syn. *C. zerumbet* Roxb.; English—Zedoary; Hindi—*Kachura*; Bengali—*Sori, kachura*; Mumbai—*Kachura*; Tamil—*Pulan-kizhaga*; Telugu—*Kachoram*; Malayalam—*Kach-chalam*; Kannada—*Kachora*; Sanskrit—*Sati, karchura*; Persian—*Kazhur*; Family—Zingiberaceae.

It is native of the East Indies and China. It is extensively cultivated in many parts of India, for its tuberous rhizomes. The plant possesses pale-yellowish or white flowers and showy purple or crimson bracts.

Uses. The zedoary possesses a bitter and strong camphoraceous taste. The odour resembles with that of ginger, but weaker unless the rhizomes are powdered, when it develops a powerful aromatic odour similar to that of cardamoms. It was formerly an important spice, and even today used for flavouring liqueurs and curries. The rhizomes possess aromatic, stimulant and carminative properties.

Gastrochilus

Gastrochilus pandurata Ridley; Family - Zingiberaceae. This is a perennial herb found in Andaman Islands and Konkan.

Uses. The rhizomes are used as spice and condiment.

Orris

Iris florentina Linn. and *I. pallida* Lam.; English—Orris, Orris root; Family—Iridaceae. The species are cultivated for their rhizomes which make the *Orris* of commerce. They are natives of Europe and Mediterranean region. In India, the species are cultivated in Kashmir.

They are perennial herbs. The rhizomes of these species resemble one another and occur in commerce as hard, cream-coloured, jointed or branched pieces, 5-10 cm. long and 2-3 cm. broad, dorsiventrally compressed and exhibiting yearly growth of enlargements alternating with constrictions. They possess an agreeable aromatic odour and bitter taste.

The plants may be propagated by seeds or by divisions of old rhizomes. They thrive best in dry and gravel soils. The rhizomes are harvested after about three years, freed from roots and aerial parts, peeled and dried. Fresh rhizomes are odourless and acrid, but during the process of drying, they lose their acridity and develop the characteristic aromatic odour. The Orris found in Indian bazaars is usually unpeeled and somewhat dark in colour. The characteristic violet-like odour of dried and aged rhizomes is due to the presence of an essential oil, Orris oil. The rhizomes also contain a flavone glucoside, iridin, sugar, starch, resin and tannins.

Uses. Orris is used for flavouring food stuffs. Orris oil is used also for flavouring soft drinks, candies and gelatin desserts. Orris possesses stimulant, cathartic and diuretic properties, and is used in bronchitis, dropsy and liver complaints.

Kaempferia

Kaempferia galanga Linn.; Hindi—*Chandramula*; Sanskrit—*Chandramulika*; Bengali—*Chandumula*; Marathi—*Kapur-kachri*; Telugu—*Kachoram*; Tamil—*Kacholam*; Kannada—*Kachchura*; Malayalam—*Katijulam*; Family—Zingiberaceae.

A common, ornamental herb, found throughout the plains of India. It is often grown for its aromatic rhizome. Leaves 2 or 3, spreading flat on the ground; petioles short; flowers white, 6-12 on a short scape; lip bilobed with purple spots.

Uses. The rhizomes are used as condiment. It is commonly used as a flavouring for rice. They are eaten along with betel leaf and arecanuts as a masticatory. The rhizomes are also used medicinally as stimulant, carminative, diuretic and expectorant. They are used in the preparation of gargles.

Cassumunar Ginger

Zingiber cassumunar Roxb.; English—Cassumunar ginger; Hindi—*Jangli adrak*; Family—Zingiberaceae.

A perennial herb. Commonly found throughout India.

Uses. The rhizomes are used as condiment. They are similar in aroma to those of ginger rhizomes.

Ginger

Zingiber officinale Rosc.; English—Ginger; Hindi—Plant = *adrak*, Dried rhizome = *sonth*, fresh rhizome = *adrak*; Bengali—Plant = *ada*, dried rhizome = *sont*, fresh root = *adrak*, *ada*; Assamese—Plant = *ada*; Nepalese—Dried rhizome = *sunt*; Oriya—Plant = *ada*; Punjabi—Plant = *ada*, *adrak*, dried rhizome = *zangzabil*, *sunt*, fresh rhizome = *zunjbel*, *adrak*; Marathi—Plant = *ale*, Gujarati—Dried rhizome = *sunt*, fresh rhizomes *adu*, *adhu*; Tamil—Dried rhizome = *shukku*, fresh rhizome = *inji*; Telugu—Plant = *allam*, dried rhizome = *sonti*, *sonthi*, *allam*, fresh

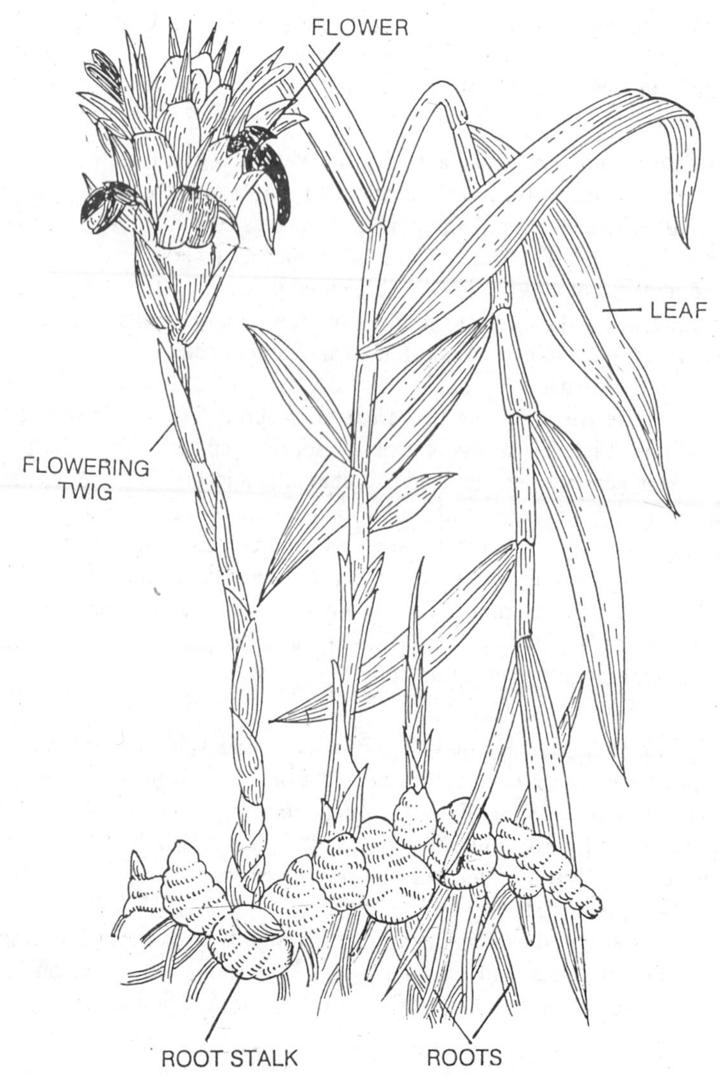

Fig. 15.5. *Zingiber officinale.* Ginger (*Adrak*). Rhizomes are used as spice and condiment.

rhizomes = *allam*; Kannanda—Dried rhizomes = *vana sunthi*, fresh rhizome = *hasisunthi*; Malayalam—Dried rhizome = *Chukka*, fresh rhizome = *inchi*; Sanskrit—Plant = *adraka, sringavera*, dried rhizome = *vishva-bhishagam, nagara, sunti, mahaushadha*, fresh rhizome = *ardrakam*; Persian—Dried rhizome = *zanjabile-khushk*, fresh rhizome = *zanjabile-tar*; Family—Zingiberaceae.

Ginger of commerce is a dry product prepared from the green underground stems or rhizomes. The most important producers of commercial dry ginger are India, Jamaica and Sierra Leone. The most important ginger growing area in India is Kerala State. The preparation of the dry ginger (sonth) of commerce is also exclusively confined to this area. In Kerala, the main producing centres lie in the Kottayam division. The other main producing areas in the West Coast are in the Malabar district.

Ginger is grown generally in areas of heavy rainfall. It requires a rich and well drained soil for its cultivation. The most suited soils are sandy or clayey loams and laterites of the Malabar Coast. Ginger is cultivated usually in small home gardens.

Cultivation. Though a perennial plant, yet it is cultivated as an annual crop. It takes 9-10 months to mature, and every year a fresh crop is taken. The plant is propagated vegetatively by means of rhizomes. The field is well ploughed, and long and narrow beds are prepared. The beds are heavily manured with cattle-manure. Seed ginger, consisting of small portions of the rhizomes, collected from the previous crop, is then planted in these beds in small pits at a distance of about 6-10 inches each way. Planting is done in May-June, and the beds are covered with mulch leaves to protect the seed from following rains. In about 2-3 weeks, the beds shoot up. The plant grows to a height of about two feet, and develop several lateral shoots in each clump. The rhizomes develop and become ready for harvest by December-January. As the crop matures, the aerial shoots

Fig. 15.6. Spices and condiments. ***Zingiber officinale*** Rosc.; Eng. Ginger (Zingiberaceae). Rhizomes.

begin to dry up and lodge on the ground. The leaves also become yellow. Harvesting is done by digging the whole crop or only portions of it.

After digging out, the rhizomes are thoroughly washed with water to remove the soil and dirt sticking to them. These underground stems are kept in the sun for a day or so, and then taken to the market for sale as green ginger (*adrak*). The preparation of dry ginger (*sonth*) is a local craft of Kerala State. For purpose of export only dry ginger is used.

In India, about 48,000 acres are under the crop, and the estimated production of cured ginger is 24,000 tonnes of which about 18,000 tonnes are produced on the west coast alone. About 25 per cent of the produce on the west coast is sold as the green ginger (*adrak*) and (*sonth*) and exported to other countries. The chief countries which import ginger from India are the U.K., U.S.A., France, Germany and Arabia.

Chemical composition. Moisture 10.5%; volatile oil 1.8%; oleoresin 6.5%; water extract 19.6%; cold alcohol extract 6.0%; starch 53.0%; crude fibre 7.17%; crude protein 12.4%; total ash 6.64%.

Uses. Ginger is used more as a condiment than as a spice. The aroma of ginger is due to essential oil, whereas the pungent taste is due to the presence of carminative and digestive stimulant. The essential oil, obtained from the rhizomes, is used for flavouring food stuffs. Ginger is exceedingly popular for flavouring beverages, such as ginger bear and ginger malt.

Zerumbet Ginger

Zingiber zerumbet (Linn.) Rosc. ex Smith; English—Zerumbet ginger; Hindi—*Mahabari bach, nar kachur*; Punjabi—*Kachur, nar kachur*; Malayalam—*Kathu-inshi-kua*; Sanskrit—*Sthula granthi*; Family—Zingiberaceae.

A perennial herb. Found throughout India.

Uses. The rhizomes are used as spice and condiment. The rhizomes has a slightly aromatic odour and possesses similar properties to those of *Zingiber officinale*, but in a minor degree.

SPICES AND CONDIMENTS OBTAINED FROM BARK

Cassia China

Cinnamomum aromaticum Nees.; English—Cassia China; Family—Lauraceae.

Cassia China is obtained from *C. aromaticum* Nees cultivated in China. The bark is the regular cassia China of commerce. Because of its brittle nature, much of the bark is broken, thus yielding the '*broken grade*'. The main grades are: 1. Whole Quills or China Rolls, 2. Selected Brokens or Canton Rolls, and 3. Extra selected brokens. Trees growing at higher altitude (180-300 metres) yield better quality bark with higher volatile oil content. The bark has been known from the earliest times. The dried immature fruits constitute the well known '*Cassia buds*', known in India as *Kala Nagkesar*. This species of Cassia is not grown in India but considerable quantities of it are imported into the country.

When the trees are about six years old, the first cut of bark is obtained. The season for barking commences in March and continues till the end of May, after which the bark loses its aroma. The bark, after its removal and while it is still moist with sap,, is then laid with the convex side downwards. The bark is left to dry for about 24 hours, and tied up in bundles of about 45 cm. diameter and sent for marketing.

Uses. It is used as a spice or condiment in curries and similar preparations. The important constituent of commercial importance is the volatile oil which finds numerous uses in the various culinary preparations, perfumery and cosmetics as well as pharmaceutical preparations.

Batavia Cassia

Cinnamomum burmannii Bl.; English—Batavia Cassia, Indonesian Cassia, Java cassia, Cassia Vera; Family—Lauraceae.

Cinnamomum burmannii Bl., grows wild throughout the Malay Archipelago and yields Batavia Cassia. The tree is found at all altitudes, *i.e.,* from sea level to 1830 metres, but grows best between 450 and 1350 metres.

The best bark is obtained from the trunk; the bigger trunk, the thicker and more valuable the bark. The dried quills are sorted into three grades - A, B and C according to their quill length, colour, thickness and aroma, tied tightly into bundles. India imports Cassia.

Uses. It is used for flavouring of bakery, confectionery and curries. It is also used in medicine and in perfumery.

Cassia

Cinnamomum inners Reinw.; English—Cassia; Hindi—*Jangli darchini*; Marathi—*Tikhi, Ranacha dalchini*; Kannada—*Adavi lavangapatte*; Tamil—*Kuttukharuvapatty*; Telugu—*Adavi lavangapatte*; Family—Lauraceae.

Cassia is a popular spice commonly used in the Indian dietary. The name *Darchini* is derived from the Arabic term *Dar-al-chini* which means the wood or bark of China. European and Arabian travellers wrote about the trade in cassia from Sri Lanka, Seychelles, China, Indonesia and West Coast of India to other countries. The demand for cassia has been so considerable that during the fifties, India used to be the second largest importer of cassia in the world, being next to the United States.

The cassia of commerce consists of layers of dried inner barks of branches of tropical trees belonging to the genus *Cinnamomum*.

It is a large evergreen tropical tree reported to occur in the evergreen forests on the Western Ghats from Mysore and Coorg to Annamalais and Kerala State. The method of preparation and curing is the same as described under cinnamon (*Cinnamomum zeylanicum*).

Uses. The bark is utilized in the similar way as that of true cinnamon (*dalchini*). It is used as a spice or condiment in curries and similar preparations. The bark contains 0.5 per cent volatile oil with the odour of cloves and musk.

Saigon Cassia

Cinnamomum laureirii Nees; English—Saigon cassia, Indo-Chinese Cassia, Royal Cinnamon, Que-thanh; Family—Lauraceae.

Saigon Cassia (*C. laureirii* Nees.) occurs in Indo-China from the upper Tonkin to Cochin China. The best bark known as *Que-thanh* or Royal Cinnamon comes from the province of Thanh-hoa in North Annam. Barking is usually carried on the standing tree, starting with the main branches and working down the trunk. The removal of the ring of bark arrests the flow of sap and renders the remainder of the bark easier for removal.

Fig. 15.7. Spices and condiments. *Cinnamomum tamala* (Buch. Ham.) T. Nees. & Eberm; Eng. Indian Cassia (Lauraceae). A flowering twig. *Tejpat* leaves are used as a spice. The bark of the tree known as Indian cassia is also used as a spice.

The bark is marketed in two forms - 1. square pieces, unscraped and unrolled, and 2. quills, which are often scraped. The former known as *Canelle plate* or *banque* is used principally in China, whereas the quills called '*Canelle roulee*' or '*dong que*' are preferred in the Western markets. The former is prepared exclusively from bark of the trunk and the main branches, the latter (quills) are obtained principally from the bark of the smaller branches. For the Chinese trade, bark has to be packed so that it retains its aroma and essential oil during prolonged storage. For export to Western markets, the quills are packed in bales 133 1/3 and 66 2/3 lbs. and are graded as *thin*, *medium* and *thick*.

Uses. The cassias are used for flavouring of bakery, confectionery, curries and other similar preparations. Also used in pharmaceuticals and cosmetics.

Indian Cassia

Cinnamomum tamala Nees. & Eberm.; English—Indian Cassia; Hindi, Bengali and Punjabi—*Tejpat*, *tejput*; Gujarati—*Tamalapatra*; Marathi—*Darchini*; Sanskrit—*Tamalaka*, *tejpatra*; Tamil—*Talishapattiri*; Telugu—*Talishapatri*; Family—Lauraceae.

A moderate-sized evergreen tree. Distributed in tropical and subtropical Himalayas (915 to 2440 metres altitude), Khasi and Jaintia hills (915-1220 metres altitude).

The bark of the tree, known in trade as Indian Cassia or Indian Cassia Lignea, is collected from trees growing at the foot of the Sikkim Himalayas. Regular plantations of *C. tamala* are grown in Khasi and Jaintia Hills, Gara Hills, Mikir Hills, Manipur and Arunachal.

Uses. The bark is aromatic. It is coarser than the bark of true cinnamon and is one of the common adulterants of true cinnamon. The essential oil from the bark is pale yellow and contains 70-85 per cent cinnamic aldehyde.

Cinnamon

Cinnamomum zeylanicum Bl.; English—Cinnamon; Hindi—*Dalchini*, *qalmi-dar-chini*; Bengali—*Dalchini*; Punjabi—*Dar-chini*, *kirfa*; Mumbai—*Taj*, *dalchini*, *tikhi*; Marathi—*Darachini*, *dalchini*; Tamil—*Karruwa*, *lavangap-pattai*; Telugu—*Sanalinga*, *lavanga-patta*; Malayalam—*Lavanga-patta*; Kannada—*Lavanga-patta*, *dalachini*; Sanskrit—*Gudatvak*; Persian—*Tali khahe*, *dar-chini*; Family—Lauraceae.

Cinnamon is one of the oldest known spices. The true cinnamon, also called as Ceylon Cinnamon is the dried bark of *Cinnamomum zeylanicum* Bl.

It is native of Sri Lanka. In India, it is grown in the Nilgiris, South Kanara, Malabar, Assam and Kumaon. It is an evergreen tree. The tree attains a height of 8 to 12 metres, but in cultivation it is coppiced or cut back to a height of about 2 metres. Its highly aromatic leaves are 12 to 17 cm. long, dark glossy green above and lighter beneath. The flowers are small, yellow and inconspicuous developing into dark purple ovoid and one seeded berries, about 1.5 to 2.5 cm. long.

Soil and climate. It is a hardy plant which can very well grow in almost all types of soils under a wide variety of tropical conditions. The quality of the bark is highly influenced by the soil and ecological factors. In Sri Lanka which is the major cinnamon growing country in the world the plants are cultivated on white sandy soil. Sandy or siliceous soils with an admixture of humus are considered to be ideal for cinnamon cultivation. Sheltered situations upto an altitude of 800 to 1000 metres receiving an annual rainfall of 200 to 250 cm. are considered to be good for the crop.

Propagation. The easiest and most widely adopted method of cinnamon propagation is by seed. It can also be propagated by planting, cutting and layers.

Cinnamon fruits ripen in July-August. The fruits fall down when fully ripe. The fleshy berries are left in heaps in shade to soften and rot. The mass is then trempled. The pulp free seeds are washed and dried in shade. They are sown without much delay as they have a short period of viability. The nursery is raised in a suitable spot in soil, rich in organic matter. The place is dug well twice or thrice. The soil is broken to powder and made loose, making it altogether free from stones, root bits, etc. The seed beds are made 1 metre wide and of suitable length with adequate provision for drainage. The seeds are sown in lines 12 cm. apart and are covered with a layer of soil to a thickness of about 2.5 cm. Germination of seed occurs in about 20 days. The beds have to be provided with artificial shade and watered regularly. The shade should continue untill the plants are about 12 cm. in height and then removed gradually. Transplanting is best done when the plants are about 10 to 12 month old.

Planting. Seedlings are transplanted when they are about 12 months old at a distance of about 2 metres between plants and rows. The planting is done in ideal weather conditions, preferably in June-July.

Harvesting and Curing. There are two regular cutting seasons in South India, which more or less synchronize with two monsoons. The appropriate time for cutting the shoots for peeling is determined with reference to the circulation of sap between the wood and the corky layer. The sap flow and the time for peeling are judged by the peelers by making a test cut on the stem with a sharp knife. If the bark separates readily, the cutting is commenced immediately. The shoots should be of at least two years growth with the bark having a uniform brown colour and should have attained a length of 1 to 1.25 m. and a thickness of not less than 1.25 cm. Shoots satisfying these requirements are cut and bundled after the leaves and the terminal shoots have been removed.

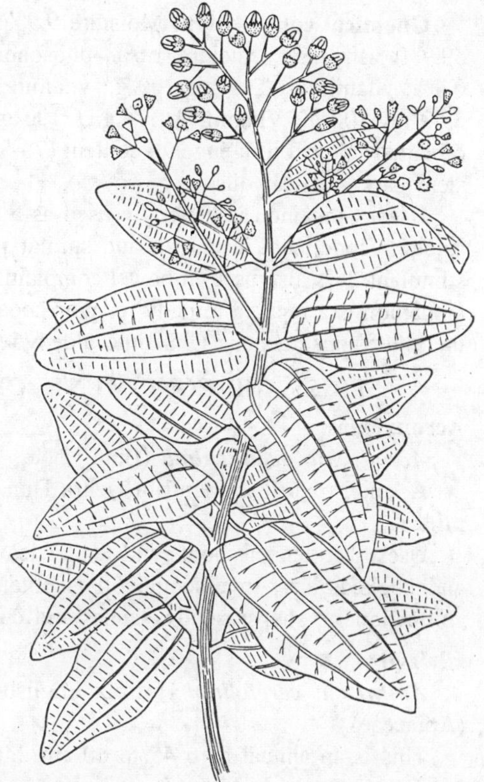

Fig. 15.8. Spices and condiments. *Cinnamomum zeylanicum* Blume; Eng. Cinnamon (Lauraceae). A flowering twig.

Cutting is followed by scraping and peeling operations. The peeling is a specialized operation peculiar to this industry which requires some skill and considerable experience. It is done by using a specially made knife which has a small and round pointed end with a projection on one side to facilitate ripping of the bark. The peeler takes the cut shoots one by one and after scraping outer skin, first makes a longitudinal slit from one end to the other. Then working the knife between the bark is ripped quickly. The twigs cut in the morning are peeled on the same day. The peeled slips are gathered and kept overnight under shade for what is called '*fermentation*'. However, while no real fermentation process develops during the course of a night, a little softening of the bark does occur with the result that peels become more easy and pliable for ensuring piping operation. The bark as it dries contracts and assumes the shape of a pipe otherwise known as quill. The smaller quills or pipes are inserted into larger ones to form compound quills. They are then arranged systematically end to end in length of about 90 cm. The quills are rolled by hand when they are soft and fresh and after rolling they are dried on mats in shade. The drying lasts for 2-5 days, depending upon the weather conditions and the type of bark. When the drying is complete the bark is collected and packed in bundles of different grades for trade.

Yield. The yield from 1 hectare of cinnamon plantation after attaining the proper yielding stage is 200 to 350 kg. of bark and 2-3 kg. of leaf oil per year on an average.

Chemical composition. Moisture 9.9%; protein 4.6% fat 2.2%; fibre 20.3%; carbohydrates 59.9%; ash 3.5%; calcium 1.6%; phosphorus 0.05%; iron 0.004%; sodium 0.01%; potassium 0.4%; vitamin B_1 0.14 mg/100 g.; vitamin B_2 0.21 mg/100 g.; niacin 1.9 mg/100 g.; vitamin C 39.8 mg/100 g.; Vitamin A 175 I.U./100 g.; calorific value 355 calories/100 g.; cinnamon leaf oil equals clove oil in eugenol content (70-95%), root bark yields 3% oil which differs from both stem bark and leaf oils.

Uses. Cinnamon is generally used as a spice or ingredients of curry powder. It is used as a spice or condiment in curries and similar preparations. It is also used in medicine as a cordial stimulant. It is also used in bowel complaints such as dyspepsia, diarrhoea and vomiting. Powder cinnamon is a reputed remedy for diarrhoea and dysentery. The bark also yields an oil of which the chief constituent is Cinnamic aldehyde.

SPICES AND CONDIMENTS OBTAINED FROM LEAVES AND TWIGS

Acronychia

Acronychia pedunculata (Linn.) Miq.; Syn. *A. laurifolia* Blume; Family—Rutaceae.

A small tree. It is found in Dehra Dun, West Bengal, the Western Ghats and in the Khasia hills.

Uses. The tender shoots and tender leaves are used as condiment for flavouring food stuffs and cooked rice. An essential oil is obtained from the leaves, which is used for flavouring, confectionery, soups, salads, sauces and curries.

Chervil

Anthriscus cerefolium Hoffm.; English—Chervil; Hindi—*Baz-atrila*; Family—Umbelliferae (Apiaceae).

This is an annual herb 45 to 60 cm. high. It is native of Europe, and may easily be grown in temperate regions. In tropics it is grown at higher elevations. The curled-leaf varieties of chervil are quite popular because of their attractive appearance. The leaves may be harvested within six to eight weeks after sowing. The plants are hardy and thrive best in winter season.

Uses. The leaves are used both for flavouring salad and as a condiment for similar prepared foods. The flavour of chervil is similar to that of parsley and aniseed. The leaves are chopped fine and sprinkled over fish before removing from the boiler. It is also used in soups, salads, sauces and egg curries. It is a great favourite in French dishes. Dried leaves are used in stuffings. Chervil vinegar is used in the preparation of salad dressing.

Celeriac

Apium graveolens var. *rapaccum* DC., English—Celeriac; Family—Umbelliferae (Apiaceae).

This is a cultivated variety. It possesses dark green foliage with less developed stalks and swollen roots, 5-6.5 cm. in diameter. Sometimes this is known as turnip - rooted celery. In India, it is cultivated to a limited extent in Uttar Pradesh and Punjab.

Uses. The leaves and stalks are used for flavouring soups and curries.

Tarragon

Artemisia dracunculus Linn.; English—Tarragon; Family—Compositae (Asteraceae).

A small perennial herb. It is native of Western Asia. In India, it is found in the Western Himalayas.

Uses. The aromatic leaves are used as a condiment for flavouring curries and other food products. It is also used for seasoning soups and salads. The essential oil is used in perfumery. It is used for flavouring *tarragon vinegar*, pickles, prepared mustard, meat dishes, egg dishes, cheeses and vegetables.

Indian Cassia

Cinnamomum tamala (Buch.-Ham.) T. Nees & Eberm.; English—Indian Cassia; Hindi—Leaves = *tejpat*, *tajpat*, bark = *taj kalami*, *dalchini*, *talispatri*, *silkanti*; Bengali—*Tejpat*; Nepali—*Chota sinkoli*; Assamese—*Dopatti*; Mumbai—*Darchini, tamala*; Gujarati—*Taj*, *tamalpatra*; Marathi—*Dalchini tiki*; Tamil—*Talisha-pattiri*; Telugu—*Tallisha-patri*; Kannada—*Lavanga patte*; Sanskrit—*Tamal*, *tespatra*; Arabic—*Zarnab*; Family—Lauraceae.

A moderate-sized evergreen tree, found on the Himalayas, between 3,000 and 7,000 feet, in Bengal, the Khasia hills and Assam.

A cinnamon known as Indian Cassia is obtained from this plant. It is a coarser and sold in larger pieces than the true cinnamon or bark of *C. zeylanicum* Bl., for which it is often used as an adulterant. The bark of the root is quite as good as the true cinnamon bark. In Manipur, the natives collect the root-bark instead of the stem-bark.

Cassia buds. These are the immature fruits of tree yielding Indian Cassia. They are something like cloves, and consist of the unexpanded flower-heads. They possess properties similar to those of the bark. Cassia buds were once employed in preparing the speed wine called *Hippocras*.

Cassia bark. The bark has a general resemblance to cinnamon, but is in simple quills, not inserted one within the other. The quills, are less straight, even and regular, and of dark-brown colour. It breaks with a short fracture.

Uses. The leaves are known as *tejpat*, and bark as *taj*. The bark and dried leaves are used to flavour dishes. The leaves are commonly used as a condiment. They are famous for their pungency and added in cooked rice (*pulao*) and curries.

Leaves are also used medicinally. They are stimulant, carminative, used in rheumatism, colic, diarrhoea, and in scorpion sting. Bark is aromatic and used in gonorrhoea.

Leaves yield an essential oil, which contains d–phellandrene and 78% eugenol.

Fig. 15.9. *Cinnamomum tamala* (Tejpat). A plant with inflorescence and leaves.

Farsetia

Farsetia jacquemontii Hook. f. & Thoms. Hindi—*Farid-buti*; Punjabi—*Mulei, farid buti, lathia, farid muli*; Family—Cruciferae (Brassicaceae).

It is found in the Rajasthan and north-western parts of India. It is native of Mediterranean region.

It is a much-branched hairy shrub having linear - lanceolate leaves and pink flowers. It has a pleasant pungent taste.

Uses. The tender twigs are used as condiment. They are eaten raw as tonic and chutney is prepared from them. It is taken as a cooling medicine after pounding. It is considered to be a specific for rheumatism in Punjab.

Fig. 15.10. Indian cassia (Tejpat). Leaves as sold in the market.

Indian Wintergreen

Gaultheria fragrantissima Wall.; English—Indian wintergreen; Verna—*Jirhap*; Nepali—*Machino*; Lepcha—*Kalomba*; Khasi hills—*Jirhap*; Family—Ericaceae.

A much-branched, evergreen aromatic shrub, about 12 feet in height with orange-brown bark; leaves alternate, oblong-lanceolate to elliptic-rhomboid, serrate, coriaceous, gland-dotted; flowers greenish white in short, pubescent axillary racemes; fruit capsular, sub-globose, completely enclosed in an enlarged, fleshy, deep blue calyx. It is grown as an ornamental plant in hills gardens. It is native of eastern North America. In India, it is common in central and eastern Himalayas, Khasi hills and hills of South India, at an altitude of 5,000-8,000 feet.

Uses. Indian Wintergreen oil is distilled on a small scale in Nilgiris from wild plants. The oil is colourless and possesses an agreeable odour and flavour. The volatile oil obtained from the carminative and antiseptic. It is applied externally in the form of ointment in rheumatism, sciatica and neuralgia. When given internally, it has vermicidal action against hookworm. The oil is an ingredient of several insecticides. The fruit is edible. Sometimes, a medicinal tea is prepared from the leaves.

Hyssop

Hyssopus officinalis Linn.; English—Hyssop; Hindi—*Zufah yabis*; Family—Labiatae (Lamiaceae).

It is native of Europe and Temperate Asia. In India it is found in Western Himalayas from Kashmir to Kumaon at 8,000-11,000 feet above sea level.

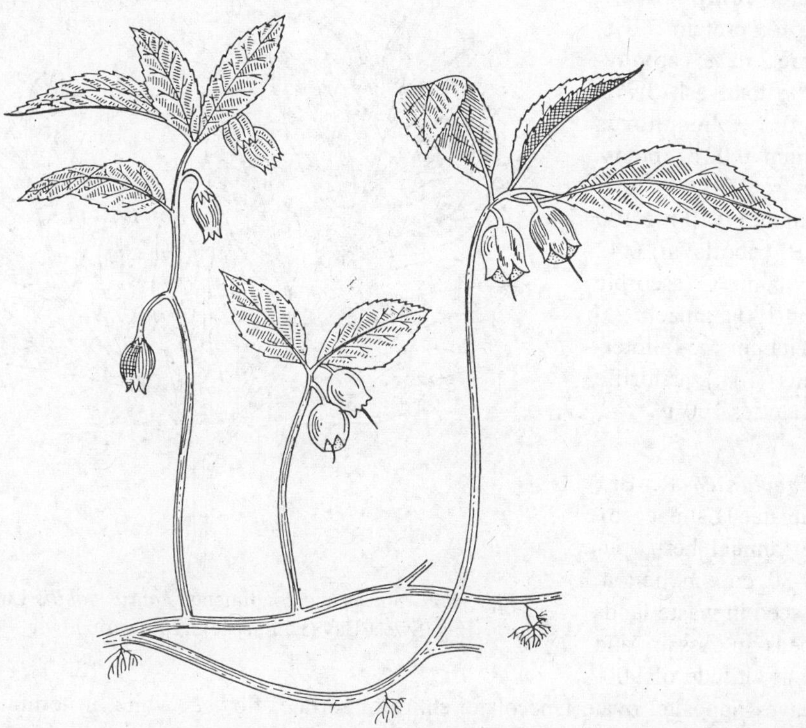

Fig. 15.11. Spices and condiments. *Gaultheria* spp.; Eng. Indian wintergreen (Ericaceae). A flowering plant. The volatile oil obtained from the leaves is used for flavouring foodstuffs. This oil is also used medicinally. It is aromatic, stimulant and carminative. It is used in rheumatism, neuralgia, as flavouring agent and as antiseptic.

An aromatic shrub, 1.2 feet high. Branches erect or diffuse; leaves sessile, linear oblong or lanceolate, entire; flowers bluish purple, in axillary tufts arranged unilaterally on terminal branches; nutlets dark brown, narrow, smooth.

In India, they are cultivated at Baramula in Kashmir.

Uses. The aromatic leaves and flowering tops are employed as flavouring for salads and soups, and also in the preparation of liqueurs. It has a pungent and bitterish taste. It is stimulant and carminative, and used in medicine. The leaf juice is employed for the expulsion of round worms.

Sweet Bay

Laurus nobilis Linn.; English—Sweet bay; Hindi—*Hab-el-ghar*; Family—Lauraceae.

It is a shrub or a small tree. It is native of Asian Minor. Sweet bay (true laurel) is a small evergreen hardy tree with stiff narrow fragrant leaves. It is widely cultivated in Europe and America as an ornamental tree. It is also grown in Indian gardens. It is propagated by cutting or by seeds. The berry is ovoid (1.5 cm. long), black coarsely wrinkled and contains a single seed. Both leaves and berries possess aromatic, stimulant and narcotic properties.

Uses. The leaves are used as a condiment and flavouring agent in food and confectionery. Commercial oil of Laurel berry is mainly used in veterinary medicine. Commercial fat (20-34%) is obtained from the whole berry by pressing or by boiling with water and skimming off the separated fat. The aromatic fat is used to some extent in pharmacy, veterinary practice and perfumery. The fresh leaves, on steam-distillation, yield 0.5% of a volatile oil with a characteristic sweet and spicy odour. This oil is employed for flavouring food stuffs.

Chemical composition. Moisture 4.5%; protein 7.6%; fat 8.8%; fibre 25.2%; carbohydrates 50.2%; total ash 3.7%; calcium 1.0%; phosphorus 0.11%; sodium 0.02%; potassium 0.6%; iron 0.53%; vitamin B_1 (thiamine) 0.10 mg/100 g.; vitamin B_2 (riboflavin) 0.42 mg/100 g.; vitamin C (ascorbic acid) 46.6 mg/100 g.; niacin 2.0 mg/100 g.; vitamin A 545 international units (I.U.); calorific value 410 calories/100 g.

Fig. 15.12. Spices and condiments. *Laurus nobilis* Linn. Eng. Sweet Bay (Lauraceae). A flowering twig.

Leucas

Leucas zeylanica R. Br.; Family—Labiatae (Lamiaceae).

An erect annual herb, pubescent, 15-30 cm. in height, found as a weed in waste lands and river beds in Assam and South India at altitude of 900-2100 m. Leaves opposite, ovate lanceolate, entire or serrate; flowers white, in terminal whorls; nutlets small, obovoid-oblong, dark-brown or black shining.

Uses. The leaves are used for flavouring curries and food stuffs.

Limnophila

Limnophila rugosa (Roth) Merrill; Bengali—*Kala karpur*; Family—Scrophulariaceae.

An erect aromatic annual herb, 1-2 feet high, found in aquatic situation and moist lands almost throughout India ascending to 1,800 m. in Himalayas. Leaves elliptic to ovate, crenate to serrate; flowers in axillary shortly pedunculate head; capsules oblong ellipsoid.

Uses. The leaves contain an essential oil. They are used for flavouring food. It is also used as hair perfume.

Lemon Verbena

Lippia citriodora H.B. & K.; English—Lemon Verbena; Family—Verbenaceae.

It is a large shrub or small tree, 3-6 m. in length. It is native of South America. In India, it is commonly grown in gardens, especially on hills. Leaves lanceolate, lemon scented, in whorls of three; flowers small, pale purple, white, in terminal panicles.

It is propagated from seeds and cuttings and thrives well on loamy soil. It is the source of true *Oil of Verbena.*

Uses. The leaves are used for flavouring beverages, desserts, fruit salads and jellies and for seasoning food. A decoction of leaves and flowers is given as febrifuge, sedative and anti-flatulent. Fresh branchlets of the plants yield, on steam distillation, 0.1-0.7% of a greenish yellow essential oil (*Oil of Verbena*) with an agreeable odour, flavouring liqueurs and non-alcoholic beverages.

Sweet Marjoram

Majorana hortensis Moench; English—Sweet marjoram; Hindi—*Murwa*; Bengali—*Murru*; Kannada—*Maruga*; Malayalam—*Maruvamu*; Tamil—*Marru*; in Kumaon it is called Bantulsi and in Dakshin Bharat *Murwa*; Family—Labiatae (Lamiaceae).

This is an aromatic, branched perennial, 1-2 feet high. It is native of mediterranean region, North Africa and Asia Minor. The herb is commonly grown in Indian gardens. It is well suited for hilly places in India. Leaves oblong-ovate; flowers small, whitish or purplish, in terminal clusters; seeds minute, oval, dark brown.

It is treated as an annual under cultivation. It thrives well in well-drained fertile garden loam. It is propagated by seeds and cuttings. Seeds are sown in pots and seedlings, when large enough, are transplanted in the field 8-10" apart in rows. The crop becomes ready for harvesting in three and half months. The tops are cut when the plants are coming into flower and dried in the shade. The volatile oil content of leaves is maximum when the plant is harvested before seed formation.

Sweet marjoram possesses a strong, spicy and pleasant odour and flavour. The dry herb contains - water, 7.61%; protein 14.31%; fixed oil 5.6%; volatile oil 1.72%; pentosans 7.68%; fibre 22.06%; and ash 9.69%. Tannin and ursolic acid are also present.

Steam distillation of the leaves and flowering heads yields a volatile oil, known in the trade as *Oil of Sweet Marjoram*. The oil is colourless or pale-yellow to yellow-green with a peculiar odour.

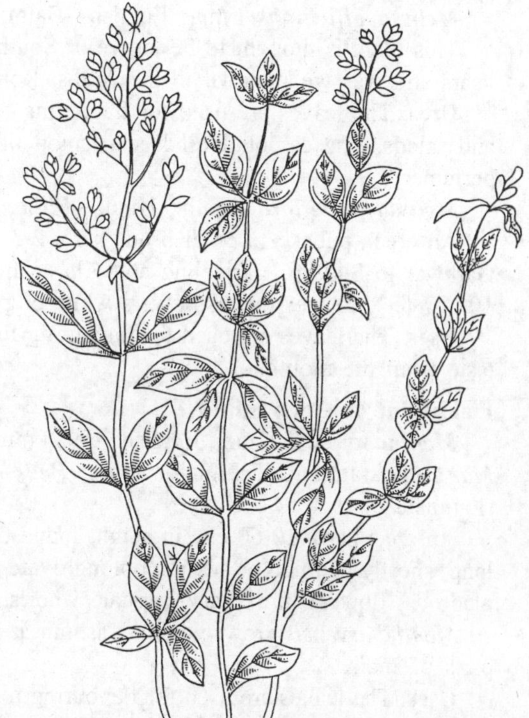

Fig. 15.13. Spices and condiments. *Majorana hortensis* Moench., Eng. Marjoram (Labiatae, Lamiaceae). The dried leaves of the herb are used as spice and flavouring agent.

Uses. The leaves of the plant are used fresh or dried and highly esteemed as a condiment for seasoning food. Fresh leaves are employed as garnish and incorporated in salad. They are used also for flavouring vinegar. The aromatic seeds are used in confectionery. It is a carminative, expectorant and tonic. Leaves and seeds are astringent. An infusion of the plant is used as stimulant, emmenagogue and gulactagogue. The oil is used in high grade flavour preparations and perfumes, and in soap and liqueur industries. It is used as an external application for sprains, bruises, paralytic limbs and toothache.

Hoarhound

Marrubium vulgare Linn.; English—Hoarhound; Hindi—*Pahari gandana*; Family—Labiatae.

It is a perennial herb, 1-4 feet high. Leaves leathery or thick, orbicular or elliptic-ovate, villous, rugose or wrinkled; flowers small, white, in dense axillary whorls; nutlets small, rugose.

It is native of Europe and Central Asia. In India, it is found in Kashmir and extending westwards at altitudes of 1,500-2,400 m. It thrives best in light calcareous, rather dry, soil and sunny situations. It is propagated by seeds, cutting and divisions. Leaves and tops are harvested just before full flowering and cured in shade to preserve the green colour. Dried leaves and flowering tops of the plant constitute the drug hoarhound, used as aromatic stomachic and expectorant.

Hoarhound, has an agreeable, somewhat musky odour which diminishes on drying and a pungent bitter, yet pleasant aromatic taste. It contains narrubiin with other bitter principles, a saponin and choline.

Uses. It is used as a condiment for flavouring food stuffs. It is a bitter tonic and possesses expectorant and diuretic properties. It is employed as domestic remedy for colds and coughs.

Balm

Melissa officinalis Linn.; English—Balm; Family—Labiatae (Lamiaceae).

This plant is thought to be native of Southern Europe. It has been cultivated for over 2,000 years and was well known to the Arabs, Romans and Greeks. This is a perennial herb.

Uses. The leaves are used for flavouring food products. The leaves are used in soups, stews and salads. The essential oil has a lemon like taste and is used to flavour beverages and in perfumery.

Melissa parviflora Benth.; Hindi—*Bililotan*; Family—Labiatae (Lamiaceae).

An erect, pubescent or glabrous herb, 2-3 feet high, found in the temperate Himalayas from Garhwal to Sikkim, Darjeeling and Khasia hills, Ala and Mishmi hills, at altitude of 4,000 to 10,000 feet. Leaves ovate; flowers white or pale-pink.

Uses. The leaves are used for flavouring food products. They are also chewed to remove bad taste from the mouth.

Field mint or Corn mint

Mentha arvensis Linn.; English—Field mint or corn mint; Hindi—*Podina*, *Pudina*; Bengali—*Mar*.; Gujarati and Telugu—*Podina*, *Pudina*; Kannada—*Chetni maragu*; Family—Labiatae (Lamiaceae).

An erect herb, 10-60 cm. in height found throughout temperate North Asia. Leaves 2.5-5 cm. long, shortly petioled or sessile, oblong ovate or lanceolate, serrate, cuneate at the base, hairy or glabrous; flowers in axillary, capitate whorls, borne on axils of leaves on upper stem.

Mentha arvensis grows wild in Kashmir at 1,500-3,000 m. It has also been reported from other parts of India.

Uses. The leaves are used for flavouring food stuffs. It is used as a stimulant and carminative. An infusion of leaves is a remedy for indigestion and rheumatism. A volatile oil is obtained from the plants by steam-distillation. The oil is used as a flavouring ingredient.

Mint

Mentha longifolia (Linn.) Huds.; English—Mint; Hindi—*Pudina*, *jungli pudina*; Punjabi—*Baburi*, *pudnakushma*; Mumbai—*Pudina*, *vartalau*; Family—Labiatae (Lamiaceae).

This is an aromatic herb, found in temperate Himalayas and in Kashmir, Garhwal, Kumaon, Punjab, Maharashtra and Uttar Pradesh.

Uses. The leaves are used as a flavouring agent. The dried leaves are used as carminative and stimulant. The essential oil obtained from the plant on steam distillation can be used as a substitute for peppermint oil to flavour confectionery.

Peppermint

Mentha piperita Linn.; English—Peppermint; Hindi—*Vilaiti pudina*, *paparaminta*, *gamathi phudina*; Family—Labiatae (Lamiaceae).

This is an aromatic herb. It is native of Europe. It is grown in Indian gardens and also cultivated in Kashmir, Nilgiris, Mysore, Delhi and Dehradun. *M. piperita* is considered to be hybrid between *M. spicata* and *M. aquatica*.

Uses. The leaves and flavouring tops are used for flavouring food stuffs. The herb is aromatic, stimulant, stomachic and carminative, and used medicinally. The herb yields a true peppermint oil on steam distillation. The oil is extensively used for flavouring food stuffs and in pharmacy. Peppermint oil is one of the most popular and widely used essential oils. It is employed for flavouring chewing gums, candies, ice creams, confectionery, mouth washes, pharmaceuticals, dental preparations and alcoholic liqueurs. It is widely employed for nausea, flatulence and vomiting. It has a refreshing odour and a persistent cooling taste. It may be taken with sugar or in the form of tablets or lozenges.

Spearmint

Mentha spicata Linn.; English—Spearmint; Hindi—*Pahari pudina, pudina*; Family—Labiatae (Lamiaceae).

A perennial aromatic herb, 1 to 3 feet high, with creeping rhizomes. It is native of temperate Europe and Asia. In India, it is cultivated in the Punjab, Uttar Pradesh and Maharashtra.

Uses. Both fresh and dried leaves are used for mint sauce and jelly and to flavour soups, stews, sauces and beverages. The leaves are the source of spearmint oil, which is used for flavouring food products. Spearmint oil is used in U.S.A. for flavouring chewing gums, confectionery and pharmaceuticals. In India, green leaves are used for making chutney and for flavouring culinary preparations, vinegars, jellies and iced drinks. The herb is considered stimulant and carminative. A soothing tea is brewed from the leaves.

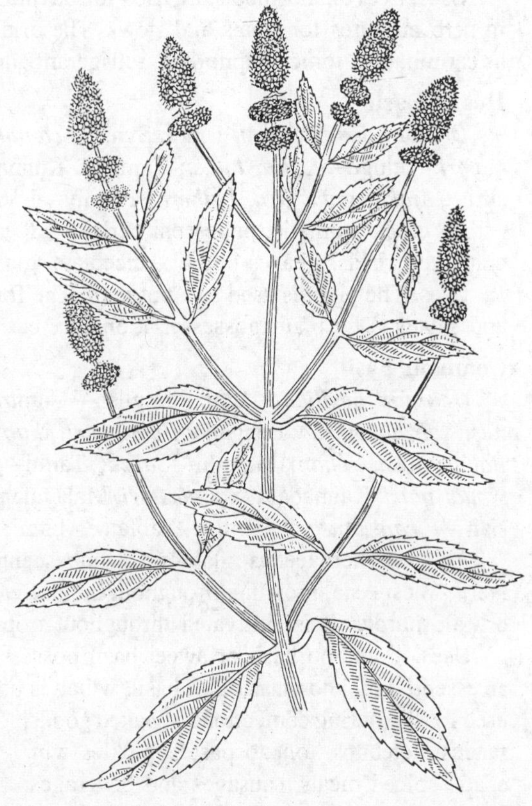

Fig. 15.14. Spices and condiments. *Mentha piperita* Linn.; Eng. Peppermint (Labiatae, Lamiaceae). A flowering twig. The leaves and flowering tops are used for flavouring purposes.

Curry Leaf Tree

Murraya koenigii (Linn.) Spreng.; English—Curry leaf tree; Hindi—*Kathnim, mitha nim*; Bengali—*Barsanga*; Marathi—*Karhinimb*; Gujarati—*Goranimb*; Telugu—*Karepaku*; Tamil—*Kattuveppilei*; Oriya—*Barsan*; Family—Rutaceae.

A large aromatic shrub or a small tree, upto 6 m. in height and 15-40 cm. in diameter, found throughout India and the Andaman Islands upto an altitude of 1,500 m. Bark dark-brown or almost black; leaves imparipinnate; leaflets 9-25, ovate, lanceolate, crenate-dentate, obtuse or acute, glabrous above, pubescent beneath, gland dotted, strongly aromatic; flowers in terminal corymbose cymes, white, fragrant, berries subglobose or ellipsoid, purplish black when ripe, 2-seeded. The plant is much cultivated for its aromatic leaves. Propagation is by seeds.

Uses. The leaves of the plant are extensively employed as flavouring agent in curries and chutneys. The leaves, root and bark are used also as tonic, stomachic and carminative.

Catmint

Nepeta cataria Linn.; English—Catnip, catmint; Family—Labiatae (Lamiaceae).

It is native of Europe. In India, it is found in western temperate Himalayas from Dalhousie to Kashmir, upto an altitude of 1500 m. This is an aromatic, erect, pubescent, perennial herb 2-3 feet in height; leaves ovate; coarsely crenate; flowers white, dotted with purple; nutlets broadly oblong, smooth, brownish black. It is grown for its scented leaves and flowering tops. It is propagated by seeds or root divisions. It thrives best on well-drained, moderately rich garden loam. Leaves and the plant possess a strong, aromatic and somewhat disagreeable odour, and a warm, bitterish and camphoraceous taste. A volatile oil (oil of catnip) is obtained by the steam distillation of the herb in 0.3% yield. The oil possesses the characteristic odour of the herb.

Uses. Leaves and shoots are used for flavouring sauces and cooked foods; dried leaves are used in herb mixtures for soups and stews. The dried leaves and flowering tops are used medicinally as carminative, tonic, diaphoretic, refrigerant and sporific. Leaves are chewed to relieve toothache.

Hoary Basil

Ocimum americanum Linn.; Syn. *O. canum* Sims; English—Hoary basil; Hindi—*Kala tulsi, mamri*; Telugu—*Kukka tulasi*; Tamil & Kannada—*Nayi tulasi*; Malayalam—*Kattu tulasi*; Sanskrit—*Ajaka, gambhira, kuthera*; Family—Labiatae (Lamiaceae).

An erect, aromatic, pubescent herb, 1-2 feet high, found throughout India in abundance near cultivated fields. Leaves elliptic-lanceolate, gland dotted; flowers small, white, pink or purplish.

Uses. The plant is used as a pot herb. The fragrant leaves are used for flavouring sauces, soups and salads. The plant possesses aromatic, carminative, diaphoretic and stimulant properties.

Common Basil

Ocimum basilicum Linn.; English—Common basil, sweet basil; Hindi—*Babui-tulsi, Kali-tulsi, sabzah, marua*; Bengali—*Babui tulsi, khub kalam, debunsha*; Oriya—*Dhala tulasi*; Punjabi—*Baburi tulsi, rehan*; Marathi—*Sabza*; Tamil—*Tirmutpatchie*; Telugu—*Bhu-tulasi, rudra jada, vibudi-patri*; Kannada—*Kam-kasturi*; Malayalam—*Tiru-nitru*; Sanskrit—*Munjariki, Varvara*; Persian—*Firanj-Mushk*; Family—Labiatae (Lamiaceae).

An erect, herbaceous, glabrous or pubescent annual, 1-3 feet high, native of Central Asia and North-West Asia, growing throughout India. Leaves ovate - lanceolate, gland dotted; flowers white or pale purple. It is cultivated throughout tropical India from the Punjab to West Bengal.

Uses. Common basil or sweet basil possesses a clove like scent and a saline taste. It yields an essential oil known as oil of basil, which is used as a flavouring agent. Oil of basil is commonly used for flavouring confectionery, baked goods, sauces, ketchups, tomato pastes, pickles, vinegars, spiced meats, sausages and beverages.

Water Dropwort

Oenanthe javanica (Bl.) DC.; English—Water dropwort; Bengali—*Pantursi*; Family—Umbelliferae (Apiaceae).

It is found in marshy places and on river banks in Northern India from Kashmir to Assam, ascending to an altitude of 1800 m. It is a perennial stoloniferous herb, 2-4 feet high. Leaves pinnate to bipinnate - segments lanceolate, serrate, sometimes lobed; flowers white in compound umbels; fruit obovoid, seed terete.

Uses. The young shoots and leaves are used as condiment. It is used for flavouring rice and other cooked food. The plant contains a flavonoid, persicarin. The fruit yields 1.5% of an essential oil containing phellandrene and myristicin.

Pot Marjoram

Origanum vulgare Linn.; English—Pot Marjoram or Common Marjoram; Hindi—*Sathra*; Telugu—*Mridu-maruvamu*; Kannada—*Maruga*; Punjabi—*Mirzanjosh*; Family—Labiatae (Lamiaceae).

Fig. 15.16. Spices and condiments. *Origanum vulgare* Linn.; Eng. Pot marjoram (Labiatae, Lamiaceae). The dried leaves of the herb make the Origanum, Oregano, Mexican Origanum or Pot marjoram, which is used as a condiment. The taste of the herb is spicy, warm, pungent and bitter. The herb contains a volatile oil, tannin and bitter principle.

It is native of Europe. In India, it is found in the temperate Himalayas from Kashmir to Sikkim at altitudes of 1,500-3,600 m. It is an aromatic, branched perennial herb, 1-3 feet in height. Leaves broadly ovate, entire or rarely toothed; flowers purple or pink, in corymbose cymes; nutlets smooth, brown.

It is commonly found in Shimla hills and Kashmir valley. It is grown in warm garden soils. It is propagated by seeds, cutting and root-division. It is commonly sown during October in the plains and during March and April in the hills.

It contains a volatile oil, tannin and a bitter principle. The oil possesses an aromatic spicy, basil like odour. It contains thymol (7%), carvacrol, free alcohols, esters, and bicyclic sesquiterpene.

Chemical composition of the leaves. Moisture 8.0%; protein 11.7%; fat 6.4%; crude fibre 11.0%; carbohydrates 53.9%; total ash 9.0%; calcium 1.7%; phosphorus 0.20%; iron 0.05%; sodium 0.02%; potassium 1.7%; vitamin A 1010 international units (I.U.)/100 g.; vitamin B_1 0.34 mg/100 g.; vitamin B_2 0.41 mg/100 g.; niacin 6.2 mg/100 g. and vitamin C 12 mg/100 g; calorific value 360 calories/100 g.

Uses. The leaves and tops of the herb cut prior to blooming are used to flavour food stuffs especially meats, egg preparations and salads. The volatile oil from herb is aromatic, carminative, stimulant, tonic, stomachic, diaphoretic, emmenagogue, rubifacient, given in colic, diarrhoea and hysteria, also applied in toothache.

Perilla

Perilla frutescens (Linn) Britton; Syn. *P. ocimoides* Linn.; Hindi—*Bhanjira*; Bengal—*Ban tulsi*; Assamese—*Angami*; Family—Labiatae (Lamiaceae).

A herb; it is native of India, China and Japan. In India, it is frequently grown in the Himalayas and Assam by the local people.

It is an aromatic, bushy annual, upto 5 feet in height. Leaves broadly ovate, acuminate, coarsely serrate or crenate; flowers small, white in axillary and terminal racemes; nutlets, rounded, pale brown with reticulate markings.

It grows well in sandy loam. The seeds are sown in rows. When the seedlings are 3 cm. in height they are thinned out 7-10 cm. apart. The plants flower after about five months and bear seeds a month later.

Uses. The leaves and flavouring tops are used as condiment. The herb possesses sedative, antispasmodic and diaphoretic properties and is recommended for uterine and cephalic troubles.

Parsley

Petroselinum crispum (Mill) Nym.; English—Parsley; Kannada—*Achu mooda*; Family—Umbelliferae (Apiaceae).

A hardy, aromatic, biennial herb. It produces a rosette of finely divided radical leaves in the first year and a flavouring stalk up to 3 feet high in second year. It is a cool weather crop, thrives best in a rich moist soil. In India the herb grows better at higher altitudes. Sowing is done March-May on the hills and August-November in the plains. The seeds are sown in nursery beds and later the seedlings are transferred to the site. The leaves are ready for harvesting in about three months.

Uses. Fresh leaves possess strong culinary odours and commonly used for garnishing and seasoning. The leaves are eaten fresh, incorporated in salads, and used as an ingredient of soups, stews and sauces. The dried leaves and roots are used as condiments. The fresh leaves are a good source of iron, calcium, carotene and ascorbic acid.

Rosemary

Rosmarinus officinalis Linn.; English—Rosemary; Hindi—*Rusmary*; Family—Labiatae (Lamiaceae).

A small shrub, native of Mediterranean region. It is found wild on dry rocky hill in the Mediterranean region. In India, it is cultivated to some extent in the temperate Himalayas and Nilgiris hills in dry to moderately moist climate.

Leaves of rosemary possess an agreeable aromatic odour and a pungent, bitter and camphoraceous taste. It yields a volatile oil known as oil of rosemary. The volatile oil is obtained by steam distillation of leaves, flavouring tops and twigs. The finest produce is being obtained from the dried leaves freed of stalks. The oil is pale-yellow or colourless liquid with the characteristic odour of the leaves and a warm camphor, borneol and bornyl acetate.

Uses. The leaves are employed as a condiment dried and powdered. They are added to cooked meats, fish, eggs, soups, stews, sauces, dressings, preserves and jams. They are mixed with sage in pork and veal stuffings and added to biscuits. Carminative, diaphoretic, astringent, diuretic, emmenagogue, stimulant and stomachic oil or rosemary is used for flavouring of meats, sausages, soups, table sauces and other food products. The oil is mainly used in perfumery. It is also used in medicinal preparations.

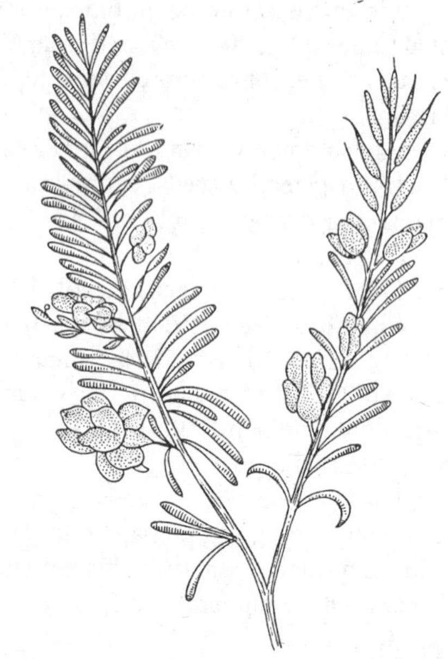

Fig. 15.16. Spices and condiments. *Rosmarinus officinalis* Linn; Eng. Rosemary (Labiatae, Lamiaceae). A flowering twig. The dried leaves of *R. officinalis* Linn., make the rosemary of commerce.

Rumex

Rumex hastatus D. Don; Hindi—*Bhilmora*; Kumaon—*Amlora*; Punjabi—*Khattimal, Katambal, amla, amlora*; Family—Polygonaceae.

An under shrub; found in the North-Western Himalayas, chiefly occurs on 8,000 feet. It is abundantly found in Almora district.

Use. The leaves are used as condiment; the leaves have a pleasant acid taste, and are eaten raw by the natives.

Garden Rue

Ruta graveolens Linn.; English—Rue; Hindi—*Sadab, Satari*; Telugu—*Aruda*; Tamil—*Aruada*; Malayalam—*Nagadhali*; Kannada—*Naga dali soppu*; Family—Rutaceae.

It is native of the Mediterranean region. In India, it is commonly cultivated in the garden.

It is a strong-scented, erect, glabrous herb 1-3 feet in height, leaves 2-3 pinnate, segments oblong—spathulate, strongly aromatic; flowers small, yellowish in corymbs; capsules small with lobes somewhat rounded.

A pale yellow or greenish volatile oil known as Rue oil, is obtained on steam distillation of the fresh plant material.

Uses. The aromatic leaves are used as condiment and for flavouring purposes. The herb is of medicinal value. It is diuretic, emmenagogue, stimulant and antispasmodic. It is given in hysteria and amenorrhoea. The juice of the herb relieves earache and toothache. Rue oil is also used medicinally. It is anthelmintic, antispasmodic, rubefacient and emmenagogue.

Garden Sage

Salvia officinalis Linn.; English—Sage, Garden sage; Hindi—*Salvia sefakuss*; Family—Labiatae (Lamiaceae).

A small shrub. It is native of Mediterranean region, cultivated as a spice and for medicinal purposes. Stems shrubby, white woolly; flavouring branches tomentose-pubescent, leaves aromatic, petiolate, entire, oblong, base narrowed; flowers, purple, blue or white, in racemes. The greyish-green hairy leaves are very aromatic.

Uses. The leaves and tender twigs are used for flavouring food products. Sage has been extensively employed in the food industry as a standard spice in making stuffing for fowl, meats and sausage. Dried and powdered aromatic leaves are mixed with cooked vegetables and sprinkled on cheese dishes, cooked meats and other curries. Fresh sage leaves are used in salads, vegetable sandwiches.

Fig. 15.17. Spices and condiments. *Salvia officinalis* Linn.; Eng. Sage, Garden sage (Labiatae, Lamiaceae). A flowering twig. The dried leaves of *S. officinalis* Linn., make the "sage" of commerce.

Chemical composition of dried leaves. Moisture 5.7%; Protein 10.2%; fat 14.1%; crude fibre 16.0%; carbohydrates 46.3%; total ash 7.7%; Calcium 1.8%; Phosphorus 0.09%; Iron 0.03%, Sodium 0.01%; Potassium 1.0%; vitamin B_1 0.75 mg/100 g.; vitamin B_2 0.34 mg/100 g.; vitamin C 39.8 mg/100 g.; niacin 5.7 mg/100 g.; vitamin A 2395 International Units (I.U.)/100 g.; calorific value 415 calories/100 g.

Summer Savory

Satureja hortensis Linn.; English—Summer savory; Family—Labiatae (Lamiaceae).

An erect, small herb with pinkish branches. It is native of the Mediterranean region. In India, it is found in the Himalayas from Kashmir to Kumaon. Leaves oblong linear or lanceolate with deep pitted glands on both sides; flowers in small axillary cymes. The plant is grown in the cooler part of the year and in the hills during spring. Tops of the plants are cut when they flower and dried in shade.

Uses. The leaves possess a strong aroma and a warm bitter taste. Fresh and dried leaves are used for flavouring meat, egg curries and vegetable dishes, and for other culinary purposes. Savory of commerce consists of dried leaves and terminal twigs, the best savory comprises only of leaves.

The plant is a source of a volatile oil, oil of savory. It is golden yellow in colour with a characteristic thyme odour. The oil is used for flavouring processed foods.

Tanacetum

Tanacetum tenuifolium Jacq.; Family—Compositae (Asteraceae).

A herb, found in Kumaon, at 14,000 feet.

Uses. The herb is used as a flavouring substance for flavouring puddings.

Thyme

Thymus vulgaris Linn.; English—Thyme; Family—Labiatae (Lamiaceae).

A flowering plant. It is a perennial garden plant. It is native of Mediterranean region. In India it is found in Himalayas from Kashmir to Kumaon.

The dried leaves and tender tops make "thyme" of commerce. A volatile oil of pleasant odour is obtained on steam distillation of herb.

Chemical composition of dried leaves. Moisture 7.1%; Protein 6.8%; fat 4.6%; crude fibre 24.3%; carbohydrates 44.0%; mineral matter 13.2%; Calcium 2.1%; Phosphorus 0.20%; Iron 0.14%, Sodium 0.08%; Potassium 0.9%; vitamin B_1 51 mg/100 g.; vitamin B_2 0.4 mg/100 g.; vitamin C 12.0 mg/100 g.; niacin 4.9 mg/100 g.; vitamin A 175 International Units (I.U.)/100 g.; calorific value 340 calories/100 g.

Uses. The dried leaves are used for dressing and seasoning meat preparations, egg curries, pork sausage, etc.

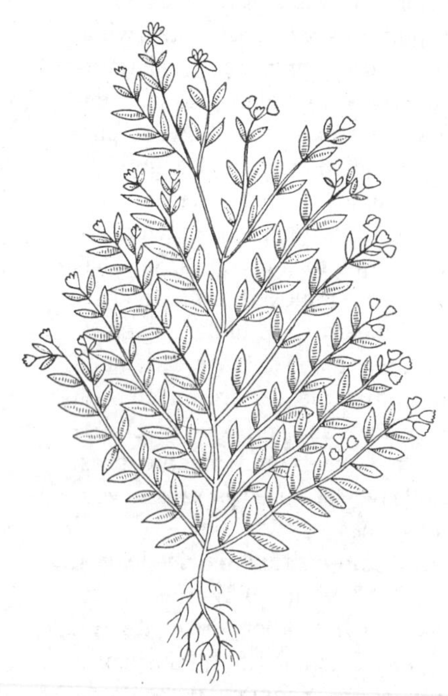

Fig. 15.18. Spices and condiments. *Thymus vulgaris* Linn.; Eng. Thyme (Labiatae, Lamiaceae). A flowering plant. The dried leaves and tender tops of *T. vulgaris* Linn., make "thyme" of commercial value. It is a perennial garden plant.

New Zealand Spinach

Tetragonia tetragonoides Kuntze; English—New Zealand Spinach; Family—Aizoaceae.

A herb, grown in Northern India, Karnataka and West Bengal.

Uses. The leaves are eaten as vegetable, and used as condiment.

SPICES AND CONDIMENTS OBTAINED FROM FLOWER BUDS, FLOWERS AND INFLORESCENCE

Edible Caper

Capparis spinosa Linn.; English—Edible caper bush, caper; Hindi—*Kabra*, *ber*; Ladak—*Kabra*; Kumaon—*Ulto kanta*; Punjabi—*Kaur*, *kiari*, *bauri*, *kakri*, *kabra*; Mumbai—*Kabar*; Arabic—*Kabar*; Persian—*Kebir*; Family—Capparidaceae.

The caper bush is a trailing spiny shrub, a few feet high. It is a native of the Mediterranean region. In India, it is found in the Punjab, Gujarat and Western Ghats. The solitary berry-like fruits are born on thick stalks. The unopened "*flower buds*" are known as *capers*. The capers are collected every morning and pickled in salt and vinegar.

Uses. The capers possess a very pungent taste and, therefore, used as condiment especially in meat, sauces and pickles.

Saffron

Crocus sativus Linn.; English—Saffron; Hindi—*Kesar*, *zafran*; Bengali—*Jafran*; Mumbai—*Safran*, *kessar*, *kecara*; Marathi—*Kecara*; Gujarati—*Keshar*; Tamil—*Kungumapu*; Telugu—*Kumkum apave*; Kashmiri—*Kong*; Sanskrit—*Kunkuma*, *kashmirajanma*, *kumkuma*; Arabic & Persian—*Zaafaran*; Family—Iridaceae.

A rhizomatous herb. It is cultivated in Kashmir at Pampur near Srinagar. To a small extent it is cultivated also at Bhersar and Chaubattia in Uttar Pradesh. The dried stigmas and tops of the styles make the saffron of commerce. The product is obtained from the stigmas of the flowers, 4,000 of which are required to produce 25 gms. of saffron.

Uses. The saffron is used as a spice and dyestuff. It possesses a pleasant aroma. Saffron is an ingredient of many Indian dishes, particularly cooked rice and sweet rice (*Kesaria chaval*) and *pulao*.

Screw Pine

Pandanus odoratissimus Linn.; English—Screw pine; Hindi—*Keura, kewda, ketki*; Bengali—*Keya, keori*; Telugu—*Mugali*; Tamil—*Tazhai*; Kannada—*Tale mara*; Malayalam—*Kaida*; Family—Pandanaceae.

It is found along the sea coasts of both the Peninsulas, Orissa, Uttar Pradesh and the Andaman Islands. It is common on the sea shore forming a belt of dense, impenetrable vegetation above the high water mark.

It is a densely branched shrub or small tree. Stem upto 6 m. high supported by aerial roots; leaves glaucous green, 3-5 feet long, caudate acuminate, coriaceous, with spines on the margins and on the midrib; spadix of male flowers, 25-50 cm. long, with numerous subsessile cylindric spikes, 5-10 cm. long enclosed in long, white, fragrant, caudate acuminate spathes; spadix of female flowers solitary, 5 cm in diameter; fruit ablong or globose, 15-25 cm in diameter, yellow or red; drupes numerous.

The male inflorescences are valued for the fragrant smell emitted by the tender white spathes covering the flowers and for the valuable attar obtained from them. The commercial exploitation of male spadices is centred mostly around Kollapalli, Meghna and Agrarum in Ganjam district of Orissa, and a few centres in Tamil Nadu and Uttar Pradesh.

The plant is generally propagated by offsets or division of the suckers. A fertile well drained soil is preferable. The tree flowers 3 to 4 years after planting. The flowering takes from July to October. The spadices mature in a fortnight. In India the male flowers are valued for their fragrance. They are used for the extraction of *kewda* attar and *kewda* water. A fully mature tree bears 30-40 spadices in a year.

Uses. *Kewda* attar and water are used for flavouring various foods, sweets, syrups and soft drinks. They are popular in Northern India, especially on festive occasions. The *kewra* water is commonly used for flavouring rasgullas, lassi and cooked rice.

Clove

Syzygium aromaticum (Linn.) Merr. & Perry; Syn. *Eugenia aromatica* O. Kuntze; *E. caryophyllata* Thunb.; English—Clove; Hindi—*Laung*; Bengali—*Lavange, langa*; Punjabi—*Laung, karanfal*; Kashmiri—*Raung*; Marathi & Gujarati—*Lawanga, lavinga*; Tamil—*Kiramber, ilavang ap-pu*; Telugu—*Lavangalu*; Malayalam—*Chanki*; Sanskrit—*Lavanga*; Persian—*Mekhak*; Family—Myrtaceae.

Clove is one of the most ancient and valuable spices of the orient and holds a unique position in the international spice trade. Native to Moluccas, the so called 'Spice Islands' in the East Indian Archipelago, this spice was first introduced in

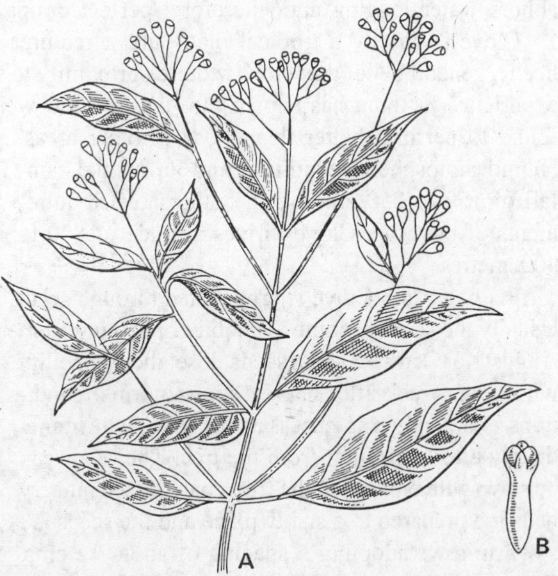

Fig. 15.19. Spices and condiments. *Syzygium aromaticum* (Linn.) Merr. & Perry, syn. *Eugenia aromatica* O. Kuntze, *E. caryophyllata* Thunb; Eng. Clove (Myrtaceae). A, a flowering twig; B, dried unopened flower bud.

India around 1800 A.D. by the East India company. The company's spice garden in Courtallam in Tamil Nadu was then established to cultivate clove and nutmeg as the principal spice crops. Induced by the success of the cultivation in Courtallam, cultivation of clove was extended during the period after 1850 A.D. to Nilgiris (Burliar) in Tamil Nadu, Southern regions of Travancore and also to Cochin State on the slope of Western Ghats. The Important clove growing regions in India now are Nilgiris, Tenkasi hills, and Kanyakumari districts of Tamil Nadu and Kottayam and Quilon districts of Kerala.

Although clove has been under cultivation in India for over about 170 years, its development has been very slow owing probably to its long pre-bearing period and lack of knowledge regarding the method and economics of its cultivation.

Clove is the dried unopened flower bud of *Syzygium aromaticum*, a medium statured, cone-shaped ever-green tree belonging to the Family—Myrtaceae. Clove tree attains a height of 10 to 12 metres. The stem is usually forked near its base with two or three main branches. Smaller branches are slender, rather brittle and covered with grey bark. The leaves appearing in pairs, are lanceolate, acute at both ends and are of dark shining green colour. The aromatic nature of the leaves is due to numerous oil glands found on their under-surfaces. The flower buds are greenish when fresh and are borne on ends, which are picked green and dried in the sun till they become dark brown, form the 'clove' of commerce. The buds have slightly cylindrical base and are surmounted by the plump ball like unopened corolla which is surmounted by the four toothed calyx. If the bud is left unpicked, the flower develops after fertilization into a fleshy, purple and one-seeded oval fruit as 'Mother of clove'. The fruit is about 2.5 cm. long and 1.25 cm. in width. The seed is oblong, rather soft in texture and grooved on one side. The leaves, unripe fruit and broken clove, including the stalk are all aromatic and yield an essential oil.

Soil and climate. Deep and rich loams with high humus content are best suited for clove cultivation. In India, clove has developed well in the open sandy loams and the laterite soils of South Kerala region. But the best growth is seen in black loams of the semi forest regions. Clove abhors water logging and, therefore, perfect drainage is essential.

Clove is strictly a tropical plant and it requires a warm humid climate. Although there has been a general belief that clove requires proximity to sea for the proper development and cropping, experience in India has shown that the trees do well in the submontane regions and have been found to perform better than those in other areas. Humid atmospheric condition and an annual rainfall of 150 to 250 cm. are the other ideal requirements of the crop. Clove thrives altitude of 800 to 900 metres.

Propagation. Clove is propagated through seed. Usually the seeds become available for sowing from August to October. The seeds lose their viability within one week after harvest under normal conditions and hence it is necessary to sow them immediately after collection from the tree. The seeds can be sown with or without the fruit coat. Raised nursery beds are prepared in a shady place and the seeds are sown in rows adopting a spacing of about 12 cms. The seeds begin to germinate in four to five weeks after sowing. The seedlings are very slender and delicate and grow very slowly. Watering is necessary throughout the nursery period. The seedlings after about six months of nursery life are transferred to baskets made of bamboo or mud pots and nurtured

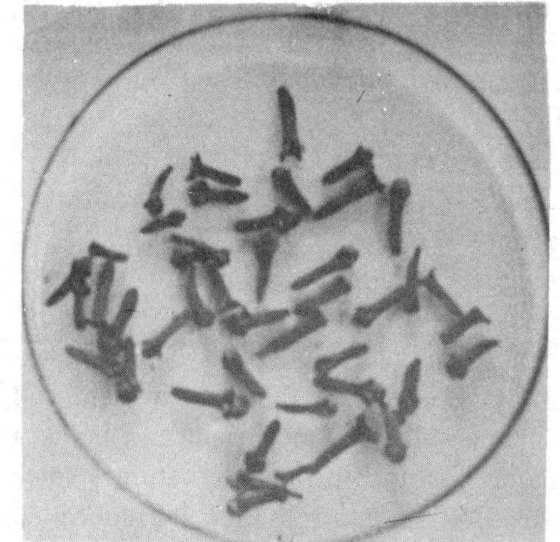

Fig. 15.20. *Syzygium aromaticum*-clove **(laung)** The dried flower buds are used as spice.

properly under the shade till they attain an age of 12 to 18 months. Clove can also be propagated vegetatively by grafting on its own stock.

Clove can conveniently be grown mixed with other commercial crops like arecanut, coconut, nutmeg, etc. The shade cast by these crops will provide enough protection to clove from the sun.

Harvesting and curing. Clove tree begins to yield from the seventh or eighth years after planting. The full bearing stage is attained after about 15 to 20 years. The flowering season is September-October in the plains and December-January in high altitudes. The buds are ready for harvest in about four months.

Just before flowering there is fresh flush of young leaves and soon after this, the flower buds begin to appear. The optimum stage for picking clove buds is indicated by the change in the colour from green to slightly pinkish tinge. The unopened clove buds are carefully picked with hand when they turn pink in colour. It is necessary to pick the buds before they open, otherwise, the value of spice will be lost to a considerable extent.

The harvested buds are spread evenly to dry in the sun either on grass-mats or on cement drying floor. During nights the buds should be stored under cover, lest they re-absorb moisture. Normally it is possible to dry the cloves in four to five days under direct sun and in about four hours when they are heated in zinc trays over a regulated fire. Fully dried buds develop the characteristic dark brown colour and crisp. If the produce is uniformly good, approximately 8,000 to 10,000 cloves would weigh one kilogram. Clove is graded according to its appearance and impurity content. Good quality clove should be brownish black in colour, with full and plump crown, somewhat rough to the touch and without wrinkles and it should not contain more than 16% moisture and 5% foreign matter. Also it should have fine aroma and flavour and should readily exude oil when the stem is pressed with the finger nail.

Yield. There is considerable variation in the yield of clove. Under favourable conditions well grown trees may yield as much 4 to 8 kg. of cloves. It is, however, common to meet with low yielding tree in clove plantations, sometimes in large numbers, resulting in low and uneconomic production, particularly in certain years. The average yield from a bearing tree in a well maintained plantation, in India is reported to be about 2.5 kg. considering that the percentage of bearing tree will be around 60, one hectare of plantation containing about 250 trees will yield about 375 kg. of cloves.

Uses. Clove is very aromatic and fine flavoured and imparts warming qualities. In all Indian homes, it is used as a culinary spice as the flavour blends well with both sweet and savoury dishes. Clove is used for flavouring pickle, curries, ketchup and sauces. It is highly valued in medicine as a carminative, aromatic and stimulant. Clove has stimulating properties and is one of the ingredients of betel chewing. In Jawa, clove is used in preparation of a special brand of cigarette for smoking. The essential oil which is obtained by distilling clove with water or steam, has even more uses. It is used medicinally in several ways. The chief constituent of the oil eugenol, is extracted and used as an imitation carnation in perfumes.

Clove requirement in India. The requirement of clove in India is being met through imports. We have been importing large quantities of clove in the past. The average annual imports of cloves was about 2,940 tonnes valuable at about Rs. 183 lakhs in the 1950's. However, owing to the restrictions imposed from time to time, it has been possible to cut down the imports considerably and in the 1960's the annual import was of the order of 500 tonnes valued at about Rs. 47 lakhs.

The chief clove producing countries are Zanzibar, which grow 90% of the total output, Indonesia, Mauritius, and the West Indies.

Chemical composition. Moisture 5.4%; protein 6.3%; volatile oil 13.2%; fat (non volatile ether extract) 15.5%; crude fibre 11.1%; carbohydrates 57.7%; mineral matter 5.0%; ash 0.24%; calcium 0.7%; phosphorus 0.11%; iron 0.01%; sodium 0.25%; potassium 1.2%; vitamin B_1 0.11 mg/100 g.; vitamin B_2 0.04 mg/100 g.; niacin 1.5 mg/100 g.; vitamin C 80.9 mg/100 g. and vitamin A 175 International Units (I.U.), calorific value 430 calories per 100 g.

SPICES AND CONDIMENTS OBTAINED FROM FRUITS

Dill

Anethum graveolens Linn.; Syn. *Peucedanum graveolens* Benth & Hook. f.; English—Dill, Hindi—*Soya, sowa, sutopsha*; Bengali—*Sulpha, sowa, sulpa*; Kumaon—*Soya*; Kashmiri—*Soi*; Punjabi—*Soya*; Gujarati—*Surva, suah*; Tamil—*Sata kuppi*; Sanskrit—*Misreya, satapushpi*; Myanmar—*Samyeit*; Arabic—*Shubit*; Family—Umbelliferae (Apiaceae).

A glabrous herb, with light-green leaves and yellow flowers. The fruits are oval, light-brown and compressed. It is native of Eurasia. In India, it is cultivated on small scale in Jammu and Kashmir. In Indian plains it is often cultivated in cold season for use as a vegetable, and for its carminative fruit (seed).

Uses. The fruits are used as spice and are also used as carminative. It is much used in soups, sauces and for other culinary purposes. Dill oil is also used for flavouring purposes. Both seeds and oil are used in medicine. The leaves are used as flavouring substance.

Fig. 15.21. Spices and condiments. *Apium graveolens* Linn. var. *dulce* DC.; Eng. Celery, Celery seed, Garden celery (Umbelliferae, Apiaceae). A, leaf; B, floral axis with typical umbels.

Celery

Apium graveolens Linn. var. *dulce* DC.; English—Celery, celery seed, garden celery; Hindi—*Ajmud, bari-ajmud, karafs*; Bengali—*Chanu, randhuni*; Mumbai—*Bori ajamoda, ajmud*; Punjabi—*Bhut jhata*; Arabic—*Karafs*; Persian—*Karasb*; Family—Umbelliferae (Apiaceae).

It is native of England and other parts of Europe. In India, it is cultivated in the North-Western Himalayas and in the hills of Uttar Pradesh, the Punjab and South India. In Indian plains, it is cultivated during the cold weather, chiefly as a garden crop in the vicinity of towns. The fruits are small and dark brown with peculiar flavour.

Uses. The herb is eaten as salad, or made into soup. The seed (fruit) is eaten as spice. The seed tied into a piece of cloth are used to flavour soup. The soil extracted from the seeds is stimulant and used chiefly for flavouring soups and curries. The seeds are also cordial, tonic, carminative, diuretic, emmenagogue and antiseptic.

Chemical composition. Moisture 5.1%; protein 18.1%; fat (ether extract) 22.8%; crude fibre 2.9%; carbohydrates 40.9%; total ash 10.2%; calcium 1.8%; phosphorus 0.55%; iron 0.45%; sodium 0.17%; potassium 1.4%; iron 0.53%; vitamin B_1 (thiamine) 0.41 mg/100 g.; vitamin B_2 (riboflavin) 0.49 mg/100 g.; vitamin C (ascorbic acid) 17.2 mg/100 g.; niacin 4.4 mg/100 g.; vitamin A 650 international unit (I.U.); calorific value 450 calories/100 g.

Black Caraway

Bunium persicum (Biss.) Fedts.; Syn. *Carum bulbocastanum* Clarke nan koch; English—Black caraway; Hindi—*Kala zira, shah-zirah*; Telugu—*Sima-jilakara*; Tamil—*Shimai-shiragam*; Malayalam—*Shima-jirakam*; Kannada—*Shine-jerage*; Mumbai—*Shiah zirah*; Persian—*Zirahe-siyah, zirahe kirmani*; Family—Umbelliferae (Apiaceae). A perennial herb. In India, it is grown in Kashmir.

Uses. The fruits are used as spice. In Western India the seeds (fruits) are used mostly by Muslims as a spice, and also to some extent by the people of Maharashtra. The starchy tubers are eaten as vegetable.

Sweet Pepper

Capsicum annuum Linn.; English—Sweet pepper, paprika, Hungarian paprika, Spanish pimento; Family—Solanaceae.

This is a non-pungent variety of chilli. The red paprikas are famous for their brilliant red colour and mild flavour. Several varieties of paprika have been successfully grown at the IARI, New Delhi and Central Food Technology Research Institute (CFTRI), Mysore. Since several varieties of *Capsicum annuum* Linn. are used to produce paprika, pods in one growing area may differ in shape and appearance of those from another. Some pods are round in shape with pointed end whereas, the others are elongated. They are medium to small and quite fleshy. They are produced on small, bushy plants. When ripe, the pods are picked and dried in the sun. It is processed into powder where it is grown. The paprika pods are supposed to be one of the richest sources of ascorbic acid (vitamin C).

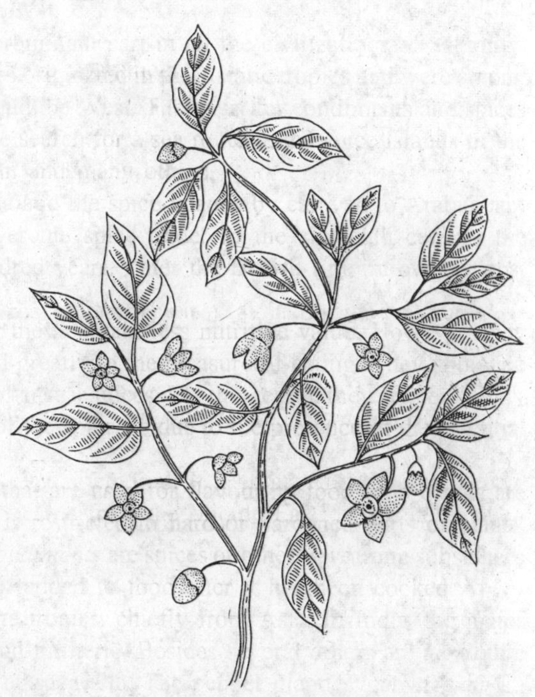

Fig. 15.22. Spices and condiments. *Capsicum annuum* Linn.; Eng. Paprika, Hungarian Paprika, Sweet pepper, Spanish pimento (Solanaceae). A flowering twig. This is a mild or non-pungent variety of chilli.

Uses. Paprika is used for its colouring and flavouring properties. It is used largely as a garnish for light coloured foods, such as eggs, fish, potatoes, salads and salad dressings. In food manufacturing, flavour is important, as this spice becomes a vital ingredient of sausages, soups, salad, salad dressings and many ready prepared foods.

Chemical composition. Moisture 7.9%; protein 13.8%; fat 10.4%; fibre 19.2%; carbohydrates 41.1%; total ash 7.6%; calcium 0.2%; phosphorus 0.3%; iron 0.23%; sodium 0.02%; potassium 2.4%; vitamin B_1 (thiamine) 0.60 mg/100 g.; vitamin B_2 (riboflavin) 1.36 mg/100 g.; vitamin C (ascorbic acid) 58.8 mg/100 g.; niacin 15.3 mg/100 g.; vitamin A 4915 International Units (I.U.)/ 100 g.; calorific value 390 calories/100 g.

Red Pepper, Chillies

Capsicum frutescens Linn.; Syn. *C. annuum* Linn.; English—Red pepper; chillies; Hindi & Punjabi—*Lal mirch*, *marcha*, *mirch*, *gachmirch*, *mattisa*, *wangru*; Bengali—*Lal-mirch*, *lanka-marich*, *gach-marich*; Kumaon—*Matisa-wangru*; Kashmiri—*Mirtz-a-vangun*, *mirch-wangum*; Gujarati—*Lal-mirch*, *marchu*; Marathi—*Mirsinga*; Tamil—*Milagay*, *mollagu*; Telugu—*Mirapakaya*; Malayalam—*Kappal-melaka*; Kannada—*Mena-sina-kayi*; Sanskrit—*Marichi-phalam*; Persian—*Filfile-surkh*, *pilpile-surkh*; Family—Solanaceae.

It is native of the West Indies and tropical America, most probably of Brazil. Commonly cultivated for its fruit throughout the plains of India, and on the lower hills such as Kashmir, and in the Chenab valley upto altitude 6,500 feet. When grown on the hills it is said to be very pungent. There are several varieties, differing chiefly in the length, shape, and colour of the fruit,

Fig. 15.23. Green fresh chillies.

some being round, others oblong, obtuse, pointed or bifid, smooth or rugose, and red, white, yellow or variegateu. The very pungent principles are present in the flesh and rind as well as seeds. The ripe fruits are dried in the sun and used whole or powdered.

It is a herb, 2 or 3 feet in height, with entire ovate leaves, white flowers with a rotate corolla, and many-seeded fruits, which are technically berries.

Cultivation. Chillies can be cultivated as a field crop over a wide range of tropical and subtropical conditions. It grows well up to an altitude of 5,000 feet a.s. 1., in tracts receiving an annual rainfall of about 25 to 50 inches. Heavy rains during the growing period lead to rotting of leaves and fruits. The crop can be grown both in the rainy season commencing from about June and in the hot weather about February. In areas of heavy rainfall the crop is taken after the monsoon period. In tracts of low rainfall, the crop is grown under irrigation.

The soils best suited for the cultivation are of the heavy loam type, from grit, gravel, and stones and well drained.

The seed is sown in the end of June or in the beginning of July in well prepared nursery beds. When the seedlings are about 40-45 days old, they are transplanted. At the nursery stage frequent irrigation is required. The flowering of the plants begin after about a month of transplanting. In another one month, the first crop of green chillies is ready for picking. From about the month of November the fruits begin to ripen and change in colour, first to a deep orange and

Fig. 15.24. Dried red chillies.

then to red. The fruits are gathered as they become red. About the middle of December the picking is in full swing. The plants continue to bear the fruit to the end of February. After being gathered, the fruits are dried in the sun. If the colour is not fully developed, they are heaped up until the colour develops fully, and then dried in the sun. They are usually trampled lightly when they are drying, so that they may be flattened and their packing in bags for storage and transport may become easy. Process of drying usually takes about a fortnight. Well irrigated crops yield 1,500 lb. to 2,500 lb. per acre.

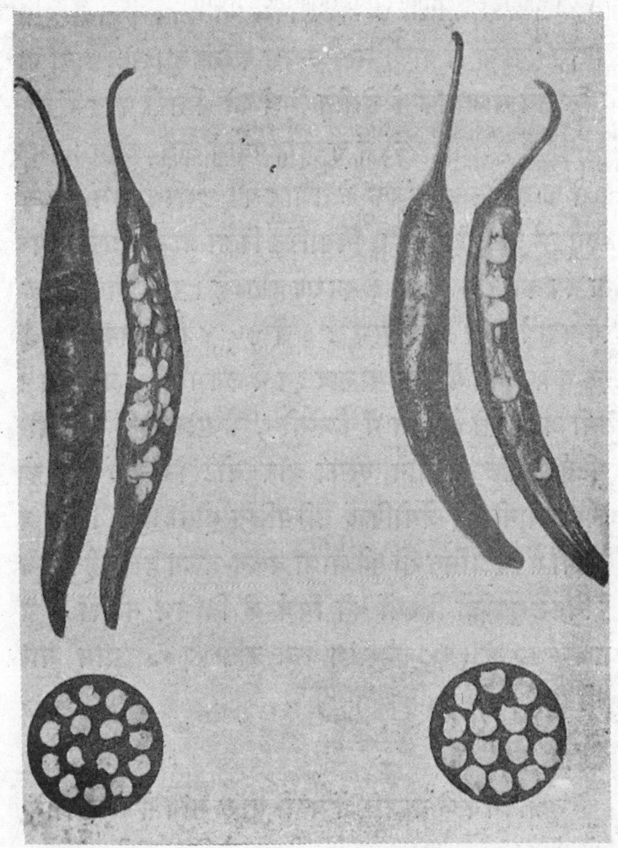

Fig. 15.25. *Capsicum frutescens* (red pepper - **Lal mirch**). Fruits and seeds.

Uses. The fruit when green is used for pickling and when ripe is mixed with tomatoes to make sauces. It is also dried and ground for use. The consumption of chillies is very great, and both rich and poor daily use them. The chillies form the principal ingredient in all chutneys and curries. They are ground into a paste, between two stones (*Silbatta*), with a little mustard oil, ginger and salt, they form the only seasoning which the millions of poor can obtain to eat with rice.

Caraway

Carum carvi Linn.; English—Caraway; Hindi—*Shiajira*, *zira*; Bengali—*Jira*; Punjabi—*Zira siyah*; Kashmiri—*Gunyun*; Ladakh—*Unbu*; Tamil—*Shimai-shombu*; Telugu—*Shimai-sapu*; Sanskrit—*Sushavi*; Persian—*Karoya*; Family—Umbelliferae (Apiaceae).

A herb; native of Europe and West Asia, now cultivated in Bihar, Orissa, the Punjab, Bengal and Andhra Pradesh, as a cold season crop. It is frequently cultivated on the hills, as a summer crop in Kashmir and Garhwal at an altitude of between 9,000 and 12,000 feet. It is a perennial with thick roots, compound leaves with linear segments and small white flowers. The fruits are brown in colour, slightly curved and tapering. The spice seems to have come into use in Europe about the thirteenth century, and it was known in England at the close of the fourteenth century.

Uses. The fruit (seed) is used parched and powdered, or raw and entire. In the former case it is employed to flavour curries; in the latter it is put in cakes. It is used in confectionery and in flavouring drinks. It also produces a spirit cordial. It is carminative and largely used in curry powder.

A valuable essential oil is obtained from the "seeds" called caraway oil. The oil is a mixture of a hydrocarbon $C_{10}H_{16}$ (*carvene*) and an oxygenated oil, $C_{10}H_{14}O$ (*carvol*). The oil possesses a strong odour and flavour of the fruit. It is used medicinally and to flavour curries.

Chemical composition of fruit of *Carum carvi* Linn. Moisture 4.5%; protein 7.6%; fat 8.8%; fibre 25.2%; carbohydrates 50.2%; total ash 3.7%; calcium 1.0%; phosphorus 0.11%; sodium 0.02%; potassium 1.9%; iron 0.09%; vitamin B_1 (thiamine) 0.38 mg/100 g.; vitamin B_2 (riboflavin) 0.38 mg/100 g.; vitamin C (ascorbic acid) 12.0 mg/100 g.; niacin 8.1 mg/100 g.; vitamin A 580 International Units (I.U.); calorific value 465 calories/100 g. of spice.

Fig. 15.26. Spices and condiments. *Carum carvi* Linn.; Eng. Caraway (Umbelliferae, Apiaceae). A flowering twig.

Coriander

Coriandrum sativum Linn.; English—Coriander; Hindi—*Dhaniya*; Bengali—*Dhane*; Sanskrit—*Dhanyaka*; Tamil—*Kotamalli*; Telugu—*Danyalu, kotimiri*; Kannada—*Kotambari*; Nepali—*Danya*; Persian—*Kushniz*; Family—Umbelliferae (Apiaceae).

It is a native of the Mediterranean region. It is grown extensively in India, Russia, Central Europe, Asia Minor and Morocco. In India, it is cultivated in all the States. The more important States for its cultivation are - Andhra Pradesh, Assam, Maharashtra, Tamil Nadu, Uttar Pradesh, Karnataka, Himachal Pradesh and Madhya Pradesh.

The plant is generally 2-3 feet in height, with white or pinkish flowers. The lower leaves have broad segments, while the upper are very narrow. The fruits are small, oval and aromatic. Technically the fruit is known as "cremocarp". Each fruit consists of two one-seeded carpels, or mericarps with numerous oil ducts (vittae).

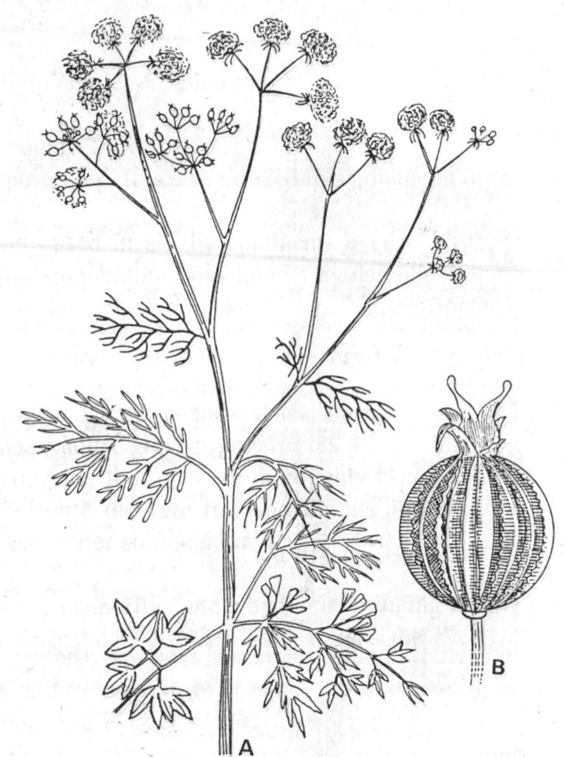

Fig. 15.27. Spices and condiments. *Coriandrum sativum* Linn.; Eng. Coriander (Umbelliferae, Apiaceae). A, a flowering twig; B, a fruit. India produces, on an average, 80,000 tons per year from about 2.5 lakh hectares.

Cultivation. It is cultivated mostly as rain-fed crop, either pure or mixed with other crops. It is also grown as an irrigated crop in some areas. The sowing time differs in different parts of India. In Northern India, the seeds are sown during winter; in Maharashtra during rainy season; in Tamil Nadu during autumn and in Karnataka and some parts of Tamil Nadu from May to August and from October to January.

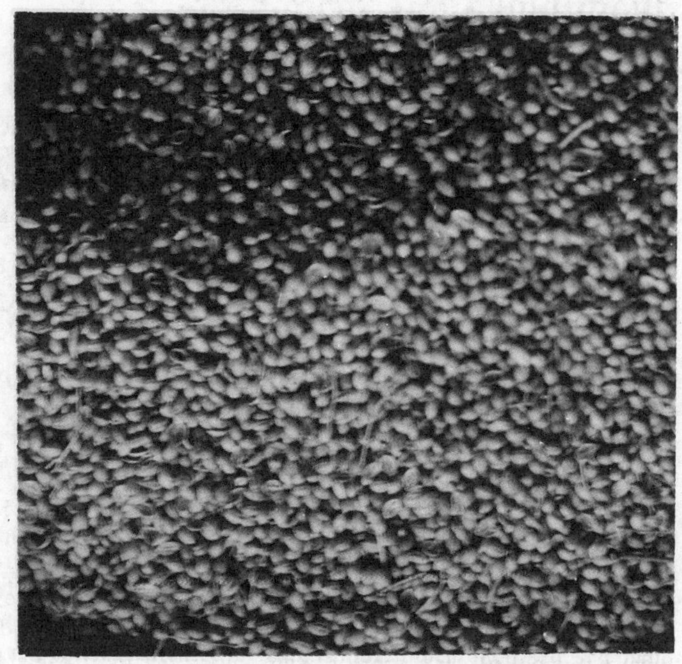

Fig. 15.28. Coriander fruits used as spice.

The seed germinates in 10-25 days after sowing. Two or three weedings are necessary for the crop. Irrigation is done when it is needed. Proper manuring increases the yield. The crop is ready in three to three and half months after sowing. The plants are pulled out along with the roots and dried. Therefore, the fruits are threshed out, further dried in the sun, winnowed and stored. The yields vary from place to place. It ranges from 300 kg. to 900 kg. per acre.

Uses. The fruits and leaves are aromatic and used as flavouring materials. The fruits are used extensively in the preparation of curry powder, pickles, sauces, soups, sausages and seasonings. They are also employed for flavouring pastries, cookies, buns and cakes and tobacco products. It is also used for flavouring liqueurs, particularly gin. The fruits are used both as spice and condiment. It is an important condiment used widely in all parts of India. A major part of the produce is used within the country. The fruits are also used as stimulant, carminative, stomachic, diuretic, antibilous, refrigerant, aphrodisiac and tonic. Oil of Coriander is used in medicine and for flavouring beverages. The Coriander leaves make one of the richest sources of vitamin C (*i.e.*, 250 mg/100 g.) and vitamin A (*i.e.*, 5200 I.U./100 g.).

Cumin

Cuminum cyminum Linn.; English—Cumin; Hindi—*Zira*; Bengali—*Jira*; Gujarati—*Jiru*, *jira-utmi*; Marathi—*Jire*; Tamil—*Shirangam*; Telugu—*Jiraka*, *Jilakarra*; Kannada—*Jiringe*, *jirage*; Malayalam—*Jirakam*; Sanskrit—*Jiraka*, *jirana*; Persian—*Zira*; Arabic—*Kamun*; Family—Umbelliferae (Apiaceae).

It is native of Mediterranean region. In India, it is cultivated mainly in the Punjab, Rajasthan, Haryana and Uttar Pradesh. The plant is a little annual herb with small pinkish flowers. The elongated oval fruits are aromatic and light brown in colour.

Cumin is a very old spice. It would appear to have been known to the ancients; at least there are names for it in most of the classical languages. During middle age it was one of the most favoured of spices. It was in frequent use, for example, in England in the thirteenth century.

Uses. The aromatic fruits are used as spice and condiment. The fruits are used in soup, curries, cake, bread, cheese and pickles. They form an ingredient of some curry powders and pickles.

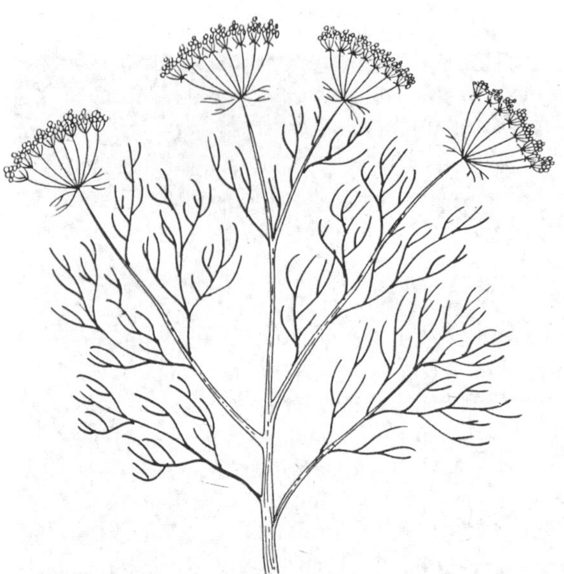

Fig. 15.29. Spices and condiments. *Cuminum cyminum* Linn.; Eng. Cumin (Umbelliferae, Apiaceae). A flowering

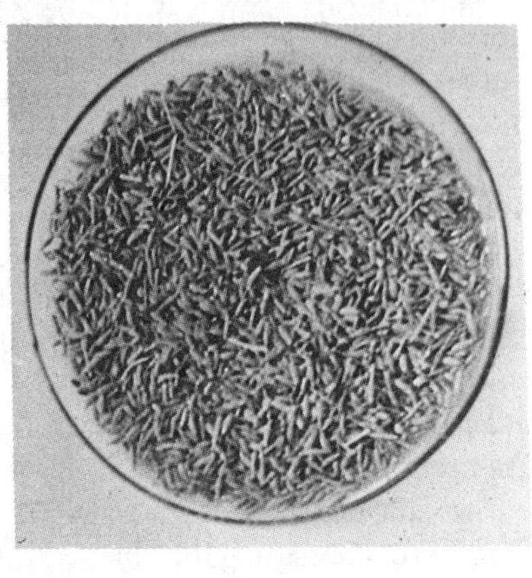

Fig. 15.30. *Cuminum cyminum* (Cumin—**Zira**). The fruits are used as spice.

The fruit contains an essential oil, which is a mixture of cymol and cuminol and other hydrocarbons. Both fruit and oil possess carminative properties. They are also aromatic, stimulant, stomachic and astringent. Their warm bitterish taste and aromatic odour reside in the volatile oil.

Chemical composition. Moisture 6.2%; protein 17.7%; fat 23.8%; crude fibre 9.1%; carbohydrates 35.5%; mineral matter 7.7%; calcium 0.9%; phosphorus 0.45%; sodium 0.16%; potassium 2.1%; iron 0.048%; vitamin B_1 (thiamine) 0.73 mg/100 g.; vitamin B_2 (riboflavin) 0.38 mg 100 g.; vitamin C (ascorbic acid) 17.2 mg/100 g.; niacin 2.5 mg/100 g.; vitamin A 175 International Units (I.U.); calorific value 460 calories/100 g.

Badrang

Fagara budrunga Roxb.; Hindi—*Badrang*; Family—Rutaceae. This is a large tree, commonly found in the Western Ghats and West Bengal.

Uses. The fruits are used as spice and the seeds are the source of an aromatic oil. The oil is used for flavouring food substances and cooked food.

Fennel

Foeniculum vulgare Mill.; English—Fennel; Hindi—*Saunf*; Bengali—*Mauri*, *pan-muhori*; Mumbai—*Bari-shopha*, *panmohuri*; Gujarati—*variari*, *Variyali*; Marathi—*Badishep*; Kannada—*Badisopu*; Tamil—*Sohikira*; Telugu—*Pedda-jila-kurra*; Sanskrit—*Madhurika*; Family—Umbelliferae (Apiaceae).

It is a stout, glabrous, aromatic herb, 5-6 feet high; leaves pinnately decompound; flowers small, yellow, in compound terminal umbels; fruit oblong, ellipsoid or cylindrical, 6-7 mm. in length, straight or slightly curved, greenish or yellowish brown; mericarp 5-ridged with prominent vittae.

Fennel is a very old flavouring substance. It was known to the ancient Hindus, Chinese and Egyptians as a culinary spice. It is native of Mediterranean region. In India, it is cultivated mainly in the Punjab, Uttar Pradesh, Assam, Maharashtra and Baroda. It is cultivated mostly as a garden or homeyard crop throughout India at all altitudes upto 6,000 feet. It requires a fairly mild climate and is cultivated as a cold weather crop in parts of Northern India.

Cultivation. It grows well in any good soil, but thrives best in rich, well-drained loam or black sandy soil containing sufficient lime. It is propagated mainly by seeds. Seeds are sown broadcast by hand or by shallow drills. The time of sowing is October-November in the plains of India and March-April on the hills. When 3-4 inches high, the plants are thinned out to about 1 feet apart. Occasional weeding and irrigation once a week during dry weather is being done.

Harvesting. The crop is harvested before the fruits are fully ripe. The stems are cut with a sickle and spread out to dry in the sun. When dry (4-5 days after cutting) the fruits are threshed out and cleaned by winnowing. The yield ranges from 2800 to 1120 lb. per acre.

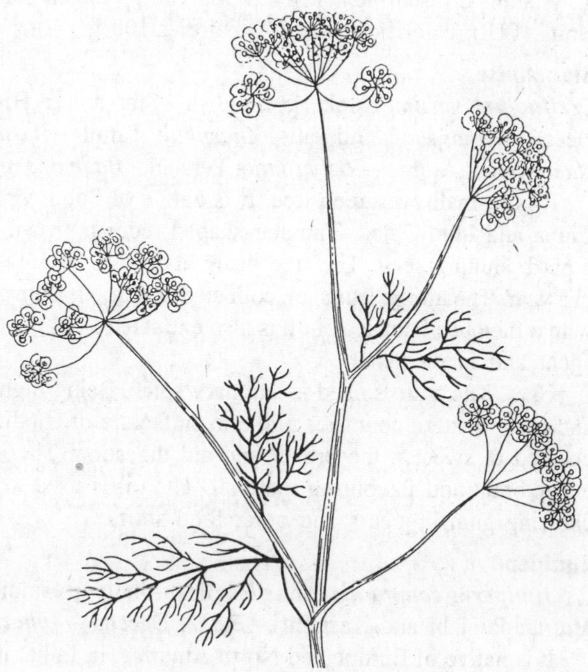

Fig. 15.31. Spices and condiments. *Foeniculum vulgare* Mill.; Eng. Fennel (Umbelliferae, Apiaceae). A flowering twig.

Uses. The leaves are used in fish sauce and for garnishing. Leaf stalks are used in salad. The leaf stalks are also used as vegetable. The leaves also possess diuretic properties. The roots are used as purgative and they possess an aromatic odour and taste.

Dried fruits of fennel have a fragrant odour and a pleasant aromatic taste. They are used for flavouring curries, soups, meat dishes, sauces, bread rolls, pastries and confectionery. They are also used for flavouring liqueurs and in the manufacturing of pickles.

The fruits are aromatic, stimulant and carminative. They are useful in the diseases of chest, spleen and kidney. It is used as one of the condiments in liquorice powder. A hot infusion of the fruits is used to increase lacteal secretion and to stimulate sweating. Fennel fruits from Lucknow are considered to be the best and are priced higher than those from other areas.

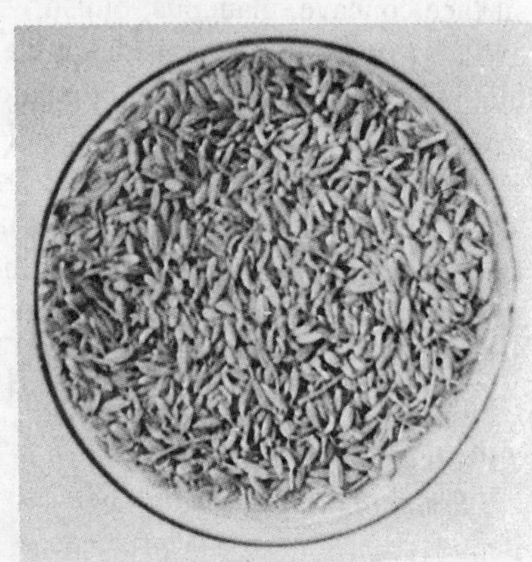

Fig. 15.32. *Foeniculum vulgare* (Fennel-**saunf**). The fruits are used as spice.

Fennel oil. The fruit of fennel also contains a volatile oil. The oil is obtained by steam distillation of crushed fruits. It is a colourless or pale yellow liquid with a characteristic taste and odour. The main constituent of the fennel oil is anethole. Oil of good quality contains 50-60% anethole. Fennel oil is largely used as a flavouring agent in culinary preparations, confectionery, cordials and liqueurs. It is a good aromatic and widely carminative. It is used in colic and flatulence of infants. It makes a good vermicide against hookworm. It is used in the preparation of fennel water. It is used in perfumery and in scenting soaps.

Chemical composition. Moisture 6.3%; protein 9.5%; fat 1.0%; crude fibre 18.5%; carbohydrates 42.3%; mineral matter 13.4%; calcium 1.3%; phosphorus 0.48%; sodium 0.09%; potassium 1.7%; iron 0.01%; vitamin B_1 (thiamine) 0.41 mg/100 g.; vitamin B_2 (riboflavin) 0.36 mg/100 g.; vitamin C (ascorbic acid) 12.0 mg/100 g.; niacin 6.0 mg/100 g.; vitamin A 1040 International Units (I.U.); calorific value 370 calories/100 g.

Star Anise

Illicium verum Hook. f.; English—Star anise; Hindi—*Anasphal*, *sonf*; Mumbai—*Badian*; Deccan—*Anasphal*; Gujarati—*Anasphal*; Tamil—*Anashuppu*; Telugu—*Anasapuvu*; Burmese—*Nanat-poen*; Arabic—*Riziya naje*; Persian—*Badiyan*, *raziyanahe-khatai*; Family—Magnoliaceae.

It is a small evergreen tree. It is native of South China. In India, its fruits are imported from China and Indo-China. The star-shaped reddish brown fruits consist of eight carpels, each with a hard, shining seed. The tree bears at from 6 to 100 years of age, often producing two crops in a year. The unripe fruits are collected and dried. Both the seeds and the fruits are highly aromatic with a flavour of anise. An oil is also extracted from the fruits. Star anise oil is used as a flavouring agent and in medicine.

Uses. The fruit is used as culinary spice. Being highly aromatic, it is in great repute in China and other Eastern countries in the manufacture of condiments and flavouring of spirits. It is often chewed to sweeten the breath and aid digestion. The oil is used in medicine as a carminative, expectorant and flavouring material. The oil is used to flavour confectionery, ice cream, candy, chewing gum, liqueurs and other food stuffs.

Juniper

Juniperus communis Linn.; English—Juniper; Hindi—*Hauber*; Bengali—*Havusha*; Marathi—*Hosha*; Punjabi and Kashmiri—*Betar*; Deccan—*Abhal*; Family—Cupressaceae, a gymnosperm.

It is native of Europe and North America. In India, it is found in the Himalayas from Kumaon westwards at altitude of 5,000 to 14,000 feet.

A dense, more or less procumbent shrub, rarely a small tree; bark reddish brown, peeling off in papery shreds; leaves in whorls of 3, linear subulate, 0.2-0.6 in. long, sharply pointed, upper surface concave, glaucous, bluish white, lower surface bluntly keeled; flowers usually dioecious, axillary; fruit sub-globose bluish black when ripe, 0.4-0.5 inch in diameter, covered with a waxy bloom; seeds usually three, elongated, ovoid. The plants flower in March-April and fruits ripen in August-September of the second year.

Uses. The fruits are employed for flavouring gin and food products. They are sometimes used as an article of food. For flavouring gin the fruits are crushed, immersed in warm water and fermented. The fermented mass is then distilled and rectified; 1,000 kg. of fruits yield 16-18 litres of beverage (containing 40-50% alcohol) and 5-6 kg. of volatile oil. The fruits and the volatile oil possess carminative, stimulant and diuretic properties. Juniper oil is largely used in compounded gin flavours, liqueurs and cordials.

Allspice

Pimenta dioica (Linn) Merrill syn. *P. officinalis* Lindl.; English—Allspice tree, Jamaica Pepper tree; Family—Myrtaceae.

It is a bushy evergreen tree, 6-9 m. high. It is native of West Indies and tropical America. It is grown in gardens in India. Leaves oblong-lanceolate, coriaceous; flowers white, in terminal and axillary trichotomous paniculate cymes; fruit a globose berry, about the size of a pea seed, black or purple, two seeded; seeds deep brown, reniform.

The unripe berries, when rapidly dried, form the Allspice, Jamaica Pepper or Pimento of commerce. In India, the plant is cultivated in gardens, especially in Bengal, Bihar, Orissa and Mysore. It grows well in Mysore and fruits heavily. The plants are propagated by seeds. Seeds are sown under shade in freshly prepared beds or in pots. The beds should be well watered before sowing, and the seeds should be thinly broadcast and lightly covered with coir waste. Seedlings are potted in large containers at the 2- and 4- leaf stage and are ready for planting in the field when they are 25-35 cm. in height. The tree begins fruiting when about seven years old and continues so upto twenty years. The berries are picked when mature, but still green. The berries

Fig. 15.33. Spices and condiments *Pimenta dioica* (Linn.) Merrill, syn. *P. officinalis* Lindl.; Eng. All spice tree, Pimento tree, Jamaica pepper tree (Myrtaceae). A flowering twig.

are quickly dried in the sun for 4-10 days; They become wrinkled like black pepper and turn reddish brown in colour and the aroma becomes more pronounced. Average yield of a tree is about 35-45 kg. of berries. Dried berries are nearly globular, 4-7 mm. in diameter, reddish brown and with a wrinkled surface. The berries possess an aromatic taste and flavour resembling a mixture of cinnamon, cloves and nutmeg, and hence the name Allspice. Allspice owes its characteristic odour to the presence of an essential oil (3.3-4.5%), found mainly in the pericarp. Allspice is available whole or ground.

Uses. The berries are used as a condiment, as a flavouring agent in sauces, soups, pickles, canned meats, sausages, etc., It is an important ingredient of whole mixed pickling spice, and spice mixtures, *viz*., curry powders, mincemeat spice, poultry dressing, pastry spice, etc., They are used for flavouring liqueurs. Pimenta berry oil is used for flavouring condiments and food products. The oil is used also as a carminative and stimulant.

Chemical composition of ground berries. Moisture 8.8%; protein 6.0%; fat 6.6%; fibre 21.6%; carbohydrates 52.8%; total ash 4.2%; calcium 0.8%; potassium 1.1%; iron 7.5%; vitamin B_1 (thiamine) 0.1 mg/100 g.; vitamin B_2 (riboflavin) 0.06 mg/100 g.; vitamin C (ascorbic acid) 39.2 mg/100 g.; niacin 2.9 mg/100 g.; vitamin A 1445 International Units (I.U.); calorific value 380 calories/100 g.

Anise

Pimpinella anisum Linn.; English—Aniseed, Hindi—*Saunf*, *badian*; Bengali—*Muhuri*, *mitha-jira*; Marathi—*Somp*; Gujarati—*Anisa*; Telugu—*Kuppi*; Tamil—*Shombu*; Kannada—*Shombu*; Oriya—*Sop*; Family—Umbelliferae (Apiaceae).

It is native of the Eastern Mediterranean region. It is widely cultivated in Europe, Russia, North Africa, Mexico and South America. In India, it is grown to a small extent. It is grown in Uttar Pradesh, the Punjab and Orissa. It is stated that a commodity commonly available in Indian market under the name *Lucknow saunf* is called Indian Aniseed.

The plant is an annual about 2 feet in height with simple based leaves and pinnately dissected steam leaves. The small fruits are greyish brown and covered with short hairs.

The plant is well grown in light, fertile or moderately rich, well drained sandy loam. It is propagated by seeds. Seeds are sown broadcast or in drills, 20-40 cm apart. It is commonly sown from the middle of October to the end of November in the plains, and from the beginning of April to the end of May on hills. When the seedlings are 5-8 cm. in height, they are thinned to stand 10-15 cm. apart in a row. The plants require, occasional weeding and watering once a week. The crop is ready to harvest in three and half months after planting when the fruits turn greyish green. The plants are tied in bundles and then stacked with the fruiting heads towards the centre. Fruits ripen in 4-5 days and are then threshed out, cleaned and bagged for the market. A yield of 445-665 kg. of fruit per hectare is expected.

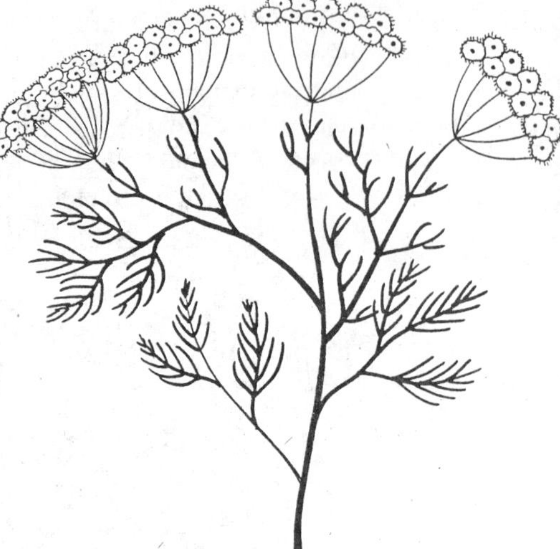

Fig. 15.34. Spices and condiments. *Pimpinella anisum* Linn.; Eng. Anise, Aniseed (Umbelliferae, Apiaceae). A flowering twig.

Aniseed on steam distillation yields an essential oil, known as oil of anise, anise oil is a colourless or pale yellow liquid having the characteristic odour and taste of the fruit.

Uses. The fruits are used as condiment. Now, oil of anise replaces the fruits for medicinal and flavouring purposes. Aniseed possesses a sweet aromatic taste and emits when crushed, a characteristic agreeable odour and is used for flavouring food, confectionery, bakery products, beverages and other liqueurs. The fruits are expectorant, stimulating, carminative, diuretic and diaphoretic.

Fresh leaves of the plant are used as a garnish and for flavouring salads.

Chemical composition of fruits. Moisture 9-13%; protein 18%; fatty oil 8-23%; essential oil 2-7%; sugars 3-5%; starch 5%; nitrogen free extract 22-28%; fibre 12-25%; ash 6-10%; choline is also present.

Indian Long Pepper

Piper Longum Linn.; English—India long pepper; Hindi—*Pipal, pipulmul, pipli, gaz pipal*; Bengali—*Pipul, pipli*; Nepali—*Popal pipal*; Punjabi—*Pipal, pipla mul*; Mumbai—*Pipli, pipla mul*; Marathi—*Pimpli*; Gujarati—*Pipli, piper*; Tamil—*Tippili, pipili, pippallu*; Telugu—*Pipili, pippali katte*; Kannada—*Yippali*; Malayalam—*Lada, mulagu*; Sanskrit—*Pippali, upukulya, videhi, ushuna*; Persian—*Pilpil, filfil-i-daraz*; Family—Piperaceae.

It is a native of India, Sri Lanka and Philippines Islands. It is a slender aromatic climber with perennial woody roots occurring in the hotter parts of India, from central Himalayas to Assam, Khasi and Mikir hills, lower hills of Bengal and evergreen forests of Western Ghats from Konkan to Travancore.

Cultivation. It is cultivated on a large scale in limestone soil, 450 to 600 m. below the Cherrapunji region. It is cultivated mainly by layering of mature branches or by suckers. The plants flower in August and September, and the fruits mature in January. The spikes are harvested in January, while still green and unripe, as they are most pungent at this stage. They are dried in the sun when they turn grey. The yield increases from 560 kg. per hectare in the first year to 1,680 kg. per hectare in the third. After third year the vines become less productive and should be replaced.

India imports 'a large' quantity of long pepper from Malaysia and Singapore.

Uses. The fruits are used as spices and also in pickles and preserves. They have a pungent pepper-like taste and produce salivation of the mouth. Long pepper, though considered inferior to black pepper, is considerably used as spice.

The dried unripe fruit and the root have long been used in medicine. They are considered heating, stimulant, carminative, alterative, laxative, and useful in cough, hoarseness, asthma, dyspepsia and paralysis.

The work on the fruit of *Piper longum* has shown the presence of the alkaloids—*piperine* and *piplartine*.

Black Pepper

Piper nigrum Linn.; English—Black pepper; Hindi—*Kalimirch*, *golmirch*, *choca mirch*; Bengali—*Kalamorich*, *golmorich*; Marathi—*Kalimirch*, *mire*; Gujarati—*Kalamari*, *kalomirich*; Tamil—*Milagu*; Telugu—*Miriyala tige*; Kannada—*Kare menasu*; Malayalam—*Kurumulaku*, *nallamulaku*; Punjabi—*Golmirich*; Kashmiri—*Martz*; Sanskrit—*Maricha*, *ushana*, *hapusha*; Family—Piperaceae.

A branching, climbing perennial shrub, mostly found cultivated in the hot and moist parts of India, Ceylon and other tropical countries. Branches stout, trailing and rooting at the nodes; leaves entire, 12.5-17.5 by 5.0-12.5 cm., very variable in breadth, sometimes glaucous beneath, base acute rounded or cordate, equal or unequal; flowers minute in spikes, usually dioecious; fruiting spikes very variable in length and robustness, rachis glabrous; fruits ovoid or globose, bright red when ripe; seeds usually globose, testa thin, albumin hard.

Pepper is one of the most ancient crops cultivated in India and has probably originated in the hills of south-western India, where it is met with in a wild state in the rain forests from North Kanara to Kanyakumari. The pepper vine in its wild state is mostly dioecious and consequently

Fig. 15.35. *Piper nigrum* (Black pepper—**Kali mirch** or **golmirch**).
The dried berries are used as black pepper.
A, a branch with fruits; B, Part of a spike.

rarely sets fruit. Under cultivated condition, however, the fruiting is very much better, since most of the cultivated types are bisexual. Almost all the types cultivated at present are selections from wild plants. Some crosses have already been evolved at the Pepper Research Station, Panniyur (Kerala). Of these, the hybrid, *Panniyur I*, a progeny from a cross between *Taliparamba No. 1*, noted for its high yield. *Panniyur I* is a good hardy climber and has long spikes with close set, large fruits.

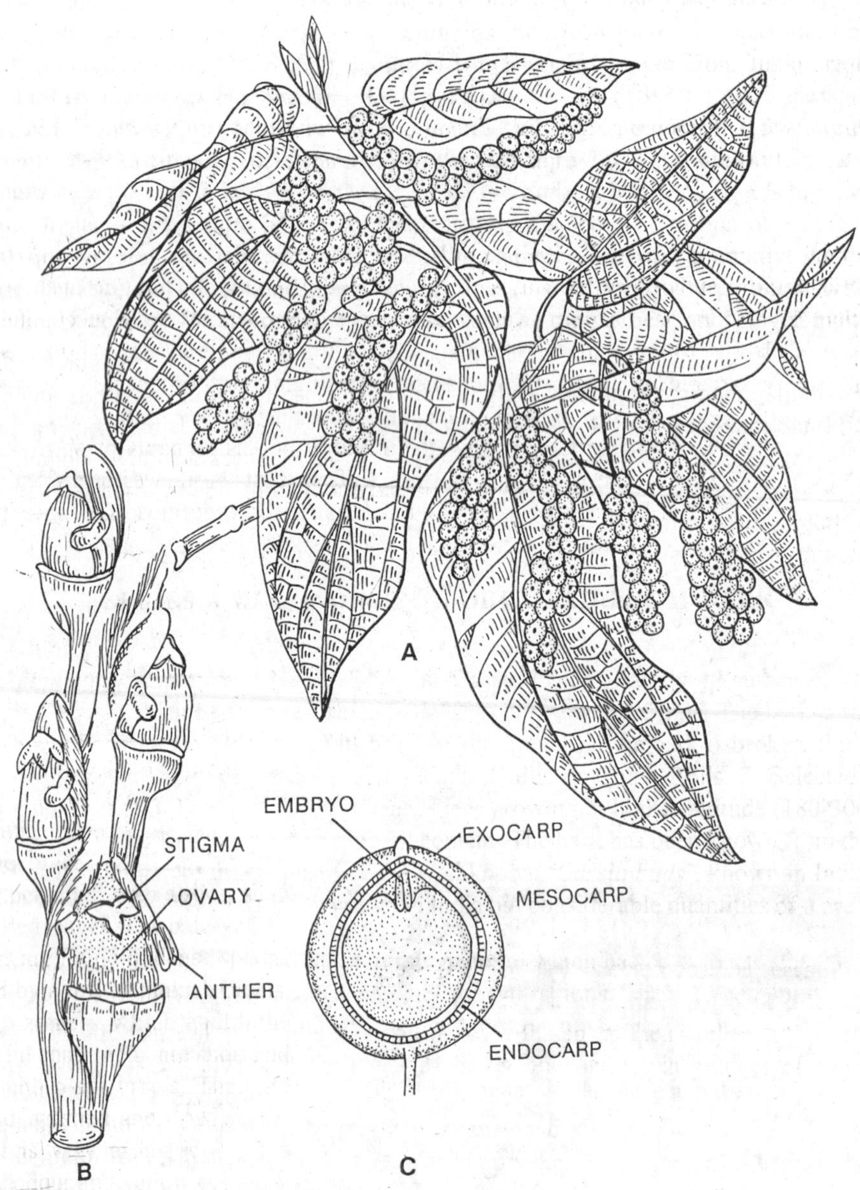

Fig. 15.36. Spices and condiments. *Piper nigrum* Linn.; Eng. Black pepper (Piperaceae). A, a branch with fruits; B, part of a spike with four flowers; C, L.S. of a berry. Black pepper are the dried mature but unripe fruits of *P. nigrum* Linn. About 24 varieties of black pepper are grown in India.

Cultivation. In India, pepper is cultivated mostly as a mixed crop in homestead gardens. Under this system, the vines are trained on to existing trees like jack, mango, coconut or arecanut. The vines receive intensive care and continue to be productive upto 60 or even 100 years. The vines are also grown as in inter - or subsidiary crop along with other plantation crops like coffee, cardamom, arecanut, coconut and orange. This practice is more common in North Kanara, Coorg and parts of Kerala. As a pure crop, pepper is grown on a plantation scale only in small areas in the sub-montane tracts and lower altitudes in South Kanara, Malabar, Wynad and Travancore. India has been a leading producer and exporter of pepper. India, Indonesia, Ceylon, Cambodia and Sarawak, Brazil and Malagasy are main producers of pepper. In India Kerala on the west coast produces nearly 95% of the total output. In India, pepper is grown mainly in Kerala, Tamil Nadu and Mysore and to a less extent in Assam.

Climate and soil. The pepper vine is found growing upto an altitude of 1,500 m., but thrives best at about 500 m. Pepper requires a warm and humid climate. It thrives in places where the annual rainfall is well over 200 cm. Pepper plants are pollinated by rain-water and consequently satisfactory yields are obtained when frequent showers of rain occur during blossoming period.

Pepper thrives best on virgin soil rich in humus and other plant nutrients. Naturally well drained, red laterite soils or alluvial soils rich in humus seem to be highly suitable for this crop. In Kerala, major part of the pepper cultivation is on laterite or sandy loam along the alluvial banks of the rivers.

Propagation. The pepper vine can be propagated either vegetatively or by seeds. Vegetative propagation is universally adopted. However, it is believed that the plants raised from seeds live longer, and give heavier crops in later years than plants raised vegetatively.

There are two types of vegetative shoots preferred for planting, *viz.,* the runners which are usually confined to the bottom portion of the vines, and the terminal shoots towards the top region. The runners form the most common material for propagation. In some areas, particularly in North Kanara, the terminal shoots of the top fruiting branches are preferred. Plants raised from such cuttings are said to yield even in the second year of planting and produce a satisfactory crop from the fourth year onwards while vines raised from runner shoots give satisfactory yields only from the sixth or seventh year.

Planting. The planting of the pepper vine is done either in July or August. The pits for planting are made preferably on the north and north-eastern sides of the standard so that the severe western sun is avoided.

Manuring. Pepper is an exhausting crop and the soil needs high level of nitrogenous manuring.

Harvesting and yield. An adult vine yields annually 0.5 kg. dried fruits under Indian conditions. The yield goes on increasing as the vines get older upto about the tenth year. Usually there are two crops in a year. One in August-September and the other in March-April. When the fruits are ready for harvest, the whole spikes are removed from the vines with the aid of a ladder. They are dried in the sun for 3-4 days, and the fruits separated from the stalks by beating the dried spikes with a stick or by trampling under foot. When completely dry, the outer·skin of the fruits become black and wrinkled. It is estimated that 100 kg. of green pepper roughly yield 33 kg. of dried pepper.

In India, the average yield of commercial black pepper has been found to vary from 110 to 335 kg. per hectare.

Black Pepper. Black pepper consists of the dried, fully developed unripe fruits. It is nearly globular in shape, about 4-5 mm in diameter with the characteristic coat with deep set wrinkles. The freshly harvested spikes are spread on cement floors or mats and dried in the sun for about a week. During drying the green or red fruits gradually change in colour to dark brown or almost black and their skin becomes tough and wrinkled. The fruits are detached from the stalks by beating the heaped up material with sticks or treading upon it barefooted. The outturn of dried black pepper obtained from green fruits varies from 26-29% depending upon the type.

White Pepper. It consists of dried ripe fruits, freed of their pericarp. In preparing white pepper the separation is done through the zone of fibro-vascular bundles, *i.e.,* only a part of the mesocarp is removed. In India, the production is limited to very small quantities, by some households in Kerala, for medicinal and domestic usage. For this purpose, the fully ripe fruits are rubbed with hand to remove the soft rind, and dried. The white fruits obtained are then washed and finally dried.

Uses. Pepper fruits are used as spice or condiment. In Kerala, fresh green pepper is sometimes used for preparing pickles. Black and white pepper make the major condiments employed for seasoning freshly cooked and prepared foods. In U.S.A. and other European countries they are used mainly for preserving meat. The whole fruits are added to pickles, certain types of sausages, etc., but the bulk of the product is generally ground before use. Black pepper is mostly used for its characteristic aroma and pungent taste. White pepper is less pungent. The aromatic odour of pepper is due to a volatile oil, while the pungent taste is caused by an *oleoresin*. An alkaloid is also present. Pepper stimulates the flow of saliva and the gastric juices and has a cooling effect. In modern Indian medicine, it is much employed as an aromatic stimulant in cholera, weakness following fevers, coma, etc., as a stomachic in dyspepsia, as an antiperiodic in malarial fever.

Chemical composition of different type of black pepper (*Piper nigrum* Linn.) varies. Moisture 8.7-14.1%; total nitrogen 1.55-2.60%; nitrogen in non-volatile ether extract 2.70-4.22%; volatile ether extract 0.3-4.2%; non-volatile ether extract 3.9-11.5%; alcohol extract 4.4-12.0%; starch 28.0-49.0%; ash 3.6-5.7%; crude fibre 8.7-18.0%; crude piperine 2.8-9.0%; piperine 1.7-7.4%; Alkaloid piperine makes the biting taste.

Ammi

Trachyspermum ammi (Linn.) Sprague; English—Ammi, Lovage; Hindi—*Ajwain, ajowan*; Bengali—*Jowan, juvani*; Gujarati—*Ajamo*; Marathi—*Owa*; Kashmiri—*Jawind*; Tamil—*Aman, oman*; Telugu—*Omami, omamu*; Kannada—*Omu, oma*; Mumbai—*Ajwan owa*; Sanskrit—*Yamani*; Persian—*Zinian*; Family—Umbelliferae (Apiaceae).

This is an aromatic herb. Cultivated extensively in India for its fruits (seeds); from the Punjab and Bengal to the South Deccan. It is first mentioned in Europe as brought from Egypt about 1549 A.D.

Uses. The fruits are used as spice. They yield an oil of distillation with water, which is used medicinally. The fruits (seeds) are much valued for their antispasmodic, stimulant, tonic and carminative properties. They are considered to combine the stimulant quality of capsicum or mustard with the bitter property of chiretta, and the antispasmodic virtues of asafoetida.

Vanilla

Vanilla planifolia Andr.; Syn. *V. fragrans* (Salisb) Ames; English—Vanilla; Family—Orchidaceae.

It is climbing vine with fleshy adventitious roots, large succulent leaves, and greenish-yellow flowers. The fruits are long, thin, pod like capsules, known as vanilla beans. It is a tropical species and thrives in hot climate with frequent rains. It is propagated by means of cuttings. It is native of tropical America. In India, it is cultivated in the Nilgiris, Wynad, Coorg and Mysore.

The flavour and aroma are not present in the pods until they have been cured. The unripe fruits are subjected to a sweating process. They are exposed to the sun in the morning and are then protected by blankets in the afternoon, while at night they are placed in airtight boxes. During curing process a glucoside is changed by enzyme action into a crystalline substance, *vanillin*, which possesses the characteristic odour.

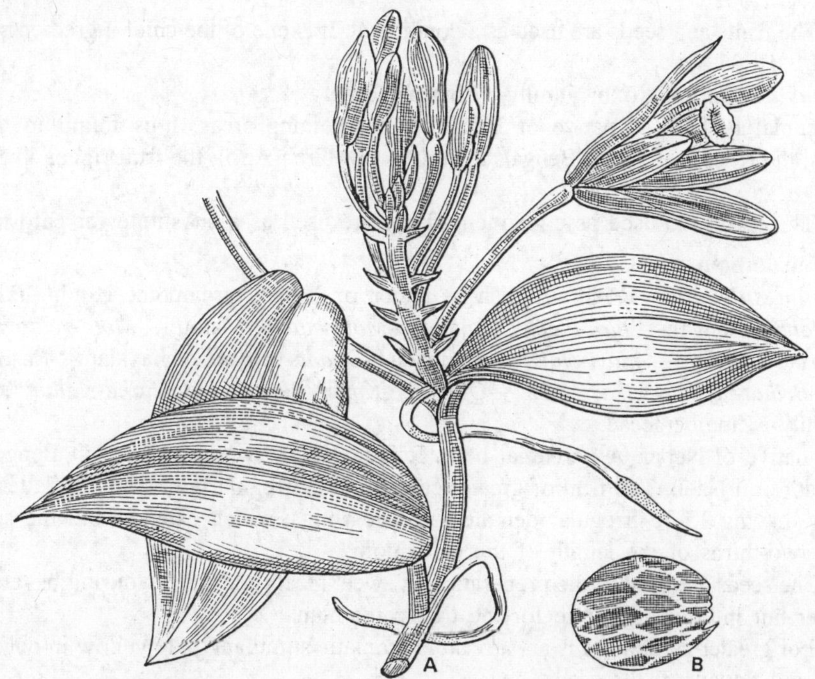

Fig. 15.37. Spices and condiments. *Vanilla planifolia* Andr.; Eng. Vanilla (Orchidaceae). A, a flowering twig; B, a seed. The vanilla pods or sticks of commerce are the cured fruits, or beans of *V. planifolia* Andr. (an orchid).

Chemical composition. Moisture 25.85-30.93%; protein 2.56-4.87%; fatty oil 1.68-6.74%; volatile oil 0.0-0.64%; nitrogen free extract 30.35-32.90%; fibre 15.27-19.6%; carbohydrates 7.1-9.1%; ash 4.5-4.7%; vanilline 1.48-2.90%; resins 1.5-2.6%; calcium 19.7%; phosphorus 9.5%; sodium 6.7%; potassium 16.2%; iron 0.3%.

Uses. The pods are used for flavouring confectionery and in perfumery. Vanilla is used in flavouring chocolate, ice creams, candy, beverages, cakes and other food products.

SPICES AND CONDIMENTS OBTAINED FROM SEEDS

Bengal Cardamom

Amomum aromaticum Roxb; English—Bengal cardamom, Nepal cardamom; Hindi—*Bari elaichi*; Marathi—*Veldode*; Family—Zingiberaceae.

It is a perennial herb, grown in North Bengal and Khasia hills. The fruit ripens in September; the capsules are then carefully gathered by the natives and sold to the druggists.

Fig. 15.38. *Amomum aromaticum* (**Bari elaichi**). The fruits and the seeds are used as spice.

Uses. The fruits and seeds are used as a condiment. It is one of the chief ingredients of '*garam masala*'.

Amomus dealbatum Roxb; Family—Zingiberaceae.

A perennial herb. It is native of Bengal and adjoining areas. It is found in the Eastern Himalayas, the Khasia hills and Bengal. It flowers in March-April; the fruit ripens in September-October.

Uses. The seeds are used as condiment. They are used as a substitute for cardamom.

Greater Cardamom

Amomum subulatum Roxb; English—Greater or Nepal cardamom; Hindi—*Bari elaichi, morang elaichi*; Bengali—*Bara elachi*; Gujarati—*Moto-elachi*; Marathi—*Mote-veldode*; Tamil—*Periya-elakkay*; Telugu—*Adavi-elakaya*; Kannada—*Dodda-yalakki*; Malayalam—*Periya-elattari*; Sanskrit—*Brihat-upakunchika*; Arabic—*Qakilahe-Kibar*; Persian—*Qakilahe-Kalan*; Myanmar—*Pala*; Family—Zingiberaceae.

It is a native of Nepal. A perennial herb. It is grown in swampy places in Bengal, Sikkim, Assam and Tamil Nadu. The fruit of greater cardamom is irregularly obcordate, flattened antero-posteriorly, having 15-20 irregular dentate-undulate wings, which extend from the apex downwards for two-thirds of the length of the cardamom.

Uses. The seeds are used in the preparation of sweet meats and for flavouring beverages. They are cheaper but inferior substitute for the true cardamom.

The oil of greater cardamom is an agreeable aromatic stimulant, pale yellow in colour, having the odour and flavour of the seeds.

White Mustard

Brassica hirta Moench; Syn. *B. alba* (Linn.) Boiss.; English—White mustard; Hindi—*Safed rai, safed rayan*; Bengali—*Dhop-rai*; Marathi—*Pandhora-mohare*; Gujarati—*Ujlo-rai*; Tamil—*Vellai-Kadugu*; Telugu—*Tella-avalu*; Malayalam—*Vella-Katuka*; Kannada—*Bili-sasave*; Sanskrit—*Siddhartha*; Persian—*Sipandane-supid*; Family—Cruciferae (Brassicaceae).

This is considered native of Europe and Western Asia. It is a herb, 2-6 feet high, with yellow flowers, hairy lobed leaves and a long-beaked pod with bristles. The small round seeds are yellow on the outside and white within. The seeds contain a glucoside, *sinalbin*. When ground seeds, are treated with water, the glucoside breaks down through enzyme action, and yields a sulphur compound with a characteristic sharp and pungent taste.

Uses. It is used as a condiment in food stuffs and pickles. The powdered seeds made into a paste with cold water act as a stimulant. It is used also in medicine. The seeds yield fatty oil.

Fig. 15.39. Spices and condiments. *Brassica* spp.; Eng. Mustard (Cruciferae, Brassicaceae). *Brassica hirta* Moench. (White mustard); *B. juncea* (Linn.) Czern. & Coss., (Indian mustard); *B. nigra* Koch (black mustard).

Indian Mustard

Brassica juncea (Linn.) Czern. & Coss; English—Indian Mustard; Hindi—*Rai*, *sarson*, *sarson-lahi*, *gohna-sarson*, *barirai*, *barlai*, *badshahi-rai*, *shahzada-rai*, *khas-rai*; Bengali—*Rai sarisha*; Kashmiri—*Asur*; Gujarati—*Rai*; Mumbai—*Rai*, *sarson*, *rajika*; Marathi—*Mohari*, *rayan*; Sanskrit—*Rajika*; Family—Cruciferae (Brassicaceae).

A tall, erect annual herb, 3-5 feet high, with bright green foliage; racemes terminal, with a rough reticulate testa. It is cultivated chiefly in the Punjab, West Bengal and Uttar Pradesh. It is sown in October-November and cut in March-April.

Uses. The seeds are used as a condiment in pickles. The seed oil is used for pickles and cooking purposes.

Black Mustard

Brassica nigra Koch; English—Black mustard; Hindi—*Kali rai*, *rai*, *tira*, *tara mira*, *lahi*, *jag rai*, *makra rai*; Bengali—*Rai sarisha*; Gujarati—*Rai*, *kalirai*; Mumbai—*Rai*, *sarson*; Tamil—*Kadagho*; Telugu—*Avalo*; Kannada—*Bile sasive*; Sanskrit—*Rajika*; Persian—*Jarshaf*; Family—Cruciferae (Brassicaceae).

It is a native of Eurasia. In India, it is cultivated mainly in the Punjab, Uttar Pradesh and Tamil Nadu. The plant is smaller than the white mustard. It has smooth pods with dark brown seeds, which are yellow inside. The seeds contain a glucoside, sinigrin. This glucoside yields on decomposition a volatile oil containing sulphur, which is responsible for the aromatic odour and pungency.

Uses. Ground black mustard seeds are used as a condiment in pickles. The seed oil is used for cooking purposes.

Chemical composition of *Brassica* seeds. Moisture 7.6%; nitrogen substances 29.1%; nitrogen free extract 19.2%; ether extract 28.2%; crude fibre 11%; ash 5%; seed oil 27-33%.

Lesser Cardamom

Elettaria cardamomum Maton; English—Lesser cardamom, cardamom; Hindi—*Chhoti elaichi*; Bengali—*Ilachi*; Gujarati—*elachi*; Punjabi—*Illachi*; Mumbai—*Malabari elachi*; Marathi—*Velloda*; Tamil—*Ellakay*, *aila-cheddi*, *elakaya*; Telugu—*Ellakay*, *elaki chettu*, *ela-kaya*; Kannada—*Yalakki*, *yelaki*, *yerakki*; Sanskrit—*Prith weeka*, *chundruvala*, *ela*, *bahoola*; Family—Zingiberaceae.

This is a tall herbaceous perennial, with branching subterranean rootstock, from which arise a number of upright leafy shoots, 5-18 feet high, bearing alternate, elliptical or lanceolate sheathing leaves, 1-3 feet long. Flowers borne in panicles 2-4 feet long, arising from the base of vegetative shoots; panicles upright throughout their length or upright at first and ultimately pendent or prostrate; flowers about 1.5 inches long, white or pale green in colour with a central lip

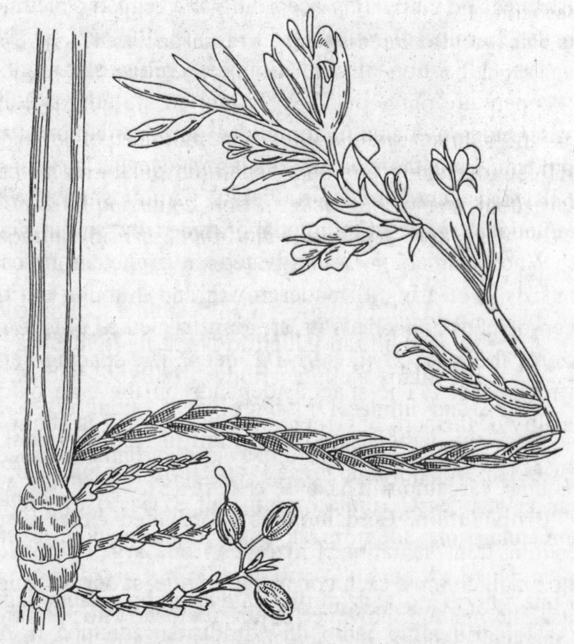

Fig. 15.40. Spices and condiments. *Elettaria cardamomum.* Maton.; Eng. Lesser cardamom, Cardamom. (Zingiberaceae). A plant with inflorescence and fruits.

streaked with violet, borne in a close series on the rachis; bisexual. Fruits trilocular capsules, fusiform to ovoid, pale green to yellow in colour, containing 15-20 hard, brownish black, angled and rugose seeds, covered by a thin mucilaginous membrane.

It is native of the moist evergreen forests of South India, growing wild in the Western Ghats, between 2,500 and 5,000 feet. It is found wherever the overhead canopy has been thinned. It is also found along stream banks, where the overhead shade is less dense.

There are two varieties: 1. *E. cardamomum* var. *major* Thw. comprising all the wild races and 2. *E. cardamomum* var. *minor* Watt. (syn. *E. Cardamomum* var. *minuscula* Burkill) comprising all the cultivated races, particularly those included under the names Malabar and Mysore cardamoms. Variety major is the more primitive variety from which the cultivated variety minor is derived. The minor variety is commonly grown in India.

Malabar cardamom is characterised by a short leafy shoot, rarely exceeding 9 feet in height, leaves 1-1.5 feet long, hairy on the under-surface; panicles 2-3 feet long and prostrate; fruits globose, rounded or ovoid and lightly ribbed. It is chiefly cultivated in Mysore and Coorg, and to some extent in Travancore.

Mysore cardamom is characterised by robust growth, leafy stem, upto 17 feet high, and large leaves with smooth underside; panicles erect; fruit fusiform, three-angled and ribbed. It is cultivated mainly in Travancore, Annamalai and Nelliampathy hills.

Distribution. It is usually cultivated in those regions which form the natural habitat of the species. They are chiefly cultivated in Kerala, Mysore, Maharashtra, Assam, Sikkim and Tamil Nadu.

Cultivation. The mode of cultivation varies considerably in different regions. The Mysore system of cultivation is the most important for commercial production and accounts for about 90% of the total cardamom crop in India. This type consists in clearing selected areas of jungle land of all undergrowth, thinning out the overhead shade, planting cardamom seedlings at regular distances and cultivating according to a regular schedule. In Mysore, cardamom is planted mostly on the margins of ravines.

The cultivation of cardamoms is confined to regions with an annual rainfall of 60-130 in., a temperature range of 50-95^{0}F. and an altitude of 2,000 to 5,000 feet. It grows on a variety of soils. Under ravine cultivation in Mysore, cardamom attains good growth in pockets of soil among boulders, even though the soil is not deep. The soil should be moist and well drained.

It thrives best under moderate natural shade. When inter-planted in coffee and arecanut plantations, shade is indirectly provided by coffee plants and areca palms.

Where cardamom is cultivated on a plantation scale in virgin forest areas, the initial work consists in clearing all undergrowth and thinning out the overhead canopy in order to obtain an even density of shade. Pits are dug in cleared land for planting cardamom seedlings. The usual size of the pit is 2' x 2' x 1-1/2' deep. The spacing between pits varies from 5 feet to 10 feet in different areas. The spacing depends on the type of cardamom grown and also on the natural fertility of the soil. In richer soils, the spacing is about 12 feet each way however, it is much less in poor soil. The spacing for Mysore cardamom ranges from 8 to 10 feet each way, while for Malabar cardamom it is 6 ft. x 6 ft.

Propagation. Cardamom is propagated either vegetatively by division of rhizomes or by seedling transplantation. Rhizomes from large clumps of growing plants are taken out, separated into small clumps, each consisting of atleast one old and one young shoot and planted in prepared pits. The method, however, is unsatisfactory for planting large areas as the number of rhizomes obtainable is limited.

Propagation by seedlings is common adopted in North Kanara, Mysore, Coorg, Tamil Nadu and Travancore-Cochin. Seedlings which come up spontaneously in jungle clearings and those which occur in old cardamom plantings are collected and planted in prepared nursery beds until they are required for planting in permanent positions. The seeds may be sown broadcast in favourable positions on cleared land and transplanted into permanent positions when they are sufficiently developed. The sowing time is usually August-October in Coorg, Mysore and North Kanara, and February-March in Tamil Nadu.

Seeds germinate in 6-7 weeks after sowing. Seed beds are prepared on raised ground and seeds are sown broadcast and covered with soft earth. Nurseries are generally opened near stream with abundant water. The beds are not shaded from the sun until after the seedlings appear. Low pandals are put up, 4-5 feet above the beds to provide shade for seedlings. Seedlings when 3-4 months old are transplanted 6-18 in. apart in nursery beds. Seedlings are usually kept in the nursery for about two years. The time of planting out seedlings is June-July, before the onset of monsoon rains. Generally two seedlings are planted in each pit, one of which is later used for filling up gaps in the plantation. Frequent weedings are necessary in the first two years to prevent young cardamom plants from being choked up.

Harvesting. Cardamom comes into bearing in three years after planting, which may be the fourth or fifth year after sowing. Flowering commences in April-May and continues till July-August; flowers may be seen almost throughout the year. The peak flowering period is May-June and as flowers appear for long, fruits ripen irregularly at intervals necessitating several pickings. Fruits are gathered at intervals of 30-40 days. Only those fruits which are just to ripe are picked. This is necessary to prevent capsules from splitting on the drying floors. If picked in the under-ripe stage, the fruits shrink on drying and have a shrivelled appearance. Individual fruits are picked with the peduncle and each clump has to be periodically visited to gather the fruits. The first crop obtained in the third or fourth year is usually small, higher and sustained yields are obtained in subsequent years upto the tenth or fifteenth year, depending upon the type cultivated, after which the plants become exhausted.

Drying and curing. The fruits are dried after harvesting either in the sun on barbecues of beaten earth or by artificial heat in drying houses. Artificial drying takes about 48 hours and cardamoms so dried retain their green colour, which is much liked in European and American markets. Sun drying, on the other hand, takes 3-5 days, but the mucilaginous coats on the seeds remain intact in sun-dried capsules and the seeds possess a characteristic sweetish aroma. A certain amount of bleaching also takes place during sun-drying and the capsules attain a straw colour. Dried capsules are hand-rubbed and winnowed to remove stalks, calyces and foreign matter, and then bagged.

The average yield of dry capsules from a well maintained cardamom estate is 100-150 lb. per acre.

Uses. Cardamom is used as a spice and masticatory, and in medicine. The seeds possess a pleasant aroma and a characteristic, warm, slightly pungent taste. It is used for flavouring curries, cakes, bread and for other culinary purposes. It is also used for flavouring liqueurs. In the Arab countries, cardamom is used for flavouring coffee and tea. In medicine, it is used as an aromatic, stimulant, carminative and flavouring agent.

The cardamom capsules contain—Moisture 20%; protein 10.2%; ether extract 2.2% mineral matter 5.4% crude fibre 20.1% carbohydrate 42.1% calcium 0.13% phosphorous 0.16% iron 5.0 mg/100 g. The seeds of cardamom contain 2-8% volatile oil. Seeds of green cardamom yield appreciably more oil than those of bleached cardamom.

Chemical composition of seeds. Moisture 8.3%; volatile oil 8.3%; total ash 3.7%; non-volatile ether extract 2.9%; crude fibre 9.2%; crude protein 10.3%; calcium 0.3%; phosphorus 0.21%; sodium 0.01%; potassium 1.2%; iron 0.012%; vitamin B_1 (thiamine) 0.18 mg/100 g.; vitamin B_2 (riboflavin) 0.23 mg/100 g.; vitamin C (ascorbic acid) 12.0 mg/100 g.; niacin 2.3 mg/100 g.; vitamin A 175 International Units (I.U.) per 100 g. of seeds.

Cardamom oil of commerce is obtained by the distillation of the whole fruits of *E. cardamomum*. The cardamom oil is a colourless or pale yellow liquid with a penetrating, camphoraceous odour and a strong pungent taste. The main constituents of the oil are - cineol, terpineol, terpinene, limonene, sabinene and terpineol in the form of formic and acetic esters. Cardamom oil is used in flavouring beverages.

The chief importing countries of Indian cardamom are - Arabia, Sweden, U.K., U.S.A., Germany and Middle East countries.

Nutmeg and Mace

Myristica fragrans Houtt.; English—Nutmeg tree; Hindi—*Jaiphal* (fruit kernel), *Japatri* (aril); Bengali—*Jaiphal* (nut), *jotri* (aril); Punjabi—*Japhal* (nut), *jauntari* (aril); Marathi—*Jaiphala* (fruit), *jayapatri* (aril); Gujarati—*Jayephal* (nut), *javantari*, *japatri* (aril); Tamil—*Jadikkay* (nut), *jadi-pattiri* (aril); Telugu—*Zevangam*, *jajikaya* (nut), *japatri* (aril); Malayalam—*Jatikka* (nut), *jati-pattiri* (aril); Sanskrit—*Jajiphalam* (nut), *jajipatri* (aril); Family—Myristicaceae.

It is native of the Moluccas Islands. In India, it is grown in the Nilgiris, Kerala, Mysore, Andhra Pradesh, Assam and West Bengal.

This is a dioecious or rarely monoecious evergreen aromatic tree, usually 30-40 feet high. The bark is greyish black and longitudinally fissured in old trees. The leaves are elliptic or oblong-lanceolate, coriaceous. The flowers are arranged in umbellate cymes. The flowers are creamy-yellow and fragrant. The fruits are yellow, broadly pyriform or globose, 6-9 cm. long, glabrous often drooping. The pericarp is fleshy, 1.25 cm. thick, splitting into two halves at maturity. The seeds are broadly ovoid, arillate, albuminous, with a shell like purplish brown testa. The aril is red and fleshy.

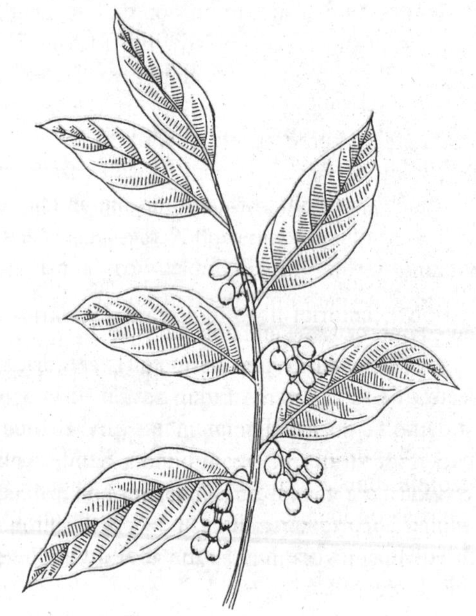

Fig. 15.41. Spices and condiments. *Myristica fragrans* Houtt.; Eng. Nutmeg (kernel) and mace (aril) (Myristicaceae). A flowering twig.

Cultivation and propagation. It requires a hot and moist climate with a rainfall of 150-300 cm. per annum. It grows best at low elevations in alluvium formed of deep friable loam with good drainage, well sheltered from high winds.

The plant is propagated from seeds. Fresh seeds with shells, collected from mature fruits and dried for a day, are sown either directly at the site or in nursery beds, sheltered from wind and strong sun. Seeds are sown 30 cm. apart, 2.5 cm. below the soil in nursery beds and take 2-3 months to germinate. The seedlings are transplanted when 60-90 cm. high to their permanent in the field during wet weather. The spacing between trees varies from 6-7.5 m., depending upon the gradient and fertility of the soil; wider spacing is adopted in rich level lands. Young plants require shade. This is provided by growing bananas in advance of planting. It may also be cultivated as a mixed crop in orchard, and in coffee, coconut, arecanut and rubber plantations.

In plantations raised from seeds there are usually more male than female trees and the sex is not determined until the plants are 6-7 years old and begin to bear flowers. In order to ensure a sufficient number of female trees in the plantation, two seedlings are planted close together and the one which has proved to be female is retained.

Harvesting and yield. Under favourable conditions, the tree starts to bear fruit from the seventh year or earlier and the yield increases upto the fifteenth year or beyond. Trees continue to bear for many years and 70-80 years old trees give good and sustained yields. Fruits are borne more or less throughout the year but the main harvesting season is June-October. The fruits are picked up every morning from the ground or are gathered from the trees by the use of a long stick to which a hook is attached. Seeds that drop out of split fruits are also gathered. The average annual yield at Burliar is 1,250-4,000 fruits per tree.

Preparation of nutmeg and mace. The harvested ripe fruit with the valves split, discloses the seed with a shell-like testa covered by a scarlet fibrous aril. After collection, the pericarp is removed and the seed separated from the aril and dried. Drying is complete when the kernel rattles in the shell. The shells are cracked off with wooden hammers and the kernels removed and sorted. Dried kernels make the nutmeg of commerce.

Mace is the dried fibrous aril covering the testa. It is obtained by separating the arils and drying in the sun. Mace of superior quality is produced by drying in specially constructed ovens. During drying the arils at first become dark red and brittle, and in about six weeks they become bright amber coloured.

Fig. 15.42. Spices and condiments. *Myristica fragrans* Houtt; Eng. Nutmeg (kernel) and mace (aril) (Myristicaceae). A twig with an opened fruit.

Nutmeg. The commercial nutmeg is ovoid, 2.0-3.5 cm. long x 1.5-2.8 cm. diameter, greyish brown in colour with minute reddish brown spots and lines and reticulately furrowed. East Indian nutmeg is of three grades - 1. Banda nutmeg is supposed to be the best which contains upto 8% essential oil; 2. Siauw nutmeg is of second grade containing 6.5% essential oil and 3. Penag nutmeg, which is suitable only for distillation purposes.

Composition. The nutmeg contains - Moisture 14.3%; protein 7.5%; ether extract 36.4%; fibre 11.6%; carbohydrates 28.5%; mineral matter 1.7%; calcium 0.12%; phosphorus 0.24%; iron 4.6 mg/100 g.; Nutmeg contains a volatile oil (6-16%), starch (14.6-24.2%); pentosans (2.25%); furfural (1.5%), and pectin (0.5-0.6%). The flavour and therapeutic action are due to the volatile oil.

Nutmeg oil. The percentage of volatile oil varies from 6-16% according to the origin and quality of the spice. Commercial oil is derived from broken and wormy nutmegs. The material is comminuted, pressed to remove fixed oil, and immediately subjected to steam distillation.

The nutmeg oil is a mobile, colourless or pale yellow liquid with a characteristic odour. The major constituents of the oil are *d*-pinene and *d*-camphene.

Nutmeg butter. Commercial nutmeg butter, a highly aromatic fat, is obtained from undersized, damaged or worm-eaten kernels which are unfit for sale as spice. The material is ground and cooked or steamed before pressing.

The nutmeg butter is a soft solid, yellow or yellowish red in colour, with the odour and taste of nutmeg.

Mace. Commercial macc consists of flattened lobed pieces, 2.5 cm. or more in length, somewhat less in breadth and 1 mm. thick. When soaked in water the lobes swell up and regain their original form. It is dull yellowish red in colour, translucent and brittle. It is nutmeg-like in odour and taste. There are three grades of mace.

1. *Banda mace* is supposed to be finest - it is bright orange in colour and possesses fine aroma; 2. *Java Estate mace* is golden yellow in colour with brilliant crimson streaks; 3. *Siauw mace* is of light colour and contains less volatile oil.

Mumbai mace is derived from *Myristica malabarica*, it is dark red in colour and consists of narrow pieces, divided into numerous lobes twisted together at the apex. It is almost devoid of aroma, and is useless as a spice.

Composition. Mace contains - Moisture 15.9%; protein 6.5%; ether extract 24.4%; fibre 3.8%; carbohydrates 47.8%; mineral matter 1.6%; calcium 0.18%; phosphorus 0.10%; iron 12.6 mg/100 g. It contains a volatile oil (4-15%; average 10%), amylodextrin (25%), reducing sugars, pectin and resinous colouring matter. The chief constituents are volatile oils and amylodextrin.

Uses. Both nutmeg and mace are used as condiment and in medicine. They are commonly used more as a drug than as condiment. Nutmeg is stimulant, carminative, astringent and aphrodisiac. It is used in tonics and forms a constituent of preparations prescribed for dysentery, flatulence, nausea, vomiting, malaria, rheumatism, sciatica and leprosy.

Nutmeg oil and mace are used for flavouring food products and liqueurs. Nutmeg butter is used as mild external stimulant in ointments, hair lotions and plasters, and forms a useful application in case of rheumatism.

Black Cumin

Nigella sativa Linn; English—Black cumin, small fennel; Hindi—*Kalonji, kalajira*; Bengali—*Mugrela, kala-jira*; Kashmiri—*Tukm-i-gandna*; Mumbai—*Kalonji, kalenjire*; Tamil—*Karunshirogam*; Telugu—*Nalla-jilakra*; Kannada—*Karijirigi, kare-jirage*; Malayalam—*Karun-chirakam*; Sanskrit—*Krishna-jirata*; Persian—*Siyah-danah*; Myanmar—*Samon-ne*; Family—Ranunculaceae.

A herb. It is a native of Southern Europe. In India, it is cultivated in the Punjab, Bengal, Assam and Bihar for its seeds. Its Sanskrit names indicate, its introduction at a very early period.

Uses. The seeds, possess a strong, pungent, aromatic taste, and therefore, are much used in curries, pickles and other dishes. They are also frequently sprinkled over the surface of bread along with sesamum seed. French cooks employ the seeds of this plant under the name of *Quatre epices* or *toute epices*, and they were formerly used as a substitute for pepper.

Poppy Seeds

Papaver somniferum Linn. var. album; English—Poppy seeds; Hindi—*Post*; Bengali—*Pasto*; Marathi—*Posta*; Gujarati—*Posta*; Family—Papaveraceae.

It is native of West Asia. In India, it is cultivated in the East Punjab, Uttar Pradesh, Rajasthan and Madhya Pradesh. The plant is an erect, rarely branched, usually glaucous annual herb, 60-120 cm. in height. Leaves ovate-oblong or linear-oblong, amplexicaul, lobed, dentate or serrate; flowers large, usually bluish white with a purplish base or white, purple or variegated; capsules large, 2.5 cm. diameter, globose, stalked; seeds white or black, very minute, reniform.

As we know poppy is cultivated also for its seeds. In India, var. *album*, with white seeds has been cultivated for many years for the production of seeds (*post*) under licence in Dehradun and Tehri Garhwal districts of Uttar Pradesh, and in Jalandhar, Kapurthala, Hoshiarpur and Patiala districts of the Punjab. In India, best seeds are obtained when the capsules have not been incised for extraction of opium. The crop yields 220-275 kg. of seeds per hectare.

Uses. Poppy seeds are considered nutritive and are used in breads, curries, sweets and confectionery. Poppy seeds contain upto 50% of an edible oil. The oil is odourless and possesses a pleasant almond - like taste. In India the oil is extracted by cold-pressing the seed in small presses. The oil is widely used for culinary purposes. It is free from narcotic properties and used as a salad oil. The cake or the meal left after extraction of the oil from the seeds is sweet and nutritious and is eaten by poor people.

Chemical composition of seeds. Moisture 4.3-5.2%; protein 22.3-24.4%; ether extract 46.5-49.1%; nitrogen free extract 11.7-14.3%; crude fibre 4.8-5.8%; ash 5.6-6.0%; calcium 1.03-1.45%; phosphorus 0.79-0.89%; iron 8.5-11.1 mg/100 g.; thiamine 740-1181 ug/100 g.; riboflavin 756-1203 ug/100 g.; nicotinic acid 800-1280 ug/100 g; iodine 6 ug/kg.; manganese 29 mg/kg.; copper 22.9 mg/kg.; magnesium 15.6 g/kg.; zinc 130 mg/kg.; lecithin 2.80%; Oxalic acid 1.62%; pentosans 3.0-3.6%; traces of narcotin, an amorphous alkaloid and enzymes—diastase, emulsin.

Fig. 15.43. Spices and condiments. *Papaver somniferum* Linn. var. *album*; Eng. Poppy seeds (Papaveraceae). A flowering twig. The poppy seeds are white and very small.

Pomegranate

Punica granatum Linn.; English—Pomegranate; Hindi—*Anardana*; Assamese—*Dalim*; Bengali—*Dalimb*; Gujarati—*Dalamb*; Kannada—*Dalimbari*; Kashmiri—*Daan*; Malayalam—*Mathalam pazham*; Marathi—*Dalimb*; Tamil—*Mathalam pazham*; Telugu—*Dannima pandu*; Punjabi—*Anardana*; Sanskrit—*Dadima*; Family—Punicaceae.

A shrub or small tree, 5-10 m. high, native of Iran, Afghanistan and Baluchistan, found growing wild in the warm valleys and outer hills of the Himalayas between 900 and 1,800 m. and cultivated throughout India. Leaves 2.0-8.0 cm. long, oblong or obvate, shining above; flowers usually scarlet red; fruits globose, crowned by persistent calyx, with a coriaceous woody rind and an interior septate with membranous walls, containing numerous seeds; seeds angular with a fleshy testa which is red, pink or whitish.

Though mostly found cultivated in many parts of India, the tree is also very common and gregarious in the gravel and boulder deposits of dry ravines in the outer Himalayas, upto about 1,800 m. In Jammu, it is found gregariously on dry limestone soils in the upper extremities of sub-tropical forests; it is also found in Chamba, Kangra and Mandi districts of Himachal Pradesh and valleys below 1,800 m. in Jaunsar and Tehri - Garhwal districts of Uttar Pradesh. As a cultivated crop, maximum area is said to be devoted to it in Maharashtra, particularly in Pune, Sholapur and Satara districts. In Gujarat, its cultivation is concentrated in Dholka taluka. It is cultivated in the districts of Almora, Aligarh, Tehri - Garhwal, Meerut and Farrukhabad in Uttar Pradesh; Uthukuli, Michaelpatti, Vellodu and Dindigul in Tamil Nadu state; Penukonda and Madakasira in Andhra Pradesh; Tumkur, Kolar, Bangalore and Mysore districts in Karnataka state.

Among the numerous types grown in India, the *Bedana* and *Kandhari* are considered the best.

The pomegranate is a sub-tropical fruit tree, growing best in semiarid climate where cool winters and hot summers prevail.

The fruit is a good source of sugars and vitamin C, and a fair source of iron but is poor in calcium. The sugar content increases with the age of the fruit and of the tree. The fruit contains 0.27% of pectin. During ripening, the insoluble pectin changes into soluble pectin. The concentration of vitamin C is said to increase with maturity and ripening of the fruit, and in most cases the maximum amount is deserved in the nearly ripe fruit.

Anardana of commerce. The seeds along with the fleshy portions are dried and commercially marketed as *anardana* and are widely used as condiment. The source of *anardana* of commerce is said to be the wild trees, covering vast tracts of hill slopes in Jammu, and parts of Chamba, Kangra and Mandi districts of Himachal Pradesh. The main areas from where *anardana* is collected are the Riasu, Udhampur, Ramban, Kishtwar and Bhadarwah Forest Divisions. In these areas, fruits are hand-picked towards the middle of October, when they are ripe and brown red in colour. Seeds with the pulp are separated by hand from the rind and are dried in the sun for about 10-15 days when the colour of the seeds turns reddish brown. The main assembling centre for *anardana* is Udhampur (J & K State) from where distribution is done all over the country.

Sesame

Sesamum indicum Linn.; English—Sesame; Hindi—*Til*; Sanskrit—*Tila*; Telugu—*Nuvvulu*; Tamil—*Ellu*; Oriya—*Khasa*; Family—Pedaliaceae.

It is one of the most ancient crops of India. Sesame seed has been an essential article in Hindu religious ceremonies. Charred remains of sesame at Harappa excavations (3600-1750 B.C.) indicate that sesame was in cultivation during Indus valley civilization.

The sesame plant is an erect, branched or unbranched annual, 2-6 feet in height; leaves 3 to 5 inches long, simple; flowers white or pink with dark coloured markings, borne in racemes in the leaf axils; fruit capsular, oblong quadrangular, slightly compressed, deeply 4-grooved, 1.5-5 cm. long; seeds black, brown or white, 2.5-3 mm. long and 1.5 mm broad.

It is cultivated throughout the plains of India and upto an altitude of 1200 m. It is mainly grown in Uttar Pradesh, Madhya Pradesh, Rajasthan, Andhra Pradesh, Tamil Nadu and Maharashtra.

Two distinct types of seeds are recognized - the white and the black. Seeds yield on expression an edible oil and the bulk of sesame seed production is utilized for the extration of oil.

Cultivation. Sesame is raised as a kharif crop under rained conditions in the States of Uttar Pradesh, Rajasthan, Maharashtra and Punjab. In these regions the crop is sown in June-July and harvested in October-November. In other states like Andhra Pradesh, Tamil Nadu, Madhya Pradesh, Maharashtra and Orissa it is raised as a *rabi* crop and as an irrigated summer crop. In these regions the crop is sown in October-November and harvested in December-January as a *rabi* crop and again planted in February-March and harvested by May-June as an irrigated crop.

The crop grows well on a light well-drained soil but thrives best on typical sandy loams. Sesame is harvested well before the plants are completely dry in order to prevent loss of seeds due to dehiscence of the capsules. The plants are cut at the base with a sickle. Now the plants are carried to the threshing floor where they are piled in stacks. The capsules are allowed to dry in this way for about a week. By inverting the bundle of plants and shaking them, most of the seeds are emptied on the threshing floor. A light beating is given with a stick if necessary to open the few capsules which have not yet opened. The seeds are collected and cleaned thoroughly.

The elements present in the seed include calcium, iron, iodine, zinc, cobalt, molybdenum and nickel. The seeds are a good source of protein for human nutrition. The main protein of sesame seed is a globulin.

Uses. Sesame seed is used for flavouring food stuffs. It is also used as a nourishing food. It is invariably dehulled for use as food. The characteristic flavour of sesame seed is developed by dry-toasting the dehulled material. Dehulled seed is an ingredient of a large variety of sweet meat and confections widely consumed in India. The seeds are used also in bread to which they impart an agreeable flavour. The seed is an ingredient of porridges, soups, confections and other food preparations. The seeds are considered emollient, diuretic, lactagogue and a nourishing tonic.

Sesame oil is widely used as an ingredient of confectionery and for making margarine.

Fenugreek

Trigonella foenum-graecum Linn.; English—Fenugreek; Hindi—*Methi*, *muthi*; Bengali—*Methi*, *methika*; Punjabi—*Methi*, *methri*; Gujarati—*Methi*, *methini*; Tamil—*Vendayam*; Telugu—

Menti kura; Kannada—*Menthya*; Malayalam—*Ventayam*; Sanskrit—*Methi*, *methika*; Persian—*Shanbalid*; Family—Papilionaceae.

It is a native of Southern Europe and Asia. It is grown mainly in Northern India. It is an annual legume with white flowers and long slender pods with a pronounced beak. The crop is sown in October-November. It is ready to harvest in April.

Uses. The seeds are chiefly used as a condiment to flavour curries made of rice, pulse, flour and meat, or as a relish with unleavened bread. The aromatic leaves are used as vegetable. The seeds are carminative, tonic and aphrodisiac.

SPICES AND CONDIMENTS OBTAINED FROM LICHENS

Parmelia

Parmelia abessinica Kremp.; Telugu—*Rathipoovvu*; Family—Parmeliaceae.

This is a crustaceous lichen found in rocky areas of Bellary, Anantpur and Guddapah districts of Andhra Pradesh. The lichen is available in large quantities in the market and used as food material and condiment. It contains atranorin (1.1%), lecanoric acid (3.3%), salazinic acid (0.1-0.5%) and isolichenin (3.4%).

Parmelia tinctorum Despr. is a foliose lichen commonly found in the plains of India and in the Himalayas on tree bark and rocks. It is commonly used in certain parts of South India as a component of various food preparations. It contains—crude protein, 13.8; ether extract, 6.0; isolichenin, 25.0; crude fibre, 13.4; and ash, 12.6%; calcium 1,728 mg.; ascorbic acid, 5.0 mg.; and riboflavin 100 mg./100 g.

Uses. The above mentioned lichens are used as spice and condiment.

INGREDIENTS USED AS GARNISHES

Cashewnut

Anacardium occidentale Linn.; English—Cashew; Hindi—*Kaju*; Gujarati—*Kaju*; Bengali—*Hijli-badam*, *kaju*; Marathi—*Kajucha-bi*, *kaju*; Tamil—*Mundiri*, *kottai*; Telugu—*Jidi-mamidi-vittu*, *jiedi pundu*; Kannada—*Jidi-vate*, *gerabija*; Malayalam—*Paranki-mava kuru*; Family—Anacardiaceae.

Mainly found in the tropics such as - India, East and West Africa, Northern Brazil, Tropical lowlands of Northern South America, Central America, Mexico. In India, *Kaju* is found in Kerala, Goa, Tamil Nadu, Karnataka, Orissa, Andhra Pradesh and Maharashtra. The cashew tree is an evergreen, of spreading, somewhat of straggly habit, low branched and reaching a height of 30 to 40 feet. The most conspicuous product of the tree is the "false fruit", the swollen peduncle or hypocarp commonly known as the 'cashew apple'.

Garnish. The cashews are used to garnish sweet meats, confectionery, cakes, pastries and many other food stuffs. The cashews are generally used to garnish marzipans, cakes, ice-creams, chocolates and so many other items of confectionery. Cashews are also eaten as salted and roasted. Chopped cashew nuts and the vanilla sugar are sprinkled on chocolates, biscuits, sweet meats (especially *Shahi Halwa*) and other confectionery items.

Cuddapah Almond

Buchanania lanzan Spreng.; Syn. *B. latifolia* Roxb.; English—Cuddapah almond; Hindi—*Chironji*, *piyala*; Bengali—*Piyal*, *chironji*; Punjabi—*Chirauli*, *chironji*; Garhwali—*Pial*, *payala*; Uriya—*Charu*; Tamil—*Mowda*, *marum*, *katma maram*, *katma parpu*; Telugu—*Chara-puppu*, *chara-pandu*; Kannada—*Nuskul*, *murkalu*; Malayalam—*Kala maram*; Gujarati—*Charoli*; Marathi—*Pyal-char*; Sanskrit—*Piyala*, *chirika*; Family—Anacardiaceae.

A tree, leafless only for a short time. Found in the sub-Himalayan tract, ascending to 2,000 feet, throughout the hotter and drier parts of India.

Garnish. *Chironji* is used as a garnish and for flavouring purposes. The kernels possess a delicious nutty flavour and are much used as ingredients of sweet meats and confectionery. The kernel is a common substitute for almonds. Its flavour is described as between that of the pistachio and the almond. It is eaten roasted with milk.

Pistachio or Green Almond

Pistacia vera Linn.; English—Pistachio, Green almond; Hindi, Bengali and Persian—*Pista*; Family—Anacardiaceae.

It is native of the eastern Mediterranean region, Iran, Afghanistan and Central Asia. It is grown on a commercial scale in Italy, Turkey, Syria, Iran, Afghanistan and West Pakistan. It is grown to a lesser extent in Northern India. *Pista* is imported in India from Afghanistan and Iran.

It is a small deciduous tree, up to 10 m. high; branches spreading; leaves imparipinnate; leaflefts 5-10 cm. long and 3-6 cm. broad, lanceolate to broadly ovate, coriaceous; flowers in panicles, minute; drupe oblong-linear to globose, laterally compressed, 10-20 mm. long and 6-12 mm. broad; outer husk variously coloured, readily separating from the greyish white, bony, keeled nut-shell (endocarp) which encloses green coloured kernel with a reddish coat.

Uses. *Pista* is used as a garnish and for flavouring purposes. The kernels possess a delicious nutty flavour and are much used as ingredients of sweet meats, confectionery and ice creams. It is also eaten as a dessert. Salted and roasted *pista* is much relished. They are highly prized for their colour and resinous flavour, and are much used in mixed nuts and as a flavouring material in ice cream and candy.

Pista kernels yield 50% of a fatty oil used to a small extent in confectionery as spice oil in medicine.

Almonds

Prunus amygdalus Batsch; English—Almonds; Hindi, Bengali, Marathi, Punjabi—*Badam*; Tamil—*Vadamkottai*; Telugu—*Badam vittulu*; Kannada—*Badami*; Malayalam—*Vatam-kotta*; Family—Rosaceae.

It is native of the East Mediterranean region. In India, it is cultivated in Himachal and Kashmir. The almond is cultivated in Kashmir at the altitude of 760-2,400 m. A tree upto 8 m. in height having oblong-lanceolate minutely serrate leaves; flowers solitary, pink or nearly white; fruit a drupe, about 3-6 cm. long, pubescent with tough flesh splitting at maturity, exposing the pitted stones; endocarp thin or thick; seed flattened, long, oval with a brownish seed coat.

The almond requires a cold and dry climate, but a fairly warm weather during its ripening period. In India, the plants are propagated by means of seedlings. The trees are planted 6-8 m. apart.

The almond crop is harvested from July to September. When the fruits ripen the husks or flesh splits open, exposing the stone. The fruits are harvested at this stage before the stones begin to fall. The harvested fruits are cleaned of the dried skin. After being cleaned the almonds are dried in the sun for a short time.

Uses. Almonds are used to garnish the sweet meats and other food stuffs. Almond kernels are eaten fresh or as dessert. The kernels are used in confectionery and for preparation of almond milk. Almond oil extracted from almonds is used in confectionery and also for medicinal preparations. Almond kernels are highly nutritious, demulcent and stimulant nervine tonic in indigenous medicine. The seeds are made into a paste to be used in making bread or cake.

INGREDIENTS USED AS SOURING AGENTS

Bilimbi

Averrhoa bilimbi Linn.; English—Bilimbi; Hindi—*Bilimbi*; Bengali—*Bilimbi*; Gujarati—*Bilimbu*; Marathi—*Bilambi*; Tamil—*Pulich-chakkay*, *bilimbikay*; Telugu—*Bili-bili-kayalu*; Malayalam—*Vilimbi*, *karichakka*; Family—Oxalidaceae.

A tree, native of S.E. Asia but cultivated throughout India. It flowers in the beginning of summer and ripens its fruit in about two or three months after. The fruit is cylindrical, about two inches long, and pulpy.

Souring agent. The fruit is sour when green, but loses some of its acidity when ripe. The fruit is generally used in pickle and in curry. The flowers are made into preserves.

Carambola

Averrhoa carambola Linn.; English—Carambola; Hindi—*Kamrakh*, *karmal*; Bengali—*kamranga*, *kamarak*; Assamese—*Kardai*; Gujarati—*Tamarak*; Marathi—*Kamarakha*; Tamil—*Tamarta*; Telugu—*Karomonga*; Malayalam—*Tamarat-tuka*; Kannada—*Kamarak*; Sanskrit—*Kamaranga*; Family—Oxalidaceae. A small tree, 15 to 20 feet in height; cultivated throughout India.

Souring agent. *Kamranga* is found in two forms in Bengal - the sweetish acid and the extremely sour. The former is cooling and useful in feverishness. It is highly cooling. The ripe fruit has a pleasant acid and sweetish taste, and is used for culinary purposes. The fruit is also used for making pickles. The fruit is used in curries. The flesh of the fruit is soft, juicy and refreshing. It is sometimes stewed in syrup with a little cinnamon, and is then very pleasant. It is also made into a jelly.

Karanda

Carissa carandas Linn.; English—Karanda; Hindi—*Karaunda*; Bengali—*Kurumia*; Marathi—*Karavanda*; Gujarati—*Karamarda*; Uriya—*Kendakeri*; Sanskrit—*Avinga*; Tamil—*Kalaka*; Telugu—*Kalivi kaya*; Kannada—*Karekai*; Family—Apocynaceae.

A spiny shrub found throughout India. It flowers from February to April, and produces a small fruit which is grape-green when young, white and pink when approaching maturity, and nearly black when ripe. The fruit is ripe in July to August.

Souring agent. The fruit is made into pickle just before it is ripe, and is also employed in tarts and puddings; for these purposes it is superior to any other fruit. The unripe fruits are used as souring agent for making chutney and curries. When ripe it makes a very good jelly.

Sour Lime

Citrus aurantiifolia (Christm.) Swing.; Syn. *Limonia aurantifolia* Christm.; *Limonia acidisima* Houtt.; *C. lima* Lam.; *C. acida* Roxb.; *C. medica* var. *acida* Hook. f.; *C. hystrix* subsp. *acida* Engl.; English—Sour lime; Hindi—*Nimbu*, *kaghzi-nimbu*; Bengali—*Kaghzi nimbu*, *nimbu*; Punjabi—*Nimbu*; Gujarati—*Khata limbu*; Tamil—*Elemitchum*; Telugu—*Nimmapandu*; Kannada—*Nimbe-hannu*; Malayalam—*Cheru-naranna*; Sanskrit—*Jambira*, *limpaka*, *nimbuka*; Family—Rutaceae.

A shrub or small tree, cultivated throughout India, especially in Andhra Pradesh, Maharashtra, Karnataka, Assam, Bihar, Uttar Pradesh, the Punjab, West Bengal, Madhya Pradesh and Rajasthan. The sour lime of India has small flowers, fruit usually small, globose or ovoid, with a thick or thin rind, pulp pale, sharply acid.

Souring agent. The fruits are cut vertically into two pieces, and the fresh juice, squeezed out with the fingers, is sprinkled on soup, *dal*, curry, to which it imparts a pleesent acid taste and agreeable flavour. A pickle of *nimbu* in its own juice and salt is a popular and effectual medicine for indigestion. The fruits are first rubbed over a stone, or their rind scraped to little so as to thin it. They are then steeped in juice obtained from other fruits of the sort, and exposed to the sun for a few days with the addition of common salt. When crisp and of a brown colour, they are preserved in porcelain vessels or glass jars. This preparation is called *zarak-nimbu* (digestive lemon). The small sour limes are used for sherbets and making lime-juice, and the larger ones for sherbets and for preserves, especially chips.

Elephant Apple

Dillenia indica Linn.; English—Elephant apple; Hindi—*Chalta*; Bengali—*Chalta*; Gujarati—*Karambel, karmal*; Telugu—*Peddakalinga, uva*; Kannada—*Betta kanigala*; Malayalam—*Chalita, punna*; Uriya—*Uva*; Family—Dilleniaceae.

A large evergreen tree, found in Bengal, Madhya Pradesh and South India. It also occurs along the base of the hills from Kumaon and Garhwal eastward, and becomes plentiful from South Kanara southwards. It is commonly seen along banks of forest streams.

Souring agent. The fruit is large, about 3 inches in diameter, and is surrounded by fleshy accrescent calyces (sepals), which are sour, used as flavouring in curries or made into jams and jellies. The acid juice sweetened with sugar is used as a cooling drink. The sepals contain (on dry wt. basis): tannin 0.37%; glucose 2.92% and malic acid 0.51%.

Wood Apple

Feronia limonia (Linn.); Swingle; Syn. *F. elephantum* Correa; English—Elephant apple, wood apple; Hindi—*Kaith, kat-bel, kavitha*; Bengali—*Kat-bel, kait*; Uriya—*Koeta*; Punjabi—*Kait*; Gujarati—*Kotha, kavit*; Tamil—*Vilam, vallanga*; Telugu—*Velaga, elaka*; Kannada—*Biwar, byala*; Malayalam—*Vilam*; Family—Rutaceae.

A medium-sized tree, 30-40 feet high and 2-4 feet in girth bearing thorny branches; leaves pinnate, with small ovate or obovate leaflets; fruit large, globose, with hard rough, woody pericarp; seeds numerous, embedded in a sweetish pulp. The plant is a native of India and Sri Lanka and is found in the Deccan, Maharashtra and Madhya Pradesh. To a considerable extent cultivated as a road-side tree near villages. It flowers in February to May, and the fruits ripen about October, and often remain a considerable time on the trees.

Souring agent. The ripe fruit, made into a sort of *Chatni*, with oil, spices, and salt, is esteemed by the natives. The natives sometimes eat the raw fruit with sugar. A jelly is prepared from the pulp of the fruit. The fruit is also used as a condiment. Wood apple is rich in mineral constituents, especially calcium and phosphorus. The acid content of the pulp varies from 7.6% in unripe fruit to 2.3% in fully ripe ones. It contains 3-5% pectin and form an excellent material for making jelly. Wood apple jelly resembles black currant jelly; it is clear of bright purple colour with agreeable odour.

Red Sorrel

Hibiscus sabdariffa Linn.; English—Roselle, Jamaica sorrel, Red sorrel; Hindi—*Lal-ambari, patwa*; Bengali—*Lal mista, patwa*; Assamese—*Chukiar*; Marathi—*Lal-ambadi, patwa*; Telugu—*Yerra gogu*; Tamil—*Pulichchai kerai, gogu*; Kannada—*Pulachakiri*; Malayalam—*Polechi*; Family—Malvaceae.

An annual erect shrub possessing red or green stem, unbranched or branched near the base; flowers large, yellow with dark crimson eye; epicalyx united at the base and adnate to calyx; calyx dark purple, fleshy. It is native of tropical Africa or Asia. In India, the edible types are found in the Punjab, Uttar Pradesh, Bihar, Bengal, Assam, Orissa, Maharashtra, Mysore, Andhra Pradesh and Tamil Nadu.

Souring agent. The edible fleshy calyces mature readily and are ready for picking within 15-20 days after blossoming. They are gathered when tender, plump, fleshy, crisp and deep red in colour. The yield of calyces averages to 3 lb. per plant.

Tender leaves and stalks are eaten as salad or used for seasoning curries. They are acid to taste and are used for the preparation of jelly, flavouring extracts, syrup and wine.

The calyces are rich in acid (3.74%, citric acid) and pectin (3.19%) and are used for making jellies, chutneys and preserves. They make a jelly of good quality with a bright red colour. They are also employed in making soups, puddings and cakes and for colouring syrups and liqueurs.

The calyces may also be dried and stored in air-tight containers. The principal water-soluble acids present in the calyces are citric acid and malic acid; tartaric acid and hibiscus acid are present.

An infusion of the calyces is used for preparing a beverage of agreeable acid taste. It is cooling and refreshing.

Kokam butter

Garicinia indica (Thou.) Choisy; English—Kokam butter, mangosteen oil, brindonia tallow; Hindi—*Kokam*; Marathi—*Bhiranal, chirand, kokam, ratambi*; Gujarati—*Kokam*; Tamil—*Murgal mora*; Kannada—*Murgala*; Goa—*Brindao*; Family—Guttiferae.

A slender tree with drooping branches. The tree is found in the tropical rain forests of Western Ghats of Konkan and Kanara, most commonly in Southern Konkan. It bears a conspicuous spherical purple fruit, the size of a small orange, which ripens about April.

Souring agent. The fruit posseses an agreeable flavour and a sweetish acid taste. By the natives, the fruit is chiefly employed in the form of a preparation called *kokam*, which is prepared as follows - when the fruits begin to ripen completely; after this it becomes soft and pulpy, the outer skin is removed and dried in the sun. The seeds and pulpy substance are then put in bamboo basket, which is kept in a boat-shaped wooden trough. The juice is allowed to trickle down the basket for some time into the trough, and when it ceases to trickle, the seed and pulp are stirred and pressed by the hand, and the whole juice is drained off into the trough. The pieces of the outer skin as they are dried are dipped into the juice, and again dried in the sun. In this way they receive three or four coatings of juice. The pieces of rind are now ready for use and are stored in bamboo baskets. Sometimes a little salt is added to the juice. *Kokam* contains 10 per cent malic acid and a little tartaric or citric acid. It is used as a garnish to give an acid flavour to curries and also for preparing cooling syrups during hot months.

Mango

Mangifera indica Linn.; English—Mango; Hindi—*Aam, amb*, unripe fruit = *amchur*; Bengali—*Am, ambra*; Assamese—*Am, ghariam*; Uriya—*Am*; Punjabi—*Am, amb, mawashi*; Marathi—*Amba*; Gujarati—*Ambo*; Tamil—*Maa, mangas, mam-marum*; Telugu—*Elamavi, mamidi, mavi*; Kannada—*Mavina, amba*; Malayalam—*Mava, mampalam*; Sanskrit—*Amra, chutu, madha-dut*; Family—Anacardiaceae.

A large glabrous tree, with a heavy dome-shaped crown; bark thick, rough dark grey; leaves linear-oblong or elliptic-lanceolate; inflorescence a large panicle; flowers small, reddish white or yellowish green, pungently odorous; staminate and hermaphrodite flowers borne in the same panicle; fruit a large drupe; fruit skin (epicarp) thick or leathery; flesh (mesocarp) whitish yellow, yellow or orange; seed solitary, encased in endocarp. Found throughout tropical India, especially in Bihar, Bengal, Orissa, Assam, Andhra Pradesh, Uttar Pradesh, Madhya Pradesh and Tamil Nadu.

Amchur, pickle (*achar*) and chutney. When unripe, the stone is extracted, the fruit cut into halves or slices and (a) put into curries; (b) made into a pickle (*achar*), with salt, mustard oil, chillies and other ingredients; (c) made into preserves (*morabba*), *chutneys* and jellies by being boiled and cooked in syrup; (d) boiled, strained, and with milk and sugar made into a custard known as, mango-food; (e) dried and made into the native *amchur*, used for adding acidity to certain curries.

The common method of preparing *amchur* is as follows: The unripe peeled mangoes are cut into slices and dried in the sun; slices may be seasoned with turmeric powder before drying. Known as *amchur* the dried material is used as such or after grinding into powder. The powder can be kept well for about 3 years if packed in air-tight containers. *Amchur* is used as a souring agent for curries, soups, chutneys and vegetables. The *amchur* contains - moisture 14.7%; tartaric acid 15.2%; glucose 3.0%; and ash 5.4%.

Tamarind

Tamarindus indica Linn.; English—Tamarind, tamarindo; Hindi—*Imli*, *amli*; Bengali—*Tintiri*, *nuli*, *ambli*; Assemese—*Teteli*; Nepali—*Titri*; Uriya—*Tentuli*; Punjabi—*Imli*; Gujarati—*Amli*; Tamil—*Puli*, *pulia*; Telugu—*Chinta*; Kannada—*Karangi*; Malayalam—*Puli*, *balam*; Sanskrit—*Tintiri*, *amlika*; Family—Caesalpiniaceae.

A large evergreen tree, reaching to a height of 80 feet, with a circumference of 25 feet. It is grown throughout India. It is frequently grown in avenues. In Uttar Pradesh, Madhya Pradesh and many parts of South India, it is found self-sown in waste and forest lands.

Souring agent. The unripe fruits are rich source of tartaric acid. Tartaric acid and its salts are extensively used in various foods, chemical and pharmaceutical industries.

It is used in a variety of ways, usually as the pulp of the ripe pod freed from the rind and the seeds, well boiled, fried, or even raw. In Andhra Pradesh, the unripe fruit ground up with the green, brittle rind, provides the relish to some dishes, and is the inspiration behind that zestful chutney that tends such a tang to boiled rice, *chintakaya-patchadi*. And though it is available only for a few days in the year, the tender leaf-buds of the tamarind, minced fine almost to a powdery consistency, are cooked with spices and chopped onion and *dhal* into a thick *sambar*. The pulp is also employed in making a pleasant cooling drink or sherbet, with sugar and water.

Fig. 15.44. *Tamarindus indica* (tamarind). A twig with leaves, flowers and lomentaceous fruits. The fruits make a good souring agent. The unripe fruits are rich source of tartaric acid. The ripe fruits are used in several south Indian dishes.

The seeds and epicarp are more or less removed by hand, and the pulp is then generally mixed with about 10 per cent of salt and trodden into a mass with the naked feet. The best qualities are free from fibre and husk, the worst contain both, as well as a large proportion of seed. Careful house-keepers prepare their own pulp by exposing it for about a week to the sun and dew.

The leaves are employed to make curries and chutneys, especially in times of scarcity, and the flowers, made into a dish called *chingar*, are also eaten.

16

Tannins and Dyes

TANNINS

Tannins are a rather heterogeneous group of complex compounds of widespread occurrence in plants. They are organic compounds, chiefly glucosidal in nature, which have an acid reaction and are very astringent. They are also known as tannic acid which is a complex glucoside of gallic acid. It is a colourless solid, dissolves in water to solutions of astringent taste. They are of importance commercially because of their property of forming an insoluble colloidal compound (leather) with the hides of animals. They also react with the salts of iron to form dark-blue or greenish-black compounds, the basis of common inks.

Tannins vary greatly in amount from one species to another. They are sometimes present in the cell sap but are of more frequent occurrence in the cell walls, often accumulating in very considerable amounts in dead tissues. Tannins are found in leaves of many plants; in the xylem, the phloem and the periderm of stems and roots; in unripe fruits; in the testa of seeds; and in pathological growths like galls. The woody tissues of many species contain tannins.

Tannins may be present in individual cells or in special containers known as tannin sacs. In the individual cells the tannins occur in the protoplast and may also impregnate the walls as in cork tissue. The tannins are the substances which protect the protoplast against dessication, decay or injury by animals.

The tannins are obtained from the different parts of the plants. A list of important tannin yielding plants is given here.

TANNINS OBTAINED FROM BARKS

Silver Wattle

Acacia decurrens Willd. var. *dealbata* F. Muell. Syn. *A. dealbata* Link; Eng. Silver wattle; Family—Mimosaceae. A tree, spreading rapidly by root sucker. Introduced from Australia, now grown in the Nilgiris. The bark is used for tanning. The bark of this variety is much thinner and greatly inferior to the Black Wattle in quality, yielding only about half the quantity of tanning principle. It is mainly employed for lighter leather.

Golden Wattle

Acacia pycnantha Benth.; Eng.—Golden wattle, Green wattle; Family—Mimosaceae. It is native of Australia, now grown in the Nilgiris. A tree, attaining a maximum height of about 30 feet.

Tan. The bark is used as a tan. The quality of tan is even sometimes superior to that of the Black wattle, but its yield is less, as the tree is smaller and bark thinner.

Kokko

Alibizia lebbeck (Linn.) Benth.; Eng.—Kokko, Lebbeck tree; Hindi—*Siris*, *siras*; Beng.—*Sirisha*; Sans.—*Pit shirish*; Tam.—*Kot vaghe*; Tel.—*Darshana*; Kan.—*Kal baghi*; Guj.—*Pilo sarshio*; Family—Mimosaceae. A common road side tree in most parts of India. Heads of flowers not panicled, 3-4 together; pod strap-shaped, yellow brown 6-12 in. long by 1-1½ in. broad, 6-10 seeded.

Tan. The bark is used in tanning leather.

White Mangrove

Avicennia officinalis Linn.; Eng.—White mangrove; Hindi—*Bina*; Beng.—*Bina*; Tel.—*Mada*, *nalla-mada*; Mal.—*Oepata*; Family—Verbenaceae. A tree of the Sundarbans, the Coromondel

Coast and the Andaman Islands. Leaves usually lanceolate and indistinctly white-tomentose beneath; flowers shortly spiked; style very short.

Tan. The bark is used as a tanning agent.

Indian Oak

Barringtonia acutangula (Linn.) Gaertn.; Eng.—Indian Oak; Hindi—*Hijjal*, *Samundar phal*; Beng.—*Hijal samundar*; Uriya—*Kinjolo*; Ass.— *Hendol*; Tel.—*Kanapa-chettu*; Kan.—*Hole kauva*; Konkan—*Nivar*; Family—Lecythidaceae. A moderate-sized, evergreen tree, found in the sub-Himalayan tract from the Jamuna eastward in Uttar Pradesh, West Bengal, Madhya Pradesh and South India. Leaves alternate, often crowded near the apices of the branches; flowers in terminal or lateral racemes.

Tan. The bark is used for tanning.

Malabar Mountain Ebony

Bauhinia malabarica Roxb.; Eng.—Malabar mountain ebony; Hindi—*Amli*; Beng.—*Karmai*; Uriya—*Gourubati*; Ass.—*Kattra*; Kan.—*Cheppura*; Tel.—*Pulla dondur*; Family—Caesalpiniaceae. A moderate-sized, bushy, deciduous tree found in the sub-Himalayan tract. It is a common tree of South India, Assam and West Bengal. Leaves 7-9 nerved, slightly cordate, deeply bifid; flowers in short, mostly simple corymbs; bracts minute; calyx-limb 5-cleft, style produced.

Tan. The bark is used as a tan.

Pink Bauhinia

Bauhinia purpurea Linn.; Eng.—Pink Bauhinia, Camel's foot tree; Hindi—*Kaliar*, *lal Kachnar*, *kandan*; Beng.—*Deva kanchan*, *rakta kanchan*; Mal.—*Kundrow*; Mar.—*Deva kanchana*; Tam.—*Mandarvi*; Tel.—*Bodanta-chettu*; Kan.—*Kanchivala*; Sans.—*Kanchan*; Family—Caesalpiniaecae. A moderate-sized deciduous tree. Found in Northern India, Assam, the Khasia hills and the Western Peninsula. Leaves 9-11 nerved, pubescent grey, pedicels short; sepals not fully distinct, exceeding the turbinate tube; petals oblanceolate, glabrous, exserted.

Tan. The bark is used for extraction of tannin.

Mountain Ebony

Bauhinia variegata Linn.; Eng.—Mountain ebony, Variegated Bauhinia; Hindi—*Kachnar*, *khairwal*; Konkan—*Kanchan*; Tam.—*Segapu-munthari*; Kan.—*Kanchivala-do*; Uriya—*Borara*; Family—Caesalpiniaceae. A moderate-sized deciduous tree. Found in the Punjab, the Western Peninsula and Assam. Leaves 9-11 nerved, pubescence grey, petioles short; flower white having red and yellow stripes; calyx-limb entire, spathaceous, equalling the cylindrical tube; petals glabrous-obovate, clawed, much exserted.

Tan. The bark is used in dyeing and tanning.

Ringworm Cassia

Cassia alata Linn.; Eng.—Ringworm Cassia; Hindi—*Dadmurdan*, *dat-ka-pat*; Beng.—*Dadmardan*, *dadmari*; Mar.—*Dadamardana*; Sans.—*Dadrughna*; Tam.—*Shimai-agati*, *vandukolli*; Tel.—*Sima avisl*; Mal.—*Shima akatti*; Kan.—*Shime-agase*; Family—Caesalpiniaceae. A small shrub, with thick, finely downy branches, found in West Bengal and Tamil Nadu. It is introduced into India from West Indies.

Tan. The bark is used for tanning.

Indian Laburnum

Cassia fistula Linn.; Eng.—Indian laburnum, Golden shower; Hindi—*Amaltas*; Punj.—*Alash*; Beng.—*Sundali*; Mal.—*Sonawir*; Uriya—*Sunari*; Guj.—*Garmal*; Tam.—*Kone*; Tel.—*Reylu*; Kan.—*Kakee*; Family—Caesalpiniaceae. A moderate-sized, deciduous tree of the sub–Himalayan tracts, and common throughout India ascending to 3,000 feet in altitude. It occurs as a small spreading tree, leafless in March, the long pendulous racemes of bright yellow flowers and fresh green leaves appearing together in April. The long, brown, pendulous, sausage like pods, 1-1½ feet in length, ripen in the cold season.

Tan. The bark is used in tanning, chiefly along with *Terminalia*. Skins after being treated with lime and cleaned, are soaked in the stringent solution prepared by pounding the bark of *Cassia fistula*, bark of *Terminalia tomentosa* and pods of *Caesalpinia digyna*, and soaking in water for 24 hours. The process of soaking is repeated three times.

Casuarina

Casuarina equisetifolia Linn.; Eng.—Casuarina, Sheoak; Hindi—*Jangli jhau*, *jangali saru*, Beng.—*Jau*; Tam.—*Chouk*, *shavuku-pattay*; Tel.—*Serva*, *chavuku-patta*; Kan.—*Sura*; Family—Casuarinaceae. A tall evergreen tree, with leafless drooping branches and branchlets, which are deciduous, and perform the function of leaves. Found on the coastal regions of both the Peninsulas. Cultivated all over India as an avenue tree. It thrives best in the sandy tracts near the sea shore.

Tan. The bark is used in tanning.

Ghoran

Ceriops roxburghiana Arn.; Hindi—*Bara goran*; Beng.—*Ghoran*; Family—Rhizophoraceae. A shrub or small tree, found in the Sundarbans. It grows abundantly on the sea shore. The bark is used in tanning leather.

Goran

Ceriops tagal (Perr.) C.B. Robins; Syn. *C. candolleana* Arn.; Hindi—*Goran*; Beng.—*Goran*; Family—Rhizophoraceae. A small, evergreen tree, found on the muddy shores and tidal creeks of the Sundarbans and the Andaman Islands.

Tan. The bark is used for tanning. It is an exceedingly valuable tan, imparting a good red colour to leather. Sole leather tanned with this is reported to be more durable than any other.

Ghogar

Garuga pinnata Roxb.; Hindi—*Ghogar*; Beng.—*Nilbhadi*; Uriya—*Mohi*; Assam—*Gendeli poma*; Punj.—*Sarota*; Mar.—*Kudak*; Guj.—*Kusimb*; Tam.—*Karvambu*; Tel.—*Garga*; Kan.—*Hala*; Family—Burseraceae. A tree, 30-40 feet high, found in the sub-Himalayan forest upto 3,000 feet altitude, also in Madhya Pradesh and South India. It flowers from February to March, and the fruit ripens in June and July.

Tan. The bark is used for tanning in many parts of the country, and is said to be good for this purpose.

Bhaulan

Hymenodictyon excelsum Wall. Hindi—*Bhaulan*, *bhamina*, *phargur*; Uriya—*Kono*; Punj.—*Manabina*; Tam.—*Sagapu*; Tel.—*Bandara*; Kan.—*Bandarayanni*; Family—Rubiaceae. A large deciduous tree, 30-40 feet high, found on the dry hills at the base of the Western Himalaya sfrom Garhwal to Nepal. Also found in Bihar, Madhya Pradesh and the Western Peninsula.

Tan. The bark possesses powerful astringent properties and is used for tanning.

Box Myrtle

Myrica nagi Thunb.; Eng.—Box myrtle, Bay-berry; Hindi—*Kaiphal*; Beng.—*Satsarila, kaiphal*; Punj.—*Kaphal*; Mar.—*Kaya-phala*; Guj.—*Kariphal*; Tam.—*Marudampattai*; Tel.—*Kaidaryamn*; Mal.—*Marutamtoli*; Sans.—*Katphala, kaidaryama*; Family—Myricaceae. An ever-green dioecious tree with an aromatic odour. It is native of China and Japan. In India, it is found in the subtropical Himalayas, the Khasia hills and Assam from 3,000 to 6,000 feet. Bark grey, rough; leaves lanceolate, oblong-obovate; flowers minute, unisexual, in axillary spikes, fruit an ellipsoid or ovoid drupe.

Tan. The bark is used as a tanning agent for fancy leather work. The tannins of the bark belong to the pyrogallol group and 73% of thc total tannins can be extracted with water. Hides tanned with box myrtle bark tend to be cracky; they are somewhat darker in colour than those tanned with wattle bark.

Jhingan

Odina wodier Roxb.; Hindi—*Jingan, jhingan, kashmala*; Beng.—*Ghadi, jiyal, Lohar*; Oriya—*Indramai*; Punj.—*Kamlai*, Mar.—*Shimti, munidi*; Tam.—*Otiyam*; Tel.—*Gumpena chettu*; Kan.—*Shimli, shimti*; Sans.—*Jingini*; Family—Anacardiaceae. A large deciduous tree, 40 to 50 feet high found in the hotter parts of India. It ascends in the sub-Himalayan tract upto 4,000 feet altitude. Also found in Assam, Tamil Nadu and the Andaman Islands.

Tan. The bark is used for tanning and dyeing. The bark contains 8% phlobatannin.

Myrobalam Emblic

Emblica officinalis Gaertn.; Syn. *Phyllanthus emblica* Linn.; Eng.—Myrobalam Emblic; Hindi—*Amla, Aonla, Aura*; Beng.—*Amla, Amlaki*; Uriya—*Amlaki*; Assam—*Amluki*; Punj.—*Ambal*; Mar.—*Avala*; Guj.—*Amla*; Tam.—*Nelli*; Tel.—*Usirika, Nelli*; Kan.—*Nilika*; Mal.—*Boa-malacca*; Sans.—*Amulki, Amalaki, Amala*; Family—Euphorbiaceae. A moderate-sized deciduous tree, found throughout tropical India. The tree possesses smooth, greenish grey, exfoliating bark; leaves feathery with small, pinnately arranged leaflets; fruits depressed globose ½ to 1 in. in diameter, fleshy and obscurely 6-lobed.

Tan. The bark is used for tanning. The bark is rich in tannin. Twig bark contains 21% and stem bark 8-9% of tannin.

Pomegranate

Punica granatum Linn.; Eng.—Pomegranate; Hindi—*Anar, Dhalim*; Beng.—*Dalim gach, Anar*; Uriya—*Dalim*; *Dalimba*; Assam—*Dalim*; Punj.—*Daru, Daruni*; Mar.—*Dalima-jhada*; Guj.—*Dadam-nu-jhada*; Tam.—*Madalai*; Tel.—*Danimma*, Kan.—*Dalimbe-gida*; Mal.—*Matalam*; Sans.—*Dadima-vrikshaha*; Family—Punicaceae. A shrub, or small tree. It is native of Iran but now cultivated throughout India. It is found growing wild in the warm valleys and outer hills of the Himalayas between 900 and 1,800 m. Bark smooth, dark grey; branchlets sometimes spiniscent; fruits globose, crowned by persistent calyx with a coriaceous woody rind and an interior septate with numerous walls, containing numerous seeds; angular with a juicy testa which is red, pink or whitish.

Tan. Tannin occurs in all parts of tree, particularly in fruit rind (up to 26% in dried rind), stem bark (10-25%), root bark (28%) and leaves (11%). The fruit and the bark of the stem and root have been widely used as tanning materials from early times in the Mediterranean countries. The stem bark was once much used for production of Morocco leather.

White Oak

Quercus incana Roxb. Eng.—White Oak; Hindi—*Banj, ban*; Kashmir—*Sila supari*; Punj.—*Ban, banj, maru*; Family—Fagaceae. It is a large, evergreen tree found on the temperate Himalayas between 3,000 and 8,000 feet altitude. It is associated with the Deodar, *Rhododendron* and the long-leaved pine. In spring it becomes of a purplish colour, owing to the appearance of a fresh burst of softly tomentose leaves.

Tan. The bark is extensively employed for tanining purposes.

Mangrove

Rhizophora mucronata Lamk.; Eng.—*Mangrove*; Beng.—*Kamo, bhora, bhara*; Uriya—*Rai*; Tel.—*Upu-poma, adavi ponna*; Andaman—*Bairada, jumuda*; Family—Rhizophoraceae. A small evergreen tree, found in the Sundarbans, along the Coromandel Coast and the Andaman Islands.

Tan. The bark is used for tanning.

Elm-leaved Sumach

Rhus cotinus Linn.; Syn. *Cotinus coggygria* Scop.; Eng.—Elm-leaved sumach; Hindi—*Tunga, tung*; Punj.—*Tittri, tung*; Kashmir—*Darengri*; Family—Anacardiaceae. A shrub or small tree of the Western Sub-tropical Himalayas, from Marri to Kumaon, found upto altitudes of 5,000 feet.

Tan. The bark and leaves are much used locally by native tanners in various parts of India. Stem bark is used as a tan.

Sal

Shorea robusta Gaertn. f.; Eng.—Sal; Hindi—*Sal, Sala, Sakhua*; Beng.—*Sal, Shal*; Uriya—*Salwa*; Punj.—*Sal, Seral*; Tel.—*Gugal*; Kan.—*Kabbu*; Family—Dipterocarpaceae. A large robust tree, found in the Punjab, Uttar Pradesh, Bihar, Bengal, Assam, Madhya Pradesh, the Coromandel Coast and Andhra Pradesh.

Tan. The bark is used as a tan. The bark extract contains 7-12% of tannin. The spray-dried aqueous extract of the bark contains 39.6% of tannins. The extract is of a pale reddish colour and the tannins are of pyrogallol type. The extract may be used locally for cheap tanning or in blend with other tanning materials. After bleaching it could be used in tanneries.

Indian Redwood

Soymida febrifuga A. Juss.; Eng.—Indian redwood; Hindi—*Rohun, rohunna, rakat rohan*; Beng.—*Rohan, rohina*; Guj.—*Rohina*; Tel.—*Sumi, sonida manu*; Tam.—*Shem, wond*; Kan.—*Suani*; Uriya—*Karwi, sohan*; Family—Meliaceae. A lotfy deciduous tree, found in the dry forests of peninsular India right upto Kerala and also occurs in Gujarat, Uttar Pradesh, Bihar and Madhya Pradesh. Bark very tough, exfoliating in large plates or scales. Leaves crowded at the ends of the branches, paripinnate; flowers greenish white in large terminal panicles; capsule black, woody with winged seeds.

Tan. The bark is a rich source of tanning material, the tannin content being 17-41%, on dry weight basis. The tannins belong to the catechol group of tannins. Trials at Madras, have shown that the higher proportion of yellow colour in the bark enables it to produce light coloured leathers which are supposed to be superior than the dark coloured ones. The leathers tanned by the bark of Indian redwood possess good colour, good fell and fullness. Indian redwood bark is said to be a potential source of an indigenous tanning material, which can be substituted for imported wattle bark, for producing high grade leathers.

Indian Jujube

Zizyphus jujuba Lamk.; Eng.—Indian jujube; Hindi—*Ber, beri*; beng.—*Kul, ber, bor*; Uriya—*Bar koli*; Punj.—*Ber, unab*; Raj.—*Ber*; Mar.—*Bora, bera*; Guj.—*Bor, bordi*; Tam.—*Elanda, elandap*; Tel.--*Regu, ganga regu*; Kan.—*Yelchi*; Mal.—*Elenta*; Sans.—*Baderi, kola, badras*; Family—Rhamnaceae. A large shrub or small spiny tree. It is native of China. In India it is found in the Punjab, Uttar Pradesh, Bihar and Rajasthan.

Tan. The bark is used for tanning.

TANNINS OBTAINED FROM WOOD

Cutch

Acacia catechu (L.f.) Willd; Eng.—Catechu, Cutch, Black cutch; Hindi—*Khair, khadira, katha*; Beng.—*Khayer, kuth*; Assam—*Khoira, koir*; Tam.—*Vodalam, vodalai, baga*; Tel.—*Podalamanu, kaviri sandra, nalla sandra*; Sans.—*Khadira*; Uriya—*Khoiru*; Kan.—*Kagli, tare*; Guj.—*Kher*; Family—Mimosaceae. Commonly found in most parts of India, extending in the sub-Himalayan tract westward to the Punjab and eastward to Sikkim ascending to altitude 5,000 feet. Also found in Madhya Pradesh, Andhra Pradesh, Uttar Pradesh and Bihar.

A moderate-sized deciduous tree, with dark brown, much cracked bark, and short hooked spines in pairs. Leaves composed of from 40 to 80 pinnae and 60 to 100 leaflets. Flowers white or pale yellow, peduncled spikes in the axils of leaves; pod straight, strap-shaped, narrow, thin dark brown.

Tan. As a tan catechu does not hold a very high position owing to the colour it imparts to the skin. It is said to contain from 45 to 55 per cent of tannin, or about 10 per cent less than dividivi pod, and 20 per cent less than gall nuts.

Mainly, the cutch obtained from the heart wood, is used as a masticatory and as a dyeing and preserving agent. It is extensively used for dyeing canvas, fishing nets and ropes.

Banda

Dendrophthoe falcata (Linn. f.); syn. *Loranthus falcatus* Linn. f.; *L. longiflorus* Desr.; Hindi—*Banda*; Beng.—*Bura manda, bara manda*; Pb.—*Panda, amut, banda*; Guj.—*Vanda*; Tel.—*Yelinga, wadinika, ippa wajna*; Sans.—*Vanda, vrikshadani, vrikshabhaksha, vriksharuha*; Family—Loranthaceae. A large bushy parasite found throughout India. It possesses grey bark, thick, usually opposite leaves, orange red or scarlet flowers and ovoid-oblong berries. It occurs on a great variety of forest and fruit trees, almost throughout India. The seeds are spread for the most part by birds.

Tan. The wood is largely used as a finishing tan-stuff in order to render leather soft. The tender shoots contain 10% tannins. The wood is prepared as follows for the purpose. The leaves are stripped from the shrub, and the sticks, laid together on a hide or mat, are bruised by two or three persons beating them with the *musal*. After the skin has been subjected to the action of various tanning materials, and the process of preservation is complete, it is roughly cleaned, served up into the form of a bag, and hung upon a tripod. The bruised wood is then placed inside and the skin is filled with water. It is allowed to remain in this state for a couple of days, and the process is then completed by rubbing one side or other with *khari* salt.

European Chestnut

Castanea sativa Mill.; Eng.—European chestnut; Family—Fagaceae. A large tree. In India it is grown in Kashmir, Darjeeling, the Khasia hills, the Punjab and Himachal Pradesh. The wood contains 30 to 40% tannin. The tannin is extracted at high temperatures from chips of wood.

TANNINS OBTAINED FROM LEAVES

Gum Ghatti

Anogeissus latifolia Wall.; Eng.—Gum ghatti; Hindi—*Dhaura, dhauri*; Raj.—*Dau, dhaukra*; Tam.—*Vellay naga, namme*; Tel.—*Chirimanu yettama*; Uriya—*Dohu*; Mal.—*Dhau*; Guj.—*Dabria*; Kan.—*Dinduga, dindiga*; Family—Combretaceae. A large tree, found in sub-Himalayan tract from the Ravi eastward, ascending to 3,000 feet in Madhya Pradesh and South India. Found also in Western Peninsula and Nilgiris. Leaves from elliptic-obtuse at both ends, peduncles one or more from the same axil, often branched, bracteoles inconspicuous; ripe fruits shining and glabrous.

Tan. The leaves are used as a tan. The leaves contain as much tannin as those of the *Sumach* tree. The leaves yield a black dye and are very useful in tanning.

Common Custard Apple

Annona reticulata Linn.; Eng.—Bullock's heart, Common custard apple; Hindi—*Ramphal, louna*; Beng.—*Nona*; Bomb.—*Ramphal*; Guj.—*Ramphal*; Kan.—*Ramphal*; Tam.—*Ramsita*; Tel.—*Rama-pandu*; Family—Annonaceae. A small tree. Native of West Indies. In India, it is grown in Bengal and South India.

Tan. The leaves and young twigs are largely used for tanning.

Guava

Psidium guajava Linn.; Eng.—Guava; Hindi—*Amrud*; Beng.—*Piyara, goaacchi-phal*; Assam—*Madhuriam*; Punj.—*Amrud, anjr zard*; Mar.—*Jamba*; Guj.—*Piyara, jamrud*; Tam.—*Segapu*; Tel.—*Jama, jam-pandu*; Kan.—*Sebe*; Mal.—*Malakkoppera*; Sans.—*Amruta-phalam, bahu-bija-phalam*; Family—Myrtaceae. A small tree. Native of Central America. In India, cultivated chiefly in Uttar Pradesh, the Punjab, Bihar, Maharashtra and Andhra Pradesh for its delicious fruit rich in vitamin C.

Tan. The bark and leaves are used for tanning. The leaves are employed for tanning either alone or along with mango or *mahua* leaves.

Dhaula

Woodfordia fruticosa (Linn.) Kurz; Syn. *W. floribunda* Salisb.; Hindi—*Dawi, thawi, dhaula*; Beng.—*Dhaura, dhainti, dawai*; Uriya—*Jatiko, harwari*; Kashmiri—*Thawi*; Guj.—*Dhavadina*; Tel.—*Serinji godari*; Sans.—*Dhataki, agnijuala*; Family—Lythraceae. A small much branched shrub, brilliantly purple in the hot season owing to the numerous flowers all along its branches; commonly found throughout India, ascending on the Himalayas to an altitude of 5,000 feet.

Tan. The leaves are one of the most commonly employed of Indian tanning materials. The flowers also contain a large quantity of tannin, sometimes used for tanning. The bark is also employed for this purpose.

Smoke Tree

Cotinus coggygria Scop.; Syn. *Rhus cotinus* Linn. Eng.—Smoke-tree; Hindi—*Tungla*; Family—Anacardiaceae. A shrub of the North-Western Himalayas. The leaves are used for tanning.

Titri

Rhus punjabensis J.L. Stew. ex-Brand; Hindi—*Titri*; Family—Anacardiaceae. A tree, found in the North-Western Himalayas from Kashmir to Kumaon. The leaves are used for tanning.

TANNINS OBTAINED FROM FRUITS

American Sumach

Caesalpinia coriaria (Jacq.) Willd.; Eng.—Divi-Divi, American sumach; Hindi—*Libi-dibi*; Maha.—*Amriki sumach*; Tam.—*Sumak*; Kan.—*Vilayati-aldekayi*; Family—Caesalpiniaceae. A small tree. It is native of South America and West Indies. It is grown mainly in South India; also cultivated in North-Western India, West Bengal and Kanara in Maharashtra.

Tan. The sinuous pods of this plant are used in tanning. The tanning material is obtained from the dried twisted seed pods.

Vakeri-mul

Caesalpinia digyna Rottler; Hindi—*Vakeri-mul*; Beng.—*Umul-kuchi*; Tel.—*Nuni gatcha*; Maharashtra—*Vakeri-chebhate*, *vakeri-mula*; Family—Caesalpiniaceae. A prickly shrub, found in the Eastern Himalayas, East and Western Peninsulas, Bengal, Assam and Andaman Islands.

Tan. The fruits, called tari, are used for tanning.

Coromandel Ebony

Diospyros melanoxylon Roxb.; Eng.—Coromoandel ebony; Hindi—*Tendu*, *kendu*, *abnus*; Beng.—*Kend*, *kiu*; Uriya—*Kendhu*; Mal.—*Kend*; Madhya Pradesh—*Tendu*; Mar.—*Timburni*; Guj.—*Tamrug*; Tam.—*Tumbi*, *tumbali*; Tel.—*Tumi*, *tumki*, *damadi*; Kan.—*Balai*; Sans.—*Kenduka*; Family—Ebenaceae. A large tree, attaining a height of 60 to 80 feet and a girth up to 7 feet, with a straight cylindrical bole of 15-20 feet. It bears coriaceous leaves varying in size and structure. It is found in the Indian Peninsula, extending northward to Bihar, Madhya Pradesh and Maharashtra, and is one of the most characteristic trees of the dry, mixed, deciduous forests in these regions. The leaves of *D. melanoxylon* are highly esteemed for wrapping bidis.

Tan. The distribution of tannins in the plant is as follows—bark 19; fruit 15; and half ripe fruit 23%. The bark and half ripe fruits are used for tanning.

Indian Gooseberry

Emblica officinalis Gaertn.; Syn. *Phyllanthus emblica* Linn.; Eng.—Emblic myrobalan, Indian gooseberry; Hindi—*Amla*, *aonla*, *amlika*; Beng.—*Amla*, *amlaki*; Guj.—*Amali*, *ambala*; Tel.—*Amalakamu*, *usirikai*; Tam.—*Nelli*; Kan.—*Amalaka*, *nelli*; Mal.—*Nelli*; Sans.—*Amalaka*, *adiphala*, *dhatri*; Family—Euphorbiaceae. A medium-sized deciduous tree, found in the mixed deciduous forests of India ascending upto 4,500 feet on the hills. It is often cultivated in gardens and homeyards.

Tan. The fruits, bark and leaves are rich in tannin. The distribution of tannin is as follows—fruit 28; twig bark 21; stem bark 8-9; and leaves 22%. The fruit contains two tannins, one giving on hydrolysis gallic acid, ellagic acid and glucose, and the other giving ellagic acid and glucose. Immature fruits are employed for tanning in combination with other tan stuffs such as myrobalans.

Belliric Myrobalan

Terminalia bellirica (Gaertn.) Roxb.; Eng.—Belliric myrobalan, Bastard myrobalan; Hindi—*Bahera, vibhitaka, behra, sagona*; Beng.—*Boyra, bohora*; Assam—*Hulluch, bauri*; Uriya—*Bahadha*; Punj.—*Behera, balela, baheri*; Mar.—*Berda, hela*; Guj.—*Behedan, beheda*; Tam.—*Tanikoi, vallai murudu*; Tel.—*Bahadha, thandi*; Kan.—*Yehela, bherda, tari*; Mal.—*Thani*; Sans.—*Vibhitaki, vibhitaka, akasha, bahira*; Family—Combretaceae. A large deciduous tree, commonly found in the plains and lower hills throughout India. Two kinds of fruits are met with, one nearly globular ½ to ¾ in. in diameter, the other ovate and much larger. Both narrow suddenly into a short stalk, are fleshy and closely covered with a fulvous tomentum, and when dried are obscurely five-angled. The stone is hard and pentagonal, and contains a sweet oily kernel, having three prominent ridges from base to apex.

Tan. The fruit is known under the name of myrobalans and is employed for dyeing and tanning. It is largely employed for dyeing, as a mordant, as a tan and also medicinally.

Indian Almond

Terminalia catappa Linn.; Eng.—Indian almond, Tropical almond; Hindi—*Janglibadam*; Beng.—*Bangla badam*; Uriya—*Badam*; Mar.—*Bengali badam*; Tam.—*Nattuvadam-kottai*; Tel.—*Badam vittlu*; Kan.—*Tari*; Sans.—*Ingudi*; Family—Combretaceae. A large, deciduous tree, attaining 80 feet height, and possesses branches in almost horizontal whorls cultivated throughout the hotter parts of India. It is raised easily from seed; and in a good light soil, well watered grows in two years more than 10 feet in height, and blossom.

Tan. The fruits and bark are used for tanning.

Chebulic Myrobalan

Terminalia chebula Retz.; Eng.—Chebulic myrobalan, yellow myrobalan; Hindi—*Harir, Harara, Haritaki*; Beng.—*Haritaki*; Assam—*Hilika*; Uriya—*Haridra*; Punj.—*Har, Harrar*; Mar.—*Hirda*; Guj.—*Pilo harle, hardi*; Tam.—*Kadaki*; Tel.—*Karaka, Karakkaya*; Kan.—*Hirda, alabkayi*; Mal.—*Katukka*; Sans.—*Haritaki*; Family—Combretaceae. A large, deciduous tree, commonly found in North India from Kumaon to West Bengal. It is found also in Maharashtra, Tamil Nadu and above the Ghats in Kanara in South India.

The dried fruit forms the *chebulic* or *black* myrobalan of commerce, one of the most valuable of Indian tanning materials. With iron salts it is employed in making country ink, and mixed with ferruginuous mud it makes a black paste employed by harness and shoe makers as well as by dyers.

Tan. The chief commercial value of chebulic myrobalan is, as a tanning material. It forms greater part of the ground myrobalans, though belleric myrobalans are occasionally mixed with it. The liquor prepared from it is not only a powerful tan, but imparts a bright colour to the leather, and therefore, is highly esteemed to mix with other tanning agents.

Asan

Terminalia tomentosa Wight and Arn.; Hindi—*Asan, saj, sain*; Beng.—*Asan, ashan*; Uriya—*Ansun, saj*; Guj.—*Hadri, sadri*; Punj.—*Sain, asun*; Mar.—*Ain, madat*; Tam.—*Karumaradu*; Tel.—*Maddi, nalla-maddi*; Kan.—*Matti, kari-matti*, Mal.—*Kara-maruta*; Sans.—*Asana*; Family—Combretanceae. A large, deciduous tree, attaining a height of 90 to 100 feet, commonly found throughout the moist parts of India. In the Siwalik tract and outer Himalayan valleys it ascends upto 4,000 feet altitude. Found also in South India. The tree thrives best in heavy binding soil, flowers in April, and the fruits ripen in February to April of the following year.

Tan. The fruit, like those of most species of *Terminalia*, is a myrobalan, but is very much inferior in tanning power to the belleric or chebulic myrobalans. It is, therefore, occasionally employed as a tanning agent.

The nuts of myrobalans (*Terminalia* spp.) have a tannin content of 30 to 40 per cent. They yield a spongy leather of light-yellow colour. They are used with goat, calf, and sheep skins, and for sole and harness leather.

Ber

Zizyphus mauritiana Lamk.; Syn. *Z. jujuba* Lamk.; Eng.—Indian jujube; Hindi—*Ber*; Beng.—*Ber, bor*; Punj.—*Unab, beri*; Raj.—*Ber*; Guj.—*Bor*; Tam.—*Elanda*; Tel.—*Rengha*; Kan.—*Yalachi*; Sans.—*Badari*; Family—Rhamnaceae. A small tree, wild and extensively cultivated throughout India. Chiefly cultivated in the Punjab, Uttar Pradesh, Bihar and Rajasthan.

Tan. The fruit, bark and leaves contain tannin. They are employed for tanning.

Kat-ber

Zizyphus xylocarpa Willd.; Hindi—*Kat-ber, goth-ber*; Uriya—*Got, gotoboro*; Mar.—*Goti*; Tel.—*Gotte*; Kan.—*Challe*; Family—Rhamnaceae. A large, straggling shrub or small tree, found in North-West India, Rajasthan and Uttar Pradesh, ascending the Himalayas to 2,000 feet. Found also in Bihar and Western Peninsula.

Tan. The berry contains a considerable amount of tannin. It is used for tanning in various parts of India.

TANNINS OBTAINED FROM FRUIT RIND AND SAP

Mangosteen

Garcinia mangostana Linn.; Eng.—Mangosteen; Hindi—*Mangustan*; Beng.—*Mangustan*; Mumbai—*Mangustan*; Mar.—*Mangostin*; Mal.—*Manggusta*; Family—Guttiferae. An evergreen tree. It is native of Malayan region now cultivated mainly in the Nilgiris, Maharashtra and Tamil Nadu.

Tan. The rind of the fruit is employed, in combination with the fruit of *Terminalia catappa* Linn., for dyeing black and is also employed for tanning.

Tamal

Garcinia morella Desr.; Eng.—Gamboge tree; Ceylon—gamboge; Hindi—*Tamal, gota ganba*; Beng.—*Tamal*; Kan.—*Arsinagurgi mara, aradal*; Mal.—*Daramba*; Family—Guttiferae. A small evergreen tree, found in the Khasia hills and the Western Ghats from South Kanara and Mysore to Travancore and the Western Peninsula.

Tan. The rind of the fruit is employed as a tan.

Akhrot

Juglans regia Linn.; Eng.—Walnut, Persian Walnut; Hindi—*Akhrot*; Beng.—*Akrut*; Assam—*Kabsing*; Kumaon—*Akhor, Kharot*; Kashmir—*Akhor, crot, dun*; Punj.—*Akhrot, dun, than than*; Mar.—*Akroda*; Guj.—*Akhrot*; Tam.—*Akrottu*; Tel.—*Akrotu*; Kan.—*Akrodu*; Sans.—*Akshota, akhoda*; Family—Juglandaceae. A large, deciduous tree. Native of Iran. In India, it is cultivated in Kashmir, Himachal Pradesh, Khasia hills, Manipur and Kumaon from 3,000 to 10,000 feet elevation. Now the tree is cultivated all over the Temperate Himalayas, especially in Kashmir. It requires a climate with neither extremes of heat nor cold, good soil, and in the young state, careful weeding.

Tan. The bark and rind of the fruit are used in tanning and dyeing.

Banana

Musa paradisiaca Linn.; Syn. *M. sapientum* Linn.; Eng.—Banana; Hindi—*Kela, Kach-kula*; Beng.—*Kala, Kach-kula*; Punj.—*Kela, Khela*; Mumbai—*Kela, Kel*; Guj.—*Kela*; Tam.—*Valei*; Tel.—*Kadali, bonta-ariti*; Kan.—*Bale*; Mal.—*Vasha*; Sans.—*Kadali, rambha*; Family—Musaceae. A perennial herb of 8 to 15 feet in height, extensively cultivated throughout India. It is native of India and Malaya. Chiefly cultivated in Assam, Madhya Pradesh, Bihar, Uttar Pradesh, Andhra Pradesh, Maharashtra, Tamil Nadu and Kerala.

Tan. The rind of the fruit is used as a tan, and for blackening leather. The sap contains a considerable amount of tannin, and stains cloth a dark, almost black colour, which is fairly permanent; and difficult to wash out; and sometimes may be employed as marking ink.

TANNINS OBTAINED FROM BRANCHES

Ephedrine

Ephedra gerardiana Wall.; Ephedrine; Hindi—*Khanda, Phok*; Punjab—*Asmania, budagur, chewa*; Ladakh—*Tse, teapat, trano*; Family—Ephedraceae. A small erect shrub. It bears dark green, cylindrical striated, often curved branches arising in whorls. Fruit ovoid, red, sweet and edible containing one or two seeds, more or less enclosed by succulent bracts. It is found in the temperate Himalayas from Kashmir to Sikkim at altitude of 7,000 to 10,000 feet and is frequently found at Chamba, Lahaul and Spiti, Simla hills, Kashmir and Ladakh. It is abundantly found on the Shalai hill north of Simla at an altitude close on 10,000 feet.

Tan. The branches yield tannin. However, the yield is only 3% of tannin, giving a whitish precipitate with gelatine and with acetate of leaf, and a greenish precipitate with acetate of iron.

Ephedra pachyclada Boiss.; Family—Ephedraceae. A tall shrub found in the drier regions of the Western Himalayas. The branches are employed in tanning the skins of goats for water-bottles (*masak*).

TANNINS OBTAINED FROM ALL PARTS OF PLANT

Masuri

Coriaria nepalensis Wall.; Hindi—*Masuri, makola*; Punjab—*Guch*; Kashmir—*Tadrelu balel*; Simla—*Raselwa, archaru*; Nepal—*Bhojinsi*; Family—Coriariaceae. A shrub, found in the temperate and sub-tropical Himalayas, Sikkim, Bhutan and Manipur. In Simla this common shrub flowers in February and March.

Tan. All parts of the plant are rich in astringent acids which might be used for tanning or for dyeing.

TANNINS OBTAINED FROM GUM

Bengal Kino

Butea monosperma (Lamk.) Taubert; Syn. *B. frondosa* Koenig ex Roxb.; Eng.—Bengal kino, flame of the forest; Hindi—*Dhak, palas, tesu, palasa*; Beng.—*Palas*; Bihar—*Paras, faras*; Nepal—*Palasi*; Uriya—*Porasu*; Guj.—*Khakara, khakhado*; Tam.—*Porasan, parasa, murukkan*; Tel.—*Moduga, mohtu, moduga chettu, kimsakamu*; Kan.—*Mutugga, thoras, muttuga gida*; Mal.—*Plach-cha*; Sans.—*Kinsuka, palasa*; Family—Papilionaceae. A moderate-sized, deciduous tree, found throughout India, extending in the North-West Himalayas as far as the Jhelum.

Tan. It yields naturally, or from artifical scars on the bark, a gum which is sold as "*Bengal kino*" or "*chunia-gond*". This occurs in the form of round tears, as large as a pea, often fragmentary, of an intense ruby colour and astringent taste. This gum may be purified by solution in water. It is translucent but with age it darkens and becomes opaque. The gum is used medicinally. It is also employed in tanning.

Indian Kino

Pterocarpus marsupium Roxb.; Eng.—Indian kino tree, Malabar kino tree; Hindi—*Bijasal, bija*; Beng.—*Pitshal*; Guj.—*Biyo, hiradakhan*; Tel.—*Yegi, peddagi*; Tam.—*Vengai*; Kan.—*Honne*; Mal.—*Venga*; Uriya—*Byasa*; Family—Papilionaceae. A moderate-sized to large deciduous tree, upto 30.0 m high and a girth of 2.5 m. with a straight clean bole. It is found commonly in hilly regions throughout the Deccan Peninsula, and extending to Gujarat, Madhya Pradesh, Uttar Pradesh, Bihar and Orissa. Bark grey, rough, longitudinally fissured and scaly; older trees exuding a blood red gum-resin.

Tan. The tree yields a gum-kino which exudes when an incision is made through the bark upto the cambium. The exudate is collected and dried in the sun or shade. Kino occurs in small, angular, glistening, brittle fragments, appearing almost black in colour. It is odourless and bitter with astringent taste. Kino contains a non-glucosidal tannin Kinotannic acid (25-80%), Kinoin ($C_{28}H_{24}O_{12}$) and Kino-red ($C_{28}H_{22}O_{11}$).

Kino finds application in tanning, dyeing and printing and is of potential use for the paper industry.

TANNINS OBTAINED FROM GALLS

East Indian Mastiche

Pistacia khinjuk Stocks; Syn. *P. integerrima* Stens ex Brand.; Eng.—East Indian mastiche; Hindi—*Kakra*, galls = *Kakrasingi*; Guj.—*Kakra*; Beng.—galls = *Kakra-sringi*; Kashmir—*Kakkar, drek, gurgu*; Punj.—*Kangar, kakar, kakra, tungu*, galls = *Kakrasingi*; Tam.—galls = *Kakkatashingi*; Tel.—galls = *Kakara shingi*; Family—Anacardiaceae.

A shrub or small tree, found in the North-Western Himalayas from the Himachal Pradesh to Kumaon, at altitudes of 1,200 to 8,000 feet. Also found in Kashmir.

Tan. The hard, rugose, hollow, irregular galls, which form in October on the leaves and petioles, are used to a small extent for dyeing and tanning. These galls sometimes attain a very large size, and from their peculiar shape well warrant the name of '*singhi*' or '*horns*'.

Terebinth

Pistacia terebinthus Linn.; Eng.—The terebinth tree; Hindi—galls = *gul-i-pista, kabuli mastaki, buzghanj*; Punj.—*Khinjak*; Family—Anacardiaceae. A small tree, 20 to 40 feet or more in height, in some countries a shrub, common on the islands and shores of the Mediterranean, as well as throughout Asia Minor. It is found in North-Western Himalayas.

Tan. The leaves are affected by a flat horse-shoe-shaped gall, that extends round the margin of the leaf; the gall is so very distinct in form, much resembling the lobe of the ear, that the leaves get their name *goshwara*, meaning ear-like, owing to this resemblance. The leaves with these galls are valued for drying and tanning.

Green Almond

Pistacia vera Linn.; Eng.—Pistachio nut, green almond; Hindi—*Pista*; Beng.—*Pista*, galls = *boz-ghanj*, Family—Anacardiaceae. A small deciduous tree, upto 30 feet in height; branches spreading; leaves imparipinnate; leaflets lanceolate to broadly ovate, leathery; flowers in panicles, minute; drupe oblong linear to globose with a variously coloured husk; it encloses a light yellow to deep green edible kernal with a reddish coat. It is native of the eastern Mediterranean region, Iran, Afghanistan and Central Asian countries. It is thought to be native of Central Asia.

Tan. The leaves of pistachio bear small, irregularly spheroid galls (Bokhara galls) which have been reported to be imported into India for dyeing and tanning purposes. Galls contain 50% tannins.

Athel Tamarisk

Tamarix aphylla (Linn.) Karst.; Syn. *T. articulata* Vahl.; Eng.—Athel tamarisk; Hindi—*Lal-jhau, farash*, galls = *chhoti-mayin*; Beng.—*Rakta-jhau*; Punj.— *Faras, farwa*, galls = *mai*; Guj.—*Lal-jhau-nu-jhada*; Tam.—*Shivappu-atru-shavukku*; Tel.—*Erra-erasaru*; Family—Tamaricaceae. A moderate-sized tree, commonly found in the Punjab and Uttar Pradesh. It is grown in Rohilkhand region. It grows well on sandy and saline soils, requires little water after it has once taken root, springs up freely from seeds, and is readily propagated by cuttings.

Tan. The small irregularly rounded tuberculate galls, often abundantly produced on the branches by the punctures of an insect, are used as a mordant in dyeing and also in tanning.

Tamarix

Tamarix dioica Roxb.; Hindi—*Jhau, jau*; Beng.—*Lal-jhau*; Punj.—*Jhau, faras, panj-pilchi*, galls = *main*; Family—Tamaricaceae. A gregarious shrub, found near rivers and on the sea coast, in Northern India, Bengal and Assam. The galls like those of other species are astringent and are employed as a mordant in dyeing, to produce various shades of grey and black with salts of iron, and in tanning.

French Tamarisk

Tamarix troupii Hole; Syn. *T. gallica* Auct; non Linn.; Eng.—French tamarisk; Hindi—*Jhau*, galls = *bari-main*; Uriya—*Jaula, jaura*; Punj.—*Pilchi, rukh, jhau, farash*, galls = *bari mahin*; Guj.—*Jhau-nu-jhada*; Tam.—*Atru-shavukku, shiru-shavukku*; Tel.—*Pakki, eru-saru, shiri-saru*; Sans.—*Jhavuka, shavaka*; Family—Tamaricaceae. A bushy shrub, found in the Punjab, Uttar Pradesh and the Western Peninsula. It is found, especially along rivers and near the sea-coast. It favours sandy or gravelly soils, and flourishes on land impregnated with salt. It is easily propagated from seeds and cuttings. The galls formed on the branches are used for tanning. The galls, which are similar in properties to those of *T. aphylla* (Linn.) Karst., are from two to three times larger, and are extensively employed in dyeing and tanning.

Terminalia

Terminalia tomentosa Wight and Arn.; Hindi—*Asan, assain*; Beng.—*Piasal, usan, asan*; Guj.—*Sag, hag, sadri*; Punj.—*Aisan, sein, asan, arjan*; Tam.—*Karra marda, anemui*; Tel.—*Maddi, nello-madu*; Kan.—*Matti, kari-matti, banapu*; Mal.—*Karu-maruta*; Sans.—*Asana*; Family—Combretaceae. A large deciduous tree, attaining a height of 80 to 100 feet, common throughout the moist regions of India. The fruits are used for dyeing and tanning.

Tan. The galls often found in the calyx of the flower-buds are used as a tan. They make a tan of high quality.

TANNIN INKS

Tannin is a complex glucoside of gallic acid. It is used for making blue-black inks. The tannins react with the salts of iron to form dark-blue or greenish-black compounds, the basis of blue-black inks. The tannin is obtained from insect galls which also contain gallic acid.

Blue-black inks. The blue-black writing inks consist chiefly of an aqueous solution of gallic acid, ferrous sulphate and a little colouring material, such as logwood or blue dye. Some gum arabic, a little sulphuric acid and a trace of phenol is also added. When the ink flows from the pen on the paper, the sulphuric acid is neutralized by ammonia of the paper. The ferrous sulphate is then rapidly oxidised to ferric sulphate and this forms a deep black precipitate with gallic acid. Hence, the ink writes blue but becomes intensely black on drying.

DYES

The dyes are coloured compounds capable of being fixed to fabrics and which do not wash out with soap and water or fade on exposure to light. Therefore, a coloured organic substance is not necessarily a dye. For example, both picric acid and trinitrotoluene are yellow in colour, but only the former can fix to a cloth and is a dye, while the latter does not fix to a cloth and is not a dye.

Natural dyes and stains, obtained from the roots, leaves, bark, fruit or wood of plants, have been used from ancient times. Today most of the available dyes are synthetic ones prepared from aromatic compounds. These latter dyes are brighter, more permanent, cheaper, easier to use and afford a wide range of colours.

About 2,000 species of plants secrete different pigments. About 150 dye-yielding plants have been used commercially and only a few have been able to compete with the artificial dyes. A list of important dye yielding plants is being given here:

DYES OBTAINED FROM BARK

Acacia

Acacia leucophloea (Roxb.) Willd.; Syn. *Mimosa leucophloea* Roxb.; Hindi—*Safed kikar, raunj, karir, rohani*; Beng.—*Safed babul*; Uriya—*Goira*; Sans.—*Shveta-barburavrikshaha*; Punj.—*Safed kikar*; Raj.—*Arinj*; Guj.—*Haribaval*; Tam.—*Vel-velam, vel-vel*; Tel.—*Tella-tuma*; Kan.—*Bili-jali, velvaila*; Family—Mimosaceae. Found in the plains of the Punjab, and in all forest tracts of Madhya Pradesh and South India, Rajasthan and Delhi State. This is a large deciduous tree, with short, straight, and white spines. Flowers in small, rounded, yellow heads, aggregated into terminal panicles, which, when fully expanded, are a food long and broad, and densely tomentose. Pod sessile, narrow ligulate-falcate, thin, flat tomentose, with straight sutures.

Dye. The leaves and bark are used in dyeing, and give a black colour. The bark is also used for dyeing and gives a red colour, but mixed with other barks gives black colour.

Mahua Tree

Madhuca indica J.F. Gmel.; Syn. *M. latifolia* (Roxb.) Macb.; *Bassia latifolia* Roxb.; Eng.—Mahua tree, Illipe butter; Hindi—*Mahua, mahwa, mowa*; Beng.—*Mahwa, mahula, maul*; Uriya—*Moha*; Guj.—*Mohura*; Tam.—*Kattuirrupai*; Tel.—*Ippi, ippa*; Kan.—*Hogne, hippe, kadu-ippe-gida*; Mal.—*Poonam, kattirippa bonam*; Sans.—*Madhuka*; Family—Sapotaceae. A large, deciduous tree, indigenous in the forests of Madhya Pradesh. It is cultivated mainly in Eastern Uttar Pradesh and Bihar.

Leaves elliptic or oblong elliptic, shortly acuminate; calyx lobes 4, rusty tomentose; corolla-tube fleshy, lobes erect; anthers 20-30, 3-seriate, subsessile. It attains a height of 40 to 60 feet. It thrives on dry, stony ground. It sheds its leaves from February to April. The cream-coloured flowers, clustering near the ends of the branches, appear in March-April. The fruits are green when unripe, and reddish yellow or orange when ripe, fleshy, one to two inches in length, with 1 to 4 seeds, which ripen about three months after the flowers have fallen.

Dye. The bark is often used as an adjunct in dyeing where dark colours or black are desired; along with the leaves it is also sometimes employed as a tan.

Casuarina

Casuarina equisetifolia Linn.; Syn. *C. muricata* Roxb.; Eng.—Beefwood of Australia, She oak, Casuarina; Hindi—*Jangli saru*; Beng.—*Jau*; Tam.—*Chouk, shavuku-maram*; Tel.—*Serva, chavuku-manu*; Kan.—*Sura*; Mal.—*Aru, chavaka-maram*; Family—Casuarinaceae. A tall tree, with leafless, drooping branches and branchlets, which are deciduous, and perform the function of leaves. Cultivated throughout India. Thrives best in the sandy tracts near the sea-shore. Introduced into the plains as an avenue tree.

Dye. The bark is used in dyeing and tanning. A brown dye is extracted from the bark. The bark contains a small quantity of colouring matter, and produces in dyeing light-reddish drab colours on the fabrics. The shades produced by this dye-stuff are very good though faint, but the dye-stuff contains too small an amount of colouring matter to be of any great value in the dye-house.

Eriobotrya

Eriobotrya bengalensis Hook.; Hindi—*Berkung*; Family—Rosaceae. A large tree found in the Eastern Himalayas, Sikkim, Bengal and the Khasia hills up to 4,000 feet.

Dye. The bark is said to be used in Nepal and Bengal for dyeing scarlet.

Spindle Wood

Euonymus tingens Wall.; Eng.—Spindle wood; Nepal—*Newar, Kasuri*; Simla—*Chopra, mer mahaul*; Hindi—*Bar-phali*; Family—Celastraceae. A small tree, found in the temperate Himalayas from Himachal Pradesh to Nepal, between 6,500 and 10,000 feet.

Dye. The bark is a source of beautiful yellow dye.—The inner bark yields a yellow dye which is reported to be used in Nepal for making *tika* on the forehead.

Cluster Fig

Ficus glomerata Roxb; Eng.—Cluster fig; Hindi—*Gular*; Beng.—*Yajnadumbar*; Nepal—*Dumri*; Punj.—*Kathgular*; Guj.—*Umbar*; Tam.—*Atti*; Tel.—*Moydi, atti*; Kan.—*Kulla-kith, atti*; Sans.—*Udumbara*; Family—Moraceae. A large tree of the salt range and Rajasthan along the sub-Himalayan tracts to Bengal, Madhya Pradesh and South India, Assam, Arunachal and Sikkim.

Dye. This tree is said to afford a dye. The bark, under the name of *goolur*, is mentioned as one of the ingredients used in *Lohardaga* in preparing a good black dye.

Peepul Tree

Ficus religicsa Linn.; Eng.—The peepul tree; Hindi—*Pipal*; Beng.—*Ashathwa, aswat*; Uriya—*Jari*; Punj.—*Pipal bhor*; Guj.—*Pipul*; Tam.—*Arasa, aswartham*; Tel.—*Rai, raig kulla ravi*; Kan.—*Rangi, basri*; Sans.—*Aswaththamu, asvattha*; Family—Moraceae. A large tree, found wild in the Sub-Himalayan forests in Bengal and Madhya Pradesh. Extensively cultivated in most states of India.

Dye and tan. The bark yields to boiling water a reddish pale-brown colouring substance which by the employment of various processes gives to tasar, mulberry silk, and woollen fabrics, faint reddish fawn colours. The amount of colouring matter in the bark is small, but it might prove a convenient dye where faint shades are required or for modifying the colours produced by other dye-stuffs. The bark of this tree is also mentioned as being used along with other barks when preparing a permanent black in Bengal. The bark is said to be sometimes used in tanning.

Garcinia

Garcinia cowa Roxb.; Syn. *G. kydịa* Roxb.; Hindi—*Cowa*; Family—Guttiferae. A tall ever-green tree, found in West Bengal, Assam, the Nilgiris and the Andaman Islands.

Dye. The bark is often employed to produce a light yellow colour, principally in the colouring of cloth for the garments of Buddhist monks. It is cut into small pieces, boiled in water and strained, the acid liquid of applewort bark being used as a mordant.

Egg Tree

Garcinia xanthochymus Hook. f.; Syn. *G. tinctoria* Dunn.; Eng.—Egg tree; Hindi—*Dampel, tamala, tumul*; Beng.—*Tamal*; Assam—*Tepor, tihur*; Mar.—*Jharambi*; Tel.—*Iwara memadi tamalamu, chitakamraku*; Sans.—*Tamala*; Family—Guttiferae. A tree, found in Orissa, Maharashtra, Karnataka, the Nilgiris, Bengal and Andaman Islands.

Dye The bark is employed in Assam for dyeing cotton. The process is as follows: Chips of the bark, with leaves of *Symplocos grandiflora* as a mordant, are boiled, and the colour produced is bright yellow. If the dye thus obtained be mixed with the blue derived from the leaves of *Strobilanthus flaccidifolia*, a green colour is produced. The dyeing property of the bark is due to the gum-resin which it contains.

Spanish Cherry

Mimusops elengi Linn.; Eng.—Spanish-cherry; Hindi—*Maulsari, mulsari, maulser*; Beng.—*Bakul-bakal, bohl*; Uriya—*Baulo*; Punj.—*Maulsari*; Guj.—*Bolsari*; Tam.—*Mogadam, magila-maram*; Tel.—*Pogada*; Kan.—*Bakal-boklu, mugali, pogada*; Mal.—*Elengi*; Sans.—*Bakul, vakula*; Family—Sapotaceae. A moderate-sized tree, cultivated in N. India, the Western Peninsula and S. India.

Dye and tan. The bark is used for dyeing shades of brown. The bark contains the small amount of brownish-red colouring matter. With cotton a light grey colour is obtained, and with silks various shades of reddish-drab, and fawn. The bark is also employed as a tan. The bark contains 3-7% tannin.

Andaman Bullet Wood

Mimusops littoralis Kurz.; Eng.—Andaman bullet wood; Hindi—*Kappali*; Family—Sapotaceae. A large ever-green tree, found in the coast forests of Andaman Islands. It forms nearly pure forests on the level lands behind the beach, and in the mangrove swamps.

Dye. The bark is used in Andaman Islands to produce a red dye.

Box Myrtle

Myrica nagi Thunb.; Eng.—Bay-berry, Box myrtle; Hindi—*Kaiphal*; Beng.—*Kaiphal, satsarila*; Nepal—*Kobusi*; Punj.—*Kaiphal, kaphal*; Guj.—*Kariphal*; Tam.—*Marudam-pattai*; Tel.—*Kaidaryamu*; Mal.—*Marutamtoli*; Sans.—*Katphala*; Family—Myricaceae. An evergreen dioecious tree with an aromatic odour, found in the subtropical Himalayas from the Ravi eastwards, at altitudes from 3,000 to 6,000 feet; also found in the Khasia hills.

Dye and tan. The bark is used in Simla district (Sirmur) for dyeing pink—*piazi*. The bark is occasionally used as a tanning agent for fancy leather work.

Jhingan

Odina wodier Roxb.; Syn. *Lannea coromandelica* (Houtt.) Merr.; Hindi—*Jhingan, kashmala, mohin*; Beng.—*Jiol, bohar, jival*; Uriya—*Mooi, indramai*; Tam.—*Otiyam, wodier*; Tel.—*Odaimanu, gumpenna chettu*; Kan.—*Shimti, punil, gojal*; Sans.—*Jingini*; Family—Anacardiaceae. A large, deciduous tree, 40 to 50 feet in height, found throughout the hotter parts of India. It ascends in the sub-Himalayan tract to an altitude of 4,000 feet; found also in Assam, Tamil Nadu, and the Andaman Islands.

Dye and tan. The bark of the tree furnishes a small amount of brownish-red dye, which produces on tasar silk, a golden or pale brown tint. It is also used in tanning.

Indian Blue Pine

Pinus wallichiana A.B. Jackson; Syn. *P. excesla* Wall.; Eng.—Indian blue pine, five-leaved pine; Hindi—*Kail, chil*; Kumaon—*Raisalla, dolchilla*; Kashmir—*Chil, laiar, yari*; Punj. *Chil, chir, darchir, tser*; Family—Pinaceae (Gymnosperms)—A lofty symmetrical gregarious tree, found on the temperate Himalayas between 6,000 and 12,500 feet, but most common upto 8,000 feet. It is readily recognized by the blue tint of its tufted foliage, the long pendulous cone, the compact greenish blue bark of the young stems and leaves remaining on the tree for three years.

Dye. The bark contains a fair amount of colouring matter. They produce various tints with different mordants. It produces good yellows on bleached Indian tasar, and a deep orange on wool. The yellow on corah silk is a very good colour. The bark possesses a small quantity of tannin which is shown by the dark colour it produces with an iron salt.

Guava

Psidium guajava Linn.; Eng.—Guava; Hindi—*Amrud*; Beng.—*Piyara*; Assam—*Madhuriam*; Guj.—*Jamrud, jamrukh*; Tam.—*Goyya-pazham, segapu*; Tel.—*Jama, jam-pandu*; Kan.—*Sebe, shibe-hannu*; Mal.—*Pela, perakka*; Sans.—*Amruta phalam, bahu-bija-phalam*; Family—Myrtaceae. A small tree; native of Central America but cultivated chiefly in Uttar Pradesh, Assam, the Punjab, Bihar, Maharashtra and Andhra Pradesh.

Dye. In Assam the bark and leaves are used for dyeing. It produces a black dye.

Indian Kino

Pterocarpus marsupium (Roxb.); Eng.—Indian Kino, Malabar Kino; Hindi—*Bijasar, pitasara*; Beng.—*Pit-shal*; Uriya—*Bijasa*; Guj.—*Bibla*; Tam.—*Vengai, vengai-maram*; Tel.—*Peddagi, yeggi*; Kan.—*Banga, bibla*; Mal.—*Venna-maram*; Family—Papilionaceae. A medium-sized, deciduous tree. Found in the Western Peninsula and South India.

Dye and tan. The bark is occasionally employed for dyeing. It contains a brownish-red colouring substance, which produces reddish fawn colours with tussar silk.

Quercus

Quercus fenestrata Roxb.; Hindi—*Kala chakma*; Assam and Manipur—*Kuhi*; Nepal—*Patle*; Khasia—*Dingjing*; Family—Fagaceae. A large evergreen tree of the Eastern Himalayas, found at altitudes between 5,000 and 8,000 feet. Occurs also on the Khasia hills.

Dye. The acorns (fruits) as well as bark are said to be used in Manipur for dyeing. A decoction of the bark is used by the Manipuris for changing the blue dye of the *rum* (*Strobilanthes flaccidifolius*) into a black.

Elm Leaved Sumach

Rhus cotinus Linn. Eng.—Elm leaved sumach; Vern.—*Tunga, tung, ami*; Punj.—*Tittri, phan, larga*; Kashmir—*Darengri*; Family—Anacardiaceae. A shrub or small tree of the Western Sub-tropical Himalayas, found upto altitudes of 5,000 feet.

Dye. The bark, leaves and wood are locally employed in dyeing.

Indian Red Wood

Soymida febrifuga A. Juss.; Eng.—Indian red-wood, Bastard cedar; Hindi—*Rohan, rakat rohan*; Beng.—*Rohan, rohina*; Guj.—*Rohina*; Tam.—*Shem, wundmaram*; Tel.—*Sumi, somida manu*; Kan.—*Swami*; Sans.—*Rohuna, Patranga*; Family—Meliaceae. A lofty, glabrous tree, found in the Western Peninsula, Rajasthan and Bihar. Also found in S. India extending southward to Karnataka and Cochin.

Dye and tan. The bark is employed in Karnataka as one of the second-rate dye substances producing the dirty browns in which cotton fabrics are often dyed. It is also used in combination with more valuable dyes such as arnatto (*Bixa orellana*). The bark is also used for tanning.

Lodh Tree

Symplocos racemosa Roxb.; Eng.—The Lodh, Lodh tree; Hindi—*Lodh*; Beng.—*Lodh*; Assam—*Kaviang*; Khasia—*Lapongdong*; Nepal—*Chamlani*; Uriya—*Ludhu*; Kumaon—*Lodh*; Guj.—*Lodar*; Tel.—*Lodduga*; Sans.—*Lodhra, margana*; Family—Symplocaceae. A small tree, of the plains and lower hills of Bengal, Assam and Khasia hills, chiefly found in dry forests upto altitudes of 2,500 feet.

Dyes. The bark and the leaves are used in dyeing, and a yellow dye is extracted from both. By itself the *lodh* bark yields a yellow dye, which is obtained by simply steeping it in hot water. *Lodh* bark is chiefly employed as an auxiliary or mordant in dyeing with *al* (*Morinda tinctoria*), *bac* (*Caesalpinia sappan*), and *paras* (*Butea monosperma*). *Lodh* bark is also an ingredient of the *abir* of red powder used during the *Holi* festival. The bark is used in calico-printing and dyeing leather as an auxiliary to other dyes, being usually pounded up and mixed with them.

Symplocas

Symplocos crataegoides Buch.—Ham. ex D. Don; Syn. *S. paniculata* Wall. ex D. Don; Hindi—*Lodh*, bark = *pathani lodh*; Family—Symplocaceae. A large shrub or small tree of the Himalayas, from Kashmir to Assam, occurring at altitudes between 3,000 and 8,000 feet. Found also in Khasia hills.

Dye. The bark and leaves are used in dyeing and yield a yellow colour. They are principally employed in combination with madder, and act probably more as a mordant than as a colouring material.

The Yew

Taxus baccata Linn.; Eng.—The yew; Hindi—*Thuno, birni*; Beng.—*Burmie, sugandh*; Khasia—*Dingsableh*; Nepal—*Tcheiray sulah*; Kumaon—*Thaner, gallu*; Kashmir—*Tung, postal, birmi*; Family—Taxaceae (Gymnosperms). An evergreen tree, found in the temperate Himalayas from 6,000 to 11,000 feet and the Khasia hills at 5,000 feet.

Dye. The inner part of the bark yields a red dye. The red juice of the bark is used in Nepal as an inferior dye.

Arjun

Terminalia arjuna (Roxb. ex DC.) Wight and Arn.; Eng.—Arjun; Hindi—*Arjun, kahu, anjan, anjani, koha*; Beng.—*Arjun, kahu*; Assam—*Orjun*; Uriya—*Arjun, hanjal*; Punj.—*Arjan, jumla*; Guj.—*Arjun sadada*; Tam.—*Vella maruthu, vella marda*; Tel.—*Yer muddi, tella maddi*; Kan.—*Billi matti*; Mal.—*Vella maruta*; Sans.—*Arjuna, Kukubha*; Family—Combretaceae. A large deciduous tree, commonly found in Madhya Pradesh, Bihar and the Western Peninsula. Found also throughout the Sub-Himalayan tracts.

Dye and tan. The bark in various localities is used in dyeing. In South India the inner bark is broken into chips and the dye extracted by boiling in water. The tint produced is a dirty brown or *khaki* colour. In Bengal, it is used to dye cotton a light brown.

Indian Almond

Terminalia catappa Linn.; Eng.—Indian almond, tropical almond; Hindi—*Janglibadam*; Beng.—*Bangla badam*; Uriya—*Badam*; Tam.—*Kottai, nattu-vadom, natvadom*, Tel.—*Badam-vittulu*; Kan.—*Tari, nat-badami*; Sans.—Fruit = *desha-vadda-mittee*; Family—Combretaceae. A large deciduous tree, which attains 80 feet in height, and has branches in almost horizontal whorls. Cultivated throughout the hotter parts of India.

Dye and tan. The bark and leaves are astringent and contain tannin. They are mixed with iron salts to form a black pigment, with which natives in certain localities colour their teeth and make Indian ink. The bark contains 9% tannin, and a small amount of colouring matter soluble in water, which, by various processes produces light brownish-yellow, light drab, golden fawn and slate colours in silk, light drab, olive and grey in cotton, and pale-fawn in wool.

Asan

Terminalia tomentosa Wight and Arn.; Hindi—*Asan, asna, sain, saj*; Beng.—*Ashan, usan, piasal*; Uriya—*Shaju, ansun*; Guj.—*Sag, hag, sadir*; Punj.—*Aisan, sain*; Tam.—*Karumardu, anemui*; Tel.—*Maddi, nella-madu*; Kan.—*Matti, banapu*; Mal.—*Karu-maruta*; Sans.—*Asana*; Family—Combretaceae. A large deciduous tree, attaining a height of 80 to 100 feet, commonly found throughout the moist parts of India. In the Siwalik tract and outer Himalayan valleys it extends west as far as the Ravi, and in places ascends to 4,000 feet.

Dye and tan. The bark is used occasionally, as a dye stuff being broken up and boiled in water to extract the dye. The resulting colour is brown or buff. In Bengal it is employed along with the bark of bakul (*Mimusops elengi*) to produce a reddish dye, used in colouring gunny bags. A mixture of the barks of *asan* and *porashi* (*Thespesia populnea*) produces a very good red dye. Also the bark is used for dyeing fishing nets. In many places it is employed, with iron salts, to obtain a black dye. It contains 16.7% tannin, and a moderate amount of brownish red colouring matter. The chief use of the bark is, however, as a tan, for which it is largely employed all over India.

DYES OBTAINED FROM WOOD

Black Cutch

Acacia catechu (Linn. f.) Willd. Eng.—Black cutch, Catechu; Hindi—*Khair, khadira, khair-babul, katha*; Beng.—*Khaiyar*; Assam—*Khaira, Kat*; Tam.—*Vodalam*; Tel.—*Podala-manu*; Sans.—*Khadera*; Uriya—*Khoiru*; Kan.—*Kagli*; Mar.—*Khaderi*; Guj.—*Kher*; Family—Mimosaceae. A tree, native of India and Myanmar. In India, it is found in the Punjab, Madhya Pradesh, Uttar Pradesh, Andhra Pradesh and Bihar. The plant serves as a host for lac insect.

KATTHA—A SMALL-SCALE INDUSTRY REPORT OF KHADI AND VILLAGE INDUSTRIES COMMISSION, MUMBAI

A cumbersome task made easy. Kattha manufacture once a painfully slow process has taken remarkable strides in its advancement of improved methods. Thanks to the introduction of

Fig. 16.1. *Acacia catechu* **(Khair)**. A tree.

improved technology and equipment into this field by the Khadi and Village Industries Commission (KVIC), the sponsoring institution. The processing which used to take as much time as eight months has now shrunk to just 2 weeks, and the yield has almost doubled.

The traditional kattha manufacturers—the Chais, the Kathiyas, the Khairooas and hill men—work for eight months in a year in kattha extraction work and for the rest 4 months, the *chaumasa* are engaged in farming, growing wheat, jowar, etc. This is thus a special category of industry we come across where agriculture is a subsidiary occupation. Mostly it is vice versa.

Source of kattha. Kattha is extracted from the heartwood of *Khair* tree. The tree belongs to Mimosaceae family and the three important species of Khair are: *Acacia catechu*, *Acacia sundra* and *Acacia catechuoides*. These species are found largely in Uttar Pradesh, Madhya Pradesh, Rajasthan, Gujarat, Himachal Pradesh, Bihar and to a lesser extent in West Bengal, Orissa, Andhra Pradesh, Maharashtra, Haryana and Karnataka.

The industry can be broadly divided into two sectors: the factory or organised sector and the indigenous sector, run by professional artisans. Even though there are a number of factories in

Fig. 16.2. Extraction of 'kattha' from khair heartwood chips by boiling them in earthen utensils.

the country, the bulk of kattha is still manufactured by the indigenous sector. The artisans engaged in this work are being exploited by big contractors to a very great extent. To safeguard their interests, Forest Labour Co-operatives are being organised which in turn undertake kattha production with the financial and technical assistance of KVIC.

Production. It is estimated that about 2000 to 3000 tonnes of kattha are produced by the indigenous sector annually. Even in this sector cutch was often allowed to drain into the soil, resulting in considerable loss in the yield. But this situation is saved now due to the improved methods.

Extraction. Khair heartwood chips are boiled in water. The decoction is further concentrated and allowed to settle. Kattha crystals are formed which are separated from the mother liquor containing cutch. This is cut into cakes and dried. Heartwood of a mature tree contains about 3-4% of kattha.

Kattha is mainly used in paan, and also in the indigenous system of medicine. It is bitter, acrid, cooling, astringent to bowels, cures itching, bronchitis, indigestion, boils, ulcers, leprosy, urinary and vaginal discharges, *kapha*, *vata*, throat diseases, leucorrhoea, piles and mouth diseases. Kattha is used in Ayurvedic preparations as *khadirasav*, *khadirvidhan*, *khadiratel*, *khadiradivati,* etc.

BY-PRODUCTS

1. *Cutch* is a by-product of Kattha Industry. It is one of the most important sources of vegetable tanning material and also used extensively for boiler descating.

2. *Kheersal.* It is found in the cavities of wood and occurs in small irregular fragments. It is used in cough treatment and sore throat.

Gums. The trees marked for kattha extraction can be tapped profitably for 2-3 years for gum collection before felling. A tree yields 1-1.5 kg. of gum per year. Khair gum gives a thicker solution than babul gum. It is very useful in medicine and textiles.

Lac. Khair tree is a host of winter brood crops. Lac brood from *Ficus* can thrive very well on these species.

Charcoal. Old and dry khair wood, which may not be useful for kattha extraction, can be converted into charcoal. This charcoal is much valued by goldsmiths.

Improved technology. Aluminium extractors, pans, containers, etc., have been introduced in place of earthen vessels by KVIC. Control of temp., and use of hydrometer for measuring density have been added to help in improving the quality of output. Filter press, screw press, cake cutting machine, etc., have replaced the age old crude methods.

As a result (*i*) the yield of kattha has increased

(*ii*) Catechin content has gone up to 68% as compared with 45% of old method.

(*iii*) time is very much less.

(*iv*) Cutch can be recovered.

ANALYSIS OF KATTHA

		Sample produced by the existing indigenous method %	*Sample produced by improved method %*	*Sample produced by the factory %*
1.	Catechin Content	44.9	68.8	66—68
2.	Cutch content	49.6	14.1	—
3.	Water insoluble non-tannin	—	9.8	—
4.	Water insoluble matter	5.5	7.3	—
5.	Ash	3.00	2.80	—

ANALYSIS OF CUTCH

1.	Catechin	21.2%
2.	Catechu tannic acid	62.4%
3.	Nontans	15.7%
4.	Water insoluble	0.7%
5.	Ash (% to above)	2.8%

The KVIC also offers free training facilities to artisans deputed by kattha units.

Dye. Cutch is obtained from the heartwood of the tree. It is used as a dyeing stuff and as a masticatory. It is extensively used for dyeing canvas fishing nets and ropes. In India, *Katha* industry is mostly located in Bareilly, Gwalior and Mumbai. Small pieces of the wood are boiled in water and the extract is evaporated down to a purplish-black, gummy, semisolid substance, which is molded into blocks for marketing.

A solution of catechu, by the action of lime or alum, changes into a dull red-colour, which constitutes a fairly good dye, and is used for the purpose in some parts of the country. With salts of copper, catechu gives a permanent bronze brown. This colour is deepened by the use of perchloride of tin, with the addition of copper nitrate.

Coral Wood

Andenanthera pavonina Linn.; Eng.—Coral wood, Bead tree, Red wood; Hindi and Beng.—*Barighumchi, rakta kanchan, rakta kambal*; Tam.—*Anai-gundumani*; Tel.—*Bandi gurivenda*; Mal.—*Manjati*; Guj.—*Bari-gumchi*; Kan.—*Manjadi*; Andaman—*Recheda*; Family—Caesalpiniaceae. A large, deciduous tree, found in Bengal, South India, Western Ghats, the Andaman Islands and the Eastern sub-Himalayan tract. Leaves compound; racemes short, peduncled, 2-6 in. long; seeds bright scarlet.

Dye. The wood is powdered and used as a red dye. The dye is used by religious Hindus for making marks on the forehead.

Jack Tree

Artocarpus heterophyllus Lamk.; Syn. *A. integra* (Thunb.) Merr.; *A. integrifolia* Linn.; Eng.—Jack tree, Jack fruit; Hindi—*Kathal*; Beng.—*Kanthal*, *kathal*; Uriya—*Panasa*; Tam.—*Pila*; Tel.—*Panasa-pandu*; Kan.—*Halsu*; Sans.—*Panasa*; Family—Moraceae. A large tree, probably indigenous to the Western Ghats, ascending to 4,000 feet. Cultivated in Uttar Pradesh, Bihar, Bengal, Andhra Pradesh and Orissa. Leaves coriaceous, smooth, shining above, rough beneath; stipules large, caducous; fruit large, hanging on short stalks, oblong, with a thick cylindrical receptacle; rind muricated; seeds reniform.

Dye. The heartwood yields a yellow dye. The wood, or its sawdust, yields on boiling a decoction used as a yellow dye. It is fixed with alum, and often intensified by a little turmeric. With indigo, it gives a green. *Kanthal* yellow is often used in dyeing silk.'

Indian Barberry

Beriberis aristata D.C.; Eng.—Indian barberry; Hindi—*Dar-hald*, *rasaut*; Family—Berberidaceae. A spiny shrub, found in the North-Western Himalayas, the Nilgiris, Kulu and Kumaon between 6,000 and 10,000 feet in altitude. Leaves evergreen, obovate or oblong entire, or with few distant spinous teeth; flowers in compound corymbose racemes; berries tapering into a short style.

Dye. A yellow dye obtained from the wood and root is used in dyeing and tanning leather. The wood is generally known as *dara-halda*; the extract as *rasot* or *rasavanti*. The colour exists chiefly in the bark and in the young wood immediately below the bark, and that in old wood the proportion is small, but much superior in quality.

Common Barberry

Berberis vulgaris Linn.; Eng.—Common barberry; Hindi—*Chatroa*, *kashmal*, *zirishk*, *chachar*; Family—Berberidaceae. Found in the North-Western Himalayas, in shady forests, above 8,000 feet.

Dye. The wood and bark are the source of a yellow dye, which is used for dyeing wool and leather.

Sappan Wood

Caesalpinia sappan Linn.; Eng.—Sappan wood; Hindi and Beng.—*Patang*, *bakam*; Uriya—*Bokomo*; Guj.—*Patang na lakaru*, *bakam*; Mar.—*Patang*; Tam.—*Patanga*, *vatteku*; Tel.—*Bakamu*, *patanga-katta*; Kan.—*Patanga chekke*; Mal.—*Chappanum*; Sans.—*Pattanga*; Family—Caesalpiniaceae. A small thorny tree. It is native of India and Malaya. Cultivated chiefly in Bengal and South India.

Dye. The heartwood yields a red dye, which is used for dyeing cottons and woollens and for preparing red ink.

Logwood

Haematoxylon campechianum Linn. Eng.—Logwood, Campeachy tree; Hindi—*Patang*; Beng.—*Bokkan*; Tel.—*Gubbi*; Kan.—*Partanga*; Family—Caesalpiniaceae. It is a small spreading tree. It is native of Central America and the West Indies. Cultivated in Indian gardens for its delicate foliage and fragrant flowers. Leaves paripinnate, leaflets obovate or obcordate; flowers small, yellow, in axillary racemes; pods flattened, lanceolate, usually 2-seeded. The tree is propagated by seed and by cuttings.

Dye. This is supposed to be one of the oldest dyestuffs. The heartwood of this plant constitutes the logwood of commerce. It yields a famous dye, or rather a series of dyes of the darker tints, such as grey, violet, blue and even black. The colouring principle is haematoxylin ($C_{16}H_{14}O_6.3H_2O$), which on mild oxidation gives haematein ($C_{16}H_{12}O_6$), a dark violet crystalline body with a green metallic lustre.

Haemotoxylin, when pure, is nearly colourless, very soluble in hot water and alcohol, but only sparingly so in cold water and ether. It is sometimes found crystallized in clefts of the wood. When exposed to the air under the influence of alkalies, it becomes red, haematein being formed.

Chipped logwood is subjected to the 'ageing' process to facilitate haematein formation. Logwood and its extracts are used for dyeing of blacks on wool and silk, and on cotton chiefly in combination with iron chromium mordants; they can also be used for dyeing rayon and nylon. It is also used for dyeing leather, fur, jute and bone and in the manufacture of inks. Haemotoxylin is used as a histological stain, mainly for cell nuclei.

Red Sandal Wood

Pterocarpus santalinus Linn. f.; Eng.—Red sandal wood, Red Sanders; Hindi—*Lalchandan, rukhto-chandan, rajat chandan*; Beng.—*Kuchunduna, ranjana, rakta-chandana, rukt-chundun*; Uriya—*Raktachandan*; Punj.—*Chandan lal*; Guj.—*Ratanjli*; Tam.—*Rakta chandan, seyapu chandanum*; Tel.—*Gerra chandan, rakta chandan, lal chandan*; Kan.—*Kempugandha-cheke, agaru*; Mal.—*Uruttah-chundanum*; Sans.—*Raktachandana*; Family—Papilionaceae. A small to medium-sized, deciduous tree, chiefly found in Cuddapah, North Arcot, and the southern portion of the Karnul district. Bark blackish brown, deeply cleft into rectangular plates, and exuding a deep red juice when cut; leaves compound, imparipinnate; flowers yellow; seeds reddish brown, smooth, leathery. It favours a dry rather rocky soil, and a hot fairly dry climate.

Dye. The wood contains 16% of a red colouring matter santalin (santalic acid, $C_{30}H_{16}O_6(OCH_3)_4$), possessing a quinonoid structure. Santalin is easily dissolved in any alkaline solution, and is used as a dye. It yields a blood red solution with alcohol, yellow with ether and violet with ammonia and caustic alkalies, but insoluble in water. The wood also contains deoxysantalin ($C_{20}H_{16}O_6$), the yellow isoflavone santal, and two unidentified pigments A and B.

The ground wood was chiefly employed for dyeing cotton, wool and leather and for staining other woods. The dye is also used for colouring pharmaceutical preparations and foodstuffs and is suitable for colouring paper pulp. A histological stain is prepared from the alcohol-soluble fraction of the heartwood.

DYES OBTAINED FROM ROOT, ROOT BARK AND ROOT STOCK

Indian Walnut

Aleurites moluccana (Linn.) Willd.; Syn. *A. triloba* Forst; Eng.—Indian walnut, candlenut, Varnish tree; Hindi—*Jangli akhrot*; Sans.—*Akshota*; Guj.—*Akhoda*; Tam.—*Nattu-akrotu-kottai*; Tel.—*Natu-akrotu-vittu*; Kan.—*Nat-akrodu*; Family—Euphorbiaceae. A handsome tree, native of Malaya and Pacific Islands but grows wild in South India and Assam.

Dye. The root of the tree affords a brown dye, which is used for dyeing coarse cloth.

Greater Galangal

Alpinia galanga (Linn.) Willd.; Syn. *Languas galanga* (Linn.) Stuntz; Eng.—Greater galangal; Hindi—*Kulinjan, bara-kulanjan*; Guj.—*Kolinjan*; Sans.—*Kulin-jana*; Tam.—*Pera rattai*; Tel.—*Pedda-dumpa-rash-trakam*; Mal.—*Pera-rattai*; Kan.—*Dumpa-rasmi*; Family—Zingiberaceae. A perennial herb. Native of Java and Sumatra, now cultivated in Bengal, Eastern Himalayas and South India.

Dye. The root stock is used in calico-printing along with myrobalans.

Indian Barberry

Berberis aristata D.C.; Eng.—Indian barberry; Hindi—*Dar-hald, rasaut*; Family—Berberidaceae. A spiny, much branched shrub, found in the North-Western Himalayas, the Nilgiris, Kulu and Kumaon, between 6,000 and 10,000 feet in altitude. Leaves obovate or oblong entire, or with few distinct spinous teeth; flowers in compound, often corymbose, racemes; berries tapering into a short style; stigma small, subglobose.

Dye. The roots and stems yield a yellow dye. The dye is used in tanning and colouring leather. Beberry is one of the best tanning dyes in India.

Common Barberry

Berberis vulgaris Linn.; Eng.—Common barberry, True barberry; Hindi—*Chatroa*, *kashmal*; Punj.—*Zirizk*, *kashmal*; Family—Berberidaceae. A deciduous thorny shrub on the Himalayas from Nepal westward, above 8,000 feet.

Dye. A yellow dye is extracted from the roots. It is used for dyeing wool and leather.

Wild Turmeric

Curcuma aromatica Salisb.; Syn. *C. zedoaria* Roxb. non Rosc.; Eng.—Wild turmeric, yellow zedoary, Cochin Turmeric; Hindi—*Jangli-haldi*, *ban-haldi*; Beng.—*Banhalud*; Guj.—*Kapur kachali*; Tam.—*Kasturi-manjal*; Tel.—*Kattu-mannal*; Mal.—*Anakuva*; Sans.—*Vanaharidra*; Kan.—*Kasturi-arishina*; Family—Zingiberaceae. Native of Asia. Cultivated chiefly in Bengal and Travancore. Flowers in summer season. The plant when in flower is highly ornamental.

Dye. The rhizome yields a yellow dye which is used in the preparation of the *Abir* powder. It is used as a dyeing agent. It gives a dirty yellow colour with the alkaline earth *chaulu*.

Turmeric

Curcuma domestica Valet; Syn. *C. longa* Auct. non-Linn.; Eng.—Turmeric; Hindi—*Haldi*; Beng.—*Halud*; Punj.—*Haldar*; Sans.—*Haridra*; Tam.—*Manjal*; Tel.—*Pasupu*; Mal.—*Mannal*; Kan.—*Arishina*; Guj.—*Halada*, Family—Zingiberaceae. Turmeric is extensively cultivated all over India. Cultivated commonly in Maharashtra, Bengal, Tamil Nadu, Andhra Pradesh, Karnataka and Orissa. A special form of turmeric is grown for the purpose of dye, namely, a harder root, much richer in the dye principle, than in the ordinary condiment form.

Dye. The above-mentioned dye rhozomes are known by separate names in various parts of India, such as—*lok-handi haldi*, *jowala-haldi* and *amba-haldi*. The colour is only deposited in the rhizome with age, and therefore in all probability, the above-mentioned forms have been obtained by a process of careful selection of stock observed to produce the colour freely. Mordants are but rarely required with the dye, as is found to attach itself readily enough to wool, silk or cotton. Alkalies deepen the colour, making it almost red. A brilliant yellow is produced, known as *basanti rang*, by mixing turmeric with *sajimati* (sodium carbonate) and lemon or lime juice. Myrobalans are sometimes employed with turmeric, but the chief compound colour in which turmeric plays an important part is the green shade formed along with indigo. The fabric is first dyed with indigo and then dipped in a solution of *haldi*. Turmeric is also often added to sharpen or brighten other colours, such as—*Singrahar* (*Nyctanthes arbortristis*), *al* (*Morinda tinctoria*), safflower (*Carthamus tinctorius*), and *toon* (*Cedrela toona*). The colouring matter of turmeric is one of the few for which cotton has naturally a strong attraction. With the use of potassium dichromate and ferrous sulphate as the mordant, the colours produced are olive or brown. The colouring matter of turmeric is very sparingly soluble in water, but alcohol, ether and fatty and essential oils dissolve it readily.

False Hemp

Datisca cannabina Linn.; Eng.—False hemp; Hindi—*Akalbir*, *bhang-jala*; Kashmir—*Waftangel*; Punj.—*Akilbir*, *drinkhari*; Family—Datiscaceae. A tall, erect herb, resembling hemp, hence the specific name. It is found in the temperate and sub-tropical Western Himalayas from Kashmir to Nepal, at altitudes from 1,000 to 6,000 feet.

Dye. The roots yield a dye, which is used for colouring wool and cotton. This is a special dye used in Kashmir to dye silk a delicate yellow colour. Throughout the Himalayas it is more or less employed, being combined, it is said, with red colours to soften the tint, and with indigo to produce *pista* green. The roots are known for producing yellow dye. In the month of August the roots are dug up (the bark peeled off), dried in the sun, and then sold. One root yields ½ kg. to 1 kg.

Fibraurea

Fibraurea trotteri Watt.; Syn. *F. manipurensis* Brace ex Diels; Manipur—*Napu, napoo*; Family—Menispermaceae. An extensive climber occurring in the forests of Manipur.

Dye. The root of this plant is reported to be used locally for dyeing cloth. The process of dyeing is as follows.

The dry root of the plant is washed and beaten into long shreds, and then soaked in water for few minutes, the water becomes yellow in colour; this water is kept aside. Now the pounded roots are taken out and re-steeped in the same quantity of fresh water for 24 hours. Wash the cloth to be dyed clean, thoroughly soak it in the first solution and take out and repeat the process in the second water, leaving the cloth to soak in it about for half an hour, then wring out and dry in the shade.

Nepal Geranium

Geranium nepalense Sweet; Eng.—Nepal Geranium, Nepalese crane's bill; Hindi—*Bhanda*; Punj.—*Bhanda*, root = *rowil, bhand*; Family—Geraniaceae. A herbaceous prostrate plant, common throughout the temperate Himalayas at altitudes of from 5,000 to 9,000 feet; found also in the Khasia hills and Nilgiris. Stems 6 to 10 in. long, pubescent; leaves palmately 3 to 5 lobed, irregularly toothed; flowers pink or purple.

Dye. The root affords an abundance of red colouring matter, which is employed in colouring medicinal oils. It makes an article of trade, being brought from the hills to the plains of the Punjab and sold as a dye, known as *rowil*, or *bhand*.

Morinda

Morinda angustifolia Roxb.; Hindi—*Daru haridra, ban hardi*; Beng.—*Daruharidra*; Assam—*Asugach, asukat*; Nepal—*Banhardi, hardi*; Family—Rubiaceae. An erect bush or small tree, found in the tropical Himalayas, wild and cultivated, from Nepal eastward to Assam, the Khasia and Naga hills. It attains an altitude of 6,000 feet on the Himalayas at Sikkim, and 4,000 feet on the Khasia hills.

Dye. The dye is prepared by pounding the bark of the root and boiling it in water; it is then strained and the water boiled over again to the required consistency. The method of dyeing in Assam is as follows—The roots are cut into very small chips, and thrown either into cold or boiling water. The cold water process gives a very bright red, while the solution in boiling water gives a very pale red. The colour varies according to the greater or less care bestowed on the dyeing, and is said to be very permanent. The thread is first dyed and then woven into cloth.

Indian Mulberry

Morinda citrifolia Linn; Eng.—Indian mulberry; Hindi—*Al, ach, ak, barra-al*; Beng.—*Ach, aich, achhu*; Mar.—*Al, surangi, bartondi*; Guj.—*Al seraoji*; Tam.—*Mina-maram*; Tel.—*Maddi, Nuna torugu*; Kan.—*Siranjikadi*; Mal.—*Ka-da pilva*; Family—Rubiaceae. A small tree, cultivated throughout India; found in the wild state or as an escape in parts of Bengal, Bihar, Orissa and Andaman Islands. Leaves broadly elliptic bright green, glabrous; flowers white, in dense ovoid heads; fruits ovoid, glossy, white when ripe; pyrenes compressed, winged on edges.

Dye. The roots yield a yellow and red dye. The roots are dug out when the plants are 3-4 years old, dried and sorted for use; roots thicker than 1-3 cm. are discarded as worthless. The colouring matter resides in the root bark and is present to the maximum, when the plants are 3-4 years old. The method for dyeing is as follows—The roots are chipped into small pieces and steeped in water to wash off the free acids. Best results are obtained from a neutral dye-bath of washed morinda root. Red, purple and chocolate shades are produced on mordanted cotton, silk or wool, the shades being fast to soap. The use of *al* dye has been totally given up since the advent of synthetic dyestuffs.

The dye contained in the root bark seems to be the best red, whereas that contained in the woody part of the roots is more yellow than red, and consequently where the wood dominates over the bark the resulting dye is reddish-yellow. *Al* dye is used principally for dyeing the thread or yarn out of which the coloured borders of the cotton garments worn by the poor are woven. It is used also for dyeing silk thread to form the borders of coarse silk fabrics known as *erendi* or *endi* cloth.

The colouring principle of morinda root is morindone ($C_{15}H_{10}O_5$), present in the bark mainly as the glucoside morindin ($C_{27}H_{30}O_{14}$); on hydrolysis morindin yields glucose, rhamnose and morindone. Morindone may be conveniently prepared in quantity by extracting the roots with sulphurous acid; the non-tinctorial yellow compounds are obtained as byproducts.

Morinda

Morinda umbellata Linn.; Hindi—*Al*, Tam.—*Nuna, kai, numa-maram*; Tel.—*Mulughudu*; Kan.—*Maddi chekhe*; Family—Rubiaceae. A diffuse shrub found in the hilly regions of Bengal, ascending to 4,000 feet in Khasia hills, also in Bihar and Deccan Peninsula, in South-West India and in Nilgiris. Leaves elliptic, oblong or lanceolate; flowerheads in terminal sessile umbels; fruit scarlet, lobed.

Dye. Like all other species of this genus the root yields dye of a brilliant yellow colour. The root bark contains the colouring principles, morindone ($C_{15}H_{10}O_5$) and morindin ($C_{17}H_{30}O_{14}$).

Indian Madder

Oldenlandia umbellata Linn.; Eng.—Chay-root, Indian madder; Hindi—*Chirval, saya*; Beng.—*Surbuli*; Uriya—*Surbuli*; Tam.—*Emburel cheddi, imbural, saya-wer*; Tel.—*Cherivelu, chiri-vera*; Family—Rubiaceae. A biennial herb, found in Orissa, Bengal and South India; chiefly occurs on sandy soils. It grows wild in the Puri district, and cultivated in many parts of Tamil Nadu, near the sea-coast. In its wild state it is a low widely spreading almost stemless plant, but under cultivation it assumes a more erect habit and grows six or eight inches high.

Dye. The root-bark of this plant (commercially known as chay-root), with alum as a mordant, gives a beautiful red dye, which is fast, and was formerly much employed in Madras for dying the handkerchiefs for which that town was once so famous. For dyeing purposes the roots of wild plants are considered preferable. They are shorter and thicker than the cultivated state, and are said to yield one-fourth part colouring matter. Roots of two years' growth are favoured. Only the bark of the root contains the dying substance, the woody portions being quite useless. Both old and new cloth may be dyed with this substance, but in old cloth the resulting colour is not so good as in the new, the freshness of the colour depends on the texture of the fabric.

Ratanjot

Onosma hispidum Wall. ex D. Don; Syn. *O. echioides* C.B. Clarke non Linn.; Hindi—*Ratanjot*; Nepal—*Newar maharangi*; Punj.—*Ratanjot, maharanga, lal jari*; Family—Boraginaceae. A hispid, biennial plant, commonly found in Western Himalayas, from Kashmir to Kumaon.

Dye. The root used in the Himachal Pradesh as a red dye for wool, being applied with ghee and the acid of apricots. It is employed by the Lambas of Lahaul-Spiti and Kanawar to stain their images. It imparts a rich red colour to medicinal oils, fats and curries. In Nepal these roots are boiled with oil and used as a dye for the hair. *O. hispidum* has been reported to be the source of Ratanjot, a red dye yielding root, commonly used for colouring footstuffs oils and medicinal preparations. The commercial product, however, is imported into India from Afghanistan.

Chab

Piper retrofractum Vahl; Syn. *Piper chaba* Hunter; Hindi—*Chab, chavi*; Beng.—*Chai, choi*; Sans.—*Chavika, chuve*; Family—Piperaceae. A glabrous rather fleshy climber, with adhesive roots, native of the Moluccas, reported to be cultivated in India. Leaves oblong, ovate or lanceolate, acuminate, base rounded, unequal fruiting spikes cylindroconic; fruits very small, bright red, globose.

Dye. The root and wood are reported to have been used in Bengal for dyeing; they give a pale brown colour on cotton if used alone and a brownish red when mixed with *Caesalpinia sappan*.

Rewand

Rheum moorcroftianum Royle; Vern.—*Rewand*; Family—Polygonaceae; A stout herb. This species occurs in the Western Himalayas, mainly in Kumaon region.

Dye. The roots, known as rewand, are powdered and steeped in cold water for two days. The decoction is then boiled and woollen stuff immersed in it, while still boiling. The result is a fleeting yellow, which can be deepened by the addition of turmeric to the dye.

Manjit

Rubia cordifolia Linn.; Eng.—Indian madder; Hindi—*Manjit*; Beng.—*Manjistha*; Assam—*Majathi*; Manipur—*Moyum*; Uriya—*Manjistha*; Kumaon—*Majethi*, *manjit*; Kashmir—*Dandu*, *fahrghas*; Punj.—*Kukarphali*, *manjit*; Tam.—*Manjitti*, *shevelli*; Tel.—*Tamravalli*, *majishta*; Kan.—*Manjushta*; Mal.—*Man-chetti*; Sans.—*Manjistha*; Family—Rubiaceae. A herbaceous creeper with perennial roots, found in the hilly regions of India from the North-West Himalayas eastwards, and South India. On the Himalayas, it occurs upto an altitude of 8,000 feet. Rootstocks perennial, roots long, cylindric flexuose with thin, red bark; stems quadrangular; leaves cordate-ovate to ovate-lanceolate; flowers small, white or greenish, or in shades of red and yellow, terminal panicles of cymes; fruits globose, dark purplish or black fleshy with two small seeds.

Dye. The Indian madder has long been employed for dying coarse cotton fabrics, blankets and carpets, various shades of scarlet and coffee brown. The colour obtained from *Rubia cordifolia* is brighter though more fleeting than that of *Rubia tinctorum* (European madder), and the colouring power of *R. cordifolia* is lesser than that of the latter. As regards the method of dyeing, the cloth is steeped in an infusion of the root and mordanted with alum.

The dye present in the roots of *R. cordifolia* is a mixture of purpurin (trihydroxy anthraquinone) and munjistin (xanthopurpurin-2-carboxylic acid). Purpurin is the main colouring principle in the roots while munjistin, an orange dye, occurs in the form of glycoside. When alum is used as mordant, purpurin dyes cotton fabrics a bright scarlet-red, and munjistin a shade of orange.

Naga Madder

Rubia sikkimensis Kurz.; Eng.—Naga madder; Manipur—*Moyam*; Family—Rubiaceae. A stout, handsome creeper, found in the Eastern Himalayas in Sikkim and Bhutan, at an altitude of from 2,000 to 5,000 feet in Upper Assam, and in Manipur and the Naga hills. Leaves equal in size, 4 in each whorl, lanceolate or elliptic lanceolate, sometimes flowers dull yellow or creamy white with greenish tinge, in terminal or axillary cymose panciles; fruits globose.

Dye. This is the main dye-yielding plant in Eastern India. Dried roots yield a brilliant red dye, which is used by the hill tribes for dyeing woollen clothes, human hair, decorations for spears and ornaments, and cane and bamboo articles.

Roots contain purpurin as the principle colouring matter munjistin, purpuroxanthin and traces of red colouring matter are also present. The shades obtained by the root extracts with various mordants are similar to those of *R. cordifolia*.

European Madder

Rubia tinctorum Linn.; Eng.—European madder; Punj.—*Bacho*; Family—Rubiaceae. A climbing herb with perennial root; leaves 4-6 in whorls, elliptic or elliptic-lanceolate; flowers greenish yellow in terminal and axillary panicles of cymes; fruit globose, red to black. Native of South Europe and Asia. In India, the plant is found in Kashmir.

Dye. The dried and ground root was formerly one of the most valued and most largely employed dye-stuffs both in Europe and Asia. It has, however, now been entirely replaced by the coal-tar derivatives—Alizarin and allied colouring matters. Until its replacement by synthetic dyes, *Rubia tinctorum* had been one of the most important dye-yielding plants, giving brilliant shades of red, known for their beauty and fastness of tint to both light and water.

The dye is found principally in the cortex of the roots. It is found in the form of glycosides (2-3.5%) which readily break down to give both alizarin and purpurin.

Madder yields the famous Turkey-red colour, on aluminium mordanted cottons, but now it has almost entirely been replaced by synthetic alizarin which is more effective and cheaper than the natural dye-stuff.

The European madder is used in calico printing and dyeing. It is preferred to synthetic alizarin for coating the printing foil; rugs are also dyed with European madder. European madder is also used to colour food products and cosmetics.

DYES OBTAINED FROM LEAVES

Safed kikar

Acacia leucophloea (Roxb). Willd.; Hindi—*Safed kikar*; Tam.—*Vel-velum*; Tel.—*Tella-tuma*; Family—Mimosaceae. A large, deciduous tree, found in the Punjab, Rajasthan and Madhya Pradesh.

Dye. The leaves and bark are used in dyeing, and give a black colour.

Arusa

Adhatoda vasica Nees; Eng.—Malabar nut; Hindi—*Arusa*, *adalsa*; Beng.—*Bakas*, *vasaka*; Guj.—*Aduso*; Kumaon—*Bashangarus*; Sans.—*Arus*, *vasaka*; Uriya—*Basung*; Tam.—*Adhatodai*; Tel.—*Adasara*; Mal.—*Atalotakam*; Nepal—*Alesi*, *kath*; Family—Acanthaceae. A small shrub, commonly found in N. India, ascending upto 4,000 feet in altitude.

Dye. A yellow dye, obtained from the leaves by boiling, is used for dyeing coarse cloth. It gives a greenish-blue when combined with indigo.

Barbados Aloe

Aloe barbadensis Mill.; Syn. *A. vera* Tourn. ex Linn.; Eng.—Barbados Aloe; Hindi—*Ghikanvar*, *ghigvar*, *kumari*; Beng.—*Ghritakumari*, *girta-kunvar*; Sans.—*Ghrita-kumari*, *kanya*; Guj.—*kunvar*, Tam.—*Kattalai*; Tel.—*Kalabanda*; Mal.—*Kattala*; Kan.—*Lola-sora*; Family—Liliaceae. A succulent shrub; native of the West Indies but grows wild in South India.

Dye. A dye, designated as "Chrysammic acid" is prepared from the succulent leaves of this plant. It is prepared by heating 8 parts of nitric acid with one part of aloe leaves. After the violent action has subsided, a second proportion of aloes is added to the mixture until the fumes of hypo nitric acid subside. The mass is then poured into water, when chrysammic flakes settle in the bottom of the vessel. These are washed several times in water. The crystals change their colour under varying circumstances, giving a purple colour to silk, black to wool, and pink to linen.

Latka

Baccaurea sapida Muell.—Arg.; Hindi—*Lutco*; Beng.—*Latka*; Nepal—*Kala bogoti*; Assam—*Latecku*; Kan.—*Koli kuki*; Family—Euphorbiaceae. A moderate sized evergreen tree, found in Bengal, Tripura and Andaman Islands.

Dye. The leaves are used in Bengal and Assam for dyeing. A green dye is extracted from the leaves by natives in Assam.

Indian Cassia

Cinnamomum tamala (Buch.—Ham.) T. Nees and Eberm.; Eng.—Indian cassia; Hindi—*Talisputar*, *talispatri*, *silkanti*, bark = *taj kalami*, *taj kalam*, leaves = *tezpat*, *tajpat*; Beng.—*Tejpat*; Nepal—*Chota sinkoli*; Assam—*Dopatti*; Guj.—*Taj*, *tamalpatra*; Tam.—*Talisha-pattiri*; Tel.—

Tallisha-patri; Kan.—*Lavanga patte*; Sans.—*Tamal, tespatra*; Family—Lauraceae. A moderate-sized evergreen tree, found in Assam, Bengal and Khasia hills, between 3,000 and 7,000 feet elevation.

Dye. The leaves are employed in calico-printing in combination with myrobalans.

Bakrelara

Hedyotis capitellata Wall.—Syn. *Oldenlandia rubioides* Miq.; Hindi—*Bakrelara* (*Pahiria*); *Kathenyok* (Lepcha); Family—Rubiaceae. A climber. This plant is found in West Bengal (Darjeeling district), Manipur and Khasia hills.

Dye. It is used by the natives as a green dye. The green leaves are put into water and infused, and the cloth to be dyed steeped in the infusion. It is thought, that it is of value more as a mordant than a real dye. The dye is chiefly used by the Lepchas.

Common Holly

Ilex aquifolium Linn.; Eng.—Common Holly; Family—Aquifoliaceae. A small tree; native of Europe; introduced in India and found in gardens within the temperate tracts. It is found in many parts of Manipur and Naga hills. Bark grey, smooth for a long time, eventually finely fissured; leaves alternate, short petioled, elliptic or oblong, spiny-toothed, shining, coriaceous; flowers in axillary clusters, white or yellow, fragrant.

Dye. The leaves contain a crystalline yellow colouring matter, ilixanthin ($C_{34}H_{22}O_{22}$), which dyes yellow with alum and green with ferric chloride.

Indigo

Indigofera tinctoria Linn.; Eng.—Indigo; Hindi—*Nil*; Beng.—*Nil*; Punj.—*Nil*; Guj.—*Nil*; Tam.—*Nilam*; Tel.—*Nili mandu*; Kan.—*Nili*; Mal.—*Nilam*; Sans.—*Nili, nilini, tuli, nilika, ranjani*; Family—Papilionaceae. A small shrub, formerly cultivated in India for its famous dye '*nil*'. Found throughout the tropical belt of India. Leaves imparipinnate bearing laterally attached hairs; stamens diadelphous; pods round, dehiscent, not jointed.

Dye. For a very long time indigo (*nil*) was known as the most important dye. Now, it has almost entirely been replaced by a synthetic product. The indigo industry was formerly widespread in India and other countries. However, the dye is not present in the plant itself. The leaves contain a soluble colourless glucoside, the *indican*, which oxidises in water to form the insoluble indigo. Indigo has been used as a dyestuff in India from earliest time. Now the dye has been substituted by a synthetic one.

Lal Bherenda

Jatropha glandulifera Roxb.; Syn. *J. glauca* Vahl.; Beng.—*Lal-bherenda*; Tam.—*Addalai*; Tel.—*Nela amudamu*; Sans.—*Nikumba*; Family—Euphorbiaceae. A shrub or a small tree. Commonly found near the villages in Bengal. Found also in the Punjab, Travancore and along the Coromandel coast.

Dye. One of the most interesting economic features of this plant is the fact that it is one of those reputed to yield a green dye. The juice of the leaves is employed for dyeing cotton, silk and wool.

Henna

Lawsonia inermis Linn.; Syn. *L. alba* Lam.; Eng.—Henna, Egyptian privet; Hindi—*Mehndi, hena*; Beng.—*Shudi, mehedi, mendi*; Uriya—*Manghati*; Punj.—*Mehndi, hinna*; Raj.—*Mehendi*; Guj.—*Medi, mendi*; Tam.—*Marithondi, aivanam*; Tel.—*Goranta, iveni*; Kan.—*Gorante*; Mal.—*Mayilanchi, ponta letshi*; Sans.—*Mendhi, mendika, sakachara*; Pers.—*Hina*; Family—Lythraceae. A small, elegant, and sweetly scented bush, cultivated commonly throughout India. It is grown as a hedge plant throughout India; as a commercial dye crop it is cultivated mainly in Haryana and Gujarat, and to a small extent in Madhya Pradesh and Rajasthan. The more important centres of production are—Faridabad in Haryana State and Bardoli and Madhi in Surat district of Gujarat

State which together account for 87% of the total production of henna leaves. The plant may be grown in any type of soil, but it thrives best on heavy soils which are retentive of moisture. Propagation is done by seeds and cuttings. Once established, the plants continue to flourish and yield successive crops of leaves for several years.

Fig. 16.3. *Lawsonia inermis*. Henna (**Mehndi**). A flowering twig. It is used for colouring palms of hands, soles of feet and finger nails. It is used also for dyeing hair, etc. Henna powder is pasted with water and applied to the part to be dyed. It is harmless and causes no irritation of skin.

Dye. Henna has long been used in India and Middle East countries for colouring palms of hands, soles of feet and finger nails. It is used also for dyeing hair, beard and eye brows. Tails and manes of horses are sometimes dyed with henna. For use as dyeing material, henna powder is pasted with water and applied to the part to be dyed. For dying hair it is applied as a pack; it imparts an orange red colour. Mixture of henna with indigo gives compound henna used by hair dressers as vegetable pack. A mixture of 1 part of henna and 2 parts of Indigo, Henna-Reng as it is called, imparts a brown tint, while henna-reng containing 1 part of henna and 3 parts of indigo gives a dark brown colour. Henna was once extensively used for dyeing silk and wool. The use of henna as a textile dye has declined since the advent of synthetic dye-stuffs.

The principal colouring matter is lawsone, 2—hydroxy—1 : 4—naphthaquinone ($C_{10}H_6O_3$) which is present in dried leaves in a concentration of 1.0-1.4%. Henna extracts are useful for dyeing in acid baths; alkalies intensify the colour but destroy the dyeing properties. Besides lawsone, other constituents present in henna are—gallic acid, glucose, mannitol, fat, resin, mucilage and traces of an alkaloid.

About 15% of the total production is used within the country in the form of powder of leaves; the rest is exported in the form of dried leaves or powder.

Three grades of henna are recognized in the trade and designated as Delhi, Gujarat and Malva. Henna of Delhi quality is sold in the form of powder and chiefly marketed in Faridabad. Gujarat henna is marketed in the form of leaves. Malva henna (produce of Rajasthan) is marketed in the form of powder, the main trading centre being Bhawani mandi near Kotah. Delhi henna is supposed to be superior than others in dye content; Malva henna and Gujarat henna, follow in order.

Iron Wood Tree

Memecylon umbellatum Burm. f.; Syn. *M. edule* Roxb.; Eng.—Iron wood tree; Mar.—*Anjan, anjana, limba*; Tam.—*Cashamaram, kasa*; Tel.—*Alli, alli topalu*; Kan.—*limbtoli*; Mal.—*Myenphoete-nyet*; Uriya—*Nirassa, bonohorono*; Assam—*Lalidimabophang*; Family—Melastomataceae. A large ornamental shrub or a small tree, found mostly in the coastal regions of Deccan Peninsula, Eastern parts of India and Andaman Islands. Bark grey with vertical furrows; leaves elliptic or ovate, shiny; flowers in umbellate cymes, bright blue; berries globose, purplish or black. Flowers during March-April and fruits appear in July-August.

Dye. The leaves yield a yellow dye which gives beautiful light shades. In combination with myrobalans and sappan wood, they produce a deep red tinge much used for dyeing and grass mats, and also good for cloth. The dye content is, however, low. The leaves contain a yellow glycosidal substance, tartaric and malic acids (1.38%), a resin (6%) and calcium oxalate (1.44%).

Smoke Tree

Rhus cotinus Linn.; Syn. *Cotinus coggygria* Scop.; Eng.—Smoke tree, Elm-leaved sumach; Hindi—*Tunga, tung*; Punj.—*Phan, tung, tittri*; Kashmir—*Darengri*; Family—Anacardiaceae. A shrub or small tree of the Western Sub-tropical Himalayas from Kashmir to Kumaon; found upto altitudes of 5,000 feet.

Dye. The leaves, bark and wood are locally employed in dyeing. The astringent leaves are used in dyeing with *kahi* to produce black and grey shades.

Pathani Lodh

Symplocos crataegoides Buch.—Ham. ex. D. Don; Syn. *S. paniculata* Wall. ex. D. Don; Kumaon—*Lodh*; Punj.—*Pathani lodh*; Family—Symplocaceae. A large shrub or small tree, found in the Himalayas from Kashmir to Assam, occurring at altitudes between 3,000 and 8,000 feet. It is found also on the Khasia hills.

Dye. The leaves and bark are used in dyeing and yield a yellow colour. They are principally employed in combination with madder, and act probably more as a mordant than as colouring material.

Lodh Tree

Symplocos racemosa Roxb.; Eng.—The lodh or lodh tree; Hindi—*Lodh*; Beng.—*Lodh*; Uriya—*Ludhu*; Tel.—*Ludduga*; Sans.—*Lodhra*, *marjana*; Family—Symplocaceae. A tree, found in the plains and lower hills of Bengal and Assam, chiefly found in dry forests, upto altitudes of 2,500 feet.

Dye. The leaves and the bark of this species are used in dyeing, and a yellow dye is said to be extracted from both.

Lodh

Symplocos spicata Roxb.; Hindi—*Bholia*, *lodh*; Beng.—*Buri*; Assam—*Bhumrati*; Family—Symplocaceae. A small tree, found in North and East India upto altitudes of 4,000 feet. It is commonly found near the base of the hills from Kumaon to Bhutan and in Assam.

Dye. In the Eastern Himalayas the leaves are dried, pounded, mixed with the fruit of a plant called *kauda*, which is also pounded, and the article to be dyed is steeped in the infusion, a yellow colour is thus obtained.

Tamarind

Tamarindus indica Linn.; Eng.—Tamarind; Hindi—*Imli*, *amli*, *teter*; Beng.—*Amli*, *ambli*, *tinturi*; Assam—*Teteli*; Nepal—*Titri*; Uriya—*Tentuli*; Punj.—*Imli*; Guj.—*Amli*, *ambli*; Tam.—*Puli*, *puliyam*; Tel.—*Chinta-pandu*; Kan.—*Hunashe-hannu*; Mal.—*Puli*, *puliyam-pazham*; Sans.—*Tintiri*, *tintili*, *amlika*; Family—Caesalpiniaceae. A large evergreen tree, which grows to a height of 80 feet, with a circumference of 25 feet. Cultivated throughout India. Grown as an avenue. In Madhya Pradesh and many parts of South India it is found self-sown in waste and forest lands.

Dye. An infusion of the leaves yields a red dye and imparts a yellow shade to cloth already dyed with indigo. The leaves, flowers and fruit contain a large proportion of acid and are much employed as auxiliaries in dyeing, especially along with safflower. They act as mordants in silk dyeing.

Teak

Tectona grandis Linn.; Eng.—Teak; Hindi—*Sagun*, *sakhu*; Beng.—*Segun*; Assam—*Chingjagu*; Uriya—*Singuru*; Punj.—*Sagun*, *sagwan*; Guj.—*Saga*; Tam.—*Tekkumaram*; Tel.—*Teku*, *pedda teku*; Kan.—*Sagwani*, *tega*; Mal.—*Tekka-maram*; Sans.—*Saka*; Family—Verbenaceae. A large tree, found in the Western Ghats, Tamil Nadu, Madhya Pradesh, Orissa, Mysore and Bihar.

Dye. The leaves yield a red or yellow dye. The leaves yield a yellowish-purple colour, used as a dye for silk and cotton. They are also used as mordants along with a species of Labiatae in dyeing black.

Bhangra

Wedelia calendulacea Less.; Hindi—*Bhanra*, *bhangra*; Beng.—*Kesraj*, *Kesaraja*, *bhimraj*; Guj.—*Pila-bungra*; Tam.—*Postaley-kaintagerai*; Sans.—*Bhringaraja*, *kesaraja*, *pitabhringi*;

Family—Compositae. A herb. Found in wet places in Bengal, Assam, the Eastern and Western Peninsulas. The plant possesses a slight camphoraceous odour.

Dye. The leaves and their juice are used in dyeing grey hair and for promoting the growth of hair.

Dhaula

Woodfordia fruticosa (Linn.) Kurz; Syn. *W. floribunda* Salisb.; Hindi—*Dhawi*, *dhauta*, *dhaula*; Beng.—*Dawa*, *dhai*, *dhaura*; Nepal—*Dahiri*; Uriya—*Jatiko*, *harwari*; Kashmir—*Thawi*, *thai*; Punj.—*Dhaur*, *tawi*, *thai*; Guj.—*Dhavadina*; Tel.—*Jargi*, *dhataki*; Sans.—*Dhataki*, *agnijvala*; Family—Lythraceae. A common ornamental shrub. Found throughout India, ascending upto 5,000 feet in Himalayas.

Dye. The leaves and twigs yield a yellow dye called *nauti*, occasionally used in calico-printing.

Sweet Indrajao

Wrightia tinctoria R. Br.; Eng.—Sweet indrajao; Hindi—*Indarjou*; Beng.—*Indrajau*; Guj.—*Indarjou*; Tam.—*Pala*, *velpala*; Tel.—*Tella pal*, *amkudu*; Kan.—*Kodmurki*; Mal.—*Kotakappala*; Sans.—*Hyamaraka*; Family—Apocynaceae. A tree, found in Rajasthan, Madhya Pradesh, Tamil Nadu and Western Peninsula.

Dye. From time immemorial the natives of South India have employed the leaves as a source of a blue dye or indigo.

DYES OBTAINED FROM FLOWERS

Hollyhock

Althaea rosea Linn.; Eng.—Hollyhock; Hindi—*Gulkhera*; Guj.—*Gul-khair*; Tam.—*Shemai-tutti*; Family—Malvaceae. An ornamental sub-bushy erect herb. Largely cultivated in Indian gardens, flourishing freely at all hill stations.

Dye. The flowers yield a red dye (anthocyanins).

Pot Marigold

Calendula officinalis Linn.; Eng.—Pot-marigold; Punj.—*Zergul*, *Akel-ul-mulk*; Family—Compositae. An ornamental herb. Found in the plains of the Punjab. It is called *zergul* in the Trans-Indus tracts, where it is wild in some parts.

Dye. An extract of the flowers is said to be used to colour butter and cheese.

Flame of the Forest

Butea monosperma (Lamk.) Taubert; Syn. *B. frondosa* Koenig ex Roxb.; Eng.—Flame of the forest, Bengal—*Kino*; Hindi—*Dhak*, *palas*, *tesu*; Beng.—*Palas*; Bihar—*Paras*, *faras*; Nepal—*Palasi*; Uriya—*Porasu*; Guj.—*Khakara*; Mar.—*Paras*, *palas*; Tam.—*Porasan*, *murukkan*; Tel.—*Moduga*, *palasamu*; Kan.—*Muttuga-gida*; Mal.—*Plach-cha*; Sans.—*Kinsuka*, *palasa*; Family—Papilionaceae. A moderate-sized, deciduous tree, found throughout India, extending in the North-West Himalayas as far as the Jhelum. This is one of the most beautiful trees of the plains and the lower hills of India.

Dye. The flowers, called *tesu*, or *palas-ke-phul*, yield a brilliant but fleeting yellow dye, used by the natives of India, especially during the *Holi* festival. The dye is extracted either by expressing the coloured sap of the fresh flowers, or as a decoction or infusion from the dried flowers. The flowers are collected in March-April, and as a rule are sun-dried. The petals are separated from the rest of the flower and preserved. Simple immersion in water will extract the colour, but in some parts of the country the dyestuff is boiled. The cloth to be dyed is sometimes boiled in the solution without the aid of any auxiliary or mordant.

The process of extraction of the dyc is as follows.—A given weight of dye is mixed thoroughly with twice as much water. After having been allowed to soak for some time, the mixture is boiled down to half its volume. It is then strained and allowed to cool. The cloth is either immersed in it before it cools, or is boiled with a required amount of dye solution. The natives prefer the fleeting but brilliant yellow colour produced without the aid of any auxiliary, especially to dye the clothes worn at the *Holi* festival; the fact of the colour being fleeting is viewed rather as an advantage than otherwise, since it can be got rid of after the festival is over. The addition of an alkali, deepens the colour into orange and makes it at the same time a little less fleeting. Alum, lime, wood ash or *sajjimati* serves this purpose.

Safflower

Carthamus tinctorius Linn.; Eng.—Safflower, African saffron, American saffron; Hindi—*Kusum*, *kasumba*; Beng.—*Kusum*, *kusamphul*; Punj.—*Kusam*, *kurtam*; Raj.—*Bundi*; Guj.—*Kusumbo*; Tam.—*Sendurgam*, *kushumba*; Kan.—*Kusumba*; Sans.—*Kusumbha*; Family—Compositae. An annual herbaceous plant, with large orange-coloured flower heads, cultivated as a dye-crop all over India. It is native of Europe and West Asia. In India, chiefly cultivated in eastern Uttar Pradesh, Bihar, Bengal, Orissa, Punjab, Andhra Pradesh, Madhya Pradesh, Karnataka and Maharashtra.

Dye. The flowers are used as dye for colouring food and cloth. The florets are picked as fast as they appear, since they lose their colour if left exposed to the sun. They are dried in the shade, when perfectly dry they are powdered and sifted. The best harvests are taken in the middle of the season. The water is sprinkled on the dried flowers while beaten. This process is continued until the water passes through quite clear. At first it is bright yellow, because of the soluble yellow colour which the florets contain in addition to their valuable red dye. The florets are repeatedly washed until they are quite freed from the yellow colour. The pulpy mass is now squeezed between the hands into small, flat, round cakes. These are known in the trade as stripped safflower. When these cakes are dried, they are ready for the market.

The quality of safflower cake is estimated by dyeing a known weight of cotton; about 125 gms. of safflower will dye 450 gms. of cotton cloth light pink; 250 gms. will dye it full rose pink; and from 375 gms. to 450 gms. will dye it a full crimson. In order to take up this quantity, the cotton must be several times dyed in fresh solutions of the colouring matter.

The florets contain three colouring principles, namely, two yellows and one red, the latter (red) being Carthamin or Carthamic acid ($C_{14}H_{16}O_7$). The first yellow colouring substance is soluble in water, and constitutes 26 to 30% of the florets, while 0.3 to 0.6% is the usual amount of Carthamin.

If the dye-stuff, after the removal of soluble yellow principle, be acidified with acetic acid, filtered, and first acetate of lead and next ammonia added, the second yellow colour will be precipitated along with lead salt. Carthamin, the valuable red colour, may be extracted in a pure form by making use of its solubility in alkaline solutions and insolubility in pure or acidified water. The *sajji-mati* (sodium carbonate) may be employed as alkali.

Carthamin is insoluble in ether, almost so in water, but soluble in alcohol, to which it imparts a cherry-red. This alcoholic tincture dyes silk immediately, no mordant being required.

Saffron

Crocus sativus Linn.; Eng.—Saffron; Hindi—*Kesar*, *zafran*; Beng.—*Jafran*; Guj.—*Keshar*; Sans.—*Kunkuma*; Tam.—*Kungumapu*; Tel.—*Kunkum apave*; Kashmir—*Kong*; Family—Iridaceae. A herb. It is cultivated in Kashmir at Pampur near Srinagar. The Indian supply is obtained from Kashmir.

Dye. The dried stigmas and tops of the styles yield a dye. It is chiefly used to colour cheese, puddings, pulao and curries. In India, it is too expensive to be used as a dye-stuff. The product (saffron) is obtained from the stigmas of the flowers, 4,000 of which are required to produce about 30 gms. of saffron.

Indian Coral Tree

Erythrina indica Lam.; Eng.—Indian coral tree; Hindi—*Pangra*, *mandara*; Beng.—*Palita mandar*; Uriya—*Paldua*; Guj.—*Panarvo*; Tam.—*Muruka*; Tel.—*Barijamu*; Kan.—*Haliwara*; Mal.—*Dudap*; Sans.—*Palitmandar*; Family—Papilionaceae. A moderate-sized, ornamental tree with straight trunk, which is usually armed with prickles when young. It occurs throughout India from the foot of the Himalayas.

Dye. The dried red flowers on being boiled yield a red dye.

Sunflower

Helianthus annuus Linn.; Eng.—Sunflower; Hindi—*Surajmukhi*; Beng.—*Surja muki*, *shuria mukti*; Tel.—*Aditya bhakti-chettu*, *podda-trim-gudda-chettu*; Sans.—*Suria-mukhi*; Family—Compositae. An ornamental annual herb; introduced from Canada. It bears large, flat, circular flower heads. Chiefly grown in gardens. In the plains flowering occurs in cold season whereas in hills it flowers during summer.

Dye. The blossoms yield a brilliant, lasting useful dye. The petals are quite rich in the amorphous resinous substance—*Xanthin*—the base of the yellow pigment from which the colour is derived. The seeds contain *helianthic acid*, which when treated with hydrochloric acid in a current of hydrogen, is resolved into glucose and a violet dye.

The colouring materials present in the flower are b- carotene, cryptoxanthin, taraxanthin, lutein and quercimeritrin.

Shoe Flower

Hibiscus rosa-sinensis Linn.; Eng.—Shoe flower, Rose of China; Hindi—*Gurhal, jasut, jasum*; Beng.—*Juwa*, *joba*, *jiwa*; Guj.—*Jasuva*; Mar.—*Jasavanda*; Tam.—*Shappat-tup-pu*; Tel.—*Java pushpamu*; Mal.—*Chempa-rattip-puva*; Kan.—*Dasavala*; Sans.—*Japa-pushpam*; Family—Malvaceae. An evergreen, ornamental bush, 5-8 feet high; leaves bright green, ovate, entire below, coarsely toothed above; flowers solitary axillary. It is native of China. Grown as an ornamental plant in gardens throughout India and often planted as a hedge plant. Numerous types with single and double flowers of red, yellow, white, cherry and stripped colours are in cultivation. Many of them are hybrids with allied species, such as *H. tiliaceus* and *H. schizopetalus*. The plant thrives best in manured and irrigated soils. It can be propagated by cuttings.

Dye. An infusion of the flowers produces a purplish dye. A red dye is obtained by rubbing the flowers on the paper. Crushed flowers yield a dark-purplish dye which was once used for blackening shoes, hence the English name of the plant. The Chinese people use the dye for colouring hair, eyebrows, foods and liquors. The flowers contain an anthocyanin pigment, cyanidin diglucoside.

Garden Balsam

Impatiens balsamina Linn.; Eng.—Garden balsam; Hindi—*Gul-mendi*; Beng.—*Dupati*; Uriya—*Haragaura*; Punj.—*Bantil*, *halu*, *tilphar*, Mar.—*Tirada*; Tam.—*Kasittumbai*; Guj.—*Gulmendi*; Mal.—*Mecchingom*; Family—Balsaminaceae. A succulent annual herb; leaves alternate, lanceolate, serrate; flowers solitary or fascicled, purple, pink, red and white, with long slender incurved spurs; fruits capsular, hairy. Found throughout tropical India.

Dye. The flowers of this plant in Garhwal are used to obtain a dye, where it is called *mujethi*. The flowers are also used for a dye in Jaintia hills by tribal people. The deep red flowers contain a monoglycosidic anthocyanin based on pelargonidin. Sometimes the flowers are used as a substitute for henna (*Lawsonia inermis* Linn.) for dyeing finger nails.

Tree of Sorrow

Nyctanthes arbor-tristis Linn.; Eng.—Tree of sorrow, Night flowering jasmine; Hindi—*Harsinghar*, *seoli*; Beng.—*Sephalika*, *harsinghar*; Uriya—*Gongo, seoli*; Punj.—*Pakura*, *laduri*, *harsingar*; Mar.—*Khurasli*; Tam.—*Manja-pu*; Tel.—*Paghada*, *karchia*; Kan.—*Harsing*; Sans.—

Parijataka, sephalika, rajanikasa; Family—Oleaceae. A large shrub, with rough leaves and sweet-scented flowers; branchlets quadrangular, strigose; leaves ovate, acuminate, entire; flowers small, 3-7 in each bunch, arranged in trichotomous cymes.

Dye. The corolla tubes are orange-coloured, and when severed from the limbs give a beautiful but fleeting orange or golden dye. The dye was formerly used for silk, sometimes in combination with turmeric. For dying purposes, the fabric was steeped in a hot or cold water decoction of the material. It imparts a beautiful, but fleeting, orange or golden colour. The addition of lime juice or alum to the dye bath makes the colour more permanent.

The bright orange corolla tubes of the flowers contain a colouring matter, nyctanthin, which occurs in the material in a concentration of 0.1% probably as a glucoside.

Toon

Toona ciliata Roem.; Syn. *Cedrela toona* Roxb. ex. Rottl. and Willd; Eng.—Cedrela tree, Red cedar, Toon, Indian mahogany; Hindi—*Toon, tuni*; Beng.—*Tuni, tun*; Uriya—*Maha limbu*; Punj.—*Drawi, Chiti-sirin, tun, gultan* (flowers); Nepal—*Tuni*; Assam—*Poma, tun*; Tam.—*Tunu-maram*; Tel.—*Nandi-chettu*; Mal.—*Arana-marani*; Kan.—*Tundu, mara*; Sans.—*Tunna, kachla*; Family—Meliaceae. A large tree, about 50 to 60 feet in height, found in the Punjab, Assam, Bihar, the Western Ghats, Maharashtra, Madhya Pradesh and the Nilgiris; ascending to 3,000 feet in the N-W. Himalayas and in Sikkim to 7,000 feet.

Dye. The flowers yield a red and a yellow dye, known as *gulnari* in Bengal. It is said to be used for dyeing cotton. The flowers are boiled to extract the colour known as *basanti*. It is fleeting and only used by the natives.

Dhaura

Woodfordia fruticosa (Linn.) Kurz; Syn. *W. floribunda* Ballab.; *Lythrum fruticosum* Linn.; Hindi—*Dhaura, dawi*; Beng.—*Dhai, dhaura*; Nepal—*Dahiri*; Uriya—*Harwari*; Kumaon—*Dhai, dhaula*; Kangra—*Guldaur*; Kashmir—*Thawi*; Punj.—*Tawi, dhaur*; Guj.—*Dhavadina*; Tel.—*Gaji, dhataki*; Sans.—*Dhataki, agnijvala*; Family—Lythraceae. A small much branched shrub, brilliantly purple in the summer season owing to the numerous flowers all along its branches. Commonly found throughout India, ascending to an altitude of 5,000 feet in the Himalayas.

Dye. The flowers are employed in dyeing either to produce a colour of themselves, or as a mordant principally with *Morinda citrifolia* (*al*). The plant flowers from February to April, during which period the flowers are gathered and dried. When used, the flowers are boiled in water, or else steeped for a considerable time in cold or hot water. To the solution thus prepared alum or lime and alum, is added as a mordant, and the material to be dyed is immersed in this solution several times until a pink colour of the required depth is obtained.

DYES OBTAINED FROM FRUITS AND FRUIT RIND

Bael

Aegle marmelos (Linn.) Corr. Eng.—Bael, Bengal quince; Hindi—*Bel, si-phal, siriphal*; Beng.—*Bela, bel, vilva*; Assam—*Bel*; Guj.—*Bil*; Sans.—*Sriphal bilva, bilvaphalam*; Tam.—*Vilva-pazham*; Tel.—*Maredu*; Mal.—*Kuvalap-pazham*; Kan.—*Belpatri*; Family—Rutaceae. A tree, cultivated throughout India. Straight, strong, axillary spines; leaves pale green of three leaflets; lateral leaflets sessile, ovate-lanceolate, 3-5 inches long, terminal long-petioled. Flowers an inch in diameter, greenish white, sweet scented.

Dye. A yellow dye is obtained from the rind of the fruit. The unripe rind is also used along with myrobalans in calico-printing.

Common Custard Apple

Annona reticulata Linn.; Eng.—Bullock's heart, common custard apple; Hindi—*Louna, Ramphal*; Beng.—*Nona*; Guj.—*Ramphal*; Tam.—*Ramsita*; Tel.—*Rama-pandu*; Family—Annonaceae. A small tree, native of the West Indies, now cultivated in Bengal and South India.

Dye. The dry unripe fruit yields black dye, and the fresh leaves a good quality of indigo.

Sappanwood

Caesalpinia sappan Linn.; Eng.—Sappanwood; Hindi—*Patang*; Uriya—*Bokmo*; Guj.—*Bakam*; Tam.—*Patanga*, *vatteku*; Tel.—*Bakamu*; Mal.—*Chappanum*; Sans.—*Pattanga*; Family—Caesalpiniaceae. A small thorny tree of the Eastern and Western Peninsulas; also cultivated in Madhya Pradesh and West Bengal.

Dye. A red dye is prepared from the pods, from the wood, from the bark or from all parts of the plant. However, root is reported to afford a yellow dye.

The pods are used with proto-sulphate of iron, to give a black colour. Sappan colour, however, is not permanent, being formed through the presence of the soluble substance *Brazilin*. Tannin and alum are used as mordants in attaching the dye to cotton, and a mixture of alum and cream of tartar in the case of wool. It is used for giving a red tint to silk, and for dyeing straw-plait for hat making.

Subali

Crozophora plicata Juss.; Hindi—*Shadevi*, *subali*; Guj.—*Okharada*; Beng.—*Khudiokra*; Sans.—*Suryavarta*; Punj.—*Nilkhānti*; Tel.—*Guruguchettu*; Family—Euphorbiaceae. Herb or shrub. Found throughout the warmer parts of India, from the Punjab to Maharashtra, Tamil Nadu and Bengal.

Dye. The fruits afford a purplish-blue dye. The cloth moistened with the juice of the green capsules, becomes blue after exposure to the open air. The fruits are used for the preparation of the dye named *turnsole*, used for giving a blue and a red tint.

Turnsole

Crozophora tinctoria Juss.; Eng.—Turnsole; Hindi—*Shadevi*, *sonballi*, *subali*; Punj.—*Tappal buti*, *nilan*, *kukronda*; Family—Euphorbiaceae. Commonly found in the Punjab and South India. A herb, possessing almost ovate-deltoid, leaves, on long thin petioles.

The plant is wild everywhere on the waste lands of India, luxuriantly found on both dry sandy tracts and river margins.

Dye. The fruits yield a purplish dye. By oxidation the dye becomes blue. The dye is called turnsole, and is obtained by grinding the plants, little herbs, to a pulp in a mill, when they yield about half their weight of a dark green coloured juice, which becomes purple by exposure to the air.

Indian Persimmon

Diospyros peregrina (Gaertn.) Gurke; Syn. *D. embryopteris* Pers.; Eng. Indian persimmon; Hindi—*Gab*, *makur-kendi*, *tendu*; Beng.—*Makarkenda*; Uriya—*Gusvakendhu*; Assam.—*Kendu*; Guj.—*Zeeberwo*; Tam.—*Tumbika*, *pani-chika*; Tel.—*Tubiki*, *tumika*; Kan.—*Kusharta*; Mal.—*Panichhi*; Sans.—*Trinduka*, *sindika*; Family—Ebenaceae. A dense evergreen small tree, with dark-green foliage and long shining leaves. Found in the Western peninsula and Madhya Pradesh. Found also in Bengal.

Dye and tan. The fruit is largely used as a tan. By simply steeping the half-ripe fruits in water a brownish liquid is obtained, which is sometimes used in dyeing a brown colour. This is made into a good black by combining with myrobalans (*Terminalia chebula*). The dried fruit and the calyx contain a small amount of colouring matter which is soluble in boiling water. The colouring matter from the fruit imparts a brown colour, dark, inclining to red, to the object to which it is applied, and also protects the timber or fibre from the action of water. The fruit is very often used for the brown dye which it yields. The fruit is simply pounded and boiled and the cloth steeped in the liquid, no auxiliary being employed. The dye is chiefly used for dyeing (usually called tanning) fishning nets and lines.

Japanese Persimmon

Diospyros kaki Linn. f.; Syn. *D. chinensis* Blume; Eng.—Japanese persimmon; Hindi—

Halwatendu, vilaiti-gab, kaki vhando; Family—Ebenaceae. A small tree, native of China but now grown in Kulu, Himachal Pradesh, Kumaon, The Nilgiris and West Bengal.

Dye. A black dye is produced from the fruit of this tree with sulphate of iron.

Tay

Diospyros pyrrhocarpa Miq.; Hindi—*Tay*; Family—Ebenaceae. An evergreen tree of the Andaman Islands.

Dye. A red dye is extracted from the fruit. The Chinese umbrellas are dyed with this substance, which has the property of rendering them waterproof.

Tikul

Garcinia pedunculata Roxb.; Beng.—*Tikul, tikur*; Assam—*Borthekra*; Manipur—*Heribung*; Family—Guttiferae. A tall tree, 50-60 feet high, found in Assam upto an altitude of 3,000 feet and in Manipur. Leaves obovate or oblanceolate, 6-12 in. long and 3-5 in. broad, leathery with a stout mid-rib; fruits subglobose, 3-4.5 in. diam., yellow with fleshy pericarp nearly one inch thick; seeds 8-10, enclosed in a pulpy aril.

Dye. The fruit is employed by the natives to deepen and render fast saffron dye.

Warrus

Flemingia grahamiana Wight and Arn. and *F. macrophylla* (Willd.) Kuntze ex. Prain; Eng.—Warrus, Wurrus, Wars, Varas; Hindi and Beng.—*Bara salpan, bhalia*; Mar.—*Dowdowdla*; Mal.—*Kamatteri*; Uriya—*Bonokandulo*; Assam—*Samnaskhat*; Nepal—*Batwasi*; Bihar—*Birbut*; Family—Pipilionaceae. An erect shrub. Native of the tropical and sub-tropical regions of the old world. *F. grahamiana*, a low erect shrub with tomentose young shoots, found in Nilgiri, Annamalai and Pulney hills of Travancore. *F. macrophylla*, an erect shrub, 4-6 feet high, with sulcate silky young branches, occurring at lower elevations throughout India and Andaman Islands.

Dye. *F. grahamiana* and *F. macrophylla* form the principal sources of the resinous powder, variously known as Warrus, Wurrus, Wars and Varas. For collecting warrus, the fruits (pods) are gathered, spread on boards or paper and dried by exposure to the sun. Dried pods are pressed or rubbed by hand over sieves. The powder collected is again sieved through fine muslin cloth. Warrus so prepared is a purplish red, granular powder without any marked odour or taste. Warrus contains 3.44, moisture; 72.83, resinous colouring matter; 8.20, albuminous matter; 9.50 fibrous matter and 6.03% ash. *F. grahamiana* yields warrus more freely than other Indian species. Pods of *F. macrophylla* also yield warrus.

Kamala

Mallotus philippinensis Muell.—Arg.; Eng.—Kamala, monkey face tree; Hindi—*Kamala, kambila, wussantha-gandha* (powder); Darjeeling—*Sinduri*; Beng.—*Kamalagundi, kamalaguri*; Uriya—*Bosonto-gund*; Assam—*puddum*; Nepal—*Sinduria*; Bundelkhand—*Rori*, Kumaon—*Roli, launti*; Kashmir—*Kaimbil*; Punj.—*Kamela, reini*; Guj.—*Kapilo*; Tam.—*Kapli, kapila*; Tel.—*Kunkuma, chendra-sinduri*; Kan.—*Kurku, rangamale, kunkuma*; Mal.—*Ponnagam*; Sans.—*Kapila, kampilla, rechanaka*; Family—Euphorbiaceae. A shrub or a small tree, with a short or often buttressed bole, found throughout tropical India; along the foot of Himalayas from Kashmir eastwards upto 5,000 feet elevation; found also in Bengal and Andaman Islands.

Dye. This tree has long been valued as the source of a dyeing material, known in the trade as *Kamala, kamala powder* or *kamala dye*, and used in the dyeing of silk and wool. The colouring material is found in the red glandular pubescence covering the ripe capsules (fruits) and is usually collected in February-March, when the fruits ripen.

The red pubescence is separated from ripe fruit by beating and shaking. It may also be obtained by stirring the fruits vigorously in water, when the dye settles down as a sediment; the sediment is collected, dried and pieces of pericarp and other refuse separated by sifting. The yield of powder is 1.4-3.7% of the weight of fresh fruits. Kamala dye is a granular, reddish brown resinous powder, almost without odour and taste. This dye is insoluble in cold water, slightly soluble in boiling water and freely soluble in alkalies, alcohol and ether, forming deep red solutions. The main colouring principles are the salmon-coloured rottlerin ($C_{30}H_{28}O_8$) and its yellow isomer, isorottlerin; together they constitute 11% of the weight of powder.

This dye was formerly extensively employed in dyeing silk and wool. It produces a white orange or flame colour which is fairly fast to soap, alkalies and acids, but fades somewhat in sunlight. A boiling alkaline bath having 4 parts of kamala dye, one part of alum and two parts of sodium bicarbonate is used for dyeing. The colour is heightened by the addition of turmeric powder. At one time this dye was exported from India, but with the advent of synthetic dye-stuff, the use of this material is practically ceased.

Kalampatti

Melastoma malabaricum Linn.; Nepal—*Choulisi*; Lepcha—*Tungbram*; Assam—*Phutuka*; Tel.—*Pattuda*; Tam.—*Nakkukaruppan*; Mal.—*Kalampatti*; Kan.—*Ankerki*; Uriya—*Gongai*; Family—Melastomataceae. A spreading shrub or a small tree found near water courses and moist places throughout India upto an elevation of 6,000 feet and in the Andaman Islands. Bark reddish brown, thin; leaves lanceolate to oblong; flowers in terminal corymbose panicles, purple coloured; fruit broadly ovoid, truncate, pulpy within.

Dye. The fruit yields a purple dye used for cotton. It is stated that an ink can be prepared from it. A pink dye is obtained from leaves and roots. The ashes of the plant are used as a dye mordant.

Common Myrtle

Myrtus communis Linn.; Eng.—Common myrtle; Hindi—*Vilayati mehndi*, *murad*; Beng.—*Sutrasowa*; Punj.—*Vilayati mehndi*, *murad*; Mar.—*Malati*; Guj.—*Makali-na-patran*; Family—Myrtaceae. An evergreen shrub, 1-3 m. in height. Leaves opposite, ovate to lanceolate, aromatic; flowers white, fragrant, axillary; berries ellipsoid, blue black with hard, kidney-shaped white seeds.

Dye. The berries are employed for dyeing. The berries contain an essential oil, citric acid, malic acid, resin and tannin.

Emblic Myrobalan

Emblica officinalis Gaertn.; Syn. *Phyllanthus emblica* Linn.; Eng.—Emblic myrobalan; Hindi—*Aonla*; Beng.—*Amlaki*, *amla*; Uriya—*Gondhona*; Assam—*Amluki*; Nepal—*Amla*; Guj.—*Ambala*, *amla*; Tam.—*Nelli*; Tel.—*Osirka*; Kan.—*Niliha*; Mal.—*Boa-malacca*; Sans.—*Amulki*, *umrita*, *dhatri*, *amala*; Family—Euphorbiaceae. A moderate-sized deciduous tree found throughout the forests of tropical India. Often cultivated in gardens and homeyards. Leaves feathery with small narrowly oblong, pinnate leaflets. Fruits depressed globose, fleshy; 6-lobed, containing 6 trigonous seeds.

Dye. The fruit, known as the *emblic myrobalan* (aonla), is used in dyeing and tanning. The dye obtained from the fruit is blackish-grey, but it is very rarely used alone, being generally employed like other myrobalans to produce a black with salts or iron, or the barks of others. It plays the part more of a colour concentrator than of a dye. Several beautiful light brown colours on silk, were obtained. It is also thought that the emblic myrobalan seems to be valuable when used alone as a dyestuff. The bark and leaves are also similarly employed, and produce the same colours. The leaves contain a small amount of colouring matter, and produce, by various processes, light drab and brownish-yellow colours. The fruit, leaves and bark all contain tannin, and are used for tanning purposes in various parts of the country.

Emetic Nut

Randia dumetorum Lamk.; Syn. *Gardenia dumetorum* Roxb.; Eng.—Emetic nut; Hindi—*Mainphal*; Beng.—*Menphal*; Uriya—*Pativa*; Assam—*Gurol*; Nepal—*Maidal*; Kumaon—*Mainphal*; Punj.—*Mindhal*; Guj.—*Medhola*; Tam.—*Madukarray*; Tel.—*Manda*; Kan.—*Kare*; Sans.—*Madana*;

Family—Rubiaceae. A deciduous thorny shrub or small tree, found in the sub-Himalayan tract, Maharashtra, Gujarat, Uttar Pradesh and Tamil Nadu. In Sikkim it occurs upto altitudes of 4,000 feet.

Dye. The fruits are used as a colour intensifier in calico-printing. In China, the fruits are used to produce a yellow dye.

Marking Nut

Semecarpus anacardium Linn.; Syn. *Anacardium orientale* Linn.; Eng.—Marking nut tree; Hindi—*Bhela, bhilawa*; Beng.—*Bhela, bhelatuki*; Uriya—*Bhallia*; Assam—*Bhola-guti*; Punj.—*Bhilawa, bhela*; Guj.—*Bhilamu*; Tam.—*Shen-kottai*; Tel.—*Jidivittulu, jidi, nellajedi*; Kan.—*Geru, kari-gheru*; Sans.—*Bhallataka*; Family—Anacardiaceae. A deciduous tree, reaching upto a height of 12-15 m. and a girth of 1.25 m. Found in the Punjab, Assam, the Khasia hills, Madhya Pradesh and the Western Peninsula. Found also in the outer Himalayas from the Sutlej to Sikkim and fairly common throughout the hotter parts of India. Bark dark brown, rough; leaves large, simple, obovate-oblong; flowers small, greenish yellow, dioecious, in terminal panicles; drupes 2.5 cm. long, obliquely ovoid, smooth and shining, black when ripe, situated on a fleshy orange coloured receptacle.

Dye and tan. The pericarp of the fruit contains a bitter and powerfully astringent principle, which is universally used in India as a substitute for marking-ink. It gives a black colour to cotton fabrics, which is insoluble in water but soluble in alcohol. The juice of the pericarp is mixed with lime water as a mordant before it is used to mark cloth. The fruits are used as a dye for cotton cloth. They are employed either alone or with alum.

The fruit is soaked in water for three days. The material to be dyed is well washed with water, and then half dry washed again in a solution of alum. When again half dried it is dipped in the fruit infusion, worked well about till the required depth of colour is obtained, then removed and dried in the sun. When quite dry it is washed frequently in fresh water to get rid of the smell of the dye-stuff. The colour produced by the use of this dye-stuff is a dark grey or greyish black.

Sterculia

Sterculia balanghas Linn.; Family—Sterculiaceae. A tree, found in South India, Maharashtra, Andaman Islands and throughout the hotter parts of India.

Dye. The capsules are burnt for the preparation of the colouring matter called by the natives *kussumbha*.

Belleric Myrobalan

Terminalia bellirica (Gaertn.) Roxb.; Syn. *Myrobalanus bellirica* Gaertn.; Eng.—Belleric myrobalan; Hindi—*Bahera*; Beng.—*Bohora*; Assam—*Bauri*; Guj.—*Beheda*; Tam.—*Vallai muruda*; Tel.—*Bahedha*; Kan.—*Bherda*; Mal.—*Thani*; Sans.—*Vibhitaki*; Family—Combretaceae. A large deciduous tree found in the plains and lower hills throughout India. The fruit ripens during the cold season, from November to January.

Dye. The fruit is known under the name of myrobalans, and is employed for dyeing and tanning. In India it is largely employed for dyeing, as a mordant, as a tan, and also medicinally. It may be used alone, in which case it gives a yellowish or brownish-yellow colour to the cloth, or with various other dye-stuffs to produce dark brown and black. The method employed for dyeing is as follows.—For each square metre of cloth take 250 gms. of *bahera* nuts. Extract and throw away the stones, and break the rind into small pieces. Put these pieces into one litre water along with 12.5 gms of pomegranate rind. Leave this to stand for one night. Now boil the infusion, allowing it to boil over three times. Then allow it to cool and strain through a coarse cloth. Wash the cloth to be dyed in water; when half dry wash again in water, in which 12.5 gms. of alum has already been dissolved. Then dip the cloth in the dye solution, working it about well so as to make the colour uniform. When the colour is deep enough, dry the cloth in the sun, and afterwards wash frequently in clear water so as to get rid of the smell of the dye. The resulting colour is a snuffy yellow.

Yellow Myrobalan

Terminalia chebula Retz.; Eng.—Yellow myrobalan, Chebulic myrobalan; Hindi—*Harara*; Beng.—*Haritaki*; Assam—*Hilika*; Guj.—*Hardi*; Tam.—*Kadakai*; Tel.—*Karaka*; Kan.—*Hirda*; Sans.—*Haritaki*; Family—Combretaceae. A large deciduous tree, commonly found in northern India from Kumaon to West Bengal.

Dye. In India the fruit is used as a dye by itself, the rind of the fruit being powdered and steeped in water. The cloth steeped in this infusion acquires a dirty grey colour. When used with alum, it gives a good permanent yellow. But the most extensive use to which *harara* is put as dye is in the production of various shades of black in combination with some salts of iron, generally the protosulphate. In some cases *gur* or molasses is mixed with dye to give depth to the colour. A dark neutral tint called *kakraiza* is obtained from *harara*, protosulphate of iron and safflower. The fruits are mixed with the pods of *Caesalpinia sappan* to produce a black dye. A mixture of the fruit and ferrous sulphate produces a *khaki* or iron-grey colour. It is commonly employed as a mordant, to concentrate the colour in dyeing with *Carthamus tinctorius* (safflower), *Morinda cetrifolia* (*al*), *Rubia cordifolia* (*Manjit*), *Curcuma domestica* (*haldi*) and *Butea monosperma* (*palas*). The fruits and galls are used for making country ink, and a black dye for staining the teeth.

Umbrella Tree

Thespesia populnea (Linn.). Soland. ex Corr.; Eng.—Portia tree, Umbrella tree or tulip tree; Hindi—*Paras-pipal*, *gajahanda*; Beng.—*Porash*; Punj.—*Paras pipal*; Guj.—*Bhindi*, *parasapiplo*; Tam.—*Purasha*; Tel.—*Gangarenu*; Kan.—*Asha*, *adavi bende*; Mal.—*Puvarasha*; Sans.—*Suparsha-vaka*; Family—Malvaceae. A moderate-sized evergreen tree, found along the coastal regions of Konkan, and the Andaman Islands. Largely cultivated along roadsides in Tamil Nadu.

Dye. The capsules (fruits) and the flowers yield a yellow dye. The dried capsules contain a small amount of yellow colouring matter soluble in water, and capable of producing faint shades of brownish yellow and light brown on silk and wool by adding suitable processes.

Harmal

Vitis setosa Wall.; Hindi—*Harmal*; Tam.—*Puli-naravi*; Tel.—*Barre bach-chali*; Family—Vitaceae. A climber, found in the Western Peninsula and Karnatak State.

Dye. The berries may be used for dyeing, staining, or colouring. A dark coloured black substance is obtained from the fruits (berries).

DYES OBTAINED FROM SEEDS

Arnatto

Bixa orellana Linn.; Eng.—Arnatto or Arnotto dye; Hindi and Beng.—*Latkan*, *latkhan*, *watkana*; Assam—*Jarat*, *jolandhar*; Uriya—*Gulbas*; Manipur—*Reipom*; Mar.—*Kesari*, *sendri*; Tel.—*Japhra-chettu*; Tam.—*Japhra-maram*; Kan.—*Kuppa-mankala*, *rangamali*; Family—Bixaceae. A shrub, with white or pinkish flowers and echinate red capsules. Native of America. It is cultivated in South India.

Dye. A red or orange dye is obtained from the pulp which surrounds the seed. The pulp gives a beautiful flesh colour, largely used in dyeing silks. It is altered by certain combinations into orange, deep orange or red, the brighter orange and red colours being obtained in combination with red powder of *Mallotus philippinensis*.

The dye may be extracted from the seeds direct, or the pulpy matter may by boiling be separated from the seeds and made into cakes. The mode in which it is obtained is by pouring hot water over the pulp and seeds, and leaving them to macerate and then separating them by pounding with a wooden pestle. The seeds are removed by straining the mass through a sieve, and the pulp being allowed to settle the water is gently poured off, and the pulp put into shallow vessels, in which it is gradually dried in the shade. After acquiring a proper consistency it is made into cylindrical rolls or balls, and placed in an airy place to dry.

The decoction of Arnatto in water has a strong peculiar odour and a disagreeable taste. Its colour is yellowish red, and it remains somewhat turbid. An alkaline solution makes it orange-yellow and more clear.

The mordant used with arnatto dye is alkali (usually crude pearl-ash); the alkali facilitates its solution, and the quantity of alkali used must be regulated according to the depth of colour required. The colour is chiefly used for dyeing silk. After dying the silk the colour may be deepened or reddened by means of vineger, alum or lemon-juice.

Cotton

Gossypium herbaceum Linn.; Eng.—Cotton; Hindi—*Kapas, Rui*; Beng.—*Kapas*, *tula*; Punj.—*Rui*; Guj.—*Ru*, *kapas*; Tam.—*Parutti*; Tel.—*Paratti*; Sans.—*Karpasi*, *karpas*; Family—Malvaceae. Cultivated mainly in Maharashtra, Gujarat, Tamil Nadu, Andhra Pradesh and Karnataka state.

Dye. Cotton seeds contain a principle which, by the action of acids, yields a blue colouring matter—*cotton seed blue*. This blue is not altered by reducing agents, but is at once decomposed by oxidising ones. It is insoluble in water. An alcoholic solution gives a fine blue colour to fabrics either mordanted or unmordanted with alum; the colour, however, fades rapidly.

Sunflower

Helianthus annuus Linn.; Eng.—Sunflower; Hindi, Beng.—*Surajmukhi*; Tel.—*Adityabhakti-chettu*; Sans.—*Suriya mukhi*; Family—Compositae. An annual herb, bearing large, flat, circular flower heads. Native of Mexico and Peru. In India, it is chiefly grown in gardens.

Dye. The seeds contain *helianthic acid*, which when treated with hydrochloric acid in a current of hydrogen, is resolved into glucose and a violet dye.

Perilla

Perilla frutescens (Linn.) Britt.; Syn. *P. ocimoides* Linn.; Eng.—Perilla; Hindi—*Bhanjira*, *bhasinda*; Kumaon—*Bhangara*, *Jhutela*; Family—Labiatae. A coarse aromatic herb, found in the tropical and temperate Himalayas, from Kashmir to Bhutan, at altitudes of 1,000 to 10,000 feet; also in the Khasia hills from 3,000 to 6,000 feet.

Dye. The seed is used as a dye auxiliary. The oil from the seeds is used for varnishes, paints, printer's ink and waterproof cloths.

Nux Vomica

Strychnos nux-vomica Linn.; Eng.—Nux vomica or Strychnine tree; Hindi—*Kuchla*; Beng.—*Kuchila*; Nepal—*Nirmali*; Uriya—*Kuchla*; Punj.—*Kuchila*; Mar.—*Kajra*; Guj.—*Kuchla*; Tam.—*Yetti-maram*; Tel.—*Musidi*; Kan.—*Kasaraka*; Sans.—*Vishamushti*; Family—Loganiaceae. A tree, attaining a height of 40 feet, found in Orissa, Bihar, Andhra Pradesh and Tamil Nadu.

Dye. The seeds yield a dye, which is used, for producing light brown shades on cotton cloth. Boiled with proto-sulphate of iron and lime they give darker shades of brown. They give light drab shades on silk, but are not well adapted for wool, on which only a faint colour can be obtained.

Crape Jasmine

Tabernaemontana divaricata (Linn.) R. Br.; Syn. *Ervatamia coronaria* Stap.; Eng.—Crape jasmine; Hindi—*Chandni*, *tagar*; Beng.—*Tagar*, *chameli*; Nepal—*Asuru*; Guj. and Mar.—*Sagar*, *tagar*; Tel.—*Grandi tagarapu*; Kan.—*Naginkada*; Sans.—*Tagara*; Family—Apocynaceae. A small, evergreen shrub, with silvery bark and glossy leaves, cultivated in gardens throughout India.

Dye. The red pulp obtained from the aril, or extra coat of the seed, gives a red colour, which is occasionally used as a dye by the hill-people.

Fenugreek

Trigonella foenum-graecum Linn.; Eng.—Fenugreek; Hindi—*Methi*; Beng.—*Methi*; Punj.—*Methi*; Guj.—*Methi*; Tam.—*Vendayam*; Tel.—*Mentula*; Kan.—*Menthya*; Mal.—*Ventayam*; Sans.—*Methika*; Family—Papilionaceae. An annual herb, native of South Europe, now grown mainly in North India.

Dye. The seeds yield a yellow dye, which enters into the composition of an imitation of carmine. The yellow decoction produces a fine permanent green with sulphate of copper.

Dharauli

Wrightia tomentosa Roem and Schult.; Syn. *Nerium tomentosum* Roxb.; Hindi—*Dharauli*; Beng.—*Dudh-koraiya*; Assam—*Atkuri*; Nepal—*Keringi*; Uriya—*Pal kurwan*; Punj.—*Dudhi*; Mar.—*Kala indarjau*; Tel.—*Tella pal*; Family—Apocynaceae. A small deciduous tree with corky bark, found in the Punjab, Rajasthan, Bihar, Assam and the Western Peninsula.

Dye. The seeds yield a yellow dye which is used for dyeing cotton cloth.

DYES OBTAINED FROM TWIGS

Rang

Peristrophe bivalis Merrill; Syn. *P. tinctoria* Nees; Beng.—*Rang*, *bet* or *batia rang*; Family—Acanthaceae. An erect spreading herb, 2 feet high, doubtfully wild, but cultivated in Bengal and Assam, especially near Midnapur (West Bengal); leaves ovate, sub-acute; flowers purple, axillary or terminal; capsules ellipsoid.

Dye. The twigs yield a dye ranging in colour from yellow orange to deep red orange. They are used in dyeing, to colour the *masland* mats of Midnapur. *Bet rang* is used exclusively in dyeing the sticks from which *masland* mats are manufactured. The sticks are made into convenient bundles, and pounded *rang* is placed upon the parts to be dyed. 1 kg. of pounded *rang* is then mixed with 20 litres of water, and the whole boiled. The part of the bundle to be dyed is dipped in this and boiled in it for three hours; then removed and dried. It has acquired a red colour. The dye is also used for dyeing fabric. The alum is used as a mordant for this purpose.

Assam Indigo Plant

Strobilanthes flaccidifolius Nees.; Eng.—Rum or Assam indigo plant; Assam—*Rum*, *rampat*; Manipur—*Khuma*, *khum*; Family—Acanthaceae. A shrub, found in Bengal, Assam and Manipur. It is often cultivated and usually occurs on the lower hills of these regions, at altitudes between 1,000 and 4,000 feet. It is propagated freely by cuttings, yields prunings twice or thrice a year, and is perennial.

Dye. In Manipur, the *Khuma* is cultivated, and the dye is extracted for home use; nearly every farmer cultivates a small plot and prepares his own dye. The process is as follows

The tops of the plants are cut twice a year, in May and October tied into bundles and immersed in a large earthen vessel containing water where they are left to steep for few days. Then the vegetable matter is taken out and thrown away, and the liquid after being thoroughly stirred up, is allowed to settle for the night. In the morning the liquid is poured off and the sediment put aside in an earthen vessel for use as required. In this state the dye, now known as *ṇamham* can be retained for six months or so without deterioration.

The dye is used in combination with turmeric to produce shades of green; with lime and turmeric, browns and almost reds; with lime alone, deep blue black; with safflower, purple and so on.

DYES OBTAINED FROM ALL PARTS OF THE PLANT

Indian Spinach

Basella rubra Linn.; Syn. *B. alba* Linn.; Eng.—Indian spinach; Hindi—*Poi*; Tam.—*Vasla-*

kire; Tel.—*Polam-bachcholi*; Mal.—*Basella-kira*; Family—Basellaceae. A herb. Cultivated in almost every part of India, especially in lower Bengal and Assam.

Dye. The herb yields a very rich purple dye.

Kalamath Weed

Hypericum perforatum Linn.; Eng.—Kalamath weed; Hindi—*Bassant, dendlu*; Punj.—*Basant, dendlu*; Family—Hypercaceae. A perennial herb, found in the Temperate Western Himalayas from Kumaon, altitude 6,000 to 9,000 feet, to Kashmir, altitude 3,000 to 6,500 feet. Rhizomatous herb upto 3 feet high. Stem 2-edged; leaves opposite, sessile, long, ovate or linear 0.3 to 1 inch long, black dotted; flowers yellow, in terminal corymbose cymes; capsule ovoid, 0.3 in. long; seeds many small. The herb possesses a characteristic balsamic odour and a bitter, resinous, somewhat astringent taste.

The principal constituents of the herb are—volatile oil, tannins, a resinous substance and a red flourescent pigment, hypericin ($C_{30}H_{14}O_{18}$).

Dye. The dried herb boiled in alum communicates a yellow or yellow red colour to wool and silk. Hypericin dyes wool and silk deep violet-red while the shade produced on mordanted cotton varies from bright green yellow and brown black to weak rose.

Knot Grass

Polygonum aviculare Linn.; Eng.—Knot-grass; Hindi—*Macihoti, nisomali, ban-natia, endrani, hunraj*; Beng.—*Machutie*; Punj.—*Tesru, banduke*; Sans.—*Miromati, nisomali*; Family—Polygonaceae. An annual herb, with prostrate stems and branches upto 2 feet high, found in the North-West Himalayas from Kashmir to Kumaon between 6,000 and 10,000 feet.

Dye. A blue dye, similar to indigo, is prepared from the plant.

Anil

Tephrosia tinctoria Pers.; Hindi—*Anil, alu-pilla*; Family—Papilionaceae. An undershrub, found in the Western Peninsula, ascending upto 5,000 feet.

Dye. A blue dye, similar to indigo, is sometimes extracted from this plant in Mysore.

DYES OBTAINED FROM GUM, RESIN, OIL AND SAP

Gumarabic Acacia

Acacia senegel (Linn.) Willd.; Eng.—Gumarabic Acacia; Hindi and Raj.—*Kumta*; Family—Mimosaceae. A low tree with grey bark; spines strong, short, sharp-hooked, often 3-nate, two lateral and one below the petiole; flowers in peduncled spikes 2-3 inches long; pod 3 by 3/4 inch, thin, grey, indehiscent, straight, strap-shaped, with a strong, fibrous, marginal midrib, and constricted between the seeds.

Dye. The gum is used for giving lustre to silk, and for thickening colours and mordants in calico-printing; for suspending tannate of iron in the manufacture of ink and blacking.

Neem Tree

Azadirachta indica A. Juss.; Syn. *Melia azadirachta* Linn.; Eng.—Margosa tree, Neem tree; Hindi—*Nim*; Beng.—*Nim, nimgachh*; Kumaon—*Betain*; Punj.—*Nim, mahanim*; Mar.—*Limba*; Guj.—*Limba*; Tam.—*Vembu*; Tel.—*Vepa, nim-bamu*; Kan.—*Heb-bavu*; Mal.—*Veppa*; Sans.—*Nimba, arishta*; Family—Meliaceae. A large tree 40 to 50 feet in height, common, wild or more often cultivated throughout the greater part of India.

Dye. The tree yields a bitter gum in abundance, which is used by the silk-dyers in the preparation of colours. The nim-oil is employed in dyeing cotton cloth; it imparts a deep yellow colour to the fabric.

Elephant Apple

Feronia limonia (Linn.) Swingle; Syn. *F. elephantum* Correa; Eng.—Elephant apple, Wood apple; Hindi—*Kaith*; Beng.—*Katbel*; Uriya—*Koeta*; Punj.—*Kait*; Mar.—*Kavatha*; Guj.—*Kotha*;

Tam.—*Vilam*; Tel.—*Velaga*; Kan.—*Bilwar*; Mal.—*Vilam*; Sans.—*Kapittha*; Family—Rutaceae. A medium-sized tree, found in the sub-Himalayan forests, being more plentiful in the moist tracts of Maharashtra, Bengal and Northern India.

Dye. The gum is used for the preparation of water colours. When an incision is made in the trunk, a transparent oily fluid exudes, which is used by the painters for mixing their colours. The wood-apple gum is used by dyers and painters; it is also employed in making ink and certain varnishes.

True Gambose

Garcinia morella Desr.; Eng.—True gambose; Hindi—*Gota ganba*, *tamal*; Beng.—*Tamal*; Mar.—*Tamal*; Tam.—*Makki*; Tel.—*Revalchini-pal*; Kan.—*Tamal*; Mal.—*Daramba*; Family—Guttiferae. A small evergreen tree, found in Bengal, Khasia hills and both the Peninsulas.

Dye. Gambose, a yellow gum resin, obtained from the pith, leaves, flowers and fruits is used for preparing water colours and gold-coloured spirit varnishes for metals. The yellow gum-resin of Burma, however, has been used as a yellow dye for the silk robes of the Buddhist priests. The yellow juice (gum-resin) of the tree under the Sanskrit name of *Tamala* has long been employed as a pigment for making sectarian marks on the forehead by the Hindus of Kanara and Karnataka.

Physic Nut

Jatropha curcas Linn.; Eng.—Physic nut; Hindi—*Safed arand*, *Bagbherenda*; Beng.—*Bagbherenda*; Uriya—*Baigab*; Nepal—*Kadam*; Punj.—*Japhrota*, seeds = *jamalgota*; Mar.—*Mogali eranda*; Guj.—*Jamalgota*; Tam.—*Kattamanakku*; Tel.—*Pepalum*; Kan.—*Maranarulle*; Mal.—*Katta-vanakka*; Sans.—*Kanana eranda*; Family—Euphorbiaceae. A shrub or small tree; native of tropical America cultivated along the Coromandel Coast and in Travancore.

Dye. The juice when dried in the sun forms a bright reddish-brown, brittle substance, which dyes linen black.

Banana

Musa paradisiaca Linn.; Syn. *M. sapientum* Linn.; Eng.—Banana; Hindi—*Kela*; Beng.—*Kala*; Punj.—*Kela*; Guj.—*Kela*; Tam.—*Pazham*; Tel.—*Kadali*; Kan.—*Bale*; Mal.—*Vasha*; Sans.—*Kadali*, *rambha*; Family—Musaceae. A perennial herb, 8-15 feet high, extensively cultivated throughout India, nearer the coast tracts than inland, chiefly for its fruit. Native of India and Malaya; cultivated in Assam, Madhya Pradesh, Bihar, Uttar Pradesh, Andhra Pradesh, Jalgaon district (Maharashtra), Tamil Nadu and Kerala.

Dye. The sap contains a considerable amount of tannin, and stains cloth a dark, almost black colour, which is fairly permanent, is very difficult to wash out, and may be employed as a substitute for marking ink.

Marking Nut Tree

Semecarpus anacardium Linn.; Syn. *Anacardium orientale* Linn.; Eng.—Marking nut tree; Hindi—*Bhilawa*; Beng.—*Bhela*; Uriya—*Bhallia*; Assam—*Bhola-guti*; Nepal—*Bhalai*; Punj.—*Bhilawa*; Mar.—*Bibha*; Guj.—*Bhilamu*; Tam.—*Shen-kottai*; Tel.—*Jidi chettu*; Kan.—*Geru*; Sans.—*Bhallataku*; Family—Anacardiaceae. A deciduous tree of the sub-Himalayan tract, from the Sutlej eastward, ascending to an altitude of 3,500 feet, and found throughout the hotter parts of India as far east as Assam.

Dye. The juice of the pericarp contains a bitter and powerfully astringent principle, which is used as a substitute for marking ink. It gives a black colour to cotton fabrics, which is insoluble in water, but soluble in alcohol. The juice of the pericarp is mixed with lime water as a mordant, before it is used to mark cloth.

Dharauli

Wrightia tomentosa Roem. and Schult.; Syn. *Nerium tomentosum* Roxb.; Hindi—*Dharauli*; Beng.—*Dudh-koraiya*; Assam—*Atkuri*; Nepal—*Karingi*; Uriya—*Pal kurwan*; Punj.—*Dudhi*; Mar.—

Kala inderjau; Tel.—*Tetta pal*; Family—Apocynaceae. A small deciduous tree, found in the Punjab, Rajasthan, Bihar, Assam, and the Western Peninsula.

Dye. Every part of the tree discharges a yellow, milky juice on being wounded. The juice yields a fairly good yellow dye when diluted with water, and the pieces of cotton so coloured retained their colour unimpaired for two years.

DYES OBTAINED FROM GALLS

China Turpentine Tree

Pistacia khinjuk Stocks; Syn. *P. integerrima* Stew. ex Brand.; Eng.—China turpentine tree; Hindi—*Kakra*, galls = *Kakrasingi*; Beng.—Galls = *Kakrasingi*; Punj.—*Kakra*, galls = *Kakra singi*; Tam.—galls = *Kakkatashingi*; Kan.—galls = *Dushtapuchattu*; Sans.—galls = *Karkata sringi*; Family—Anacardiaceae. A tall, nearly glabrous tree, 40 feet high, found in the North-Western Himalayas and the Punjab, and on the hot slopes of the Western Himalayas from the Indus to Kumaon, at altitudes of 1,200 feet to 8,000 feet.

Dye and tan. The hard, rugose, hollow, irregular galls, which are formed in October on the leaves and petioles, are used to a small extent for dyeing and tanning. These galls sometimes attain a very large size, and from their peculiar shape well warrant the name of *singhi* or "horns".

Green Almond

Pistacia vera Linn.; Eng.—Pistachio, green almond; Hindi—*Pista*; Beng.—*Pista*; Family—Anacardiaceae. A small tree. Native of W. Asia; cultivated in N. India.

Dye, tan and mordant. The leaves are very frequently affected by galls, which are irregularly-shaped spheroids, borne on a short stalk and usually growing from the surface of the leaf. The galls, with the pericarp of the fruit, and the unfertilized fruit-like ovaries are used locally for dyeing silk. The galls are known as *gul-i-pista*, *baz-ganj* or *boza-ganj*.

The galls contain 45% of tannin allied to gallo-tannic acid, also gallic acid, and 7% of an oleo-resin to which the odour is due.

Grey Oak

Quercus incana Roxb.; Eng.—Grey oak, white oak; Hindi—*Banj*; Kumaon—*Ban*; Kashmir—*Sila supari*; Punj.—*Ban*, *banj*, *shindar*; Family—Fagaceae. A large, gregarious, evergreen tree, found on the temperate Himalayas, generally between altitudes of 3,000 and 8,000 feet

Dye. The galls are used in the Punjab for dyeing the hair.

Dyer's Oak

Quercus infectoria Oliver.; Eng.—The gall or Dyer's Oak; Hindi—*Majuphul*, *mazuphal*; Beng.—*Majuphal*; Tam.—*Machakai*; Mal.—*Majakani*; Sans.—*Majuphul*; Family—Fagaceae. A middle-sized tree or shrub. This is not indigenous to India.

Dye. It is reported, that some of the gall-nuts are obtained from Oak trees found in the forests of Kumaon and Garhwal. In dyeing, the galls are boiled in water in the proportion of 60 gms. of the former to one litre of latter, till three-fourth of the water is evaporated. The cloth is then dipped in the decoction. The galls contain gallic acid and gallo-tannic acid.

Dingrittiang

Quercus serrata Thunb.; Assam—*Dingrittiang*; Family—Fagaceae. A moderate-sized, deciduous tree of the Eastern temperate Himalayas, at altitudes between 5,000 and 6,000 feet. It is found in Manipur and on the Khasia hills at altitudes between 3,000 and 5,500 feet.

Dye. A black dye is produced from the galls of thin species, by the addition of the sulphate of iron.

Pilu

Salvadora oleoides Dcne.; Hindi—*Pilu*, *jhal*; Punj.—*Kubbur*, *pil*, *pilu*; Mar.—*Pilu*, *khakhan*; Tam.—*Ughai*, *koku*; Family—Salvadoraceae. A large evergreen shrub or tree, of the arid tracts of the Punjab, Haryana and Rajasthan, often forming the greater part of the vegetation of desert.

Dye. The galls found upon the plant are used in dyeing.

Athel Tamarisk

Tamarix aphylla (Linn.) Karst.; Syn. *T. articulata* Vahl; Eng.—Athel tamarisk; Hindi—*Lal jhau*, galls = *Chhoti-mayin*; Beng.—*Rakta-jhau*; Punj.—*Faras*, *rukh*, *khagal*, galls = *mai vari*, *mai chhoti*; Guj.—*Lal jhau-nu-jhada*; Tam.—*Shivappu-atru-shavukku*; Tel.—*Erra-erusaru*, galls = *habbul asle*; Family—Tamaricaceae. A moderate-sized tree, found in the Punjab and Uttar Pradesh. It grows well on sandy and saline soils, and is readily propagated by cuttings.

Dye and tan. The small irregularly rounded tuberculate galls, often abundantly produced on the branches by the punctures of an insect, are used as a mordant in dyeing and also in tanning.

Jhau

Tamarix dioica Roxb.; Hindi—*Jhau*; Beng.—*Lal jhau*; Punj.—*Jhau*, galls = *main*, Family—Tamaricaceae. A gregarious shrub, found near rivers and on the sea-coast throughout India. Commonly found in Northern India, Bengal and Assam.

Dye and tan. The galls like those of other species are astringent and are employed as a mordant in dyeing, to produce various shades of grey and black with salts of iron, and in tanning.

Tamarisk

Tamarix troupii Hole; Syn. *T. gallica* auct; non Linn.; Eng.—Tamarisk; Hindi and Beng.—*Jhau*, galls = *bari main*; Uriya—*Jaula*; Punj.—*Jhau*, galls—*mahim*, *bari mahin*; Guj.—*Jhau-nu-Jhada*; Tam.—*Shiru-shavukku*; Tel.—*Shiri-saru*; Sans.—*Jha-vaka*, *shavaka*; Family—Tamaricaceae. A shrub or small tree found especially along rivers and near the sea-coast, throughout India. Found in the Punjab, Uttar Pradesh and Western Peninsula. It favours sandy soils, and flourishes on land impregnated with salt.

Dye and tan. The galls which are similar in properties to those of *Tamarix aphylla*, are from two or three times larger, and are extensively employed in dyeing and tanning.

DYES AND MORDANTS OBTAINED FROM ASHES

Prickly Chaff Flower

Achyranthes aspera Linn.; Eng.—Prickly chaff-flower; Hindi—*Chirchitta*; Beng.—*Apang*; Assam—*Apang*; Mar.—*Aghada*; Sans.—*Apamarga*; Tel.—*Utta-reni*; Tam.—*Na-yurivi*; Punj.—*Kutri*; Mal.—*Katalati*; Kan.—*Uttarane*; Guj.—*Aghedo*; Family—Amarantaceae. A shrub, 3-4 feet high; found all over India, ascending to 3,000 feet in altitude. A troublesome weed in gardens. Stem erect; leaves ovate-obtuse, petiole short, pubescent.

Mordant. The ashes of this plant are used as an alkali in dyeing.

Milk Hedge

Euphorbia tirucalli Linn.; Eng.—Milk-hedge, Spurge; Hindi—*Sehund*; Beng.—*Lanka sij*; Uriya—*Seju*; Mar.—*Shera*; Guj.—*Thordandalio*; Tam.—*Tirukali*; Tel.—*Jemudu*; Kan.—*Bontakalli*; Mal.—*Tirukali*; Family—Euphorbiaceae. A shrub or small tree, with round stems and smooth branches. Native of Africa; found mainly in Bengal and South India; elsewhere cultivated for hedges.

Dye and mordant. The ashes are employed in South India as a mordant. It is burnt, and the ashes form an ingredient of the red dye with *Oldenlandia* root.

Pongam

Pongamia pinnata (Linn.) Pierre; Syn. *P. glabra* Vent.; Eng.—Pongam; Hindi—*Karanja*; Beng.—*Karanja*; Uriya—*Karansa*; Kumaon—*Papar*; Punj.—*Paphri*; Raj.—*Charr*; Mar.—*Karanj*; Guj.—*Kanaji*; Tam.—*Pungam-maram*; Tel.—*Kanuga*; Kan.—*Honge*; Mal.—*Unna-maram*; Sans.—*Karanja*; Family—Papilionaceae. A tall erect tree, commonly found near banks of streams in both the Peninsulas, West Bengal and Travancore.

Dye. The ash of wood is employed for dyeing.

Sal

Shorea robusta Gaertn.; Eng.—Sal tree; Hindi—*Sal*; Beng.—*Sal*; Nepal—*Sakwa*; Uriya—*Salwa*; Punj.—*Sal*; Guj.—resin = *ral*; Tam.—resin = *Kungiliyam*; Tel.—*Guggilamu*; Kan.—resin = *guggala*; Sans.—*Sala*, resin = *asvakarna*; Family—Dipterocarpaceae. A large gregarious tree, found in the Punjab, Uttar Pradesh, Bihar, Bengal, Assam, Madhya Pradesh, the Coromandel Coast and Andhra Pradesh.

Dye. The ashes of the wood are used in dyeing.

DYES OBTAINED FROM LICHENS

Table: Dye yielding Indian Lichens: Distribution and chemical components

Name of lichen	*Distribution*	*Chemical components*	*Dye*
Alectoria jubata (Linn.) Ach. (Usneaceae)	Himalayas		pale, green and brown dye
A. virens Tavl.	Himalayas	Vulpinic acid, virensic acid and *d*—arabitol	pale, green and brown dye
Caloplaca murorum (Hoffm.) Th. Fr. (Caloplacaceae)	Himalayas	Physcion (0.7%)	yellow dye
Candelariella vatellina (Ehrh.) Muell.—Arg. (Lecanoraceae)	Himalayas	Calycin and pulvic anhydrided	yellow dye
Cetraria fahlunensis (Linn.) Schaer.	Himalayas	Cetraric acid	Red brown dye
C. pinastri (Scop.) Rohl.	N.W. Himalayas	Pinastric acid, usnic acid and vulpinic acid	Green dye
Cadonia fimbriata (Linn.) Willd. (Cladoniaceae)	Himalayas		Red-purple dye
Diposchistes scruposus Norm. (Diploschistaceae)	Himalayas	Diploschistesic acid	Brown dye; for calico printing
Gyrophora cylindrica (Linn.) Ach. (Gyrophoraceae)	Darjeeling		Green-brown dye
G. lecanocarpoides Th. Fr.	Himalayas		Red-brown dye
G. papillosa Nyl.	Himalayas		Red-brown dye
Laçanora calcarea (Linn.) Nyl. (Lecanoraceae)	Kumaon		Red-brown dye
Lepraria chlorina (DC.) Ach. (Imperfect lichen)	Dharamshala	Leprapinic acid and its methyl ether, calycin, arabitol and mannitol	Brown dye
Lobaria pulmonaria (Linn.) Hoffm. (Stictaceae)	Himalayas and Assam	Arabitol, mannitol and *nor*—stictic acid	Orange and brown dye
L. scrobiculata (Scop.) Ach.	Simla		Brown dye
Ochrolechia parella (Linn.) Mass. (Lecanoraceae)	Scarce in India		Violet dye

Name of lichen	*Distribution*	*Chemical components*	*Dye*
Parmelia caperata (Linn.) Ach. (Parmeliaceae)	Sub-tropical and temperate Himalayas	Arbitol and Mannitol	Brown-orange to lemon-yellow dye
P. conspersa (Ehrh.) Ach.	Kashmir and Darjeeling	Salazinic acid, arabitol and mannitol	Red-brown dye
P. olivacea (Linn.) Nyl.	Himalayas		Brown-dye used in calico printing
P. physodes (Linn.) Ach.	N.W. Himalayas	Atranorin, capraric acid, physodic acid, arbitol and mannitol in traces	Brown dye
P. saxatilis (Linn.) Ach.	N.W. Himalayas		Orange, yellow and red-brown dye
Physcia pulverulenta (Schreb.) Hampe. (Physciaceae)	Himalayas		Yellow dye
Ramalina calicaris (Linn.) Rohl. (Usneaceae)	West Bengal, Nepal, Chaubattia and Nainital	*d*-Usnic acid, sekikaic acid, *d*-arbito and lichenin	Yellow-red dye
R. farinacea (Linn.) Ach.	Himalayas	Usnic acid, sekikaic acid, *nor*-stictic acid and mannitol	Light brown dye
R. fraxinea (Linn.) Ach.	N.W. Himalayas		Grey-white dye
Roccela tinctoria Lam. & D.C. (Roccellaceae)	Tamil Nadu	Lecanoric acid, roccellic acid and erythrin	Orchil and cudbear
Solorina crocea (Linn.) Ach. (Peltigeraceae)	N.W. Himalayas	Solorinic acid	Yellow dye
Stereocaulon paschale (Linn.) Hoffm. (Cladoniaceae)	Himalayas	Friedelin	Ash-green dye
Sticta crocata (Linn.) Ach. (Stictaceae)	Himalayas		Brown dye
Teloschistes flavicans (Swartz.) Norm. (Teloschistaceae)	Nilgiris	Physcion, teloschistin, fallacinal and vicanicin	Yellow dye
Umbilicaria pustulata var. *populosa* Tuck (Gyrophoraceae)	N.W. Himalayas	Gyrophoric acid	Red-brown dye
Usnea florida Wiggs (Usneaceae)	Himalayas	Usnic acid, salazinic acid and stictic acid	Green-yellow, red-brown dye

Name of lichen	Distribution	Chemical components	Dye
Xanthoria parietina (Linn.) Th. Fr. (Teloschistaceae)	Kashmir	Physcion (1.2%), mannitol, lichenin, isolichenin and parietinic acid	Yellow dye

INK AND MARKING NUTS

Various substances are used by the natives of India in making ink, the usual process being to mix some astringent principle such as galls or myrobalans with one of the iron salts or oxides. The following plants are known as adjuncts in the formation of inks:

1. ***Alnus nepalensis*** D. Don. (Betulaceae)—Bark forms an ingredient in native red inks.
2. ***Cordia dichotoma*** Forst. f.; Syn. *C. myxa* Linn. (Ethretiaceae)—The unripe fruit is used as a marking nut, though its colour is less enduring, than that from *Semecarpus*.
3. ***Emblica officinalis*** Gaertn.; Syn. *Phyllanthus emblica* Linn. (Euphorbiaceae)—Fruits largely employed in making black ink.
4. ***Semecarpus anacardium*** Linn. f.; Syn. *Anacardium orientale*, Linn. (Anacardiaceae)—The marking nut is used by all the Indian washermen to mark the linen given them to wash. The juice of the unripe pericarp with lime forms the ink. Without the addition of lime it is often employed as ordinary writing ink.
5. ***Terminalia bellirica*** (Gaertn.) Roxb. and ***T. chebula*** Retz. (Combretaceae). The unripe fruit, is combined with iron in making ink.

17

Gums and Resins

Among the forest products of India, *gums* and *resins* occupy an honourable position. They have been known to mankind from time immemorial. Gum arabic was known and used by the people of Egypt as early as 2,000 B.C. Balsams, myrrh and frankincense have been frequently referred to in the Old Testament. These products have been in use in India and China for religious purposes from immemorial times. These products were greatly used in ancient times by Egyptians for embalming the dead bodies and for use in cosmetics. These balsamiferous products of the Middle East and North Africa were carried to European countries over the deserts through camel caravans and over the seas through Arab merchants. The ancient philosophers and botanists like Theophrastus, Celsus and Dioscorides have mentioned the products like gum tragacanth and gum arabic in their famous works.

These products are exuded by plants, partly as a normal phenomenon and partly as the results of wound or injury to the bark or wood. The gums and resins arise chiefly from the stem of the plant but sometimes they also exude from the roots, leaves and other parts of the plants.

Ordinarily, a gum is a substance of a more or less sticky nature. The gums are insoluble in alcohol or other organic solvents but soluble in water, readily swells up in it, and makes a viscous mass. On the other hand, a resin, though it resembles a gum in appearance, neither dissolves nor softens in water; it is more or less soluble in organic solvents.

Closely related to true gums are the gum-resins which are also the plant products. However, they differ from true gums in that they also contain some resin and therefore, they do not dissolve in water completely. Assafoetida, frankincense, myrrh and other such fragrant substances are well-known gum-resins. On the other hand, when the resins remain mixed with a high percentage of essential oils, they are known as oleo-resins (*e.g.*, the resins from several species of *Pinus*).

When the oleo-resins include some gum in them, they are called gum oleo-resins (*e.g.*, the exudate from *Boswellia serrata*).

CLASSIFICATION OF INDIAN GUMS AND RESINS

Category	*Typical product in world trade*	*Source of typical Indian products*
True gums	Gum arabic Gum tragacanth	*Acacia nilotica* *A. catechu* *A. modesta* *A. senegel* *Anogeissus latifolia* *Bauhinia retusa* *Cochlospermum religiosum* *Lannea coromandelica* *Pterocarpus marsupium* *Sterculia urens* *S. villosa*
Hard resins	Copal Dammar	*Canarium strictum* *Hopea odorata* *Shorea robusta* *Vateria indica*
	Amber Lacquer	

Category	Typical product in world trade	Source of typical Indian products
	Shellac	
	Sandarac	
	Mastic	
Oleo-resins	Turpentines	*Pinus roxburghii* and some other species of *Pinus*.
	Balsams of peru of Tolu of Styrax or storax	*Boswellia serrata*
	Other Oleo-resins	*Dipterocarpus turbinatus*
	Copaiba	*Kingiodendron pinnatum*
	Elemi	
Gum-resins	Gamboge	*Garcinia morella*
	Assafoetida	
	Galbanum	
	Myrrh	
	Olibanum, or Frankincense	*Commiphora mukul*

Some plants yield only gum, some others only resin, and still others both gum and resin, the gum and resin in the latter case being exuded either separately from different parts of the same plant or together as an emulsion or mixture. In many plants the resins remain dissolved in a volatile oil, and exude in a liquid state, and soon after exudation they solidify through the evaporation of part of the volatile oil. Some resins possess aromatic acids in them, while the other known as balsams are made up of certain liquid organic compounds of these aromatic acids with a solid resin dissolved in them. It is a fact that almost all gums and resins come from wild plants and not from cultivated ones.

Gums

The true gums are formed as a result of the disintegration of internal tissues, for the most part from the decomposition of cellulose, through a process known as gummosis. The gums are insoluble in alcohol but soluble in water, readily swell up in it, and form a viscous mass. On heating they decompose completely without melting, usually showing charring. They are colloidal in nature, they contain a large amount of sugar and are closely allied to the pectins. The bulk of the natural gums of commerce exudate from plants as a liquid, which on exposure dries up into translucent, amorphous tear-shaped bodies or flakes. The exudation of gums is particularly seen after a long dry hot season, especially from diseased or wounded trees, and are usually copious on the trunks of trees charred by a forest fire. Artificial incisions are also made on tree trunks to increase the yield of gum. Intensive tapping for gums, however, causes considerable damage to the trees, and sometimes it leads to the damage and destruction of trees. Harvesting of gums is done mostly by hand-picking by tribal people living in the neighbourhood of forests. The larger lumps are broken with a wooden mallet, and the tears are made clean by hand before grading. The gum that reaches to the market requires careful grading and sorting before it is sold in the market or exported. Grading is based on colour size and transparency of the tears of gum. The gums are utilized as adhesives, and are also used in printing and finishing textiles, as a sizing for paper, in the paint and candy industries, and is a glaze in painting. In medicine it is used as an emulsifying agent and as a demulcent.

The more important gums come largely from tropical and subtropical species of the Leguminosae. They are gathered in the dry seasons and brought to the market in the shape of tears. The commercial gums come to the market in the form of dried exudations and are used in a variety of ways in industry. The varieties with least colour and highest adhesive power are the most valuable. The finer grades of gums are utilized in clarification of liqueurs, finishing of silk and making of quality water colours. Intermediate grades are used in confectionery, pharmaceuticals and printing inks, in sizing and finishing textile fabrics, and in dyeing. The cheaper grades are utilized in calico-printing, paint industry, sizing of paper and as adhesives. Many gums are utilized in cosmetic and ice-cream industry.

Important Gums in World Trade

Two kinds of gums, viz., *gum arabic* and *gum tragacanth* play an important role in world trade. Gum arabic is obtained from certain species of the genus *Acacia*. However, the true gum arabic is obtained from *Acacia senegal*. On the other hand gum tragacanth of commerce is obtained from *Astragalus gummifer* and other species of the same genus.

Important Indian Gums

A large number of trees exude Indian gums. The more important gums are known as *Acacia gums*, largely known as *gum arabic*, and are obtained from several species of genus *Acacia*. The gum available in the market is often a mixture of the gums of *Acacia* species and small quantities of the gums of *Anogeissus latifolia*, *Feronia limonia*, etc. The important Indian species of *Acacia* that yield gum are—*Acacia catechu, A. modesta, Acacia nilotica, Acacia senegal*. The other gum yielding trees of India are—*Anogeissus latifolia*, *Bauhinia retusa*, *Cochlospermum religiosum*, *Lannea coromandelica, Pterocarpus marsupium, Sterculia urens*, etc. The minor gum-yielding Indian trees are—*Albizia chinensis*, *A. lebbeck, A. odoratissima, A. procera, Anacardium occidentale, Azadirachta indica, Bauhinia purpurea, B. racemosa, B. variegata, Chloroxylon swietenia, Elaedendron glaucum, Feronia limonia, Mangifera indica, Pithecolobium dulce, Spondias pinnata, Terminalia bellirica, T. tomentosa,* etc.

Resins

The resins occur in special cavities or passages in a wide variety of plants. They originate through reduction and polymerization of carbohydrates. They normally ooze out through the bark and harden on exposure. Unlike gums, resins are insoluble in water, but more or less soluble in ether, alcohol and turpentine. They are brittle, amorphous and transparent. They possess lustre, and burn with a smoky flame when ignited. A few families of plants are commercially important as source of resin, they are—Anacardiaceae, Burseraceae, Dipterocarpaceae, Guttiferae, Hamamelidaceae, Leguminosae, Liliaceae, Pinaceae, Styracaceae and Umbelliferae (Apiaceae). The plants belonging to other families are of minor importance, and not exploited commercially.

The resins obtained from the Dipterocarpaceae are usually called dammars, while resins from Pinaceae, which contain essential oil are known as oleo-resins. The resins obtained from Caesalpiniaceae are known as copal. Highly aromatic resins are termed balsams. The resins that contain gums are called gum-resins. The resins formed thousands of years ago are termed as fossil-resins.

The resins may be classified into three groups. 1. hard resins, 2. oleo-resins and 3. gum-resins. The important hard resins are-copals (from Caesalpiniaceae), Dammar (from Dipterocarpaceae and Burseraceae), Amber (a fossil resin; principal source was the now extinct *Pinus succinifera*), Lacquer (from *Rhus verniciflua*, *Melanorrhoea usitata*), Sandarac (from *Callitris* spp.), Mastic (from *Pistacia* spp.)

The main oleo-resins are—Turpentines (exclusively from coniferous trees,) Balsams (Balsam of Peru from *Myroxylon pereirae*; Balsam of Tolu from *Myroxylon balsamum*; Styrax —a liquid balsam from *Liquidamber orientalis*). The other oleo-resins are—copaiba (from *Copaifera* spp.) Elemi (from trees of Burseraceae such as *Protium* and *Canarium*).

The important gum-resins are—Gamboge (from *Garcinia* spp.), Assafoetida (from *Ferula assafoetida*), Galbanum (from *Ferula galbaniflua*), Myrrh (from *Commiphora* spp.), Frankincense (from *Boswellia* spp.)

The natural resins are utilized in several ways. They are applied to a great variety of industrial purposes. The most important use of resins is the manufacture of varnishes and lacquers. Thus the resins may be divided into two classes—1. those resins which after melting can be combined with linseed oil or turpentine to form an 'oil varnish' and 2. the resins that dissolve in alcohol, turpentine or other volatile solvents to form a 'spirit varnish'. The amber and copals belong to class 1, while the class 2, include rosin, dammar, sandarac, mastic and elemi. Resins dissolve readily in alkali and therefore utilized for making soaps. They are also used in pharmaceutical, for sizing paper, for making sealing wax and other such products. World trade in natural resins is estimated a million tonnes annually.

GUMS

The gums are formed in various kinds of plants. They are found in many phanerogamic plants, and are of various kinds. *Acacia senegal* (Linn). Willd., yields the best gum-arabic of commerce. The true gums are formed as the result of the disintegration of internal tissues, for the most part from the decomposition of cellulose. The gums are insoluble in alcohol but soluble in water, readily swell up in it, and form a viscous mass. They are colloidal in nature. They contain a large amount of sugar and are closely allied to the pectins. They exude naturally from the stems, or in response to wounding. The commercial gums are the dried exudations of the plants. Gums are commonly found in the plants of dry regions. The gums are utilized as adhesives, and are also used in printing and finishing textiles, as a sizing for paper in the paint and candy industries, and as drugs. A list of important gum-yielding plants is being given here.

Acacia senegal (Linn.) Willd.; Eng. *Gumarabic Acacia*; Hindi—*Kumta*; Family—Mimosaceae.

Found in the Punjab and Rajasthan. It is native of north Africa and abundantly found in West Africa, north of the river Senegal.

It is a low tree with grey bark with flexuose branches; *spines* strong, short, sharp–hooked, often 3-nate, two lateral and one below the petiole. *Leaves* with 6-10 pinnae and 16-28 leaflets. *Flowers* in peduncled spikes 2-3 inches long and not crowded. *Corolla* yellow, twice the length of the campanulate, glabrous, deeply toothed *calyx*. *Pod* 3 by ¾ inch, thin, grey indehiscent, straight, strap-

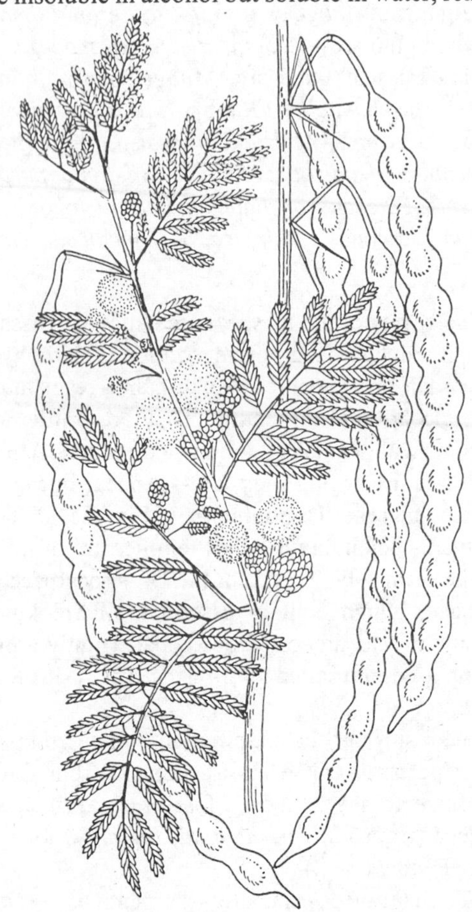

Fig. 17.1. *Acacia senegel.* A twig with leaves, inflorescences and fruits (lomentum). The plant yields a gum of excellent quality.

shaped with a strong, fibrous, marginal midrib, and constricted between the seeds.

Gum. The plant yields a gum of excellent quality. The gum, obtained from the stem, bark, is used in the textile, mucilage, paste, polish and confectionery industries, and as a glaze in painting. The *Kordofan* form of the gum from *Acacia senegal* is the purest and most valuable, and is the Gum Arabic of Pharmacy.

The Persian and Arabic writers described the *gum arabic* under the name of *samgh-i-arabi*. From the very ancient times gum was known to the Egyptians. It is frequently mentioned by the ancient writers, and there are numerous representations both of the plant and of the gum itself. Gum was used by the Arabian physicians, but in the 12th century it was apparently unknown in Europe; it first reached Europe in 1340 A.D., through Italian merchants trading with Egypt and Turkey.

Gum arabic is a type of gum found in the juice of various plants, in the genus *Acacia*. Chemically gum arabic consists of *Arabic acid* (C_{12} H_{22} O_{11}) in combination with lime, potash and magnesia.

Gum dissolves slowly in an equal weight of water, and forms a thick glutinous liquid which possesses a distinctly acid reaction. Gum is insoluble in alcohol and most other liquids.

Large quantities of gum arabic are used for giving lustre to crape and silk, and for thickening colours and mordants in calico-printing, for suspending tannate of iron in the manufacture of ink and blacking.

The gum is used in medicine as a demulcent and emollient. The powdered gum is used for checking haemorrhage from leech-bites.

Acacia nilotica Desf. Syn. *A. arabica* Willd.; English—Indian Gum Arabic Tree; Hindi—Bengali and Punjabi *Babul, babur, kikar*; Sanskrit—*Vabbula*; Tamil—*karuvelum*; Telugu—*Nella tuma*; Karnatak—Gobli; Kannada—*Jali*; Marathi—*Babhula*, *Kali—kikar*, Gujarati—*Baval;* Santal—*Gabur*, Malayalam—*Babola*; Arabic—*Ummughilan*; Persian—*Khare—mughilan*; Family—Mimosaceae.

Found throughout India. It is very commonly found in South India.

An erect tree, with straight spines. *Leaves* composed of from 6-12 pinnae and 20-40 leaflets. *Flowers* in rounded heads, axillary, supported upon the short peduncles with the bracts above the middle. *Pod* stalked, straight, sub-indehiscent, sutures deeply indented between the seeds, *seeds* 8-12.

Gum. Hindi—Babul ki gond, kikar ki gond; Bengali—Babla–ata, babla–gond; Tamil—pishin; Telugu and Malayalam—Pasha; Arabic and Persian—Samgh-i-Arabi.

The gum consists of tears or broken pieces of a light dusky-brown tint. When dry it is permeated by cracks and is very brittle. It is perfectly soluble in water.

The gum is extensively used by the calico–printers. For certain colours a mixture of this gum, with that obtained from *Anogeissus latifolia* (of Combretaceae) the *dhawa* is regarded as most serviceable and this mixture is likewise used to stiffen dyed fabrics and to give them a polish.

The true Gum Acacia is used in the preparation of mucilage. It is used as a demulcent and also in lozenges. Gum Acacia (*babul–ki–gond*) is also administered to recently-delivered women as a tonic. The gum is used in the form of mucilage in diarrhoea and dysentery. It is quite useful for diabetics. The powdered gum is useful, combined with quinine, in fever cases complicated with diarrhoea and dysentery.

Acacia catechu Willd.; English—Black cutch, catechu; Hindi—*Khair, katha*; Bengali—*Khayer, kuth*; Assamese—*Khoira koir, kat*; Tamil—*Wothalay*; Telugu—*Podala—manu, nalla sandra*; Sanskrit—*Khadira*; Uriya—*Khoiru*; Kannada—*Kagli*; Marathi—*Khaderi*, *khaira*; Gujarati—*Kher*, Family—Mimosaceae.

The *khadira* tree is mentioned in the *Vedas*, where it is used as a simile for "*strength*".

A tree. It is commonly found in most parts of India, extending in the sub-Himalayan tract west ward to the Punjab and east ward to Sikkim, ascending to altitude 5,000 feet. It is commonly met with in the Punjab, Madhya Pradesh, Uttar Pradesh, Andhra Pradesh, Maharashtra, Gujarat and Bihar.

It is a moderate-sized, deciduous tree, with dark brown, much cracked bark, and short-hooked spines in pairs. *Leaves* composed of from 40 to 80 pinnae. *Flowers* white or pale yellow, peduncled, spikes in the axils of leaves; *rachis* downy. *Corolla* 2 to 3 times the tomentose *calyx*. *Pod* straight, strap-shaped, narrow, thin, dark brown.

Gum. The tree yields a pale-yellow gum, often occurring in tears one inch in diameter, generally less than half an inch in size. It is sweet and soluble in water. It forms a strong mucilage. The tears are mostly bright coloured, varying to dark amber. They occur chiefly in broken pieces, the fragments being cracked and granular.

Acacia farnesiana Willd.; English—The Cassie Flower; Hindi—*Vilayati kikar*, *vilayati babul*, *gand–babul*; Bengali—*Guya babula*; Marathi—*Gubabhul*; Gujarati—*Talbaval*; Tamil—*Veddavala*, *puj–velam*; Telugu—*Kusturi*, *murki tumma*; Kannada—*Jali*; Family—Mimosaceae.

A small tree. Native of the West Indies. It is found in all parts of India. Its strong, scented, yellow flower-heads, perfume the atmosphere very pleasantly.

It possesses straight spines and flowers in the cold season. *Flowers* in rounded heads, axillary, fragrant bright yellow, supported upon *peduncles* which are crowded in the nodes of the leaves, and possessing a whorl of bracts like an involucre at the apex. *Pod* thick, swollen or fleshy, cylindrical, more or less curved or hooked, glabrous and possessing straight sutures. *Seeds* biserial.

Gum. The gum exudes from the trunk in considerable quantity. Sometimes it is considered superior to gum arabic in the arts and as a medicine. Sometimes it is used to adulterate gum arabic. It is found as dark, conchoidal masses, translucent and transparent at the edges. Some pieces are much whiter. The gum is generally used as a substitute for gum arabic.

Acacia leucophloea Willd.; Hindi—*safed kikar*, *raunj*, *jhind*; Bengali—*safed babul*, Uriya—*Goira*; Sanskrit—*shveta–barbura-vrikshaha*; Punjabi—*safed kikar*, Rajasthani—*Arinj*; Marathi—*Hewar*; Gujarati—*Hari baval*; Tamil—*Vel-velam*; Telugu—*Tella-tuma*; Kannada—*Bili-jali*, *vel-vaila*; Family—Mimosaceae.

Found in the Punjab, Rajasthan and Madhya Pradesh. A large, deciduous tree possessing short, straight and white spines. *Leaves* composed of 12-24 pinnae and 30-60 leaflets. *Flowers* in small, rounded, yellow heads, aggregated into terminal *panicles*, which when fully expanded, are a foot long and broad, and densely tomentose. *Pod* sessile, narrow ligulate falcate, thin, flat tomentose, with straight sutures. The inflorescence of this species is most characteristic.

Gum. The plant yields a gum. The gum yielded by this plant is used in medicine by natives. It resembles to some extent the gum arabic.

Acacia modesta Wall.; Punjabi—*Phulahi*; Gujarati—*Kantosariyo*; Family—Mimosaceae.

Found in the sub-Himalayan tract and the plains of the Punjab. It is one of the characteristic trees of the Punjab.

A moderate-sized tree; *spines* in pairs, short hooked. *Leaves* with 4-6 pinnae and 6-8 leaflets. *Flowers* in peduncled spikes 2-3 inches long, not very dense. Corolla greenish coloured, twice the length of the glabrous campanulate *calyx*. *Pod* 2-3 inches long by 3/8, glabrous, glossy, venulose, straight strap-shaped, narrowed into a short peduncle.

Gum. It yields a gum which occurs in the form of small, round, smooth, subtranslucent and very characteristic tears. The gum is used in calico-printing. It is quite tasteless. It is restorative and used in native medicine.

Adenanthera pavonia Linn.; English—Coral Wood, Red wood; Bengali—*Rakta kanchan*; Assamese—*Chandan*; Tamil—*Anai-gundumani*; Telugu—*Bandi gurivenda*; Malayalam—*Manjati*; Gujarati—*Bari-gumchi*; Kannada—*Manjadi*; Family—Caesalpiniaceae.

A large deciduous tree. Found in Bengal and South India. Leaves compound, with 8-12 pinnae and 12-18 obtuse leaflets. Racemes short, peduncled, 2-6 inches long; seeds bright scarlet.

Gum. The plant yields a gum known as *madatia.*

Aegle marmelos (Linn.) Corr; English—Bael, Bengal quince; Hindi—*Bel*, *bilva*, *siriphal*; Bengali—*Bela, bel, vilva*; Assamese—*Bel*; Marathi—*Bel*; Gujarati—*Bil*; Sanskrit—*Sriphal, bilva; Tamil—Vilva-pazham*; Telugu—*Bilvapandu*; Malyalam—*Kuvalap-pazham*; Kannada—*Belpatri*; Family—Rutaceae.

A tree, found throughout India. A small, deciduous, glabrous tree, with straight, strong, axillary spines. *Leaves* pale green, of tree leaflets; lateral leaflets sessile, ovate–lanceolate, 3-5 inches long, terminal long petioled. *Flowers* an inch in diamter, greenish white, sweetly scented.

Gum. The stem yields a good gum, occurring in tears like gum arabic, or in fragmentary pieces resembling coarse brown sugar. From the seeds a mucous fluid is secreted within the cells of the fruits which hardens into a transparent, tasteless, gummy substance. This mucilaginous substance, is used as a cement and is also employed as a varnish.

Ailanthus excelsa Roxb; English—Tree of Heaven; Hindi—*Maharukha*; Marathi—*Mahanimb*; Uriya—*Mahanim*; Gujarati—*Motoaduso*; Tamil—*Perumaruttu*; Telugu—*Peddamanu putta*; Malayalam—*Perumarum*; Sanskrit—*Madala*, *aralu*; Family—Simarubaceae.

A Small tree; probably introduced into India. Commonly found in Bihar and Karnataka. Occasionally found in Rajasthan. Leaves 1-2 feet long, glandularly hairy; *leaflets* very coarsely toothed. *Stamens* with the filaments about half the length of the anther. *Samara* 2 inches by 1/2 inch, red, twisted.

Gum. A red gum is obtained from this tree. It resembles Moringa gum, and consists of large rounded tears of a deep vinous red.

Albizzia stipulata Boivin; Hindi—*Siran*, *samsundra*; Bengali—*Chakua, amluki*; Punjabi—*Sirin, shirsha*; Nepalese—*Kala siris;* Tamil—*Katturanji*; Telugu—*Chindaga*; Marathi—*Udala*; Kannada—*Kal baghi*; Family—Mimosaceae.

A large, deciduous tree, found in the sub-Himalayan tract; in Uttar Pradesh, Assam, Bengal and South India.

Pinnae 12-40 with many glands on the rachis; leaflets 40-80; *stipules* and *bracts* large, membranous and persistent; *flower heads* panicled, terminal racemes densely pubescent. *Pod* 5-6 inches by 1 inch, pale brown, thin indehiscent, sub-sessile, 8-10 seeded.

Gum. It yields a gum which exudes copiously from the stem, and is used by the Nepalese for sizing their *Daphne* paper.

Anacardium occidentale Linn.; English—Cashewnut; Hindi—*Kaju*; Gujarati—*Kaju*; Bengali—*Hijli–badam*, *Kaju*; Marathi—*Kaju*; Tamil—*Mundiri*; Telugu—*Jidi-mamidivittu*; Kannada—*Jidi vate*; Malayalam—*Paranki–mava kuru*; Family—Anacardiaceae.

A small tree; native of Brazil beat now cultivated in Malabar, Kerala, Karnataka, Tamil Nadu and Andhra Pradesh.

Leaves alternate, simple; *panicles* terminal; stamens 8-10, all or only a few fertile; *nut* kidney-shaped, seated on a large pyriform fleshy body (cashew apple); *seed* kidney–shaped; *cotyledons* semilunar

Gum. The bark of this plant yields a gum. This gum occurs in large stalactitic pieces; it is yellow or reddish, and only slightly soluble in water. It is obnoxious to insects.

The juice exuding from incisions in the bark is used as an indelible marking ink. The astringent juice is used by native workmen as a flux for soldering metals.

Anogeissus latifolia Wall.; English—Gum ghatti; Hindi—*Daura, bakla*; Rajasthani—*Golra, dhauk*; Tamil—*Vellay naga*; Telugu—*Chirimanu*; Uriya—*Dohu*; Malayalam—*Dhau*; Gujarati—*Dhavdo*; Marathi—*Dhamora*; Kannada—*Dinduga*; Family—Combretaceae.

A large, decidous tree, occurring in the drier parts of India. Found in the sub-Himalyan tract, Madhya Pradesh, the Western Peninsula and the Nilgiris. Very plentiful in Melghat; common in the upper Godavari; not found in the trans-Gangetic Peninsula.

Leaves broad elliptic-obtuse at both ends peduncles one or more from the same axil, often branched; ripe fruits shining, glabrous, the beak as long as the main part or longer.

Gum. It yields a gum, known as *ghati gum*. This gum is used in calico–printing. The gum occurs in clear straw-coloured, elongated tears, adhering into masses, sometimes honey–coloured or even brown from impurities. As an adhesive gum it is inferior to gum arabic.

The gum is used by the natives in the treatment of cholera.

Astragalus heratensis Bunge; English—Indian tragacanth; Hindi—*Anjira*; Marathi—*Gujar*, Family—Papilionaceae.

A large shrub. The stem of the plant yields gum.

Gum. Tragacanth is a gum obtained from several species of *Astragalus*. It is of a dull white colour, translucent, inodorous and tasteless. The gum is officinal being emollient and demulcent, useful in irritation of the mucous membranes. Tragacanth or *Katira gabina* is a valuable medicine in gonorrhoea. During the hot season it is given to horses for its demulcent and refrigerant qualities.

Azadirachta indica A. Juss; Syn. *Melia azadirachta* Linn; English—Margosa tree, Neem tree; Hindi—*Nim, nimb*; Bengali—*Nim, nimgachh*; Kumaon—*Betain*; Punjabi—*Nim*; Mumbai—*Nim, bal-nimb*; Marathi—*Limba*; Gujarati—*Limba, limb*; Tamil—*Vembu, veppam*; Telugu—*Vepa, nim-bamu*; Kannada—*Heb-bavu*; Malayalam—*Veppa*; Sanskrit—*Nimba, arishta, nimba–vrikshaha*; Persian—*Nib*; Family—Meliceae.

A large tree 40 to 50 feet in height, common, wild or more often cultivated, throughout the greater part of India. It is native of Myanmar.

Gum. The bark exudes a clean, bright amber-coloured gum, which is collected in small tears and fragments. It is said to form a portion of the commercial *gum gatti* and of *East India gum*. It is considerably esteemed medicinally as a stimulant. It is bitter and fully soluble in cold water. The silk-dyers use the gum for the preparation of their colours. The gum is said to be demulcent tonic.

Bombax ceiba Linn.; English—Silk cotton tree; Hindi—*Sermul, semal, semur, ragatsemal*; Bengali—*Rokto–simul, simul*; Uriya—*Bouro*; Lepcha—*Sunglu*; Marathi—*Samar, savara*; Gujarati—*Shemolo*; Telugu--*Mundla—buraja—chettu*; Tamil—*Pula, mulilavu*; Malayalam—*Pulamaram*; Kannada—*Mullu–buragamara*, Sanskrit—*Salmali, mocha*; Family—Bombacaceae.

A large, deciduous tree, with branches in whorls, spreading horizontally, and the stem with large thorny buttresses. Found throughout the hotter parts of India. It is abundantly found on the eastern side of India, ascending the mountains to 4,000 feet in altitude. It is the largest and most characteristic tree of east Rajasthan.

Trunk and branches remain covered with large corky prickles; *leaflets* 5-7, entire, cuspidate, base tapering, filaments ligulate, half the length of the petals; capsules oblong-obtuse.

Gum. A brown, astringent gum-like substance is commonly known as *mochras* in Indian markets. It occurs in the form of light or dark brown tears, which are often hollow, much resembling galls. The formation of the gum is due to some functional disease, and commences below the bark like a large swelling. Dry *mochras* when soaked in water swells up, and resumes very much the appearance of the fresh exudation. The taste is purely astringent like tannin.

Mochras is collected by Bheels and wandering tribes in Western India. It is used in medicine and sold by the indigenous druggists.

The gum or dried juice, *mochras*, which the tree yields, is used as an aphrodisiac. This gum contains a large proportion of tannic and gallic, acids, and may be successfully employed in cases requiring astringents. It has also tonic and alterative properties; it is regarded as a styptic, and is used in diarrhoea, dysentery, and menorrhagia. In Gujarat, this gum is known as *kamarkas*; it is ground to powder and drunk in milk as a tonic. The gum of the *semul* tree, is given to children as a laxative.

Borassus flabellifer Linn.; English—Palmyra palm; Hindi—*Tar, tari, tal*; Bengali—*Tal*; Gujarati—*Tad*; Marathi—*Tada*; Telugu—*Tati–chettu*; Tamil—*Panai–maram*; Malayalam—*Pana*; Kannada—*Tale*; Sanskrit—*Tala*; Family—Palmae.

A tall palm, covered with dry leaves, or with the lower part of the petioles, while the old stems are marked with the hard, black, long and narrow scars of the fallen petioles. Found throughout tropical India, Bihar, coastal areas of Bengal, the Western and Eastern Peninsulas.

Gum. A gum is obtained from this palm. It is black and possesses shining fracture. It is used medicinally.

Buchanania latifolia Roxb.; Syn. *B. lanzan* Spreng; English—Cuddapah almond; Hindi—*Piyala, chironji*; Bengali—*Chironji, piyal*, Uriya—*Charu*; Tamil—*Mowda*; Telugu—*Chara, charumamudi*; Kannada—*Nuskul*; Malayalam—*Kalamaram*; Gujarati—*Charoli*; Marathi—*Pyal–char*, Sanskrit—*Piyala, chara*; Family—Anacardiaceae.

A tree, leafless only for a short time. Found in the sub-Himalayan tract from the Sutlej eastward, ascending to 2,000 feet; throughout India except in the arid regions of N.W. India.

Panicles terminal and axillary, crowded. *Drupe* small, flesh scanty. *Seed* acute at one end; *cotyledons* thick.

Gum. A pellucid gum exudes from wounds on the stem. The gum is said to be half, soluble in water. It occurs in irregular broken fragments, brittle, pale, horn-coloured, tinged with brown, tasteless. It possesses good adhesive properties similar to the inferior kinds of gum arabic, and suitable for dressing textiles. The gum is used medicinally. It is said to be administered in diarrhoea.

Butea monosperma (Lamk.) Taubert; Syn. *B. frondosa* Koenig ex Roxb.; English—*Bengal kino*, Flame of the forest; Hindi—*Dhak, palas, tesu, chichra*; Bihar—*Paras, faras*; Uriya—*Porasu*; Gujarati—*Khakara*; Marathi—*Paras, palas*; Tamil—*Porasa*; Telugu—*Palas; moduga, tella modugu*; Kannada—*Muttuga*; Malayalam—*Plach–cha*; Sanskrit—*Palasa, kinsuka*; Family—Papilionaceae.

A moderate-sized, deciduous tree, found throughout India, extending in the North-West Himalayas as far as the Jhelum. This is one of the most beautiful trees of the plains and lower hills of India. The leaves drop when the flowers appear, the top and outer branches stand out like sprays of unbroken scarlet.

Gum. It yields naturally, or from artificial scars on the bark, a gum which is sold as "Bengal kino" or *chuniya–gond*. This occurs in the form of round tears, as large as pea, often fragmentary, of an intense ruby colour and astringent taste. This gum may be purified by solution in water. It is translucent, but with age it darkens and becomes opaque. It is brittle. It is generally known as "Bengal kino", *palas–ki gond* or *chuniya–gond*. In native medicine, Bengal kino is used as an astringent—a substitute for true kino. It is also employed in tanning.

Calophyllum apetalum Willd.; English—Poonspar of Travancore; Hindi—*Bobbi*; Kannada—*Kalpun*; Tamil—*Cheru, pinnay*; Malayalam—*Tsirou–panna*; Marathi—*Irai*; Family—Guttiferae.

An evergreen tree, found in Maharashtra, Karnataka, the Western Ghats from Konkan to Travancore.

Gum. The gum occurs in large, translucent, irregular lumps of a yellowish colour, it is of horny texture, somewhat brittle, without odour; the taste is soapy. When placed in water it gradually softens, and finally disintegrates into a fine granular matter which floats in the form of flacky particles of a dirty-white colour, and numerous oil-globules which gradually collect upon the surface; the water dissolves a small portion and becomes slightly viscid.

The gum-resin is used medicinally. It acts as a vulnerary, resolutive, and anodyne.

Calophyllum elatum Bedd.; English—Poonspar tree, sirpoon tree; Hindi—*Bobi sirpon*; Malayalam—*Pungu*; Tamil—*Pongu*; Kannada—*Sirepune kuve*; Marathi—*Nagani*; Family—Guttiferae.

A large, evergreen tree, often growing to a height of 150 feet, found in the evergreen forests of Western Ghats from North Kanara to Travancore.

Gum. The tree yields a black opaque gum, which in the market, occurs much mixed with pieces of bark; it is mildly astringent, and very soluble in cold water. The solution of brownish yellow, showing a strong blue fluorescence.

The gum is used medicinally.

Calophyllum inophyllum Linn.; English—Indian laurel, laurelwood; Hindi—*Sultan champa, surpan, surpunka*; Bengali—*Sultana champa, punnag*; Uriya—*Polang*, Marathi—*Undi, nag–champa, pumag*; Malayalam—*Bintangor, punna*; Tamil—*Pinnay, punagam*; Telugu—*Puna, panna–chettu*; Kannada—*Wuma, pinna bija*; Sanskrit—*Punnaga*; Family—Guttiferae.

Grown near the sea-coast throughout India as an ornamental tree; indigenous to the Western Peninsula, Orissa, South India and the Andaman Islands.

Gum. The tree yields a black resinous gum. The gum is yellowish green and translucent. The gum is also obtained from the fruit in small quantity, in the form of very small tears. It is soft and entirely soluble in rectified spirit.

The gum–resin is used medicinally as a remedy for indolent ulcers. The tears which are obtained from the tree and its fruit are emetic and purgative.

Ceiba pentandra (Linn) Gaertn.; Syn. *Eriodendron anfractuosum* DC.; English—Kapok; Hindi—*Safed simal, hattian*; Bengali—*Shwet–simul*; Tamil—*Illavam*; Telugu—*Buruga*; Malayalam—*Paniyala*; Marathi—*Pandhari*; Kannada—*Bili barga*; Family—Bombacaceae.

A tall tree with straight trunk, prickly when young; branches horizontal and whorled. Flowers dirty white, with staminal tube splitting into five portions, each with two anthers. Found in the forests throughout the hotter parts of India. It is met with in South Western India and in the Western Ghats. It is often found planted round villages and temples.

Gum. The gum obtained from the bark of this tree is of a dark-red colour and almost opaque. It is generally known as *hattian-ke-gond*, and by Europeans it is thought to be one of hog-gums or pseudo-gums. It is insoluble in water and forms a pasty mass like *katera* gum. This gum is astringent and employed medicinally in bowel complaints. A solution of this gum is given in conjunction with spices in certain stages of bowel complaints.

Chickrassia tabularis A Juss.; English—Bastard cedar; Hindi—*Chikrasi*; Bengali—*Chikrassi, dalmara*; Assamese—*Boga poma*; Marathi—*Pabba*; Tamil—*Aglay, agal*; Telugu—*Chittagong karru*; Malayalam—*Dovedah*; Kannada—*Dalmara, devdari*; Andamanese—*Arrodah*; Family—Meliaceae.

A large tree, native of the hills of Eastern Bengal and South India. Found in the Western Ghats, Andhra Pradesh and Andaman Islands.

Gum. It yields a transparent, amber–coloured gum. The gum consists of irregular tears, amber–coloured or light brown, somewhat transparent, but not brittle. The gum is used as an adhesive.

Cochlospermum religiosum (Linn.) Alston; English—White silk cotton; Hindi—*Pili kapas, kumbi, galgal*; Uriya—*Kontopalas*; Punjabi—*Kumbi*; Telugu—*Gunju; gondu–gogu*; Tamil—*Tanaku*; Kannada—*Betta tovare*; Malayalam—*Chima-punji*; Marathi—*Ganeri, kathalya gonda*; Persian—*Katira-i-Hindi*; Family—Cochlospermaceae.

Hindi—*Hindi katera*; Tamil—*Tanaku—pishin*; Telugu—*Konda-gogu-pisunu*; Malayalam–*Shima—Pangi-pasha*.

A small deciduous tree, with short thick spreading branches; found commonly in Andhra Pradesh, Bihar and Madhya Pradesh. When the tree is devoid of leaves (March-April) it bursts into its handsome large yellow flowers, its pendulous, pear shaped fruits ripening before the new leaves appear.

Gum. The gum is often sold in the market as *katira* or *kathira*. The gum obtained from *Cochlospermum religiosum* is known as pseduo-gum. It is insoluble in water, but swells and forms a pasty mass. The gum is obtained freely on the trees being tapped. It occurs in striated and twisted pieces, of a pale semi-transparent colour, transversely fissured with a tendency to split up into flat scales. While not soluble it readily becomes diffused in minute particles through a large quantity of water.

The gum is used as a mild demulcent in coughs. The larger lumps being of a dullish red. It is used for making cheap varnishes. It may be applied as water proofing material.

Delonix regia Rafin.; Syn. *Poinciana regia* Bojer; Hindi—*Gulmohar*, Family—Caesalpiniaceae.

A tree, native of Madagascar, which was introduced into India, and now grown as an ornamental tree, all over India.

Gum. The tree yields a gum plentifully. It has irregular, granular or warty tears of a yellowish or reddish-brown colour, soluble in water, forming a thick opalescent mucilage. The surface of some of tears is an opaque yellow colour; this portion consists largely of beautiful sphaero–crystals of oxalate of lime. On moistening this gum with water a cloud of small crystals often separates and the sphaero–crystals, arrange into bundles of acicular crystals.

Dichopsis polyantha Benth, & Hook.; Syn. *Bassia polyantha* Wall, *Isonandra polyantha* Kurz; Bengali—*Tali;* Assamese—*Sill kurta*; Bihar—*Thainban*; Family—Sapotaceae.

A moderate-sized ever-green tree found in Assam.

Gum. It produces a good quality of Gutta–percha in large quantities. The milky substance is obtained on tapping the trees. The gutta–percha yielded from the milk after processing is about one third of its weight.

Elaeodendron glaucum Pers.; Hindi—*Chauri jamrasi*; Lepcha—*Chikyeng*; Kumaon—*Shauriya*; Punjabi—*Mirandu*; Marathi—*Aran, tamruj*; Tamil—*Karkava, siri*; Telugu—*Nerasi, kanemi*; Kannada—*Tha–maroja*; Family—Celastraceae.

A moderate-sized tree; found throughout hotter parts of India. Along the outer Himalayas, it ascends upto 6,000 feet.

Gum. It is supposed to yield the gum called *jumrasi*, which occurs in roundish tears about half an inch in diameter, rough or cracked on the surface. It is tasteless, and forms a sherry–coloured solution with water.

The gum is used medicinally.

Eucalyptus globulus Labill.; English-Blue Gum-tree, Tasmanian blue gum; Vernacular—*Karpura maram*; Family—Myrtaceae.

A lofty tree; native of Australia, now cultivated chiefly in the Nilgiris, the Anamalais, the Pulney hills, Simla hills and at Shillong in Assam.

Gum. The bark of this tree exudes an astringent gum. It is known in trade as *Australian* or *Eucalyptus kino*. A kino of better quality is obtained from other species of *Eucalyptus*, such as *E. corymbosa* and *E. citriodora*.

This gum is used for tanning and dyeing.

Feronia limonia (Linn.) Swingle; Syn. *F. elephantum*; English—Elephant apple, Wood apple; Hindi—*Kaith, bilin, kat bel, kavitha*; Bengali—*Kath–bel, kait*; Uriya—*Koeta*, Punjabi—*Kait, bilin*; Marathi—*Kawat, kavith*; Gujarati—*Kotha, kavit*; Tamil—*Vilam, kavit*; Telugu—*Velaga, kapidh*; Kannada—*Bilwar, byala*; Malayalam—*Vilam*; Sanskrit—*Kapittha*; Persian—*Kabit*; Family—Rutaceae

A moderate-sized tree with short, erect, cylindrical stem, bearing thorny branches; leaves pinnate, with small ovate leaflets; fruit large, globose with hard, rough, woody pericarp. The plant is a native of India and Sri Lanka. Found in the sub-Himalayan forests, throughout the greater part of the plains of India. Abundantly found in moist tracts of Maharashtra, Tamil Nadu and Bengal.

Gum. It is known as *Feronia gum*. The stem and branches of the tree exude a gum resembling gum arabic in properties. The exudation is profuse after rainy season. It occurs in irregular, semitransparent tears varying in colour from reddish brown to pale yellow or colourless. The gum dissolves in water forming a tasteless mucilage, not less adhesive than that of gum arabic. It is a better gum for mixing with colours than gum arabic. The gum is also used as an ingredient of writing inks.

Gardenia turgida Roxb.; Hindi—*Thanella, khuriari, ghurga*; Uriya—*Bamenia*; Marathi—*Khurphendra, phanda*; Telugu—*Manjunda*; Kannada—*Bongeri*; Malayalam—*Malankara*; Rajasthan—*Karumba*; Madhya Pradesh—*Karhar*; Family—Rubiaceae.

A small deciduous tree found in the sub-Himalayan tracts from Nepal to the Jumna, ascending to 4,000 feet; also in Rajasthan, Madhya Pradesh and South India.

Gum. The tree yields a yellow gum with a pleasant odour, almost completely soluble in water. The gum exudes from the cuts made on the upper part of the stem. The gum is used medicinally.

Garuga pinnata Roxb.; Hindi—*Ghogar, kaikar*, Bengali—*Jum, tum kharpat*; Uriya—*Mohi*; Assamese—*Gendeli poma*; Nepal—*Dabdabbi*; Lepcha—*Maldit*; Malayalam—*Kasramba*; Kumaon—*Kilmira*; Punjabi—*Kharpat*; Marathi—*Kuruk*; Gujarati—*Kusimb*; Tamil—*Karre vembu*; Telgu—*Garugo;* Kannada—*Hala, balage*; Family—Burseraceae.

A tree, attaining the height of 30 to 40 feet, found in the sub-Himayalan forest, ascending upto 3,000 feet, also in Madhya Pradesh and South India. Flowers from February to March, and the fruit ripens in June-July.

Gum. This tree yields a greenish yellow, translucent exudation in small masses. It possesses a wild terebinathinate odour and taste. By many workers, it has been regarded a true gum.

The gum is employed medicinally. The juice of the stem is dropped into eyes, to cure opacities of the conjuctiva.

Lannea coromandelica (Houtt.) Merr.; Syn. *L. grandis* (Dennst.) Engl.; *Odina wodier* Roxb.; Hindi—*Jingan, kaimil, kashmala*; Bengali—*Jiyal, jival*. Uriya—*Mooi, indramai*; Punjabi—*Kamali, kiamil*; Marathi—*Moi, moja*; Tamil—*Wodier, odiyamaram*; Telugu—*Gumpini, dumpini*; Kannada—*Shimti, punil*; Sanskrit—*Jingini*; Family—Anacardiaceae.

A large deciduous tree, 40 to 50 feet in height, found throughout the hotter parts of India, ascending in the sub-Himalayan tract to an altitude of 4,000 feet; found also in Assam, Tamil Nadu and the Andaman Islands.

Gum. The gum exudes from wounds and cracks in the bark of this tree. At first it is yellowish white in colour (*Kanne-ki-gond*), which takes on a brownish tinge, and afterwards, if it falls on the ground, it becomes blackish (*jingan-ki-gond*). The gum usually exudes in October and occurs partly in tears of a yellowish tinge, partly in gum arabic. It is not astringent and possesses a disagreeable taste; about one half is completely soluble in water; the remaining portion forms a slimy mucilage, but not gelatinous.

It is used in calico–printing and for sizing of paper. Sometimes it is used mixed with lime in white washing. It is used also as the basis of an inferior varnish.

Macaranga peltata Muell. Arg.; Marathi—*Chanda*; Uriya—*Piania, gondaguria*; Tamil—*Vattakanni*; Telugu—*Boddi kondatamara*; Kannada—*Chandakanne*; Malayalam—*Uppila*; Family—Euphorbiaceae.

A moderate sized tree, found in Bengal, Bihar, Orissa and the Deccan Peninsula, mostly in the hills.

Gum. A reddish gum kino exudes from the cut branches, bases of petioles, young shoots and fruits. It is found in the form of hard tears or agglutinated masses with a shiny lustre and little or no taste. It is partly soluble in water. The gum contains: 17.1% moisture, 63.4% gum, 15.00% tannin and 2.1 ash.

The gum is used for sizing paper and for taking impression of leaves, coins, etc. A paste of the kino is used as an application for venereal sores.

Mangifera indica Linn.; English—Mango; Hindi—*Am*; Family—*Anacardiaceae.*

A large evergreen tree, found throughout the hotter parts of India.

Gum. The bark yields a gum, which is frequently sold in the markets as gum arabic. The gum is partly soluble in cold water. The gum-resin mixed with lime-juice or oil is used in scabies and cutaneous affections.

Moringa oleifera Lam.; Syn. *M.pterygosperma* Gaertn.; English *Drumstick tree*, *Horse Radish tree*; Hindi—*Shajna*; Bengali—*Sojna, Sajina*; Uriya—*Munigha, sajina*; Punjabi—*Senjna*; Marathi—*Shevgi Shevaga*; Gujarati—*Saragavo*; Tamil—*Morunga*; Telugu—*Munaga*; Kannada—*Nugge–gida*; Malayalam—*Murinna*; Sanskrit—*Sobhanjana*; Family—Moringaceae.

A medium-sized tree, found wild in the sub-Himalayan tract, from Chenab eastwards to eastern Uttar Pradesh, and cultivated all over the plains of India.

Gum. The tree yields a gum which is white when it exudes, but gradually turns to a mahogany colour on the surface. It is little soluble in water and swells in contact with it giving a highly viscous solution. It is polyuronide consisting of arabinose, galactose and glucuronic acid; rhamnose is present in traces. The gum is locally used in calico-printing.

Pithecellobium dulce (Roxb.) Benth.; Syn. *Mimosa dulcis* Roxb.; *Inga dulcis* Willd.; English—Manila tamarind; Hindi—*Vilaiti imli, jangal jalebi*; Mumbai—*Vilayti amli, deccan babul*; Marathi—*Hatichinch*; Tamil—*Karkapilli*; Telugu—*Sima chinta*; Kannada—*Sime hunase*; Family—Mimosaceae.

A tree, introduced from Mexico and now cultivated throughout India, especially along railway lines in Tamil Nadu, flowers during cold season, pods 4-5 inches in length and ½ inch in breadth with 6-8 seeds.

Gum. This tree yields a gum, usually in spheroidal tears, about half an inch in diameter, of a deep reddish–brown colour, transparent and with a polished surface. It is freely soluble in water, forming a thick brown mucilage.

Prosopis chilensis Stuntz; Syn. *P. juliflora* DC; English—Mesquite; Hindi—*Vilayati kikkar, vilayati babul, kabuli kikkar, vilayati khejra*; Family—Mimosaceae.

An evergreen tree, with drooping branches, found in the drier parts of India. The pods have a tough pericarp, consisting of a waxy exocarp, a spongy mesocarp and a cartilaginous endocarp, enclosing the seeds in lomentaceous segments.

Gum. The tree exudes a gum, consisting of nearly smooth, light yellowish brown, more or less opaque tears, which are translucent and glassy when fractured. The gum forms a somewhat adhesive mucilage and can be used as an emulsifying agent. It also finds use in confectionery and is sometimes employed for mending pottery. The mesquite gum is used as an adulterant and substitute for gum arabic. It is inferior to gum arabic.

Prosopis cineraria Druce; Syn. *P. spicigera* Linn; Hindi—*Jhand, jand*; Bengali—*Shami, somi*; Uriya—*Shami, savandal*; Punjabi—*Jand, khunda, khar*; Rajasthan—*Khejra*, *sangri*; Marathi—*Shemi, saunder*; Gujarati—*Semru, hamra*; Tamil—*Perumbe*, *vanni*; Telugu—*Chani, shumi, jammi*; Kannada—*Perumbe, vunne*; Family—Mimosaceae.

A moderate-sized, deciduous, thorny tree, found in the arid dry zones of the Punjab, Rajasthan, Gujarat, Madhya Pradesh, Bundelkhand and South India. Pods cylindric, torulose or flattish with coriaceous exocarp.

Gum. A gum exudes similar to gum arabic, from the stumps of pruned branches and other scars of tree. The gum is unusually friable, occurs in small angular fragments of a yellowish colour, more or less deep sometimes in large ovoid tears about two inches long, of an amber colour internally, but having a frosted or candied appearance externally from the presence of numerous minute cracks which cause the tears to crumble under pressure. The gum forms with water a dark coloured tasteless mucilage of about the same viscosity as that of gum arabic. It is used like mesquite gum.

Prunus amygdalus Batsch; Syn. *P. communis* Fritsch; *Amygdalus communis* Linn; English—*Almond*; Hindi and Punjabi—*Badam*; Bengali—*Bilati–badam*; Marathi and Gujarati—*Badam*; Tamil—*Vadam–kottai*; Telugu—*Badam–vittulu*; Kannada—*Badami*; Malayalam—*Badam*; Sanskrit—*Badamitte*; Persian—*Badam*; Family—Rosaceae.

A Tree, native of central and western Asia. In India it is cultivated in Kashmir and Himachal Pradesh between 3,000-8,000 feet elevations.

Gum. This tree exudes a gum, known as "*Badam* or Hog tragacanth", which has been employed in place of tragacanth. It is obtained mostly from the trunk and larger branches of old trees. Almond gum hydrolyses into L-arabinose (4 parts), D-xylose (2 parts); D-galactose (3 parts) and D-glucuronic acid (1 part).

Prunus armeniaca Linn; English—Common apricot; Hindi—*Khubani, zardalu, chuari*; Kumaon—*Chuaru, zard–aru*; Kashmir—*Gurdalu, cherkush*; Punjabi—*Hari, cheroli, sari*; Persian—*Mishmish*; Family—Rosaceae.

A moderate-sized deciduous tree, cultivated in the north-western Himalayas, chiefly in the valleys of Kashmir, Kulu and in Simla hills at altitudes upto nearly 10,000 feet.

Gum. The tree yields a gum similar to tragacanth, which along with the gums from other members of this genus, is known commercially as *Cherry gum*. It is used medicinally.

Pterocarpus marsupium Roxb.; English—Indian kino; Hindi—*Bija, bijasal*; Bengali—*Pitsal*; Uriya—*Piasal*; Marathi—*Dhorbeula, asan*; Gujarati—*Bibla*; Tamil—*Vengai*; Telugu—*Peddagi*; Kannada—*Benga, honne*; Malayalam—*Venna*; Hindi—Gum = *Hira dokhi, rang barat*; Persian—*Khune-siyavushane–hindi*; Family—Papilionaceae.

A large deciduous tree, found commonly in hilly regions throughout the Deccan Peninsula, and extending to Gujarat, Madhya Pradesh, Uttar Pradesh, Bihar and Orissa.

Gum. The tree yields the gum kino. The gum as it exudes has the appearance of red currant jelly, but hardens in a few hours after exposure to the air. The gum is obtained by making a perpendicular incision with lateral ones leading into it is made in the trunk, at the foot of which is placed a vessel to receive the outflowing juice. The exudate is collected and dried in the sun or shade and yield of dried gum is reported to be 340 g. per tree. Kino occurs in small, angular, glistening, brittle fragments, appearing, almost black in colour. It is odourless and bitter with astringent taste and colours saliva pink when masticated, kino contain kinotannic acid (25-80%), Kinoin $C_{28}H_{24}O_{12}$ and kino-red $C_{28}H_{22}O_{11}$.

Kino is a powerful astringent and was formerly used in the treatment of diarrhoea and dysentery. It is locally applied in lecorrhoea and in passive haemorrhages. It is also used in toothache. It is employed in dyeing, tanning and printing.

Sterculia urens Roxb.; English—Kateera gum; Hindi—*Gulu, tabsi*; Assam—*Odla*; Uriya—*Kavili*; Marathi—*Pandruka, Kavali*; Gujarati—*Karai*; Tamil—*Vellay putali*; Telugu—*Tabsu, kavalee*; Kannada—*Penari*; Family—Sterculiaceae.

A soft-wooded tree, found commonly in Uttar Pradesh, Madhya Pradesh, Rajasthan and Bihar.

Gum. The stem yields a gum known as *kateera* gum. It is used as a substitute for gum tragacanth and is used in calico-printing.

The gum is completely soluble in cold water, forming an almost colourless solution. In mass it is slightly opalescent. Thirty grains dissolved in about 500 ml. water forms a thick tasteless mucilage, which entirely passes through a filter paper.

The gum is used medicinally as a substitute for tragacanth. Santals consider it a useful medicine in throat affections.

Terminalia bellirica (Gaertn.) Roxb.; English—Belleric myrobalan, Bastard myrobalan; Hindi—*Bahera*; Bengali—*Bohera*; Assam—*Bauri*; Lepcha—*Kanom*; Uriya—*Bahara*; Punjabi—*Bahera*; Marathi—*Bherda*; Gujarati—*Beheda*; Tamil—*Thani*; Telugu—*Tani*; Kannada—*Tanikayi*; Malayalam—*Thani*; Sanskrit—*Vibhitaki*; Persian—*Babla*; Family—Combretaceae.

A large, deciduous tree, commonly found in the plains and lower hills throughout India.

Gum. Large quantities of an insipid gum exudes from the wounds in the bark of the tree. It resembles to some extent the gum arabic. It is soluble in water, and burns away in the flame of a candle, with little smell, into black gritty ashes. The gum is found in vermicular pieces, about the thickness of a finger, of the colour of inferior gum arabic.

In water, it swells up and forms a bulky gelatinous mass; its taste is insipid. It is collected and mixed with soluble gums for sale as country gum.

Terminalia tomentosa Wight a Arn.; Hindi and Bengali—*Asan*; Assam—*Amari*; Lepcha—*Taksor*; Uriya—*Sahaju*; Punjabi—*Sain*; Marathi—*Asna*; Gujarati—*Ain*; Tamil—*Karra marda*; Telugu—*Maddi*; Kannada—*Matti, aini*; Sanskrit—*Asana*; Family—Combretaceae.

A large, deciduous tree, 80 to 100 feet in height, common throughout the moister regions of India.

Gum. The tree yields a red gum, black outside the pieces. Sometimes, the gum is found in rounded dull brown tears, soft, readily being agglutinated, and capable of being sliced with a knife. It has a better disagreeable taste, and partly soluble in water. The gum is used as an incense and as a cosmetic. It is eaten by the natives.

Zizyphus sativa Gaertn.; Syn. *Z. vulgaris* Lamk; English—Common jujube; Hindi—*Kandiari*; Kashmir—*Phitni*; Punjabi—*Phitniamni*; Persian—*Kunar*; Family—Rhamnaceae.

A shrub, or small tree, found wild and cultivated in the Punjab and Himachal Pradesh up to 6,500 feet, occasionally cultivated in Bengal.

Gum. A gum is obtained from this species, which is used for dyeing and as a drug. It is found in irregular masses and broken pieces of a dull dark brown colour, with a lustrous fracture, soluble in water, but forming a dark–coloured mucilage very like that of coarse dark *Babul gum*.

The gum is used medicinally. It is used in certain affections of the eyes.

RESINS

They represent oxidation products of various essential oils and are very complex and varied in their chemical composition. The resins are found chiefly in the plants of the phanerogamic families—Anacardiaceae, Burseraceae, Dipterocarpaceae, Guttiferae, Hamamelidaceae, Leguminosae, Liliaceae, Pinaceae, Styracaceae, and Umbelliferae. They are mostly found in the stems of the trees, and occur in abundance in special canals or ducts. They are yellowish solids, insoluble in water but soluble in alcohol, turpentine and spirit. The essential ingredients of every resin are the resin acids, volatile oils, gums and often cinnamic and benzoic acids as well as the ordinary components of plant tissues such as cellulose, tannin, etc. When present in the wood, resins add to the strength and durability of it. These occur associated with a small quantity of turpentine which is removed by distillation, and the residue is pure resin. It normally oozes out through the bark and hardens on exposure to the air. Usually tapping is necessary in order to obtain a large amount to be of commercial value. In the coniferous-pine forest of Himalayan region Northern India the resin is collected on commercial scale. Resin and Turpentine are a source of immense revenue to India. A list of resin yielding plants is being given here.

Abies excelsa DC.; English—Spruce Fir; Family—*Pinaceae.*

A noble tree found in the mountains of Central Europe, in Norway, Sweden and Russia introduced into India.

Resin. It yields a resin. A yellowish—brown opaque substance, which exudes from the bark, hard and brittle when cold, strongly adhesive, possesses an agreeable, aromatic odour. The ole-resinous exudations known as *turpentine* consist of an amorphous resin $C_{44} H_{62} O_4$ mixed with an essential oil or hydrocarbon of the composition $C_{10}H_{16}$.

Abies webbiana Lindl.; English—Himalayan Silver Fir.; Punjabi—*Paludar*, Kashmiri—*Badar tung*; Kulu—*Tos*; Garhwal—*Chilrao, chiliragha*; Kumaon—*Ragha*; Nepalese—*Gobria sulah*; Bhutia—*Dumshing*; Family—Pinaceae.

A lofty ever green tree found in the Himalayas from Himachal Pradesh to Sikkim; in the North-West Himalayas, between 7,000 and 13,000 feet; in the inner ranges of Sikkim, between 9,000 and 13,000 feet; in the outer ranges not below 10,000 feet.

The *cones* are lateral, erect, 4-6 inches long, solitary or clustered, dark blue when young; *scales* deciduous. *Leaves* flat, narrow, linear, spirally arranged.

Resin. It yields a white, resin, which is used medicinally. The resin is mixed with oil of roses, and used externally for headache, neuralgia, etc.; when taken internally produces intoxication.

Ailanthus malabarica DC.; Tamil—*Peru-marattup-pattai, maddi-pal*; Malayalam—*Peru-morat-toli, mattippal*; Marathi—*Guggula—dhupi*; Kannada—*Dhup, gogul-dhup*; Family—Simarubaceae.

A large, deciduous tree, of the evergreen tropical forests abundant in the Western Ghats. Often planted in India for ornamental purposes.

Leaves very large; *leaflets* distant, almost entire, nearly glabrous. *Stamens* exerted, upon filaments many times longer than the anthers. *Samara* large, rounded at both ends, not twisted.

Resin. On incision the bark yields a dark-coloured soft resin known as *Mattipal*, which, after some time, hardens into a brittle resin with a strong balsamic odour. The pure resin is very soft, and it is always mixed with fragments of wood, which make it more easy to handle. Alcohol readily dissolves the resin. When burned it gives out a fragrance, and therefore, it is used for incense.

The resin known as *mattipal* is used medicinally especially in dysentery.

Altingia excelsa Noronha; Hindi—*Silaras*; Assamese—*Jutili*; Tamil—*Neriuriship-pal*; Telugu—*Shila-rasam*; Malayalam—*Rasa-mala*; Gujarati—*Silaras*; Marathi—*Shilaras*; Persian—*Zak*; Family—Hamamelidaceae.

A magnificent tree of the tropical evergreen forests. Found in Assam and Sikkim.

Leaves alternate, persistent petioled; *flowers* in dense heads; *male* heads a mass of stamens; *female* heads of 12-20 flowers; *fruiting head* globose, many capsuled; *seeds* imperfect, the lowest winged and fertile, the upper without wings and sterile.

Resin. It yields in small quantity an odorous resin, known in Europe under the name *storax*, which is obtained by incisions in the trunk. It is said to afford a fragrant balsam of two varieties one pellucid and of a light-yellowish colour, obtained by simple incision; and the other, dark, opaque, and of a terebinthinous odour, procured by boring the stem and applying fire around the trunk. *Storax* has long formed an important article of medicine in India.

Boswellia serrata Roxb.; English—Indian Olibanum Tree; Hindi—*Luban, salai, sel-gond, kundur*, Bengali—*Luban, salai, kundro*; Kumaon—*Guggar*, Marathi—*Salaphali*; Gujarati—*Dhup, mukul salai*; Tamil—*Kungli, gugulu*; Telugu—*Anduga—pisum, andu*; Malayalam—*Vella-kundirukkam*; Kannada—*Chittu*; Sanskrit—*Kunduru guggulu*; Family—Burseraceae.

A moderate sized gregarious tree, found in sub-Himalayan tract from the Sutlej to Nepal, the forests of Madhya Pradesh, Rajasthan and South India.

Bark frequently papyraceous; the leaflets being sessile, pubescent, coarsely crenate-serrate; racemes axillary, shorter than the leaves.

Gum-resin. The gum-resin, *Salai gugul*, occurs as a transparent golden yellow, semi-fluid substance, which slowly hardens with lime. When it occurs in the soft massive form it is known as *Gandah ferozah*; in tears (true olibanum) it is known as *kundur*. It is pungent having a slightly aromatic taste and balsamic resinous odour. It becomes opaque when immersed in alcohol; the proportion of resin to gum being much smaller. The opaque, soft, whitish mass produced by water when rubbed in a mortar forms an emulsion. A sweet-scented gum burnt in religious ceremonies. The gum-resin is also made into ointment. It is regarded as a demulcent, aperient, alterative and a purifier of the blood. The gum is used in rheumatism, nervous diseases, scrofulous affections, urinary disorders and skin diseases, and is generally combined with aromatics. It is regarded as a diaphoretic and astringent, and is used in the preparation of ointment for sores. It is also prescribed with clarified butter in syphilitic diseases; with coconut oil for sores, and as a stimulant in pulmonary disease.

Canarium strictum Roxb.; English—Black Dammar tree; Hindi, Bengali and Gujarati—*Kala dammar*; Marathi—*Dhup raldhup*; Tamil—*Karapu kongiliam, karapuammar*; Telugu—*Nalla-rojan*; Kannada—*Manda-dhup, raldhupada*, Malayalam—*Thelli*; Family—Burseraceae.

A tall tree, found in south India. Common about Courtallum in the Tinnevelly district and in Kanara.

Resin. The tree yields a brilliant resin called the *Black Dammar* of South India. It is obtained by making vertical cuts in the bark and setting fire to the bottom of the stem. The dammar exudes from the stem as high as the flames reached commencing about after two years of the above operation. The flow continues for ten years, between the months of April and November, and resin is collected in January.

The resin occurs in stalactitic masses of a bright shining colour. It is homogeneous, with a vitreous fracture; partially soluble in boiling alcohol, and completely so in oil turpentine.

It makes an ordinary varnish. The resin is used medicinally; also used as liniment in rheumatic pains.

Cannabis sativa Linn; English—Hemp; Hindi—*Ganja, bhang, charas*; Bengali—*Ganja, bhang, sidhi*; Punjabi—*Bhangi, charas, kas*; Kashmiri—*Bangi*; Gujarati—*Ganja*; Marathi—*Bhangacha-jhada*; Tamil—*Ganja-chedi, ganja-phal, ganja-rasham*; Malayalam—*Kannchava-chetti*; Kannada—*Bhangi-gida*; Sanskrit—*Ganjika, jaya, hursini*; Persian—*Darakte-bang*; Family—Cannabinaceae.

An undershrub; native of Asia, now cultivated in Uttar Pradesh, Bengal, Maharashtra, Madhya Pradesh, Tamil Nadu and Orissa.

Resin. The female flowering tops, leaves, stems and fruits of the hemp yield a resin, commonly known as *charas*. *Charas* is collected in Madhya Pradesh by causing men to run through the hemp fields. They are generally clad in leathern aprons to which the resin adheres. The *charas* thus collected is scraped off and made into the cakes in which it is sold.

Charas is smoked. The drug is stimulant, anodyne, sedative and antispasmodic. It is also said to be narcotic, diuretic and parturifacient.

The resin is a brown amorphous solid, burning with a bright white flame and leaving no ash. It has a very potent action when taken internally, two-thirds of a grain acting as a powerful narcotic, and one grain producing complete intoxication.

Carica papaya Linn; English—The papaw or Papaya tree; Hindi—*Papita*; Bengali—*Pappaiya*, Punjabi—*Arandkharbuza*; Marathi—*Papaya*; Gujarati—*Papia, eranda kakdi*; Tamil—*Pappali*; Telugu—*Bappayi*; Kannada—*Perangi*; Malayalam—*Pappaya*; Persian—*Aanabahe-hindi*; Family—Caricaceae.

A sub-herbaceous almost branchless tree. Native of the West Indies and central America but cultivated chiefly in Uttar Pradesh, the Punjab, Rajasthan, Gujarat, Maharashtra, Tamil Nadu and the Nilgiris.

Resin. Exudes a white resin. The juice (white resin) of the fruit is regarded as medicinal. The milky juice possesses anthelmintic properties. 'Papain' may be prepared from the juice of the green fruit of papaya by adding alcohol, which precipitates papain. This precipitate is dried and powdered, and is then quite ready for use. It is an invaluable remedy in the dyspepsia of those recovering from attacks of cholera. The juice of papaya is a favourite domestic remedy for the expulsion of tape worm. Milky resin is also used in spleenic and hepatic enlargements. The unripe fruit is scraped longitudinally and the milky juice is collected. This is then put on a sand bath. After 24 hours, a dull white powder is formed. This powder is given (one or two grains only) with sugar or milk after meals to adults.

A few drops of juice added to tough meat render it quite tender and fit for immediate cooking.

The juice called *papaine* has digestive ferment properties and will remove thickened skin, as in eczema. It is also said to be a certain remedy in cases of scorpion sting.

The juice is also used in preparing chewing gum.

Cedrus deodara Loudon; English—Deodar—Hindi—*Deodar, kilan*; Punjabi—*Dewdar, kalain, kaiwal*; Kashmiri, Garhwali, Kumaon—*Diar, deodar, delu, kilar*; Gujarati—*Deodar*; Marathi—*Dewdar;* Tamil—*Devadari—chedi*; Telugu—*Devadari-chethi*; Malayalam—*Devataram*; Kannada—*Devdari–mari*; Sanskrit—*Devadara*; Persian—*Darakte–devdar*; Family—Pinaceae.

A large and tall, tree found in the North-Western Himalayas from Kashmir to Garhwal, between 4,000 and 10,000 feet.

Oleo-resin. It yields a true oleo-resin, called *kelon-ka-tel*. An oleo-resin is dark coloured, thick and resembles crude turpentine. It is used for anointing the inflated skins which are used for crossing rivers. It is also used as a remedy for ulcers and eruptions. The oleo-resin is also used by natives in the treatment of leprosy.

Commiphora mukul (Hook. ex Stocks); Syn. *Balsamodendron mukul* Hook. ex Stocks; Engl. English—Indian bdellium tree; Hindi—*Guggul*; Gujarati—*Gugal*; Kannada—*Guggala*; Tamil—*Gukkal*; Telugu—*Gugal*; Sanskrit—*Guggulu*; Family—Burseraceae.

A small tree, found in the arid zones, Haryana, Rajasthan, Gujarat and South India.

Gum-resin. The tree yields the gum-resin known as *Gugul* or as an *Indian Bdellium*. It occurs in vermicular or stalactitic pieces. It is of a brown or dull green colour, and possesses a bitter, acid taste. It exudes from incisions on the bark made in the cold season. It swells when heated, diffusing a disagreeable odour.

Indian Bdellium (*gugul or mukul*) is used in native medicine as a demulcent, aperient, carminative and alterative; especially useful in leprosy, rheumatism and syphilitic disorders. It is also used in nervous diseases, urinary disorders and skin diseases. It is found useful in the form of an ointment in cleansing and stimulating indolent ulcers. A preparation of *gugul* called *yogaraja gugul* is given internally for muscular rheumastism. The fumes of gugal are believed to be disinfectant. In combination with other medicines (black pepper and colchicum) the gum is given in the form of confection in cases of rheumatism, haemorrhoids, and flatulent dyspepsia.

Mixed with mortar the gum forms an excellent cement.

Daemonorops kurzianus Hook. f.; English—East Indian Dragon's Blood; Hindi—*Aprang, hiradukhi*; Marathi and Gujarati—*Hiradakhan*; Tamil—*Kondamurae rattam*; Family—Palmae.

It is a lofty climber found in South Andaman Islands.

Resin. The dark reddish brown resin occurs as small granules on the scaly fruits. It is used chiefly in the manufacture of red spirit varnishes for metals and in making fine line engravings. It is also used for colouring lacquers and varnishes.

Diospyros peregrina (Gaertn.) Gurke; Syn. *D. embryopteris* Pers; English—Indian persimmon; Hindi—*Kendu, gab, tendu*; Uriya—*Gusvakendhui*; Assam—*Kendu*; Marathi—*Temburni*; Gujarati—*Zeeberwo*; Tamil—*Tumbika*; Telugu—*Tinduki*; Kannada—*Kusharta*; Sanskrit—*Tinkuka*; Family—Ebenaceae.

A dense evergreen small tree with dark-green foliage and long shining leaves; commonly found throughout India, except the arid and dry zones in the Punjab. Frequently found in the Western Peninsula and Madhya Pradesh.

Gum-resin. The fruit of this tree yields a gum-resin. It is used for caulking the small canoes. The Assamese fishermen colour their nets with it, and thereby render them more durable in wear and tear.

The fruits are boiled and the juice thus obtained boiled down to a tarry or resinous consistency.

Dipterocarpus turbinatus Gaertn.f. English—Gurjan oil tree; Hindi—*Gurjan*; Bengali—*Garjan, tihya gurjun*; Gujarati—*Gurjun*; Kannada—*Challani*; Family—Dipterocarpaceae.

An evergreen tree; found in Bengal and Andaman Islands. It is said to be one of the loftiest of Indian trees, sometimes seen 250 feet in height.

Oleo-resin. The oleo-resin is obtained by cutting two or three pyramidal hollows near the foot of the tree and by applying fire to the upper cut surfaces. The oleo-resin (wood-oil) then collects at the bottom of the hollow, which is emptied every three or four days. Fire is applied every time the oleo-resin is removed, and the upper surfaces of the hollow are rechipped three or four times during the season.

The oleo-resins consist of a volatile oil holding in solution a resin and are generally classed under the head of balsams. It is stated that the oil if set aside for a time subsides into two substances, *viz.*, a clear thin liquid, floating above a thick mass known as *guad*. The oleo-resin is reported to act as a solvent to caoutchouc. The oleo-resin becomes gelatinous at 130°C, and on cooling does not recover its fluidity.

The oleo-resin is used for making good varnishes. It is also used in water proofing. In very limited cases, it is used medicinally. Gurjan oil is used internally for the treatment of gonorrhoea, gleet and similar affections of the urinary organs by the native physicians. It is also used externally as a stimulating application to indolent ulcers. It is employed both internally and externally, in the treatment of leprosy.

Erythroxylon monogynum Roxb.; English—Bastard sandal, Red cedar; Tamil—*Davadaram, devadari, Nat-ka-deodar, sammanathi, thasadaram*; Kannada—*Kuruvakumara*; Telugu—*Devadari, adavigoranta, gathiri*; Family—Erythroxylaceae.

A shrub or small tree, found in South India. It occurs plentifully in Tanjore and Tinnevelly, ascending the ghats to an elevation of 2,500 feet, and in the Nilgiris.

Oleo-resin. The wood yields an oleo-resin (oil) used as a preservative for native boats. This oleo-resin resembling tar, is known as *dummele*. It is extracted by packing pieces of the wood in an earthen pot, inverted over a similar empty one and surrounded by fire. The tar thus distilled is soluble in ether, alcohol and turpentine, and is an excellent preservative of timber.

Ferula assafoetida Linn; English—Asafetida; Hindi—*Hing*; Kashmiri—*Anjudan*; Gujarati—*Hing*; Tamil—*Kyam*; Sanskrit—*Hingu*; Persian—*Anguza*; Family—Umbelliferae (Apiaceae).

A perennial herb, commonly grown in the Punjab, Himachal Pradesh and Kashmir. It grows to a height of 2-4 feet.

Gum-resin. The gum-resin is collected as follows: When the stem begins to grow it is cut down, and the upper part of the root being wounded, a small quantity of very choice gum is collected. Afterwards, a slice of the root, about 1/4 inch thick, is removed every two or three days with exudation adhering to it, until the root is exhausted. The collected mass, consisting of alternate layers of root and gum-resin, when packed in a skin, forms the *Hing* of Indian commerce.

The gum-resin, consists of a blackish brown, originally translucent, brittle mass of extremely foetid alliaceous odour. The term '*hira-hing*' is applied to liquid of treacly consistency, often found in the centre of the bales, which is squeezed out and sold at a high price.

Hing is very much used in India, and has, from the ancient times, been held in great esteem by Indian *Vaids*. It is a carminative and antispasmodic, and used in colic, cholera, etc.,; and when taken daily it is said to ward off attacks of malarial fever. It is used as a carminative in flatulent dyspepsia; as an anthelmintic in cases of round worm, and as an emetic.

The gum-resin (hing) is employed as a condiment, and is especially prized by the vegetarians. It is mixed in various ways with *dal* and *chawal.*

Ferula galbaniflua Boiss & Buhse; English—Galbanum; Hindi—*Gandha—biroza*; Persian—*Jawashir*, Family—Umbelliferae (Apiaceae).

Found in North Western India. The young leaves on the tops of the perennial stems appear like cushions of mass.

Gum-resin. The *Jawashir* is a yellow or greenish semi-fluid resin, generally mixed with the stems, flowers and fruits of the plant.

The stem, on injury from its earliest stage of growth, yield, an orange-yellow gummy fluid, which very slowly consolidates, on the stem. It possesses a strong, celery-like odour. The gum is commonly found adhering to the lower portions of the stem.

It is used medicinally as an attenuant, detergent, antispasmodic, and expectorant; prescribed in paralytic affections, hysteria and chronic bronchitis.

Ferula narthex Boiss.; English—Narthex asafetida; Hindi—*Hing, hingra*; Sanskrit—*Hingu.*

A herb, with a circular mass of foliage commonly found in Kashmir.

Gum-resin. It forms the drug of commerce known in Europe as asafetida and in India as *hingra.*

After cutting the plants through, above the root, three or four incisions are made in the stumps. The operation of incision is repeated every three or four days, so long as the sap continues to exude.

It consists of resin, gum and essential oil in various proportions. It is used for flavouring food products and medicinally. It is used as an antispasmodic and stimulant.

Garcinia morella Desr.; English—Indian Gamboge tree; Hindi, Bengali and Marathi—*Tamal*; Tamil—*Makki, solaippuli*; Telugu—*Pasupuvarne*; Kannada—*Hardala devanabuli*; Malayalam—*Chigiri, daramba*; Arabic and Persian—*Farfiran*; Family—Guttiferae.

A small evergreen tree; found in the evergreen forests of Assam, Khasi and Jaintia hills, Bengal and in Western Ghats from North Kanara South wards to Travancore, upto an altitude of 3,000 feet.

Gum-resin. *Garcinia morella* yields 'gamboge' of commerce. Gamboge occurs as a yellowish emulsion in the cortex, pith, leaves, flowers and fruits of the tree. Gamboge is obtained by making a spiral incision on the bark of the tree, during the rainy season and collecting the exudate in bamboo cups. The juice is allowed to harden, which takes about a month, after which the bamboo container is heated and the gamboge taken out in the form of cylindrical sticks.

Gamboge is reddish, yellow or brownish orange in colour with a smooth, uniform conchoidal fracture. It forms a yellow emulsion with water. It is almost completely dissolved by successive additions of alcohol and water. The resin is precipitated from alkaline solutions by acids.

Gamboge is used as a pigment on account of the brilliancy of its colour. It is used in the preparation of water colours and gold-coloured spirit varnishes for metals. It is used in Burma for dyeing silk robes of Buddhist priests. It is used for preparing a golden yellow ink for writing on black paper.

Gamboge is also used medicinally. It is powerful hydragogue cathartic causing, in large doses, nausea, vomiting and griping. It is used in dropsy and cerebral congestion when it is desired to lower the blood pressure rapidly.

Gardenia gummifera Linn, F.; Hindi—*Dikamli, dikmali*; Marathi—*Dikamali*; Gujarati—*Kamarri, dikamali*; Tamil—*Kumbai, dika—malli*; Telugu—*Chittamatta, tella–manga, chinaka-ringuva*; Kannada—*Dikke–malli*; Uriya—*Gurudu*; Sanskrit—*Pindava*; Family—Rubiaceae.

A large shrub or a small tree found in Madhya Pradesh and South India, from the Satpura range southwards. Also found in Bundelkhand region and parts of Bihar.

Gum-resin. The leaf buds and the young shoots of this species, as well as of *Gardenia lucida*, yield a resinous exudation, known in commerce as *Dikamali* or *Cumbi gum*. The resin is secreted in the form of tears. The shoots and buds are broken off with the drops of gum-resin attached, and exposed for sale either in this form, or after agglutination into cakes or irregular masses. The gum-resin is transparent, greenish yellow, with a sharp pungent taste and a peculiar offensive odour like that of cat's urine. The drug is considered anti-spasmodic, carminative, and when applied externally, antiseptic and stimulating. It is employed by the natives, in cases of hysteria, flatulent dyspepsia, and nervous disorders due to dentition in children, also externally as an application to foul and callous ulcers, and extensively to keep away flies from sores mainly in veterinary practice. Commercial sample of *dikamali* contains 89.9% resin; 0.1% steam-volatile oil and 10.0% plant impurities.

Guaiacum officinale Linn.; English—Gum Guaiacum (resin); Family—Zygophyllaceae.

A small to middle-sized tree up to 50 feet high occasionally cultivated in Indian gardens. Introduced in India from West Indies.

Guaiac resin. Gum Guaiacum or Guaiac Resin is the resin present as a filling in the tissues of the wood. It is obtained either as a natural exudation from the tree or by burning logs of wood at one end after making incisions in the middle, when the resin flows out and is collected. In the other process the wood is boiled after reducing it to chips or saw-dust in a solution of common salt or in sea water when the resin melts and is collected from the surface. The resin occurs in large compact masses or occasionally in rounded or ovoid tears. It is brownish black in colour and acquires a greenish tinge on long exposure. It is brittle with glassy fracture. It possesses a balsamic odour.

Uses. Guaiacum resin is used as fat stabilizer. A tincture of the resin is used as a reagent in the detection of blood stains. The resin is used in paints and varnishes.

Gum Guaiacum is a mild laxative and used in the treatment of chronic rheumatism and gout. In the form of lozenges, it is used for the treatment of tonsillitis and pharyngitis.

Kingidendron pinnatum (Roxb.) Harms; Syn. *Hardwickia pinnata* Roxb.; Tamil—*Matayen samprani*; Kannada—*Enne, yenne–mara*; Malayalam—*Kolavu, shurali*; Family—***Caesalpiniaceae.***

A large, evergreen tree, attaining a height of 100 feet, found in the Western Ghats, Travancore and Karnataka.

Oleo-resin. The tree yields, on tapping, a dark or reddish brown oleo-resin, resembling copaiba balsam (from *Copaifera* spp.) in odour and taste. The tapping is done by boring a hole, 3/4 inch in diameter, in trees 5 feet or more in girth. The hole which reaches the pith is placed 3 feet above the ground and slopes downward from the pith to the bark. A lip is fixed below the hole and the exudate is collected in a tin. When the flow ceases, the hole is plugged with a piece of wood. The tree is rested for about ten years before tapping again. A tree of 8 feet girth yields about 12 gallons of the oleo-resin. It is inferior to copaiba balsam.

It is used as varnish after thinning with turpentine. It is used medicinally in the treatment of gonorrhoea.

The oleo-resin on steam-distillation yields a colourless volatile oil. The resin left after the distillation is a hard, brittle mass, greenish yellow in thin layers and dark brown in lumps. It is completely soluble in rectified spirit and is suitable for preparing spirit or oil varnishes.

Hopea odorata Roxb.; English—Rock damar; Tamil—*Urappuppicin*; Kannada—*Bilitirupu*; Malayalam—*Urappimpasa;* Andamans—*Thingan*; Trade name—*Thingan*; Family—Dipterocarpaceae.

A large evergreen tree upto 150 feet in height, found in Andaman Islands.

Resin. This species is the principal source of the copalline resin known as *Rock dammer* in commerce. The resin derived from this source occurs in nodules about as large as a walnut, rounded, of a pale straw colour, sometimes almost colourless, brittle, with a shining resinoid fracture. It is completely soluble in turpentine oil and partially in alcohol. The resin is used in the preparation of varnishes for indoor decorative work. It is also used for painting pictures, caulking boats, and mounting microscopic objects. A composition prepared by mixing the resin with beeswax and red ochre is used for fastening spear and arrowheads. The resin possesses antiseptic properties and used as an ointment for wounds and sores.

Juniperus communis Linn.; Hindi—*Aaraar*, Kumaon—*Chichia*; Kashmir—*Nuch, bentha*; Punjabi—*Dhup, gugil*; Bengali—*Havusha*; Marathi—*Hosha*; Family—Cupressaceae.

A large shrub, found in the North–West Himalayas from Kumaon Westwards, at altitudes from 5,500 to 14,000 feet.

Resin. All parts of the plant contain volatile oil. A terebinthinate juice exudes from the tree and hardens on the bark. Terminal twigs and needles yield a bright yellow oil. Juniper oil is obtained by stem distillation of ripe fruits. In overripe fruits, the oil changes into a resin. On storing the oil turns viscous and acquires a turpentinic odour. It is colourless or pale yellow and possesses a strong odour.

It is carminative, stimulant and diuretic.

Liquidambar orientalis Mill.; English—Styrax, storax, Oriental sweet gum; Hindi—*Silaras, meith-silai*; Bengali—*Silha, silaras*; Marathi—*Silarasa*; Gujarati—*Meih-sila, selaras*; Tamil—*Neri–arishippal*; Telugu—*Shila-rasam*; Malayalam—*Rasamalla*; Sanskrit—*Silhaka*; Persian—*Meih–Silla*; Family—Hamamelidaceae.

A handsome tree, reaching to the height of 34-40 feet, and forming forests in the extreme south-western part of Asia Minor. Not found in India. Storax is imported into India from France, U.K. and U.S.A.

Resin. The collection of storax commences when the tree is 3-4 years old. The bark is bruised or injured by beating and the balsam soon exudes into the inner bark. The outer bark is then peeled and discarded. The inner bark, saturated with balsam, is stripped off and boiled with water which causes the balsam to separate and float to the top, so that it can be skimmed off. Crude storax thus obtained is poured into barrels, casks or cans for shipment.

The storax is a soft, viscid resin, usually of the consistency of honey, heavier than water and greyish brown in colour which deposits, on standing, a dark brown oleoresin. The purified storax, also known as *Prepared storax* or *Styrax Preparatus*, occurs as a yellowish brown viscous mass, transparent in thin layers and possessing an agreeable balsamic odour and taste. Prepared storax is completely soluble in rectified spirit. Purified storax is composed principally of an alcoholic resin, known as *storesin* (33-50%). Stem distillation of storax yields a pale yellow to dark brown volatile oil, *Oil of Storax*, with a pleasant but peculiar odour.

Storax is used medicinally as well as in cosmetics. It is used as a stimulating expectorant and antiseptic, and in ointments for scabies and other parasitic skin diseases. It is used for scenting soaps and cosmetics, as a fixative for heavy perfumes, in the perparation of adhesives, lacquers and incense.

Melanorrhoea usitata Wall.; English—Myanmar Lacquer Tree; Manipur—*Kheu*; Trade name—*Thitsi*; Family—Anacardiaceae.

A large deciduous tree, frequent in the open forests, rare in the dry forests in Manipur. In Manipur tree attains very large dimensions and forms extensive forests from the top of the Kabo Valley for many miles in a northerly and north-easterly direction towards the Chinese frontier.

Oleo-resin. Every part of the tree abounds in a thick, viscid, greyish, terebinthinate fluid, which soon assumes a black colour on exposure to the air. This natural varnish, known as Myanmar lacquer or *thitsi*, is obtained by tapping trees through V-shaped incisions in the bark. A bamboo joint closed at the outer end is chiselled and inserted into the incision to receive the exuding varnish. After about ten days, when the flow almost ceases, a second cut is made along each side of the incision and the exudation collected. The wounds are freshened up or fresh incisions made to derive further quantities of the oleo-resin. As many as 40 to 50 taps are made in one tree, extending from the base to a height of 30 feet. The tapping season is June to February. However, maximum yields are obtained during July-October. A good tree yields 2-8 kg. of oleoresin annually.

The oleo-resin is thick at ordinary temperatures, and of a dull grey colour, but as soon as it comes in contact with the air it assumes, in a very short duration, a shining black surface. It possesses an aromatic odour. It is soluble in alcohol, turpentine and benzene. The oleoresin contains: 86.24% urushiol (urushic acid), 3.08% gum and 0.53% oil.

It is widely used in Manipur, as a water-proofing paint for boats, household vessels, paper and cloth, as size or glue for gilding and as a non-fouling preservative paint for wood, metalware and leather. It is also used for palm leaf inscriptions. Linseed oil is used for thinning the lacquer and various pigments are incorporated to obtain coloured finishes.

Black varnish is used as an anthelmintic in cases of round worms.

Pinus insularis Endl.; Syn. *P. khasya* Royle; English—Khasi pine; Khasia—*Dingsa*; Bengali—*Saral*; Family—*Pinaceae*.

This is native of Khasia mountains, and found between 3,000 to 7,000 feet elevation. Found in the Khasi, Jaintia, Lushai, Manipur, and Arunachal Pradesh. It has also been introduced into the hills of North Bengal.

Oleo-resin. It exudes an oleoresin of excellent quality. In Assam it is collected, and crude oil of turpentine is made from the resinous wood. The local people tap the trees by a traditional crude method. A hole is cut near the base of the trunk and a blaze, 4 feet long and 12 inches wide, is made above this notch. The resin exudes copiously from the blaze and is collected. The wood, at least in the vicinity of the blaze, becomes soaked with oleo-resin. It has been estimated that a full-grown tree yields about 30 kg. of crude resin from one of these wounds and that the resin encrusted wood contains another 16 per cent, of its total weight of crude resin.

Fig. 17.2. Making incision for tapping resin on the lower part of the trunk of pine tree. F.R.I., Dehradun.

Liquid turpentine is extracted from this wood as follows: The wood is cut into chips, and placed crosswise in layers in an ordinary earthenware cooking pot, the mouth of which is closed by large smooth leaves perforated with small holes. The vessel is then inverted on a support over another receptacle and fire is made on or above it, the vessel being surrounded with hot ashes and burning charcoal. As a result the turpentine exudes from the wood and drops through the holes in the leaf-covering into the pot below. Oil of turpentine from *P. insularis* is a colourless mobile liquid with a characteristic odour. The turpentine oil is supposed to be of a good quality. The use of this oil as a solvent for the extraction of quinine from cinchona bark with low-alkaloid content has been suggested.

The rosin obtained is a transparent golden yellow solid substance.

Fig. 17.3. Making incision for tapping resin on the upper part of the trunk of pine tree. F.R.I., Dehradun.

Pinus roxburghii Sarg.; Syn. *P. longifolia* Roxb.; English—Pine, Himalayan Long—leaved pine; Hindi—*Salla chir, saral*, oleo–resin = *ganda biroza, chir-ka-gond*; Nepal—*Dhup, sala*; Lepcha—*Gniet*; Kumaon—*Chir, salla*; Kashmir—*Sall, sarl*; Pubjabi—*Chir, chil*, oleo—resin = *gand biroza*, purified oleo—resin - *birza, sat biroza*; Sanskrit—*Sarala*, oleo—resin + *sarala drava*; Persian—Oleo–resin = *Birozeh*; Family—Pinaceae.

A large tree of the outer and drier Himalayan slopes, from Kashmir to Bhutan, found as low down as 1,500 feet and ascending to 7,000 feet; it comes up tolerably well in the plains also.

Oleo-resin. The oleo-resin of this species is more largely collected and used than that of any other Himalayan conifer. The oleo-resin is procured by incision. The product is valued chiefly for its resin, which is obtained by exposing the oleo-resin to heat.

Turpentine oil. Oleoresin from *P. roxburghii* is the main source of oil is turpentine in India. There are two methods of tapping 1. *light tapping*, which consists of making a moderate number of blazes in the trees not to be felled in the near future; and 2. *heavy tapping*, also known as *tapping to death*. In heavy tapping as many blazes as possible are made on the trees which will be felled within five years. In the light-tapping cycle, the trees are tapped for a period of four years, followed by a rest of eight years. The yield of resin varies greatly; it may be estimated at about 187 kg. per 100 blazes. Normally the tapping commences in March and continues till November.

The oil of turpentine is extracted as follows: The crude oleoresin is first purified by melting in stem-jacketed containers, provided with spiral mixers, and removing the impurities such as chips and lighter particles which float to the top, and sand and silt which settle at the bottom. The clarified resin is then distilled in steam-jacketed or vacuum stills. The turpentine oil which collects in the top layer of the distillate is drawn off, passed through lime water to remove rosin acids. The oil is then dehydrated with common salt and anhydrous sodium sulphate. The dehydrated product is further stored in tanks where any water still remaining over settles to the bottom and is drained off.

About 40 kilograms of oleoresin yields on the average about eight litres of first grade turpentine, 1-1.5 litres of other grades and 29-30 kilograms of rosin. This means that the average yield of total turpentine oil is about 22% and that of rosin about 75% of the oleo-resin.

Oil of turpentine is a clear, transparent liquid with a pungent and somewhat bitter taste. The Indian turpentine contains low pinene and high carene contents. American turpentine consists almost entirely of α and β pinenes.

The Indian turpentine oil is mainly used as a solvent, especially for thinning paints and varnishes. It is also used in pharmaceutical preparations, perfumery industry and in the manufacture of synthetic pine oil, disinfectants, insecticides and denaturants. The turpentine oil makes an important basic raw material for the synthesis of terpene chemicals, which are used in several industries, such as adhesives, lubrication additives, synthetic resins, solvents, plasticizers, paints, varnishes, soaps, perfumery, cosmetics and paper and rubber chemicals.

The turpentine oil is of much value in medicine. In Indian Pharmaceutical Codex, it is known as *Oleum terebinthinae*. It is used in chronic bronchitis. It is recommended also in the treatment of gangrene of lungs. It is used as a carminative in flatulent colic.

Externally it is used as a rubefacient in various rheumatic affections.

Turpentine oil producing units. There are three units which produce turpentine oil in India 1. Indian Turpentine and Rosin Co. Ltd., Bareilly; 2. Himachal Rosin and Turpentine Factory, Nahan and 3. The Government Rosin and Turpentine Factory, Miran Sahib, Jammu. The distilled oil of turpentine and the pine tar are produced only in one unit namely Camphor & Allied Products Ltd., Bareilly, which was set up in 1964. These items are obtained as by-products in the manufacture of synthetic camphor.

Rosin. The rosin is obtained as the solid residue in the distillation of the turpentine oil from the oleo-resin. The yield of rosin is about 75% of the quantity of oleo-resin distilled.

The rosin is faintly aromatic and occurs in the form of transparent or slightly translucent brittle lumps with a glassy structure. It is soluble in ether, chloroform, light petroleum, alcohol, acetone, and most volatile and fixed oils. It is one of the cheapest resins available in the market.

The rosin is chiefly used in paper, soap, cosmetics, paint, varnish, rubber and polish industries. It is used for making printing inks, casein glues, and as a binder in plastics, dry battery and insulating compositions. It is also used in linoleum and roofing cements. It is employed in the manufacture of fireworks, match compositions, shell explosives, insecticides, and disinfectants. It is also used in lubricating, hair fixing and nail polishing preparations. It is also used as a flux in soldering and tinplating. It is utilized as a source of rosin oil and rosin spirit.

On destructive distillation the rosin yields 3-10% *rosin spirit* or *pinoline*, and 80-85% *rosin oil*. The pale yellow rosin spirit is used as an illuminant and as a substitute for turpentine oil in the varnish industry. The rosin oil with lime forms rosin grease, which is used as a lubricant in wagon and trolley wheels, and in various other types of machinery. It is also used in the manufacture of printing ink, varnishes and antiseptics.

Fig. 17.4 Collection of rosin in a metalled cup, from pine tree. F.R.I., Dehradun.

Pinus wallichiana A.B. Jackson; Syn. *P. excelsa* Wall. ex. D. Don; English—Blue pine, Five-leaved pine, Bhutan pine; Kashmir—*Yiro, Kaiar, kail*; Himachal Pradesh—*Lim*; Kumaon—*Raisalla, lamshing, byans, dolchilla*; Lepcha—*Neet-kung*; Bhutan—*Tongschi, lamshing*; Trade—Kail, blue pine.

A lofty symmetrical gregarious tree, with spreading or drooping branches, found in the Himalayas from Kashmir to Bhutan at altitudes between 6,000 and 12,500 feet, but most common upto 8,000 feet; it also occurs in Balipara tract of Assam. It is readily recognised by the blue tint of its tufted foliage. The long pendulous cone, the compact greenish blue bark of the young stems and leaves remaining on the tree for three years.

Oleo-resin and turpentine oil. The wood is highly resinous and affords turpentine and tar. However, the yields of oleoresin from the blue pine is low, being about half of that from the chir pine, but the turpentine oil obtained is of superior quality. On distillation the crude oleoresin gives 27.5% of turpentine oil which possesses high pinene content (88%). The amount of resin recovered is about 68% of the oleoresin. Rosin oil is obtained in a yield of 75% when the rosin is subjected to destructive distillation.

Pistacia lentiscus Linn.; English—The Mastic tree, mastiche; Hindi—resin - *rumi mastike, kundur-rumi*; Bengali—*Rumi mastungi, kundur-rumi*; Persian—*Kundar-i-rumi, mastake-i-rumi*; Family—Anacardiaceae.

A dioecious evergreen shrub or a small tree, upto 15 feet in height, found chiefly in the Mediterranean region. It yields *mastic resin* which is imported into India. The main source of supply of mastic is the island of Chios in Aegean Sea, where the mastic industry is well organized.

Resin. The resin exudes naturally from the bark but for commercial purposes, it is obtained by making small vertical incisions in it and picking off the hardened product about three weeks later. Average annual yield of resin per tree is 3.6-5.4 kg.

The *resin, mastic*, occurs in small irregular, yellowish tears, brittle and of a vitreous fracture, but soft and ductile when chewed. It has a faint agreeable odour, which is increased by the application of heat or friction.

The mastic consists of two resins called respectively, α-masticoresene and β-masticoresene.

Mastic has been used, as a masticatory to sweeten the breath and to preserve teeth and gums. It is used in the preparation of chewing gum. It has been used also to flavour alcoholic beverages and cordials. It is considered carminative, stimulant and diuretic. It is also used as a microscopic mountant. The chief use of mastic is in the manufacture of high grade transparent varnishes employed for coating valuable art paintings and metals, for lithographic processes and retouching negatives.

Rhus verniciflua Stokes; Syn. *R. vernicifera* DC.; English—Lacquer tree; Family—Anacardiaceae.

A tree, upto 50 feet in height. It is native of China, found in Kumaon and Garhwal at altitudes of 3,000 to 6,000 feet.

Resin. The tree on injury exudes an acrid, grey-brown juice used as lacquer (Japanese lacquer). It consists principally of urushiol; small amounts of an enzyme laccase, rubber, albumin, and gum have also been reported. The trees are tapped, the exudation is a milky liquid which darkens and thickens rapidly on exposure. It can be kept unchanged, however, for long periods by storing in closed containers. When applied as a varnish, the thin film rapidly hardens in a moist atmosphere, owing in part to oxidation. The lacquer forms a remarkably protective and durable coating, which takes a high polish, and is not affected by acids, alkalies, alcohol or heat upto 160^{0}F. The latex from the lacquer-tree is mixed with pigments and when applied in multiple layers on a base of fabric or wood becomes a hard material of great strength, capable of being polished and carved. The lacquer can be coloured red with vermilion, black with lampblack, and yellow with orpiment. It is also used for waterproofing of wood, and for coating leather.

Shorea robusta Gaertn.; English—Sal tree; Hindi and Bengali—*Sal*; Nepal—*Sakwa*; Lepcha—*Teturl*; Uriya—*Salwa*; Punjabi—*Sal*, resin = *ral zard*; Marathi—resin - *rala guggilu*; Gujarati—*resin* = *ra*; Tamil—resin = *Kungiliyam*; Telugu—*Gugal*, resin = *guggilamu*; Kannada—*Kabbu*, resin = *guggala*; Sanskrit—*Sala,* resin = *rala*, *guggilam*; Persian—*Lale-moab bari*; Family—Dipterocarpaceae.

A large gregarious tree, found along the base of the Tropical Himalayas from the Sutlej to Assam, in the eastern districts of Madhya Pradesh and on the Western Bengal hills. In Chota Nagpur it is found abundantly.

Oleo-resin. On tapping, the tree exudes large quantities of an aromatic *Oleoresin* known as Sal Dammar or Bengal Dammar. An annual yield of 4-5 kg. of resin per tree is obtained. The method of tapping usually employed is to cut out from three to five narrow strips of the bark, about 3 or 4 feet from the ground. This is generally done in the month of July. In about twelve days these groves fill up with resin which is gathered, and the grooves left to fill again. The trees give three yields. The first is the best in quality. A second yield in October and a third in January are also obtained from the same wounds, but small in quantity and inferior in quality.

The resin usually occurs in small rough pieces, nearly opaque and very brittle, pale creamy-yellow in colour and having a faint resinous-balsamic odour.

Sal dammar is widely used as an incense, especially as an ingredient of *havan samagri* which is burnt in religious ceremonies and cremation rites. It is also used in inferior quality paints and varnishes and for caulking boats. It is employed in the manufacture of shoe polishes, carbon papers, typewriter ribbons, etc. It has also been used as a plastering medium for walls and roofs and as a cementing material for plywood, asbestos sheets, etc. It is used medicinally. It is an astringent and detergent and is given in diarrhoea and dysentery. Also used in the preparation of ointments for skin diseases and in ear troubles.

On dry distillation the resin yields an essential oil, known as *chua* oil. The oil is light brownish yellow in colour and possesses an agreeable odour. The oil is used as a fixative in heavy perfumes and for flavouring chewing and smoking tobacco. It is used in indigenous medicine as an antiseptic for skin diseases and ear troubles.

Styrax benzoin Dryand.; English—Benzoin tree; Hindi—*Luban*; Bengali—*Luban*; Punjabi—*Luban*; Gujarati—*Loban*; Tamil—*Shambirani*; Malayalam—*Kaminian*; Persian—*Hussi-luban, kamkam*; Family—Styracaceae.

A small tree, native of the Malay Archipelago.

Resin. The tree yields the true *Benzoin* or *Gum Benjamin* of commerce. When the trees are six or seven years old and have trunks 6 to 8 inches in diameter. They are judged capable of yielding the resin, and incisions are then made in the stems from which exudes a thick whitish resinous juice. This soon hardens by exposure to the air and is then carefully scraped off with a knife. The tree continues to yield about 1.4 kg. per annum for ten or twelve years, at the end of yield about 1.4 kg. per annum for ten or twelve years, at the end of which time it is cut down.

Benzoin is imported into India in cubic blocks. It consists most frequently of a compact mass of rich, amber-brown, translucent resin, containing a number of white tears. In some cases the tears of white resin are very small, and the whole mass has the appearance of reddish-brown granite. There is always a certain amount of admixture of pieces of wood, bark and other impurities. It is brittle, and possesses a delicate balsamic odour, but very little taste. When heated it evolves a powerful fragrance with the fumes of benzoic acid.

Benzoin is used as an internal remedy in phthisis and asthma. It is also given as an aphrodisiac in small doses. Benzoin preparations are used externally as disinfectants and stimulants. It is burnt as an incense. On account of its disinfectant properties, the smoke it gives out is employed to drive away mosquitoes and house-flies.

Vateria indica Linn.; English—White dammar, Piney varnish, Indian copal; Hindi—*Barasal, safed dammar, ral*; Bengali—*Chundrus*; Tamil—*Vellai damar, vellai–kungiliyam*; Telugu—*Tella damaru*; Kannada—*Dupa maram, dhupa*; Malayalam—*Payana*; Family—Dipterocarpaceae.

A large evergreen tree, found in the Western Ghats and South India, ascending to 4,000 feet.

Resin. The tree yields a true resin of considerable value, known as *white damar* or *Piney resin*. The resin occurs in *three* forms: 1. *Compact piney Resin*, in lumps of all shapes which varies in colour, on the outside, from bright orange to a dull yellow. It has a bright vitreous fracture, and internally presents all shades of colour from a light green to a light yellow. 2. *Cellular piney Resin*, occurs either in small lumps or in large masses. It has a shining appearance and a balsamic odour. It has a distinctly cellular structure. Notches are cut in the stem of the tree sloping downwards and inwards, the resin collects in the cavity and is allowed to dry in *situ*, or is collected and dried by heat. It varies in colour from light green to yellow or white. It may be transparent or opaque. 3. *Dark coloured Piney Resin*, is obtained on splitting open old and decayed trees.

The finest specimens of Piney resin are obtained by making incisions in the tree, and are in pale green translucent piecesf of considerable size.

The resin is used as an incense, and is also used in paints and varnishes. The varnish thus prepared is valuable for coating carriages, furniture, and other work requiring a complete and finely finished protection to the coat of paint. It is also made into candles which diffuse an agreeable fragrance, and give a clear light and little smoke.

The resin yields, on distillation, 82% of a volatile oil of agreeable odour.

Fine shavings of the resin are taken internally to check diarrhoea.

18

Non-Alcoholic Beverages and Beverage Plants

The beverages containing caffeine are used all over the world for their stimulating and refreshing qualities. Caffeine is an alkaloid, which has definite medicinal values and acts as a diuretic and nerve stimulant. The most important non-alcoholic beverages are—*tea, coffee* and *cocoa*. Caffeine is harmful in large quantities, it is present in these beverages in very small amounts, not exceeding two per cent. The other beverages which do not contain alcohol are commonly known as *soft drinks*. They contain high sugar content and make a good source of energy. The fruit juices are the simplest soft drinks.

INDIAN TEA

Tea

Camellia sinensis (Linn.) O. Kuntze; Eng.—Tea; Hindi—*Cha*; Family—Theaceae.

History. "According to Japanese folklore, the discovery of tea is attributed to an Indian Buddhist monk *Daruma* who lived hundreds of years ago. He left India to preach the message of the Enlightened One—The Lord Buddha—in China. Arriving at Peking, he vowed not to sleep for nine years till he finished his prayers. But, during his meditation he felt tired and his eyes closed. Waking up in disgust, he tore off his eyelids and threw them away. Where the latter fell there sprang up an evergreen bush. Surprised, *Daruma* plucked the leaves, boiled them and drank the liquid. Soon, he felt refreshed and was able to continue his prayers. *Daruma* eventually left China and went to Japan where he died."

In China the people have been drinking tea probably since the 4th century A.D. and the people of Upper Burma, along the Indo-Burmese border and Upper Assam have been used to tea for just as long a period.

Tea was brought to the notice of Europe in 1559 by Ramusio, the noted Venetian writer on voyages and travels, long after it was accepted beverage of the people of Asia. Half a century after Ramusio's notice of tea, the enterprising Dutch began to import it into Europe. England was strangely casual to the beverage during the first half of the 17th century. Tea then was far beyond the reach of the average person. The 2 lbs, 2 oz. tea, brought by the Governors of the East India Company in 1664, as a gift to King Charles II, cost them 85 shillings. Two years later the Company gave him a more ample present of *thea,* consisting of 22¾ lbs., costing 50 shillings per lb. Half a century later the retail price of tea in London was between 20-30 shillings per lb. for *Bohea* (black tea) and 12-30 shillings for the finest *Hyson* (green tea). It took another hundred years before consumption began to spread to the rural areas. According to Prof. Northcote Parkinson, it was presumably between 1784 and 1789 that tea "may literally be said to have descended from the palace to the cottage". About that time it became the breakfast beverage of practically the whole population. It was also drunk at other times which led Raynal (the French historian. 1713-96) to exclaim that tea had done more for English sobriety than any laws, sermons or moral treatises could have achieved.

Tea was sold publicly in London in 1657 and two years later, on 14th November 1659, Mercurius Politicus carried the news that tea was sold "almost in every street" of London. The first order by the East India Company for 100 lb. of the best tea was sent to its agent at Bantam in Java in 1667. In 1669 import of tea into England from Continental Europe was prohibited, thus making absolute the East Indian Company's monopoly in tea.

The British imported China tea from Bantam (Java) when they were dispossessed in 1684. From this period Madras and Surat became centres for the import of "secondhand" Chinese tea. Though, direct British trade with China did not weaken till the 1780 when the merchants of Canton decided to organise the conduct of their country's foreign tea trade.

Meanwhile, the Company's officials in India had realised the possibilities of raising this valuable commodity of trade in India. In 1780, Warren Hastings, the Governor-General, passed on some seeds of *Hyson* for propagation in Bhutan. A tea enthusiast Col. Robert Kyd, who later became Superintendent of the Botanical Gardens in Calcutta, similarly experimented with tea cultivation. In 1788, Sir Joseph Banks tried to interest the East India Company in the possibilities of growing tea in India; but the Company was not agreed to the suggestion as such a venture was against the monopoly of its tea trade with China. However, after five years, Lord Macartney's Embassy was making arrangements to send seeds of the tea locally cultivated to Calcutta. Although imported tea seeds gave the idea and the sustenance for the commercial cultivation of tea in India, the rise of the Tea Industry, as it is today, owes origin to the momentous discovery of the indigenous tea plant in Assam in India.

Many have claimed the discovery of the indigenous Indian tea plant which eventually became the foundation of the world's Tea Trade and Industry during the next century. There were some who believed that Moneram Dewan was the discoverer who led Major Robert Bruce to it; but it was Charles Bruce, his brother, who played a significant role in the development of tea culture in India during the ensuing twenty years.

As we have realised, the East India Company at first was averse to the idea of starting any tea plantation in India; but its officials in the country were eager and the discovery of the indigenous growth of tea in Assam acted as a catalyst in the great movement to develop the industry in the country. If there was difficulty in the early stages, it came largely from the application of Chinese tea labour since the general belief persisted that only such imported methods of tea manufacture were the safe ones to follow. This belief, however, lasted for a while till experience with Assam tea and the response to the indigenous tea in London overcame existing influences and the foundation of a truly Indian Tea Industry was laid.

An index of the steady rise in Britain's consumption of Indian tea can be seen from the figures of her tea imports. In 1836, the import of tea into Britain reached 49 million lbs. and all of it was China tea. Two years later, India sent her first consignment of Assam tea to London to be sold there, on January 10, 1839. In 1859, the last year of the East India Company before the crown took over, Britain imported 69 million lbs. of which only 2 million or 3 per cent came from India. By the close of the century Britain was importing about 250 million lbs. of tea of which 55 per cent came from India, 37 per cent from Ceylon and only 5 per cent from China.

In 1900, Indian production reached 170.5 million lbs., with exports at 164.6 million lbs. and out of which 154 million lbs. went to Great Britain. Within another five years Indian exports reached 214 million lbs., and India became the leading tea supplier to the world and she has retained that position.

Today the total output of this industry is over 779 million lbs.—176 million lbs. in South India and the rest in North-East India. It accounts for about 20% of the country's total exports in value.

Tea is the queen of beverages, and no other beverage except water is consumed by so many. Tea, a treasure of the world, is the most important non-alcoholic beverage; it is a pure, safe and helpful stimulant and one of the chief joys of life. A billion cups of tea all over the world add daily to the zest of life. They bring good cheer with health, stimulation without intoxication, and are refreshing and energising without any exacting demand on the family budget.

The tea plant is considered to be a native of Assam and the adjoining areas of Upper Burma, some regard it to be a native of southern Yunnan and Upper Indo-China. The major areas of production are situated in the northeast region comprising the Brahmaputra and Surma valleys of Assam, and Darjeeling and Jalpaiguri districts of West Bengal. These areas are the Terai, the Dooars and the Brahmaputra valley where tea is grown on the north and south banks of the river Brahmaputra. In addition, considerable quantities of tea are sown in the Cachar district in Assam and in Tripura. In North-West India a fair amount of tea is grown in Kangra, Mandi and a smaller quantity in Dehradun. Down South tea is grown on the Nilgiris, on the Kanan Devan or the High Range, Anamallais and on the slopes of the mountain stretching down to the plains of Kerala.

Brahmaputra valley. Tea is grown mostly on flat land, between 160 to 400 feet above sea level, on both banks of the river Brahmaputra. Soils are alluvial and the mostly sandy loam with some red clay. The average annual rainfall is 80–120 inches and it is well distributed. The cropping season is from April to December, May/June and October/November.

Teas of orthodox manufacture are black, clean, flinty well-twisted leaf with a good show of golden tip in the best grades, throwing creamy, strong, pungent, full-bodied top quality liquors Teas of C.T.C. manufacture from this region are brown, rich, grainy, small leaf grades throwing exceptionally strong, coloury, pungent liquors. These teas are in great demand almost all over the world and are particularly suited for blending with flavoury teas to improve strength and colour.

The Terai and Dooars. Tea is grown on flat land and hill slopes bordering the Himalayan foot-hills, from 100–1000 feet above sea level. The average rainfall is 120–200 inches but it is not very evenly distributed. The cropping season is from March to November, with the vintage periods in spring and autumn.

The Terai and Dooars produce coloury, full creamy liquoring teas. The orthodox teas are black, stylish and well-twisted, while the C.T.C. tea are hard, grainy and even. A number of gardens produce very small brown leaf Leg-cut teas which throw bright colour liquors. These teas are in great demand by blenders for mixing with flavoury teas to contribute to colour and body.

Cachar and Tripura. Tea is mostly grown on hillocks peculiar to the region and to some extent on the flat lands surrounding them. Soils are red sand and clay with some areas of peat. The annual rainfall varies from 120–200 inches but distribution is uneven and severe winter droughts are common. The cropping season is from March to November.

The districts of Cachar and Tripura produce black, well-made teas throwing colour, sweet liquors very popular with blenders.

Fig. 18.1. *Camellia sinensis* : Tea. The garden looks best, brightest and of course, busiest during the plucking season.

Darjeeling Tea

The town of Darjeeling in India is situated in the extreme north of the State of West Bengal, at a height of 7,000 feet, surrounded by the lower hills and valleys of the Himalayas and overlooks the State of Sikkim.

The town is clustered in a bowl among the hills, looking north toward Kanchenjungha, a mere 35 miles away and a very magnificent sight when it is to be seen. Most of Darjeeling's livelihood is derived from the tea gardens which surrounded the town on all sides, at lower levels.

Early in the nineteenth century tea seed was imported from China and planted both on the plains of N.E. India and also in these hills. But it was soon discovered than an indigenous tea plant was to be found in Assam. This proved to be more vigorous, in 'plains' conditions; it gave a large crop and the leaf was more easily harvested, and so the old 'China' bushes were gradually uprooted and replaced. The original bushed plantes in Darjeeling have still remained, they are up to 100 years old and still yielding good crops.

We have continued to use these bushes because they, and they alone, can produce the flavour for which Darjeeling tea is renowned. The crop yield per acre is small and the leaf is small in size and expensive to harvest. And yet, it is this climate and high altitude, the slow growth of leaf, which help to give to Darjeeling teas the flavour for which they are known throughout the world.

In South India, Sri Lanka, Indonesia and Africa the tea bush produces leaf throughout the whole year, but in N.E. India it hibernates in winter. Growth starts again in the spring when the tea bushes of Darjeeling are covered with a mass of small soft bushes. This is the first 'flush' when the tea made is of good quality. When this 'flush' is finished there is a lull in growth until the second 'flush' appears; this is usually small in appearance, the quantity of leaf on the bushes is not great, but this is the vintage period of the whole year.

The monsoon reaches Darjeeling about mid June and from July onwards the quality of tea is of standard character until the rains end in October. Then for a short period until the end of the season in November a small crop of 'autumnal' leaf is harvested which has a flavoury but 'thin' liquor.

The quantity of fine Darjeeling tea produced each year is comparatively small. Of the total crop of 16,000,000 lbs. from this district, the first and second flushes and the autumnal teas are highly prized. Most of these good, flavoury teas are used for blending with teas from other places; the latter add strength and colour; while the lighter liquor of the Darjeeling tea supplies the flavour to the blended mixture.

Public demand in many countries requires a 'strong colour' tea which the 'Assam' variety of bush contains; but for those who appreciate flavour they will find this in full measure in a pure Darjeeling tea.

In countries where tea is drunk without milk or sugar—as for example, in Russia—tea is appreciated for its flavour, and indeed it is true that the delicate flavour of a Darjeeling tea can best be appreciated by drinking the liquor 'straight'.

Good Darjeeling teas are indeed the champagne of teas. There is no other area in the world which can produce the 'muscatel' flavour of these fine teas produced on the mountain sides of this small district.

Darjeeling—Tea is grown on the mountain slopes of the Himalayas at elevations ranging from 1000 to 6000 feet above sea level. Soils are red loams and sandy loams. Annual rainfall varies between 100 and 200 inches; the winter season being fairly cold and dry. The cropping season is from April to November and vintage teas are produced in April/May and October.

Teas with exquisite aroma come from the Darjeeling district. Stylish, large leaf teas are the main feature with golden tips in the top grades. The liquors are famous for their flavour derived from the high elevation at which the teas are grown.

Kangra and Mandi. The tea plantations are on the mountain slopes of the Kangra Valley in the Himalayas, at elevations from 2000-5000 feet above sea level. The average annual rainfall but it is not evenly distributed; the winter and early summer being dry and droughty. The cropping season is from April to November.

Kangra and Mandi mostly produce green teas which are in demand in many parts of the world.

South India. Tea is grown on hill slopes in the States of Tamil Nadu, Kerala and Karnataka, the main tea districats being the Nilgiris, Nilgiri—Wynaad, High Range and Central Kerala. The plantations are at elevations ranging from 1,000 to 7,000 feet above sea level. Soils are red and yellow loams and are lateritic in origin. Most tea plantations in South India get two monsoons and the annual rainfall ranges from 50–275 inches and it is fairly well distributed throughout the year. The cropping season lasts throughout the year, and vintage teas are produced in December/ January.

Fine flavoury teas are produced in the Nilgiris—stylish leaf throwing very bright, brisk, quality liquors with pungency and pronounced flavour; cup character is the dominating feature of these teas. Flavour is derived from the high elevation and prevails throughout the year in varying degrees.

A namallai produces good, black, heavy, flinty well-made teas with a sprinkling of tips in the top grades during the tipping season and are the pride of this district. Liquors are colouty heavy, full-bodied, brisk, malty with quite useful strength and character, with cup quality during December/January.

Kerala teas are black or blackish, clean, heavy well-made grades, throwing exceptionally coloury, strong liquors with briskness and character and occasional quality. These are very useful for blending with flavoury teas of comparatively lighter liquors, and with others of different liquoring properties.

Karnataka produces black, well-made teas with malty medium liquors.

Nilgiri Tea

The Nilgiris, which literally means the "Blue Mountains" are situated in the state of Tamil Nadu in South India. The place derives its name from the blue haze that covers its wooded slopes. This range, with its half English air and rolling down is undoubtedly one of the most beautiful hill resorts in the world.

Apart from its scenic charms the region constitutes one of the most important tea-growing districts in India. It is the home of famed 'Nilgiri tea'. The Nilgiri district has about 23,000 hectares of land under tea cultivation producing nearly 34 to 35 million kg. of tea every year. The other main districts in the South are Coimbatore and the Anamallai hills in Tamil Nadu producing about 18.50 million kg. of tea and the Idukki high ranges in Kerala producing about 32 million kg. Nilgiri teas are, however, a class by themselves. They have a flavour unique to the high grown teas of South India.

The plantations are spread out over every slope, valley and plateau at elevations ranging from 30 to 2,000 metres above sea level on a soil that is predominantly red and yellow loam and lateritic in origin.

Unlike the gardens in Northern India, where tea growth slides off during a period of "winter dormancy", the gardens of the South enjoy a steady growth throughout the year. The two monsoons are a determining factor in facilitating throughout the year plucking. There are, however, two big flushing periods the first from April to May when about 25% of the annual crop is gathered, and the second from September to December fetching about 32 to 40 per cent of the annual crop. Most of the manufacturing is done in the `Orthodox' process.

No other tea has the unique Nilgiri flavour. The characteristic is a gift of nature—a result of various natural stimuli on the plant. It almost appears as if the leaves take their special qualities from the breath-taking beauty that surrounds them. This assumption is not far from the truth. The flavour prevails throughout the year in varying degrees. Nilgiri teas are famous for their stylish leaves that produce very bright, brisk and quality liquors having body and strength. Cup character is the dominant feature of these teas.

The importance of South India as a tea-producing region is tremendous, Cochin serving as its nearest auction centre and port. The largest tea-packing factory in India belonging to the Tata-Finlay group is situated in Bangalore. Besides, both the instant tea manufacturing units are in Munar and Cholari in the South. The United Planters Association of South India (UPASI) Tea Research Station in the Anamallai hills in Coimbatore, is yet another factor contributing towards the phenomenal growth in the tea industry in this region.

Nilgiri teas a blender's dream, and a large portion of these teas find their way into overseas markets in Russia, U.K., the Middle East, U.S.A., Canada and Australia to meet the ever growing demand for the Nilgiri flavour. With the emergence of India as a packet tea exporter, packets of fresh high grown Nilgiri teas are available for consumers all over the world.

The Story of Indian Tea

The fascinating story of tea from the seed to your cup is one filled with colourful legend and romance of modern organisation and commerce in India. It is end product of the patient toil of tens of thousands of tea-growers, the ingenuity of scientists, the expert skill of the blenders, business acumen of traders, care and promptness of shippers and the service and courtesy of retailers. It has passed through many phases, thousands of hands, and across the oceans of the world to reach you and to give you satisfaction.

The tea plant is an evergreen of the *Camellia* genus and is known as *Camellia sinensis* and flourishes in warm tropical and subtropical rainy regions. The highest yield will be obtained from tea which is grown in tropical climate. Paradoxically, the finest quality teas are produced from leaves grown in the cooler altitudes of 1,000 to 2,000 metres. The great 'Darjeeling teas' are grown against the background of *Mount Kanchenjungha* of the Himalayas; nowhere else in the world has the 'Darjeeling flavour' been duplicated. From here come the most delicate, the most fragrant teas ever produced.

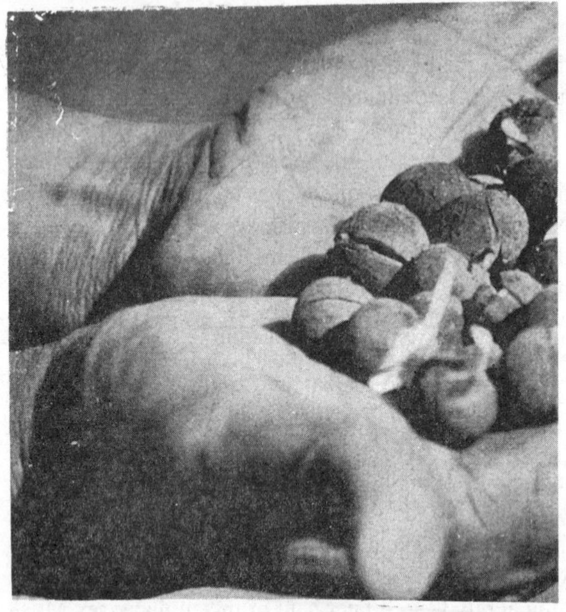

Fig. 18.2. *Camellia sinensis*. The story starts from the tea seed.

Our story starts with the tea seeds, round and hard like brown marbles. The seed itself is taken from the tea trees. Many estates have a plot of selected trees which are allowed to grow to their full size of 5 metres to 10 metres. During July to October the trees blossom and tea flowers appear. But only after a year will the seed be ripe. When the shells crack the seeds are removed and planted in the nursery beds. Within a month they germinate into seedlings and green shoots start to grow. In 9 to 12 months the young plants are ready for removal to the tea garden—their permanent home. For another four years the plants must be expertly pruned before they are ready for the first plucking. By this time

Fig. 18.3. *Camellia sinensis*. Indian tea. The cured leaves are used as a beverage.

they have become wide and flat-topped like a waist-high table to make the plucking easier. If tea plants were left unpruned, they would grow as high as 30 feet.

Plucking. Plucking takes place over 7 to 14 days and the deft fingers of the expert plucker (mostly women) can gather as much as 60 lb. of leaf in one day, or enough to make 15 lb. of black tea. Only two small leaves and a bud are gathered from each shoot.

In North India, during March or April the spring or 'First Flush' materialises; the 'Second Flush' comes a month later in May. After the 'Second Flush', the other flushes are not so well marked and the bush gives more or less a continuous supply of green leaf until November.

In India, plucking is carried on throughout the year. Even then, there are two big flushing periods, from April to May when about 25 per cent of the annual crop is gathered, and a second flush in September to December when about 35 to 40 per cent of the annual crop is harvested.

Commercially, tea may be divided into three basic classes:

(1) Black or fermented (2) Green or unfermented and (3) Oblong or semi-fermented.

Black tea is most important commercially and the production of major world supplies—viz., India, Ceylon, Indonesia and recently East Africa—is mostly of this type.

There are four principal operations in the preparation of black tea: (1) withering, (2) rolling; (3) fermenting, and (4) firing.

Withering. About 75% of the weight of a fresh leaf consists of water even on a dry day and nearly half of this has to be removed before the fibre of the leaf and stalk will stand the strain of rolling, without breaking up. The leaf is spread thinly over withering racks arranged one above the other horizontally, and allowed to remain there for 12 to 18 hours to lose its excessive moisture. Sometimes, heated air is forced over these racks if the atmosphere around is humid. The leaf slowly and evenly becomes soft and flaccid like soft leather and is ready for rolling.

Rolling. It imparts the characteristic twist to the leaf, breaks the leaf cells, exposes the juices to the air for fermentation to set in. After half-an-hour's rolling the leaf is removed in aluminium trolleys to a sifter and ball breaker. This machine consists of a long and flat metal sheet with perforations fixed on a frame which makes reciprocating motion. The movement causes the broken leaf and fine particles to fall below and the rest is taken out after sieving. The latter is again rolled for the second time with increased pressure. After the second roll, all the leaf still more or less green and quite flaccid removed to the fermenting room.

Fermentation. During fermentation, the tannin in tea is partly oxidised and the leaf changes colour and turns bright coppery-red. The rolled leaf is spread on tiles or aluminium or even glass sheets and the oxidation which has commenced during rolling continues, the leaf coming into contact with air. The period of fermentation generally extends from 3 to 3 1/2 hours, but this includes rolling time as well. As a general rule, the *shorter* the fermentation the more pungent the liquor: and the *longer* the fermentation the softer the liquor and deeper the colour.

Drying or firing. After the leaf has changed its colour, comes the final operation; namely *drying* or *firing*. The essential function of this process is to arrest further oxidation of the leaf and to remove all moisture excepting a small amount of roughly 3 to 5 per cent.

The driving machine is the largest machine required in a tea estate factory. The automatic tea drier is a large steel oven inside which the leaves, spread on tray, travel slowly from top to bottom while hot dry air is continuously forced into the oven. Careful regulation of temperature is essential as excessive heat will scorch the leaves while lack of it will result in improper drying.

Generally the process is known as first firing, removes only three-fourths of the total moisture of the fermented leaves. After a period of cooling, they are fed into another drier. The latter known as second firing removes the remaining moisture. The tea from this second drier is the finished product. There now remains only sorting or grading and packing before the tea is ready to be sent from the factory.

Fig. 18.4. *Camellia sinensis* (Tea—**cha**). The manufacturing process being now complete the different grades of tea that come from the sorters are bulked in plywood tea chests.

As the leaves come out of the driers, large and small broken and unbroken leaves are mixed together. Now they are sent through sieves, with graduated mesh, and the different sizes of tea are removed carefully so that final packages are of uniform-sized leaves or particles.

After sorting, the tea is graded and given various names such as *Orange Pekoe*, *Pekoe*, *Fannings* and *Dust*; but these are only indicative of size and not of quality.

The manufacturing process being now complete, the different grades of tea that come from the sorters are bulked, packed in plywood tea chests with linings of aluminium-foil and paper, and are ready for their adventurous journey to all corners of the world.

The process of making tea already described is generally known as the 'orthodox' method and is in vogue in most gardens in India. Recently, however, some import variations have been introduced in the process of manufacturing especially in North India. The variations consist of the use of a machine named C.T.C. or, to give its full name a *Crushing*, *Tearing* and *Curling* machine, all in one. Here, the rollers are either wholly or partly eliminated. The leaf after withering is lightly rolled without any pressure. Then the fine leaf is separated and the coarse leaf is fed into this machine two, three or even four times.

Leg-cut teas. Manufacture is vogue in the Dooars, Terai and in some Cachar gardens of North East India. This process eliminates withering, which is often difficult in these highly humid areas, and also produces a type of tea which is quick brewing and gives an instant liquor of good strength and colour.

Green tea. Most of the teas made in China and Japan are of this type. A small quantity is made in North India also but mainly for sale in Afghanistan, Iran and some for the American market. The essential difference between *green* and *black* varieties of tea is in the fermentation—green tea being completely unfermented. The leaf meant for the manufacture of green tea is generally plucked without stalk.

Oolong tea. It can be described as a cross between *black* and *green*. The fresh leaf is slightly withered before panning; during the process a light ferment is allowed to develop, the leaf is then rolled and fired. Its colour is slightly changed and the resultant teas are of greenish-brown colour. The Oolongs which have long enjoyed a special market in America, are now almost exclusively manufactured in Formosa.

Leaf Grades

These are made up of larger leaves after the broken grades have been sifted out. In brewing, flavour and colour come out of these grades more slowly but, given the necessary brewing time, they produce flavoury liquor. 'Orange Pekoe', 'Pekoe', and 'Pekoe Souchong' are the three generally used grades in this category.

Orange Pekoe (*O.P.*): This consists of long, thin, wiry leaves containing tip or bud leaf. These teas generally come from the finer portions of the shoot that is the bud, first leaf and softer parts of the stalk. The liquors are generally light or pale in colour and, it produced from high grown teas, usually flavoury.

Pekoe (*P*): The leaves are slightly shorter and so not wiry as those of O.P., the buds also are present. The liquors, however, have more colour and this grade is generally quicker brewing than the O.P.

Pekoe Souchong (P.S.): A bold and round leaf with pale liquors.

Broken Grades

These are of smaller and broken leaves and are roughly 80% of the crop including the dust varieties, which are also strictly to be classified under broken grades. There are 4 main grades in this category.

Broken Orange Pekoe (*B.O.P.*): They are much smaller than any of the leaf grades and usually contain tips, hence the word *orange* is added. The liquors have good colour and strength and this is one of the chief attractions.

Broken Pekoe (*B.P*): They are slightly larger than B.O.P. but tips or buds will be absent. The liquors are paler and are less coloury in the cup. These are mainly used as 'fillers'.

Broken Pekoe Souchong (*B.P.S.*): They are a little large or bolder than B.P. and in consequence lighter in the cup, but useful as 'fillers'.

Fannings or Pekoe Fannings: These are much smaller than B.P.S. and are quick brewing and give good coloured liquors.

Dusts

Finally, the smallest particles excluding 'fluff' and 'stalk' are graded as Dusts. These teas are useful in quick brewing, the liquors produced have both strength and colour. They are in a good demand for catering purposes.

Tea Board of India

The Tea Board's major functions at present are: promotion of the tea habit; helping the development of the Industry, more especially of its marginal units; mediation to secure essential goods and services; provision of educational facilities for the dependants of garden workers. For the buyers from abroad it is enabled by the Act to render effective assistance in ensuring the standard of tea shipments from India. But, the Tea Board is more important than its functions would signify. For it embodies the organised knowledge and experience of all tea interests, reflects the attitudes of the industry as a whole and is the only statutorily recognised link between the Industry and the Government.

India continues to be world's largest exporter of tea, and more cups of tea are drunk throughout the world than any other beverage—except water—because it is a good tranquilliser and at the same time acts as a mild stimulant. To relax over a cup of tea is a virtue worth cultivating before the human system, through stress and tension, is allowed to reach a stage when life seems distorted and the mind refuses to adjust itself. Tea's contribution to reduce tension and instil a sense of general comfort has been recognised today by eminent medical experts, though it is by no means a new claim for this little plant whose written origin dates back to some 5000 years. From the distant past an ancient philosopher has left this image of tea to posterity:

"*When I drink tea, I am conscious of peace,*
The cool breath of heaven rises in my sleeves
And blows my cares away."

Five Golden Steps to a Good Cup of Tea

1. Take fresh water from the cold tap and boil.
2. Warm teapot by rinsing out with hot water.
3. Put into teapot one teaspoonful of tea leaf for each cup.
4. Pour boiling water into the teapot. Cover and wait for three minutes. Give more time for bigger leaf teas.
5. Pour liquid tea from teapot into cup. Add milk or a slice of lemon and sugar to taste. Enjoy the taste and flavour of Indian tea.

Fig. 18.5. *Coffee berries*. There was a record production of coffee in India, one lakh and twenty thousand tonnes valued at Rs. 89 crores were exported in 1977-78.

INDIAN COFFEE

History. It has been stated that the coffee plant of commerce is native of Abyssinia, where it is called *bun* or *boun*, and coffee must have been used in that country from very early times.

There is very fair evidence in Arabic literature that the use of coffee was introduced into Aden by a certain Sheikh Shihabuddin Dhabhani, sometime about the middle of the fifteenth century. From Yemen it spread to Mecca in 1511, then to Cairo to Damascus and Aleppo, and to Constantinople, where the first coffee house was established in 1554. The first European mention of coffee of Aleppo is in 1573. The first mention of a coffee shop in Great Britain occurs in 1652. Coffee is spoken of as being in use in France in 1640, and the first public cafe was opened in Paris in 1669. Shortly after it became general throughout Europe.

It was introduced into Ceylon by the Arabs, prior to the invasion of this island by the Portuguese. Its systematic cultivation about 1690 was undertaken by the Dutch but on the cession of their territory its cultivation was continued by the natives of Ceylon (Sri Lanka).

As regards the history of its introduction into India, most writers agree that it was brought to Mysore (now Karnataka) some three centuries ago by a Muslim pilgrim named Baba Budan, who on his return from Mecca, brought seven seeds with him. George Watt writes in his 'Dictionary of the Economic Products of India': "Coffee planting in India at the present day is concentrated in the Madras Presidency, and as a European Industry it may be said to date from Mr. Cannon's plantation at Chikmuglur in Mysore. This was established in 1830, but as a curiosity Major Bevan grew coffee in the Wynaad in 1882. It was cultivated by Mr. Cockburn on the Shevaroys in 1830; Mr. Glasson formed a plantation at Manantoddy in 1840; the plant was taken to the Nilgiris in 1846 and to Darjeeling in 1856."

Fig. 18.6. Indian Coffee. Plucking Beans.

Location and area. Coffee is grown in India on the sunny slopes of the Western Ghats and it spurs extending eastwards where the ecological conditions are ideal for the growth of firstclass coffee. The F.A.O. has listed India as one of the eight countries in the world possessing optimum conditions for the growing of coffee.

The total area under coffee in India is 3,20,570 acres (or 1,29,730 hectares). About 58% of this coffee area is in the Karnataka State, 20% is in the Tamil Nadu State and 21% in the Kerala State. The remaining 1% is scattered, in small pockets in Andhra Pradesh, Maharashtra, Orissa, Assam and West Bengal States, besides the Andamans.

India produces about 100,000 metric tonnes of coffee every year. About 60,000 metric tonnes are exported to different destinations like U.S.A., Russia, West Germany, France, Italy, Netherlands, Sweden, U.K., Belgium and Canada. Indian Coffee in all grades—is among the best in the world and is used to blend with coffees of other origin to improve the quality.

Two varieties. Two main varieties of coffee are grown in India. They are *Coffea arabica* Linn.; and *Coffea robusta* Linden.

Indian Coffee is processed in two ways, *i.e.,* by wet process (washed coffee) and by dry process (unwashed coffee). The washed coffee is known as Plantation or Parchment coffee. The unwashed coffee is called cherry coffee. Thus according to the process, coffee under arabica and robusta is described as under:

	ARABICA	ROBUSTA
Wet Process (*i.e.,* washed coffee)	Plantation	Parchment
Dry Process (*i.e.,* unwashed coffee)	Cherry	Cherry

Fig. 18.7. Indian coffee. Coffee beans of *Coffea arabica.*

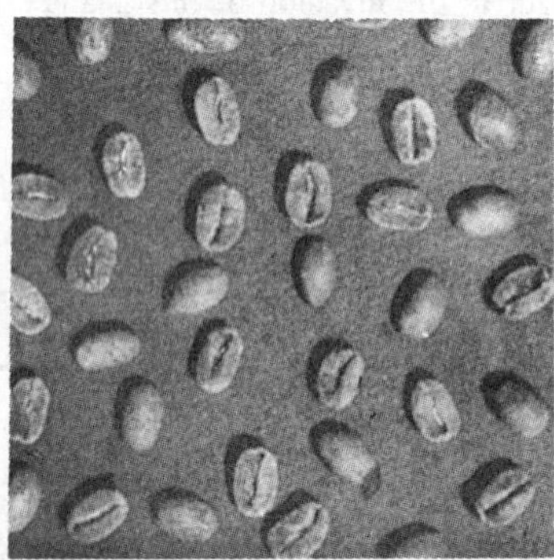

Fig. 18.8. Indian coffee. Coffee beans of *Coffea robusta.*

Plantation coffee. Arabica coffee processed by the wet or washed method is called Plantation coffee. This is an elaborate and time-consuming process in which the outer skin of the fruit and the mucilaginous matter beneath it are removed, by pulping and subsequent washing with fresh water, channelled from the springs in coffee-growing areas. The pulping and washing release the coffee beans from the coffee berries. But each bean is still encased in a soft parchment covering. The parchment-covered coffee is then dried in the sun on barbecues in the estates. Coffee in this form is known as Parchment coffee.

Curing. Parchment coffee is sent to curing works where it is further dried in the sun, if necessary, to a moisture content of not more than 10 percent. The well-dried parchment coffee is fed through the peeling machine to remove the thin brittle parchment cover.

The beans obtained are then graded according to shape and size, and carefully garbled by hand, or using electronic sorters and pneumatic separators to remove blemished and defective beans. The final product of these processes gives the Plantation coffee of commerce.

Cherry coffee. The other method of preparing Arabica coffee for the market is simpler. The berries are harvested, dried in the sun till their outer skin, the parchment cover and pulpy matter in between, dry up to form a thick husk leaving the beans loose inside. The dried cherry is then decorticated by passing through a huller, graded and garbled according to prescribed standards. The final produce of this process is known as Arabica cherry coffee or simply cherry coffee.

Robusta coffee. Robusta coffee is also processed by both wet and dry methods, and the resultant beans are known in trade parlance as Robusta Parchment or washed Robusta and Robusta cherry respectively.

Fig. 18.9. Indian Coffee—Flowering shrub.

Characteristics. The two types of coffee have certain characteristics:

Variety	*Type*	*Characteristics*
Arabica	Plantation	Long and wide beans, greyish or bluish in colour.
	Cherry	Same as above, but colour golden brown to brown.
Robusta	Plantation	Broad oval beans, smaller in size than (Parchment) Arabica, greyish in colour.
	Cherry	Same as above, but colour golden brown to brown.

Grading. Both Arabica and Robusta can have round beans, called PEABERRY (PB) or flat seeds called FLATS. Classification into A or B or AB is done according to the size of seeds.

The main grades of coffee released for export are Plantation A, B and Triage; Arabica Cherry AB and Triage; Robusta Cherry AB and Triage; Robusta Parchment AB. The main grades are tabulated below:

Arabica Plantation		*Arabica Cherry*		*Robusta Parchment*		*Robusta Cherry*	
Grade	Sieve Size (mm)	Grade	Sieve Size (mm)	Grade	Sieve Size (mm)	Grade	Sieve Size (mm)
PB	—	PB	—	PB	—	PB	—
A	6.65 and above	AB	6.00	AB	6.00	AB	6.00
B	6.00						
T	5.50	T	5.50	T	5.50	T	5.50

Fig. 18.10. Indian Coffee. Grading of coffee beans according to shape and size.

Coffee board. The production, internal marketing, export marketing, quality control, research, etc., are controlled by the Coffee Board which is a Statutory Body under the Ministry of Commerce of the Government of India. All coffee produced in India is delivered to the Coffee Board, Bangalore, a statutory organization, comprising representatives of coffee growers, curers, traders, workers, consumers and others connected with the industry.

Marketing. One of the major functions of the Coffee Board is the marketing of Indian coffee, through a common pool (as prescribed by the Indian Coffee Act). At the beginning of each season the coffee crop is estimated and after making provision for the internal market, the export allotment is made, so that the coffee trade may operate on the assurance of known quantities of coffee available for export. This quantity is subsequently increased, if warranted by actual crop realisations.

Exports. Coffee for export is sold in open auctions in which exporters who have registered with the Board, bid for the lots. Foreign firms can participate in these export sales by registering their names with the Coffee Board, through their Agents or Branch offices in India.

Export auctions are held once in two to three weeks. The dates of the sales are widely notified in advance. Facilities are provided for drawing the samples of coffee offered for sale and the despatch of the coffee to prospective buyers abroad.

Coffee is normally exported from the West Coast ports of Mangalore and Cochin. The normal packing is in new gunnies of 60 kg. net. Prices are normally quoted in rupees per 50 kg., f.o.b. New Mangalore/Cochin.

There was a record production of coffee, one lakh and twenty thousand tonnes in 1977-78. Thirty-six thousand tonnes of coffee worth Rs. 89 crores were exported in 1977-78.

Quality. Many steps are taken by the Coffee Board to ensure the export of only the best quality of Indian coffee. The measures are aimed at maintaining and improving the quality of coffee right from the stage when the coffee is a tender plant in plantations to the time it is fully cured and exported. Use of improved strains of coffee, scientific control of pests, application of balanced nutrition and extension of coffee research results to the field ensure growth of quality coffees.

Curing of the coffee is done so as to conform to rigid standards set for curing. The cured coffee is again examined by a screening panel before being made available for exports. Production of quality coffee on estates is encouraged by awarding premium points for good quality coffee. A cup testing unit has been established by the Coffee Board to select the best coffee for exports, on the basis of cup test.

Instant Coffee. India can also offer instant coffee of the highest standards in 50 gms., 100 gms. and 25 kg. bulk packings. This instant coffee is manufactured from Green Coffees of other origin to raise the quality of the latter.

Monsooned coffee. A speciality of Indian coffee is the Monsooned coffee, which is in good demand in several countries of Europe, particularly the Scandinavian countries.

Monsooning of coffee is done by coffee curers on the West Coast of India. Monsooned coffee had its origin in the days of sailing ships, when it took about six months for unwashed coffee to be sent to Europe. During this period of voyage, the coffee in the damp holds of the ship lost its original colour and acquired a special flavour which has been liked by consumers in Norway, France and Switzerland. With the opening of the Suez and speedy transport of steamships, consumers in Europe complained that they missed the special flavour of the unwashed Indian coffee ! The curers on the West Coast rose to the occasion and perfected the process known as 'monsooning', by which coffee acquired the special flavour. Indian monsooned coffee is coffee with a golden colour.

With the outbreak of the South-West Monsoon, late in May or early in June, unwashed coffee (Arabica Cherry or Robusta Cherry coffees) is evenly spread, about 4 to 6 inches thick, in airy godowns, open on all sides. It is racked, from time to time, for about four or five days. It is packed loose in gunny bags and stacked in piles with sufficient space between rows, to enable the monsoon air to circulate freely around each bag.

The coffee beans are bulked and repacked, once a week, or they are poured from one leaf to another, prevent their getting mouldy and ensure even monsooning. In about a month and a half, the coffee assumes golden colour, when it is considered to be fully monsooned. At this stage, the coffee beans are carefully hand-garbled to remove those beans that are not up to the required standard. After mild fumigation treatment to prevent attacks by weevils, the monsooned coffee is ready for shipment.

Monsooned coffee is generally sold for export during October to December, every year. The grades offered for sale are *Monsooned Malabar* AA, *Monsooned Basanally* and *Monsooned Robusta* AB.

Why Indian Coffee is one of the Finest? Lovers of good coffee often wonder how Indian coffee is distinctive, delicious and distinguishable. To comprehend fully the principal reasons for Indian coffee

Fig. 18.11. Indian Coffee. Indian coffee has a high reputation in the world market. Shipment of coffee.

being one of the world's finest coffees is to understand pretty well a whole range of coffee facts and operations in the country.

Nature with its generous Indian climate and rich soil, pampers coffee's growth. Careful scientific cultivation of the coffee on hill slopes, programmed picking of the coffee fruits by deft human hands and the gentle painstaking manner in which they are processed, unlock the full flavour latent in every bean. Rigid assessment of quality leads to the export of only the finest beans accounting for the superiority and fine aroma of India's high grown mild coffees.

Many of the different growths of Indian coffee, like Biligiris, Bababudans, other Mysores, Coorgs, Nilgiris, Naidubattams, Shevaroys, Anamallais, Pulneys, Nelliampathis, and Nilagiris Wynaad are familiar to importers of coffee in different countries of the world, as being synonymous, with supreme quality coffee. This is because the Indian Coffee Industry has to its credit three hundred years of continuous service in growing and making available high grown, good quality, mild coffee.

Indian coffee has a high reputation in the world market. This is because of the high quality of Indian coffee and the infinite care taken by the Coffee Board to maintain high standard at each stage of processing Indian coffee from the seed to the cup and special attention given to the screening and selection of only the best coffee for export, so that coffee connoisseurs around the world can always enjoy the superior and fine aroma of India's high-grown mild coffees.

COCOA AND CHOCOLATE

Cocoa

Theobroma cacao Linn.; Eng. Cocoa; Family—Sterculiaceae.

A small tree. It is native of tropical America, but now cultivated along the Malabar Coast and in the Nilgiris and Pulney hills.

It is grown throughout tropical South and Central America, in the West Indies, and in many other parts of world. The use of cocoa and chocolate by other than the native peoples is of recent origin. Chocolate was the chief drink of the Aztees and other native American peoples. It is the most nutritious of all beverages.

Cultivation. It is a tropical crop. It is sensitive to drought and wind, and so needs shelter from the direct rays of the sun and protection from strong winds. The permanent shade trees are usually grown with cocoa. A deep rich alluvial soil with moisture and suitable drainage is also nec-

Fig. 18.12. *Theobroma cacao* (Cocoa). The fruits of cocoa. The seeds are the source of a non-alcoholic beverage, and are also used as confectionery.

essary. It cannot be grown satisfactorily at altitudes above 2500 feet, and it is injured by temperatures below 60⁰F. The crop is raised from seed or trans-planted seedlings.

The tree is small, from 12 to 25 feet in height, with numerous branches. The leaves are ovate and often 1 ft. in length. The flowers and fruits are short peduncled and borne directly on the trunk and large branches. The fruits are podlike capsules 6 to 9 inches long and 3 to 4 inches thick, with tapering ends. They contain a mucilaginous pulp and 40 to 60 seeds. The fruits ripen to a reddish purple or yellow.

Preparation of cocoa and chocolate. The mature pods are carefully cut off with sharp knives and are then split open. The pulp and seeds are taken out, cured and usually fermented. The process of fermentation may be carried on by piling the seeds in heaps for several days and then spreading them out to dry. Specially made boxes are used, which allow the liquids from the disintegrating pulp to run off and also afford protection from rain. The beans in these 'sweating boxes' are constantly stirred. During the process of fermentation, which completes in about a week, beans become brownish red in colour, lose their bitter taste, and develop an aroma. They are then washed and dried.

For the preparation of commercial cocoa and chocolate the beans are first cleaned to remove any impurities and are then sorted. They are next roasted at a temperature from 257⁰ to 284⁰ F in iron drums. This decreases the amount of tannin, develops the flavour and increases the fat and protein content. The shells become dry and brittle and the seeds easier to grind. The beans are now passed between corrugated rollers which break the shells into small pieces. These are removed in a winnowing machine. The seeds also known as 'niles' are finally ground to an oily paste, constituting the 'liquor' or bitter chocolate. On cooling and hardening this material is known as the bitter chocolate of commerce. Sweet chocolate is made by adding sugar and various spices and flavouring materials. Milk chocolate contains milk, sugar, spices and flavouring materials. Cocoa is prepared by removing about two-thirds of the fatty oil in hydraulic presses and powdering the residue.

Uses—Chocolate is used for making confectionery. Cocoa is a nutritious beverage. The seeds contain less than 1 percent of an alkaloid, theobromine, which, with a few traces of caffeine, is responsible for the stimulating properties.

Fig. 18.13. Cocoa raised as intercrop in coconut. A closer view of cocoa fruits.

They also contain a fatty oil, starch and protein. A volatile oil develops during the roasting process.

Kinds of cocoa. There are several varieties of cocoa in cultivation. However, the most important are—(1) Criollo and (2) Forastero. In the Criollo type the fruit is soft and thin skinned, with a rough surface and pointed ends. The seeds are plump, pale in colour and white within. They are supposed to be finest beans for flavouring purposes. The Forastero type is of hybrid origin. They possess thick-shelled pods with seeds of a pale to deep-purple colour. Most of the commercial crop is obtained from this type.

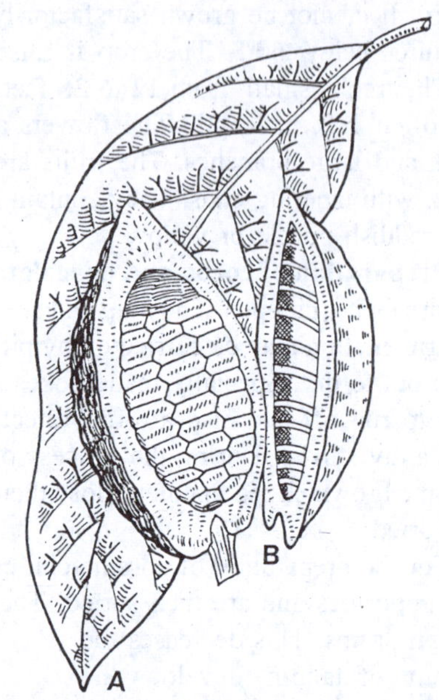

Fig. 18.14. *Theobroma cacao* (Cocoa). A, a leaf; B, longitudinally cut fruit. The seeds are used for making chocolate and other soft beverages.

COLA

The seeds of *Cola nitida*, are used in African and other countries for beverage purposes. The drink is prepared by powdering the cola nuts (seeds). When needed the powder is boiled in water for a few minutes. Cola contains two per cent caffeine, as well as other ingredients. It is quite refreshing. The cola seeds are used in various soft drinks.

FRUIT AND VEGETABLE JUICES

They make the simplest kind of soft drinks. They may consist of the extracted juice alone, or with sugar and water added. The most popular fruit juices are—*Santara* juice (*Citrus reticulata* Blanco), *Musambi* juice (*Citrus sinensis* (L.) Osbeck), Sweet lime juice (*Citrus limettioides* Tanaka), Lemon juice (*Citrus aurantiifolia* (Christm.) Swing.), Grape fruit juice (*Citrus paradisi* Macf.), Tomato juice (*Lycopersicon esculentum* Mill.), Pineapple juice (*Ananas comosus* (L.) Merr.), Pomegranate juice (*Punica granatum* Linn.), Grape juice (*Vitis vinifera* L.) and apple juice (*Malus sylvestris* (L.) Mill.). Sherbets made from strawberries and raspberries are common soft drinks. Lemon juice added with sugar and water is a common drink in India. Grape juice is made by expressing the fresh fruit. The fresh juice obtained from the grapes is pasteurized to avoid fermentation. The expressed juices of apples (sweet cidar) and pears (perry) contain wild yeasts and, therefore, they are pasteurized to kill the yeast organisms. Many tropical fruits are also used for the extraction of soft beverages. The mixed juice of tomato, carrot, ginger and beet root is becoming popular day by day in cities of India.

19

Alcoholic Beverages

'One of the most important and best known industrial fermentation is that in which ethyl alcohol is produced from sugars by yeasts. The chemical manufacturer, the brewer, the distiller, the baker, the vinegar manufacturer, the scientist, the housewife and many others depend in one way or another on the ability of the yeast to convert sugars to alcohol, carbon dioxide, and other end products.' From the time immemorial the man has exploited the natural process of fermentation, and used its products for his own pleasure. In all ages, he has celebrated the occasions with alcoholic beverages. Today the alcoholic beverages are consumed all over the world. Alcohol is a poison, and when taken to excess, produces hazardous effects on the human system. The alcoholic beverages bring about cerebral excitation, followed by depression, and may produce the complete, though temporary, suppression of the functions. The evils of excessive drinking are known all over the world.

Manufacture of ethyl alcohol. Generally the ethyl alcohol is manufactured from molasses. The molasses mash is adjusted to the desired sugar concentration and temperature by the addition of water and to the desired pH by the addition of a measured quantity of acid. A yeast 'starter' is mixed with the mash in the fermentation tank, which is usually covered. Streams of the adjusted mash and the starter flowing simultaneously into the fermentor may be caused to converge on a baffle board located in the upper part of the tank. The mash and starter become well mixed as they spatter and fall to the bottom of the tank. The fermentation rapidly becomes vigorous with the evolution of large quantities of carbon dioxide. In the modern plant, this gas is collected, purified and used for the manufacture of dry ice or for other purposes. Within 50 hours or less the fermentation is usually complete. The fermented molasses, is distilled in a continuous still to separate the alcohol and other volatile constituents from the mash. The alcohol is purified by means of rectifying columns and then stored in a bonded warehouse.

There are two distinct categories of alcoholic beverages : 1. The fermented beverages, where the alcohol is formed by the fermentation of sugar and 2. The distilled beverages, which are obtained by the distillation of some alcoholic liquor.

FERMENTED BEVERAGES

BEER

Brewing. 'Brewing' or the production of malt beverages, is the name given to the combined process of preparing beverages from infusion of grains that have undergone sprouting (malting), and the fermenting of the sugary solution by yeast, whereby a portion of the carbohydrate is changed to alcohol and carbon dioxide. It is an ancient industry and was probably invented by the Egyptians.

Composition of beer. The substances found in a beer depend largely upon the nature of the quality of the raw materials, the treatment of the sprouted grain or malt used in mashing, and the character of the ensuing fermentation, but storage and finishing operations affect the final composition. In normal beer carbohydrates—such as dextrin, maltose and glucose—and protein derivatives—such as peptones, amino acids and amides are present. The products are produced mainly as the result of the action of the enzymes of the malt. Hops contribute bitter substances—such as resins, essential oils and tannins. As a result of alcoholic fermentation, the sugars of the wort are being converted, in part to ethyl alcohol and carbon dioxide. Some of the amino acids are being transformed to higher alcohols and acids. Salts and traces of oil are always found. The finished beer contain 82 to 92 per cent water by volume.

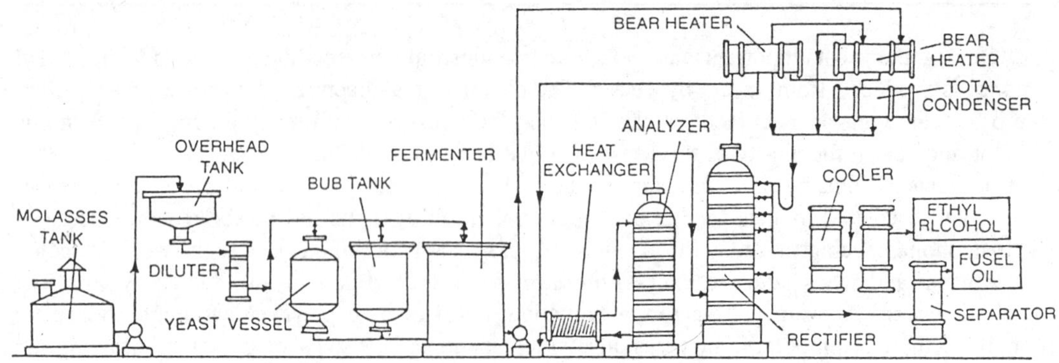

Fig. 19.1. The alcohol industry. Production of alcohol from molasses

The brewing process. The process of brewing alcoholic beverages from cereals is very old. Millet was perhaps the first to be so used, and it is still fermented in India. Rice, maize and rye have been used on a small scale but barley has always been the chief source. The important steps in the manufacture of beer or ale include the following: cleaning, grinding, and weighing of malt and processing of the malt adjuncts; filtering of the mashed materials to separate the soluble material (the wort) from the insoluble material (the spent grain); boiling of the wort with hops; separation of the wort from hop residues and protein precipitates; cooling of the wort; fermentation of the wort; clarification, attenuation, and maturing of the malt beverage during storage; chillproofing and carbonation; filtration packing; pasteurizing; labeling; inspecting; and stamping of the cases.

The commercial manufacture of beer involves two distinct processes—*malting* and *brewing*. The malting is done for the conversion of the starch present in the grains into sugar. This is brought about through the agency of an enzyme, diastase, which is produced during the process of germination. Large, fresh, perfect, light-coloured barley grains are used. The barley grains are steeped in water from one to four days. During this time the grains soak the water. The grains are then placed in heaps or layers 6 inches deep until germination starts. Thereafter they are spread out on malting floor in a temperature of 50 to 60°F, and are constantly turned over. The germinated barley grains are then kiln-dried for twelve hours. The colour of the dried product, the malt, depends on the degree of heat.

The subsequent process is brewing. Now the malt is crushed or coarsely ground in a roller mill and is mixed with water heated to 170°F. The sugar dissolves out and the infusion or wort is drawn off. This process of mashing is repeated several times. The wort is then boiled with hops for two hours. The hops impart the bitter flavour and tonic properties. The liquid is then cooled rapidly and yeast is added to bring about the fermentation of the sugar. Precaution must be taken to keep an optimum temperature for enzyme action. The beer is then drained off and strained and allowed to cool in wooden casks. A slow fermentation goes on, that increases the alcoholic content and results in the formation of carbon dioxide which is responsible for the foaming of the beer. Beer contains 3-8 per cent alcohol.

Kinds of beer. *Lager beer:* It means literally, the stored beer. Larger beer is produced by bottom fermentation and is rather high in alcohol and extract with a relatively low proportion of hops.

Bock beer: This is a heavy beer, dark in colour and high in alcohol. It is brewed for consumption in early spring.

Ale: It is produced by top fermentation. It is pale in colour, tart in taste, and high in alcohol and contains more hops than does beer.

Porter: It is a dark ale, high in extract and sweeter than the usual ale in taste. It is brewed from dark or black malt (malt roasted at a high temperature) to produce a wort of high extract. The flavour of hops is less distinct than that of normal ale.

Stout: It is a strong porter that is high in alcohol and extract. It is dark in colour and possesses a sweet taste and strong flavour of malt. The flavour of hops is ore pronounced than that of porter.

Weiss beer: A beer made mainly from wheat, is produced by top fermentation. It is rather light, possesses a distinct flavour of malt and hop, is tart and contains a large quantity of natural fermentation gas. It is somewhat turbid in appearance.

Cereal beverage: This beer contains less than 0.5 per cent alcohol. It is sometimes known as "near beer".

OTHER FERMENTED ALCOHOLIC BEVERAGES

Fermented alcoholic beverages are consumed all over the world. In some countries the use of a particular beverage has been passed down from ancient times, for example, *Kuass* in Russia, *Pulque* in Mexico, *Sake* in Japan, *Taette* in Scandinavia and *Sorgho* in Manchuria. Some beverages are enlisted here.

Kuass: It is prepared by mixing equal parts of barley malt, rye malt and rye flour, adding boiling water, and then inoculating with yeast and permitting fermentation to take place. Peppermint is added to the fermented product for flavouring. This beverage is quite common in Russia.

Pulque: It is a common fermented alcoholic beverage in Mexico. It is prepared by allowing the sweet juice of agave to undergo fermentation which usually completes in one day. Yeasts produce alcohol from the sugars.

Hard cidar: The fresh apple juice is allowed to ferment for 24 hours. It gradually increases in alcoholic content and hard cider is resulted. Instead of apple juice pear juice may also be fermented.

Taette: It is a common fermented alcoholic beverage in Scandinavia. It is an alcoholic beverage prepared from milk. Yeasts cause the characteristic changes in flavour. It has a pleasant acid taste.

Sake: It is the widely used alcoholic beverage of the Japanese. It is a yellow rice wine containing 14 to 24 per cent of alcohol. It is prepared by fermenting rice. No hops are used. It has been used in Japan for 2600 years.

Pombe: This is an alcoholic beverage made by permitting millet grain to sprout and undergo conversion of the starch to sugars and by allowing a spontaneous fermentation of the starch water.

Ginger beer: It is characterized by its distinctly acid nature, the ginger flavour, and presence of a small amount of alcohol.

Palm wine: The fermented juice obtained from the inflorescences of many palms is a beverage of great antiquity. Palm wine or toddy (*Tari*) was known to Herodotus as early as 420 B. C. The important species of palms which yield toddy are *Phoenix dactylifera*, *Borassus flabellifer*, *Cocos nucifera, etc.* When toddy is distilled, it yields arrack.

Chicha: This is a common fermented beverage of Peru and Bolivia. It is prepared from maize by fermentation.

Root beer: It consists of an infusion of various roots, barks and herbs, with the addition of sugar and yeast. The herbs commonly used for the purpose are ginger, sarsaparilla and wintergreen. Fermentation sets and the beverage becomes charged with carbon dioxide.

Mead: It is a fermented beverage of great antiquity. It is fermented from honey and water and possesses a winelike flavour.

WINE

Wine is the product made by the normal alcoholic fermentation of the juice of sound, ripe grapes (*Vitis vinifera*). Relatively small amounts of wine are made from apples, raisins, berries, peaches, cherries, oranges, currants, apricots, grapefruit; pomegranates, raspberries, pears, honey and strawberries. The wine made from the fruit is named after the fruit, for example, 'apple wine'.

Wine-making areas. A large part of the world's wine is made in the countries located near the Mediterranean sea. France leads the world in wine making, followed by Italy, Spain, Algeria, Portugal, Rumania, Argentina, Russia, Hungary, Yugoslavia, United States, Chile, Greece, Bulgaria, South Africa, Germany and other countries.

In France, the region around Bordeaux produces most of the wine. This district is the most outstanding single vine growing area in the world. Burgundy wines are produced in the hilly country of the Cote d' Or in east central France. Champagnes are produced in the vicinity of Reims and Epernay. Only wines made in this Champagne region have a right to the name. Black and red grapes are used and the manufacture involves a series of elaborate processes which extend over a period of six or seven years.

Making of red wine. Selected grapes of the proper maturity are crushed and stemmed; treated with sulphur dioxide, or a sulfite, or pasteurized; and inoculated with a starter containing a pure culture of yeast. After a short fermentation period the wine is drawn off, placed in storage tanks for further fermentation, racked, stored for aging and packaged.

Wines vary considerably in their characteristics. The alcoholic content varies from 7 to 16 per cent. The sugar content of the grapes is from 12 to 18 per cent. Fermentation of the fruit juice is carried on in vats, usually with the aid of selected yeasts (strains of *Saccharomyces cerevisiae*). The optimum temperature is 68°F. The aroma and flavours are due to various aromatic principles present in the fruit. The characteristic bouquet develops only after the wine has been aged from four or five years to several decades.

White wines are made from white grapes, or expressed juice. Red wines are made from coloured grapes and derive their own colour from the pigments present in the skins of the fruits. In dry wines or sour wines, the fermentation is stopped before all the sugar is being converted, and at least 1 per cent is still present. In sparkling wines, the wine is bottled before fermentation is complete. The still wines have a higher alcoholic content due to the addition of wine, brandy or alcohol.

Other important wines are: The dry *Rhine wines* of Germany, *Chianti*, *Asti* and other wines of Italy; *tokay*, a golden yellow wine of Hungary; *sherry*, a dry wine fortified with brandy, of Spain; *port* of Portugal; *champagne* of France and *California sherry* of California.

DISTILLED BEVERAGES

Whisky. Whisky is an alcohol distillate from a fermented mash of grains. Whisky is obtained by distillation from a fermented mash of malted or unmalted cereals or potatoes. After several distillations of the mash the "low wines" are resulted. Further distillation yield the "high wines". A mixture of water and high wines makes straight whisky. At first several principles are present which make whisky harsh and unpalatable. It must be aged to allow these principles to disappear. The whiskies are aged in charred oak containers. At first the whisky is colourless, the colour develops during the aging process. A continued distillation of high wines results in the formation of 'neutral spirits' which are used in blended whiskies and cordials.

The Scotch whisky is prepared from barley malt. Irish whisky is made from malt or unmalted grains of barley, oats and maize. The Russian Vodka is made from fermented wheat mash. The Vodka is not aged and bottled immediately after distillation and therefore, it remains colourless.

Rum. Rum is an alcoholic distillate from the fermented juice of sugarcane, sugarcane syrup, sugarcane molasses or other sugarcane by-products. Rum is manufactured in general in those countries which grow sugarcane or import molasses or other sugarcane products. It possesses a characteristic flavour, aroma, and colour. The flavour and aroma, improve with aging. Rum contains about 40 percent alcohol. Rum is usually aged in charred white-oak barrels. Rum may be used in the preparation of ice cream, candies and mincemeat; in the curing of tobacco; as a beverage; and as a medicinal.

Brandy. Brandy is distilled only from wine. It is also distilled from the fermented juice of various fruits. The best brandy is made in France in the Charente district. Only this product is known as *cognac*. The other French brandies are known as *armagnac*. The finest grades of brandy are made from white wines. The brown colour of brandy develops when it is stored in wooden casks. Sometimes the brandy is coloured with caramel. It contains about 65 to 70 percent alcohol. Apple brandy is known as applejack.

Gin. It is obtained by distillation from a fermented mash of malt or raw grain. The finest gin is distilled from the malt of barley and rye. It requires several distillations. The flavour of gin and any medicinal value are due to oil of juniper.

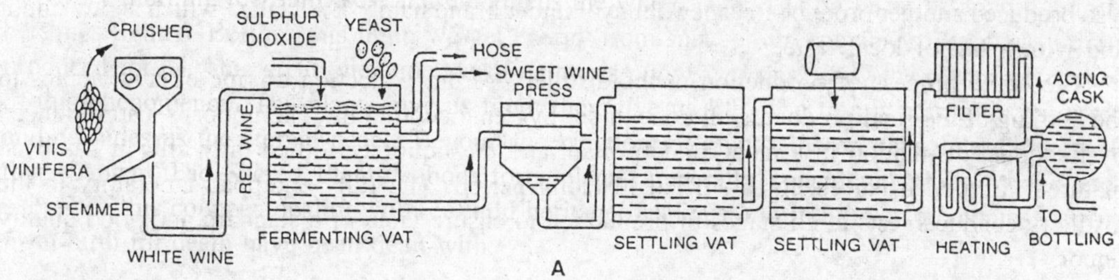

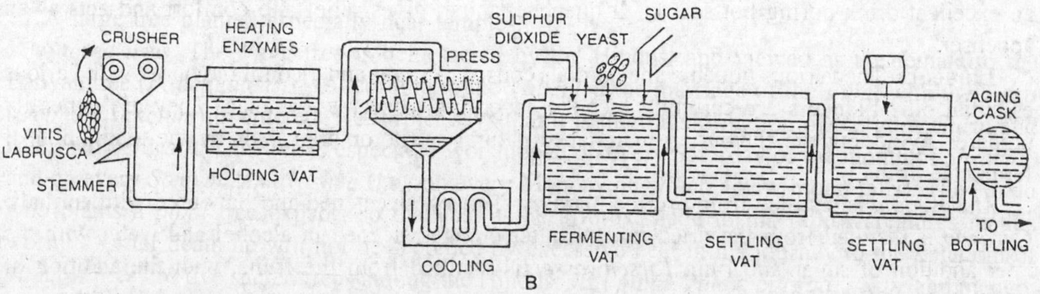

Fig. 19.2. The wine industry—A. Western U.S. method of producing red wine: B, Eastern U.S. method of producing red wine.

CASHEW APPLE LIQUOR OF GOA (FENNY)

Cashew apple (*Anacardium occidentale*) liquor is a unique liquor obtained from the excessive distillation of cashew apple juice. It is a liquor perfect by itself without the addition of any extraneous ingredients and having its own unique flavour.

Cashew apple was discovered during the fifteenth century by the Portuguese missionaries who brought it to Goa, which was then, under the Portuguese occupation. Goa is the only place, in India, where cashew liquor is being produced for the past four centuries. In Goa cashew liquor is commonly known as '*Fenny*' the word '*Fen*' being derived from the Konkani language which means froth. In fact, good cashew liquor when shaken in a bottle produces little froth.

Cashew apple grows only seasonally for about 2-2½ months in a year mostly in hilly regions where iron contents are more in soil. Cashew apple contains sugar in the form of glucose, proteins, malic acid, tannin, calcium, phosphorus pentoxide and vitamin 'C' and from nutritional and medicinal point of view it is considered as one of the best fruits, each fruit giving about 35 calories, Bharat Goa Fruit Distillery (BGFD), is the first modern distillery established in Goa with the consultancy of the Central Food Technological Research Institute producing for the first name in India a quality cashew liquor under the brand name 'NANORA' wherein the strong smell and stringent taste, normally found in the local liquor, have been successfully removed. They have also produced another product—cashew fenny—under brand name 'BARDEZ' which is raw liquor and with a tint of local taste.

The entire process of production in the distillery—from collection of ripe cashew apples to the bottling of the final product is all done under hygienic conditions. The liquor is also produced by fractional distillation in a most modern plant and the liquor is bottled only after it has been matured in wooden barrels for a sufficiently long period. The fully equipped laboratory in the distillery conducts tests at all stages of production to ensure yield of a standard and good quality liquor.

In Goa, traditionally, cashew liquor is used for various medicinal purposes. With people over middle age, men and women, a little quantity of cashew liquor is a must before dinner, which is expected to keep them fit and young in health for longer.

Cashew liquor is normally taken with aerated water and ice. It can also be taken with black coffee. A half peg of cashew liquor with fresh lemon juice, sugar, ice and aerated water makes an excellent drink during hot season. It quenches thirst, gives relief and comfort and acts as an appetiser.

Liqueurs. The various liqueurs and cordials consist of sugar and alcohol flavoured with various essential oils. Liqueurs are carefully blended according to secret formulae. They are generally prepared by adding the flavouring material to natural spirits or brandy or by the distillation of fermented fruits.

Maraschino is distilled from bruised marasca cherries, sweetened and flavoured with cordials. *Curacao* is distilled from the dried rind of bitter oranges steeped in alcohol and water with the later addition of sugar and rum. *Kirschwasser* is distilled from the fruits, with the addition of sugar and alcohol later on.

Bitters. They are prepared by steeping various herbs with bitter principles in water or alcohol. As soon as the principles are dissolved, the infusion is strained and alcohol is added to prevent decomposition. *Angostura* bitters contain quinine and aromatics. *Orange bitters* are also typical of this group.

Aperitifs. These alcoholic preparations are used as appetizers and tonics. *Vermouth*, the best known aperitif, is a light bitter wine slightly sweetened and flavoured with an infusion of bitter and aromatic herbs.

20

Useful Aspects of Lower Plants

THE ALGAE

Food for sea animals and fishes. The algae are used as a direct source of food by several sea animals and fishes. The marine algae are rich in iodine and several other important minerals. This makes the fundamental source of food for all marine animals, and in this respect sea is the richest food producing area. Marine plank–tonic diatomes together with Dinoflagellata are of fundamental biological importance since all life of sea is dependent upon them.

Mineral contents. High mineral content, up to five percent of the wet material, in which all the mineral elements important in human and animal physiology are found, makes sea weeds a unique supplement for a well balanced diet. Potassium, sodium and chloride are found in the ionic form in sea weeds (Pillai, 1956). According to Black (1953), the significance of iodine, as a constituent of foods, is that besides being present in organic combination it is also available in part in the readily available form of the precursor of thyroxine, and hence this source of iodine surpasses mineral iodine in drinking water and iodized table salt. The other micronutrients besides iodine which are important in human metabolism are iron, copper, manganese and zinc, and all of them are present as the trace elements of sea weeds. The highest copper content was found in *Sarconema furcellatum* and *Acanthophora spicigera*.

At the present time iodine is produced from brown seaweeds in Japan, France, Norway and Java. However, in Russia it is obtained from a red alga of the Black Sea, *Phyllophora nervosa* which contains from 0.2 to 0.5 per cent on the dry matter. About 80 per cent of the world's supply comes from the nitrate mines in Chile.

Direct Use of Algae as Food for Man. Since the prehistoric times several sea weeds have been used as direct source of food to human beings. Several fresh water algae have also been utilised in the preparation of various kinds of vitaminized food. As we know well that the fundamental food of sea living stock is algae and they are used as food by human beings. Since the algae are rich in vitamins and minerals; all the deficiencies are overrun by the use of algae as food. The algae (sea weeds) form the most important part of the diet of Japan and China. And some people think that the artistic taste and cultural development of the people of Japan is because of the use of the sea weeds as food. In our country a few species of *Spirogyra* and *Oedogonium* are utilised as food in South India.

Several preparations of algae are used in various countries. Of course, Japan tops the list. **Suimono** is a Japanese preparation of dried fish and several sea weeds. **Mitsu** is another Japanese preparation which contains sea weeds, fruits, sugars and dried kidney beans. **Dulse** is an English dish of algae prepared from *Rhodymenia*. **Seatron** is prepared from *Nereocystis* in United States. **Laver** or **Nori** is prepared in Japan from *Porphyra*. **Green Laver** is prepared in South India from *Spirogyra* and *Oedogonium*. **Kompu** is the product of many Laminariales.

Apart from China and Japan, in Malaya, Indonesia, Burma and Thailand, the sea weeds are used as food. *Ulva lactuca* was used in Scotland for the preparation of salad and soups. *Laminaria saccharina, Rhodymenia palmata* have been used as food in parts of Scotland and Ireland and *Porphyra* is considered to be a tasteful culinary dish in many parts of England.

Among the more important algal food industries may be mentioned **carageen**. This is a product of several sea weeds but principally *Chondrus crispus*. Carrageen is used by soaking it in water and mixing it with milk. Carrageen possesses the properties of gelling which is one of the factors which enhances its usefulness. Fruit juice may also be mixed with it forming fruit jelly. It is also employed in the preparation of ice creams and in the confectionery industry.

The sea weeds are also used as food in the regions of Far East and Australia. The inhabitants of the Hawaii island consume large quantities of sea weeds. The indigenous people of Chile use large quantities of *Durvillea antarctica* and some species of *Ulva*. The natives of New Zealand use certain green sea weeds in preparation of salads and soups.

The people of China and Japan consume the sea weeds on large scale. The people living on the sea coasts in these countries commonly use fresh sea weeds as food. In Japan *Porphyra tenera* happens to be one of the most important edible algae and a product by the name of **amanori** and **Asakusa-Nori** are made from it.

As a Source of Vitamins. The marine algae are the richest source of vitamins. The vitamins A, B and E are found abundantly in sea weeds. The vitamin B essentially required for the development of human body is found in great abundance in almost all Phaeophyceae. The cod liver oil is the rich source of vitamin A, which is acquired from sea weeds. Vitamin E is equally important for human beings which is found in many marine algae.

According to Lundin and Ericson (1956), in the sea weeds of Sweden maximum amount of vitamin B_{12} and folic acid are found in spring and summer and niacin and folinic acid in winter. Vitamin B_{12} content and also that of B_{12} are higher in green and red algae than in brown algae and that the niacin and Vitamin C content appear to be about the same in the above three groups of marine algae. Several vitamins except ascorbic acid have been reported from *Chlorella*. The vitamins found in *Chlorella* are—thiomin, riboflavin, niacin, pyridoxine, pantothenic acid, choline, biotin, vitamin B_{12} and lipoic acid.

As a Source of Agar. The best agar is manufactured from *Gelidium* of Rhodophyceae, which is also called vegetable agar. Japan produces the largest quantity of agar. It produces 95% of the world production. Agar is also obtained from several other marine algae the yield of agar, setting temperature and gel strength of the product from ten species belonging to *Gelidium*, *Sarconema*, *Hypnea* and *Gracilaria* were obtained by Thivy (1951). Japan is the chief agar producing country. According to Ohmi in 1958, Japan produced about 3.5 million lb. of agar.

The agar is used in several ways. It is employed in the preparation of ice cream, jellies, desserts, etc.; in sizing the textiles and clearing many liquids. It is also used in preparing shaving creams, cosmetics and shoe polishes. The agar has constantly been used in biological laboratories as artificial media.

In India, agar resources, as annual yield of dry sea weeds, of Chilka Lake has been estimated by Mitra (1946) to be about 4.06-5.08 metric tons; of Cape Comorin by Koshy and John (1948), Thivy (1957) about one metric ton, and of the Pamban area as estimated by Thivy (1957) about seven metric tons. Other large quantities are found in Kathiawar Peninsula and estuaries, the resources of the Andamans are believed to be considerable.

Medicines and Minerals. Several diseases caused by vitamin deficiency such as vitex, asthma, tooth decay, etc., may be eradicated if floor of the sea weeds is added to the food. According to Dr. W. Weston, iodine is the most important element to enable the thyroid glands to secrete the thyrosin which contains 60% iodine. It controls the general development of the animal. Sea weeds are the best source of iodine for human beings. Several important sea weed medicinal preparations are prepared in various countries, e.g., Kelpeck is prepared from kelps in Chicago; Burbank Vegetable tablets are sea weed preparation from United States. Kelpamalt is a sea weed medicinal preparation from New York (U.S.A.); Isokelp is prepared in California; Parakelp and Manamar are other medicinal sea weed American preparations. An antibiotic drug Chlorellum is also obtained from algae.

About fortyfive elements are found in a weed *Macrocystis pyrifera*. In addition to these elements vitamins are also found. No other food contains such a great abundance of minerals and vitamins.

Marine algae are specially rich in vitamins A, B, C and E. The diatom *Nitzschia* is very rich in vitamin A and possibly this is the main source of vitamin A that is found in liver oils of many fishes. Vitamin B is also found in certain algae, as for example, *Ulva, Porphyra,* etc., and *Alaria valida* is rich in vitamin C, while some Fucoids and *Porphyra* are even richer.

Besides these several other marine forms are quite rich in minerals and vitamins. Several tribes using the sea weeds directly as food have their people handsome and sturdy.

Manufacture of Iodine. The world's iodine supply is fulfilled from the sea weeds. Since hundred years back the iodine manufacture is in progress. Iodine is used in several ways.

Alginic Acid, Algin and Mannitol. The alginic acid is manufactured from the cell wall of Phaeophyceae. It is insoluble in water and hard when dry. Sodium alginate is used, in sizing material for water proof material dyes, buttons, handles, combs and many other such things. This is also used as a sterilizer in daily use.

The algin is found in the form of calcium alginate and alginic acid. The Fucaceae are the chief source of algin in India. Yields of algin varying from 15.6 to 19.2 per cent on air dry matter were estimated for Fucaceae and 10.4 percent for *Padina*.

A yield of 9.4 per cent of mannitol from *Sargassum tenerrimum* and 7.3% from *S. wightii* have been reported.

Manufacture of Soaps and Alums. By burning sea weeds on the sea coasts the alkalies are prepared from sea weed ashes. These alkalies are employed in the manufacture of soaps and alums.

As a Fodder for Hens and Milk Cattle. By feeding the milk cattle and hens with sea weeds, the iodine quantity of the milk and eggs may sufficiently be increased. In Scotland, Newzealand, Norway, France and United States several sea weeds are utilized for feeding hens and young livestock.

Some countries have even factories to process sea weeds into suitable cattle feed. The manufacture of cattle feed from sea weeds is made principally from brown algae and the processed food is fed to cattle, poultry and even pigs. It has been recorded that dried sea weeds served as cattle food have enhanced the milk-yielding and egg-laying capacity of cattle and poultry respectively.

Manufacture of Potash. Species of *Macrocystis* and *Nereocystis* (Phaeophyceae) possess 30% of potash in their dry weight.

Used as Fertilizers. Due to the presence of potassium chloride (KCl) in sea weeds, they are used as fertilizers in many countries, such as Japan, France, United States, England and South India.

Sea weeds are a store-house of the important potash, ionic sulphate, trace elements and growth substances besides having every other elements and radical required by plants. Sea weed manure seems to increase resistance to disease. Most of the nutrients including nitrogen compounds are in ionic form and a quick absorption by crops takes place and relatively little is left to be broken down by soil microflora, thus preventing acid conditions of the soil arising from the fermentation. In general the minerals diffuse out from the sea weed thallus rapidly. Yet another feature is that sea weed manure holds water and air at the same time and improves the soil in both respects. Like other manures sea weeds have a similar role, but also contribute the required potassium, sulphur, phosphrous and calcium.

Manufacture of Light Weight Buildings. Germany has discovered a process, in which the sea weeds are mixed with cement to make building light in weight and good heat resistant.

Manufacture of Paper. It is probably thought that a rough quality of paper may be manufactured from sea weeds, as yet it is not practised.

Ornamental Uses. Some algae like *Botrydium* and *Spirogyra* are grown in the garden ponds for thier good looking habit.

Diatomin Earth. In the beds of oceans diatomin earth is found. During millions of years immense rocks of diatomaceous earth have been formed. At Lompoc and California the deposits are several hundred feet deep, *i.e.*, 700 feet. At St. Maria oil fields California, the deposits are 3,000 feet deep and miles and miles in length. Diatomaceous earth is very important commercial product and used in several ways. America is the largest producer of diatomaceous earth. This is employed as an absorbent in the manufacture of dynamite. It is used in the filtration of liquid in sugar industry. This earth can resist very high temperature up to 1,500°C, and used in the manufacture of the fire bricks. These bricks are used in the manufacture of the fire bricks. These bricks are used in the inner lining of the blast furnaces. This is also added in the cement to increase its cementing power. It is used as a mild abrasive for metal polishes. It is also used in the manufacture of tooth pastes and car polishes. It acts as a catalyst carrier in the hydrogenation of vegetable oil. In the manufacture of paints, lipsticks and other cosmetics, it is used in many ways.

Sir John Murray is of the opinion that the sea is as important as the land in the production of food materials. Marine Biology is a new field of study and may lead to the scientists to new science 'mariculture' parallel to 'agriculture'. Certainly we have not realized importance of sea vegetation as yet.

THE FUNGI

Edible Fungi

Some fungi have been esteemed as articles of food from early times. Edible fungi have delicate flavours and are often valued as delicacies. Several such fungi grow wild in India, appearing usually during the rainy season. They are found in mountainous and submontane forest areas on dead and decaying vegetable or animal remains. Some delicious species are found in the hills under snow, *e.g.*, morels, a few species are common in the plains and in deserts. The culture of edible mushrooms is not on large scale in India as in Europe and America. A few of the edible mushrooms occurring in our country are described here.

Agaricus campestris Linn., Syn. *Psalliota campestris* (Linn.) Fr. (Verna. *Khumbi*; Eng. Field mushroom). It is distributed in many parts of India, particularly on the hills and plains of northern and eastern India. It grows during the rainy season on dead organic matter, *e.g.*, rotting leaves and manure. It has a white, smooth, solid stalk with a thin ring attached at the middle or towards the top. The cap is rounded in the young stage; later becomes convex and finally more or less flat. The gills are whitish in the button stage, as the cap opens the gills become pink and later dark, purplish brown or black.

The field mushroom is adapted for culture under artificial conditions. The most suitable time for cultivation is August to March in the plains and March to October on the hills. Spores are sown on nutrient beds and when they develop, the white cottony growth, called the spawn, is taken out and planted on suitable beds. Horse dung collected from stables in which abundant wheat straw has been used, is considered to be the best material for the bed. Fermentation is allowed to proceed in the bed for 12-15 days and when the temperature has cooled down to 90° or 80°F., chunks of spawn, cut into small squares, are planted in the composted manure which is covered with loam soil. The beds are watered lightly every other day. Mushroom fructifications (button stage) appear 4-10 weeks later. The crop is prolific when the beds are fresh, but in course of time bearing capacity falls off. They may be stimulated into vigour by applying liquid manure twice a week and productivity maintained for 6-8 months. Temperature control is essential, a temperature of about 50 to 55°F, should be maintained.

Field mushroom is cooked fresh or after dehydration. Fruit bodies are gathered before they attain maturity, caps and stalks, washed in water, cut into pieces and dried in the sun. It is an excellent source of vitamins of B complex and the vitamins are well retained during cooking and

Fig. 20.1. *Agaricus bisporus*. Mushrooms, which are highly prized for their rich protein and mineral contents.

processing. It also contains vitamin C, vitamin K and vitamin D. It also contains amylase, maltase, glycogenase, protease, catalase, tyrosinase, two phospho-monoesterases, two polyphosphatases and polyphenoloxidase. Dehydropeptidases are also present. Mushroom juice containing proteolytic enzymes may be used for tenderizing meat. It can also be used as a source of protein.

Cantharellus cibarius Fr. (Eng. Chanterelle; Punjabi—*Dhingri*). It is a small yellowish fungus, 2-3 inch across and about the same height, found in the submontane regions of northern and eastern India. It is mostly found on sandy or loose soils of decaying vegetable matter during the autumn. The cap is at first convex then becomes flat and finally depressed or funnel-shaped with an irregular wavy margin.

It is cooked fresh or dry and is relished for its cartilaginous tissue. It contains several vitamins such as thiamin, riboflavin, ascorbic acid and vitamin D. Fresh chanterelle also contains iodine.

Volvaria diplasia Berk. & Br. (Eng. Straw mushroom). This is an edible mushroom found in Bengal and Tamil Nadu. Stalk 6-12.5 cm. long, white, slightly attenuated upwards, with a bulbous base; cap campanulate in the young condition, later expanded, umbrella like with serrate margin; gills free, crowded, white, later pale.

Straw mushroom has a soft texture and a pleasant taste. It grows on rotten straw during rainy season. Its cultivation under artificial conditions has been tried in Tamil Nadu and grown successfully.

Lycoperdon bovista Fr. (Eng. Giant Puffball; Punjabi—*Bhoenphal*). It is a large creamy white, flattened, oval or pumpkin-like fungus grooved or folded at the bottom, found in the plains of northern India. It is rooted in the soil by a cluster of thick, cord-like strands of mycelium.

It is eaten only in the young stage when the inner part is white. Other edible puff balls recorded in India are—*L. gemmatum* Batsch, *L. piriforme* Schaeff., and *L. saccatum* Vahl.

Morchella esculenta (Linn.) Pers. (Eng. Morel; Punjabi—*Gucchhe*). This is an ascomycetous fungus found in Kashmir, parts of Himachal Pradesh and Naini Tal. Its stalk is white, thick, erect, tapering, bearing rounded or conical pileus covered with a number of prominet ridges. Its colour is yellowish brown to olive. It appears on decaying vegetable matter in hill forests from May to August after the snow has melted. Large quantities of dried morel are exported from Kashmir to other parts of India.

This is eaten both fresh and dry. It is sold in the market in the dried state.

Tuber spp. (Eng. Truffles). They also belong to Ascomycetes. *Tuber cibarium* Sibth has been reported from Kashmir and Kangra, where it is eaten as a curry. Truffles of white and black types are found at Bankura in West Bengal and are dug out from the soil below sal trees.

Other edible mushrooms found in India are—*Hymenomycetes*: *Armillaria mellea* (Vahl.) Fr.; *Boletus* spp.; *Clavaria fusiformis* Sowerby, *C. stricta* Pers.; *Collybia albuminosa* (Berk.) Petch; *Coprinus comatus* Fr.; *Entoloma microcarpum* Berk.; & Br.; *Fomes pachyphlaeus* Pat.; *Hydnum coralloides* Scop.; *H. repandum* Linn.; *Hypoxylum vernicosum* Schw.; *Lactarius deliciosus* S.F. Gray; *Lepiota mastoidea* Fr.; *Pleurotus cretaceus* Massee; *Polyporus grammocephalus* Berk; *Polystictus sanguineus* (Linn.) Mey.; *Schizophyllum commune Fr.*

Gasteromycetes; *Bovista gigantea*; *Geaster* spp.; *Melanogaster durissimus* Cke.; *Scleroderma aurantium* Pers.

Fungi in Medicine

The subject of fungi in medicine has attained great importance since 1929 when Alexandar Fleming reported his observation on lysis of *Streptococci* in agar plate cultures contaminated by *Penicillium notatum*. Westling, Chain *et al* showed in 1940 that an active anti-bacterial substance could be extracted from culture media in which *Penicillium notatum* was grown and that it could be preserved for appreciable periods. Since then progress has been rapid and several active principles, antibiotics as they are called, have been isolated from specific fungi and some of them have been used in the treatment of diseases.

Penicillin: It is now widely employed in the treatment of a variety of bacterial diseases in man. It is the metabolic product of a number of moulds generally referred to as *Pencillium notalum—chrysogenum* series. High yielding strains of *Penicillium* adapted for submerged culture and producing specific penicillins or increased general yield of penicillin, are selected for commercial production of this antibiotic. They are cultured in suitable media under aeration.

Penicillins are bacteriostatic and, in sufficient concentrations, bactericidal against a wide range of common pathogenic bacteria. They are non-toxic and are effective mostly against Gram-positive bacteria. A few Gram-negative bacteria are also susceptible. Among the conditions which have been successfully treated with penicillins are infections of wounds and fractures, boils, carbuncles, abscesses, acute tonsillitis, pneumonia, peritonitis, pericarditis, subacute bacterial endocarditis, acute osteomyelitis, otitis media, mastoiditis, meningitis, ocular infections, gonorrhoea and syphilis. Penicillins are administered orally or through intramuscular, intravenous or thoracic injections.

Polyporin: This is obtained from the culture filtrates of *Polystictus sanguineus* Fr. The crude antibiotic is active against typhoid, cholera and dysentery organisms (*i.e. Staphylococcus aureus and Escherichia coli*). It is completely non-toxic to animals and is also stable.

Other fungal antibiotics: *Clitocybine* is extracted from *Clitocybe gigantea* Fr. It has not been extracted in pure form. It is highly bacteriostatic. It prevents the growth of a number of Gram-negative, Gram-positive and acid-fast organisms. It inhibits the growth of *Mycobacterium tuberculosis*.

The other fungal antibiotics which have been studied in some detail are as follows—Aspergillic acid from *Aspergilus flavus*; avenacein from *Fusarium avenaceum* Sacc; candidulin from *Aspergillus candidus* Link; chaetomin from *Chaetomiun cochlioides* Palliser; citrinin from *Penicillium citrinum* and other species of *Penicilium* and *Aspergillus*; clavacin from *Aspergillus clavatus* and *Penicillium* spp; fumigacin from *Aspergillus fumigatus*; fuscin from *Oidiodendron fuscum* Robak; geodin from *Aspergillus terreus*; gladiolic acid from *Penicillium gladiolic* Machacek; javanicin from *Fusarium javanicum* Koord; lactaroviolin from *lactarius deliciosus* Fr. lycomerasmin from *Fusarium* spp; oregonensin from *Ganoderma oregonense* Murrill; penicilic acid from *Penicillium* spp. and *Aspergillus* spp; spinulosin from *Penicillium spinulosum* Thom. and *Aspergillus fumigatus*.

Other fungal drugs. Ergot: This is the sclerotium of the fungus *Claviceps purpurea* (Fr.) Tul. It is extensively employed for stimulating uterine contractions.

Fomes officinalis (Vill ex Fr.) Lloyd (Eng. Purging Agaric, Verna. *Gharikun*). This yields a drug long held in repute cheifly for its use in the treatment of night sweats in phthisis. It is diuretic, laxative and expectorant. In large doses it acts as a purgative. It is used as an ingredient of pills of asthma.

Polyporus anthelminticus Berk. is used as an anthelmintic.

Several fungal enzyme preparations have been used in digestive diseases and other disorders. The diastase obtained from *Aspergillus flavus oryzae* group is effective in salivary deficiencies. Fungal diastase preparations are also recommended for gastrointestinal disturbances. Patients suffering from arterial hypertension respond favourably to subcutaneous injections of tyrosinase obtained from the abstracts of *Agaricus campestris*.

Fungi in Industry other than Yeasts

Mould enzymes. The mould enzymes play a part in the ripening of the cheese and impart particular flavours and textures. The Camembert type of soft cheese ows its texture to the proteolytic activity of *Penicillium camemberti* Thom. *P. caseicolum* Bainier produces a cheese of somewhat different texture. The ripening agent in the green Roquefort cheese is *P. roqueforti* Thom.

Production of Acids. *Citric acid*: Oxidation of sugars to citric acid is brough about by a number of moulds, including *P. luteum* Zukal, *P. citrinum* Thom, *Mucor pyriformis* Fischer, and *Ustulina vulgaris* Tul. *Aspergillus niger* is the mould commonly employed in industry. Oxalic acid is also produced alongwith citric acid and it is possible to reduce or prevent its formation by adjusting the pH, temperature and composition of the medium.

Gluconic acid: Certain strains of *Aspergillus* and *Penicillium* have been utilized for the commercial production of gluconic acid from glucose.

d-Lactic acid: It is obtained by the fermentation of sucrose by *Rhizopus oryzae* Went & Painsen Gearlings.

Gallic acid: It is prepared by the fermentation of tannin extracts with *Aspergillus gallomyces* Calmette.

Other organic acids produced by mould fermentation are fumaric, succinic, malic, formic, pyruvic and itaconic.

Members of the genus *Fusarium* produce alcohol in good yields. Some moulds produce mannitol.

Production of Enzymes. The mould enzymes which have attained commercial importance include diastase, invertase, protease and pectinase. Several members of the *Aspergillus flavus oryzae* group have been employed for the production of enzymes.

Diastase: Fungal diastase is chiefly used for dextrinization of starch and desizing of textiles. Commercial preparations are obtained mainly from *Aspergillus oryzae* (Ahlburg) Cohn. Diastase producing moulds have long been employed for the saccharification of starch which is an important stage in the production of alcoholic beverages from starchy raw material. In the production of Japanseses *Sake*, from rice, *Aspergillus oryzae* is the active agent.

Invertase: Species of *Aspergillus*, *Mucor* and *Penicillium* produce invertase when cultivated in a medium containing sucrose. Invertase finds use in the candy industry and in the manufacture of syrups.

Proteases: Several species of *Aspergillus*, *Mucor* and *Penicillium* produce protein–hydrolysing enzymes when grown in media containing proteins. A strain of *Aspergillus oryzae* grown in liquid medium produces a highly active protease which has been obtained in a partially purified crystalline form. Fungal proteases are used for clarifying and maturing beers and for improving their stability. They are also used in the bakery industry. Cultures of *A. flavus* Link and *A. oryzae* may be used for dehairing hides and skins.

Pectinase: *Penicillium notatum* Westling grown on a medium containing lucerne yields pectinase, employed for the clarification of fruit juice.

Vitamins. Ergosterol, the precursor of vitamin D, is synthesized by a number of moulds, especially *Aspergillus* and *Penicillium* spp. Vitamin C has been reported to be a metabolic product of a strain of *Aspergillus niger*. Some fungi require specific vitamins for their normal growth and development and may be used therefore, for the estimation of vitamin concentrations. *Phycomyces blackesleeanus* Burgeff grows only in the presence of vitamin B_1 and therefore, it has been employed for the bioassary of this vitamin.

THE LICHENS

Lichens constitute a class of small perennial plants, which are unique in that they are actually combination of two organisms—a fungus and an alga-growing together in symbiotic association. The fungal element lives up to the food elaborated by the algae, while the algal element gets its moisture and mineral nutrients from the fungus; the fungus also protects the alga against desiccation and injury.

Lichens are widely distributed from the arctic to the tropics and are found on soil, barren rocks and tree trunks. In India, the genus *Parmelia* and members of the family Usneaceae are widely distributed. Series Pyrenocarpeae is of little economic value. A species, *Dermatocarpon moulinsii* (family Dermatocarpaceae) reported to occur in India finds some minor uses. In the forests of Ooty lichens (especially *Usnea* and *Parmelia*) grow wildly.

Uses. *As food*: Lichens have been used from the earliest times as animal feeds and also as food for human consumption in times of scarcity; a few are consumed as delicacies. *Cetraria is landica* (Iceland Moss), *Evernia prunastri* and *Lecanora esculenta* Evers, are instances of lichen foods. Species of *Umbilicaria* and *Gyrophora*, have been consumed by arctic explorers. *Gyrophora esculenta* Miyosh is prized as a food adjunct in China and Japan. Some species of *Parmelia*, *Peltigera*, *Ramalina* and *Usnea* are reported to possesess food value. Several lichens are used as fodder for reindeer, cattle, swine and goats. In Orissa *Usnea* and some species of *Parmelia* are used alongwith the spices and condiments.

Dyes: Lichens were once employed for dyeing wool and silk, and colouring materials derived from them were highly valued before the advent of aniline dyes. Certain species of *Roccella* and *Lecanora* yield purple dyes, known in commerce as Orchill, Cudbear and Litmus. In India species of *Caloplaca, Candelariella, Cetraria, Gyrophora, Lecanora, Lepraria, Ochrolechia, Parmelia, Physcia, Ramalina, Roccella, Solorina, Sticta, Umbilicaria, Usnea etc.* yield several dyes.

Alcohols and other Chemicals: Lichens were formerly used as source of fermentable sugars for the production of ethyl alcohol. *Cetraria islandica* and *Cladonia rangiferina* yield up to 60% polysaccharides readily hydrolysed to glucose. Some species of *Parmelia*, *Evernia* and *Ramalina* yield gums which found application in calico printing and in the production of parchment and cardboard. Many crustaceous lichens contain oxalic acid; *Pertusaria* spp. is reported to be used in France in the manufacture of oxalic acid.

Medicines and Antibiotics: Several lichens possess medicinal properties. *Cetraria islandica* is considered useful in chest complaints. Some lichens belonging to Cladoniaceae, Parmeliaceae and Usneaceae yield substances which are active against Gram-positive bacteria. Usnic acid is active against *Streptococcus haemolyticus* and *Pneumococcus* spp. and inhibits the growth of tubercle bacillus. Barbatic acid and diabasic rocellic acid, in the form of its half esters and half amides, possess antitubercular activity. *Cetraria islandica* (Iceland Moss) is almost odourless and has a bitter mucilaginous taste. It possesses demulcent and laxative properties and is given in decoction as a bitter tonic and nutrient in chronic catarrh, chronic bronchitis and consumption. Iceland moss may be used also as a source of glucose and for the production of alcohol.

Spices and condiments: *Parmelia abessinica* Kremp.; Telugu—*Rathipoovvu*; Family—Parmeliaceae. This is a crustose lichen found in rocky areas of Bellary, Anantpur and Cuddapah districts of Andhra Pradesh. The lichen is available in large quantities in the market and used as spice and condiment. It contains atranorin (1.1%), lecanoric acid (3.3%), salazinic acid (0.1-0.5%) and isolichenin (3.4%).

Parmelia tinctorum Despr., is a foliose lichen commonly found in the plains of India and in the Himalayas on tree bark and rocks. It is commonly used in certain parts of South India as a component of various food preparations. It contains—crude protein, 13.8%; ether extract 6.0%; isolichenin 25.0%; crude fibre 13.4% and ash, 12.6%; calcium 1728 mg.; ascorbic acid 5.0 gm.; and riboflavin 100 mg/100 g.

21

Methods of Crop Improvement

The subject of plant-breeding has been scientifically developed to such an extent in recent years that it is now recognized as the best practical method for the improvement of the various crops—cereals, millets, pulses, vegetables, fruits, oil-seeds, industrial plants, fodder crops and many other plants of economic value and of aesthetic value as regards their productiveness, quality, colour, size, yield and other useful characters. It has been found possible by the application of this method to combine the desired characters of parent forms and evolve new types of better than the original types in many respects. The subject of plant-breeding has been considered of great importance in agriculture and horticulture. By adopting practical methods of plant-breeding considerable improvements have been achieved in the last few decades; for example, rust resistant and high-yielding types of wheat plants have been evolved, and the milling and bread-making qualities of wheat grains considerably improved. Several new varieties of rice with higher yield and better quality have also been evolved. In tobacco the number of leaves per plant and their size and quality have been enhanced. Pea, bean, vegetable, pulses, oilseeds and condiments have been considerably improved. Fruits like mango, apple, grapes, pomegranate, pecan, pear, banana, plum, orange, litchi, strawberry, etc., have similarly been much improved. In fibre-yielding plants like cotton, jute, flax, hemp, etc., better quality of the fibre, *i.e.,* length, strength and fitness, and also higher yield have already been achieved. In India a considerable amount of work has been done on rice, wheat, maize, millets, sugarcane, pulses, oilseeds, cotton, tobacco, jute, flax, hemp, fruits, vegetables, spices and condiments.

The main methods of crop improvement are as follows:

1. **Selection.**
2. **Hybridization.**
3. **Plant introduction and acclimatization, and**
4. **Mutation breeding.**

SELECTION

Selection is the choice of certain individuals of propagation from a mixed population where the individuals vary in character. It is the oldest breeding method and is the basis of all crop improvements. Even to-day it is the most common method of crop improvement among the cultivators. There are two main types of selection, (1) natural selection and (2) artificial selection.

Natural selection. This is a rule in the nature and results in evolution. Here, the fittest can survive and rest wipe out. All the local varieties of crops resulted because of such selection. It remains always operating in nature. It is one of the natural factors which creates variations in the already existing varieties of crops.

Artificial selection. '*Artificial selection is to choose certain individual plants for the purpose of having better crop from a mixed population where the individuals differ in characters*'. There are three main types of artificial selection: (1) mass selection, (2) pure line selection, and (3) clonal selection.

Mass selection. '*The selection of a number of plants, heads of seeds phenotypically superior in desired traits from the field population harvesting and bulking their produce together for sowing the next year's crop; and repeating this process till desired improvement is achieved, is known as mass selection*'.

The mass selection is followed generally in cross-pollinated crops for producing a new variety. It consists of selection and collection of the similarly appearing best and most vigorous plants from the mixed population of a crop. The selected plants are thrashed together and a mixture of seeds is obtained. The mixture so obtained is sown for raising the new crop from which the selection is made similarly in the next year. This practice of selection is continued till the plants show uniformity in the desired characters and they constitute a new variety. In already uniform population, the mass selection cannot be practised. It means that the mass of selection is based upon the presence of variability in the population. Greater the variability better are the results of selection. The plants are selected on the basis of phenotypical characters. It is the simplest, easiest and quickest method of crop improvement. It is only the method for improving the local varieties to meet the immediate needs of farmers. Mass selection is the most common method of crop improvement among farmers. All the cultivated crops and their local strains are the results of mass selection operating in the nature automatically.

Pure-line selection. '*The process of isolating a desirable homozygous individual from the mixed population and multiplying the same without contamination to release as a new variety is known as pure-line selection*'. It consists of the progeny of single self-fertilized homozygous plant and is used for developing variety. It is used in the mixed population of self-pollinated crops and consists of testing the progeny of single individual plants separately. The usual method is to select a large number of single plants, keep them separate, compare their progeny in field trials, and to save the single most valuable progeny as a new variety. In this method the fewer plants are selected and each selected plant is tested separately. The variety developed is genetically more and more durable.

Clonal selection. This type of selection is practised in vegetatively propagated crops such as potato, sugarcane, banana, sweet-potato, citrus, mango, etc. '*All the progenies of a single plant obtained vegetatively are known as a clone*' and the method of developing varieties from the clones is known as clonal selection. *Clonal selection is the method of selection of desirable clones from the mixed population of vegetatively propagated crops*. The superior clones are selected on the basis of their phenotypic characters. The selection is always between the clones and not within the clone. The selected clones are multiplied vegetatively and compared with normal variety.

HYBRIDIZATION

'*Hybridization is the method of producing new crop varieties in which two or more plants of unlike genetical constitution are crossed together*'.

Kinds of hybridization. The plants which are crossed together may belong to the same species, different species or different genera. According to this relationship between parental plants, the hybridization may be—(1) intravarietal, (2) intervarietal, (3) interspecific and (4) intergeneric.

1. **Intravarietal hybridization**. The crosses are made between the plants of same variety. Such crosses are useful only in self-pollinated crops. In cross-pollinated crops, where the varieties are genetically heterogeneous. The controlled crossing within the varieties is only useful to maintain and improve the individual varieties.

2. **Intervarietal hybridization.** In such crosses desired characters can easily be combined. The crosses are made between the plants of two different varieties of the same species and is also known as *intraspecific hybridization*. This makes the basis of improving self-pollinated crops as well as certain cross pollinated crops. In wheat, tobacco, cotton, tomato, etc., intervarietal crosses have been done to achieve good yield, resistance to diseases, resistance to drought, good quality, nutritive value *etc*. Most of the hybrid varieties of cereals, have been evolved by this type of hybridization. Hybrid maize is an outstanding practical achievement in plant breeding of this type of hybridization.

3. **Interspecific hybridization**. The plants of two different species belonging to the same genus are crossed together. This type of hybridization is between the species and within the same genus and is sometimes known as *intrageneric hybridization*. All the disease, pest and drought resistant varieties in wheat, tomato, sugarcane, etc., have been evolved by this method.

4. **Intergeneric hybridization.** The crosses are made between the plants belonging to two different genera. Raphanobrassica, Rabbage, Triticale, sugarcane-sorghum, etc., are some examples of this type of crossing. Such crosses are generally used for transferring the characters like disease, pest and drought resistance from wild genera into the cultivated plants. Such crosses are difficult to achieve and are of scientific interest.

Hybridization Procedure. The steps of the procedure of hybridization are as follows:

Selection of parents. The first step is to select the plants which are to be used as parents with all the desired important characters which are not found in good standard variety. The plants must be selected in such a manner, so that the chances of selecting desirable genotypes become more. In the selected variety also the weak and undesirable plants must be rejected, and the healthy and vigorous ones are kept as parents. All the important characters to be combined must be kept in mind and the plants chosen as parents tested repeatedly for these characters.

Selfing of parents. The second step consists of artificial self-pollination of parents, which is also known as *artificial self-pollination* and is essentially needed for eliminating undesirable characters and obtaining inbreds. The selfing of parents is done to obtain homozygosity in desired characters so that they may easily be combined together.

Emasculation. This is the third step in hybridization. It is defined as '*the removal of stamens from female parent before they burst and have shed their pollens*'. The purpose of emasculation is only to prevent self-pollinating, and therefore, it is adopted only in those crops where there is self-pollination, may be complete or up to some extent and never in unisexual crops.

Bagging, tagging and labelling. The fourth step is the bagging, tagging and labelling of males as well as females. The females are bagged to prevent natural cross-pollination and males to prevent the contamination of pollens with foreign pollens and to collect the pollens for crossing. The bagging is usually done in the evening of previous day crossing. After the emasculation the flowers are enclosed in a muslin, paper, plastic, glass, celophane, polyethylene or parchment bags so that foreign pollen may not come in contact with stigma. A tag (tagging) recording the name of the female and male parent, date and the name of emasculator (labelling) is tied to the flower. The bags are kept on the females as such till seed setting is complete while in males they are removed as soon as the crossing is over.

Crossing. The fifth step is the crossing. It can be defined as '*the artificial cross-pollination between the genetically unlike plants*'. In this step the pollens from already bagged males are collected and dusted on bagged females. The crossed flowers are properly tagged and labelled. The labelling is done either on the bag itself or on the labels specially designed for this purpose.

Harvesting hybrid seeds and raising F_1 generation. The sixth step is the collection of seeds from the crossed plants after maturity. Such seeds are maintained separately and sown in the next season to raise F_1 generation. The plants so obtained are known as hybrids and may be defined as the '*progenies of cross*'. All the plants of F_1, although heterozygous, have similar genetical constitution and look exactly alike. They may or may not show hybrid vigour.

Hybridization methods. The seventh step consists of handling of F_1 and subsequent generations by different selection methods of hybridization which are different for self- and cross-pollinated crops. The methods are as follows:

Hybridization methods in self-pollinated crops. The well known self-pollinated crops are—wheat, oat, barley, rice, cotton, potato, peas and beans, etc., Every selection method has its application, under some specific circumstances for definite advantages. They are as follows:

(*i*) *Pedigree method.* Here the individual plants are selected from the F_1 population. The plants are selected on the basis of desired characters. The number of plants to be selected depends upon the number of important characters involved, the importance of breeding problem and the facilities available. These plants are harvested and threshed separately, and sown in the separate rows in the next year to raise the F_2 generation. This process is continued up to F_3, F_4 or F_5 generations. At F_6 due to successive self-pollination, most of the lines become homozygous and fairly uniform. The plants uniform in desired characters are harvested and bulked together to constitute a variety. The whole process takes about 10 to 13 years.

This method is well suited to those crops where the characters to be combined in crosses can be easily seen and recognized. The pedigree method is the quickest one and a new variety can be selected and well tested by the time it reaches its ninth generation after cross.

(*ii*) *Bulk method.* Here the F_3 plants are not maintained separately but are bulked together to form a single F_3 population. In F_3 again the suitable plants are selected, collected and bulked together. This is done for six generations. In F_6 the desired individuals are selected and harvested separately. The produce of each plant is kept separate. The best performers are released as new varieties. It is a simple, convenient and inexpensive method. It takes much time but minimizes the labour as the plants are not paid any individual attention.

(*iii*) *Back-cross method.* In the back-cross method the desirable variety, called as *recurrent* or *recipient parent*, is crossed to an undesirable variety, called as *donor parent*, possessing a character which lacks in the good parents. F_1 plants, instead of permitting to self-pollinate as in pedigree or bulk method, are crossed with the recurrent parent and, therefore, it is known as back-cross' method. The purpose of backcrossing is to recover the genotype of recurrent parent with disease resistance. The progenies of first backcross designated as BC_1 and containing the good character from donor parent are again backcrossed to the recurrent parent. This process of selection of plants with desired characters and then backcrossing is continued for several generations usually up to BC_5 or BC_6. In each backcross generation, the plants possessing desirable characters of non-recurrent parent are primarily selected. At the end of BC_6, the selected plants are self-pollinated to make the population homozygous in combined characters. The greatest advantage has been in the production of disease-resistant varieties.

(*iv*) *Multiple or composite cross method.* It consists in crossing of several pure lines together. The selected pure-lines are first combined into crosses as A x B, C x D, E x F, G x H and so on. F_1 of these single crosses are then combined into double crosses as (A x B) x (C x D) and (E x F) x (G x H). Finally the F_1s of double crosses are crossed with each other to produce the hybrids [(A x B) x (C x D)] x [(E x F) x (G x H)]. This cross is known as multiple cross and further breeding in these hybrids is carried out according to either pedigree or bulk method. Multiple cross in self-pollinated crop is used when three or four monogenic characters scattered in three or four different varieties are to be combined into one as the resistance to all three rusts of wheat into N.P. 809 variety from different strains.

Hybridization Methods in Cross-pollinated crops. The inbreds are combined in any of the following crosses and released as improved strains.

(*i*) *Single cross method.* (A x B), this is the cross between two inbreds such as A x B or C x D. Its hybrid seed, which is the first generation product, is used for raising the crop.

(*ii*) *Three-way cross method.* (A x B) x C, this is a cross between a single cross used as female and an inbred used as male, *i.e.,* it involves three inbreds. The advantage of a three-way cross is the use of vigorous hybrid of first generation as female in order to maximize the yield of hybrid seed as well as to obtain seeds of normal kernel size.

(*iii*) *Double cross method.* (A x B) x (C x D). This is the cross between two single crosses involving four different inbreds. An inbred involved in one single cross must not be included in the another single cross of the double cross parents. A double cross is produced by alternate planting of two single cross plants in an isolated plot. Double crosses are the most widely used commercial hybrids. All the hybrid seeds of maize distributed to farmers for cultivation are double crosses. They give very high yields in small land without any increase in the cost of production.

INTRODUCTION AND ACCLIMATIZATION

'Plant introduction and acclimatization' is the easiest and most rapid method of crop improvement where the acclimatization follows the introduction and both the processes go side by side. "The process of introducing new plants into a locality or introducing the plants from their growing place to a new locality with different climate is known as *plant introduction* and their adjustment under this changed climate is termed as *acclimatization*. Both, new crops and new varieties of crops, may be introduced either in the form of seeds or cuttings. In vegetatively propagated crops the cuttings are imported and in sexually propagated crops the seeds are imported. In the latter case, the cross-pollinated crops are generally introduced since in them there is greater frequency of gene recombinations owing to the frequent cross pollinations and some of the recombinations may be more favourably adapted in the new environment. The crops may be introduced into a locality either from outside the country or from different regions within the country. Introductions within the country are very convenient but for introductions from another country there is a definite procedure. The material desired is demanded from the concerned authority or agency of the foreign country through the plant introduction organization of the country. The material is packed properly and sent by sea. Before entering the country, it is inspected at sea port by the plant protection authorities and, if found fit according to quarantine rules, is permitted to enter the country. In the country, it is handed over to the concerned workers or institutions. There it is grown under local conditions and tested for the acclimatization and the presence of desired characters. If found suitable for both, it is either utilized as such after selection or utilized in hybridization for transferring the desired characters into the local varieties.

This method of crop improvement has been proved of great importance, especially in recent years, as a source of resistance material to some crop diseases.

MUTATION BREEDING

'*Mutations are the sudden heritable variations in the plants other than those due to Mendelian segregation or recombination*'. These changes are due to rearrangement of genes (*gene mutation* or *point mutation*), changes in chromosome size and structure (*chromosomal mutation*), changes in the chromosome number (*polyploidy*) and changes in the body of the plant (*somatic mutation*). They cover up all types of hereditary changes or changes in genotypes of plants and give rise to inexhaustible variations which make the basis of selection and production of new crop varieties in the plant breeding. Mutations may occur spontaneously in nature or may be produced artificially in crop plants with the help of mutagens such as atomic radiations of various types, sudden heat shocks and chemicals. They are X-rays, ultraviolet rays, neutrons and radiations from radioactive isotopes, extreme temperatures and colchicine and other chemicals. With their help the plant breeder can change the genotype and phenotype of the plants according to the needs for production of new superior varieties is known as *mutation breeding*. This is the latest method of crop improvement and has been responsible for increased rust resistance and baking quality in wheat, higher yield and powdery mildew resistance in barley, superior fibre content and increased yield in cotton, and increased yield of seed and oil in white mustard. In vegetatively propagated plants, selection from somatic mutations has directed to improvement, as in chrysanthemums and dahlias, crop plants like sugar-cane, potato and fruit crops.

IMPROVED VARIETIES OF FIELD CROPS

CEREALS

Wheat. N.P. 710; N.P. 718; N.P. 799; N.P. 809; N.P. 823; N.P. 836; N.P. 862 (N.P. 709 & 800 series of I.A.R.I); C. 591 of Punjab; R.S. 31—1 of Rajasthan; *Hyb.* 65 of M.P.; and *Kenphad* of Mumbai; *Lerma Rojo*; *Sonora* 64; *Sonalika* (S. 308) *Safed Larma* and *Sharbati Sonora*; R.S. 31-1, Durgapura-65, C—306, C. 281, Kharchia-65, A. 9-39-1, Kalyana Sona, Lal Bahadur (Dwarf trigenic variety), Raj. 911 for Rajasthan; N.P. 404, Narmada 4, *Hyb.* 65; N.P. 839, *Narmada* 112, *Mukta Kalyan Sona* (H.D.M. 1593), *Sonalika* (H.D.M. 1553) and *Raj* 911. for Madhya Pradesh.

Rice. *Ambemohar*-157, *Andrewsail*, *Basmati*-307, B.R. 41, B.R. 46. Co. 4, Co. 25, Co. 26, FR. 13-A., GEB. 24, Hr. 19, Hr. 35, *Keresail Mtu.* 1, *Patnai* S. 1092, SR. 26 B., T. 3, T.T. 1421; 2, T.E.B. 1, Taichung Native—1. I.R. 8 and *Tainan* 3; *Jaya*, *Sona*, *Ratna*, I.E.T. 1983 and *Culture* (*Chambal*) 8054 for Rajasthan.

Maize. T. 41. *Jaunpuri* U.S. 13, *Dixie* 11, *Dixie* 22, *Dixie* 33 and N.C. 27; *Ganga* 101, *Ganga* 1, *Ranjit*, *Deccan*, *Ganga Safed* 2, *Ganga* 3, *Hi-starch*, *Himalayan*-123 and VL. 54; *Composite varieties—Amber*, *Jawahar*, *Kissan*, *Sona*, *Vijay* and *Vikram*; *Meerut yellow*, *Meerut* 3, K. 19 and K. 41 for U.P.; *Pusa* 5 and *Jonpur* for Bihar; *Udaipur Selected* and *Bassi Selected* for Rajasthan; *Punjab Hy*. No. 1; U.S. 13 and III. 1656 for Punjab; N.C. 27, *Tex*. 26 and *Dixie* 18; *Ganga White*-2, *Ganga White*-4, *Ganga Yellow*-5, *Composite maize*, *Maharana Composite Maize* for Rajasthan.

Barley. R.S. 17. R.S. 6, R.D.B. 1, R.D. 31, T. 4, T. 5. N.P. 13 for Rajasthan and Madhya Pradesh; K. 12, K. 251 and 300 A for U.P.; BR. 22, BR. 24, BR. 27, BR. 30 and N.P. 21 for Bihar; N.P. 21 for Orissa and W.B.; other varieties: N.P. 21, T. 20, K. 74, K. 84, K. 85, K. 86, K. 285 and K. 259.

MILLETS

Jowar. T. 4, T. 22, T. 8-B, *Early-Hegari*, M. 35—1, M. 47—3, Co. 20, K. 2, PJ. 4-K, PJ. 8-K, PJ. 24-K, G. 2 and N.D. 15; CSH-1, C.S.H.-2; S. 405; C.S.H. 5, 604, *Java* for Rajasthan.

Bajra. Co. 1, Co. 2, Co. 3, Co. 4, Co. 5, PT. 17, PT. 248, PT. 728, X-1 & X-2 of Madras; AKP. 1 & AKP. 2 of A.P.; RSJ and RSK of Rajasthan; AY1/3, T. 55, T. 5 and S. 305 of Punjab; B. 207 and B. 28—15 of Gujarat; *Gwalior* 2 of M.P.; and *Improved Ghana* and *Pusa Moti* of I.A.R.I.; H.B. 1 (PAU Ludhiana); *New Hyb*. Bajra-5, P.H.B. 14, *New Hyb*. *Bajra*-3, T. 5 and P.H.B. 10 for Rajasthan; AF 3, B-207, 32 C and 14. D for Maharashtra; T. 55, A. 1/3, S. 350 & S. 530 for Punjab; *Local Selections* for U.P.: AKP. 2 for A.P.; *Local Selections* and *Transval Sajja* for Mysore; T. 55 for Delhi.

SUGARCANE

Sugarcane. Co. 312, Co. 313, Co. 419, Co. 421, Co. 449, Co. 453, Co. 513, Co. 527, Co. 617, Co. 630, Co. 644, Co. 659, Co. 686, Co. 785, Co. 975, Co. 1158, Co. S. 109, Co. S. 443, Co. S. 510, Co. L. 9, B.O. 24, Co. K. 30, and POJ. 2878. Co. 351 and Co. 357; Co. L. 19, Co. 1253.

LEGUMES

Pulses. N.P. 15, N.P.C. 15, T. 21, and S. 103 in *arhar*; N.P. 25, N.P. 58, R.S. 10, R.S. 11, B.R. 65, Pb. 7 and Co. 1 in *gram*; BR. 2, *Bonneville* N.P. 29, and *sylvia* in *pea*; T. 36, *Pusa hybrid* 1 and WB. 81 in lentil; N.P. 28, Pb. 54, R.S. 4, R.S. 37, *Jawahar* 45, and *Pusa Baisakhi* in *mung*; N.P. 4, N.P. 14, T. 9. BR. 68, *Krishna* and Pb. 48 in *urd*; Acme in *soyabean*; and *Do-fasli* in *Cowpea*.

Gram. T—87, Type—1 for U.P., T—7, G—24 for Punjab; A—18, T-192, T-10 and EP—28 for M.P.; R.S.—10 for Rajasthan; *Chafa* (Strain No. 816) for Maharashtra; B.R. 17, R.P. 65 and B.R. 77 for Bihar.

FATTY OILS AND FATS

Groundnut. AK. 10, AH. 25, HG. 1, HG. 7, HG. 9, *Big Japan*, *Faizapur* 1—5, *Improved Spanish*, *Karad* 4—11; *Kopergaon* 1 and 3, *Pondicherry* 8, R.S., 1, *Spanish Improved*, T. 25, T. 32, T. 100, T.M.V. 1, T.M.V. 2, T.M.V. 3, T.M.V. 4, and *Asiriya mutwanda*; R.S.B. 87, A.K. 12—24 for Rajasthan.

Sunflower. E.C. 68414 for Rajasthan.

Mustard. M. 3, M. 18, M. 27, and B.S.G. of brown *sarson*; Y.S. 10, Y.S. 151, and Y.S. BBI. IB of yellow *sarson*, BR. 23, BR 32, BR. 36, Sel. A and B. 54 of *toria*; R.B. 13, B.R. 40, *Faridkot selection*, RT. 1, T. 16, *Laha* 101 and B. 85 of *rai*; *Raya Durgamani* and *Raya* R.L. 18 of *Sarson* for Rajasthan.

Sesame. M. 3—2 and M. 3—3 for Bihar; No. 41 and 128 for M.P.; TMV. 1, 2 and 3 for Madras; No. 8, 85 and 128 for Maharashtra; TS. 12—24 for Punjab; No. 10, T. 4 and 12 for U.P.; and B. 14 for West Bengal; *Pratap* for Rajasthan T.M.V. 1, T.M.V. 2 and T.M.V. 3 for Tamil Nadu; No. 10 for U.P.; N.P. 3, N.P. 7, N.P. 29 for Delhi; T. 5 for Punjab: *Improved Berhampore* No. 9, *Improved Berhampore* No. 4 for West Bengal.

Soya bean. *Palmetto*, *Monetta*, *Clemson*, *Creole* and *Charlee*; *Punjab* No. 1, *Type* 33, N—49 and *Burette*.

Coconut. *Tall* and *Dwarf*.

Castor. Co. 1, *Castar* 20, EB. 9, EB. 16, EB. 31, HC. 1 and 6, MC. 1, *Mauthner's dwarf*, N. 20, *Rosy*, *Tarai* 4, *T*. 3, TMV. 1 and TMV. 2, and TMV. 3.

Linseed. B. 37, B. 67, BR. 1, BR. 12, BR. 24, BR. 40, K. 2. No. 3, No. 55, N.P. 5, N.P. 11, N.P. 45, N.P. 262, N.P. 439, N.P. 440, *Sholapur* 36, T. 1, and T. 26.

FIBRES

Cotton. *Virnar*, *Jarila*, *Pratap*, *Maljari*, *Bhog*, *Daulat*, *Molvi-9*, *Cocanadas* 2, *Gaonrani* 6, K. 2, K. 5, 231 R, R. 18, H. 420 and C. 520 in *Gossypium arboreum*; *Kalyan*, *Vijay*, *Digvijay*, *Suyog*, *Vijalpa*, *Wagod* MCU. 8. *Jayawant*, *Jayadhar*, S. 69 and H. 1 in *G. herbaceum*; Co. 4, 2, *Indore—2*, *Laxmi*, L.S.S., *Buri* 0394, 320 F and H 14 in *G. hirsutum* and *Andrews in G. barbadense*; 320 F, R.S. 89, G—1 (Ganganagar—1) 134 C.O. 2M, C. Indore—1, Virnar 197—3, Digvijai, Hyb—4, and M.C.U.—5 (E.L. 812—3E) for Rajasthan.

Jute. The improved varieties of *Corchorus capsularis* are—*Fanduk*, *Dhaleshwari*, *Fuleswari*, *Desihat*, *Kakya*, *Bombay*, D. 154 and R. 85, and the principal *C. olitorius* varieties are—*Desi*, *Tosah*, R. 26 and *Chinsurah Green* (C.G). The other promising varieties are—C. 13, C. 206, C 212, C. 321, C. 412, C. 918, C. 5854, C. 6382, C. 8429, C. 9527 (JRC series) and EC. 4142 of *C. capsularis*; and O. 620, O. 632 and O. 753 (JRO series, of *C. olitorius*).

Hemp. N.P. 3, N.P. 6, A. 10, A. 14, A. 17, A. 43 and A. 77 of *Hibiscus cannabinus*; N.P. sab. 5 of *H. sabdariffa*.

UNDERGROUND VEGETABLES

Potato. *Phulwa*, *Darjeeling Red Round*, *Up-to-date*, *Kufri Kuber*, *Kufri Kundan*, *Kufri Safed*, *Kufri Red* and *K. Chandramukhi*; *Kufri Alankar*, *Kufri Chamatkar*, *Kufri Shitman* and *Kufri Sinduri* for Rajasthan; *Kufri Lavkar* for Deccan Plateau; *Kufri Jyoti* for hills and plains of N. India; *Kufri Muthu* for Nilgiris & S. India. G. 2524 for Kumaon hills; SLB/Z—405 for hills and plains for N. India; *Kufri Deva* for Tarai regions of Western U.P.

Sweet Potato. *Pusa Sunehari*, *Pusa saffaid*, *Pusa Lal*, F. 17 *Ranger*, FB. 4004 and TST (*Tie Shin Tun*).

Colocasia (Arvi). *Bangali Banda*, *Desi Banda*, *Faizabadi*.

FUMITORY

Tobacco. T. 12, T. 49, T. 238, C. 302, N.P. 214, N.P. 219, N.P. 220, N.P. 222, N.P. 18 for hookah and snuffing; N.P. 28, N.P. 35, N.P. 63, N.P. 70, P. 4, D.P. 401 and S. 57 for chewing; *Keliu* 20, *Keliu* 49, *Surati* 20, *Delcrest*, *Virginia Gold*, *Chatham*, *Harrison*, *Special*, *Puteha*, *Amarelo* 5 for cigarettes; *Lanka*-27 (D.R.I.) for cheroot; and *Dixie shade* for wrapper.

SPICES AND CONDIMENTS

Cardamom. *Malabar*, *Mysore*, No. 71 and No. 81.

Chillies. N.P. 34, N.P. 41, N.P. 46, N.P. 51, *Hybrid* 5—1—5, *Hybrid* 17—1—1, R. 449, G. 1, G. 2, and G. 1402

Pepper. *Kaluvalli*, *Balankota*, *Cheriakodi*, *Uthirankota*, *Morata* and *Mundi*.

Turmeric. *Lokhandi*, *Soni*, *Chinna*, *Mudga*; 324, 225 and 326 of A.P.

VEGETABLES

Cabbage. *Pride of India*, *Pusa Drum Head*.

Cauliflower. *Pusa Katki*, *Snow Ball*.

Knolkhol. *Early White*, and *Purple Vienna*.

Radish. S. No. 8, and *Rashmi*.

Turnip. *Pusa Kanchan*.

Carrot. *Pusa Kesar*.

Onion. *Pusa Red*. N. 404.

Pea. N.P. 29, *Bonneville*, *Sylvia* and *Asauji*.

Guar. *Pusa Naubahar*, *Pusa Sadabahar*, *Pusa Mausmi*, *Durgapura Safed*, *Durgapura Chikni*, B. 19—1—55.

Cowpea. *Pusa Barsati*, *Pusa Phalguni* and *Dofasli*.

Tomato. *Pusa Ruby*, *Pusa Red Plum* and S. No. 120.

Brinjal. *Pusa Purple Long*, *Pusa Purple Red* and *Black Beauty*.

Bhindi. *Pusa Sawani*, *Pusa Makhmali*, *Lucknow Dwarf*, *Punjab 13*, *Hissar Selected*.

Chillies. *Chinese Giant* and *World Beater*.

Spinach. *All Green*; *Palak* 51—16, *Pusa all green*, *Pusa Jyoti*, *Raj Jobner*.

Fenugreek. *Pusa Early Bunching*; R.S. 1 for Rajasthan.

Bottlegourd. *Pusa Summer Prolific*.

Tinda. *Bikaneri*, *Hissar Selection*.

Luffa. *Pusa Chitni*, *Pusa Nasdar*, *Satputia*, *Kalyanpur Ghia Tora*i 5235.

Squash. *Early Yellow Prolific*.

Watermelon. *Asahi Yamato*, *New Hampshire Midget*, *Sugar Baby*, *Durgapura Mitha* and *Durgapura* for Rajasthan.

Green bean. *Pusa Baisakhi*.

French bean. *Contendor*.

Sitaphal. *Large red*, *Large round*, *Yellow flesh* and *Red flesh*.

FRUITS

Mango. *Langra*, *Lucknow*, *Safeda*, *Dasehari* and *Fazli* of U.P.; *Mumbai*, *Zardalu*, *Sipia* of Bihar; *Alphonso* and *Pairi* of Mumbai; *Fernandin* and *Mankurad* of Goa; *Ambalvi* of A.P.; *Neelam* and *Totapuri* of Madras; *Himayudin* of South India; *Suvaranrekha* and *Banganpalle* of the Circars and Rayalaseema (A.P.); *Kesar Jamadar* and *Salebhai-ni-Amdi* of Saurashtra; *Mulgoa* of Hyderabad-Deccan; *Shendriya* of Pune; *Rajapuri* of Gujarat; *Borsha* of East Khandesh; *Dalimbya* of S. Gujarat; Big-sized *Hathi-Jhool* of U.P. and *Jehangir* of Madras.

Pomegranate (Anar). *Kandhari*, *Dholka*, *Paper Shell*, *Spanish Ruby* and *Muskat red*.

Grape. *Black Prince*, *Bedana*, *Kandhari*, *Dakh* and *Foster's Seedling* for Northern plains; *Bangalore blue*, *Pachadraksha* and *Anab-e-Shahi* for Southern India; *Thomson seedless*, *Pusa Seedless*, *Sultana* and *Kishmish white* for dry and temperate regions.

Lemon. *Jamburi*.

Mandarin (santra). *Nagpur Santra*, *Khasi Santra*, *Coorg Santra*.

Sweet orange. *Mosambi*, *Blood Red*, *Jaffa*, *Suthgudi*.

Banana. *Rajeli*, *Velchi*, *Basrai*, *Rajapuri* of Mumbai, *Poovan* of Madras; *Champa* and *Mortaman* of Assam, *Amrit sagar* of Bengal; *Harichal*, *Singapur* and *Kothia* of Bihar; *Nendran*; *Rastali*, *Sirumalai*, *Kadali* and *Pacha Nadan* of South India; *Murtban* and *Pankela* of North India.

Date palm. *Hillawi* and *Khaudrawi*.

Fig. *Poona Fig.*

Guava. *Lucknow-49*, *Allahabad Safeda* and *Seedless*.

Litchi. *Bedana*, *Gulabi*, *Kalkatia* and *Muzaffarpur*.

Loquat. *Improved Golden Yellow*, *Large Agra*, *Tanaka*.

Papaya. *Honey Dew* and Co. 1.

Pear. *Baghu Ghosa* and *Nashpati*.

Pineapple. *Giant Kew* and *Mauritius*, *Queen*; *Jaldhup* and *Lakhat* of Assam; *Simhachalam* of A.P.; *Charlotte Rothschild* for S. India.

Plum. *Hale* and *Alu Bokhara*.

Sapota. *Kalipatti* and *Pot*.

22

Food Adulteration

There are three basic necessities of man, *i.e.*, (1) food, (2) shelter and (3) clothing. Considering the importance of food as one of the three basic needs, the food which we take in should be healthy and nutritious one. Further it should contain all the required components. But it was true only in ancient times when there was plenty of healthy nutrition. In those times even poor people could afford healthy and nutritious food. But now the situation has been changed, and there is a problem to get healthy and nutritious food. This seems to be due to exploding population. Due to increase in population, the supply of common foodstuff has fallen short. To keep pace with the increasing needs of food, the production has to be increased, but it is not always possible. So some traders resort to food adulteration without considering its hazardous effects. Although it is a crime but lack of public awareness has allowed it to continue. Due to scarcity of food the price index has also gone up. This has further made, well balanced food, out of reach of common people.

Survey of Food Adulteration in Uttar Pradesh. Survey for this problem has been made in U.P. during recent years. For the purpose, the samples of different foodstuff were taken from different parts of U.P. The samples were analysed at the Government Public Laboratory, Lucknow. The results found were rather disappointing and disgusting.

About 4,09,200 samples were taken for testing but amazingly 1,03,044 samples were found adulterated in some way or other. So, approximately 25% adulteration was there.

Now the Central Food and Technological Research Institute, Mysore (CFTRI, Mysore) is engaged in giving information about the food adulterations.

Danger Signal to Health. Thus we see that the food articles on an average have 25% adulteration which we consume daily in one form or other. These adulterations are responsible for many diseases. When these adulterated foodstuffs reach in our body, they affect the metabolism. The chemicals used for checking the skimming-curding of milk in summer, are responsible for many diseases of colon and deudenum.

Pigeonpea adulterated with *Lathyrus sativus* when taken causes **Lathyrism**, which is paralytic disease. Both human beings and livestock are susceptable to lathyrism and severe epidemics have been reported in the past from Bihar, M.P. and some eastern districts of U.P. Lathyrism is characterized by partial or complete paralysis of the lower limbs and degenerative changes in spinal cord. The onset of the disease is sudden. In the early stages the patient finds difficulty in walking, the gait becomes jerky and the legs may have to be crossed while walking. High and calf muscles become rigid and lower limbs are paralysed. The toxic principle has been identified as anicnopopionitrite ($C_8H_{13}O_3N_3$) which is present in concentration varying from 58 to 160 mg/100 gm and selanium which interferes with methionine metabolism.

1. *Cotton Seed Oil*—Cotton seed oil is detected when present in oil mixture with other oils by the *Halphen cotton test*. The test is carried out as follows:

(*i*) 1—3 c.c. of the oil is dissolved in an equal volume of aryl alcohol.

(*ii*) 1—3 c.c. of 1% solution of flowers of sulphur in carbon disulphide is added and the mixture is heated for 2 hours.

(*iii*) Red colour indicates the presence of cotton seed oil.

2. *Milk*—Milk is usually adulterated with water or separated milk. In summer it is usually adulterated with calcium hydroxide, sodium carbonate and sodium bicarbonate chemicals to prevent it from skimming (without cream).

The specific gravity of the milk should be detected with the help of Lactometer for detecting any adulteration. Pure milk shows the specific gravity 1.030—1.034.

3. *Ghee*—Ghee obtained from animal sources (*i.e.*, Buffalo, Cow. etc.) is usually an adulterant with vegetable ghee or other cheap oils. The adulterated ghee is detected by the following test:

(*i*) 1 spoon sugar is mixed with 10 c.c. HCl.

(*ii*) (Sugar + HCl) mixture is added to 10 c.c. of melted sample ghee.

(*iii*) Keep for 10 minutes — Red colouration develops over the surface.

4. *Brassica or Mustard Oil*—The oil is usually adulterated with the oil of seeds of *Argemone mexicana* or the chemical alayl isothiocynate which is cheaper than mustard oil. The adulteration can be detected by the following tests—

(1) HCl + Ferric chloride + oil.

(2) Brownish red precipitate develops. This shows the adulteration of oil of seeds of *Argemone mexicana*.

5. *Boora* (*Powdered Sugar Khandsari*)—This sugary white article is adulterated with sodium bicarbonate which is cheaper than *Boora*. To detect the adulteration the following test is performed:

(1) 5 gm. Boora + 10 c.c. of water → Red litmus paper is applied.

6. *Coffee*—Coffee is adulterated with powder of roasted wheat, gram, sunflower seeds, etc. The purity of coffee is tested as follows:

Coffee sample is mixed with water and allowed to settle; pure coffee powder floats over the surface of water, while adulterants become submerged and settle at the bottom.

7. *Tea leaves*—Tea leaves are adulterated with leaves of other plant sources, finely powdered and treated with yellow, pink or red non-permitted colours which are cheap enough to be mixed in fresh tea leaves or in used dried tea leaves. The purity of tea leaves is tested as follows:

These adulterants can be detected by keeping the adulterated tea leaves on wet white paper sheet. The wet paper receives the colouration (pink, yellow or red) which shows that the tea sample is adulterated.

8. *Turmeric*—This spice is used in every household of India either poor or rich and thus is the source of bigger adulteration by many adulterants.

(1) If adulterated with wheat or sorghum flour—dissolve 1 gm. of turmeric powder in 3 c.c. water and add 1 c.c. iodine solution — Blue colour is produced (starch).

(2) If adulterated with colours

1 gm. turmeric powder + 3 c.c. water + 3 c.c. HCl → colouration develops. (If no colouration is produced then the turmeric is not adulterated with colours).

(3) If adulterated with *geru*:

Burn spoonful of turmeric powder → Brown ash more than half in quantity than of the sample.

9. *Cut Areca Nuts and Powdered Chillies*—These two articles are usually adulterated with fine sieved saw dust. To observe adulteration the sample of powdered chillies or cut areca nuts are added to the water. The saw dust will float at the surface of the water while areca nut or chillies will settle down in the bottom. If they are mixed with colours also the water becomes coloured.

10. *Asafoetida* (*Hing*)—Asafoetida is mixed and adulterated with cheap resins and gums. This type of adulteration can be judged by the following process. A spoonful of *Hing* is dissolved in 1/2 glass of water, pure *Hing* changes the water into milky water.

11. *Cumin* (*Zeera*)—Cumin is adulterated by charcoal powder or some other seed powders which are easily available. *Zeera* powder is pressed keeping it on palm with the help of thumb—Tip of thumb turns black, due to the colour of adulterants.

12. *Black Pepper* (*Piper nigrum*; **Kali Mirch**)—This is adulterated with papaya (*Carica papaya*) seeds as such or powdered. Test—This sample is mixed with water; the pepper powder settles at the bottom while the papaya seeds float at the surface of the water.

13. *Crocus sativus* (*Kesar*)—*Kesar*, the stigmas of *Crocus sativus*, adulterated with the stamens and stigmas of other plants coloured with non-permitted colouring agents and coloured starch threads. The purity is tested as follows:

(*i*) Dissolve the *Kesar* into water, the pure *Kesar* dissolves rapidly in water while other substances are not dissolved easily.

(*ii*) A glass rod dipped in sulphuric acid placed over *Kesar* vapour — the glass rod first changes to blue, then to violet and then to brown — the *Kesar* is pure.

14. *Cardamom* (*Chotti Ilaichi*)—Cardamom fruits are often adulterated with orange seeds or unroasted coffee grains. The adulterants of seeds are small pebbles and seeds of *Amomum* spp., powdered seeds are adulterated with the powder of hulls.

15. *Pigeon Pea* (*Cajanus cajan—Arhar*)—*Arhar* is usually adulterated with *Khesari dal* (*Lathyrus sativus*) which is cheap and easy to cultivate. Consuming *Khesari* as the principal diet are known to be affected by a paralytic disease known as *lathyrism*. Prohibited colour methanil yellow is used to colour these pulses. If these coloured pulses are kept in water for some time, the water becomes coloured. Sometimes, lead chromate is used instead of methanil yellow.

16. *Fennel* (*Faeniculum vulgare*; **Saunf**)—*Saunf* is adulterated with exhausted or partially exhausted fruits, or with undeveloped or mold attacked fruits, or with the seeds of anise (*Pimpinella anisum*). Pure fennel passes easily through 54 mesh screen and has distinct features and flavours. Pure seeds contain 1.4% volatile oil and 1.5% acid insoluble ash.

17. *Licorice* (*Glycyrhiza glabra*; **Mulhati**)—Licorice which is used as expectorant, demulcent, laxative and for allaying coughs and in irritable conditions of mucous membrane has been found adulterated with rods of *Rati* (*Abrus precatorius*) and with *G. uralensis* by CFTRI, Mysore. Licorice is imported into India from Asia Minor, Iraq, Persia and other Central Asian countries. Pure licorice is soft, flexible and fibrous, internally of a light yellow colour with a characteristic sweet pleasant taste, while adulterants contain only a small percentage of sugar and gives a rather pungent taste.

18. *Surti and Pan Masala*—Adulterated with other leaves without *Nicotiana* leaves and prohibited colours, *i.e.*, lead chromate and methanil yellow.

19. *Khoya Sweets*—Adulterated with starch and prohibited colours. Test: Khoya + 1 c.c. Iodine solution → Blue colour (starch is adulterant).

20. *Honey*—Adulterated with sugar, saccharine and prohibited colours. The honey should be examined under the microscope for pollen and other natural crystals.

Oils adulterated with *Argemone mexicana* causes the disease epidemic dropsy which causes swelling of feet fingers and heart. Vegetable ghee industry uses nickel as a catalyst in hydrogenation process. When vegetable ghee is yellow in colour the nickel is in excess. The nickel in excess quantity causes respiratory diseases.

21. *Perfumes*—Perfume oils such as *Cananga odorata*, *Citronella* and *Palmarasa* have been found adulterated with coconut oil. It may be detected by the fact that oils, so adulterated will solidify wholly in freezing mixture. Coconut oil melts at 68–82°F; its sp. gr. at 212°F is 0.868 to 0.874 and its saponification value from 209 to 229.

22. *Tomato Ketchup*—In the manufacture of tomato ketchup the tomato solid is adulterated by carrot, gourd, pumpkin, sweet potato, papaya and apple pulp. The adulteration can be detected by determining the lycopene content of the ketchup. Tomato ketchup contains 475 hg/gm or 7.8 mg/100 gm. of lycopene. Lycopene is the colouring matter of tomato when adulterated with other solids artificial colouring matter is added.

23. *Almond Oil*—The almond oil is adulterated with the fatty oils from peach, apricot and plum seeds which resemble it to a great extent. The adulteration is determined by examining the specific gravity of almond oil which is 1.049—1.058 at 25°C; saponification value is 208.

24. *Orange Juice*—Pressed juice from carrot is adulterated in orange juice (carrot has sp. gr. 1.3; total acids as citric acid 1.3 and sucrose 17.8%) and the adulteration can be detected by analysis.

25. *Til Oil* (*Sesamum Oil*)—*Carthamus tinctorius* seed oil which is straw coloured. Nitrogen 3.19 sp. gr.; 0.9224 at 15°C is generally added to *Til oil* as adulterant.

26. *Animal Ghee* (*Clarified Butter*)—The chief articles used in the adulteration of ghee are vegetable oils such as—coconut, groundnut, cotton seed, sunflower, poppy and *sesamum oils*. These are harmless, several cheaper but injurious oils are also used as adulterants, especially *Salvadora* oil, castor oil, *Madhuca latifolia* oil and animal fats (especially mutton and meat are largely utilized). Starches such as that of rice, bajra, potato and corn are frequently resorted to in order to thicken oily compositions. Adulterated ghee is also often remade with milk or curd to render detection difficult. The simplest method of ascertaining adulteration is to boil a given quantity of ghee; and when it is in a state of complete ebullition add cold water on it. The oil will rise to the surface.

To face the challenge of food adulteration following measures should be taken:

(1) Consumers' councils should be organised at 1000 population level, taking the co-operation of active citizens and consumers.

(2) Organization of anti-adulteration board at district level with legal powers to deal with the adulterators.

(3) Consumers' council members should have the powers of drawing the sample for analysis.

(4) Effective enforcements of prevention of food adulteration. (Rules—Ministry of Health, Government of India 1959 by the administration).

(5) More heavy punishment should be awarded to the offenders (Even capital punishment should not be spared for serious offenders).

(6) Number of analysing laboratories should be increased to facilitate effective and early analysis of samples (at present in U.P. only one lab. exists).

(7) Number of persons concerning food and drug administration should be increased in proportion to the number of shops and population.

(8) At present administration charges Rs. 35/- per sample from citizen for analysing any sample. The charges should be moderate, so that poor section of the society may also be benefitted.

(9) Consumers should be taught about the common adulterants and the simple test to detect them. If these measures are taken effectively, we can meet the challenge of food adulteration.

23

Alcoholism

Today's popular concept that "Alcoholism is a disease" inhibits new ideas on this field of preventive medicine. The concept is based only on the act which itself starts the individual down one day road to alcoholism.

However, if we consider that alcoholism is a self-infected injury, this is a type of diagnosis that can bring us face to face with reality in considering alcoholism. Scholars in the medical, social and religious profession could then meet with victims of alcohol, their relatives and other concerned persons to draw up workable programmes to cope with this seemingly insurmountable problem.

Let us first analyse the act of taking that first drink. Now one could dispute the statement that one takes that drink of his own volition.

Effect on Body. Ethyl alcohol is consumed in the various alcoholic drinks such as—Gin, Whisky, Brandy, Rum and various other indigenous cheap liquors which contain 35 to 50% of alcohol. Bear has 3.6% alcohol.

When ingested it is observed unaltered from the gastrointestinal tract about 80% from the intestines and the remainder from stomach. Food slows absorption especially the fatty foods and milks. Once absorbed alcohol is distributed throughout all organs and fluid of body fairly uniformly except in brain and spinal fluid where concentration rises and falls slowly. Its presence in blood may be detected within 5 minutes after ingestion and the maximum concentration is reached in half to two hours. Elimination of alcohol is accomplished chiefly by oxidation less than 2% being lost through lungs, skin and kidney. The energy liberated by the oxidation of alcohol (7 cals per gm.) can be utilized completely but it is important to emphasize that alcohol cannot be stored in the body or used in the replacement of destroyed tissues. During the process of oxidation carried out mainly in liver alcohol is utilized first to acetaldehyde then the acetic acid and finally to CO_2 and H_2O. There seems to be a definite relationship between the oxidation of alcohol and oxidation of glucose, so that insulin and glucose when given to patient who has ingested alcohol, helps in rapid metabolism of the latter in the body. Thiamine and Nicatinic acid may also be necessary for the rapid oxidation of alcohol. On the other hand, the rate of oxidation of acetaldehyde depends on its concentration in the tissues. Definite tolerance to alcohol usually develops on repeated areas. This may be due to the decreased absorption, increased oxidation, increased cell tolerance and increased excretion.

Alcohol affects all tissues of body. It is now accepted that alcohol is not a stimulant of central nervous system but a depressant. The early effects seemingly of stimulation are due to the depression of central inhibition. It impairs the efficiency of mental function, faculty of judgement, discrimination association. All mental performances are slowed down and made inaccurate and random in character. The pulse rate usually increases and vasodilation occurs especially the cutaneous vessels with resultant heat loss and fall in body temperate. The effect on the alimentary tract is usually increased, salivary secretion and increased gastric secretion of high acidity and low pepsin with higher concentration, however, gastric secretion may be inhibited and long continued use results in gastritis. Alcohol acts as diuretic due to transient suppression of anti-diuretic hormone. The sex desire is increased because of abolition of inhibition but the sexual potency is usually depressed. The prolonged use of alcohol increases susceptibility of infection, semi-starvation and hypo or avitaminosis.

Alcoholism is the term applied to the clinical abnormalities that result from the excessive consumption of alcohol as well as for the state of addiction to alcohol. The various manifestations shall be discussed under 1. Acute Alcoholism 2. Chronic alcoholism. 3. Dipsomania.

1. **Acute Alcoholism**—A person is said to suffer from acute alcoholism when as a result of alcohol he is unable to function with safety to himself or to others.

Etiology (A science dealing with the causation of disease)—A scale related to the various degrees of chemical intoxication to the blood alcohol level has been constructed. It is believed that at blood concentration of 0.2% mild or moderate intoxication occurs, concentration of 0.3% or more indicates marked intoxication and fatal concentration ranges between 0.5 to 0.8%. The effect of alcohol, however, varies in different individuals and depends not only on the amount taken but also on the physical state and mental stability of the person. Effect of alcohol is enhanced by taking some other substances simultaneously, *e.g.,* tobacco, hushish, acids of badly prepared wines, states of debility, exhaustion or hunger also make alcohol more dangerous.

Pathology—The nervous system and digestive system are most markedly affected. In death from acute alcoholism there is evidence of mucasal catarrh of stomach; evidence of right heart failure, ıncreased cerebrospinal fluid and oedema of the cortex.

Clinical Features—A characteristic progression of symptoms is seen during the process of acute alcoholic intoxication. To start with the person experiences a general feeling of well being, mild relaxation and euphoria with loud and boisterous speech. There is, however, tumultuousness and increasing clumsiness of movement. As the patient gets really intoxicated, he is *verbose*, *staggering*, clumsy, *dysarthric*, *elated* or sad and may feel inordinate sense of power. He may become angry, argumentative and may start *shouting* or weeping. The pupils are *dilated*, pulse is dull and rapid and the face is flushed. *Nausea* and vomiting may occur. Increasing intoxication leads to *stupors* with heavy breathing and latter as the *coma* becomes dangerously deep. The pupils contract, the pulse weakens and respiration becomes shallow, circulatory *collapse* may occur and death may result rapidly or after hours of deep coma. In some cases instead of coma automatic behaviour may occur and upon recovery there is *amnesia* of the entire events. During coma pneumonia may result either from exposure or from aspiration of the food material.

After acute alcoholic episode the patient feels severe headache, *lessitude* malaise, nausea and vomiting; the next day he may show clinical picture of mild *melancholia* with *gloomy* thoughts, poor *volition* and excessive irritability.

Diagnosis—A history of alcohol consumption, odour of alcohol with determination of blood level of alcohol facilitate the diagnosis. It is, however, often difficult for an exact diagnosis to be made until the patient has been kept under observations. Other causes of coma should be excluded as they may be superimposed upon a case who has taken alcoholic drink.

2. **Chronic Alcoholism** (lasting for long time)—Chronic alcoholic is a person who habitually relies on alcohol and cannot carry on his ordinary life without alcohol.

Etiology—Chronic alcoholism is a mental disorder. The alcoholic almost invariably is seeking escape from worries and anxiety or from unconscious *conflicts* within his own personality which he does not recognize. He is usually insecure, self-centered, *narcissistic*, immature personality driven by unrecognised sexual conflicts.

Clinical Features—A chronic alcoholic is typically a *plethoric* person with plum coloured complexion, *termlousness* of lips and hands, restless, personality. His eyes are described as glassy, watery and the conjunctiva are red.

The most common symptoms of chronic alcoholic are gastrointestinal. Nausea and vomiting occur in arising and may be spontaneous or *provoked* by food or cigarettes. Abdominal distension, epigastric distress, belching or even peptic ulcer symptoms are complained of. The incidence of peplic ulceration of stomach or duedenum is exceptionally high in alcoholic population. *Laceration* of cardiac oesophageal junction may cause profound or fatal *haemorrhage* (Malloory-Weiss syndrome). The excessive ingestion of alcohol with concomitant semi-starvation may resu.t in fatty liver. The incidence of *cirrhosis* of liver among alcoholics is quite high and they suffer from attacks of hepato-cellular failure with ascites (free fluid in peritoneal cavity) and haemetemesis. The relative importance of alcohol or deficient diet in the etiology of alcoholic cirrhosis is still disputed but it is likely that the two factors act together. *Pancreatitis* is complication of chronic alcoholism and acute haemorrhagic necrosis of pancreas may follow the ingestion of large amount of alcohol.

The neuro-psychological manifestations of chronic alcoholism are not only of clinical importance but make the patient a problem of society. The various symptoms may be precipitated by a large drink or sudden *abstruence* from alcohol. *Concomitant* nutritional deficiency may result in the neurological manifestations. A chronic drunkard suffers from dementing demoralisation, crude or brutality of conduct, superficiality of thoughts, liability of mood and may develop delusion especially of jealously. The various clinical syndromes can be described as follows—

(*a*) *Alcoholic Epilepsy*—It is observed that the frequency of attacks of ideopathic epilepsy tends to increase in heavy drunkards.

(*b*) *Alcoholic hallucinations*—This is an alcoholic psychosis characterized by vivid auditory hallucinations with an otherwise clear sensorium, coupled with this is commonly a paranoid reaction of suspicion, distrust and ideas of persecution. This disorder may be momentary or may last for days, weeks or months.

(*c*) *Delirium tremens*—This is an acute, serious and sometimes fatal syndrome during periods of withdrawal or after particularly heavy bouts of drinking. It is characterized by disorientation of time and place and delusions of horror. The patient has muscular tremors, restlessness, loss of memory and has a tendency to commit suicide, homicide or violent assault. He may imagine insects crawling under the skin, there may be fever, furred tongue and signs of taxaemia. He may die in a few days.

(*d*) *Neuropathy*—Association of malnutrition with vitamin deficiencies especially B, may result in parenthesis in hands and feet, pain and tenderness in muscles or motor manifestation in the form of paralysis involving the feet, legs and arms. When there is associated sluggish papillary reflex it is referred as alcoholic *pseudotabes* and where mental symptoms predominate it may be called "*alcoholic psudoparesis*". Paraplegic symptoms and wenernicks encephalopathy may occur. In some cases there may be clinical manifestation of cerebellar or corpus callosum degeneration, the latter being termed as "*Marchiafava Bigname disease*".

Chronic alcoholics have enhanced degenerative tendencies such as atheroma contract or osteoarthritis.

(3) **Dipsomania**—It is a form of compulsive neurosis in which the patient is seized with an irresistible desire to take large quantities of alcohol. This is period and the frequency varies in different individuals and in the same individual under the different stresses. Usually the patient takes no alcohol in between the attacks. The patient needs a psychiatrist care for the treatment.

Conclusion—A person comes under the influence of alcohol, his body and brain changes, the degree of change increases in a proportion to the damage done to the organs of body by the alcohol, which varies in direct proportion to the amount consumed. Time has shown that this degree is determinative not only to drinkers but also to those closely related to them. It also affects millions of other people, mentally, spiritually and financially.

Drinking is the nucleus of many of our social evils. There is a direct connection between drinking and mental health. Alcohol plays an important role in illegitimacy, divorce, murder, crime of all kinds. The percentage of alcoholics in tuberculosis hospitals is increasing significantly. There is a direct connection between heart diseases and the consumption of alcohol.

24

Forests and Action Plans

Forests are Nature's most bountiful and versatile renewable resources, providing simultaneously a wide range of economic, social, environmental and culture benefits and services. The world wide demands for their numerous functions and outputs increasing with the expanding population, while the global forest resource is shrinking either as a result of over harvesting, deforestation and permanent conversion of other forms of land use, in many tropical regions, or as a consequence of forest decline associated with air borne pollutants in temperate regions.

The forests of India occupy an area of about 75 million hectares, *i.e.,* 22 per cent of the land area with a great variety of species. During the British Period, these forests were treated as a source of revenue and to supply the demands of people for timber, fuelwood, and minor forest products such as cane, resins and lac. Nearly a 100 million poor people including tribals live in and around forests. The forests supply, almost free of cost their food requirements water, fodder for livestock, firewood, timber of agricultural implement and house construction as well as medicinal plants. In addition these people make their living by collecting and selling minor forest products. They have a symbiotic relationship with the forest since ages. Like indigenous people all over the world, the tribal communities that live in forests respect the trees, and the birds and animals that give them sustenance. Strategies are being developed for a greater degree of people's participation in forest management. The role of forest communities in restoring and conserving forests is now being increasingly recognized. The challenge is to integrate modern knowledge and skills in the area of forestry with the traditional knowledge and experience of the local communities, and to evolve more effective strategies for the joint management of forests.

Nature has endowed India with a large variety of plants, about 45,000 species in number, of which 15,000 are flowering plants, 1,676 algae, 1,940 lichens, 12,480 fungi, 64 gymnosperms, 2,843 bryophytes and 1,012 pteridophytes. About 5,000 species are endemic occurring only in India. Also 28 species are now extinct, 399 are endangered and 1500 species are threatened.

India can be divided into eight floristic regions, namely, (1) Western Himalayas, (2) Eastern Himalayas, (3) Assam, (4) Indus Plain, (5) Ganaga Plain, (6) Deccan, (7) Malabar and (8) Andamans.

The temperate zone of Western Himalayan region is rich in forests of chir pine (*Pinus roxburghii*), other conifers and broad leaved temperate trees. Higher up, forests of deodar (*Cedrus deodara*), blue pine (*Pinus wallichiana*), spruce (*Picea smithiana*) and silver fir occur. The alpine zone extends from the upper limit of the temperate zone to about 4,750 metres or even higher. The characteristic trees of this zone are the higher level silver fir, silver birch and junipers which are characteristic of this high altitude area.

The temperate zone of the Eastern Himalayan region has forests of Oak (*Quercus spp.*) laurel (*Calophyllum spp.*), maple (*Acer spp.*) rhododendron, alder (*Alnus spp.*) and birch (*Betula spp.*), many conifers, junipers and dwarf willows.

The Assam region comprises evergreen forests, occasional thick clumps of bamboo and tall grasses.

Some small areas of the Ganga Plain support forests of widely different types.

The Deccan region comprises the entire tableland of the Indian Peninsula and supports vegetations of various kinds from scrub jungles to mixed deciduous forests.

The Malabar region besides being rich in forest vegetation produces important commercial crops, such as coconut (*Cocos nucifera*), betelnut (*Areca catechu*), pepper (*Piper nigrum*), coffee (*Coffe arabica*) and tea (*Camellia sinensis*). Rubber (*Hevea brasiliensis*), cashewnut (*Anacardium occidentale*) and eucalyptus trees have also been successfully grown in some parts of this region.

The Andaman region abounds in evergreen, semi-evergreen, mangrove, beach and diluvial forests.

In India, forest resources are an important factor of socio–economic development, more specially of rural development, and it is the responsibility of our generation to safeguard these for future generations. The importance of the renewable goods and services provided by trees and forests, including the positive role they play in water and carbon cycles, soil protection, and the conservation of biodiversity, is well understood, yet forests face an onslaught due to the growing demand for construction materials, fuel, food, fodder, recreation areas and the like. Techniques are available for the sustainable management of trees and forests which can ensure the permanence, and even lead to increasing their ability to provide goods, and services, though these are too often ignored. Long term planning for the management of forest resources, however, can help avoid irreversible damage to the biosphere.

The real challenge is to reconcile the economic use of natural resources with the protection of the environment, through integrated and sustainable development. Any solution of forest problem requires common efforts to reduce poverty, to increase agricultural productivity, to guarantee food security and energy supplies, and to promote development.

Thus, sustainable development is an integrated approach to all round development, keeping mind its effect on the environment. What needs to be done is to increase the productivity of the "**producer systems**" and the efficiency of the "**user systems**".

GENETIC DIVERSITY OF ECONOMIC PLANTS

India is endowed with a rich genetic variety of cereals and pulses. The cereal basket includes rich (*Oryza sativa*), wheat (*Triticum aestivum*), barley (*Hordeum vulgare*), oat (*Avena sativa*), rye (*Secale cereale*) and maize (*Zea mays*); the millets include jower (*Sorghum vulgare*), bajra (*Pennisetum typhoides*), ragi (*Elucine coracana*), fox-tail millet (*Setaria italica*), *etc*., and also pseudocereals like amaranth (*Amaranthus* spp.), Buck wheat or Kutu (*Fagophyrum esculentuam*), *etc*. The pulses include various types of gram such as green gram (moong), black gram (urad) and gram (chana), and different types of pea (matar). It has also a wealth of oilseeds both edible and non-edible.

India's fruit basket is fabulously rich and diverse. It consists of a variety of mangoes (*Mangifera indica*), bananas (*Musa paradiciaca*) and various types of citrus fruits like mandarin (*Citrus reticulata*), sweet orange (*Citrus sinensis*) and lime (*Citrus aurantiefolia*). Other fruits are papaya (*Carica papaya*), guava (*Psidium guajava*), apple (*Malus sylvestris*), grapes (*Vitis vinifera*), pinapple (*Ananas comosus*), *etc*.

The oriental spices are well known. India produces a wealth of spices such as ginger (*Zingiber officinale*), cardamom (*Elettaria cardamomum*), black pepper (*Piper nigrum*), red pepper (*Capsicum frutescens*), clove (*Syzygium aromaticum*) and nutmeg (*Myristica fragrans*).

The plant wealth of India also includes a variety of aromatic and medicinal plants. Some of these are the sandal (*Santalum album*), lemon-grass (*Cymbopogon flexuosus*), palm-rose (*Cymbopogon martini*), eucalyptus (*Eucalyptus globulus*) and keora (*Pandanus tectorius*).

ACTION PLANS

The concern for environmental conservation has long been an integral part of Indian thought and social process and is also reflected in the Constitution of India. The Government has enunciated policy statements on forestry and on abatement of pollution and formulated a National Conservation Strategy. In addition, there are laws for the protection of environment, the Wild life Protection Act, 1972, the Forest Conservation Act, 1980, and the Environment Protection, Act 1986. Significant amendments have been made in the various acts from time to time to make them more effective.

The priorities before us are conservation and sustainable utilization of natural resources, afforestation, wastelands development, control and management of industrial and related pollution.

NATIONAL FORESTRY ACTION PLAN

This programme envisages a review of the forestry sector covering critical issues affecting forestry development in the country, a perspective action programme for the long, mid and short term development of forestry at national and state levels. The focus of the National Forestry Action Plan would be the increase in the sustainable management of forest and tree resources contributing to biodiversity and conservational and climatic needs.

NATIONAL AFFORESTATION AND ECO-DEVELOPMENT BOARD (NAEB)

The programme of wasteland development through afforestation and tree planting in 1985 laid emphasis on people's participation, harnessing inputs of science and technology and achieving interdisciplinary coordination in programme planning and implementation.

The National Afforestation and Eco-development Board (NAEB), set up in the ministry of Environment and Forests, Government of India in August, 1992, is responsible for promoting afforestation, tree-planting, ecological restoration and eco-development activities in the country.

The National Wastelands Development Programme on a large scale was launched in the country in May, 1985, to deal with the ecological crisis caused by land degradation. The National Wastelands Development Board (NWDB) established in 1985 in the Ministry of Environment and Forests was the forerunner of NAEB.

The Integrated Afforestation and Eco-development Project Scheme (IAEPS) promotes an integrated approach for the development of mini-watershed by creating fuel wood and fodder plantation and treating the lands against soil and moisture losses.

The Area-Oriented Fuelwood and Fodder Projects Scheme (AOFFPS) aims at creation of plantation of fuelwood and fodder. The Minor Forest Produce including Medicinal Plant Scheme provides assistance to State Governments for increasing the production of non-timber forest produce including medicinal plants with a special focus on tribal population for whom non-timber forest produce is an important source of livelihood. There are also schemes for seed development, aerial seedling and wasteland mapping.

WILDLIFE CONSERVATION

The threat to biological diversity due to over exploitation and habitat destruction is a major challenge. Realising the fact that many species of plants and animals have become endangered, several initiatives have been taken in India for their conservation. Hunting of animals is banned throughout the country.

NATIONAL PARKS AND SANCTUARIES

Today India has a wide network of 75 National Parks, 421 Wildlife Sanctuaries, 21 Project Tiger areas and eight Biosphere Reserves covering an area of 140,675.46 sq. kms., which is approximately 4.3% of the total land area. Years of protection and scientific management have pulled out many species from the brink of extinction. Not only have individual species been saved, but also their habitats and entire ecosystems have been restored to health.

BIOSPHERE RESERVES

Biosphere Reserves would afford conservation of large ecosystems representing a number of rare and endangered species of animals that exist in these habitats.

The setting up of Biosphere Reserves for conserving the biological diversity that exists in representative ecosystems is a concept evolved under the Man and Biosphere Programme by the UNESCO. The major objectives of the Biosphere Reserve are (*i*) To conserve diversity and genetic integrity of plants, animals and micro-organisms in their totality as part of the natural ecosystems, so as to ensure their self-perpetuation and unhindered revolution of the living resources.

(ii) To promote research on ecological conservation, both within and in areas adjacent to these reserves.

(iii) To provide facilities for international cooperation and people-environment interaction in various bioclimatic and biogeographical situations of the biosphere.

(iv) To provide opportunities for people's participation in achieving these objectives.

In India 14 representative ecosystems have been identified for setting up of Biosphere Reserves of which eight have already been set up at Nilgiris, Nanda Devi, Nokrek, Great Nikobar, Gulf of Mannar, Manas, the Sunderbans and Simlipal. Others that the proposed to be set up are the Kanha, Uttarakhand, Thar Desert, Kaziranga, the little Rann of Kutch, and North Andaman.

BOTANICAL SURVEY OF INDIA

The Botanical Survey of India (BSI) was established in 1890 with the objective of surveying and identifying the plant resources of the country. The survey has its headquarters at Calcutta and nine circles located in different phytogeographical regions of the country.

The Botanical Survey of India has been serving the nation through inventorisation of the plant resources and advising on all matters related to plants. About 65% of the total area of the country has been surveyed and the herbaria of the BSI hold about three million plant specimens including about 15,000 type specimens. The research wing has flora, ecology, cryptogams, plant chemistry and pharmacognosy units.

The schemes of setting up of many Botanic Gardens and Field Centres have also been taken up during 1991-92 to augment the activities for conservation and propagation of plant genetic resources in different regions of the country through a network of these centres. Under this scheme, a one time non-recurring financial assistance is provided to botanic gardens in different phytogeographic regions of the country for strengthening their existing facilities for conservation and propagation of threatened and endangered endemic plant species of that region, and education and public awareness of endemic plant species.

The Indian Botanic Garden in Howrah is the largest Botanic Garden in the country and is used for study, introduction and conservation of flora, educational and secretarial purposes. It is a living repository of about 15,000 trees and shrubs distributed over 2,000 species. The greatest attraction of this Garden are the Great Banyan Tree, Giant Water Lilies and the large Palm House in addition to a number of interesting botanical specimens from different parts of the India and other countries.

FOREST SURVEY OF INDIA

The Forest Survey of India with its headquarters at Dehradun and four zonal offices has been set up with the following objectives:—

(*i*) To assess the extent of forest cover and monitor on two–year cycle the broad changes in forest vegetation cover of the country by using multisatellite data.

(ii) To prepare thematic maps through the use of remote sensing data with maximum essential ground with verification on a 10–year cycle.

(iii) To collect, store and retrieve necessary forestry related data for national and state level planning and to create a computer based National Basic Forestry Inventory System.

(iv) To design methodologies relating to forest surveys and subsequent updating.

(v) To support and oversee techniques/inventory work undertaken by State/Union Territory Forest departments.

INDIAN COUNCIL FOR FORESTRY RESEARCH AND EDUCATION (ICFRE)

The ICFRE in Dehredun, an autonomous body of the Ministry of Environment and Forests holds the mandate to organise, direct and manage research and education in the field of forestry. It is also responsible for framing the overall Forestry Research Policy of the country and ensuring the best method of application of all sources of scientific knowledge to the solution of problems facing the forestry sector.

CONVENTION OF BIOLOGICAL DIVERSITY

India along with 170 other countries signed the Convention on Biological Diversity during the Earth Summit held at Rio de Janeiro, Brazil in June, 1992. The Convention has come into force with effect from December 29, 1993. India has ratified the Convention in February, 1994.

The Convention through its 42 articles establishes commitments on conservation, access to genetic resources, transfer of technology, biotechnology and biosafety, benefit sharing and finance, besides establishing sovereign rights of the state on their biological resources.

25

The Sacred Plants

From the time immemorial the plants have been utilized, praised and worshipped, by human kind. Millions of Hindus for centuries have worshipped several trees, of which '*Asvattha*' (peepal tree)—*Ficus religiosa* has been supposed to be supreme. In the Tenth Chapter, Twenty-sixth *shloka* of *Shrimadbhagwadgita* Lord Krishna says: "Of all trees (I am) *Asvattha*". Further in Fifteenth Chapter—shlokas 1-3, Lord Krishna says: "They speak of the imperishable *asvattham* (peepal tree) as having its root above and branches below. Its leaves are the Vedas and he who knows this is the knower of the Vedas". Lord Krishna further says, 15, 2: " Its branches extend below and above, nourished by the modes, with sense objects for its twigs and below in the world of men stretch forth the roots resulting in action". In 15, 3: He says: "Its real form is not thus perceived here, nor its end nor beginning nor its foundation. Having cut off this firm-rooted *Asvattham* (peepal tree) with the strong sword of non attachment".

The plants are ever worshipped. This is the biggest creation of god for human beings; the gift of nature for us; static *Brahmarishi*, the ideal of donors, what he takes, takes from the mother, mother the land; mother the sole donor; sucks the nectar from the mother and gives to us.

Every morning the Sun rises, and enlightens the whole land. The darkness disappears and all becomes lighted. Besides this light, we get energy from sun. All greens absorb light and food is produced for us. This great phenomenon of nature is *photosynthesis*. Only the plants are capable of this phenomenon. The plants give us food; they give us life. The tree is eternal in its origin, and provides us the eternal peace. Beyond the kingdom of plants, there is absolute darkness. In this chapter "The Sacred Plants" about forty plants, generally considered pious by the great people of this country, dealing with cultural heritage, festivals and religious ceremonies, starting from birth till death have been dealt with.

Pipal

Ficus religiosa Linn.; Eng.—Peepul; Hindi—*Pipal*; Bengali—*Ashthwa*, *aswat*; Uriya—*Jari*; Nepali—*Pipli*; Punjabi—*Pipal*, *bhor*; Marathi—*Pimpala*; Gujarati—*Pipul*; Tamil—*Aswartham*; Telugu—*Kulla*, *ravi*; Kannada—*Rangi*, *asvalta*; Sanskrit—*Aswaththamu*, *asvattha*; Family—Moraceae.

A large tree planted especially near temples.

Sacred uses. The *Pipal* tree is held sacred by the Hindus, and viewed as the female to the Banyan tree (*F. bengalensis*). According to the *Valkhilya* the marriage of the *Pipal* with the *Tulsa* (*Ocimum sanctum*) is ordered. He further says that it is the transformation of the gods *Guru*, and is termed *Ashwath*. This tree is especially worshipped on every Saturday of the month of *Shravan*, and on every *Somvati Amavashya* (*i.e.*, on every Monday on which a new moon falls). The Hindu who plants a *pipal* tree expects so that just as he affords shade to his fellow-creatures in this world, so after death he will not be scorched by excessive heat in his journey to the kingdom of *Yama*. There are five sacred trees among the Hindus, *viz.*, *pipal*, *gular*, *bargad*, *pakar* and mango, but of these the first is held most sacred. A good Hindu (*Sanskari Hindu*) who on a journey sees a *pipal* tree will take off his shoes and walk five minutes round the tree from right to left (*pradakshana*). While doing so he repeats the verse which may be translated, "The roots are *Brahma*, the bark *Vishnu*, the branches the *Mahadeos*. In the bark lives the Ganges (*Ganga*), the leaves are the minor deities. Hail to thee, king of trees".

The *pipal* is believed to be inhabited by the sacred triad, *Brahma*, *Vishnu* and *Shiv*. It is used at the thread investiture and at the laying of the foundation of a building. Vows are made to it and it is worshipped; male offspring is entreated for under its shade. Pious women moving round its trunk 108 times. Of its wood the spoons are made with which to pour *ghi* on the sacred fire.

Bar

Ficus bengalensis Linn.; Eng.—Banyan; Hindi—*Bar, bargat*; Bengali—*Bar, but*; Uriya—*Boru*; Assamese—*Bot*; Nepali—*Borhar*; Punjabi—*Bor, bargad*; Marathi—*War, vada*; Tamil—*Ala*; Telugu—*Mari*; Kannada—*Alada*; Malayalam—*Peralu*; Sanskrit—*Vata*; Family—Moraceae.

A large tree, planted throughout India.

Sacred uses. Hindu mythology says that *Brahma* was transformed into a *Vada* tree. According to Hindu mythology the Banyan tree is viewed, as the male to the *Peepul*. It is considered as a sin to destroy either of these trees, but more especially the male. It is meritorious to plant a young male close to the female and this is done with a ceremony somewhat similar to that of marriage. It is customary to place a silver coin under the roots of young Banyan tree. The dry twigs of the tree are used as *Samidhas* for producing sacred fire. The leaves are employed as one of the *Panch pallavas* or platters, and also for pouring libations. In the *Vratraj* the women are ordered to worship this tree on *Jesht shudh* 15th (May or June), to water it, to wind a thread round it, and to worship it with *genda* (Indian marigold) flowers. They are further ordered to make *Pradakshanas* (*i.e.*, to go round it a certain number of times, to praise it, and to pray to it for the survival of their husbands and for the fulfilment of their wishes). They are told that by worshipping this tree they attain one of the heavens—*Shivlok*. They are encouraged to this worship by the tradition that *Savitri*, the wife of *Satyawan* got back her deceased husband through the adoration of this tree. They are recommended to perform the thread ceremony of this tree and its marriage with the *Durva* grass (*Cynodon dactylon*).

Charm. The Santals and other tribes wind the young thin aerial roots of this tree around the neck as a charm to ensure conception.

Bel

Aegle marmelos Corr.; Eng.—The Bael or Bel Fruit Tree, The Bengal Quince; Hindi—*Bel, siriphal*; Bengali—*Bil, vilva*; Gujarati—*Bil*; Sanskrit—*Sriphal, bilva, bilvaphalam*; Arabic and Persian—*Safarjale-hindi, shul*; Tamil—*Vilva-pazham*; Telugu—*Bilvapandu*; Malayalam—*Kuvalap-pazham*; Kannada—*Bel-patri*; Family—Rutaceae.

A tree, cultivated all over India, often sending up off-shoots from the roots, which later on become trees. Also found wild in Sub-Himalayan forests and South India.

Sacred value. This is one of the most sacred trees of India. It is generally cultivated near temples and dedicated to Lord *Shiva*, whose worship cannot be completed without the leaves of this tree. It is incumbent upon all *Hindus* to cultivate and cherish this tree, and it is sacrilege to cut it down.

The tree is sacred to the *Trimurti*, being a representative of *Shiva*. It is also sacred to the *Parvati* and is the *Vilva rupra*, one of the *Patricas*, or nine forms of *Kali*. The planting of the trees by the waysides gives long life. The leaves are used in enchantments.

Nim

Azadirachta indica Juss. (*Syn. Melia azadirachta* Linn.); Eng.—Margosa tree, Neem tree; Hindi—*Nim*; Bengali—*Nim, nimgachh*; Punjabi—*Nim*; Marathi—*Limba*; Gujarati—*Limba*; Tamil—*Vembu, veppam*; Telugu—*Vepa*; Kannada—*Heb-vebu*; Malayalam—*Veppa*; Sanskrit—*Nimba*; Persian—*Azad-darakhte-hindi*; Family—Meliaceae.

A large tree, 40 to 50 feet in height; found throughout India; extensively planted as an avenue tree.

Sacred uses. The tree is held sacred by the Hindus and is used in many religious ceremonies. It is said that when nectar was being taken to heaven from the world below for the use of gods, a few drops fell on the *nim*. Hence, on New Year's Day of *Shak Sanvat*, Hindus eat its leaves in the hope that they will acquire freedom from disease.

Buchanan writes, "Once in two or three years the Coramas of a village of Karnatak State make a collection among themselves, and purchase a brass pot, in which they put five branches of *nim* and coconut. This is covered with flowers and sprinkled with sandal-wood water. It is kept in a small temporary shed for three days, during which time the people feast and drink, sacrificing lambs and fowls to *Marima*, the daughter of *Shiva*; at the end of the three days they throw the pot into the water."

The comments of Pharmacographia Indica on the above mentioned practice are—"This practice is known in other parts of India as *Ghatasthapan*, and is considered to avert ill-luck and disease."

Among certain castes of the Hindus the leaves of the *nim* are placed in the mouth as an amblem of grief returning from funerals.

Palas

Butea monosperma (Lamk.) Taubert; Eng.—Flame of the forest, Bengal—Kino; Hindi—*Dhak, palas*; Bengali—*Palas*; Nepalese—*Palasi*; Uriya—*Porasu*; Gujarati—*Khakara*; Marathi—*Paras, palas*; Tamil—*Parasa, palasham*; Telugu—*Tella modugu, palasamu*; Kannada—*Muttuga-gida*; Malayalam—*Plach-cha, murukka-maram*; Sanskrit—*Kinsuka, palasa*; Persian—*Palah*; Family—Papilionaceae.

A moderate-sized deciduous tree found throughout India, extending in the North–West Himalayas as far as the Jhelum.

Sacred value. The flowers called *tesu, kesu, kesuda* or *palas-ke-phul*, yield a brilliant but fleeting yellow dye, much used in India, especially during the *Holi* festival.

This beautiful tree is sacred to *Soma* (Moon); the word is sacrificial, and is frequently mentioned in the Vedas. The flowers are offered to the gods. The *palas* is sometimes represented as a sacred tree of Buddhists. The *dhak* tree is supposed to be imbued with the immortalising *Soma*, the beverage of the gods. "This tree is supposed to have sprung from the feather of a falcon imbued with the *Soma*". "The *palasa* was much employed by the Hindus in religious ceremonies, particularly in one connected with the blessing of calves to ensure them proving good milkers". "The leaves of this plant are trifoliate, the middle leaf-let is supposed to represent Vishnu, the left Brahma, and the right Shiva: hence its worship is enjoined in *Chaturmas Mahatma*. Hence also its use in the following three great ceremonies:

1. The leaves are used as platters on the occasion of the investiture of the sacred thread, when a particular part of the ceremony, called *chewul* (that is, when the barber removes the last tuft of hair from the head of the child to be invested), is being performed.

2. The dry twigs, under the designation of *Samidhas*, are used for the feeding of *hom*, or sacred fire, in the ceremony which goes under the name of *nava grahas*, celebrated to secure the pacification of the nine planets (nava = nine, grahas = planets) on the occasion of *vastu shanti*, *i.e.,* entrance into a newly built house, or one acquired from a non-Hindu.

3. The stem is used as a staff on the day of *sodmunj*, a part of the thread ceremony.

Bhojpattra

Betula bhojpattra Wall.; Eng.—The Indian Birch Tree, Indian Paper Birch; Hindi—*Bhujpattra*; Punjabi—*Bhuj*; Bhutia—*Takpa*; Nepalese—*Phuspat*; Gujarati—*Bojapatra*; Sanskrit—*Bhurjputra*; Telugu—*Bhujapatri Chettu*; Family—Betulaceae.

A moderate-sized deciduous tree found in the higher ranges of the Himalayas, forming the upper edge of arborescent vegetation, and ascending to 14,000 feet.

Sacred value. The bark is well known as the material upon which the ancient Sanskrit manuscripts of Northern India are written. "The use of the birch–bark for literary purposes is attested by the earliest classical Sanskrit writers. **Kalidasa** mentioned it in his dramas and epics; **Susruta**,

Varahmihira (500-550 A.D.), knew it likewise". All ancient religious books were written on birch-bark.

Nariyal

Cocos nucifera Linn.; Eng.—Coconut; Hindi—*Nariyal*; Bengali—*Narikel*; Gujarati—*Nariyela*; Marathi—*Narela*; Tamil—*Tenna, tenga*; Telugu—*Nari kadam, kobbari chettu*; Kannada—*Tengina*; Malayalam—*Tenga, teena*; Sanskrit—*Nari-kela*; Arabic—*Narjil*; Persian—Darakhte-nargil; Family—Palmaceae.

A pinnate leaved tall plam, with a straight or curved stem, marked by annular scars; cultivated chiefly in Kerala, Tamil Nadu and Karnataka. In South India the palm thrives at altitudes upto 3,000 feet above sea level.

Sacred uses. Coconuts are largely employed as offerings to the gods by the Hindus, and coconut day (the full moon in the month of August—or *Bhadon ki Purnima*) is celebrated throughout the country. The nuts are broken on the opening ceremonies (*muhurat*). The dried *kernel* (*copra*) is sometimes cut into ornaments such as flowers or garlands; these are worn by Hindu women on religious ceremonies.

Charm. "Over his couch when born and over his grave when buried a branch of coconut blossoms is hung to charm away evil spirits."

The coconuts are offered by the higher classes of Hindus (Brahmins) to appease the sea on the coconut fair day (*i.e.,* full noon day in August). At weddings the bridegroom and bride carry the nut in their hands.

Amla

Emblica officinalis Gaertn. (Syn. *Phyllanthus emblica*) Linn.; Eng.—Emblic Myrobalan; Hindi—*Amla, aonla*; Benagli—*Amla, amlaki*; Uriya—*Amlaki*; Santal—*Meral*; Punjabi—*Amla*; Marathi—*Aonli*; Gujarati—*Amla*; Tamil—*Nelli-kai*; Telugu—*Usiriki*; Kannada—*Nelli, nilika*; Malayalam—*Boa-malacca*; Sanskrit—*Amulki, amalaki*, amala, *umrita*; Arabic—*Amlaj*; Persian—*Amelah*; Family—Euphorbiaceae.

A moderate-sized deciduous tree, wild or planted throughout the hotter parts of India.

Sacred value. This is one of the sacred trees of India. "*Kartik Mahatma* orders the worship of this tree, and that a Brahmin couple should feed under it, whereby all their sins are washed off." In *Vrat Kaumudi* also, the *vrat* (fast) and worship of the tree are ordered.

Am

Mangifera indica Linn.; Eng.—Mango; Hindi—*Am*; Bengali—*Ambra*; Assamese—*Am*; Uriya—*Am*; Punjabi—*Amb*; Marathi—*Amba*; Gujarati—*Ambo*; Tamil—*Maa, mangas*; Telugu—*Mamadi*; Kannada—*Mavina, amba*; Malayalam—*Mampalam*; Sanskrit—*Amra*; Persian—*Amba*; Family—Anacardiaceae.

A tree, cultivated chiefly in Uttar Pradesh, the Punjab, Maharashtra, Andhra Pradesh, West Bengal and Tamil Nadu.

Sacred uses. The mango is held sacred by the Hindus and is connected with many mythological legends and folk lore. The mango (Sanskrit = *Amra*) is said to be a transformation of Prajapati (lord of creatures). The tree provides one of the *panch–pallavas*. The flowers (*baur*) are used in Shiva worship on the Shivaratri. The flower is invoked in Shakuntalam as one of the five arrows of Kamadeva.

"In the Indian mythological story of Surya Bai the daughter of the sun is represented as persecuted by a sorceress, to escape from whom she became a golden lotus. The king fell in love with the flower, which was then burnt by the sorceress. From its ashes grew a mango tree, and the king fell in love first with its flower, and then with its fruit; when ripe the fruit fell to the ground, and from it emerged the daughter of the sun (Surya Bai), who was recognized by the prince as his lost wife."

The leaves and twigs of this tree are employed in adorning *mandaps* and houses on occasions of various religious ceremonies.

The twigs are used as *samidhas* and offered to sacred fire.

Khejra

Prosopis spicigera Linn.; Hindi—*Jand*; Bengali—*Shami*; Uriya—*Shami*; Punjabi—*Jhand, khar*; Rajasthani—*Khejra*; Marathi—*Shemi*; Gujarati—*Semru*; Tamil—*Perumbe*; Telugu—*Chami*; Kannada—*Perumbe*; Family—Mimosaceae.

A moderate sized, deciduous, thorn tree, found in the arid dry zones of the Punjab, Haryana, Rajasthan, Uttar Pradesh, Gujarat, Madhya Pradesh and South India.

Sacred. In Rajasthan, the tree is worshipped at certain times of the year when large crowds of natives form themselves into a procession headed by the *Pradhan* or *Mukhia* or head *Thakur* of the place, and make their way to some particular tree set apart for worship. All Hindus held the tree sacred. Its worship is ordered in the *Vratraj* to be performed on the tenth of *Ashwin Sudhapaksha*—the *Dasera* festival because on this tree five *pandavas* hung up their arms when they entered *Virat Nagari* in disguise. On the tree the arms turned to snakes and remained untouched till the owners returned. It is also believed that the tree is a transformation of the goddess that pleased Ram. It is worshipped to obtain pardon for sins, success over enemies, and the realisation of the devotee's wishes. The dry twigs are used as *samidhas* in feeding the sacred fire, and the leaves (*patri*) are offered in the worship of *Ganpati*.

Putra-jiva

Putranjiva roxburghii Wall.; Eng.—Life of the child; Hindi—*Putra-jiva*; Bengali—*Putranjiva*; Punjabi—*Jiyaputra*; Marathi—*Jewan putr*; Tamil—*Karupale*; Telugu—*Kadrajuvi, putrajivi*; Malayalam—*Pongalam*; Sanskrit—*Putranjiva*; Family—Euphorbiaceae.

A moderate-sized evergreen tree, found throughout tropical India.

Charm. The nuts are made into rosaries by Hindu *fakirs*, Brahmans, and by parents to put round the necks of their children. They are supposed to preserve the wearer from harm, and the tree is known as "child-life tree".

Chandan

Santalum album Linn.; Eng.—White sandal wood; Hindi—*Chandan, sandal*; Bengali—*Sufaid chandan, chandan, srikhanda*; Punjabi—*Chandan*; Marathi—*Chandan*; Gujarati—*Suket*; Tamil—*Sundel*; Telugu—*Hari chandanam*; Kannada—*Gandha*; Malayalam—*Chandna-mutti*; Sanskrit—*Chandana, pitachandana, srikhanda*; Arabic—*Sandal-abiyaz*; Persian—*Sandal suped*; Family—Santalaceae.

A small, evergreen tree, found in the western Peninsula, Karnataka and Tamil Nadu, generally at elevations of from 2,000 to 3,000 feet.

Sacred uses. The paste obtained by rubbing the wood on a stone with a little water is used for painting the body after bathing and is employed for making the *shardana* or caste-marks of the natives, especially in South India.

The wood of *chandan* is largely employed in the religious ceremonies of the Hindus. Idols are carved in it. An emulsion of the wood is offered to the gods and an incense made of the wood is burned before them. Large quantities of wood are used by the Parsis in their fire temples. Rich people sometimes employ sandal-wood for cremating their dead relatives, and all both rich and poor, add at least one piece of the wood to the funeral pile (Hindu Materia Medica, U.C. Dutt)

Sal

Shorea robusta Gaertn.; Eng.—Sal; Hindi—*Sal, Shal*; Santal—*Sarjom*; Nepali—*Sakwa*; Uriya—*Salwa*; Punjabi—*Tal*; Marathi—*Rala*; Gujarati—*Ral*; Tamil—*Kungiliyam*; Telugu—*Gugal*; Kannada—*Kabbu*; Sanskrit—*Sala, rala, guggilam*; Arabic—*Kaikahr*; Persian—*Lab-moab bari*; Family—Dipterocarpaceae.

A large, gregarious tree, found in the Punjab, Uttar Pradesh, Bihar, Bengal, Assam, Madhya Pradesh and Andhra Pradesh.

Sacred use. It is mentioned in Pharmacographia Indica—"The *sal* tree, called in Sanskrit *Sala* and *Asvakarna*, is of interest from a mythological point of view, since the mother of Buddha is represented as holding a branch of the tree in her hand when Buddha was born, and it was under the shade of a *Sala* tree that Buddha passed the last night of his life on earth. The small branches of the *Sala* are used by Indian villagers to detect witches; they write upon branches the name of every woman over twelve years of age in the village; the branches are then placed in water and left for four and half hours; if any woman's branch withers she is the witch".

Luban

Styrax benzoin Dryand.; Eng.—Benzoin tree; Hindi—*Luban*; Bengali—*Luban*; Punjabi—*Luban*; Gujarati—*Loban*; Tamil—*Shambirani*; Malayalam—*Kaminian*; Arabic—*Loban*; Persian—*Hussi-Luban*; Family—Styracaceae.

A small tree, native of Malyan Archipelago.

Sacred. The *Luban* is burnt as an incense by the Buddhists and Hindus in their worship, and in Europe by the Roman Catholics and members of the Greek Church.

Jamun

Syzygium cumini (Linn.) Skeels (Syn. *Eugenia jambolana* Lomk.); Eng.—Java plum, Jambolana; Hindi—*Jamun, phalanda*; Bengali—*Kala-jam*; Uriya—*Jamo*; Assamese—*Jamu*; Malayalam—*Jam*; Marathi—*Jambul*; Gujarati—*Jambu*; Tamil—*Nawal, naga*; Telugu—*Nasedu*; Kannada—*Narela*; Sanskrit—*Jambu, jambula*; Family—Myrtaceae.

A large-sized tree, cultivated throughout India for the edible fruits. It ascends to 3,000 feet in Himachal Pradesh and 5,000 feet in Kumaon.

Sacred uses. It is a sacred tree of Hindus. It is said that the god *Megh* has been transformed into a *jambul* tree. The colour of the fruit being dark like that of Krishna, this tree is very dear to him. The plant is worshipped and *Brahmins* are fed under it. The leaves are used as platters or *panch pallavas* and for pouring libations.

Birmi

Taxus baccata Linn.; Eng.—Yew; Hindi—*Birmi*; Bengali—*Burmie*; Nepali—*Teheiray sulah*; Lepcha—*Cheongbu*; Ladakh—*Pung cha*; Kashmiri—*Birmi*; Punjabi—*Birmi*; Family—Taxaceae (Gymnosperms).

An evergreen tree, found on the Temperate Himalayas from 6,000 to 10,000 feet.

Charm. A twig is worn by young unmarried Naga girls as a charm to prevent pregnancy, chastity being an exceptional virtue amongst them (G. Watt).

Sacred. The wood is burned as incense, the branches are carried in processions in Kumaon, and in Nepal are used to decorate houses at religious festivals. In certain localities of Himalayas and Khasia hills the tree is held sacred and known as *deodar* (God's tree).

Ber

Zizyphus mauritiana Lamk. (Syn. *Z. jujuba* Lamk.); Eng.—Indian jujube, Chinese date; Hindi—*Ber, beri*; Bengali—*Bor*; Uriya—*Bar, koli*; Punjabi—*Beri, ber*; Marathi—*Bhora bera*; Gujarati—*Bor*; Tamil—*Elandap*; Telugu—*Regu*; Malayalam—*Elentha*; Sanskrit—*Badari, badara*; Arabic—*Sidr*; Persian—*Kundr*; Family—Rhamnaceae.

A small tree, wild and cultivated in the Punjab, Uttar Pradesh, Bihar and Rajasthan.

Sacred. The *ber* fruits are offered to *Siva* by the Hindus on *Mahashivaratri*.

"According to the *Purana*, there was, in former times, a celebrated place of pilgrimage called Badarica Srama (now Badarinath), which abounded with the *badari* or jujube trees, and the devotees or sages of those times lived upon its fruits."

Akanda

Calotropis gigantea R. Br.; Hindi—*Madar*, *safed ak*; Bengali—*Swet akond*; Nepali—*Auk*; Marathi—*Akanda*; Gujarati—*Akado*; Tamil—*Erukku*; Telugu—*Yekka*, *nella-jilledu*; Kannada—*Yekka*; Malayalam—*Erukku*; Arabic—*Ushar*; Persian—*Khark*; Sanskrit—*Svaytaurkum*; Family—Asclepiadaceae.

An erect spreading perennial shrub, chiefly found in wastelands. It ascends to 3,000 feet on the Himalayas and extends from the Punjab to South India and Assam.

Sacred value. The ancient name of the plant, which occurs in the Vedic literature, was *Arka* (wedge), alluding to the form of leaves which were used in sacrificial rites.

The flowers are used in the worship of *Mahadeo* and *Hanuman*. In Bengal the sections of the bluish corona of the flower are carefully picked from the corolla and strung into garlands which are worn at certain religious ceremonies.

In *Chaturmas Mahatma*, in the narration of *Gallava Rushi*, taken from *Skand Puran*, this tree is mentioned to be the transformation of *Surya* (Sun). The leaves are used as *patri*, in the worship of *Ganpati*, *Haritalika*, *Pitchri*, etc. They are also employed in *shusti pujan* by women. When a Hindu is to marry a third time, it is believed that the third wife will soon die—in order to avoid such a calamity, the man is first married to this tree, which is then cut down. This ceremony is believed to ensure the longevity of the fourth, but really the third wife whom he now marries.

It is ordered in the *Shravan Mahatma* to worship *Maruti* (*Hanuman*), on every Saturday, with a garland of the flowers of tree, which are then offered to him. The twigs are also ordered to be used as substitutes for tooth-brushes in the *Smritisar Granth*. The twigs are also employed as *Samidhas* for the feeding of scared fires (*Hawan*).

Bhang

Cannabis Sativa Linn.; Eng.—Hemp; Hindi—*Bhang*, *ganja*, *charas*; Bengali—*Ganja*, *bhanj*; Punjabi—*Bengi*, *bhang*; Kashmiri—*Bangi*; Gujarati—*Ganja*; Marathi—*Bhangacha-jhada*; Tamil—*Bangi-ilai*, *Kalpam*; Telugu—*Bangi-aku*, Kalpam-*chettu*; Malayalam—*Kanchava-chetti*; Kannada—*Bhangi-gida*; Sanskrit—*Vijaya*, *jaya*; Arabic—*Kinnab*; Persian—*Bang*; Family—Cannabinaceae.

An undershrub; native of Asian, now cultivated in Uttar Pradesh, Bengal, Maharashtra, Madhya Pradesh, Tamil Nadu and Orissa. It is wild on the Western Himalayas and Kashmir and it is acclimatized on the plains of India generally.

Sacred. A curious story has been told in the Hindu mythology about the origin of this plant. "It is said to have been produced in the shape of nectar while the gods were churning the ocean with the mountain called **Mandara**. It is the favourite drink of **Indra** the king of gods, and is called *vijaya*, because it gives success to its votaries. The gods, through compassion on the human race, sent it to this earth, so that mankind by using it habitually may attain delight, lose all fear and have their sexual desires excited. On the last day of the Durga Pooja, after the idols are thrown into water it is customary for the Hindus to see their friends and relatives and embrace them. After the ceremony is over it is incumbent on the owner of the house to offer his visitors a cup of *Bhang* and sweetmeats." This is a favourite drink of Hindus on the eve of Holi festival.

Khira

Cucumis sativus Linn.; Eng.—Cucumber; Hindi—*Khira*; Uriya—*Kaknai*; Marathi—*Kakdi*; Gujarati—*Kakari*; Tamil—*Muhevehri*; Telugu—*Doza-kara*; Kannada—*Santi Kayi*; Sanskrit—*Trapusha*, *sukasa*; Family—Cucurbitaceae.

A trailing herb.

Sacred uses. In *Vratraj* it is related that *Suth* told the *Kushis* and *Shiv* told his wife *Parwati* to worship the plant, as by doing so females do not lose their husbands, or that these survive them. The fruit is cut into thin slices and employed in the worship of snakes on *Shravan shudh* 5th, (*Nagpanchmi day*). It is like-wise employed in the worship of many other gods.

Kaddu

Cucurbita pepo Linn.; Eng.—Pumpkin, Vegetable marrow; Hindi and Bengali—*Kumra, safed kaddu*; Marathi—*Kohala*; Kannada—*Kumbala kagi*; Sanskrit—*Kurkaru*; Telugu—*Potti gummadi*; Uriya—*Pani-kakharu*; Family—Cucurbitaceae.

A trailing herb; commonly cultivated in Northern India.

Sacred uses. The *Vrat Kaumudi* recommends the worship of this plant, considering it a goddess. "*Dharamraj* tells *Krishna*, and *Narad* priest of the gods tells king *Chandrasen* to observe, the *Vrat* of this cucurbitaceous plant. Its fruit is also cut with some ceremony, called *kohala muhurt*, a day or two before a marriage".

Haldi

Curcuma domestica Valet. (Syn. *C. longa* Auct. non L.); Eng.—Turmeric; Hindi—*Haldi*; Bengali—*Halud*; Punjabi—*Haldar*; Sanskrit—*Haridra*; Arabic—*Kurkum, zarsud*; Telugu—*Pasupu*; Malayalam—*Mannal*; Kannada—*Arishina*; Marathi—*Halede*; Gujarati—*Halada*; Family—Zingiberaceae.

Turmeric is cultivated throughout India, chiefly in Maharashtra, Bengal, Tamil Nadu, Andhra Pradesh and Orissa.

Sacred uses. Turmeric yields a yellow dye of a fleeting character. "Formerly on festive occasions an infusion of turmeric was used in dyeing garments." The rubbing of turmeric and oil is an essential part of the Hindu marriage festival, as well as of some religious ceremonies. "The root (rhizome) enters into many of the religious ceremonies of the Hindus. The entire, or the corners, of every new article of dress, whether of man or woman, are stained before wearing it with a paste made of the root (rhizome) and water. Mixed with lime, it forms the liquid used in the *Arati* ceremony for warding off the evil eye. Clothes dyed with it are deemed a protection against fever." "With it, in conjunction with lime-juice, the Hindus of the sect of Vishnu prepare their yellow *tiruchurnum*, with which they make the peculiar mark on their foreheads."

Dub

Cynodon dactylon (Linn.) Pers.; Eng.—Bermuda grass, Bahamas grass; Hindi, Punjabi and Bengali—*Dub, durba*; Sanskrit—*Durva*; Marathi—*Durva*; Tamil—*Hariyali*; Telugu—*Ghericha, haryali*; Family—Poaceae (Gramineae).

A perennial creeping grass, found throughout India.

Sacred value. In the *Atharwa Veda* it is said—"May *Durba*, which rose from the water of life, which has a hundred roots and hundred stems, efface a hundred of my sins, and prolong my existence on earth for a hundred years." "This elegant and most useful vegetable has a niche in the temple of Hindu religion."

Dab

Eragrostis cynosuroides R. and S.; Hindi—*Dab, durva*; Bengali—*Kusha*; Punjabi—*Dib, dab, kusa*; Marathi—*Darbha*; Telugu—*Darbha, kusa-dharbha*; Sanskrit—*Kusha, puvitrung*; Family—Poaceae (Gramineae).

A strong coarse, perennial grass, with thick far-creeping rhizomes, commonly found in the barren grounds and sandy soil in the Punjab, Haryana, Uttar Pradesh, Bihar and West Bengal. It grows luxuriantly also on the low-lying portions of the *usar* lands in the Northern India.

Sacred uses. The *dab* grass is needed in the funeral ceremonies of Hindus, and the chief mourner wears a ring of the grass upon his finger; it is also placed beneath the *pindas*. It is mentioned in *Chaturmas Mahatma*, "that this plant is a transformation of *Ketu*. *Shravan Puran* orders that these *Darbhas* should be pulled out of the ground on *Pithori Amavashya*, and that unless this is done the plants are not considered fit for use in sacred ceremonies."

"Some Hindu legends make *Garuda* the offspring of *Kashyapa* and *Diti*. This dame laid an egg, which it was predicted would produce her a deliverer from some great affliction. After a lapse of five hundred years *Garuda* sprang from the egg, flew to the abode of Indira, extinguished the fire that surrounded it, conquered its guards, the *devata*, and bore off the *amrita* (ambrosia) which enabled him to liberate his captive mother. A few drops of this immortal beverage falling on the *Kusa*, it became eternally consecrated; and the serpents, greedily licking it up, so lacerated their tongues with the sharp grass, that they have ever since remained forked; but the boon of eternity was ensured to them by their thus partaking of the imperishable fluid. This cause of snakes having forked tongues is still in the popular tales of India, attributed to the above greediness. At the Ganges (*Ganga*) bathing places for pilgrims, the *Brahman* guides usually present the pilgrim with blades of this grass."

Kotu

Fagopyrum esculentum Moench., (Syn. *Polygonum fagopyrum* Linn.); Eng.—Buckwheat; Hindi—*Kotu*; Assamese—*Doron*; Nepali—*Titaphapur*; Punjabi—*Phaphar*; Kulu—*Kathu*; Family—Polygonaceae.

Cultivated in the Himalayan tracts of Northern India. Chiefly cultivated in the Khasia hills, Sikkim, Manipur and the Nilgiris. On the Himalayas between 4,000 and 10,000 feet buckwheat (*Kotu*) is a rainy season crop, sown in July and harvested in October.

Sacred uses. The flour of *Kotu* grains (pseudograins) is eaten by the Hindus during their fasts (*bart*), being one of the *phalahars* or food-grains lawful for fast-days.

Jau

Hordeum vulgare Linn.; Eng.—Barley; Hindi and Bengali—*Jau*; Nepali—*Tosa*; Punjabi—*Jau*; Marathi—*Java*, *jav*; Gujarati—*Jau*, *jav*; Tamil—*Barli-arisi*; Telugu—*Yava*, *pachcha yava*; Kannada—*Jovegodhi*; Sanskrit—*Yava*; Persian—*Jao*; Arabic—*Shaair*; Family—Poaceae (Gramineae).

An annual herb, producing many stems, 2 to 3 feet high, from a single grain. Chiefly cultivated in Uttar Pradesh, the Punjab, Rajasthan, Madhya Pradesh, Bihar and West Bengal.

Sacred uses. Barley is required for many religious ceremonies of the Hindus. It is considered a symbol of wealth and abundance. It is claimed by astrologers as a "notable plant of Saturn". It is particularly associated with the God Indra. It is especially introduced in the ceremonies attending the birth of an infant, weddings, funerals and at certain sacrifices.

Dhan

Oryza sativa Linn.; Eng.—Rice, Paddy; Hindi—*Dhan*, *Chaval*; Bengali—*Dhan*, *Chanvol*; Uriya—*Dhan*, *Chaul*; Kashmiri—*Dein*, *tani*; Punjabi—*Dhan*, *munji*, *shali*; Marathi—*Tandula*, *Bhat*; Gujarati—*Chokha*; Tamil—*Arishi*; Telugu—*Biyam*, *vudhu*; Kannada—*Akki*; Malayalam—*Ari*; Sanskrit—*Dhanya*; Arabic—*Arruz*; Persian—*Biranj*; Family—Poaceae. (Gramineae).

A herb, grown all over India as a food crop.

Sacred uses. The Sanskrit name *Dhanya* means "the supporter or nourisher of mankind". U.C. Dutt says (in *Materia Medica of the Hindus*)—"By the Hindus it is regarded as the emblem of wealth or fortune. On a Thursday in the month of *Pausha* (December-January), after the new paddy has been reaped, a rattan-made grain measure, called *rek* (in Bengali), is filled with new paddy, pieces of gold, silver and copper coins, and some shells called *cauries*, and these are worshipped as the representative of the goddess of fortune. This apparatus is preserved in a clean earthen pot and brought out for worship on one Thursday in each of the following Hindu months, namely *Chaitra*, *Shrawana*, *Kartika*. Such is the form of domestic goddess of wealth of any agricultural people living chiefly on rice."

"The three principal classes of rice are *Sali*, or that reaped in the cold season, *Vrihi*, or that ripening in the rainy season, and *Shashtika*, or that grown in the hot weather in low lands".

"Certain rice are used as votive offerings at many religious ceremonies. Young girls desiring husbands offer dressed rice to the gods. It is used at the observances after birth of a male child and at the consecration of a Brashmanic disciple. The Brahmans, when performing the marriage rites, after having recited prayers, consecrate the union by throwing rice flour coloured with saffron on the newly-married couple. The Sanskrit word *syala* denotes the custom of the bride's brother scattering fried grains at the marriage ceremony. And later in life women, who desire male children, present offerings of rice and saffron at the temples".

Pan

Piper betle Linn.; Eng.—Betel pepper; Hindi—*Pan*; Bengali—*Pan*; Marathi—*Videcha-pana*; Gujarati—*Pan*; Tamil—*Vettilai*; Telugu—*Tamalapaku*; Kannada—*Vile-dele*; Malayalam—*Vetta*; Sanskrit—*Tambula, nagavalli*; Arabic—*Tanbol*; Persian—*Tambol*; Family—Piperaceae.

A climbing shrub grown for its leaves.

Sacred. The *pan* leaves are chewed on many religious and festive occasions of the Hindus. The *tambula* leaves are offered to Lord *Vishnu* and other gods.

Mainphal

Randia dumetorum Lamk.; Eng.—Emetic nut; Hindi—*Mainphal*; Bengali—*Menphal*; Uriya—*Pativa*; Assamese—*Gurd*; Nepali—*Maida phul*; Punjabi—*Mindhal*; Marathi—*Gelaphala*; Gujarati—*Medhola*; Tamil—*Madu-Karray*; Telugu—*Manda*; Kannada—*Kare*; Sanskrit—*Madana*; Persian—*Jauz-ul-kuch*; Family—Rubiaceae.

A deciduous thorny shrub or small tree, found in the sub-Himalayan tract, Maharashtra, Gujarat, Uttar Pradesh and Tamil Nadu.

Sacred. The Hindus consider this plant sacred to god *Shiva*, and at the marriage ceremonies of *Vaishyas*, the fruit is tied upon the wrists of both bride and bridegroom.

Kela

Musa paradisiaca Linn. (Syn. *M. sapientum* Linn.); Eng.—Banana; Hindi—*Kela*; Bengali—*Kela*; Punjabi—*Kela*; Marathi—*Kel*; Gujarati—*Kela*; Tamil—*Valei*; Telugu—*Amti*; Kannada—*Bale*; Malayalam—*Vasha*; Sanskrit—*Kadali, rambha*; Arabic and Persian—*Mouz*; Family—Musaceae.

The perennial herb is native of India and Malaya; chiefly cultivated in Assam, Madhya Pradesh, Bihar, Uttar Pradesh, Andhra Pradesh, Maharashtra, Tamil Nadu and Kerala.

Sacred uses. It is ordered in the *Vratraj* that the women should worship the plant on the 4th of *Kartik shudh*, whereby their husbands are said to survive them, and their life is lengthened. It is also worshipped on the 3rd of *Shravan*. The bunches of fruit are much used in certain festivals and ceremonies, and are generally placed by Hindus at the entrance of their houses on such occasions, as appropriate emblems of plenty and fertility. The fruits are always offered to gods. The plantain called *Kathali* is considered sacred in Tamil Nadu, and offered to the gods.

Kamal

Nelumbo nucifera Gaertn. (Syn. *Nelumbium speciosum* Willd.); Eng.—The Sacred Lotus; Hindi—*Kamal, kanwal*; Bengali—Padma; Uriya—*Padam*; Punjabi—*Pamposh*; Kannada—*Tavarigadde*; Tamil—*Ambal*; Telugu—*Erra-tamaraveru*; Malayalam—*Tamara*; Sanskrit—*Padma, nalina, rajiva*; Arabic—*Nilufer*; Persian—*Nilufer*; Family—Nymphaeaceae.

A large aquatic herb with peltate leaves, and handsome rosy, red or white flowers, found all over India.

Sacred uses. According to U.C. Dutt (mentioned in *Hindu Materia Medica*)—"These beautiful plants have attracted the attention of the ancient Hindus from a very remote period, and have obtained a place in their religious ceremonies and mythological fables hence they are described in great detail by Sanskrit writers. The flowers of *Kamal* are sacred to *Lakshmi*, the goddess of wealth and prosperity".

The wicks made from the spiral fibres of the leaf-stalks are burned by the Hindus before their idols, and the leaves are used as plates on which offerings are placed.

Dr. Dymock gives the following account of the mythology of the Indian Lotus in his famous paper entitled. "The Flowers of the Hindu Poets".

"The queen of Indian flowers (now the national flower) is the lotus; the Hindus compare the newly created world to a lotus flower floating upon the waters, and it thus becomes symbolical of spontaneous generation. It is the *Padmamani* of the Buddhists. The golden lotus of Brahminic and Buddhistic mythology is the sun, which floats in the waters above the firmament, like an earthly lotus in the deep blue stream below; from it distils the *Amrita*, the first manifestation of Vishnu. Brahma and Buddha (the supreme intelligence) were born of this heavenly lotus. Lakshmi, the wife of Vishnu, the Indian Venus, is called Padmavati, because she is represented sitting on this flower. The Hindus see in the form of the lotus the mysterious symbol Svastika, which signifies the *Ban* and *Shaluka* combined. The allusions to this flower by Indian poets are innumerable; no praise is too extravagant for it; it is the chaste flower, and its various synonyms are bestowed as names upon women. In the Koka Shastra the best class of women is called Padmini. The lotus is supposed to calm the pangs of love, but in the case of Sagarika in the Ratnavali, so violent was her malady that even this remedy was applied in vain. `Take it all away' she says to Susamyata. `I shall never obtain the object of my desire'. The red lotus is said by the poets to be dyed with the blood of Shiva, that flowed from the wound made by the arrow of Kama, the Indian Cupid".

"There are three varieties of lotus or Water Lily: The white, called *Pundarika*; the red, *Kokanada*; and the blue, *Indivara*. The entire plant is called *Padmini*; the fruit, *Karmikara*; and the honey formed in the flowers, *Makaranda*. The stalks of the leaves are called *Mrinala*. The face of a beautiful women is compared by the poets to a lotus blossom, the eyes to lotus buds, and the arms to its filaments, in Sanskrit *Kinjalka* or *Padmakesara*. The bee is represented as enamoured of the lotus. Hari likens the eyes of Radha to a blue lotus, and when sorrowful, to a red one; the face of Hari is to his Radhika like of full bloomed lotus".

Harsinghar

Nyctanthes arbor-tristis Linn. Eng.—Tree of sorrow, Night flowering jasmine; Hindi—*Harsinghar*; Bengali—*Sephalika*; Uria—*Seoli*; Punjabi—*Harsingar*; Marathi—*Khurasli*; Tamil—*Manja-pu*; Telugu—*Paghada*; Kannada—*Harsing*; Sanskrit—*Parijataka*, *sephalica*; Family—Oleaceae.

A large shrub with sweet scented flowers; commonly grown in gardens; the flowers open in the night, and fall to the ground on the following morning.

Sacred uses. The flowers are used by the Hindus in worship and as votive offerings. According to Hindu Mythology, the plant is supposed to have been brought from heaven by Lord Krishna for his wife Satyabhama. As written in *Vishnu Purana*, "this shrub was a king's daughter, named *Parijataka*. She fell in love with the sun, who soon deserted her, on which she killed herself, and was burnt. This shrub arose from her ashes. Hence it casts its flowers in the morning, as it cannot bear the sight of the sun".

Mogra

Jasminum sambac (Linn.) Ait.; Eng.—Arabian jasmine; Hindi—*Motia*, *bela*, *mogra*; Bengali—*Mogra*, *banmallika*; Punjabi—*Mugra*; Marathi—*Mogra*; Gujarati—*Mogro*; Tamil—*Mallippu*; Telugu—*Nava malika*; Malayalam—*Pun mulla*; Sanskrit—*Navamalika*; Arabic—*Suman*; Persian—*Gule-suped*; Family—Oleaceae.

A fragrant climbing shrub, cultivated throughout India.

Sacred value. The flowers are white and highly fragrant. The flowers are held sacred to Lord

Vishnu, and are used in several Hindu religious ceremonies. The flowers are supposed to form one of the darts of Kama Deva, the God of Love of Hindus.

Tulsi

Ocimum sanctum Linn; Eng.—The Sacred Basil; Hindi—*Tulsi*, *vranda*; Bengali—*Tulsi*, *Kala Tulsi*; Punjabi—*Tulsi*; Marathi—*Tulasa*; Gujarati—*Talasi*; Tamil—*Tulasi*; Telugu—*Tulasi*, *Krushna-tulaṣi*; Kannada—*Tulashi-gida*; Malayalam—*Krishna-tulsi*; Sanskrit—*Tulasi*, *tulashi*, *manjarika*; Family—Lamiaceae (Labiatae).

A somewhat shrubby, herbaceous aromatic herbaceous plant, found throughout India.

Sacred uses. The *tulsi* is the most sacred plant in the Hindu religion; it is found in or near almost every Hindu house throughout India. It is said by Sanskrit writers that it protects from misfortune and sanctifies and guides to heaven all who cultivate it. The Brahmins hold it sacred to the gods Krishna and Vishnu. The mythological story narrates that this plant is the transformed nymph Tulasi, beloved of Krishna, and for this reason near every Hindu house it is cultivated in pots or on brick or earthen pillars with hollows at the top in which earth is deposited; it is daily watered and worshipped by all the members of the family. The wood of the stem is used for making beads for the rosaries used by Hindus on which they count the number of recitations of their God's name.

In the *Vrat kauumudi*, a ceremony called the *tulashi laksha vrat*, is ordered to be performed, when a vow is made, which consists in offering a *lac* of the leaves one by one to Krishna, the performer fasting till the ceremony is complete. *Nayavad*, another ceremonial sacrifice among the Hindus, consists in taking a brash dish containing some cooked food, and placing it before the god in a square (*chauk*) previously marked out on the ground with the fingers dipped in water. The worshipper then squats on a low stool, and taking two leaves of the *tulsi* in his right hand he closes his eyes with his left, dips the leaves in water, and throws one upon the food, and the other, after five peculiar motions of the hand, on the god.

The leaves are also used in the funeral ceremonies of the Hindus.

The leaves make one of the constituents of *charnamrit* offered to Lord Vishnu.

Ikh

Saccharum officinarum Linn.; Eng.—Sugarcane; Hindi—*Ganna*, *ikh*; Bengali—*Ganna*, *ik*; Nepali—*Uk*, *chaku*; Uriya—*Aku*; Punjabi—*Ganna*, *ikh*; Marathi—*Usa*; Gujarati—*Sherdi*, *nai-sakar*; Tamil—*Karumbu*; Telugu—*Cherku*; Kannada—*Khabbu*; Malayalam—*Karinpa*; Sanskrit—*Ikshu*, *rusala*, *pundra*, *pundarika*; Arabic—*Kasib-shakar*; Persian—*Nai-shakar*; Family—Poaceae (Gramineae).

A strong cane-stemmed grom, grown chiefly in Uttar Pradesh, Bihar, the Punjab and South India.

Sacred uses. In the unrefined state sugar is used as a votive offering by the Hindus at the shrine of gods and other places of worship. Every operation in cultivation and manufacture of *gur* and unrefined sugar is governed by very pronounced religious observances, and the ultimate product (*gur* or unrefined sugar) holds a high place in the esteem of the Hindu. The bow of *Kamadeva* is sometimes represented as made of sugarcane.

The Institutes of Manu (*Manusmriti*) make undoubted allusion to sugarcane.

"The flowering of the cane was pronounced a very ominous occurrence. It was a funeral flower, foreboding death to whomever might chance to look on it".

As mentioned the sugarcane is most sacred to the Hindus, it may be observed in the following paragraph:

"The sacred appellation of the cane amongst the *ryots* is *Nagbele*, and hence..., the immediate owners of the cane plantations sedulously refrain from repairing to or even beholding them, during the continuance of the *Piterputch*. On the 26th, of *Kartik*, termed by the *ryots* (the natives) *Deouthan*, they proceed to fields, and having sacrified to *Nagbele*, a few canes are afterwards cut and distributed to the Brahmins. Until these ceremonies are performed according to the rules of established usage and custom, no persuasion or inducement can prevail upon any of them to taste the cane or to make any use whatever of it".

"Rites and sacrifices are performed on the germination of the cuttings, at the *Naudurga* festival in *Aswin* (October), and in the following month, to avert a disease (*sundi*) which affects the crop. But the most important ceremony connected with its growth is the *Deothan* in the end of *Kartik* (October–November). This, which celebrates the awaking of Vishnu after his slumber in the infernal regions, is to sugar-cane what the *Arwan* is to other crops–a sort of harvest-home. Before this day no Hindu will eat the cane, and even jackals are said to avoid it. But on the *Deothan* several stalks are cut, five being reserved by the owner of the crop, and five each distributed to the village priests and craftsmen. On a board named the *saligram* are daubed, with cowdung and clarified butter (*ghi*), the figures of Vishnu and his consort. On the same receptacle are set *urd* (*Phaseolus mungo*), cotton, and other vegetable offerings; while around it, tied together by their tops, the farmer places his five cane stalks. A burnt sacrifice and prayers are followed by the elevation of the *saligram*." During this last process the women of the household repeat five times the following incantation.

"Arise Oh God! Be seated, Oh Lord!
Spread thy carpets, God of Gaya Gajadhar:
Sit on them, highest Rama of Kampil.
Arise, God, a thousand times arise".

All present then move round the *saligram*. The tops (*juri*) of the five cane stalks around it are severed, hung up to the roof-tree, and burnt on the arrival of the *Holi* festival some months later. At the moment declared auspicious by the presiding Brahman the reaping of the crop begins. "The whole village is a scene of festivity, and dancing and singing go on frantically. Houses are set in order, and marriages which have been suspended during the rains recommence".

Kans

Saccharum spontaneum Linn.; Eng.—Thatch grass; Hindi—*Kans*, *kus*; Bengali—*Kash*, *kas*; Punjabi—*Kans*, *kahi*, *sarkara*; Rajasthani—*Kans*, *kas*; Marathi—*Kagara*; Telugu—*Rellugaddi*; Sanskrit—*Kasha*; Family—Poaceae (Gramineae).

A coarse, perennial grass, found throughout India and upto 6,000 feet on the Himalayas.

Sacred uses. "This beautiful and superb grass is highly celebrated in the *Puranas*, the Indian God of war having been born in a grove of it which burst into a flame. It is often described with praise by the Hindu poets for the whiteness of its blossoms, which gives a large plain at some distance the appearance of a broad river". The *kans* grass is employed in several religious ceremonies of the Hindus by their priests. The mats of this grass are viewed most sacred, and are used by saints and holy men.

Til

Sesamum indicum Linn.; Eng.—Sesame; Hindi—*Til*; Bengali—*Til*, *kala til*, *samsum*, *krishna til*; Uriya—*Rasi*; Nepali—*Til*; Punjabi—*Til*, *tili*; Marathi—*Til*; Gujarati—*Tal*, *til*; Tamil—*Ellu*; Telugu—*Nuvvu*; Kannada—*Wollelu*; Malayalam—*Schit-elu*; Sanskrit—*Vila*, *snehaphala*, *tilaha*; Arabic—*Sim-sim*; Persian—*Kunjad*; Family—Pedaliaceae.

An annual, commonly cultivated in Uttar Pradesh, Madhya Pradesh, Rajasthan, Andhra Pradesh, Tamil Nadu and Maharashtra.

Sacred uses. According to U.C. Dutt (in *Hindu Materia Medica*). The Sesame seeds "form an essential article of certain religious ceremonies of the Hindus, and have therefore received the names of *homadhanya* or the sacrificial grain, *pitritarpana* or the grain that is offered as an oblation to deceased ancestors".

"At the festival of *Sakat*, held in the month of *Magh*, the Hindus eat a composition of *gur* and *til*, which they call *tilkut*".

It may be pointed out that *tila* is also that name of a chapter in the *Purana-sarva sva*; *Tila-ganji-tirtha* is the name of a place mentioned in the *Kasika-ramana*; and *Tiladhenudan* is the title of a chapter in the *Varaha-Purana*.

As exhibiting the important place which *Tila*, seeds and, oil, took in Hindu mythology, it need only be necessary to cite the passage regarding it in the *Institutes of Manu* (*Manusmriti*). In the third lecture it is repeatedly mentioned. The peculiar form in which it should be offered to Brahmans is given in detail. It is supposed to be an offering that secures prosperity and confers offspring, while it delights the manes for a month. It is advised not to eat anything mixed with sesamum seeds after sunset. The punishment of an unlearned man who accepts an offering of *Tila* is indicated, as also the peculiar transmigration that will fall to the lot of the thief of this seed.

Gehun. ***Triticum aestivum*** Linn. (Syn. *T. vulgare* Vill.); Eng.—Wheat; Hindi—*Gehun*; Bengali—*Giun*; Punjabi—*Gehun, kanak*; Marathi—*Gahung*; Gujarati—*Ghavum*; Tamil—*Godumai*; Telugu—*Godumulu*; Kannada—*Godhi*; Malayalam—*Gendum*; Sanskrit—*Godhuma, mahagodhuma, madhuli, mihsuki*; Arabic—*Hintah*; Persian—*Gandum*; Family—Poaceae (Gramineae).

A herb, cultivated as a food grain mainly in Uttar Pradesh, the Punjab, Madhya Pradesh, Maharashtra, Bihar and Rajasthan.

Sacred. The wheat grain is used either as whole or its flour on every festive and religious occasion of the Hindus in Northern India.

26

An Introduction to Ethnobotany

The term **Ethnobotany** was first coined by Harshberger in 1895. Ethnobotany deals with the direct relationship of plants with man. The term has often been considered synonymous with either economic botany or with traditional medicine. However, ethnobotany is not synonymous with traditional medicine. On the other hand, ethnobotany includes study of foods, fibres, dyes, tans, other useful and harmful plants, taboos, avoidances and even magico-religious beliefs about plants (Jain, 1967; Ford, 1978).

This man-plant relationship can be broadly classified into two groups, *viz.,* (a) Abstract and (b) Concrete.

Abstract relationship. The abstract relationship of man with plants includes faith in the good or bad powers of plants, taboos, avoidances, sacred plants, worship and folklore

Concrete relationship. The concrete relationship includes mainly the material use such as in food, medicine, house-building, agriculture operations, other domestic uses, trade or barter, plants in fine arts and culture like paintings, carvings, and house decoration and the acts of domestication, conservation, improvement or destruction of plants.

Ethnobotanical studies cover a variety of approaches, such as : (a) Ethnobotany of certain large or small geographic regions, *e.g.,* Koelz (1979) on Lahul, Shah and Joshi (1971) on Kumaon, and Jain (1981) on Central India (b) Ethnobotany of selected primitive or otherwise ethnobotanically interesting human societies, *e.g.,* Jain and Borthakur (1980) on Mikir, Maheshwari, *et al* (1981) on Tharus of India (c) Ethnobotany of certain plant groups or individual plant species, *e.g.,* Wasson (1969) on the legendary 'Soma', Merlin (1984), on Poppy, Plowman (1969) on Aroids, and Mehra (1967) on Sesame. (d) Ethnobotany of certain specific utility groups like food, medicine, dyes, etc., (Borthakur, 1981).

TERMINOLOGY OF ETHNOBOTANY

A number of terms are used in varied areas of ethnobotanical research, such as ethnotaxonomy, ethnomycology, ethnoecology, ethnopharmacology, ethnomedicine, ethnotoxicology, archaoethnobotany, palaeoethnobotany, ethnonarcotics, ethnogynaecology, *etc.*

Ethnotaxonomy. This deals with the naming and classification of plants and their cultivars, by human societies in their language.

Ethnomycology. This deals with the origin and antiquity of the use of fungi for human beings, such as mushrooms, yeasts, truffles, ergot, *etc*, for food, medicine or for the preparation of beverages, etc.

Ethnoecology. This deals with the study of the past and present interrelationships between human societies, and their living and non-living environment. The importance of ethnoecological research is now keenly felt, as indigenous knowledge systems could usefully be incorporated into development planning and environmental management.

Ethnopharmacology. This makes an interdisciplinary field of research that deals with the indentification, description, observations and experimental investigations of the ingredients used in various recipes prepared by aborigines and the effects of indigenous drugs on animals and man.

Ethnomedicine. This is a field of research that deals with medicines derived from plants, animals, minerals, etc, and used in the treatment of various diseases and ailments, based on indigenous pharmacopoea, folk lore, and herbal charms.

Ethnotoxicology. This is the study of the use of various toxic plants as fish poison (ichthyotoxic), arrow poisons, etc, in human societies.

Ethnomusicology. This deals with the study of music of tribals and aborigines, its documentation, forms and contents.

Archaeoethnobotany. This type of study is the identification of plant materials from archaeological sites for studies on migration of human cultures, and origin, dispersal and domestication of crops, etc.

Palaeoethnobotany. This deals with the identification of fossilized plant materials and remains for studies an ancient plant economy, palaeobotanical history of crops and changing patterns on the use of plant life by human cultures.

Ethnogynaecology. Such type of study deals with various diseases among women in tribal societies, related to sterility, conception, abortion, *etc.*, and the use of abortifacients.

Ethnonarcotics. This is the study of the use of narcotics, snuffs, hallucinogens, *etc*, in primitive human societies.

SOME WILD PLANTS USED FOR FOOD

Tribal people through their hereditary traditional knowledge infer what to eat, and what not to eat. Actually the foods, habits of people are developed on the basis of experience and survival through successive generations. This is not known, how many humans from the beginning to the present day were poisoned by mushrooms, night-shade or deceptive berries, *etc.* Primal man learnt, that certain nuts, fruits, berries, leaves, roots, grains, mushrooms, *etc.*, could cause many ailments, and even death.

Earliest food gathering man gathered fruits, nuts, moss, tubers, mushrooms, morels and stems in season. Now they are thoroughly acquainted with the methods of excluding the harmful substances from the wild plants and preparing recipes for their meagre meals. A list of some wild plants used as food is given below:

1. *Ampelocissus tomentosa* Verna. **Datromabili**; Family—Ampelidaceae. The fruit is edible.
2. *Annona squamosa*. Verna. **Sharifa**; Family—Annonaceae. The fresh flowers are eaten; the ripe fruits are eaten, and the under-ripe fruits are roasted and eaten.
3. *Antidesma diandrum.* Verna. **Anti, Mata ara**; Family—Euphorbiaceae. The young leaves are used as salad, and small red to black drupes are eaten.
4. *Bauhinia purpurea; B. variegata* and *B. diffusa* Verna. **Kachnar**; Family—Caesalpiniaceae. The tender leaves, buds and flowers are eaten as vegetable.
5. *Bombax ceiba*. Verna. **Semal**; Family—Bombacaceae. The flowers and young fruits are eaten as vegetable.
6. *Cassia fistula*. Verna. **Amaltas**; Family Caesalpiniaceae. The flower buds and the flowers are used as vegetable by tribals.
7. *Cordia myxa*. Verna. **Lasora**; Family—Boraginaceae. The tender leaves, and the unripe fruits are used as vegetable.
8. *Emblica officinalis*. Verna. **Amla**; Family—Euphorbiaceae. The fruits are eaten raw or cooked.
9. *Ficus religiosa*. Verna. **Peepal**; Family—Moraceae. The leaf buds are used as vegetable.
10. *Grewia tiliaefolia*. Verna. **Dhaman kehla**; Family—Tiliaceae. The young tender leaves are eaten as vegetable.
11. *Holostemnia annulare*. Verna. **Dudhi**; Family—Asclepiadaceae. The leaves are used with pulses to make curry.
12. *Hymenodictyon excelsum*; Verna. **Kukurkar**; Family—Rubiaceae. The leaves are eaten as vegetable.
13. *Indigofera pulchella*. Verna. **Jirhul**; Family—Papilionaceae. The pink flowers are eaten as vegetable.
14. *Leucas cephalotes*. Verna. **Durup**; Family—Lamiaceae. The leaves are used as vegetable.

15. *Madhuca latifolia*. Verna. **Mahua**. Family—Sapotaceae. The flowers are eaten fresh and dry. The fruits are eaten as vegetable. A spirit prepared from flowers is considered as tonic and nutritive.

16. *Melothria perpusilia*. Verna. **Bankundri**; Family—Cucurbitaceae. The fruits are eaten as vegetable. The roots are pounded and used for making cold drink.

17. *Moringa oleifera*. Verna. **Sainjana**; Family—Moringaceae. The pods and flowers are used as vegetable.

18. *Polygonum plebejum*. Verna. **Sukuripota**; Family—Polygonaceae. The leaves are used as potherb.

19. *Randia dumetorum*. Verna. **Maurea**; Family—Rubiaceae. The leaves are used as vegetable, and the ripe seeds are edible.

20. *Sacciolepis interrupta*. Verna. **Horthilid,** Family—Poaceae. The seeds are used as grain.

21. *Schleichera trijuga*. Verna. **Kusum, Pasra, Baru**; Family—Sapindaceae. The fruits and nuts are edible; the young shoots are also eaten.

22. *Sesbania grandiflora*. Verna. **Agoti**; Family—Papilionaceae. The flowers are eaten as potherb.

23. *Shorea robusta*. Verna **Sal, Sakna, Sarjom, Daru**; Family—Dipterocarpaceae. The seeds are eaten by the poor as a famine food.

24. *Terminalia cremulate*. Verna. **Asan**; Family—Combretaceae. The hard gumming exudate from the stem is called 'asan-latha' is eaten and considered delicious.

25. *Xylia xylocarpa*. Verna. **Tangan**; Family—Mimosaceae. The seeds are edible.

26. *Dioscorea angunia*. Verna. **Kuikuch samga**; Family—Dioscoreaceae. Bulbs and aerial bulbs are eaten as vegetable.

27. *D. belophylla*. Verna. **Pitharu kanda**—The root is eaten raw. It makes the most prized of the forest yams.

28. *D. bulbifera*. Verna. **Gethi kanda**. The yam is cut into slices, boiled and kept in running water and eaten.

29. *D. daemona*; *D glabra* and *D. pentaphylla*. The yams are boiled and eaten as vegetable.

27

Indian Raw Materials and Industrial Products

The economic botany deals with the direct relationship of plants with man. The use of processed, improved, or otherwise modified plant products and their commerce by man primarily is **economic botany**. The primary necessities of man are threefold—**food**, **clothing** and **shelter**. This food primarily comes from plants in the form of cereals, millets, pulses, vegetables and fruits. For clothing again plants are indispensable. The plants that yield fibres are second only to food plants. From the time immemorial, shelter from the wheather inclemencies and as a protection against natural enemy, has been felt too much. In this respect, the forest products have been of service to mankind from the very beginning of his history. The most familiar and most important of these products is wood, which is used in all types of construction work. With the advancement of civilization, the need of man also ever on the increase. To meet his requirements he has tried to tap plants as sources for his comforts and varied uses by exploiting his scientific knowledge and has been successful in this direction to a great extent. Other useful plant products include drugs, tannins, dyes, paper, sugar, starch gum, resin, rubber, vegetable fats, fatty oils, essential oils, tea, coffee, cocoa, tobacco, spices, condiments, cork, *etc.* It becomes quiet evident that a knowledge of botany and its proper application led to well-being of humanity in different ways.

A list of the genera of Indian raw-materials is undermentioned alphabetically.

BEVERAGES

Anacardium (Cashewnut tree), *Ananas* (Pine apple), *Arenga* (Malacca sugar palm), *Borassus* (Sweet toddy and toddy), *Camellia* (Tea), *Caryota* (Sago palm), *Cichorium* (Chicory, endive), *Coffea* (Coffee), *Cola* (Kola), *Elaëis* (African oil palm), *Gaultheria* (Indian winter green), *Hemidesmus* (Indian sarsaparilla), *Hordeum* (barley), *Ilex* (Paraguay tea), *Musa* (Banana, plantain), *Oryza* (Rice), *Phoenix* (Date palm), *Theobroma* (Cocoa), *Vitis* (Grape), *Zingiber* (Ginger).

DYES AND TANS

Acacia (Babul, Cutch, Wattle), *Bixa* (Annatta), *Bruguiera*, *Butea* (Flame of the forest), *Caesalpinia* (Divi-divi), *Cassia* (Avaram), *Ceriops* (Cochineal), *Cotinus* (Indian sumach), *Crocos* (Saffron), *Curcuma* (Turmeric), *Datisca* (Akalbir), *Diospyros* (Persimmon), *Emblica* (Emblic Myrobalan), *Flemingia* (Warrus), *Galium* (Cheese Rennet), *Garcinia* (Gomboge), *Haematoxylon* (Logwood), *Hedyotis* (Chayroot), *Heritiera* (Sundri), *Indigofera* (Common Indigo), *Juglans* (walnut), *Kandelia, Lawsonia* (Henna), *Lichens*, *Mallotus* (Kamala), *Myrica* (Box Myrtle), *Onosma, Oroxylum, Paeonia* (Peony), *Peltophorum* (Copper pod), *Pistacia* (Galls of pistachio),

Pterocarpus (Red sanders, Malay Padauk, Andaman Padauk, Indian or Malabar kino tree), *Quercus* (Oak), *Punica* (Pomegranate), *Rhizophora* (Mangrove), *Rhus, Rubia* (Madder), *Semecarpus* (Marking-nut tree), *Sonneratia*, *Tagetes* (Marigold), *Tamarix, Terminalia, Uncaria* (Gambir), *Wagatea, Xylia, Xylocarpa.*

ESSENTIAL OILS

Alpinia (Galangal), *Amomum* (Greater Cardamom), *Anethum* (Dill), *Apium* (Celery), *Artemisia* (Oil of Taragon), *Bursera* (Linaloe), *Cananga* (Ylang-ylang), *Carum* (Caraway), *Cedrus* (Cedar wood), *Cinnamomum* (Cinnamon bark oil, camphor), *Citrus* (Lime, Lemon and Oranges), *Coriandrum* (Coriander), *Cuminum* (Cumin), *Cymbopogon* (Lemon grass), *Dianthus* (Carnation), *Elettaria* (Cardamom), *Eucalyptus, Foeniculum* (Fennel), *Gaultheria* (Indian Wintergreen), *Hi-*

biscus (Ambrette seed), *Illicium* (Star Anise), *Iris* (Orris), *Jasminum* (Jasmine), *Juniperus* (Juniper), *Lavandula* (Lavender), *Melaleuca* (Cajuput), *Mentha* (Mint), *Murraya* (Curry leaf), *Narcissus* (Jonquil), *Nyctanthes* (Night Jasmine), *Ocimum* (Basil), *Pandanus* (Screw pine), *Pavonia* (Hina), *Pelargonium* (Geranium), *Pogostemon* (Patchouli), *Polianthes* (Tuberose), *Rosa* (Rose), *Rosmarinus* (Rosemary), *Salvia* (Sage), *Santalum* (Sandal wood oil), *Saussurea* (Costus), *Tagetes* (Marigold), *Tanacetum* (Tansy), *Thalictrum, Thuja, Thymus* (Thyme), *Vanilla, Vetiveria, Viola Zanthoxylum.*

FATS AND OILS

Aleurites (Tung, wood oil tree), *Arachis* (Groundnut), *Brassica* (Mustard, Rape), *Carthamus* (Saffflower), *Cocos* (Coconut), *Croton* (Oleum Crotonis, Oleum Tigli), *Cucumis* (Wild Musk-melon), *Cucurbita* (Pumpkin), *Diploknema* (Phulwara, Indian Butter Tree), *Elaeis* (African oil Palm), *Eruca* (Taramira), *Glycine* (Soyabean), *Gossypium* (Cotton seed), *Helianthus* (Sunflower), *Hydnocarpus* (Chaulmoogra), *Juglans* (Walnut), *Linum* (Linseed), *Madhuca* (Mahua), *Mesua* (Nagkesar), *Olea* (Olive), *Pongamia* (Pongam oil), *Prunus* (Almonds oil), *Ricinus* (Castor), *Salvadora*, *Sapindus* (Soapnut tree), *Schleichera* (Lac Tree), *Sesamum* (Sesame), *Shorea* (Sal Butter), *Telfairia* (Oysternut), *Theobroma* (Cocoa), *Xanthium, Ximenia.*

FIBRES AND PULPS

Agave (Sisal), *Aloe* (Indian Aloe), *Boehmeria* (Ramie), *Borassus* (Palmyra, Bassine), *Calotropis*, *Cannabis* (True Hemp), *Caryota* (Kittul), *Ceiba* (Kapok, white silkcotton), *Cocos* (Coir), *Corchorus* (Jute), *Corypha* (Fan palm), *Crotaloria* (Sunnhemp), *Eriophorum* (False Bhabar), *Erythrina* (Indian Coral Tree), *Furcraea* (Mauritius Hemp), *Girardinia* (Nettle), *Gossypium* (Cotton), *Helicteres* (Isora Fibre), *Hibiscus* (Kenaf, Mesta), *Linum* (Flax), *Lonicera* (Honeysuckle), *Luffa* (Ridged Gourd), *Musa* (Banana), *Pandanus, Pergularia, Phoenix* (Date-palm), *Phormium* (New Zealand Flax or Hemp), *Sabal* (Cabbage-Palm), *Salmalia* (Silk-cotton Tree), *Sansevieria, Tillandsia, Triumfetta, Typha, Urena, Urtica, Yucca.*

FOOD AND FODDER

Avena (Oat), *Bothriochloa* (Sour Grass), *Brachiaria* (Crab Grass), *Cajanus* (Redgram), *Caryota* (Sago Palm), *Cicer* (Bengal-gram), *Cyamopsis* (Cluster Bean), *Cynodon* (Dub), *Dioscroea* (Yams), *Dolichos* (Horsebean, Indian butter bean, Field bean), *Echinochloa* (Barnyard Millet), *Eleusine* (Finger Millet), *Eragrostis* (Teff Grass), *Fagopyrum* (Buck wheat), *Fungi* (Edible Mushrooms), *Glycine* (Soyabean), *Hordeum* (Barley), *Hymenachne* (Dal Grass), *Ipomoea* (Sweet Potato), Lens (Lentil), *Manihot* (Tapioca), *Maranta* (West Indian Arrowroot), *Melilotus* (Sweet clover), *Mucuna* (Common Cowitch), *Musa* (Plantain), *Oryza* (Rice), *Pachyrrhizus* (Yan-bean), *Panicum* (Guinea grass, Proso or Hog Millet), *Pennisetum* (Pearl Millet, Napier , or Elephant grass), *Phoenix* (Date-palm), *Pueraria* (Kudzu), *Saccharum* (Sugarcane), *Secale* (Rye), *Sesbania* (Agathi, Sesban), *Setaria* (Italian or Fox-tail Millet), *Sorghum* (Jowar), *Theobroma* (Cocoa), *Trapa* (Water-chestnut), *Trifolium* (Clover), *Triticum* (Wheat), *Vaccinium* (Blueberry), *Vicia, Vigna* (Cowpea, Black and Greengram), Yeast.

FRUITS AND NUTS

Achras (Sapota), *Aegle* (Bael Tree), *Anacardium* (Cashewnut), *Ananas* (Pineapple), *Annona* (Custard Apple), *Areca* (Betelnut), *Artocarpus* (Jack Fruit), *Averrhoa* (Carambola), *Borassus* (Palmyra), *Carica* (Papaya), *Carissa* (Karaunda), *Cicca Citrullus* (Watermelon), *Citrus* (Lime, Lemon and Orange), *Cocos* (Coconut), *Corylus* (Hazel nut), *Cucumis* (Melon), *Cydonia* (Quince), *Diospyros* (Persimmon), *Durio* (Durian Civet Fruit), *Emblica* (Indian Gooseberry), *Eriobotrya* (Loquat), *Ficus* (Fig), *Fragaria* (Strawbarry), *Garcinia* (Mangosteen), *Grewia* (Phalsa), *Hyphaene* (Egyptian Doum Palm), *Juglans* (Walnut), *Litchi* (Litchi), *Malus* (Apple), *Mangifera* (Mango), *Morus* (Mullberry), *Musa* (Banana), *Passiflora* (Passion Fruit), *Persea* (Avocado), *Phoenix* (Dates),

Physalis (Cape Goseberry), *Pinus* (Chilgoza pine), *Pistacia* (Pistachio nut), *Prunus* (Almond, Apricot, Cherry, Plum), *Psidium* (Guava), *Punica* (Pomegranate), *Pyrus* (Pears), *Ribes* (Black Currant, Red Currant), *Rubus* (Raspberry, Blackberry), *Sorbus* (Rowan Tree), *Spondias* (Hogplum), *Telfairia* (Oysternut), *Vitis* (Grape), *Ziziphus* (Ber).

GUMS, LATEX AND RESINS

Acacia (Gum Arabic), *Algae* (Agar-agar), *Aquilaria* (Agar), *Astragalus* (Tragacanth), *Boswellia* (Indian olibanum), *Butea* (Bengal kino), *Conarium* (Black Dammar), *Cannabis* (Hemp Resin), *Ceratonia* (Carob), *Cochlospermum* (Kaitira Gum), *Commiphora* (Indian Bedellium), *Cyamposis* (Guar Gum), *Daemonorops* (Dragon's blood), *Dipterocarpus* (Gurjan Balsam), *Excoecaria* (Agallocha), *Ferula* (Asafoetida), *Garcinia* (Gamboge), *Guaiacum* (Guaiacum Resin), *Hevea* (Para Rubber), *Hopea* (Rock Dammar), *Kingiodendron* (Piney), *Liquidamber* (Storax), *Macarange* (Gum kino), *Melanorrhoea* (Burmese Lacquer), *Myroxylon* (Tolu), *Palaquium* (Gutta Percha), *Pterocarpus* (Indian or Malabar kino Tree), *Salmalia* (Semal Gum), *Schinus* (Mastic-Tree), *Semecarpus* (Markingnut Tree), *Shorea* (Dammar Gum), *Sterculia, Stereospermum, Styrax, Tamarindus* (Tamarind), *Taraxacum, Tieghemopanax, Trachylobium, Vateria* (Dammar).

MEDICINAL PLANTS

Aconitum (Aconites), *Acorus* (Sweet Flag), *Adhatoda, Aloe* (Indian Aloe), *Alstonia* (Ditabark), *Artemisia* (Absinthe, Wormseed, Santonica), *Atropa* (Belladonna), *Azadirachta* (Neem, Margosa), *Cannabis* (Hemp Drug), *Carica* (Papaya), *Cassia* (Senna), *Centella* (Indian Pennywort), *Cephalia* (Ipecacuanha), *Cinchona* (Quinine), *Cinnamomum* (Cinnamon, Camphor), *Citrullus* (Colocynith), *Claviceps* (Ergot), *Croton* (Purging-Nut, Croton Oil), *Datura* (Stramonium), *Derris* (Tuba), *Digitalis, Dryopteris* (Filix-Mas), *Emblica* (Indian Goose-berry, Emblic, Myrobalan), *Ephedra, Erythroxylum* (Cocaine), *Eucalyptus, Euphorbia, Fumaria* (Fumitory), *Fungi* (*Penicillium, etc.*), *Gentiana* (Gentian), *Glycyrrhiza* (Liquorice), *Gymnema, Hemidesmus* (Indian Sarsaparilla), *Holarrhena* (Kurchi), *Hydnocarpus* (Chaulmoogra), *Hyoscyamus* (Henbane), *Ipomoea* (Kaladana), *Jateorhiza* (Calumba), *Matricaria* (Chamomile), *Mentha* (Mint), *Myristica* (Nutmeg), *Nardostachys* (Indian Nard), *Nerium* (Indian Oleander), *Nigella* (Black Cumin), *Onosma, Pimpinella* (Anise), *Plantago* (Isafgol), *Polygala* (Senega), *Psoralea* (Bebchi), *Quassia, Randia* (Emetic Nut), *Rauvolfia* (Sarpagandha), *Rheum* (Rhubarb), *Ruta* (Garden Rue, Common Rue), *Sambucus* (European or Black Elder), *Saraca* (Asoka), *Sassafras* (Ague Tree), *Saussurea* (Costus), *Scilla* (South Indian Squill), *Sida* (Country Mallow), *Smilax* (China Root), *Strophanthus, Strychnos* (Kuchla), *Svertica, Tanacetum* (Tansy), *Tephrosia, Thalictrum, Thevetia* (Yellow Oleander), *Thymus* (Thyme), *Trachyspermum* (Carum), *Tylophora* (Antamul), *Ulex, Urtica, Valeriana, Vinca* (Periwinkle, Sadabahar), *Withania* (Asgandh), *Xanthium.*

SPICES AND FLAVOURINGS

Allium (Garlic), *Anethum* (Dill), *Artemisia* (Davanamu), *Brassica* (Mustard), *Capsicum* (Chillies), *Carcum* (Caraway), *Cinnamomum* (Cinnamon), *Coriandrum* (Coriander), *Crocus* (Saffron), *Cuminum* (Cumin), *Curcuma* (Turmeric), *Elettaria* (Cardamom), *Ferula* (Asafoetida), *Foeniculum* (Fennel), *Glycyrrhiza* (Liquorice), *Hyssopus* (Hyssop), *Illicium* (Star Anise), *Iris* (Orris), *Laurus* (True Laurel), *Mentha* (Mint), *Murraya* (Curry leaf), *Myristica* (Nutmeg and Mace), *Pastinaca* (Parsnip), *Origanum* (Wild or Common Marjoram), *Ocimum* (Basil), *Piper* (Pepper), *Pimenta* (Allspice Tree), *Rosemarinus* (Rosemary), *Salvia* (Sage), *Syzygium* (Laung), *Tamarindus* (Tamarind), *Tetragonia* (New Zealand spinach), *Tanacetum* (Tansy), *Thymus* (Thyme), *Trachyspermum* (Carum), *Trigonella* (Fenugreek), *Vanilla*, *Zingiber* (Ginger).

TIMBER AND FOREST PRODUCTS

Abies (Fir), *Albizzia* (Siris Tree, East Indian Walnut), *Anogeissus* (Axle Wood), *Bambusa* (Bamboos), *Betula* (Birch), *Boswellia* (Salai), *Buxus* (Boxwood), *Calamus* (Cane), *Calophyllum*

(Poonspar, Alexandrian Laurel), *Canarium* (Black Dammar, Dhup), *Cedrela* (Toon), *Cedrus* (Deodar), *Chloroxylon* (Indian Satinwood), *Chukrasia* (Chickrassy), *Cupressus* (Cypress), *Dalbergia* (Sissoo, Indian Rosewood), *Dendrocalamus* (Male or Solid Bamboo), *Diospyros* (Ebony), *Dipterocarpus* (Gurjun), *Duabanga* (Lampati), *Dysoxylum* (White Cedar), *Eucalyptus* (Safeda), *Eulaliopsis* (Sabai Grass), *Gardenia, Gmelina* (Gumhar), *Grewia* (Dhaman), *Hardwickia* (Anjan), *Heritiera* (Sundri), *Holoptella* (Indian Elm), *Juglans* (Walnut), *Juniperus* (Juniper, Pencil Cedar), *Melia* (Bakain), *Mesua* (Mesua), *Millingtonia* (Indian Cork Tree), *Mimusops* (Bulletwood), *Morus* (Mulberry), *Oroxylum, Ougeinia* (Sandan), *Pinus* (Chir), *Populus* (Poplar), *Pseudotsuga* (Green Douglas Fir), *Quercus* (Oak), *Robinia* (False-Acacia, Black-Locust), *Salix* (Willow), *Salmalia* (Silk-cotton Tree), *Santalum* (Sandal Tree), *Schleichera* (Lac Tree), *Sequoia* (Red Wood), *Shorea* (Sal Tree), *Soymida* (Indian Red Wood), *Swietenia, Syzygium* (Jaman Wood), *Tectona* (Teak), *Terminalia* (Dhaman), *Tetrapanax, Tsuga, Xylia.*

NARCOTICS, FUMITORIES AND MASTICATORIES

Areca (Arecanut, Betelnut), *Cannabis* (Hemp Resin), *Cola* (Kola), *Datura* (Stramonium), *Erythroxylum* (Coca Leaves), *Gaultheria* (Indian Wintergreen), *Glycyrrhiza* (Liquorice), Hyoscyamus (Henbane), *Illicium, Myristica* (Nutmeg), *Nicotiana* (Tobacco), *Papaver* (Opium), *Piper* (Betel), *Pistacia* (Mastic Tree), *Rivea* (Ololiuqui, Snake Plant), *Vernonia.*

VEGETABLES

Allium (Onion, Leek, Garlic), *Amaranthus* (Chaulai), *Amorphophallus* (Corm), *Apium* (Celery), *Artocarpus* (Jack Tree or Bread Fruit), *Beta* (Beetroot), *Brassica* (Cabbage, Cauliflower, Knol-khol, Turnip), *Canavalia* (Jack or Sword Bean), *Cichorium* (Chicory, Endive), *Citrus* (Citrus Fruits), *Colocasia* (Taro), *Cucumis* (Cucumber), *Cucurbita* (Pumpkin, Squash), *Cyamopsis* (Cluster Bean), *Cynara* (Globe Artichoke), *Daucus* (Carrot), *Dioscorea* (Yams), *Dolichos* (Lablab Bean), *Entada* (Mackay Bean), *Feronia* (Elephant Apple or Woodapple), *Fungi* (Edible Mushrooms), *Glycine* (Soyabean), *Hibiscus* (Lady's Finger), *Ipomoea* (Sweet potato), *Lactuca* (Lettuce), *Lagenaria* (Bottlegourd), *Lapidum* (Cress), *Luffa* (Ridged Gourd, Ghia Tori), *Lycopersicon* (Tomato), *Momordica* (Bitter Gourd), *Moringa* (Drumstick), *Manihot* (Cassava, Tapioca), *Musa* (Plantain), *Mentha* (Mint), *Metroxylon* (Sago Palm), *Pachyrrizus* (Yam Bean), *Pastinaca* (Parsnip), *Petroselinum* (Parsely), *Phaseolus* (Beans), *Pisum* (Peas), *Portulaca* (Common Purslanes), *Psophocarpus* (Goa Bean), *Raphanus* (Radish), *Rumex* (Sorrel), *Sechium* (Chow-Chow, Chayote), *Sesbania* (Agathi), *Solanum* (Potato, Brinjal), *Spinacia* (Spinach), *Thlaspi* (Common Pennycross), *Tragopogon* (Salisfy), *Trianthema, Trichoxanthes* (Snake Gourd), *Trigonella* (Fenugreek), *Vaccinium, Vicia* (Bakhla), *Vigna*, (Pulses), *Vitex, Xanthosoma.*

INDUSTRIAL PRODUCTS FROM PLANTS

MAJOR INDUSTRIES

Coaltar, cotton mill, Dyestuffs, Jute Mill, Paints and Varnishes, Paper and Paper boards, Pharmaceuticals, Plywood and other wood panel products, Rayon, Rice-milling, Rubber industry, Soaps, Sugar, Synthetic fibres, Tea, Textile industry, Tobacco products, Vanaspati, Vegetable oils industry.

MEDIUM AND SMALL SCALE INDUSTRIES

Alcohol, Bakery, Beer and breweries, Caffeine, Camphor, Carpentry, and cabinet work, Cigarettes, Cigars and cheroots, Cocoa and chocolate, Coffee, Confectionery, Cordages and ropes, Cotton-ginning and baling, Distilled liquors, Essential oils, Flax manufactures, Flour milling, Glucose, Glue and gelatine, Hosiery, Ink, Jute pressing and baling, Lac, Malt and malt products,

Opium and opium alkaloids, perfumery, processed fruits and vegetables, quinine and quinine products, Railway Sleepers, Starches, Vinegar, Wine, Wood-carving, Wood seasoning and preservation, Yeast.

COTTAGE INDUSTRIES

Agarbatties, Bidis, Carts, Coir, Handloom, Gur, Match, Pencils, Sago, Saw-milling, Wooden Toys.

28

Natural Resources and Vegetation Types of India

Man lives in nature and depends on the resources of nature. The progress of human beings depends upon the exploitation of different natural resources. The utilisation of soil, water, coal, electricity, oil, gas and nuclear energy is important for the development of a nation. Any component of the natural environment that can be utilised by man to promote his welfare is considered as a **natural resource**.

Food, shelter and clothing are the primary requirements of the man. Today, man's dependence on the environment is greater than that of other organisms because he is more than a mere biological creature. Modern man has developed an obsession for greater comfort and security. To meet his requirements, he consumes larger amounts of material and energy than any other organism. He has developed a new kind of environment–the **socio-cultural environment**–within the natural environment. Such environment consists of things developed by man through his tools, skills, efforts and social institutions. It depends on and draws sustenance from the natural environment.

Human civilisation has evolved gradually and passed through several distinct stages. The hunting, fishing and food-gathering stages of the primitive man were followed by the agricultural and the industrial stages. With these events the biosphere became transformed into a human-dominated environment or **noosphere** (*noo* = mind; *sphere* = domain).

Due to increase in human population, India is facing an ecological crisis and is depleting her natural resources day by day. Now the shortage of natural resources is a matter of international concern. There is increasing deficiency of energy, metals, coal, non-fuel and non-metallic materials. In the subsequent paragraphs, the natural resources, causes of their degradation and conservation will be discussed.

CLASSIFICATION OF NATURAL RESOURCES

Natural Resources

The word 'resource' means a source of supply or support generally held in reserve.

In other words, natural resources are the components of the atmosphere, hydrosphere and lithosphere, which can be drawn upon for supporting life.

The 'natural resources' include energy, air, water, land, minerals, plants and animals.

For human-beings resources are those materials and sources of energy which are needed for their survival and prosperity.

The nature of resources varies from society to society depending on a number of factors, such as culture, the level of development and the nature of work of that particular society. For example, uranium, gold or silver are of no use for the Onge tribals of Andamans.

Natural resources vary greatly in their location, quantity and quality. For example, a particular type of forest may occur only in certain countries.

Classification

Natural resources can be classified in different ways. A most convenient classification of resources is based upon their exhaustibility and renewability. They are most commonly classified as (*i*) **inexhaustible**, and (*ii*) **exhaustible** on the basis of their abundance and availability.

Inexhaustible Resources

Inexhaustible resources are available in unlimited quantities on the earth.

Some inexhaustible resources remain unaffected by human activities, while others may show changes in their quality, but their quantity is not affected.

Inexhaustible resources like solar energy, wind power, tide power, rainfall and even atomic energy cannot be exhausted significantly at global level due to human activities.

However, they may sometimes be affected at local level by human activities, *e.g.*, pollution may change the quality of air.

Exhaustible Resources

A large number of natural resources are exhaustible, *i.e.*, they have limited supply on the earth and can be exhausted if used indiscriminately. For example, the availability of water in lakes and rivers and water stream energy can be diminished or degraded if used improperly.

Exhaustible resources are classified as (*i*) **renewable**, and (*ii*) **non-renewable**.

Renewable resources. Resources that have the inherent capacity to reappear or replenish themselves by quick recycling, reproduction and replacement within a reasonable time and maintain themselves. The growth and reproduction of such resources can be successfully managed so that these resources are continuously regenerated. Soil, water and living organisms are the main renewable resources. The rate at which their renewal occurs varies.

However, if the consumption of renewable resources continues to exceed their rate of renewal, not only their quality becomes affected, they may even get totally exhausted.

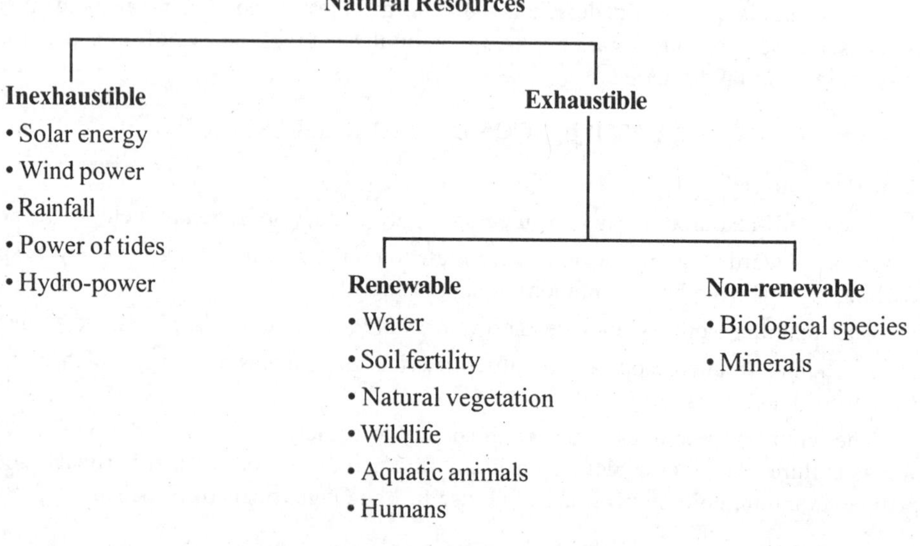

Fig. 28.1. Basic types of natural resources

Some important ecosystems and their renewable resources are :

(*i*) **Forests.** They yield timber and other plant products.

(*ii*) **Rangelands.** They sustain grazing animals for milk, meat and wool production.

(*iii*) **Wild life.** They maintain food chain.

(*iv*) **Agricultural systems.** They yield food, fibre, spices, etc.

(*v*) **Marine and fresh water systems.** They provide various products from plants and animals.

(*vi*) **Soil and water.** They are other renewable resources of great importance.

Non-renewable resources. Resources that lack the ability for recycling and replacement are called non-renewable resources. Substances with a very long recycling time are also regarded as non-renewable resources. For example, fossil fuels, like coal and petroleum, and minerals are non-renewable resources. Biological species, which have evolved in nature during the course of millions of years, are considered non-renewable. Once a biological species becomes extinct, it cannot be recreated again by man.

Resources such as underground water, forests and wildlife are renewable resources but can become non-renewable if used too rapidly by improper management.

Principal Natural Resources, that contribute significantly to human welfare are :

(*i*) Soil, (*ii*) Water, (*iii*) Land, (*iv*) Energy, (*v*) Marine, and (*vi*) Mineral.

SOIL RESOURCES

Land is a major constituent of the lithosphere and the source of many minerals essential to man and other organisms. It forms about one-fifth of the earth's surface, covering about 13, 393 million hectares. About 36.6 per cent of the land area is occupied by human dwellings and factories, roads and railways, deserts and dunes, glaciers, polar ice marshes, rocks and mountains. About 30 per cent of the total land mass is under forests, approximately 22 per cent is occupied by meadows and pastures and only 11 per cent is arable land (*i.e.*, suitable for growing crops).

The surface layer of the land is called soil. About four-fifths of the land area is covered by soil. Soil is composed of inorganic particles, organic matter, air, water and a variety of organisms. It takes hundreds of years for the development of soil horizons having different physico-chemical properties.

The physical properties of soil depend on its texture, structure, bulk, density, porosity and water retentivity. The salt content, pH, and organic and inorganic nutrients like nitrogen, phosphorus and potassium determine its chemical properties. The topography, climate and biotic factors control the conditions of the soil.

Human activities often create worldwide problems. Soil erosion, salinisation, water-logging, and acidification and alkalisation of soil by erroneous human activities are the great dangers to the land resources.

Construction of dams, roads, and railways, urban encroachment and industrialisation and mining activities are responsible for causing depletion of productive land.

Soil Erosion

The word 'erosion' literally means 'to wearing away'. In soil erosion, fertile soil surfaces are detached and removed from their original places and are deposited at some other places. According to Fox (1950), the soil erosion covers a wide range of physical and chemical actions, such as removal of soluble matters, chemical changes, disintegration by frost or by rapid changes of temperature, attrition by dust charged wind, scouring by silt laden currents, alternate impact and succession by storm waves, land slides and so on. Thus, soil erosion is the removal of top soil by movement of water and air.

Human activities accelerate soil erosion by removing natural plant cover. From croplands in India, millions of tons of top soil are eroded into sea each year. Deforestation, undesirable forest biota, and mechanical practices by man are important factors which cause soil erosion. Erosion causes a significant loss of soil fertility by transporting organic matter and nutrients that are essential part of the soil.

The eroded soil, which gets into streams, rivers and lakes in the form of sediments, affects water quality and the habitats of aquatic organisms. There are several serious effects of soil erosion, which are as follows :

1. Due to uprooting of trees, scarcity of timber and fuelwood.
2. Loss of soil stability and fertility.
3. Shortage of fodder.
4. Destruction of cropable land in plains.
5. Formation of sand dunes.
6. Greater frequency of floods and threat to communication channels.
7. Silting of river beds, lakes and dams.
8. Higher temperature and lower rainfall.

However, abundant plant cover significantly reduces soil erosion.

Depletion of Soil Fertility

If the rate of removal or loss of components is greater than the rate of addition, the soil will naturally less fertile. Leaching is an important factor which makes the soil poor in its resources. In this process minerals and organic substances are removed from the top layer of soil by rain water. Biological agencies are also active in causing loss of resources from the soil. Cultivation of crops regularly year after year makes the soil less productive. In cultivation there are little chances for the compensation of lost nutrients of the soil by way of death and decay of the vegetation which grows on it.

On the other hand, when natural vegetation is removed to develop agricultural systems, as in many parts of India, not only the nutrients stored in vegetation are removed, the organic matter and nutrients accumulated in the soil are also lost. From agricultural systems, nutrients are exported through crop harvest, thus depleting soil fertility.

Leguminous plants, however, compensate the loss of nitrogenous compounds because bacteria (rhizobia, etc.) inhabiting in their root nodules fix considerable amount of free atmospheric nitrogen into its compounds, such as ammonia, nitrites, nitrates, etc.

Soil Conservation

The main aims of soil conservation are : (*i*) to protect the soil from erosion, and (*ii*) to maintain the productive capacity of the soil.

Both engineering and biological methods have been used to check the soil erosion but it is still without a plausible check. In India, we are aware of advancing deserts of Rajasthan, and erosional losses, floods, etc., in other parts of the country. The problem has received the attention of forest ecologists, soil scientists and engineers these days. With the result several soil and crop management practices have been evolved which can minimise erosion and reduce nutrient depletion of agricultural soils. Such practices include (*i*) conservation tillage, (*ii*) organic farming, (*iii*) crop rotation (especially cereals with legumes), (*iv*) contour ploughing, and (*v*) strip-cropping terraces, etc.

(*i*) **Tillage conservation.** Many researches support the view that in dry areas, shallow ploughing gives comparatively good crop yields. Shallow ploughing removes the weeds and enables the soil to absorb water. Deep ploughing often leads to soil erosion but in the areas where rainfall is sufficiently high, deep ploughing (upto 15 to 30 cm deep) is effective in removing weeds and increasing crop yields.

In contrast to conventional tillage, conservational tillage incorporates residues from previous crops into the soil, thus increasing the organic matter which in turn improves soil moisture and nutrients. There are two kinds of conservation tillage : (i) reduced tillage, and (ii) no-tillage.

The efforts to improve erosion-affected soils are summarised in two steps. They are :

(*i*) stabilising the soil to prevent further erosion, and

(*ii*) restoration of soil fertility.

For soil stabilization, seeding of base ground with plants that can survive adverse conditions is required. For this purpose usually drought resistant grasses are grown. Such plants eventually establish vegetation cover on the soil, preventing further erosion.

(*ii*) **Organic farming.** With increasing addition of detritus, the soil organic matter, nutrient and moisture levels improve. Application of biofertilizers is also useful for enhancement of soil fertility. Various organic farming measures that provide increased organic input to soil have long-term beneficial effects on soil fertility.

Humus is formed by the accumulation of partially decayed and partially synthesized organic materials. Humus helps in making the soil granular. It causes air spaces to be formed in the clay soil and increases the water holding capacity of sandy soils. Humus is rich in nutrients and therefore, enhances plant growth. Being black humus absorbs heat and warms up the soil.

WATER RESOURCES

Water is the major constituent of the hydrosphere and covers four-fifths of the earth's surface. In fact, water is all around us. Every cubic millimetre of air, even over dry deserts, has water vapour. Water is present in the soil and also hidden in underground pools.

The total volume of water in the hydrosphere is 1.4 billion cubic kilometres (km^3), about 97.5 per cent of the earth's water is found in oceans, which is strongly saline. The rest 2.5 per cent is fresh water, and all of this is not available for direct human use. Most of the fresh water is frozen as polar or glacial ice (*i.e.*, 1.97 per cent). Remaining fresh water occurs as ground water (*i.e.*; 0.5 per cent), and water in lakes and rivers (0.02 per cent), soil (0.01 per cent) and atmosphere (0.001 per cent). Thus, only a small fraction of fresh water is available for human consumption.

About 84 per cent of the total global evaporation occurs from ocean surface and 16 per cent from land surface.

At any given time the amount of moisture in the air is only enough to meet a total rainfall requirement of ten days. Thus, there is a fast movement of water from ocean and land into the atmosphere, and an average stay time of water in the air is only about ten days.

About 77 per cent of the total rainfall on earth is received on the sea surface (as against 84 per cent evaporation from ocean surface) and 23 per cent on land (as against 16 per cent from land surface). Thus, there is net gain of 7 per cent rainfall water on land, and this excess is returned to the oceans by surface run off through rivers and sub-surface water flows.

However, on global basis, the hydrological cycle is perfectly balanced as the total annual evaporation matches with annual rainfall.

Global water is unequally distributed. Precipitation is seasonal and, therefore, the amount of water in inland bodies is variable. Irregularity in the duration and intensity of rainfall often causes floods or droughts. Scarcity of fresh water results in serious regional disparities. Arid regions suffer perennially from water shortage.

Table 28.1. Global Distribution of Fresh Water

	Water : Cubic km (km^3)
1. Water in snow caps, ice sheets, glaciers, etc.	24, 000, 000
2. Surface ponds, lakes and reservoirs	280, 000
3. Water in streams and rivers	1, 200
4. Water present as soil moisture	85, 000
5. Ground water	60, 000, 000
Total amount of fresh water on our planet	84, 366, 200

The global distribution of fresh water on our planet is only about 84.4 million cubic km. The global distribution of fresh water on earth's crust including ground water and water present as vapours in the atmosphere is given in table 18.1.

Table 28.2. Annual Water Budget of our Planet

	Water in cubic km (km^3)
1. Evaporation from sea surface	452, 600
2. Evaporation from land surface	72, 500
3. Precipitation on ocean surface	411, 600
4. Precipitation on land surface	113, 500
5. Surface and ground water run-off	41, 000
Total evaporation from land and sea surface	525, 100
Total precipitation on land and sea surface	525, 100

The excess water received by land surface, about 41, 000 cubic km has to flow back to sea, it cannot be retained on earth's surface ordinarily.

Water Use

On global basis, the water use has increased 4 to 8 per cent per year since 1950, and the consumption rate varies among different countries.

Agriculture sector is the biggest consumer of freshwater. Worldwide, about 70 per cent of total water use is accounted by agriculture.

Following agriculture, power generation (6.2 per cent), and industries (5.7 per cent) are the biggest consumers of fresh water.

Domestic requirement and livestock management taken together consume only 4.3 per cent of the total water drawn.

However, only about 1.1 per cent of fresh water is used for domestic and municipal supplies. The rest of water is consumed by various industries, such as cement, mining, pharmaceutical, detergent and leather industry, etc.

Fisheries, hydro-electric power generation, recreational activities, etc., also require a huge quantity of water, much of which flows down to the sea.

Problems Related with Water Resources

Today in almost all spheres of human activity more water is drawn than what is actually needed. Due to carelessness much of it is wasted and allowed to flow out in an impure state. At many places enough clean water is no longer available. The rapid rise in demand for fresh water is naturally a result of an equally rapid growth in the number of consumers. But there are many other causes as well which have contributed significantly to the wastage and degeneration of fresh water.

For example, about 40 per cent of the world's population lives in arid or semi-arid regions. These people spend lot of energy, time and effort in obtaining water for their domestic and agricultural uses.

To meet the needs of huge population, surface waters, *i.e.*, ponds, lakes, rivers, streams, etc., are overdrawn. Due to over use of surface water, the nearby wetlands may dry up.

When more ground water is removed for human use than can be recycled by rainfall or snow-melt, the ground water may also dry out.

However, everywhere we tend to use more water than is actually needed. A little care can result in substantial savings and longer life for our resource base or else, the water thus saved can be diverted to regions where there is shortage.

Excessive irrigation in arid and semi-arid regions can cause salt accumulation in the soil, due to which crop productivity will decline.

On the other hand, the continuous depletion of ground water along the coastal regions often leads to the movement of saline sea water into fresh water wells, spoiling the quality of water. Estuaries become more saline and consequently less productive when surface waters are over drawn.

Addition of wastes and sewage causes the water body to become exceedingly rich in plant nutrients. Blooms of algae and other organisms appear and make the water useless for human purposes. The entire biomass may suddenly die and start decomposing causing several other problems. Organically rich waters also support a population of many pathogenic organisms and vectors which transfer diseases from one individual to another.

Heavy rainfall results in rapid run-off from areas having exposed soil, particularly on mountain slopes. This causes soil erosion which results in heavy floods. Uncontrolled soil erosion is the main cause of sedimentation of water ways which causes harm to fisheries.

Now a days, pressure of demand on underground water resourses has gone up considerably. Every year more and more water is drawn up from sub-surface layers whereas recharging of underground waters has been slowed down. Massive deforestation has caused disappearance of plant cover over a large area of land surface. In absence of plant cover most of the rain water flows down quickly in streams and rivers. Little of it percolates down to sub-surface layers to recharge the ground water stock. With rapidly flowing waters considerable amount of top soil is lost with torrential flow of water. In a number of localities where underground water is regularly drawn, the water table is receding deeper, thus output exceeds the input.

Conservation and Management of Water

As we know, water is the basic need of living organisms. An acute crisis of water is expected to follow in some regions of the world, particularly in our country. The shortage of water shall make many localities barren, dry and devoid of life. Fertile land will convert into deserts. Conservation of fresh water is therefore, an absolute necessity of today. A number of steps are being undertaken to minimise wastage of fresh water resources and to make more efficient use of the available water. Main approaches for conservation of water are as follows :

(*i*) Reducing agricultural water wastage by increasing efficiency of irrigation.

(*ii*) Reducing water wastage in industry by recycling the used water.

(*iii*) Reducing domestic water wastage by constructing waste water treatment plants and recycling the treated water.

(*iv*) Rainwater harvesting by using practices to store rainwater and recharge groundwater.

(*v*) Afforestation and protection of water sheds to improve water economy.

(*vi*) Reducing pollution and recycling of water.

(*vii*) By development of an efficient distribution system of water.

(*viii*) By enhancement of surface water storage capacity.

(*ix*) By improvement of underground water storage capacity.

(*x*) By adopting water economy.

Main approaches for management of water are as follows :

(*i*) Construction of dams and reservoirs to regulate year-round supply of water. They are also helpful in controlling flood and generating electricity.

(*ii*) Desalinisation of sea water and saline groundwater, making it fit for drinking and agricultural purposes. Desalinisation plants are already under operation in Middle East countries.

(*iii*) Diversion of water and bodies (*e.g.*, through canals) to increase the natural supply of water to a particular area.

(*iv*) Regular dredging and desiltation of water bodies.

Swajaldhara. Reforms in the Rural Drinking Water Sector were adopted in 1999 and also a few projects were launched as pilots. They were intended to be implemented during the Ninth Five Year Plan and with the experience gained there on, the reform initiatives were to be firmed up and scaled up during the Tenth Plan period for adopting the demand responsive strategy and also for institutionalising community participation for the sustainability of drinking water supply system and sources in rural areas. On 15.12.2002, the reform initiative in the Rural Drinking Water Sector were scaled up throughout the country by launching the **Swajaldhara** by the Hon'ble Prime Minister of India.

LAND RESOURCES

Land is a major constituent of the lithosphere and is the source of many materials essential to man and other organisms. Earth's one fourth area is formed by land which is largely covered with natural forests, grasslands, wetlands and man made urban and rural settlements along with agriculture. About 36.6 per cent of the land area is occupied by human dwellings and factories, roads and railways, deserts and dunes, glaciers, polar ice marshes, rocks and mountains. About 30 per cent of the total land mass is under forests, approximately 22 per cent is occupied by meadows and pastures and only 11 per cent for agriculture. Low-lying areas covered with shallow water are called wetlands. The wetlands make the transitional zones between terrestrial and aquatic areas.

FORESTS

The forest is 'a plant community predominantly of trees and other woody vegetation, usually with a closed canopy'. The word **forest** has been derived from the Latin '*foris*' meaning outside, the reference being to a village boundary or fence, and it must have included all uncultivated and uninhabited land. Today a forest is any land managed for the diverse purpose of forestry, whether covered with trees, shrubs, climbers, lianes, etc., or not. The Indian word 'jungle' has been adopted in the English language to describe a collection of trees, shrubs, etc., that are not grown in a regular manner. Forests are a very striking feature of the land surface. They vary greatly in composition and density, and stand in marked contrast with meadows and pastures. Certain forests are evergreen, like the Deodar forests of Kashmir and Himachal Pradesh, while others are deciduous, becoming leafless either before the advent of winter when vegetative activity almost ceases, such as the oak forests of the Himalayas, or else just before the onset of intense dry summer, to reduce transpiration to the minimum, like the teak forests of Madhya Pradesh.

Importance of Forests

The forests of a country make a natural asset of immense value. Unlike its mineral resources, including fossil fuels, which in course of time will either get exhausted or their utilisation will become uneconomic due to increased costs for obtaining and processing them, the forests, if of adequate extent ideally dispersed, scientifically managed, and judiciously utilised can be kept perpetually productive and useful, conferring many benefits, direct and indirect, on the people. Directly, forests meet the needs of timber wood, fuel, bamboos and a variety of other products, including fodder which are indispensable requirements of the people living in close proximity of the forests. The forests provide facility for the grazing of livestock, a yield and variety of products, of commercial and industrial value such as structural timber, charcoal, raw materials for making paper, newsprint, panel products, **bidi** leaves, gums, resin, tanstuff and a number of other economic products including medicinal drugs. Indirectly, forests preserve the physical features, check soil runoff, mitigate floods and make the stream flow perennially and thus help agriculture. Forests make the climate equable and have a considerable hygienic and strategic value. They also provide shelter to the wildlife.

In developing countries the heaviest demand on forests is for fuel wood. More than 1500 million people depend on wood for cooking and other purposes. The world consumption of wood for fuel is estimated to be more than 1,000 million m^3. This is well over 80 per cent of the total use. About 58 per cent of the total energy used in Africa and 42 per cent in South East Asia is obtained from fuel wood.

Bamboos are tall, perennial, arborescent grasses. More than 100 species of them are found in the Indian forests. Bamboos are characterised by woody stems, called culms. The strength of culms, their straightness, hardness, and the facility with which they can be split make them suitable for a variety of purposes. Indeed bamboos are the poor man's timber, used as rafters, scaffolding, roofing, walling, flooring, matting, basketry, carthoods, cordage, etc. Tender shoots are eaten and pickled. Bamboos are the only indigenous plant of wide occurrence in the Indian forests which has a comparatively long fibre suitable for making paper and rayon. Canes, that come from family Palmaceae, and occur in the forests of Andamans, Assam, Kerala, Tamil Nadu, Maharashtra, Bihar, Orissa and Uttar Pradesh, are used as plaiting material, as ropes and cables of suspension bridges in the forests, for furniture, walking sticks, umbrella handles, and sports goods.

Essential oils are obtained from a variety of forest plants. These are used in the manufacture of soaps, cosmetics, pharmaceuticals, confectionery, tobacco flavouring and incense. Important from the commercial point of view are the oils of *Eucalyptus*, used in medicine, rusa grass, khus and sandalwood. Tannins, dyes, gums, resins and oleoresins are other important forest products. Babul bark (*Acacia nilotica*) is an important tanning material used in northern India, the main centre being Kanpur. Gums and resins are exuded by plants. The gum is used in the textile trade, cosmetics, dentrifices, cigar and food industry. The chief producing centres of gum are Madhya Pradesh, Uttar Pradesh, Rajasthan and Gujarat. The most important resin is that obtained by tapping the chir pine (*Pinus roxburghii*). It is a flourishing forest industry. Natural camphor is also a forest product. Drugs, spices, poisons and insecticides are other important forest products. They are derived from trees, shrubs, climbers, herbs, ferns, mosses, lichens and consist of fruits, flowers, leaves, bark, stem or root. **Sarpagandha**, a famous drug is obtained from the roots of *Rauwolfia* *serpentina*, whereas quinine, a drug used against malaria fever, is obtained from the *bark of Cinchona*. **Kala zira** is the seed of *Carum carvi*, a much valued spice. Several forest fruits, flowers and even leaves and tuberous roots are eatern, such as **bel, ber, phalsa, jamun, khirni** and **tendu**. The parts of many plants are used as pickles or vegetables. Such important trees and shrubs are—**amla**, **anar**, **imli**, **karaunda**, **kokam**, **kachnar** and **kaith**. Mushrooms and morels are found in wild state in Kashmir and Himachal Pradesh. They are edible and used as delicacy. Palmyra palm yields a fermented drink **nira**. **Mahua** corollas are fermented to get a liquor. **Phulwara** butter is obtained from the seeds of Indian butter tree (*Diploknema butyracea*). **Shahtoot** fruit is eaten or made into a sherbet. **Bhilama** fruits are salted and pickled.

The leaves of **tendu** (*Diospyros melanoxylon*) are used as wrappers of tobacco to make **bidis**. Leaves of the woody climber *Bauhinia vahlii* found in deciduous forests are collected in large quantities and made into plates and leaf cups used for feasts and other festive occasions.

Miscellaneous forest products include leaves of certain species used for various purposes, soap substitutes, such as **ritha** (seeds of *Sapindus emarginatus*), pods of *Acacia concinna* called **shikakai**, sola pith, ornamental seeds, such as **gumchi** and **rudraksha** (the drupes of *Elaeocarpus ganitrus*).

Lac is the most important animal product obtained from forests. It is a resinous secretion, formed as a protective covering on the bodies of larvae of a minute insect *Laccifer lacca*. The lac insects live on the hosts, such as **palas** (*Butea monosperma*) and **ber**. India produces about 85 per cent of the world production. Lac has been mentioned in Vedas, and known from ancient times in India. Other animal products from forests are honey and wax. **Tussar** or **kosa** silk moths, horns and hides of dead animals are also the products of forests. Antlers of deer and ivory are also animal forest products

The forests play an important role in the life of tribal people living in forests. The forests provide them food, fruits, meat, medicines, hides, skins and other products of commercial use.

Forests make large biotic communities, and they give shelter to diverse species of plants, animals and micro-organisms.

Forests protect the environment. A forest and its environment are always interacting. The most important environmental factors are micro-climate soil characteristics, moisture availability and action of animals and insects. Micro-climate is governed by solar radiation, rainfall, wind, humidity and temperature of the air and the soil. In a dense forest temperature range is narrow, and although the air temperature in and outside a forest is more or less the same, the humidity is greater inside. Soil temperature is also influenced by the forests, which in turn affects the biological activity in the top layer of the soil. Heavy winds may cause a serious erosion if the soil is dry and not covered with vegetation. Extensive forests condense low clouds and thus to some extent increase precipitation. When the tree canopy is open there is little humus on the forest floor and, therefore, rain strikes the bare soil with full fury and brings about soil erosion, which leads to denudation. In rocky tracts, roots of trees penetrate deep into the soil and thus bind it together.

The need for adopting soil conservation measures is the practice of forestry. A well managed and properly stocked forest reduces soil runoff. Rain water is absorbed by the humus and then it seeps into the sub-soil and finally moves underground, thus keeping the streams flowing perennially. Forests thus mitigate floods which can otherwise do immense damage to agricultural land in the lower region. The forest soil absorbs water during rains, and does not allow it to evaporate or run off swiftly. This ensures a perennial supply of water in springs and wells. The hill slopes with forest cover conserve water and gradually release it into streams, rivers and sub-soil springs. Forests help to increase the humidity of the atmosphere by drawing subsoil water and sending it out in the air by means of transpiration. The increased humidity of the forest helps plants and animals for their survival in a warm season by making it cool and pleasant. Forests reduce atmospheric pollution by collecting the suspended particulate matter and by absorbing carbon dioxide from atmosphere.

Forests have a great aesthetic value. All people appreciate the beauty and tranquillity of forests. In ancient times, to arrest devastation of forests, which was adversely affecting the life of people, some wise ancestors and **rishis** declared cutting of trees a sin and planting and protecting them an act of piety. Several useful species were thus saved from extinction, such as the banyan, the **pipal** and **bel**. Forests are of great educative value. The forests are supposed to be nature's laboratories which have contributed to the study and research in various branches of natural sciences. Indian culture was born in **ashramas** situated in deep forests, where the **rishis** sang—"May the gods, the water, the plants and the forest trees, accept our prayers, and may their blessings protect us for ever, and ever".

The forests are storehouse of biodiversity and provide important environmental services to mankind. These services originate from the undermentioned key functions of forests :

(*i*) **Productive functions.** They include production of wood, fruits and a wide variety of compounds, such as resins, alkaloids, essential oils, latex, pharmaceuticals, etc.

(*ii*) **Protective functions.** They include conservation of soil and water; prevention of drought, shelter against wind cold, radiation, noise, sights and smells, etc.

(*iii*) **Regulative functions.** They include absorption, storage and release of gases (CO_2, O_2), water, mineral elements and radiant energy.

Forests effectively regulate floods and drought and the global biogeochemical cycles, particularly carbon cycle.

The regulative functions also improve atmospheric and temperature conditions. They enhance the economic and environmental value of the landscape.

Forest Area in India

At the beginning of the twentieth century, about 30 per cent of land in India was covered with forests.

However, by the end of the twentieth century, the forest cover was .educed to 19.4 per cent.

This is less than the optimum 33 per cent area recommended by the National Forest Policy (1988) for the plains, and 67 per cent for the hills.

Of the existing forests, less than two-third are dense forests, and the rest are open degraded forests.

Today, per capita forest area avaiable in India is 0.06 ha, comparatively, which is much below the average for the world, *i.e.*, 0.64 ha per capita.

Table 28.3. Forest Cover in India, 1999

Class	*Area* (km^2)	*Percentage-Geographic area*
Dense forest	3,77,358	11.5
Open degraded forest	2,55,064	7.8
Mangrove	4,871	0.1
Scrub forest	5,896	1.6
Non-forest (other land use)	25,98,074	79.0
Total	32,87,263	100.0

Forests in India

Palaeobotanical studies have shown that there were dense forests in India in the Permian period, 250 million years ago. Man was evolved in the beginning of the Pleistocene Age, only a million years ago. At this time India had thick forests except in Rajasthan and parts of Punjab which lay buried under a swamp. Archaeological evidence shows that the Rajasthan swamps existed till as late as 4000 B.C., when Mohenjodaro culture flourished. During **Ramayana era** there were dense forests in Naimisharanya, Chitrakoot, Dandakaranya and Panchwati which abounded in wild life. The chronicles of Chinese pilgrims mention dense Indian forests in the birth place of Lord Krishna. Records relating to the invasion of Alexander in 327 B.C. mention the existence of almost dark and deep forests along the Indus. Later in Kautilya's time, protection of forests, planting of new species of trees and preservation of wild life were considered desirable. By the time of Samrat Ashoka, the inroads were into the forests. Ashoka also encouraged the cultivation of exotic, medicinal plants. Emperor Jahangir introduced the famous **chinar** tree in Kashmir valley. Over-population and unplanned felling of trees have drastically reduced the forest area and the quality of forests in India today.

The area under forests in India is about 75 million ha and consists of about 23 per cent of the total land area. The National Forest Policy lays down that India should maintain one-third of its land under forests. The forests of India may be classified into three classes : (*i*) reserved forests, (*ii*) protected forests, and (*iii*) unclassed forests, forming 47.1, 30.3 and 22.6 per cent respectively of the total forest area. The forests in India are largely owned (about 96 per cent) or managed by the states.

Forest types. The distribution of natural forests in India is governed principally by rainfall which may vary from 12 cm to 125 cm annually in different parts of the country resulting in a wide variety of vegetation, from xerophytic scrub forests in the arid zones in Rajasthan to evergreen rain forests in the tropics as in the Western Ghats, Assam, North Bengal, Andamans and conifers in the temperate region. In between the extremes, there are many intermediate types and sub-types. The major forest types are : (*i*) moist tropical forests, (*ii*) dry tropical forests, (*iii*) montane sub-tropical forests, (*iv*) montane temperate forests, (*v*) sub-alpine forests, and (*vi*) alpine scrub forests.

Moist tropical forests occur in Western Ghats, Andamans and Assam. They consist of many evergreen species, such as dipterocarps, hopea, irul, toon, terminalias and several others.

Dry tropical forests occur in areas with rainfall 75 cm to 125 cm containing thorn forests. Tropical thorn forests are found in the desert and arid zones of Rajasthan, Punjab, Haryana, Gujarat, Western U.P. and Madhya Pradesh. They consist mainly of acacias.

Montane sub-tropical forests occur at elevations of 750 to 1800 m in the Himalayas and also at the similar elevations in the Nilgiris in South India. The forest flora is represented by pines.

Montane temperate forests occur at elevation of 1800 to 2700 m and with a rainfall of about 125 cm to 200 cm. Deodar, blue pine, spruce and fir and the important conifers in the Western Himalayas mixed with oaks, maples, walnut and *Ulmus*.

Sub-alpine forests occur between 2900 to 3500 m a.s.l. The vegetation consists of a dense growth of small crooked trees or large shrubs with patches of conifers like *Abies* and blue pine. Birch and *Rhododendron* are the common broadleaved trees.

Alpine scrub forests occur above 3500 m along the entire length of the Himalayas. The vegetation consists of low evergreen forests mostly of *Rhododendron* with birch and other deciduous trees forming a dense growth with short branchy stems.

Deforestation. Deforestation is a serious threat to the economy, quality of life and future of the environment of this country. The forests in India have declined from about 7,000 million hectares in 1900 to 2890 million hectares in 1975. It is estimated that by the year 2000 only 2370 million hectares will be left. Tropical rain forests have been reduced from 1,600 million hectares to 938 million hectares.

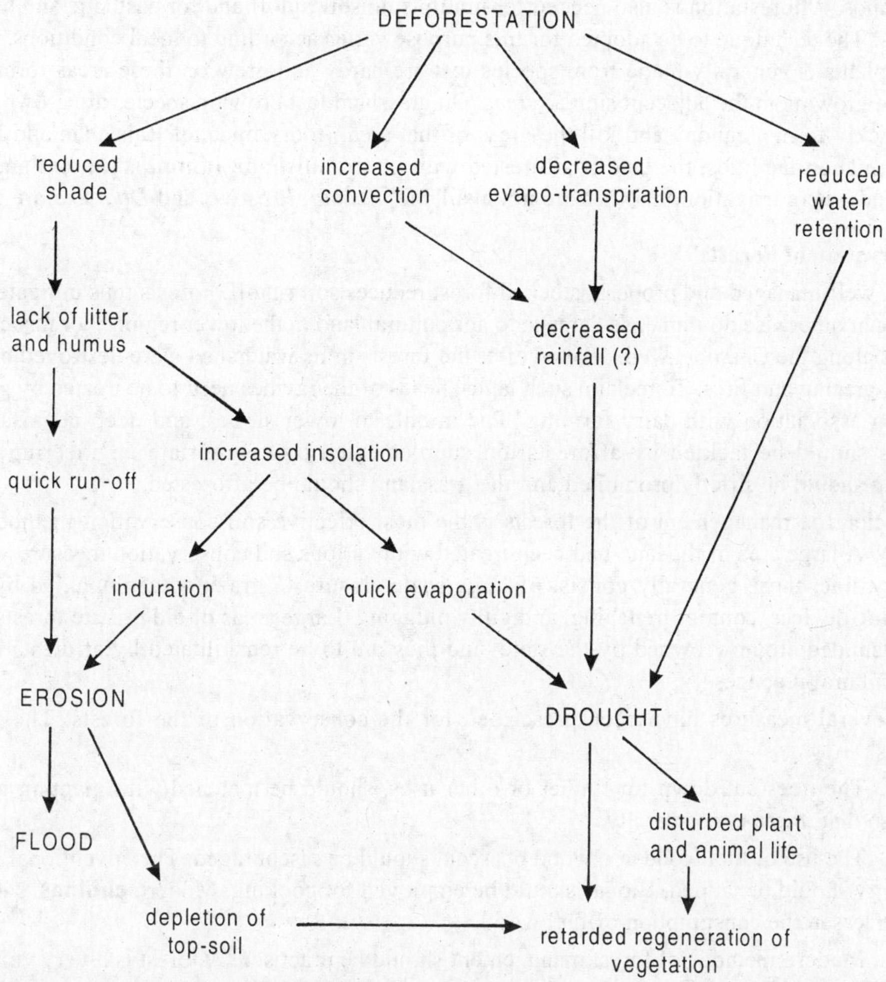

Fig. 28.2. Deforestation.

In India, the main causes of deforestation are : explosion of human and livestock population, increased requirement of timber and fuel wood, expansion of croplands for farming and enhanced grazing by cattle. Construction of roads in the hilly terrain is another important factor for deforestation. The total length of such roads is about 30,000 km in ecologically fragile areas. Development of industries, extension work for mining, quarrying, irrigation and agriculture are other causes of deforestation. Increased demand for fuel wood, timber, wooden crates, paper, newsprint and paper board led to felling of trees on large scale.

Deforestation is the main cause of soil erosion. The floods, drought and loss of precious wild life are also due to deforestation. Thus the economy of the country and quality of life of weaker sections have been much deteriorated. In this country, particularly in the Himalayan region, shortage of fire wood and water has been causing great misery to the womenfolk. They have to spend the whole day in the collection of fuel wood for cooking the meals of the day. It is thought that time is not far when the cost of food would be lesser than the cost of fuel to cook it.

Afforestation. As there is a general shortage of forests and as wood is a bulky commodity which cannot be transported to a long distance, it would be desirable to develop forests on all the wastelands. Afforestation is also needed for minimising soil runoff and for aesthetic and hygienic reasons. The technique to be adopted for this purpose varies according to local conditions. Choice of the plants is generally made from species that are hardy and grew on these areas formerly or they are growing in the adjacent similar areas. On grassland fast growing species are grown, which can quickly form a canopy and kill the grass of that area. In dry tracts of Rajasthan and Punjab, with irrigation facilities, the land is afforested with **khair**, **shisham**, **jamun**, **shahtoot** and **safed siris**, and where irrigation facilities are not available, *Acacia*, *Prosopis* and *Dodonea* are grown.

Conservation of Forests

A well-managed and properly stocked forest reduces soil runoff. Forests thus mitigate floods which can otherwise do immense damage to agricultural land in the lower region. For instance, the ravines along the Chambal were formed after the forests in its watershed were destroyed by over-felling, grazing and fires. To reclaim such lands, heads of the ravines need to be treated by growing grass in association with dairy farming. The middle to lower slopes, and deep depressions of ravines should be tackled by afforestation supplemented by appropriate engineering works. Grazing should be strictly prohibited and the grassland should be afforested.

Scientific management of the forests is the most effective soil conservation method in the country. A large area of the land had been treated with various soil conservation measures. On the forestry side, these generally consist of large-scale closure to grazing accompanied by water disposing devices, contour trenching, and gully plugging. Large areas of old private forests which were denuded are now owned by the state, and they are to be rehabilitated by afforestation and other cultural practices.

Several measures have been prescribed for the conservation of the forests. They are as follows :

1. The trees cut down for timber or other uses, should be matched with planting of more trees, so that, there is no scarcity.

2. The use of fuel wood and wood charcoal should be discouraged. The coventional sources of energy should be tapped. Biogas should be employed for cooking. Modern **chulhas**, should be used to lessen the consumption of fuel wood.

3. Modern methods of forest management should be practised. A forest is a very vulnerable asset. The slight neglect or unscientific working can rapidly deplete and may even annihilate it after which reforesting of the area may not be economically feasible.

4. The silvicultural practices should be adopted. Cultivation of forest crops consists of regeneration, that is, renewal of forest crop, and tending, that is carrying on operations from time to time for the benefit of the growing forest crop during its lifetime. The silvicultural practices include use of irrigation, fertilisers, bacterial and mycorrhizal inoculations, disease and pest management, control of weeds, cleaning, thinning, breeding of elite trees and use of tissue culture techniques.

5. The Social Forestry Programme (1976) should be encouraged. Under this programme, the so called waste lands should be utilised to produce fire wood, fodder and small timber for the use of agricultural implements. This programme includes raising, planting and protecting trees for various purposes. This programme is a boon for rural community.

6. The annual deforestation must be followed by annual reforestation of the deforested areas. Thus the forest is conserved. There should be intensification of afforestation or reforestation rates to nearly three times of the existing rate.

7. The agroforestry programme should be encouraged. Under this programme the attempt should be made to use the same land for farming, forestry and animal husbandry.

8. The urban forestry programme should also be encouraged. Under this programme, the small gardens and house compounds should be well maintained by planting flower and fruit trees. The trees of aesthetic value should also be planted along the road side. The ornamental trees should be grown in the parks and other wasteland. Such forestry programme will give nice look to the town, and mitigate the foul environment.

According to the opinion of experts in this field, it is believed that economy of this country would be sound if at least one-third of country's land comes under forest. To achieve this, there should be prompt action at the levels of individuals, community and the government. In India, several such movements have been taken by the people. For instance, *Chipko* movement of Tehri-Garwal region, and the public agitation against the construction of the hydroelectric project in silent valley, are current examples. India became independent in 1947, and thereafter, attention of the government was drawn to the need for remodelling the management of Indian forests so that they could play an increasingly useful role in promoting national welfare. As a first step towards making the people forest-conscious, a nation-wide celebration of an annual tree-planting festival, called **Van-Mahotsava** was conceived. All the waste-lands, road sides, railway track sides, parks and barren lands should be planted to regenerate forests.

Forest conservation and management programmes may be summarised as follows :

(*i*) They should ensure sustainable supply of tree products and services to people and industry.

(*ii*) They should ensure the maintenance of long-term ecological balance through protection, restoration and conservation of forest cover.

Extensive programmes of **afforestation** are needed to save the diminishing forest cover. The goals can be achieved by undermentioned practices :

(*i*) **Conservation forestry.** Also known as **protection forestry**. This involves protection of degraded forests to allow recoupment of their flora and fauna.

In such forestry practices, forest areas designated as national parks and sanctuaries are protected from human interference.

On the other hand, well-stocked forests are managed scientifically for producing timber and other forest products without causing any negative environmental impact on the forest.

(*ii*) **Production forestry.** Also known as **commercial forestry**. This aims to fulfil the commercial demand, without causing denudation of natural forests, through intensive plantation in available land. Social forestry and agro-forestry programmes are also included in this category.

Social forestry. This programme aims to plant trees and shrubs on all unused and fallow land to provide fuel wood, fodder, etc., thus, reducing pressure on existing forests. For example, unused farmland, usar land, community land, road and rail sides, etc., are planted with suitable tree species.

Agro-forestry. This includes a variety of land uses, where woody species are grown in combination with herbaceous crops, either at the same time or in time sequence.

Taungya system involves growing agricultural crops between rows of planted trees, such as sal, teak, poplars, etc.

Jhum system is well known shifting cultivation. This is a traditional agro-forestry system usually practiced in the north-eastern region of our country. This system involves felling and burning of forests, followed by cultivation of crops for few years and then gain abondoning cultivation to allow forest regrowth.

FORESTS AND WILDLIFE LAWS

National Forest Policy (1988). The National Forest Policy, 1988, emphasizes on scientific forestry research toward attainment of policy objectives which aim at restoring ecological balance and conservation of country's natural heritage by evolving socially acceptable, technically viable and ecologically sustainable management practices. Some priority areas of research broadly indicated in the policy are :

(*i*) Increasing forest productivity per unit area and per unit time.

(*ii*) Conservation eco-restoration and management of existing forest resources.

(*iii*) Social forestry in rural/tribal development.

(*iv*) Efficient utilisation of forest produce and substitutes for wood

(*v*) Formulate a meaningful curriculum and training model for forestry education to ensure professional competence.

(*vi*) Creating awakening amongst masses for increasing tree cover and adopting agroforestry models using suitable extension methodologies.

To enhance the effectiveness of management, conservation, sustainable development and utilisation of forest resources it becomes expedient to build up technical skills and improve the quality of human resources, research capability and support, forestry extension, information dissemination system and public education.

The principal aim of Forest Policy is to ensure environmental stability and maintenance of ecological balance, including atmospheric equilibrium, which are vital for all life forms, human, animal and plant.

Wildlife (Protection) Act 1972 (Amended 1991). This Act aims to protect wild animals, birds and plants. The main objectives of this Act are :

(*i*) Restriction and prohibition on hunting of animals.

(*ii*) Protection of specified plants ;

(*iii*) Setting up and managing sanctuaries and national parks;

(*iv*) Empowering zoo authority with control of zoos and captive breeding; and

(*v*) Control of trade and commerce in wildlife, wildlife products and trophies.

Forest Act 1927. This Act aims to consolidate the law relating to forests. The basic objectives of this act are :

(*i*) Setting up and managing reserved forests, protected forests and village forests ;

(*ii*) Protection of non-government forests and forest land;

(*iii*) Control of movement of forest produce ; and

(*iv*) Control of cattle grazing.

WWF – INDIA

The Word Wide Fund for Nature-India (WWF – India) has been working to promote harmony between humankind and nature for almost three decades. Today it is recognised as a premier NGO (non-government organisation) in the country dealing with conservation and development issues.

At a time when the Web of Life has come under increasing threats, WWF – India's attempts have been to find and implement solutions so that human beings can live in harmony with nature, and leave for future generations a world rich in natural resources and natural wonders.

Formerly known as the World Wildlife Fund, WWF – India was established as a Charitable Trust in 1969. With its network of State/Divisional and Field offices spread across the country to implement its programmes. WWF – India is the largest and one of the most experienced conservation organisations in the country. The Secretariat functions from New Delhi. The organisation is part of the WWF – India family with 27 independent National Organisations. The coordinating body, the WWF – International, is located at Gland in Switzerland.

WWF – India Mission

'The promotion of nature conservation and environmental protection as the basis for sustainable and equitable development'

To achieve this Mission WWF – India undertakes the following activities :

(*i*) **Tiger and Wildlife conservation**

(*ii*) **Freshwater and Wetland Conservation Programme**

(*iii*) **Educational Awareness**

(*iv*) **Centre for Environmental Law**

(*v*) **Oceans and Coasts Programme**

(*vi*) **Library and Documentation Services**

(*vii*) **Communication and Fund Raising**

(*viii*) **Indira Gandhi Conservation Monitoring Centre (IGCMC)**

WWF – India, through its biodiversity conservation, ecorestoration and environment education programmes has been extending its support to the conservation efforts and trying to maintain the ecological balance.

Table 28.4. Environmental Calendar

Month	*Date*	*Event*
January	14 – 30	Animal Welfare Fortnight
February	2	World Wetlands Day
February	28	National Science Day
March	21	World Forestry Day
March	22	World Water Day
April	22	World Earth Day
April	28	World Heritage Day
June	5	World Environment Day
June	17	World Day to Combat Desertification
July	1	Vanmahotsava Day
July	11	World Population Day
September	16	World Ozone Day
October	1 – 7	Worldlife Week
October	2	International National Disaster Prevention Day
October	5	World Habitat Day
November	1	World Ecology Day
November	10	Forest Martyr's Day
November to December	19 to 18	National Environment Month
December	29	International Day for Biological Diversity

GRASSLANDS

The **grasslands**, also known as **rangeland**, provide forage and habitat to domestic animals and wildlife. Natural grasslands occur where rainfall is intermediate between that of desert lands and forest lands. In the temperate zone this generally means an annual precipitation between 10 and 30 inches, depending on temperature, seasonal distribution of the rainfall, and the water-holding capacity of the soil. Tropical grasslands may receive up to 60 inches concentrated in a wet season that alternates with a prolonged dry season. Grasslands are one of the most important of terrestrial ecosystem types.

A well-developed grassland community contains species with different temperature adaptations, one group growing in the cool part of the season (*i.e.*, spring and winter) and another in the hot part (summer). Forbs (*i.e.*, nongrassy herbs) are often important components, and woody plants (trees and shrubs) also occur in grasslands either as scattered individuals (savanna) or in belts or groups along streams and rivers.

In India, the area under various kinds of grass cover, including fallow and waste lands is about 18 per cent of the total land area. While forest cover is about 19 per cent of total land. This means, about 37 per cent of land can be said to be available for grazing.

Uses. In rural areas of our country, dried hay obtained from grasslands, particularly from tall grasses, is used as fuel or thatching material.

Grass cover is extremely effective in binding soil particles with the help of highly branched fibrous root system, thus significantly reducing soil erosion.

The annual average production of dry grass or hay in India is about 250 million tons.

Large herbivores are a characteristic feature of grasslands.

When man uses grasslands as natural pastures he usually replaces the native grazers with its domestic animals, *i.e.*, cattle, sheep and goats.

Since grasslands are adapted to heavy energy flow along the grazing food chain, such a switch is ecologically sound.

Degradation of Grassland

However, man has had a persistent history of misuse of grassland resources by virtue of allowing overgrazing and overplowing. The result is that many grasslands are now man-made deserts. The conversion of grassland (or forest) to desert is called **desertification**.

Degradation of destruction of grassland is mainly related to overpopulation. To enhance food production, grasslands possessing fertile soils are ploughed and converted to agricultural lands.

In India, grassland areas are frequently overgrazed. For example, the number of animals grazing in the arid and semi-arid regions has been found to be 2 – 10 times greater than the capacity of the grassland to feed the animals.

The lack of plant cover due to overgrazing causes soil erosion. When overgrazing occurs in combination with drought, the deserts are resulted.

Grassland Management

The important measures of grassland management are :

(*i*) **Protection from grazing.** This allows recovery of severely damaged vegetation.

(*ii*) **Use of rotational grazing.** While some areas are closed to grazing, allowing the plant cover to recover, grazing is permitted in other selected areas.

(*iii*) **Removal of unwanted plants.** This includes removal of woody bushes or shrubs and weeds, which usually adversely affect the productivity of grasses.

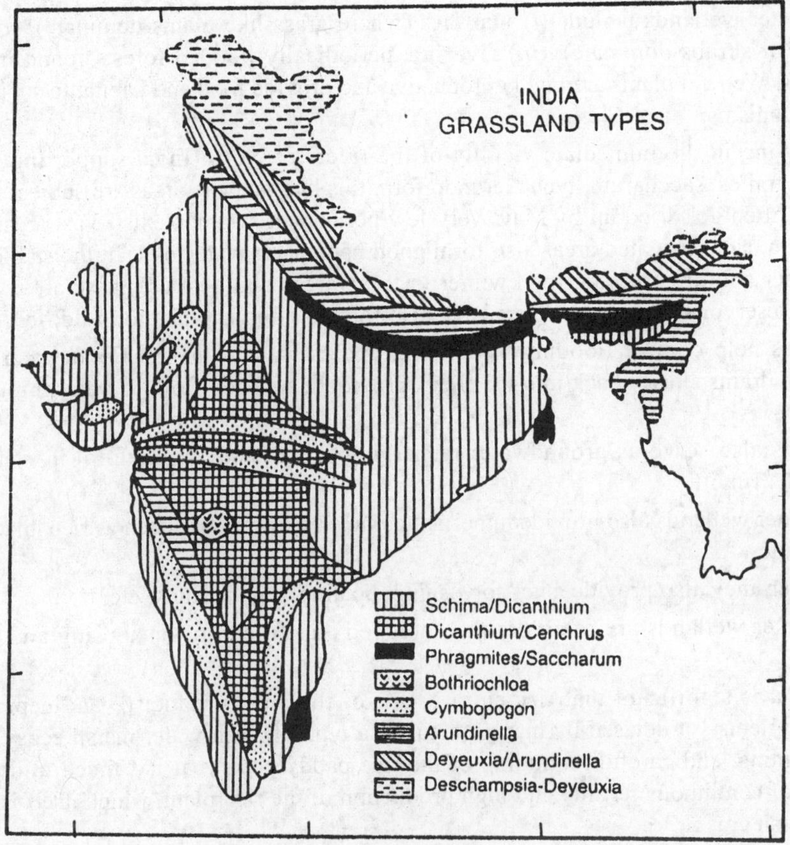

Fig. 28.3. Major grassland types of India.

(*iv*) **Conservation of soil and water.** This may be done by reducing loss of soil and water from the grassland.

(*v*) **Use of controlled burning.** This promotes recycling of nutrients stored in dried mulch and to reduce woody species invasion.

Grassland Types of India

The tropical grasslands of India may be classified as follows :

(*i*) **Xerophilous grasslands.** They are found in dry regions of North West India under semiarid conditions. The common examples are : *Andropogon contortus*, *Cenchrus ciliaris*, *C. barbatus*, etc.

(*ii*) **Mesophilous grasslands.** Also known as 'savannahs' they are extensive grassy plains typically occurring in moist deciduous forests of Uttaranchal. The dominant species are : *Saccharum munja*, *Saccharum narenga*, etc.

(*iii*) **Hygrophilous grasslands.** These are called wet 'savannahs' adapted to wet soil.

WETLANDS

Wetlands are low-lying areas, usually covered by shallow water and have characteristic soil and water tolerant vegetation. Wetlands may be either freshwater or salt water.

Wetlands occupy almost 6 per cent of the world's land surface and provide crucial environmental services. Now a days the wetlands are increasingly threatened by agriculture, pollution and construction of dams, etc.

Freshwater Wetlands

Freshwater wetlands include (*i*) **marshes** (where grass-like plants dominate), (*ii*) **swamps** (where trees or shrubs dominate), (*iii*) **riverine**, periodically flooded forests found in lowlands along streams. Wetland plants are highly productive and provide food and habitat to support a wide variety of organisms.

Areas lying in the immediate vicinity of the rivers of North India supporting tamarisks, acacias, phragmites, saccharum, typha, arundo form this habitat. The areas are subject to summer flooding but often get dried up by May. Very few permanent swamps exist. Vast areas rendered waterlogged in canal irrigated areas also form good habitat for waterfowl. In the stabilised areas occur hog deer and nilgai. Waterfowl winter in these areas and also use these areas as staging points; large reservoirs created on rivers have also formed good habitats for wildlife.

Wetlands help control flooding by holding excess water, and the flood water stored in wetlands then drains slowly back into the rivers providing a steady flow of water throughout the year.

Wetlands also serve as groundwater recharging areas. They are helpful in cleaning and purifying water runoff.

Freshwater wetlands also provide important commercial products, such as rice, black berries, blue berries, etc.

Moreover, they also provide, sites for fishing, boating, nature study, etc.

Freshwater wetlands are often drained, dredged and filled up for housing and industrial purposes.

Finally, it is significant that rice culture, one of the most productive and dependable of agricultural systems yet devised by man, is actually a type of fresh-water marsh ecosystem. The flooding, draining, and careful rebuilding of the rice paddy each year has much to do with the maintenance of continuous fertility and high production of the rice plant, which itself is a kind of cultivated marsh grass.

Saltwater Wetlands

Coastal wetlands are also known as saltwater wetlands. They include highly productive estuaries which provide food and habitat for a large number of marine organisms. Between the seas and the continents lie a belt of diverse ecosystems that are not just transition zones but have ecological characteristics of their own. Whereas physical factors such as salinity and temperature are much more variable near shore than in the sea itself, food conditions are so much better that the region is packed with life.

Along the shore live thousands of adapted species that are not to be found in the open sea, on land, or in freshwater. The word 'estuary' (Latin, *aestus* = tide) refers to a river mouth or coastal bay where the salinity is intermediate between the sea and freshwater and where tidal action is an important physical regulator. Estuaries and inshore marine waters are amongst the most naturally fertile in the world.

Tidal action promotes a rapid circulation of nutrients and food, and aids in the rapid removal of the waste products of metabolism. A diversity of plant species and life forms provides a continuous photosynthetic carpet despite variable physical conditions.

Mangrove swamps are coastal wetlands in tropical regions which contain certain trees and shrubs growing best in the intertidal zone. As mangroves expand into the ocean, other plants colonise the soil left behind. Mangrove roots provide habitat for oysters, crabs and other marine organisms. However, the coastal wetlands are also being destroyed for coastal development and agricultural lands.

Organisms have evolved many adaptations to cope. with tidal cycles, thereby enabling them to exploit the many advantages of living in an estuary. Some animals such as fiddler crabs, have internal 'biological crops' that help to time their feeding activities to the most favourable part of the tidal cycle.

Wetland Conservation

Conservationists and trained marine environmental engineers have become alarmed by the needless destruction of coastal resources. Ultimately it may be necessary to set up some kind of conservation plan so that the use of such areas by man can be placed on a sound ecological basis. Some wetland conservation programmes are as follows :

(*i*) Preparation of wetland inventories

(*ii*) Identification of wetlands of critical importance for their protection.

(*iii*) Checking waste disposal in wetlands

(*iv*) Reduction of excessive inflow of nutrients and silt into wetlands from surrounding uplands by keeping them under plant cover.

ENERGY RESOURCES

The energy crisis is a global problem today. The survival of the man will be difficult if the energy problem is not solved on the priority basis. Future energy needs of rapidly expanding human population will demand the exploitation of most energy resources. The energy consumption is maximum in developed countries as compared to developing nations.

The nineteenth century industrial societies used fossil fuels, and the daily per capita energy utilisation jumped to 70,000 kcal. Today we require energy for agriculture, industry, transport, communication, comfort and defence. The per capita energy consumption per day in the USA has reached about 2,50,000 kcal. People in other countries use far less energy. It is as low as 10,000 kcal in several developing countries. However, it is increasing fast with rapid industrialisation.

Today the world's energy resources have reached a critical stage. The reasons are many. First, the world's almost total dependence on the fossil fuels-coal, petroleum and natural gas has caused such depletion that these fuels may last only another few centuries.

Today, 30 per cent of the world's population, living in industrialised countries, consumes about 80 per cent of the global energy. On the other hand the poorer nations rely more on firewood for cooking and heating and animal power for transport.

(*i*) **Non-renewable energy resources.** Such resources include various fossil fuels and nuclear energy. Fossil fuels include petroleum products, natural gas and coal. While nuclear energy is mainly obtained from the nuclear fission of the uranium.

The world reserves of fossil fuels and uranium are limited and will eventually be depleted. It is estimated that the stock of the mineral oil will be depleted in twenty first century, if it continues to be used at present rate.

However burning fossil fuels for energy has negative environmental consequences, such as global warming, air pollution, acid rain and oil spills. Now, this has become necessary to minimise use of non-renewable energy resources, and to replace them with renewable resources.

(*ii*) **Renewable energy resources.** Such energy resources are regenerated by natural processes so that they could bc used indefinitely.

Renewable energy generally causes much less negative environmental impact than fossil fuels or nuclear energy. However, the generation of renewable energy is often more expensive than energy produced by fossil fuels or nuclear energy. With the advancement of technology, the costs of renewable energy is expected to decrease.

The most important renewable energy is solar energy, while other resources of renewable energy are hydropower, wind geothermal energy, ocean waves and tidal energy.

Solar Energy

There are two types of solar energy :

(*i*) Direct solar energy, and

(*ii*) Indirect solar energy.

Direct solar energy is the radiant energy, while **indirect solar energy** is obtained from materials that have previously incorporated the sun's radiant energy.

Solar energy can be used for direct heating and alternatively the heat converted into electricity, *i.e.*, thermal electric generation. Solar thermal pumps have been developed and are being tested under actual field conditions. Photovoltaic cells convert direct solar energy into electricity. However, a back-up system is required to store and generate electricity when solar power is not operative at night or during cloudy days. Solar photovoltaic panels, cookers, heaters and solar battery driven cars are being made and are in use.

Biomass energy. Among various energy resources, where solar energy is utilised indirectly, biomass energy is most important one. This is detained from those materials whose origin can be traced to photosynthesis, *e.g.*, live plant material and dried residues, freshwater and marine algae, agricultural and forest residues, such as straw, husks, corn cobs, bark, sawdust, roots, animal wastes, etc. Biomass also includes biodegradable organic wastes from industries, such as sugar mills, breweries, etc. About half of the world's population depends upon biomass as their main source of energy for domestic use.

In India, fuel wood is still a major source of energy for domestic purposes, at least in rural areas. Biomass fuel, may be solid, liquid or gas. This is burned to release its energy. Solid biomass includes wood, charcoal, animal dung and peat.

Biomass can be converted to liquid fuels, such as methanol (methyl alcohol) and ethanol (ethyl alcohol), which can be used in internal combustion engines of automobiles. Gasoline mixed with 10–20 per cent ethanol can be used in conventional gasoline engines.

Biogas. Biomass in the form of animal waste and hydrophytes, such as *Eichhornia* can be converted into biogas in biogas digesters by using the process of anaerobic microbial decomposition. This is a clean anaerobic fuel whose combustion produces less pollutants than other combustible energy resources. The biogas is composed of 60 per cent methane and 40 per cent carbon dioxide.

Energy Plantations

Production of biomass for energy requires sufficient area of land and water. Fuel wood has been the primary energy source for mankind from the beginning of civilisation and still continues to be the main source of energy in the developing countries, particularly in rural areas. Plant based energy is obtained through energy plantations which can produce biomass from slected species of trees in the shortest possible time at a low cost. These plants yield solid, liquid or gaseous fuel through burning, gasification, digestion, etc. Such plantations can be raised in both hills and plains particularly on marginal land. It has been calculated that about four tons of dry matter can be obtained in 0.25 ha land, which is sufficient for an average family.

Some important plants for energy plantation, which are extensively grown in India are *Leucaena leucocephala*, *Casuarina*, poplar tree and *Eucalyptus*.

Petroplants. Also called **hydrocarbon plants** are known to yield liquid hydrocarbons the substitute for liquid fuels. The hydrocarbons present in such plants can be converted into petroleum hydrocarbons of high molecular weight. *Jatropha curcas* tree yields 2 kg of seed oil per plant/year.

The oil obtained from this plant is expected to replace the conventional diesel fuel. These plants can easily be grown in arid regions and other waste lands.

Other Renewable Resources of Energy

Among other renewable resources of energy some are important which are mentioned here

(*i*) **Hydropower.** Hydroelectric power is the most important and widely used resource of renewable energy. This is generated from water. Water falling from a height turns turbines at the bottom of dams to generate electricity. Hydropower produces approximately one-fourth of the worlds electricity, and is usually cheaper than electricity produced by thermal power plants. India has big potential for hydroelectric power generation. The energy obtained from hydropower is widely used in agriculture, industry, transport, domestic sectors, etc.

However, building a dam to hold the water results in several environmental problems, such as submergence of plant and animal habitats and displacement of people.

(*ii*) **Wind energy.** The wind can also be exploited as a source of energy in different ways. When fans are rotated by the action of wind, its energy can be used for generation of electricity. National Aeronautics Limitd, Bangalore is engaged in research and development of power generation through wind mills. In many states wind mills have been set up for irrgation purposes. However, harvesting wind energy is possible only in the areas that receive fairly continual winds, such as islands, coastal areas and mountain passes. Wind energy may be converted into mechanical and electrical energy. Now a days, wind energy is being used for pumping water is rural areas. According to an estimate, about 20,000 mw electricity can be generated in India from wind alone.

(*iii*) **Tidal energy.** The difference in the level of water between high tide and low tide can be used to generate electricity.

(*iv*) **Geothermal energy.** Heated groundwater flowing upward as hot water or steam, or as hot springs, can be used to turn turbines and generate electricity is geothermal power plants.

(*v*) **Ocean waves energy.** Ocean waves, produced by winds, possess the potential to turn a turbine and generate electricity.

The present critical energy position demands an organised effort at all levels from individual practices to international action. Considerable amount of energy can be saved by reducing wastage and using energy efficient devices. Some of such measures are :

(*i*) Reduction in man's dependence on fossil fuels and development of newer alternative sources of energy.

(*ii*) Preparation of smokeless and efficient chulhas (wood stoves).

(*iii*) Improvement and expansion of sources of solar energy. Solar photovoltaic panels, cookers, heaters, and solar-battery driven cars need to be improved technically and made cost-effective.

(*iv*) The use of biogas plants must be encouraged so that agricultural and animal wastes can be used to produce both energy and fertiliser.

(*v*) Improvement in design, manufacture and maintenance of biogas plants, engines and pumps.

(*vi*) Generation of hydroelectric, wind and tidal power has to be explored more extensively.

(*vii*) Programmes for growing fuel wood trees and shrubs under the control and maintenance of local communities have to be implemented in the rural areas of the developing countries.

MARINE RESOURCES

The oceans cover nearly three-fourth of earth's surface. Ocean has been the source of many needs of man from the time immemorial. The rapid growth of human population and the advancement of industrialisation have exerted great pressure on the resources and the environment of oceans. Land resources are depleting at tremendous rate and in view of this man has started thinking to exploit the oceanic resources.

Marine resources can be broadly divided into two major categories : (*i*) **Living resources**, such as algae and the animals of the sea, (*ii*) **Non-living resources**, *e.g.*, various kinds of minerals.

Alga Resources

Marine algae vary greatly in form and range from one-called microscopic phytoplanktons to giant helps, which attain a length of 100 to 150 metres, *e.g.*, *Nereocystis*, *Macrocystis*, etc.

Green, blue, red and brown algae are commonly found in oceans. From time immemorial, algae have been widely used as human food. The algae are used as direct source of food by several sea animals and fishes. Marine planktonic diatoms are of fundamental biological importance since all life of sea is dependent upon them. Since the prehistoric times several sea weeds have been used as direct source of food to human beings. The sea weeds form the most important part of the diet of Japan and China. **Suimono** is a Japanese preparation of dried fish and several sea weeds. **Mitsu** is another Japanese preparation which contains sea weeds, fruits, sugars and dried kidney beans.

In many countries, animals are still regularly fed on fresh or processed sea weed (*e.g.*, *Laminaria* and *Fucus*).

Marine algae have been used as a manure in many countries because of their high nutrient content. Balanced fertiliser can be made by mixing sheep manure, fish and shells with seaweeds. Seaweeds are a store-house of the important potash, ionic sulphate, microelements and growth substances, besides having every other element and radical required by plants. Seaweed manure seems to increase disease resistanc to plants.

Agar extraction. Mainly red algae, *e.g.*, Gracilaria, Gelidium, etc., are used for the extraction of commercially important agar. Japan produces the largest quantity of agar, and exports to most of the countries of the world. The agar is used in several ways. It is employed in the preparation of ice creams, jellies, desserts, pharmaceuticals, photography, metal plating, lithography, metallurgy, and also in the manufacture of explosives, detergents, pesticides, in sizing the textiles and clearing many liquids. It is also used in preparing shaving creams, cosmetics and shoe polishes. The agar has constantly been used in biological laboratories for media preparation for fungal and bacteriological cultures.

Animal Resources

Fish, molluscs, crustaceans and many mammals found in sea waters are used as human food. Oceans provide an excellent habitat for above mentioned organisms.

Fish. The bulk of fish and other aquatic animals come from marine habitats. Marine fish provide considerable amount of food throughout the world. Fish are also used for the manufacture of several edible products, such as fish glue, fish meal, fish oil, fish proteins and vitamins.

Economically important fish are generally divided into two categories : (*i*) **demersal fish** and (*ii*) **pelagic fish**. The demersal fish are found at the sea-bottom, while pelagic fish floating free in the water column.

Fishing activity near the shore-line is carried on by small boats and vessels. This is called **coastal fishing**. On the other hand a large part of the total annual catch of fish comes from deep waters, and called **deep sea fishing**. This type of fishing is done by large boats or ships. In developed countries, where fishing is an organised industry, large boats and vessels are used for the purpose.

Molluscs. They are soft-bodied and shelled animals, (Latin, molluscus = soft bodied). Important molluscs from commercial point of view are the mussel, oyster, clam, etc. Many types of molluscs are used as food. However, pearl oyster is commercially important. The pearl worn as jewellery is secreted by certain oysters (bivalves). The inner calcareous surface of several bivalve shells is iridescent and pearl-like (nacreous) in appearance.

Crustaceans. They include crabs and prawns. They are the dominant arthropods of the sea. Prawns, lobsters and crabs are important economically. They make valuable items of food. India ranks first among the prawn producing countries of the world.

Phytoplankton-feeding, tiny crustaceans form an important link in the food chain of oceans.

Mammals. Whales, dolphins and porpoises are the economically important mammals found in the sea. The blue whales are the largest animals, some 30 m in length. Whales may be filter-feeders, living on planktons, or may feed on fish. Whales provide many valuable primary products, such as meat, ivory, skin, and frozen glands. They also produce many secondary products, such as oils, liver oil, etc. Fresh meet of all cetaceans, *i.e.*, whales, dolphins, porpoises, etc., has been used for human consumption.

Minerals in Sea

The sea is a store house of many valuable minerals. Most abundant elements in sea water are sodium, chlorine, magnesium, and bromine, that are commercially extracted from sea water. Phosphorite nodules are also found in sea, which can meet the shortage of phosphate fertilizers.

The sea weeds make an unique supplement for a well balanced diet. Potassium, sodium and chloride are found in the ionic form in sea weeds. The significance of iodine, as a constituent of food, is that besides being present in organic combination it is also available in part in the readily available form of the precursor of thyroxine and hence this source of iodine surpasses mineral iodine in drinking water and iodised table salt. The another micronutrients besides iodine which are important in human metabolism are iron, copper, manganese and zinc and all of them present as the trace elements of sea weeds. The highest copper content is found is *Sarconema* and *Acanthophora*. Today, iodine is produced from brown algae (seaweeds) in Japan, France, Norway and Jawa.

MINERAL RESOURCES

Some mineral elements are essential for the formation and functioning of the living body of all organisms, including man. Most of the essential elements enter animals through food chains and food webs. However, minerals are essential to our industrialised society and daily life. Mineral resources are non-renewable. The geographical distribution of essential minerals is unequal. Industry, transport, agriculture and defence preparation are making ever-increasing demands on the limited mineral deposits of the world. Depletion of almost all known and easily accessible deposits are anticipated within a few decades. Moreover, there may be shortage of some crucial elements, such as mercury, silver, gold, platinum, copper, tungsten within next 20 to 100 years. The limited resources of phosphorus which is an essential component of fertilisers may also cause retardation of agricultural growth.

Minerals can be divided into two groups : (*i*) metallic, *e.g.*, iron, copper, gold, silver, platinum, etc., and (*ii*) non-metallic, *e.g.*, sand, stone, salt, phosphates, asbestos, sulphur, corundum, etc.

Mining. This includes extraction, processing and disposal of minerals. However, mining has adverse effects on environment. This not only disturbs and damages the land, but also pollutes the soil, water and air.

The land that has been destroyed due to mining is known as **mine spoil**. Such destroyed lands can be reclaimed to a semi-natural condition by revegetation to prevent further degradation.

Conservation of Minerals

Efforts are being made to check the wasteful and injudicious use of minerals by recycling and adopting more efficient technologies, exploiting untapped deposits and by deep sea-mining. In process of recycling, used and discarded items are collected, remelted and reprocessed into new products, such as iron scraps, aluminium cans, etc.

The problem of high mineral consumption can be minimised by scientific and technological developments. Search for new production processes, designing smaller equipment, use of new raw materials and restrained use of materials have to be intensified. Finding new uses for glass, plastics, ceramics and synthetic fibres and using them as substitutes for exhaustible minerals is in progress.

Some minerals present in products can be recycled, e.g., gold, silver, lead, copper, nickel, steel, zinc, aluminium, etc. However, minerals in several other products are lost through normal use, such as paints containing lead, zinc, chromium, etc.

Table 28.5. Some Important Minerals and Their Uses

Minerals	*Uses*
1. Metallic	
Uranium	Nuclear bombs, electricity, tinting glass
Iron	Steel
Manganese	Alloy steel, disinfectants
Chromium	Metallurgy, refractory, chemicals
Nickel	Over 3000 alloys, coins, metal plating
Copper	Electrical products, alloys, gold jewellery, silver, etc.
Lead	Batteries, gasoline, paints, pipes, etc.
Tin	Tin plate, solder, chemicals, cans, containers, alloys.
Zinc	Galvanising, solder, die-casting, chemicals, brass, electrodes
Aluminum	Air crafts, rockets, building materials, electrical wiring, utensils
Gold	Monetary purposes, jewellery, dentistry, alloys
Silver	Jewellery, photography, utensils, alloys
Potassium	Fertiliser, glass, photography
Platinum	Jewellery, equipments, industrial catalyst
Radium	Medical and industrial uses, radiography
2. Non-metallic	
Phosphorus	Fertilisers, chemicals, medicine, detergents
Sulphur	Fertilisers, acid, iron and steel industries, insecticides, medicines
Corundum	Abrasives
3. Metallic liquid	
Mercury	Thermometers, dentistry, electric switches, etc.

Recycling and reusing not only renew the mineral resources, but also help in : (*i*) saving land from disruption of mining, (*ii*) reducing the amount of solid waste that is disposeable, and (*iii*) reducing pollution and consumption of energy.

However, to maintain the extended supply of minerals for a longer time, consumers must decrease their mineral consumption.

Durable and repairable products should be encouraged to be used again. Manufacturing industries should use the waste products of one manufacturing process as the raw materials for mother industry.

ENVIRONMENTAL ETHICS AND RESOURCE USE

Every morning the sun rises, and enlightens the whole land. The darkness disappears, and all becomes lighted. Besides light, we get energy from sun. All greens absorb light and food is produced for us. This great phenomenon of nature is **photosynthesis**. Only the plants are capable of this phenomenon. The plants give us food; they give us life. The plants are producers while rest consumers. The plants are biggest creation of god for us ; they are ideal donors. They only give, and from the time immemorial we worship them. The tree is eternal in its origin, and provides us the eternal peace. Beyond the kingdom of plants there is absolute darkness. Hundreds of plants are being considered sacred by the great people of this country, dealing with cultural heritage, festivals and religious ceremonies, starting from birth till death.

Lord Krishna says (Gita X, 26), "of all trees (I am) *Asvattha* (peepal tree)".

"They speak of the imperishable *asvattham* (peepal tree) as having its root above and branches below. Its leaves are the *Vedas* and he who knows this is the knower of the *Vedas*". (Gita XV, 1).

"Its branches extend below and above, nourished by the modes, with sense objects for its twigs and below in the world of men stretch forth the roots resulting in action". (Gita XV, 2).

"Its real form is not thus perceived here, nor its end nor beginning nor its foundation. Having cut off this firm-rooted *asvattham* (peepal tree) with the strong sword of non-attachment". (Gita XV, 3).

It is said in *Taittiriya Upnishad*, I, 10

"I am the originator of the world tree"

The cosmic tree is supreme–to know the cosmic tree is true knowledge, and the knower is complete in wisdom.

Plato says, "As regards the most lordly part of our soul, we must conceive of it in this wise ; we declare that God has given to each of us, as his demon, that kind of soul which is housed in the top of our body and which raises us-seeing that we are not an earthly but a heavenly plant-up from earth towards our kindred in heaven". (*Timaeus*, 90–A).

Dr. Sarvapallai Radhakrishnan says in his famous work '*The Bhagavadgita*' : 'As the tree originates in god, it is said to have its roots 'above' ; as it extends into the world, its branches are said to go downwards'.

He further says; "According to ancient belief, the vedic sacrificial cult is said to sustain the world and so the hymns are said to be the leaves which keep the tree with its trunk and branches alive".

"India is a unique country with a great cultural diversity associated with all kinds of climates and rich flora and fauna. The human societies in our country have evolved within magnificent environments and reverence to nature is inherent in our cultural ethos". I quote the lines: "The vast sky; and so the vast land; The vast ocean and the great Mountains; The great people and the majestic animals; The beautiful birds and the colourful butterflies; All the insects and all the moths; The creator has created them. Why ? Because they are need of creation; They are components of creation". One can enjoy nature by giving up greed.

Here I quote 'the philosophy I ever dreamt': "Where I find, no concrete walls, and no cement pillars; No chimneys and no factories; Neither transport nor poisonous gases, nothing like pollution, Only nature, the vast nature; Nature from horizon to horizon; great Nature in full grandeur ; Only huts and hamlets and their inhabitants; the innocense and serenity; the philosophy I ever dreamt; This is life; This is God; The 'beauty' and 'love' at the same time; great elements *khsiti* (soil), *Jal* (water), *Pavak* (fire) *gagan* (ether) and *Samira* (air), are components of great Nature; Where air develops from ether and water from air; where fire bakes our bread and earth makes base to stand; Then survival of all creatures and survival of whole 'global family'; The philosophy I ever dreamt".

Through history, the Indian people have not been exploiters but utilisers of nature. We belive in 'Live and let live'. We must recognise our responsibility to conserve earth resources for future generations. We must grow trees to maximum to protect the environment from poisonous gases. "In this hour of peril; Every hand is on trigger, and waits for order to fire; where does this race end, they do not know; Humanity is on brink of disaster; War mongers are roaming to convert the world in flames and ashes; Whole bio-kingdom is under threat; Count down has begun; By order or by mistake; All have to face doom's day; Oh contractors of humanity; Thou trust in God; Then why thou do not know; what His sons and daughters need; I tell thee, they need 'plain drinking water' ; Neither *agni* nor fire.

To conclude, I may say that the plant is the biggest creation of God; the gift of nature for us; static *Brahmarishi*, the ideal of donors, what he takes, takes from the mother, mother the land, mother the sole donor; sucks the nectar from the mother and gives to us. Grow more plants; protect the earth; protect the environment.

I worship the plants; I bow my head to plant kingdom.

29

Biodiversity and Biogeographical Regions of India

'Biological diversity' encompasses all species of plants, animals, and microorganisms and the ecosystems and ecological processes of which they are parts. It is an umbrella term for the degree of nature's variety including both the number and frequency of ecosystems, species, or genes in a given assemblage. Thus, the term '**biodiversity**' refers to the totality of 'genes, species and ecosystems' of a region.

It is usually considered at three different levels: (*i*) genetic diversity, (*ii*) species diversity and (*iii*) ecosystem diversity.

Genetic diversity is the sum total of genetic information, contained in the genes of individuals of plants, animals and microorganisms that inhabit the earth.

Species diversity refers to the variety of living organisms on earth and has been variously estimated to be between 5 and 50 million or more, though only about 1.4 million have actually been described.

Ecosystem diversity relates to the variety of habitats, biotic communities, and ecological processes in the biosphere, as well as the tremendous diversity within ecosystems in terms of habitat differences and the variety of ecological processes.

Biodiversity is defined as "the variability among living organisms from all sources, including terrestrial, marine and other aquatic ecosystems and the ecological complexes of which they are a part; this includes diversity within species, between species and of ecosystems."

Conservation and sustainable use of biodiversity is fundamental to ecologically sustainable development. Biodiversity is part of our daily lives and livelihood, and constitutes resources upon which families, communities, nations and future generations depend.

Every country has the responsibility to conserve, restore and sustainably use the biological diversity within its jurisdiction. Biological diversity is fundamental to the fulfilment of human needs. An environment rich in biological diversity offers the broadest array of options for sustainable economic ability, for sustaining human welfare and for adapting to change. Loss of biodiversity has serious economic and social costs for any country.

The experience of the past few decades has shown that as industrialisation and economic development in the classical sense takes place, patterns of consumption, production and needs, change, straining, altering and even destroying ecosystems.

India, a megabiodiversity country, while following the path of development, has been sensitive to needs of conservation and hence is still rich in biological resources. Ethos of conservation and harmonious living with nature is very much ingrained in the lifestyles of India's people.

MAGNITUDE OF BIODIVERSITY

The foundation for assessing the importance of biodiversity is an inventory of how many species exist and which species exist where. At the global level the plants and vertebrates are relatively well known. Erwin (1982), for example, suggests as many as 30 million species in total, with most undescribed species living in tropical forests. The known and described number of species of all organisms on the earth is between 1.7 and 1.8 million which is fewer than 15 per cent of the actual number'. About 61 per cent of the known species are insects. Only 4650 species of mammals are known to biological science. A large number of plant species (2,70,000) and vertebrates are known. But the fact remains that basic knowledge of the organisms that make up most ecosystems, especially in the tropics is inadequate. Information about bacteria,viruses, protists and Archaea is only fragmentary.

For convenience, many assume that about 10 million species exist, though the final figure is likely to be 30-50 million.

Given these limitations, the following table 29.1, represents a summary of the current state of knowledge (from Wilson, 1988):

Table 29.1. How many species exist

Group	Number of described species
Bacteria and blue-green algae: (*i.e.*, cyanobacteria)	4760
Fungi:	46,983
Algae:	26,900
Bryophytes (mosses and liverworms):	17,000 (WCMC, 1988)
Gymnosperms (conifers):	750 (Raven *et al.*, 1986)
Angiosperms (flowering plants):	250,000 (Raven *et al.*, 1986)
Protozoans:	30,800
Sponges:	5,000
Corals and Jellyfish:	9,000
Roundworms and earthworms:	24,000
Crustaceans:	38,000
Insects:	751,000
Other arthropods and minor invertebrates:	132,461
Molluscs:	50,000
Starfish:	6,100
Fishes (Teleosts):	19,056
Amphibians:	4,184
Reptiles:	6,300
Birds:	9,198 (Clements, 1981)
Mammals:	4,170 (Honacki *et al.*, 1982)
Total	**1,435,662**

However, new species are being discovered faster than ever before due to the efforts of Global Biodiversity Information Facility and the Species, 2000. The approximate number of species is being mentioned in the table 292.

Table 29.2. Approximate Numbers of Species.

Group	*Number of described species*
Higher plants	2,70,000
Algae	40,000
Fungi	72,000
Bacteria and cyanobacteria	4,000
Viruses	1,550
Mammals	4,650
Birds	9,700
Reptiles	7,150
Fish	26,959
Amphibians	4,780
Insects	10,25,000
Crustaceans	43,000
Molluscs	70,000
Nematodes and worms	25,000
Protozoa	40,000
Others	1,10,000

The number of plant and animal species in different groups recorded in India are given in table 29.3.

Table 29.3. Number of Species of Different Taxonomic Groups

Plants		*Animals*	
Group	*No. of species*	*Group*	*No. of species*
Angiosperms:	17,500	Mammalia:	390
Gymnosperms:	64	Aves:	1232
Pteridophytes:	1100	Reptilia:	456
Bryophytes:	2850	Amphibia:	209
Lichens:	200	Pisces:	2546
Fungi:	14,500	Protochordata:	119
Algae:	6,500	Other invertebrates:	8329
Bacteria:	850	Arthropoda:	68389
		Mollusca:	5070
		Protozoa:	2577

India has 47,000 species of flowering and non-flowering plants representing about 12 per cent of the recorded world's flora. Out of 47,000 species of plants, 5,150 are endemic and 2532 species are found in the Himalayas and adjoining regions and 1,782 in the peninsular India. India is also rich in the number of endemic faunal species it possesses.

Table 29.4. Comparative statement of recorded number of animal species in India and the world

Taxa	*INDIA*			*World*	*Percentage of India to the world*
	Species	*Endemic species*	*Threatened species*		
Protista	2577			31,259	8.24
Mollusca	5070	967		66,535	7.62
Arthropoda	68389	16214 (insects)		987,949	6.90
Other Invertebrates	8329		22	87,121	9.56
Protochordata	119			2106	5.65
Pisces	2546		4	21,723	11.72
Amphibia	209	110	3	5150	4.06
Reptilia	456	214	16	5817	7.84
Aves	1232	69	73	9026	13.66
Mammalia	390	38	75	4629	8.42

Source: MoEF 1999; Baillie 1996.

LEVELS OF BIODIVERSITY

Biodiversity is usually considered at three different hierarchical levels:

(*i*) Genetic diversity,

(*ii*) Species diversity, and

(*iii*) Community and ecosystem diversity.

Though these levels of biodiversity are interrelated yet distinct and can be studied separately to understand the interrelationships that sustain life on earth.

Genetic Diversity

Genetic diversity is the sum total of genetic information, contained in the genes of individuals of plants, animals, and microorganisms that inhabit the earth. Each species, stores an immense amount of genetic information.

For example, the number of genes is about 450-700 in Mycoplasma; 4000 in *Escherichia coli*; 13,000 in *Drosophila melanogaster*; 32,000-50,000 in *Oryza sativa* and 35,000 to 45,000 in *Homo sapiens*.

The characteristics of genetic diversity are as follows:

Genetic diversity refers to the variation of genes within species. The differences could be in alleles, *i.e.*, different variants of same genes; in entire genes, *i.e.*, the traits that determine particular characteristics or in chromosomal structures.

The genetic diversity enables a population to adapt to its environment and to respond to natural selection. On the other hand, if a species has more genetic diversity, it can adapt better to the changed environmental conditions.

Lower diversity in a species leads to uniformity, for example, in large monocultures of genetically similar crop plants. This may be an advantage when increased crop production is considered, whereas this may be a problem when a pest or pathogen attacks the crop field, and the whole crop is threatened.

The amount of genetic variation is the basis of **speciation**, *i.e.*, evolution of new species. It has a key role to play in the maintenance of diversity at species and community levels.

The total genetic diversity of a community will be greater if there are many species, as compared to an area with few species. Genetic diversity within a species often increases with environmental variability.

Species Diversity

Species diversity refers to the variety of living organisms on earth and has been variously estimated to be between 5 and 50 million or more, though only about 1.4 million have actually been described.

Species are distinct units of diversity, and each unit plays a specific role in an ecosystem. Thus, loss of species has consequences for the whole ecosystem.

Species diversity refers to the variety of species within a region.

Species richness or the number of species per unit area, denotes the measure of species diversity. The number of species increases with the area of the site, that is greater the species richness, greater is the species diversity. On the other hand, number of individuals among the

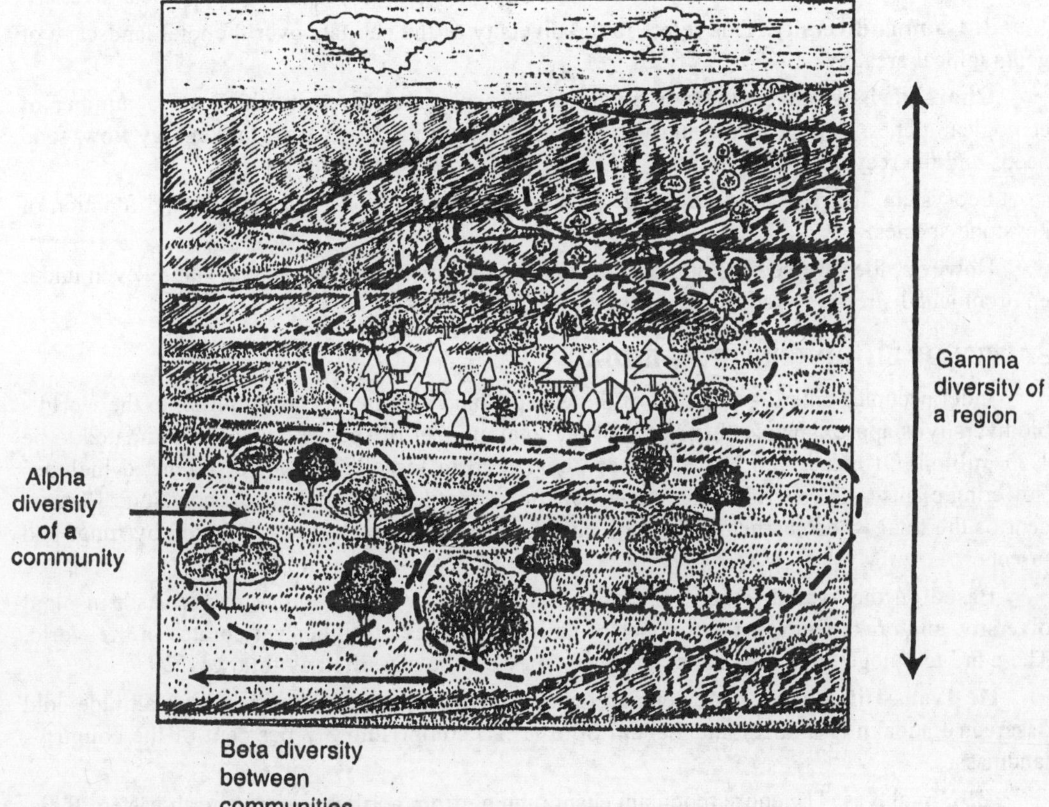

Fig. 29.1. Alpha, beta and gamma diversity.

species may also vary, resulting into differences in **evenness**, or **equitability** and consequently in diversity.

This example shows, there are equal number of species in each sample area, *i.e.*, samples 1, 2 and 3, but having varying number of individuals per species.

However, in nature, both the number and kind of species, as well as the number of individuals per species vary,which result in greater diversity.

Community and Ecosystem Diversity

Ecosystem diversity is related to the variety of habitats, biotic communities, and ecological processes in the biosphere, as well as the tremendous diversity within ecosystems in terms of habitat differences and the variety of ecological processes.

Diversity at the level of community and ecosystem has been divided into three categories:

1. Alpha diversity (= within community diversity). This refers to the diversity of organisms sharing the same community or habitat.

A combination of species richness and equitability or evenness is used to represent diversity within a community or habitat. Species frequently change when habitat or community changes.

2. Beta diversity (= between-community diversity). This refers the rate of replacement of species along a gradient of habitats or communities. In beta diversity, there are differences in species composition of communities along environmental gradients, *e.g.*, altitudinal gradient, moisture gradient, etc. Greater the dissimilarity between communities, higher is the beta diversity.

3. Gamma diversity. This refers to the diversity of the habitats over the total landscape or geographical area.

Characteristics of ecosystem diversity. In ecosystem diversity there are a number of ecological niches, trophic levels and various ecological processes that sustain energy flow, food needs and the recycling of nutrients.

Ecosystem diversity has a focus on various biotic interactions and the role and function of keystone species.

However, the diverse communities are functionally more productive and stable, even under environmental stresses, such as dry or drought conditions.

STATUS OF BIODIVERSITY IN INDIA

India occupies only 2.4 per cent of the world's land area but its contribution to the world's biodiversity is approximately 8 per cent of the total number of species, which is estimated to be 1.75 million. Of these, 126,188 have been described in India. The species recorded includes flowering plants (angiosperms), mammals, fish, birds, reptiles and amphibians, constitute 17.3 per cent of the total wheareas nearly 60 per cent of India's bio-wealth is contributed by fungi and insects.

Based on the available data, India ranks tenth in the world and fourth in Asia in plant diversity, and ranks tenth in the number of endemic species of higher vertebrates in the world. There are ten biogeographical zones in India. They can be classified as follows:

1. Trans-Himalayas. An extension of the Tibetan Plateau, harbouring high-altitude cold desert in Laddakh (J and K) and Lahaul Spiti (H.P.) comprising 5.7 per cent of the country's landmass.

2. Himalayas. The entire mountain chain running from north-western to north-eastern India, comprising a diverse range of biotic provinces and biomes, 7.2 per cent of the country's landmass.

3. Desert. The extremely arid area west of the Aravalli hill range, comprising both the salty desert of Gujarat and the sand desert of Rajasthan. 6.9 per cent of the country's landmass.

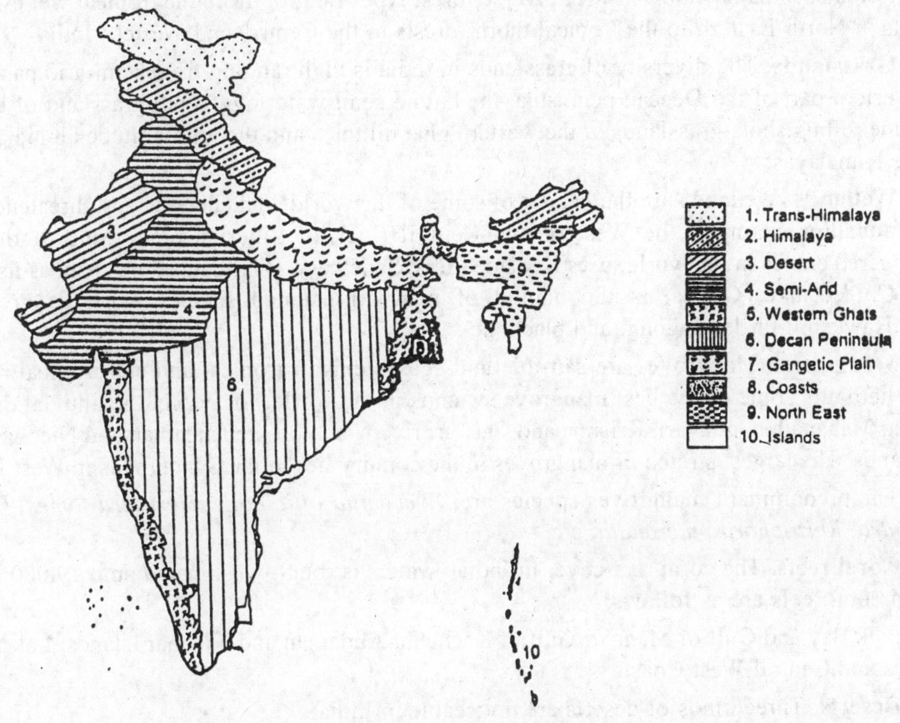

Fig. 29.2. Biogeographical regions of India (source WII Dehradun).

4. Semi-arid. The zone between the desert and the Deccan plateau, including the Aravalli hill range. 15.6 per cent of the country's landmass.

5. Western ghats. The hill ranges and plains running along the western coast line, south of the Tapti river, covering an extremely diverse range of biotic provinces and biomes. 5.8 per cent of the country's landmass.

6. Deccan peninsula. The largest of the zones, covering much of the southern and south-central plateau with a predominantly deciduous vegetation. 4.3 per cent of the country's landmass.

7. Gangetic plain. Defined by the Gangas river system, these plains are relatively homogenous. 11 per cent of the country's landmass.

8. North-east India. The plains and non-Himalayan hill ranges of north eastern India, with a wide variation of vegetation. 5.2 per cent of the country's landmass.

9. Islands. The Andaman and Nicobar Islands in the Bay of Bengal, with a highly diverse set of biomes. 0.03 per cent of the country's landmass.

10. Coasts. A large coastline distributed both to the west and east, with distinct differences between the two; Lakshadeep Islands are included in this with the percent area being negligible.

Apart from the above mentioned biogeographic classification, ecosystems can also be demarcated on the basis of purely geographical or geological features, such as mountains, islands, valleys, plateau, oceans, etc., on the basis of vegetative cover like forests, grasslands, mangroves and deserts; on the basis of climatic conditions, such as arid and semi-arid areas, permanently snow-covered areas, high rainfall areas; on the basis of soil characteristic and other such criteria.

The main natural biomes/ecosystems are: forests, grasslands, wetlands, mangroves,coral reefs, deserts, etc.

Forests. India is endowed with diverse forest types ranging from the tropical wet evergreen forests in North-Eastern to the tropical thorn forests in the Central and Western India.

Grasslands. The diversity of grasslands in India is high ranging from semi-arid pastures of the western part of the Deccan peninsula, the humid semi-waterlogged tall grassland of theTerai belt, the rolling shola grasslands of the western ghat hilltops, and the high-altitude alpine pastures of the Himalayas.

Wetlands. Wetlands are the habitat of some of the world's endangered and threatened flora and fauna.For example, the Western and Central flock of Siberian crane, one of the most endangered cranes in the world, uses Keoladeo Bird Sanctuary, Bharatpur, Rajasthan, as its winter site. Chilka lake, Orissa, is the habitat of many threatened species, such as green sea turtle,Hawksbill turtle, dugong, and blackbuck.

Mangroves. Mangroves are salt-tolerand ecosystems in tropical and sub-tropical regions. India harbours some of the best mangrove swamps in the world, located in the alluvial deltas of Ganga, Mahanadi, Godavari, Krishna and Cauveri rivers, and on the Andaman and Nicobar group of Islands. The largest stretch of mangroves in the country lies in the Sunderbans in West Bengal.

The predominant mangroves species are; *Avecennia officinalis*, *Heritiera fomes*, *Ceriops decandra*, *Rhizophora mucronata*, etc.

Coral reefs. The coral reef cover in Indian waters is roughly estimated upto 19,000 sq.Km. Indian coral reefs are as follows:

Palk Bay and Gulf of Mannar; Gulf of Kachchh; Andaman and Nicobar Islands; Lakshadeep Islands and Central West Coast.

Deserts. Three kinds of deserts are noticeable in India:

1. The sand desert of western Rajasthan and neighbouring areas.
2. The vast salt desert of Gujarat
3. The high-altitude cold desert of Jammu and Kashmir and Himachal Pradesh.

Desert fox, desert cat, houbara bustard and some sandgrouse species are restricted to Thar Desert, Rajasthan.

In the Rann of Kachchh, lies the nesting ground of Flamingoes and the only known population of Asiatic wild ass.

The distinctive animals of cold desert are:

Kiang (relative of wild ass of Kachchh), snow leopard, Yak, Tibetan antelope, Ibex, Blue sheep, Tibetan gazelle, Woolly hare, etc.

GRADIENTS OF BIODIVERSITY

Biodiversity varies with change in latitude or altitude.When one moves from high to low latitudes (*i.e.*, from the poles to the equator), the biodiversiy increases.

In the temperate region, the climate is severe having short growing period for plants. On the other hand, in tropical rain forest the conditions are favourable for growth of plants throughout the year. Favourable conditions are helpful for speciation, and large number of species develop and grow.

The mean number of species of vascular plants per 0.1 ha sample area in tropical rain forests varies from 118-236, while in temperate zones, this range is 21-48 species.

Such a correlation between diversity and latitude is also found for a variety of several other taxa, such as ants, birds, butterflies, moths, etc.

In the similar way, there is a decrease in species diversity from lower to higher altitudes on a mountain. Increase of 1000 metres in altitude results in a temperature drop of about 6.5° C. Such

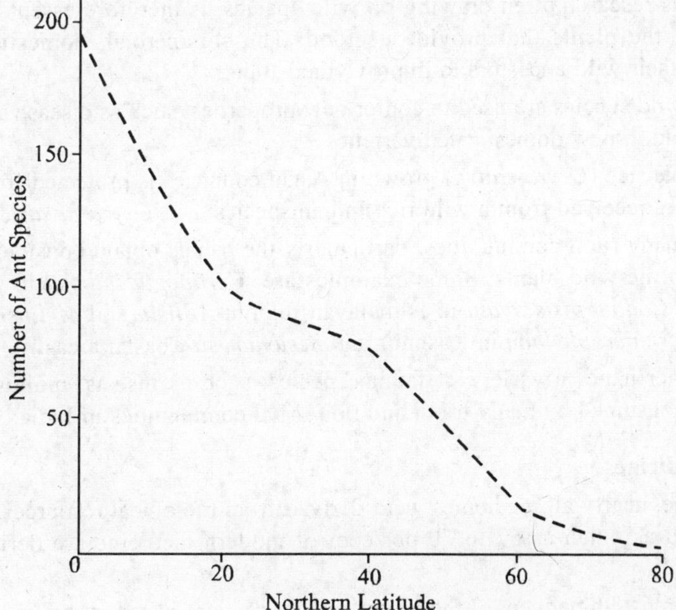

Fig. 29.3. Latitudinal gradient of biodiversity.

temperature drop and seasonal variability at higher altitudes are responsible for lowering the diversity.

The **latitudinal** and **altitudinal** gradients of species diversity are two master gradients. To some extent regional and taxa-related gradients also affect biodiversity. It is expected that prevalence of more complex and heterogeneous environment makes the flora and fauna more diverse and complex.

Uses of Biodiversity

Humans derive many direct and indirect benifits from biological diversity. All our food comes from wild species brought into domestication. Most of our medicines, pharmaceutical, fibres, rubber and timber come from biological resources. The biodiversity also provides many ecological services free of charge that are responsible for maintaining ecosystems. Our water is supplied by one of nature's most important processes, called the 'hydrological cycle'. Forested watersheds provide clear,high-quality water for domestic or agricultural use,while healthy rivers provide water, transport and fish.

Source of Food and Improved Varieties

Biodiversity has direct consumptive value in agriculture. In modern agriculture, biodiversity is used as follows:

(*i*) As a source of new crops
(*ii*) As a source material for breeding improved varieties, and
(*iii*) As a source new biodegradable pesticides.

Approximately 80,000 edible plants have been used at one time or another in human history, of which only about 150 have even been cultivated on a large scale. Today a mere 10 to 20 species provide 80 to 90 per cent requirements of the world. Over half of human nutrition is provided by the three major carbohydrate crops, *i.e.*, rice, wheat and maize. Fats, oils, fibres, etc., are other uses for which many new species are to be investigated.

Continuous research,often drawing on wild species, is therefore essential to maintain the productivity of the plants that provide us food. The commercial, domesticated species are crossbred with their wild ancestors to improve their traits.

Genes of wild species are used to confer new properties, such as disease and pest resistance or increased yield in new domesticated varieties.

For example rice (*Oryza sativa*) grown in Asian countries is protected from the four major diseases by genes received from a wild rice Indian species, *i.e.*, *Oryza nivara*.

In India, many rural communities, particularly the tribals obtain considerable part of their daily food from the wild plants. Some examples are: *Ceropegia bubosa* in Central India and Western Ghats; *Codonopisis ovata* in Himalayan region; *Ardisia* and *Meliosma pinnata* in the North-east; *Cicer microphyllum* in Kashmir and *Sesuvium* in Coastal areas.

On the other hand, a variety of faunal species, such as insects, molluscs, spiders, wild herbivores are consumed by many tribal and non-tribal communities in India.

Drugs and Medicines

At one time, nearly all medicines were derived from biological resources. Even today they remain vital and as much as 67 to 70 per cent of modern medicine are derived from natural products.

In India, almost 95 per cent of the prescriptions are plant-based in the traditional systems of Ayurveda, Unani and Sidha. Many indigenous medicines also utilise medicines and their parts or extracts as remedies for various diseases.

Examples of plant-derived substances developed into valuable drugs are: Morphine obtained from *Papaver somniferum* and used as an analgesic; Quinine obtained from *Cinchona ledgeriana*, used for treatment of Malaria, and Taxol, an anticancer drug obtained from the bark of yew tree (*Taxas baccata*, *T.brevifolia*).

Botanochemicals. Many plants are also used for the manufacture of hundreds of synthetic products, which are called 'botanochemicals'.

Aesthetic and Cultural Benefits

Biodiversity has also great aesthetic value. Diverse habitats and species have non-consumptive use-value. Examples of aesthetic benefits include ecotourism, bird watching, wildlife, pet keeping, gardening, recreation and scientific research.

Many plants, animals and their parts are used in rituals all over the country. In a majority of Indian villages and towns, plants like *Ocimum sanctum* (Tulsi), *Ficus religiosa* (Pipal), *F. bengalensis* (Bargad), etc., are considered sacred and worshipped by the people. Flowers of *Hibiscus*, *Datura* and *Euphorbia*, leaves of *Aegle marmelos* (Bel), *Eragrostis cynasuroides* (Kusa grass), rice, til, chenopods sacred, odorous roots of *Dolomiaea macrocephala* (Dhup) are used in rituals. Besides, sacred values are attached to entire ecosystems, for example, patches of forests were believed to be the abode of demigods and are used only for prayers and rituals. Many 'sacred groves' exist in different parts of our country.

Several birds, snakes and other animals are treated sacred and worshipped.

Ecosystem Services

The indirect use-value of biodiversity includes ecosystem process of biological diversity, which provides valuable ecological services to the biosphere.

For example, the ecosystem's ability to absorb pollution, maintain soil fertility and microclimates, recharge groundwater and other invaluable services. These services include maintenance of gaseous composition of the atmosphere, climate control by forests and oceanic systems, natural pest control, pollination of plants by insects and birds, formation and protection of soil, conservation and purification of water, and nutrient cycling, etc.

The above mentioned ecosystem services have been valued in the range of 16 to 54 trillion US dollars per year.

Threats to Biodiversity

Major biodiversity threats are as follows:

(*i*) Habitat destruction and fragmentation
(*ii*) Extention of agriculture
(*iii*) Filling up of wetlands
(*iv*) Conversion of rich biodiversity site for luman settlement and industrial development
(*v*) Destruction of coastal areas,
(*vi*) Uncontrolled commercial exploitation
(*vii*) Disturbance and pollution, and
(*viii*) Introduction of non-native (exotic) species

Habitat Loss and Fragmentation

The destruction of habitats is the primary reason for the loss of biodiversity. This erosion of biodiversity is largely due to habitat loss caused by the expansion of various development projects, such as mines, dams, roads and canal construction. When people cut down trees, fill a wetland, plough a grassland or burn a forest the natural habitat of a species is changed or destroyed. Such changes can kill or force out many plants, animals and microorganisms. These changes also disrupt complex interactions among the species.

Habitat loss leads to the fragmentation of continuous stretches of land and consequently fragments wildlife populations inhabiting them. These small populations are increasingly vulnerable to inbreeding depression, high infant mortality, susceptibility to environmental stochasticity and in the long run possibly to extinction.

Apart from the primary loss of habitats, there are many other problems contributing to the loss and endangered status of several plant and animal species. Habitat degradation can lead to declines in primary food species for wildlife.

Poaching is another threat as one of the primary reasons for extinction of species such as the tiger.

Population pressures, such as collection of fuelwood and fodder, and grazing in forests by local communities also take their toll on the forests and consequently its biodiversity.

Other minor factors include fire, and natural calamities like droughts, diseases, cyclones and land slides.

It is estimated that, after independence, the country has lost 4,696 million ha of forest land to non-forestry purposes.

Disturbance and Pollution

Communities are affected by natural disturbances, such as fire, tree fall, landslides and defoliation by insects.

Anthropogenic disturbances differ from natural disturbances in intensity, rate and spatial extent. For example, man by using fire more frequently may change species richness of a community.

The new anthropogenic impacts may be large number of synthetic compounds, massive releases of radiations by nuclear projects, and spill over of oil in sea. These impacts may lead to a change in the habitat quality.

Over explcitation of natural resourses, such as fishing by boats, mechanical catching of other animal species, etc., are serious threats to wildlife. Disturbance in migratory routes of several fishes, due to construction of dams, etc., hence they are not able to reach their spawning grounds and face extinction.

Pollution is also responsible for reduction and elimination of populations of sensitive species. For example, pesticide linked decline of fish-eating birds and falcons.

Lead poisoning of water is also a cause of mortality of many species, such as ducks, swans and cranes, as they take in the spent shotgun pellets that fell into lakes and marshes. Eutrophication, *i.e.*, nutrient enrichment of water bodies is also responsible for reduction of species diversity.

Introduction of Exotic Species

New species that enter a geographical region are called **exotic** or alien species. The invasion of such species may cause disappearance of native species through changed biotic interactions. The invasion of exotic species is considered second one to habitat destruction as a major threat to extinction of species.

Exotic species have more effective impact in island ecosystems, which give shelter to the world's threatened biological diversity. Examples are as follows:

(*i*) An exotic species *Lantana camara* has invaded many forest lands in different parts of our country, and strongly competes with the native species.

(*ii*) On the other hand water hyacinth (*Eichhornia crassipes*), an exotic aquatic species, clogs rivers, lakes and village ponds in different parts of India and several other tropical countries, which threaten, the survival of several native aquatic species.

(*iii*) Nile perch, an exotic predatory fish has been introduced into Lake Victoria of South Africa, where this has threatened the entire ecosystem of the lake, and eliminated several native species of small fish, *e.g.*, Cichild fish species.

(*iv*) *Periplaneta americana* (American cockroach) has threatened the existence of native cockroach, *i.e.*, *Blatta orientalis*.

(*v*) *Parthenium hysterophorus* has threatened several native herbs and shrubs of this country.

(*vi*) Goats and rabbits introduced in the islands of Indian and Pacific oceans are destroying the habitats of several plants, birds and reptiles.

Extinction of Species

Extinction has been a fact of life since life first emerged. The present few million species are the modern-day survivers of the estimated several billion species that have ever existed. All past extinctions have occurred by natural processes, but today humans are the main cause of extinction.

The rapid destruction of the world's most diverse ecosystems, especially in the tropics,has led most experts to conclude that perhaps a quarter of the earth's total biological diversity is at serious risk of extinction during the next 20-30 years. By many indications, the world is already experiencing extinction rates of greater scale and impact than at any previous time in the earth's history.

More species than even before are threatened with extinction, with thousands, mostly insects, disappearing each year. There are three main types of extinction processes:

1. Natural extinction. With the change in environmental conditions, some species become extinct, while others, which are more adapted to changed conditions,take their place.This type of extinction of species which took place in the geological past at a very slow rate, is called **natural extinction**.

2. Mass extinction. As we know, there have been several periods in the geological history of earth when large number of species disappeared because of catastrophes. This kind of extinction is called '**mass extinction**'. Such mass extinctions occurred in millions of years.

3. Anthropogenic extinction. All past extinctions occurred either by natural or mass extinctions, but today humans are the main cause of extinctions. Today an alarming number of species are disappearing from the surface of earth due to irresponsible human activities. This man-made extinction of species represents a very severe depletion of biodiversity, particularly because this occurs within a short period of time.

It has been estimated at World Conservation Monitoring Centre (WCMC), that 533 species of animals (mostly vertebrates) and 384 species of plants (mostly angiosperms) have become extinct since 1600 AD. More species have gone extinct from the islands than from the mainland or oceans. The current rate of extinction of species is 1000 to 10,000 times higher than the background rate of extinction. Some interesting observations about the current rate extinction of species are as follows:

1. From ten high-diversity localities in tropical forests which cover about 300,000 sq. km, some17,000 endemic plant species and 350,000 endemic animal species are threatened and could be lost in near future.

2. The tropical forests alone are losing about 14,000 to 40,000 species per year,which comes about 2-5 species per hour.

3. It is estimated that by the end of twenty first century, if the current rate continues, the earth will lose up to fifty per cent of the species.

Susceptibility to Extinction

The characteristics of species that make them susceptible for extinction are as follows:

1. Large body size. Many animals are large-sized, *e.g.*, Bengal tiger, lion and elephant.

2. Small population size and low reproductive rate. This characteristic of species is responsible for susceptibility to extinction, *e.g.*, Blue whale and Giant panda.

3. Feeding at high trophic levels in the food chain. This characteristic also makes the animals susceptible to extinction. The examples are: Bengal tiger and Bald eagle.

4. Fixed migratory routes and habitat. Such routes cannot be changed, if any interruption occurs that leads towards extinction. The examples are; Blue whale and Whooping crane.

5. Localised and narrow range of distribution. This characteristic also makes the species susceptible to extinction, *e.g.*, woodland caribon and many island species.

The Iucn Red List Categories

The World Conservation Union (WCU), formerly known as International Union for the Conservation of Nature and Natural Resources (IUCN), has recognised eight Red List Categories of species. They are as follows:

1. Extinct. A taxon is **Extinct** when there is no reasonable doubt that the last individual has died.

2. Extinct in the Wild. A taxon is **Extinct in the Wild** when exhaustive surveys, in known and/or expected habitats, have failed to record an individual.

3. Critically Endangered. A taxon is **Critically Endangered** when it is facing an extremely high risk of extinction in the wild in the immediate future.

4. Endangered. A taxon is **Endangered** when it is not Critically Endangered, but is facing a very high risk of extinction in the wild in the near future.

5. Vulnerable. A taxon is **Vulnerable** when it is not Critically Endangered or Endangered, but is facing a high risk of extinction in the wild in the medium-term future.

6. Lower Risk. A taxon is **Lower Risk** when it has been evaluated and does not satisfy the criteria for Critically Endangered, Endangered or Vulnerable.

7. Data Deficient. A taxon is **Data Deficient** when there is inadequate information to make a direct, or indirect, assessment of its risk of extinction.

8. Not Evaluated. A taxon is **Not Evaluated** when it has not yet been assessed against the above criteria.

The IUCN Red List is a catalogue of those taxa which are facing the risk of extinction.

The uses of the Red List are:

(*i*) To develop awareness about the importance of threatened biological diversity.

(*ii*) Identification and documentation of endangered species.

(*iii*) Providing a global index of the decline of biodiversity.

(*iv*) Defining conservation priorities at the local level and guiding conservation action.

The species which are threatened with extinction are included in **Vulnerable**, **Endangered** or **Critically Endangered** categories.

Rare species. Taxa with small world populations that are not at present 'Endangered' or 'Vulnerable' but are at risk.

In practice, 'Endangered' and 'Vulnerable' categories may include, temporarily, taxa whose populations are beginning to recover as a result of remedial action, but whose recovery is insufficient to justify that transfer to another category. These taxa are usually localised within restricted geographical areas or habitats or are thinly scattered over a more extensive range.

Indeterminate. Taxa known to be 'Endangered', 'Vulnerable', or 'Rare', but where there is not enough information to say which of the three categories is appropriate.

The IUCN Red List System was initiated in 1963, and since then evaluation of the conservation status of species and subspecies is continuing on a global scale.

The 2000 IUCN Red List is the most comprehensive inventory of the global conservation status both of plant and animal species.

The 2000 IUCN Red List contains assessments of more than 18,000 species, of which 11,000 are threatened.

Status of Threatened Species. According to 2000 IUCN Red List, there are 11,046 species listed as threatened, *i.e.*, Critically Endangered, Endangered, or Vulnerable. Of these 5,485 are animals and 5,611 are plants. 1939 species are listed as Critically Endangered, of which 925 are animals and 1,014 are plants.

The percentage of threatened species of Angiosperms and four verteberate groups of animals are as follows :

Table 29.5. Percentage of Threatened Species

Categories	*Angiosperms*	*Amphibians*	*Reptiles*	*Birds*	*Mammals*
1. Critically Endangered	16%	14%	15%	9.0%	10%
2. Endangered	19%	22%	21%	17%	19%
3. Vulnerable	51%	48%	43%	48%	34%
4. Lower Risk	14%	16%	21%	16%	37%

According to 2000 IUCN Red List, in India 44 species of plants are critically endangered; 113 endangered and 87 vulnerable. While amongst animals 18 species are critically endangered, 54 endangered and 143 species, vulnerable.

Table 29.6. The Number of Plant and Animal Species of Various Threat Categories in India

Categories	*Plants*	*Animals*
1. Critically Endangered	44	18
2. Endangered	113	54
3. Vulnerable	87	143
4. Lower Risk	73	109
5. Data Deficient	14	31

Table 29.7. Examples of Threatened Species in India

Categories	*Plants*	*Animals*
1. Critically Endangered	*Berberis nilghiriensis*	*Sus salvanius* (Pigmy hog)
2. Endangered	*Bentinckia nicobarica*	*Ailurus fulgens* (Red Panda)
3. Vulnerable	*Cupressus cashmeriana*	*Antilope cerbicapra* (Black buck)

CONSERVATION OF BIODIVERSITY

The complex threats to biological diversity call for a wide range of responses across a large number of private and public sectors. All are necessary, with the mix of responses adjusted to the local conditions. As government policies are often responsible for depleting biological resources, it stands to reason that policy changes are often a necessary first step toward conservation.

National policies dealing directly with wildlands management or forestry, or influencing biodiversity indirectly through land tenure, rural development, family planning, and subsidies for food, pesticides, or energy can have significant impacts on the conservation of biodiversity.

Protecting species can best be done through protecting habitats. Most national governments have established legal means for protecting habitats that are important for conserving biological resources. These can include: national parks and other categories of reserves; local laws protecting particular forests, reefs or wetlands; planning restrictions on certain types of land; and customary laws protecting sacred groves or other special sites. The responsibility for such management is often spread widely among public and private institutions.

The Ministry of Environment and Forests (MOEF) is the main agency in the Government of India for planning, promotion, coordination and overseeing the implementation of the environmental and forestry programmes. The Mandates of the MOEF included survey of flora, fauna, forests and wildlife, and conservation of natural resources. A number of institutions affiliated with the Ministry are involved in the work related to various aspects of biological diversity.

Today, the ecosystems are undergoing change due to pollution, invasive species, anthropogenic over exploitation and climate change. Biodiversity at all the three levels (*i.e.*, genetic, species and ecosystem or community biodiversity) is important and needs to be conserved. The most effective and efficient mechanism for conserving biodiversity is to prevent further destruction or degradation of habitats by us.

People form the foundation for the sustainable use of biological resources. Local communities need to be more or involved in the management of biological resources, and to benefit from their sustainable use.

There are two basic strategies of biodiversity conservation, (*i*) *in situ* (on site) and (*ii*) *ex situ* (off site).

***In situ* Conservation (Within natural habitat) Strategies**

Protected areas. The areas of land and/or sea especially for protection and maintenance of biodiversity, and of natural and associated cultural resources. These areas are managed through legal or other effective means, *e.g.*, National Parks and Wildlife Sanctuaries. The earliest national parks are: The Yellowstone National Park in USA and the Royal National Park near Sydney, Australia. These parks were chosen because of their scenic beauty and recreational values.

Today, many such protected areas throughout the world protect rare species. World Conservation Monitoring Centre (WCMC) has recognised 37,000 protected areas around the world.

In India, some important measures are taken. They are as follows :

Approximately 4.7 per cent of the total geographical area of the country has been earmarked for extensive *in situ* conservation of habitats and ecosystems. A protected area network of 89 National Parks and 492 Wildlife Sanctuaries have been created (MOEF, 2002). The results of this network have been significant in restoring viable population of large mammals, such as tiger, lion, rhinoceros, crocodiles, elephants, etc.

The Jim Corbett National Park, Nainital, Uttaranchal, was the first National Park, in India.

The Indian Council of Forestry Research (ICFRE) has identified 309 forest preservation plots of representative forest types for conservation of viable and representative areas of biodiversity. 187 of these plots are in natural forests and 112 in plantations covering a total area of 8,500 hectares.

Eco-development. A programme entitled 'eco-development' for *in situ* conservation of biological diversity involving local communities has been initiated in recent years. The concept of 'eco-development' includes the ecological and economic parameters for sustained conservation of ecosystems by involving the local communities with the maintenance of earmarked regions surrounding protected areas. The economic needs of the local communities are taken care of under this programme through provision of alternative sources of income and a steady availability of forest and related produce.

The main benefits of protected areas are:

1. To maintain viable populations of all native species and subspecies.
2. To maintain the number and distribution of communities and habitats. Conservation of the genetic diversity of all the existing species.
3. To prevent human caused introductions of alien species.
4. To make it possible for species and habitats and shift in response to environmental changes.

Biosphere reserve programme. Biosphere reserves are a special category of protected areas of land and/or coastal environments, wherein people are an integral component of the system.

The biosphere reserves are representative examples of natural biomes and contain unique biological communities.

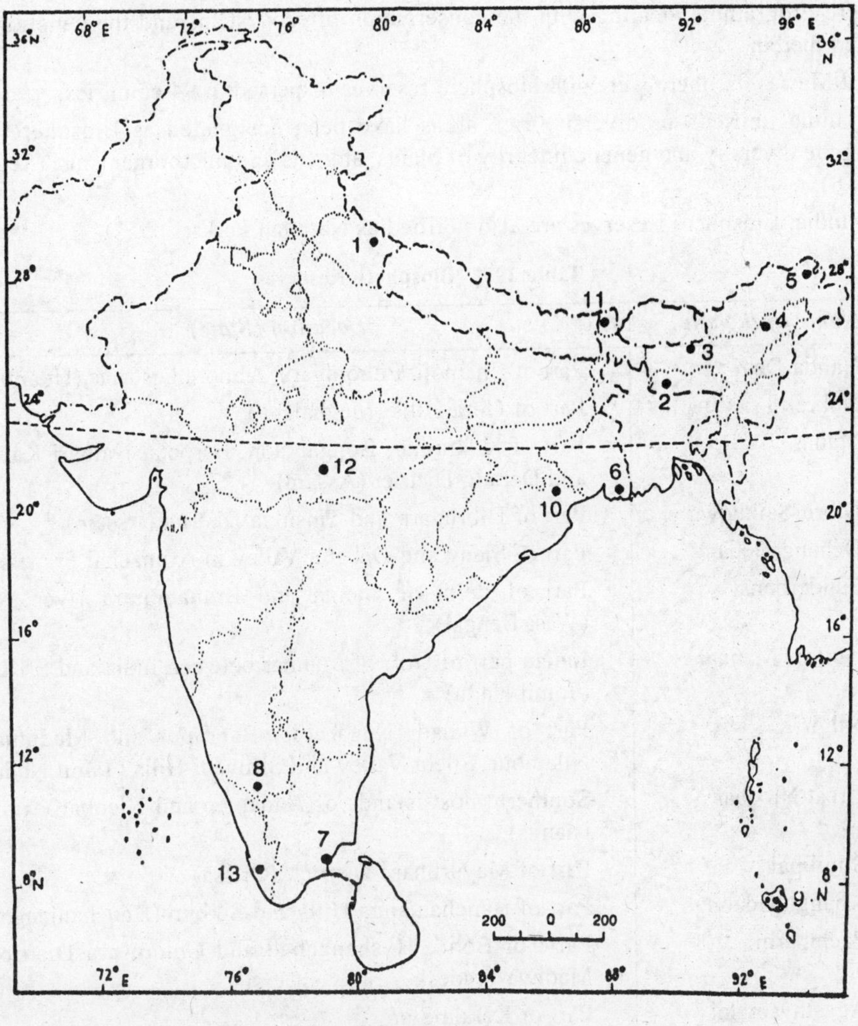

Fig. 29.4. The Biosphere Reserves in India.

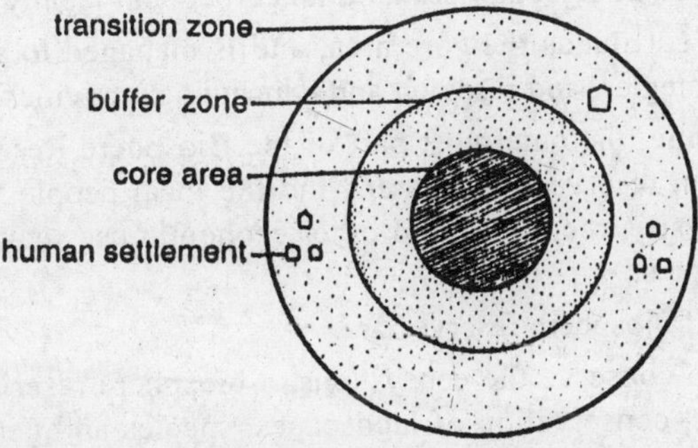

Fig. 29.5. Zonation in a terrestrial biosphere reserve.

The concept of Biosphere Reserves was launched in 1975 as a, part of UNESCO's 'Man and Biosphere Programme, dealing with the conservation of ecosystems and the genetic resources contained therein.

Till May 2002, there were 408 biosphere reserves dispersed in 94 countries.

In India, thirteen biodiversity rich areas have been designated as Biosphere Reserves applying the diversity and genetic integrity of plants, animals and microorganisms. (See map and table 29.8).

In India, Biosphere Reserves are also notified as National Parks.

Table 29.8. Biosphere Reserves

Name of the site	***Location (State)***
1. Nanda Devi	Part of Chamoli, Pithoragarh, Almora Districts (Uttaranchal)
2. Nokrek	Part of Gora Hills (Meghalaya)
3. Manas	Part of Kokrajhar, Bongaigaon, Barpeta, Nalbari Kamprup and Darang District (Assam)
4. Dibru-Saikhowa	Part of Dibrugarh and Tinsukia District (Assam)
5. Dehang Debang	Part of Siang and Debang Valley in Arunachal Pradesh
6. Sunderbans	Part of delta of Ganga and Brahamputra river system (West Bengal)
7. Gulf of Mannar	Indian part of Gulf of Mannar between India and Sri Lanka (Tamil Nadu)
8. Nilgiri	Part of Wynad, Nagarhole, Bandipur and Madhumalai, Nilambur, Silent Valley and Siruvani Hills (Tamil Nadu)
9. Great Nicobar	Southern most islands of Andaman and Nicobar (A and N islands)
10. Similipal	Part of Mayurbhanj District, (Orissa)
11. Khanghendzonga	Part of Kanchanjanga Hills and Sikkim (Kanchanjanga)
12. Pachmarhi	Parts of Betul, Hoshangabad and Chindwara Districts of Madhya Pradesh
13. Agasthyamalai	Part of Kerala State

Zonatism of a terrestrial biosphere reserve. A terrestrial biosphere reserve consists of **core**, **buffer** and **transition** zones.

(*i*) The **natural** or **core zone** comprises an undisturbed and legally protected ecosystem.

(*ii*) The **buffer zone** surrounαs the core area, and is managed to accommodate a greater variety of resource use strategies, and research and educational activities.

(*iii*) The **transition zone**, the outermost part of the Biosphere Reserve. This is an area of active cooperation between reserve management and the local people, wherein activities like settlements, cropping, forestry, recreation and other economic uses continue in harmony with people and conservation goals.

The main functions of biosphere reserves are:

(*i*) **Conservation.** To conserve the ecosystems, a biosphere reserve programme is being implemented, for example, conservation of landscapes, species and genetic resources. It also encourages traditional resource use.

(*ii*) **Eco-development.** The concept of eco-development integrates the ecological and economic parameters for sustained conservation of ecosystems by involving the local people with the maintenance of earmarked regions. Biosphere reserves are also used to promote economic development which is culturally, socially and ecologically sustainable.

(*iii*) **Scientific research programme.** Programmes have also been launched for scientific management and wise use of fragile ecosystem. Specific programmes for management and conservation of wetlands, mangroves and coral reef systems are also being implemented. Under this programme, 21 wetlands,15 mangrove areas and 4 coral reef areas have been identified for management. National and sub-national level committees oversee and guide these programmes to ensure strong policy and strategic support.

Sacred forests and sacred lakes. In India and some other Asian countries, a traditional strategy for the protection of biodiversity has been in practice in the form of **sacred forests** or **groves**. These forest patches of varying dimensions are protected by local people due to their religious sanctity. Generally, they are most undisturbed forests without any human impact. In India, sacred forests are located in several parts, such as Karnataka, Maharashtra, Kerala, Meghalaya, Uttaranchal, Uttar Pradesh, etc., and serve as refuge for a number of rare and endangered taxa.

Similarly several water bodies are declared sacred by the people, *e.g.*, Khecheopalri lake in Sikkim. Such water bodies protect aquatic flora and fauna.

Ramsar sites. Six internationally significant wetlands of India have been declared as **Ramsar Sites** under the Ramsar Convention. To focus attention on urban wetlands threatened by pollution and other *anthropogenic* activities, state Governments were requested to identify lakes that could be include the National Lake Conservation Plan (NLCP).

World heritage sites. Under the World Heritage Convention, five natural sites have been declared as 'World Heritage Sites'.

Table 29.9. World Heritage Sites

Site	*Location*
1. Kaziranga National Park	Assam
2. Keoladeo Ghana National Park	Rajasthan
3. Manas Wildlife Sanctuary	Assam
4. Nanda Devi National Park	Uttaranchal
5. Sunderban National Park	West Bengal

Five natural World Heritage Sites are as follows:

1. The Tura Range in Gora Hills of Meghalaya is a gene sanctuary for preserving the rich native diversity of wild *Citrus* and *Musa* species.

2. Sanctuaries for rhodendrous and orchids have been established in Sikkim.

3. **Project tiger.** A potential example of an highly endangered species in the Indian Tiger (*Panthera tigris*). It is estimated that India had about 40,000 tigers in 1900, and the number declined to a mere about 1,800 in 1972. Hence project tiger was launched in 1973.

At present these are 25 Tiger Reserves spreading over in 14 states and covering an area of about 33875 sq. km and the tiger population has more than doubled now due to total ban on hunting and trading tiger products at national and international levels.

4. **Project elephant.** This project was launched in 1991-92 to assist states having free ranging population of wild elephants to ensure long term survival of elephants in their natural habitats.

5. **Rhinos.** Rhinos have been given special atention in selected sanctuaries and national parks in the North East and North West India.

All these programmes, though focussed on a single species, have a wider impact as they conserve habitats and a variety of other species in those habitats.

Ex-situ Conservation (outside natural habitats) Strategies

The *ex situ* conservation strategies include: botanical gardens, zoological gardens, conservation stands and gene, pollen, seed, seedling, tissue culture and DNA banks.

Seed gene banks make the easiest way to store germplasm of wild and cultivated plants at low temperature.

While in **field gene banks**, preservation of genetic resources is being done under normal growing conditions.

Table 29.10. In Situ and Ex-situ Biodiversity Conservation

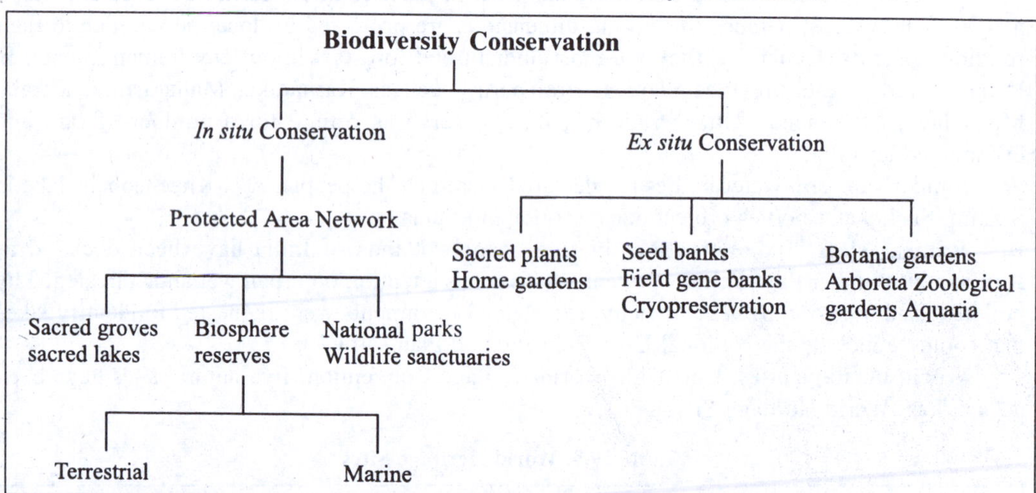

Cryopreservation. This type of *in vitro* conservation is done in liquid nitrogen at a temperature of –196°C. This is particularly useful for conserving vegetatively propagated crops, *e.g.*, potato.

Cryopreservation is the storage of material at ultra-low temperature (*i.e.*, –196° C) either by very rapid cooling, as used for storing seeds, or by gradual cooling and simultaneous dehydration, as being done in tissue culture.

In cryopreservation, the material can be stored for a considerable long period of time in compact low maintenance refrigeration units.

Botanical gardens. According to currently available survey, Central Government and State Governments together run and manage 33 Botanical Gardens, while Universities have their own botanic gardens.

An scheme entitled Assistance to Botanic Gardens provides one-time assistance to botainc gardens to strengthen and institute measure for *ex situ* conservation of threatened and endangered species in their respective regions.

There are more than 1500 botanic gardens and arboreta in the world containing more than 80,000 species. Many of these botanic gardens now have seed banks, tissue culture facilities and other *ex situ* technologies.

Zoological gardens (zoos). In India, there are 275 zoos, deer parks, safari parks, aquaria, etc. A Central Zoo Authority was set up to secure better management of zoos.

There are more than 800 professionally managed zoos around the world with about 3000 species of mammals, birds, reptiles and amphibians.

Many of these zoos have well developed captive breeding programmes.

Conservation of wild species. The conservation of wild relatives of crop plants, animals or cultures of microorganisms provides breeders and genetic engineers with a ready source of genetic material.

India has 47,000 species of flowering and non-flowering plants representing about 12 per cent of the recorded world's flora. Out of 47,000 species of plants, 5150 are endemic and 2532 species are found in the Himalayas and adjoining regions, and 1782 in peninsular India.

India is also rich in the number of endemic faunal species it possesses, while its record in agro-biodiversity is very impressive as well.

There are 166 crop species and 320 wild relatives along with numerous wild relatives of domesticated animals. Overall India ranks seventh in terms of contribution to world agriculture.

Table 29.11. Wild Relatives of Some Crops and Medicinal Plants

Crop	*Number of wild relatives*
Millets	51
Fruits	104
Spices and condiments	27
Vegetables and pulses	55
Fibre crops	24
Oil seeds, tea, coffee, tobacco and sugarcane	12
Medicinal plants	3000
Source MOEE, 1999	

Table 29.12. Wild Relatives of Domesticated Animals

Group	*Numbers*
Cattle	27
Sheep	40
Goats	22
Camels	8
Horses	6
Donkeys	2
Poultry	18
Buffalo	8
Source MOEF, 1999	

HOT SPOTS OF BIODIVERSITY

Norman Myers developed the **hot spots** concept in 1988 to designate priority areas for in situ conservation.

Biodiversity hot spots are areas that are unusually rich in species, most of which are endemic, and are under a constant threat of being overexploited.

The main criteria for determining a hot spot are:

(*i*) Number of endemic species (*i.e.*, the species confined to that area, and not found elsewhere).

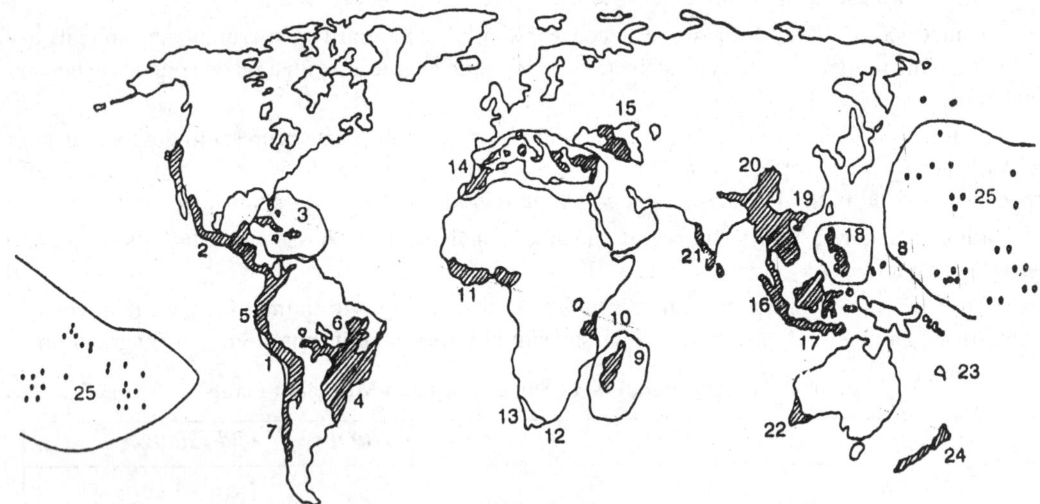

1. Tropical Andes, 2. Mesoamerica, 3. Caribbean, 4. Brazil's Atlantic Forests, 5. Choco/Darien/Western Ecuador, 6.Brazil's Cerrado, 7. Central Chile, 8. California Floristic Province, 9. Madagascar, 10. Eastern Arc & Coastal Forests of Tanzania/Kenya, 11. West African Forests, 12. Cape Floristic Province, 13. Succulent Karoo, 14. Mediterranean Basin, 15. Causcasus, 16. Sundland, 17. Wallacea, 18. Philippines, 19. Indo-burna, 20, South-Central China, 21. Western Ghats Sri Lanka, 22. Southwest Australia, 23. New Caledonia, 24. New Zealand, 25. Polynesia/Micronesia.

Fig. 29.6. The terrestrial biodiversity worldwide hot spots.

(*ii*) Degree of threat, which is measured in terms of habitat loss.

Among the 25 hot spots in the world, two are found in India.

These are two distinct areas:

(*i*) The Eastern Himalayas, and

(*ii*) The Western Ghats.

These hot spots extend into the neighbouring countries also. These areas are rich in angiosperms (flowering plants), also in reptiles, amphibians, swallow-tailed butterflies and some mammals. Most of the species are endemic.

Together these 25 sites of the world contain approximately 49,955 endemic plant species or 20 per cent of the world's recorded plant species, in only 746,400 sq km or 0.5 per cent of the earth's land surface.

1. **Eastern Himalayas.** Phytogeographically, the Eastern Himalayas forms a distinct floral region and comprises Nepal, Bhutan, neighbouring states of east and north-east India, and a part of southwestern China.

The temperate forests are found at altitudes of 1780 to 3500 metres a.s.l. In the whole of Eastern Himalayas, there are an estimated 9000 plant species, with 3500, *i.e.*, 39 per cent of them are endemic.

In India's sector of the area, there occur 5800 plant species, approximately 2000, *i.e.*, 36 per cent of them are endemic.

At least 55 flowering plants endemic to this area are recognised as rare, *e.g.*, the pitcher plant (*Nepenthes khasiana*). The species of numerous primitive angiosperm families such as Magnoliaceae and Winteraceae, and primitive genera of plants, like *Magnolia* and *Betula* are found in Eastern Himalayas.

The area is also rich in wild relatives of plants of economic significance, such as rice, banana, citrus, ginger, chilli, jute and sugarcane. The rigion is regarded as the centre of origin and diversification of commercially important palms, like coconut, arecanut, palmyra palm, sugar palm and wild date-palm.

As regards faunal diversity, 63 per cent of the genera of land mammals in India are known from this area, *e.g.*, golden langur from Assam Bhutan region, and Namdapha flying squirrel from Arunachal Pradesh, indicating the species richness in the region.

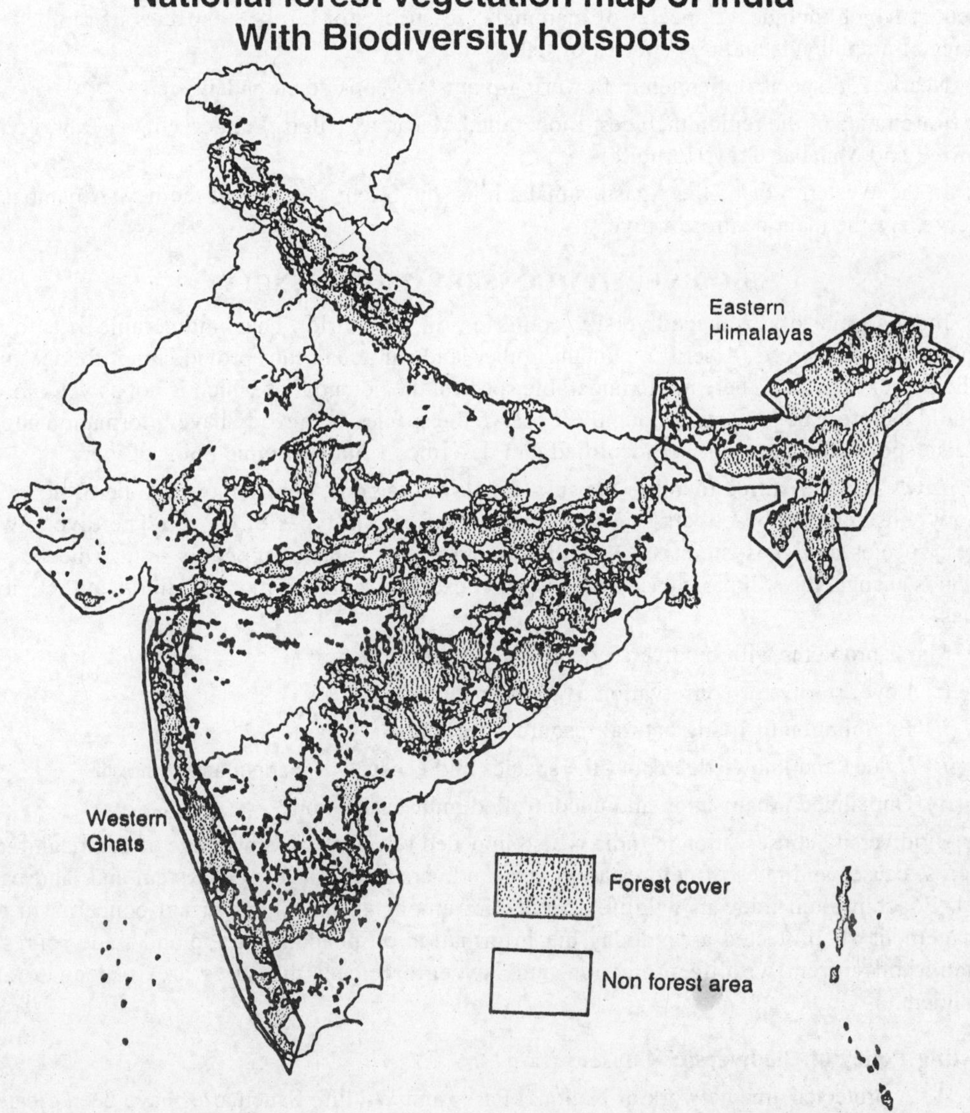

Fig. 29.7. National forest vegetation map of India with biodiversity hotspots.

2. **The Western Ghats.** The Western Ghats region is considered as one of the most important biogeographic zones of India, as it is one of the richest centres of endemic species. This region lies parallel to the western coast of Indian peninsula for almost 1600 km in Maharashtra,

Karnataka, Tamil Nadu and Kerala. The forest at low elevation (500 m a.s.l.) are mostly evergreen, while those found at 500 to 1500 metres a.s.l. are generally semi-evergreen forests.

Due to varied topography and micro-climate, some areas within the region are considered to be active zones of speciation.

The region has 490 arborescent taxa, of which 308 species are endemic. About 1500 endemic species of dicotyledonous plants are reported from the Western Ghats. 245 species of orchids (monocotyledons) belonging to 75 genera are found in this region, of which 112 species in ten genera are endemic to this region.

As regards the fauna, as many as 315 species of vertebrates belonging to 22 genera are endemic, which include 12 species of mammals, 13 species of birds, 89 species of reptiles, 87 species of amphibians and 104 species of fish.

Nearly 235 species of endemic flowering plants are considered endangered.

Rare fauna of the region includes: Lion Tailed Macaque, Nilgiri Langur, Nilgiri Tahr, Flying Squirrel, and Malabar Gray Hornbill.

In the Western Ghats, the Agasthyamalai hills, the Silent Valley, and the new Amambalam Reserve, are the main centres of diversity.

BIODIVERSITY CONSERVATION IN INDIA

India is one of 12 megadiversity countries of the world. The innumerable life forms harboured by the forests, deserts, mountains, other land, air and oceans provide food, fodder, fuel, medicine, textiles, etc. There are innumerable species, the potential of which is not as yet known. It would therefore be prudent to not only conserve the species we already have information about, but also species we have not get identified and described from economic point of view.

Taxus baccata, a tree found in the sub-Himalayan regions, once believed to be of no value is now considered to be effective in the treatment of certain types of cancer. The diversity of genes, species and ecosystem is a valuable resource that can be tapped as human needs and demands change, the still more basic reasons for conservation are the moral, cultural and religious values.

Major problems with biodiversity conservation in India are:

(*i*) Low priority for conservation of living natural resources.

(*ii*) Exploitation of living natural resources for monetary gain.

(*iii*) Values and knowledge about the species and ecosystem inadequately known.

(*iv*) Unplanned urbanisation and uncontrolled industrialisation.

Biodiversity conservation in India is also impeded by a lack of knowledge of the magnitude, patterns, causes and rates of deforestation and biodiversity loss at the ecosystem and landscape level. Poaching and trade in wildlife species are among the most important concerns in the management of protected areas today but information on poaching, trade and trade routes is sketchy and current wildlife protection and law enforcement measures are inadequate and inefficient.

Existing Policy of Biodiversity Conservation

1. A protected area network of National Parks and Wildlife Sanctuaries have been created.

2. The Indian Council of Forestry Research and Education (ICFRE) has identified forest preservation plots of representative forest types for conservation of viable and representative areas of biodiversity.

3. A programme entitled '**ecodevelopment**' for *in situ* conservation of biological diversity involving local communities has been initiated.

4. To conserve the respective ecosystems, a **Biosphere Reserve Programme** is being implemented.

5. Programmes have also been launched for scientific management and wise use of fragile ecosystem.

6. Specific programmes for management and conservation of wetlands, mangroves and coral reef systems are also being implemented.

7. Six internationally significant wetlands of India have been declared as '**Ramsar Sites**' under the Ramsar Convention.

8. Wildlife Protection Act is in the final stage of revision and provisions have been made for conservation reserves and community reserves to allow restrictive use to make it more people oriented. There will also be State Biodiversity Boards to control access to domestic consumers.

9. Under the World Heritage Convention, five natural sites in India, have been declared as **World Heritage Sites.**

10. Project Tiger and Project Elephant have been launched to protect the wildlife. Rhinos have been given special attention in selected sanctuaries and national parks.

11. The Ministry of Environment and Forests (MOEF) constituted the National Afforestation and Ecodevelopment Board (NAEB) in August 1992, which has evolved specific schemes for promoting afforestation and management strategies and ecodevelopment packages for promoting biomass production through a participatory planning process of Joint Forest Management (JFM) and microplanning.

Certain *Ex situ* conservation strategies are as follows:

Central Government and State Government together run and manage Botanical Gardens, zoos, deer parks, safari parks, aquaria, etc. A Central Zoo Authority has been set up to secure better management of zoos. A scheme entitled Assistance to Botanic Gardens provides asstistance to botanic gardens.

State Governments have already received the funds for the preparation of management action plans for Pongdam in Himachal Pradesh, Wullar lake in Kashmir, Loktak in Manipur, Rudrasagar in Tripura and Kolleru in Andhra Pradesh.

The National Bureau of Plant, Animal and Fish Genetic Resources has a number of programmes to collect and conserve the germplasm of plants and animals in seed gene banks, and field gene banks *in vitro* conservation.

The diverse food and medicinal plants are also being conserved successfully by the tribal people. The women particularly have an important role in the conservation of agrobiodiversity. The contribution of natural and agricultural biodiversity in terms of crops, live stock, fisheries, etc., is very substantial in terms of commercial value.

Biodiversity Contribution To Indian Economy

Biodiversity products have obtained a commercial value and have been increasingly exchanged in the markets having a monetary value, from which their share in the national economy can be judged.

The contribution of natural and agricultural biodiversity has a major contribution to the Indian GDP (gross domestic product).

The large economic implication of biodiversity in its wild and domesticated forms is the rice improvement programme. Rice accounts for 22 per cent of the total cropped area and 39 per cent of the total area under cereals, which reflects its importance in the country's struggle to attain self sufficiency in food.

When the rice crop was doomed in the 1970s, one single gene from the wild strain of rice, namely *Oryza nivara* from Uttar Pradesh, showed resistance to this virus and proved vital in the fight against the virus.

With respect to the commercial value of the plant species of medicinal value, the world trade is of several billion dollars and this is growing. The export market for medicinal plants has also increased. India's foreign exchange reserves from horticultural products are from high yielding varieties. Increased production of oilseeds also helped in saving large amounts of foreign exchange spent on edible oil import.

International Efforts For Conserving Biodiversity

It is apparent that action plans and strategies, when designed appropriately and implemented, can make important contributions to conservation. A collaborate effort of the World Resources Institute, IUCN (International Union for Conservation of Nature and Natural Resources), and UNEP (United Nations Environment Programme), working with other institutions, is leading to the preparation of a Global Strategy for the Conservation of Biodiversity, as a companion to the new version of the **World Conservation Strategy** now being prepared. The aim of the Strategy is to provide a comprehensive framework to stimulate urgent, positive, innovative and coordinated action to stem the loss and degradation of the world's biological resources and enhance the contribution of these resources to human well-being. The strategy will be developed by and for national governments, NGOs (Non-government organisations), resource managers, scientists, international institutions, multilateral banks, and bilateral aid agencies. The development of the Global Strategy will be centred around a series of regional workshops in Asia, Africa, Europe, Latin America and North America. The Global strategy will include considerations of a variety of factors influencing biological resource conservation, such as international financing, international cooperation, research, education, training, public awareness, and ecological restoration.

The Earth Summit held in 1992 at Rio de Janeiro, Brazil resulted into a Convention on Biodiversity, which came into force on December, 29, 1993. The Convention has three main objectives:

(*i*) Conservation of biological diversity

(*ii*) Sustainable use of biodiversity, and

(*iii*) Fair and equitable sharing of benefits that arise by the utilisation of genetic resources.

The World Conservation Strategy (WTS) and the World Wide Fund for Nature (WWF) support projects worldwide to promote conservation of biological diversity.

WILDLIFE CONSERVATION

When we hear the term "wildlife", it generally refers to large ferocious animals living in jungles and forests such as tigers, lions, elephants, wolves, etc. But in fact, "wildlife" implies to any living organism in its natural habitat which includes all plants, animals and microorganisms except cultivated plants and domesticated animals. From ecological view point, wildlife is a renewable resource.

Necessity for Wildlife Conservation

The conservation of wildlife is required for the following benefits:

1. The wildlife helps us in maintaining the '**balance of nature**'. Once this equilibrium is disturbed it leads to many problems. The destruction of carnivores or insectivores often leads to the increase of herbivores which in turn affects the forest vegetation or crops.

2. The wildlife can be used **commercially** to earn money. It can increase our earning of foreign exchange, if linked with tourism.

3. The preservation of wildlife helps many naturalists and behaviour biologists to study morphology, anatomy, physiology, ecology, behaviour biology of the wild animals under their natural surroundings.

4. The wildlife provides best means of **sports** and **recreation**.

5. The wildlife of India is our **cultural asset** and has deep-rooted effect on Indian art, sculpture, literature and religion.

Reasons for Depletion of Wildlife

Many wild animals have become extinct due to various human and natural activities:

1. Absence of cover or shelter to wild animals.

2. Due to deforestation for cultivation or for urbanisation, reduction in area for free movement of wild animals which retard reproductive capacity of certain wild animals like deer, bison, rhino, tiger, etc.

3. Destruction of wild plants of forests for timber, charcoal and fire wood often deprives wild animals their most palatable food and affects their survival.

4. Noise pollution by different transporting media (trucks, buses, rails, aeroplanes, etc.) and polluting river water have adversely affected wild animals.

5. Various natural calamities such as floods, droughts, fires, epidemics,etc., have also caused great destruction of wild animals.

6. Hunting methods of all kinds and for any purpose (*i.e.*, for food, recreation, hide, fur, plumage, musk, tusk, horn, etc.) have caused destruction of wildlife.

SOME IMPORTANT WILDLIFE OF INDIA

Indian subcontinent is unique in having immense natural beauty in its different regions (like Himalayas, Gangetic Plain and Plateau of Deccan, etc.) and also in possessing a rich and diverse wildlife fauna. However, some of the mammalian wildlife representatives of our country are:

1. **The Majestic Elephants** (*Elephas maximus*). It is the largest terrestrial mammal and is confined to the Terai and foot-hills because of its dependence on succulent grass, bamboo and plenty of water. It is gregarious and its herds move constantly in search of new feeding grounds; it cannot tolerate high temperature. It is an intelligent animal and can be domesticated easily. Its lifespan is about 70 years. Its population is hunted out for tusk and, therefore, its number is decreasing slowly.

2. **The great one-horned Rhinoceros** (*Rhinoceros unicornis*). It is found in the grasslands and jungle areas of the foot hills of the Himalayas and also in the plains of West Bengal and Assam. It prefers swamps and open savannah, covered with tall elephant grass. It has a single horn near the tip of its snout; the horn is made of keratinized skin and matted hair and it is said to have aphrodisiac and other medicinal values. It is, therefore, hunted for this purpose and is at the verge of extinction.

3. **The Sloth Bear** (*Melursus ursinus*). It is a stout carnivore which is found in the hilly forests like Mount Abu and Erinpura ranges. It lives mostly on fruits, honey and insects, it sucks up white ants from their hills which are broken by its claws with great speed. It is rapidly vanishing due to ruthless hunting and it needs conservation. **Brown Bears** (*Ursus arctus*) are found on mountains upto 3,000 metres height and **Black Bears** (*Selenarctos tibetanus*) at 1500–3000 metres height.

4. **The lion** (*Panthera leo*). It is a rare mammal and was found in Thar desert, in arid plains of Sind, Rajasthan and Punjab but now it is restricted to Gir forest of Gujarat. It is somewhat nocturnal and referred to as the **King of Animals**. Male possesses mane which is not found in

female. This animal is said to possess enormous strength. The lion is the emblem of the Government of India. Its extermination is well known.

5. **Indian Panther** (*Panthera paradus fusca*). It is smaller than tiger and is spotted. It can live in all types of forests. Once it was found all over India but now its number has been drastically reduced. Now, it is under protection in Mount Abu sanctuary. It is commonly called the **Leopard**. The **snow leopard** (*Panthera unica*) is found in the high Himalayas from Kashmir to Sikkim near snow line. It has a creamy grey coat with large black rings. The leopards, especially the snow leopards are greatly hunted for their fur. It is an endangered species.

6. **The Indian Tiger** (*Panthera tigris*). It is gracious carnivore and distributed in Uttar Pradesh from the Himalayas to the Vindhya forests in the South. It lives in a variety of habitats, from thorn forests to the dense terai forests. It is carnivorous and nocturnal. Its number, once estimated to be nearly 25,000 in 1948, has gone considerably down nearly 1540 in 1970. Realizing this fact, Government of India has launched a project commonly called **Project Tiger** to save the tigers from extinction. This project was launched in 1972; due to this effort the population of tiger has improved considerably.

The **White Tiger** is a mutant variety of Tiger which was first noticed in the forests of Gidhaur in Monghyr District of Bihar State.

7. **The Cheetah** (*Acinonyx jubatus*). It was found in Central India,Deccan region and in Thar desert upto Jaipur in Rajasthan. Now the Cheetah, one of the most beautiful, agile and fastest animals, is regarded to be extinct from India.

8. **The Clouded Leopard** (*Neofelis nebulosa*). It is found in the evergreen forests of Assam. It is black in colour and lives on trees, *i.e.*, arboreal in habit. It feeds on birds and small animals.

9. **Striped Hyaena** (*Hyaena striata*). It is found generally roaming near jungles throughout our country. It is terrestrial and carnivorous; it feeds on goat, sheep, etc.

10. **Wolf** (*Canis lupus*). It is found in the plains throughout our country. It preys on goat, sheep and deer.

11. **Wild Ass** (*Equus hemionus khur*). It had a fairly wide distribution in the dry regions of North-West India (Jaisalmer and Bikaner), but its distribution has become restricted to the southern part of the Thar desert. Its number is estimated to be nearly 870. Its population is severely reduced due to hunting, catching for breeding mules and surra disease, etc. Since, it is the only true **wild ass** of the world, hence, it needs protection and conservation.

12. **Wild Boar** (*Sus cristatus*). The wild boar or pig has a long mobile snout with terminal nostrils The canines of both jaws grow continuously and form long triangular and up curved tusks for defence and digging roots; the tusks are better developed in males. It is omnivorous, it feeds on carrion, snakes, insects, roots, tubers and cultivated crops. It is gregarious living in families in grass or bush of marshy places. It is now rare animal and is mostly found in small numbers along the Aravalli ranges in the eastern part of Thar desert where it inhabits rocky slopes.

13. **Wild goat and sheep.** These are restricted only to the Northern part of our country. The **Nilgiri Tahr** (*Hemitragus hylocrius*) is a variety of wild sheep; the only variety found in south (in Nilgiris, Tamil Nadu). It lives in groups and grazes on grasses. This species is greatly endangered due to habitat disturbance, predation by black panther, wild dogs and hunting by poachers. The **Himalayan Ibex** is a variety of wild goat, *Capara siberica*.

14. **Nilgai or Blue Bull** (*Boselaphus tragocamelus*). It lives in plains and Rajasthan deserts in groups of 4 to 10 in number. It prefers to live near cultivated lands and is also found in desert grasslands.Males have small cone-shaped curved horns. Both sexes have a short dark mane on the neck. It is hunted under Crop-Protection Act, because herds of Nilgai inflict severe damage to crops. In Thar desert it is regarded sacred like 'cow', so is spared from hunting.

15. **Wild-Buffalo** (*Bas bubalis*). It is large-sized, robust animal having streamlined body. It prefers tall grassy forests close to marshy areas in the vicinity of rivers and lakes. It was once distributed over the grass jungles and rivers in forests of Gangetic plain and in the foot-hills of Assam, Madhya Pradesh and Nepal. Its number has also decreased rapidly due to great hunting.

16. **Indian Bison** (*Bos gaurus*). It is related to bull and is found in Andhra Pradesh, Orissa, Madhya Pradesh and Bihar. Due to various reasons, it is now confined only to Bandipur of Karnataka. Its number has also reduced down to a considerable extent.

17. **Indian Gazelle** (*Gazella bennetii*). It is also called **Chinkara** and is mostly found through out desert region of Southern part of Uttar Pradesh and Rajasthan. It lives in herds and adapted to live in extremely arid conditions. It prefers to live in open grass-lands with thick bushes.

18. **Four-horned Antelope** (*Tetracerus quadricornis*). It is also called **Chausingha** and has two pairs of horns. It is found in the foot-hills of Himalayas and in Bundelkhand region. It generally leads solitary life.

19. **The Black Buck** (*Antelope cervicapra*). It has four subspecies in India. It is a graceful and fast running animal, found in the grassy plains of India and avoids forests. Males have spirally ringed, unbranched long and permanent horns. It has been hunted out from the most parts of India and now it is confined only in certain localities in Rajasthan.

20. **The Musk Deer** (*Moschus moschiferus*). It is found in Himachal Pradesh, Northern Uttar Pradesh, Nepal and Sikkim. It is a shy animal and lives a solitary life. Male possesses a musk gland for which it is hunted. The secretion of this gland is a valuable product used as perfume fixative, and as an important ingredient of several Aurvedic and Homeopathic drugs.

21. **The Barking Deer** (*Cervulus muntjac*). It is found in the foot-hills and plain. It has suborbital glands below eyes, antlers and non-protruding canines. It is diurnal and lives in pairs.

22. **The Dancing Deer** (*Cervus unicolor*). It is also called **Sambhar**. It is a rare beautiful deer, lives in herds and found in the forests all over the country. It has very long antlers. It comes out on plains only at mid-night. It is mostly hunted for its skin and its flesh is used as food by many people.

23. **The Spotted Deer** (*Axis axis*). It is also called **Cheetal**. It is beautiful deer with white spots on its brown coat. It has long slender antlers, each having three branches which shed annually. It is mostly found in the forests of Indian plains and is also reported from Assam, Punjab and Rajasthan. It is also hunted greatly for skin and meat.

24. **The Hog Deer** (*Axis porcinus*). It is smaller and stout in build and has hog-like appearance. It lives either solitary or in pair and prefers to live in grass patches at the borders of the forests. It is diurnal and comes out to feed only during morning and evening.

25. **The Nilgiri Langur** (*Cercopithecus johni*).
26. **The Lion-tailed Monkey** (*Macaca silensus*).
27. **The Hanuman Monkey** (*Semnopithecus entellus*).
28. **The Rhesus Monkey** (*Macaca mulatta*).
29. **The Indian Giant Squirrel** (*Rotufa indica*).
30. **The Porcupine** (*Hystrix lencura*).
31. **The Pangolin** (*Manis crassicaudata*).

The above mentioned list is related to mammals only. Likewise, a large number of birds, reptiles, etc., are also among wildlife of India.

PROTECTED SPECIES OF INDIAN WILDLIFE

The Government of India has declared followings species as protected:

1. Tiger,
2. White tiger,
3. Lion,
4. Panther,
5. Snow Leopard,
6. Clouded Leopard,
7. Leopard cat,
8. Kashmir stag,
9. Brown Antlered Deer,
10. Black Buck,
11. Chital,
12. Indian swamp deer,
13. Four-horned Antelope,
14. Golden Cat,
15. Indian Bison,
16. Indian Rhinoceros,
17. Wild Ass,
18. Barking Deer,
19. Slender Loris,
20. Pigmy hog,
21. Musk deer,
22. Indian Gazelle,
23. Gibbons,
24. Nilgiri Langur,
25. Slow Loris,
26. Golden Langur,
27. Lion-tailed Monkey,
28. Great Indian Bustard,
29. Mountain Quail
30. Grey Jungle Fowl,
31. White Winged Wood Duck,
32. Pea Fowl,
33. Horned Owl,
34. Indian Gharials,
35. Crocodiles.

Measures for Wildlife Conservation in India

In India, following measures have been undertaken for wildlife conservation:

(*i*) **Protection of natural habitats.** Jungles and forests are the natural habitats of many wild animals. Hence, these natural habitats should not be destroyed and for this law should be framed to protect such areas from any damage and hunting should be prohibited. However, realizing the importance of natural resources, several national and international organisations have been established to look after such problems.

An international organisation **IUCN** was established in Switzerland in 1948. This organisation has framed many projects to overcome such problems for which this organisation arranges a meeting once a year in any country. Its 10th general meeting was held in New Delhi in 1969. For the normal and effective functioning of this organisation, an additional organisation referred to as **WWF** (World Wildlife Fund) has been established. The **WWF** is related to the collection of funds from its member countries, of course, India is also a member, which are 24 in number. This fund is utilized to run several projects dealing with the conservation of natural resources. The **Tiger Project** in India has been financed by **WWF** and working efficiently since 1972.

(*ii*) **Protection by Law.** India was probably the first country to enact a Wildlife Protection Act. The **Wild birds and Animals Protection Act** was passed in 1887 and repealed in 1912. For game protection in the states the **Forest Act XVI** was enacted in 1927. Then **Indian Board of Wildlife** was established in 1952 which is followed by setting up of Wildlife Boards in different states. In 1972, new **Wildlife Protection Act** was passed. Under this Act, possession, trapping, shooting of wild animals alive or dead; serving their meat for eating, their transport and export and all controlled and watched by special staff (Chief Wildlife Warden and anthorized officers). This Act prohibits hunting of females and young ones under this Act, threatened species are absolutely protected and the rest and provided graded protection according to the condition of their population size. However, in India every year we observe a **Wildlife Week** from 1st October to 8th October.

(*iii*) **Establishment of sanctuaries and national parks.** In India, nearly 200 sanctuaries and national parks have been established for wildlife management. These are meant for providing natural habitat to wild animals.

A **sanctuary** is a place or area with natural environment having optimum living conditions and protection for wild animals; in fact, in a sanctuary wild indigenous animals are kept for protection. Shooting and hunting is prohibited in a sanctuary. A sanctuary is established by **State Forest Department** by Notification and it can be abolished by similar procedures. Contrary to a sanctuary, **a national park** is a permanent place or area established by **Central Legislation** for preserving flora, fauna, landscapes and historic objects of a particular area. However, shooting and hunting is prohibited in such areas also.

Conservation Strategies

For the judicious use of resources, the scientists from nearly 100 countries of the world have formulated some conservation strategies. Few of the steps proposed to save the existing species of the wildlife are given below:

1. Every effort should be made to preserve the endangered species; these species should be given priority over a vulnerable one, a vulnerable species over a rare one and a rare species over other categories.

2. Proper management and planning should be made to preserve wildlife in their natural habitat (*in situ*) and in zoos and national parks (*ex situ*).

3. A wide range of varieties of livestock, animals for aquaculture and their wild varieties should be preserved, and priority should be given to those that are most threatened and needed for national and international breeding programme.

4. Each country should locate the habitats of wild relatives of the economically valuable and useful animals and ensure their preservation in protected areas.

5. The feeding, breeding, nursery and resting areas of species should be safeguarded.

6. In case of migratory animals, a network of protected areas should be established to preserve their habitats. For those species which migrate from one national jurisdiction to another, bilateral or multilateral agreements should be made to meet out the required network.

7. Unique ecosystems should be protected as a matter of priority.

8. The national protection programmes should be coordinated with the international programmes like UNESCO's Man and Biosphere Project 8 and national parks and protected areas of International Union for Conservation of Nature and Natural Resources. This would safeguard the genetic diversity and their continuing evolution.

9. The productive capacities of species should be determined, so that, their utilisation should not exceed those capacities.

10. Proper legislative and administrative measures should be taken to regulate the International trade of wild animals.

Protected Areas in India

As referred to, Indian subcontinent is very rich in possessing natural beauty. Its biogeographical range consists of the coldest Ladakh and Spiti, the hot deserts of Thar, the temperate forest in Himalayas to the lush green tropical rain forests of the low lands; Indian subcontinent also possesses large fresh water bodies such as Wular and the Manasbal lakes in Kashmir, the Chilka in Orissa and Kolleru lake in Andhra Pradesh and the ruggest and rich coastline and coral reefs of the Deccan.

Government of India passed the Wildlife Protection Act in 1972, under which national parks and sanctuaries could be created.

However, out of 434 national parks and sanctuaries in India, 17 have been selected as "Project Tiger" areas.

During the past twenty years UNESCO has evolved **Biosphere Reserves** under its Man and Biosphere (MAB) Programme. Accordingly, India has located 13 areas as Biosphere Reserves; from these Nilgiri Biosphere Reserve, including parts of Karnataka, Kerala and Tamil Nadu was declared in 1986 and the Nanda Devi Biosphere Reserve in 1988. Probably two more Biosphere Reserves, one at Uttaranchal (including the Valley of Flowers in North-Western Himalayas) and another at Nokrek (North-Eastern Himalaya), has also been included in the list of Biosphere Reserve.

In a Biosphere Reserve, the land is designated into different zones; these are the **core zone** (where no human activity is allowed), the **buffer zone** (where limited human activity is allowed) and the **manipulation zone** (where a large number of human activity may go on). In Biosphere Reserves, wild population as well as traditional life styles are protected.

Some Important Wildlife Sanctuaries

1. **Annamalai Sanctuary.** This sanctuary is situated in the southern part of Coimbatore District in Tamil Nadu and extends in a vast area of 958 sq km. This sanctuary has rich fauna of animals like elephant, gaur, sambhar, spotted deer, barking deer, Nilgiri tahr, lion-tailed monkeys, tiger, panther, sloth bear, langur, porcupine and pangolin.

2. **Kaziranga Wildlife Sanctuary.** This sanctuary is situated on the south bank of the river Brahmaputra in Sibsagar district of Jorhat subdivision of Assam. It consists of 430 sq km of forest grasslands and swamps and supports a fauna of nearly 700 rhinoeros. Besides, it also has elephant, wild buffalo, bison, tiger, leopard, sloth bear, sambhar, swamp deer, hog deer, barking deer, wild boar, gibbon, and birds like pelicon, stork and ring-tailed fishing eagles. This is, in fact, a **Rhino Sanctuary**.

3. **Jaldapara Wildlife Sanctuary.** This sanctuary is situated in Jalpaiguri district of West Bengal and is extended in a 65 sq km stretch of grassland. It is known for rhinoceros and, hence, referred to as a **Rhino Sanctuary**. Besides, it also has gaur, elephant, tiger, leopard, deer, and a variety of birds and reptilian fauna.

4. **Manas Wildlife Sanctuary and Tiger Reserve.** This sanctuary is situated in Kamrup district in Assam and is extended in an area of 540 sq km. River Manas passes through it. It contains the following wild animals: tiger, panther, wild dog, wild boar, rhinoceros, gaur, wild buffalo, sambhar, swamp deer and golden langur.

5. **Periyar Wildlife Sanctuary.** This sanctuary is situated in Kerala and has an area of 777 sq km. It has the fauna of wild elephants, gaur, leopards, sloth bear, sambhar, barking deer, wild dogs, wild boars, black Nilgiri Langur and some birds.

6. **Mudumalai Wildlife Sanctuary.** This sanctuary is situated in north-western part of Nilgiris in Tamil Nadu and extends in an area of 22 sq km. It has diverse fauna of wild elephant, gaur, sambhar, chital, barking deer, mouse deer, four-horned antelope, tiger, panther, bonnet monkey, langur, giant squirrel, flying squirrel, wild dog, jackal, wild cat, sloth bear, porcupine, pangolin, flying lizard, monitor lizard, rat snake, python and various birds.

7. **Bandipur Wildlife Sanctuary.** This sanctuary is situated in south of Mysore city of Karnataka state and extends in an area of 874 sq km. Its fauna includes elephant, leopard, sloth bear, wild dog, chital, panther, barking deer, porcupine and langur.

8. **Sesan Gir Sanctuary.** This sanctuary is situated near Ahmedabad in Gujarat state and extends in a vast area of 1295 sq km. It is the largest sanctuary of our country. It is known for Asiatic lions and other fauna includes spotted dear, blue bull, four-horned antelope, chinkara, striped hyaena, wild boar, porcupine, langur, python, crocodiles and some birds.

9. **Dachigam Wildlife Sanctuary.** This sanctuary is situated in Kashmir and extends in an area of 89 sq km. Its fauna includes Kashmir stag or hungal, musk deer, leopard, black bear, brown bear and baboon.

10. **Bharatpur Bird Sanctuary.** It is also named as Koeldeo Ghana Sanctuary. This sanctuary is situated at Bharatpur of Rajasthan state and extends in an area of 29 sq. km. It is a'sanctuary meant for various bird fauna.

In addition to these, the mention of the following sanctuaries may also be made:

(*i*) **Point Calimera Wildlife Sanctuary** in Tamil Nadu and includes the fauna of Chital, wild boar and black buck.
(*ii*) **Sariska Sanctuary** near Alwar in Rajasthan and includes tiger, leopard, spotted deer, four-horned antelope, etc.
(*iii*) **Chilka Lake Sanctuary** in Orissa and has variety of water birds.
(*iv*) **Shivpuri Sanctuary** in Madhya Pradesh and is known for its tiger fauna.
(*v*) **Gautam Buddha Sanctuary** in Gaya of Bihar state and has tiger, leopard, chital, sambhar, etc.
(*vi*) **Bhimbadh Sanctuary** of Monghyr district in Bihar state and has wolf, tiger, leopard, etc.
(*vii*) **Valmikinagar Sanctuary** in Champaran of Bihar state and has tiger, leopard, sambhar, chital, etc.

Some Important National Parks

1. **Simlipal National Park.** This national park is situated in Mayurbhanj district in Orissa state and is extended in an area of 2750 sq km. It is covered over by dense sal forests and is chosen for **project tiger**. Its fauna includes tiger, elephant, deer, pea fowl, talking mynas, chital, sambhar, panther, gaur, hyaena and sloth bear. It is the largest national park of our country.

2. **Corbett National Park.** This national park is situated between Nainital and Garhwal districts of Uttar Pradesh and extends in an area of 525 sq km. It supports rich and diverse fauna of tiger, panther, sloth bear, elephant, blue bull, swamp deer, barking deer, Indian antelope, porcupine, and many birds and reptiles. This national park is one of the most prestigious wildlife parks of our country.

3. **Hailey National Park.** This national park is situated in Ramnagar in the valley of Nainital in Uttar Pradesh and covers an area of 200 sq km. It supports the fauna of elephant, sambhar, chital, langur, wild fowls, pea fowls, etc.

4. **Palamu or Betla National Park.** This national park is situated in Daltonganj district of Bihar state and covers an area of 345 sq km. It supports the fauna of tiger, panther, sloth bear, elephant, chital, gaur, nilgai, chinkara, mouse deer, etc.

5. **Hazaribagh National Park.** This national park is situated in Hazaribagh district of Bihar state and covers an area of 184 sq km. It supports the fauna of wild boar, sambhar, nilgai, tiger, leopard, sloth bear, hyaena and gaur.

6. **Kanha National Park.** This national park is situated in Mandala district of Madhya Pradesh and extends in an area of 160 sq km. It supports the fauna of tiger, chital, deer, black buck, sambhar, blue bull, bison, panther, wild dogs, hyaena, etc.

7. **Tandoba National Park.** This national park is situated in Chandrapur district of Maharashtra state and covers an area of 116 sq km. It supports the fauna of tiger, sambhar, sloth bear, bison, chital, chinkara, barking deer, blue bull, four-horned deer, langur, pea fowl, and few crocodiles.

Zoological Gardens

The **zoological gardens** are somewhat different from sanctuaries and wildlife national parks as the animals are kept in cages in zoological gardens for show. The animals, thus, do not get natural habitat in zoological gardens. There are various zoological gardens in our country but the zoological garden (Zoo) of Delhi is worth mentioning

30

Global Environmental Change

An outcome of the atmospheric pollution is warming of the earth's environment. This is a serious problem and in twenty-first century this may be a major global problem. The temperature of the earth has increased to some extent during the past fifty years has been exhibited in the following fig. 30.1. This increase in temperature of atmosphere is dangerous for living organisms. The human activities are mainly responsible for global warming. Though there are several causes for global warming, however, some important ones are mentioned here : They include deforestation, industrialisation, energy production from fossil fuels, such as diesel, petroleum, coal, etc., and urbanisation. The conversion of a forest to a grazing land or a cropland through deforestation causes loss of carbon stored in soil and vegetation to the atmosphere, and affects the global carbon cycle.

Biomass burning associated with agricultural practices releases a large amount of CO_2 into the atmosphere. Besides, due to domestic and industrial coal burning, huge amount of carbon dioxide is also released in the atmosphere.

Besides carbondioxide, some other gases like methane (CH_4), nitrous oxide (N_2O) and chlorofluorocarbons (CFCs) are also responsible for increasing the temperature of the earth and its environment.

These gases CO_2, CH_4, N_2O and CFCs are **radiatively active gases**, also called **greenhouse gases** as they can absorb long wave infrared radiation.

The increased amounts of greenhouse gases in the atmosphere affect the global climate, and this phenomenon is called **global change**.

Greenhouse Gases (Greenhouse Effect)

The gaseous mantle around the globe allows a considerable portion of solar radiations to enter right upto the surface of earth which absorbs it and radiates back infra-red and heat waves. This heat is transferred to layers above, as warm layer rises and in turn passes on to higher and higher layers. The outgoing longwave infrared radiation is absorbed by the greenhouse gases normally present in the atmosphere. The atmosphere radiates part of this energy

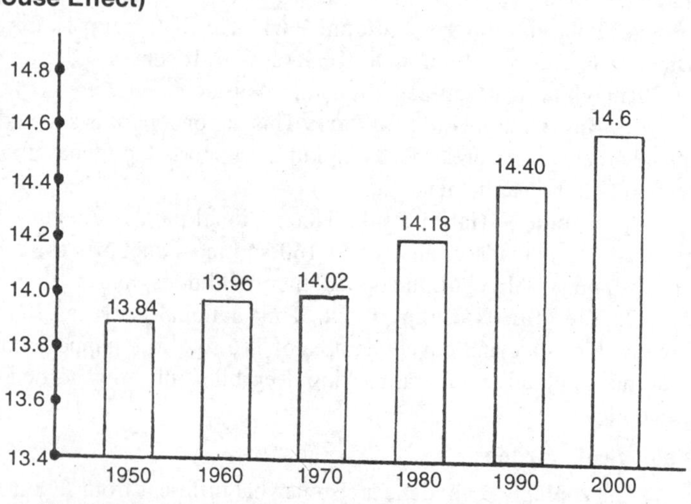

Fig. 30.1. Average temperature of earth recorded in different years.

back to the earth. This downward flux of radiation, known as **greenhouse flux**, keeps the earth warm. Thus, the greenhouse gases of the atmosphere form a blanket over the earth that controls the escape of heat from the surface of earth to outer space so as to keep it warm and cosy. This phenomenon is called **greenhouse effect**.

The name, greenhouse effect is derived from the fact that inside a glass enclosed **green house**, temperature is warmer than outside. The mean annual temperature of the earth is about 15°C. If there would have no greenhouse gases in the atmosphere, the earth's mean temperature would go sharply down to about—20°C.

This capacity of the atmosphere to keep the earth warm depends upon the concentration of the greenhouse gases. The excessive increase in concentrations of greenhouse gases in the atmosphere would retain more and more of the infrared radiation, resulting in enhanced **green house effect**. The consequent increase in the global mean temperature is referred to as **global warming**.

The rise in temperature of earth's atmosphere caused due to greenhouse effect depends on the amount of CO_2 present in the atmosphere. At normal CO_2 concentration (0.03 per cent) in the atmosphere the surface temperature of the earth remains constant due to energy balance of the sun rays which strike on the earth, heat it and then radiate back into the space. This is called **energy budget**. But when there is increase in CO_2 concentration in the atmosphere, the thick layer of CO_2 prevents the heat from being radiated out into space. This layer of CO_2 thus functions as a glass pane of green house which allows the sunlight to filter through it but prevents heat from being radiated back into the outer space.

However, an international panel on climate change periodically makes an assessment of the atmospheric abundance of greenhouse gases and its possible impact on climate and other such issues.

The trends in the increase in concentrations of greenhouse gases since pre-industrial times, are as follows :

Carbon dioxide. CO_2 is the most abundant greenhouse gas in the atmosphere. This is about sixty per cent of the greenhouse gases. The level of the concentration of carbon dioxide in the atmosphere has increased from the pre-industrial level of 280 ppm (parts per million) to about 368

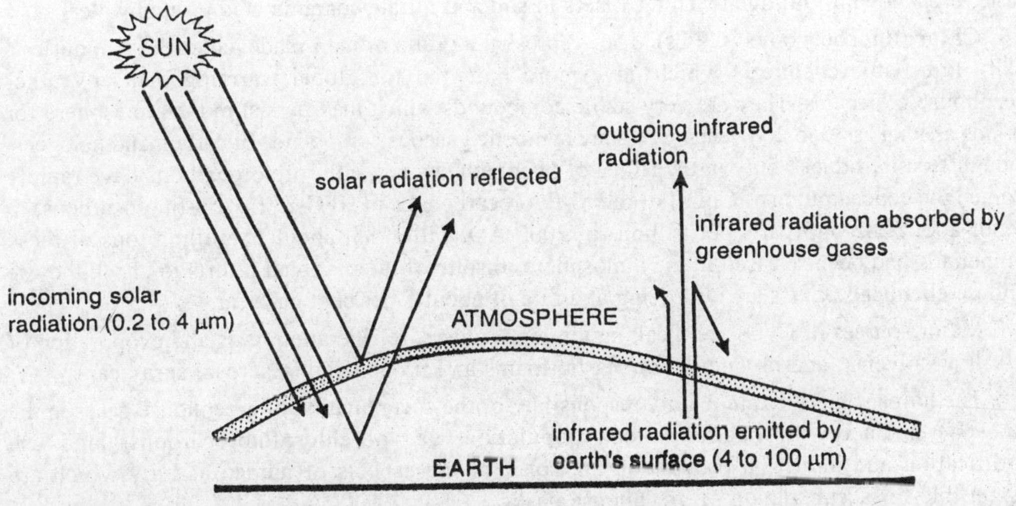

Fig. 30.2. The greenhouse effect.

ppm in 2000 AD. Here, pre-industrial time is considered 1750 AD. This increase in concentration is mainly due to fossil fuel burning, deforestation and change in land use.

Methane. CH_4 is another greenhouse gas which is produced when organic matter decays under anaerobic conditions. Methane concentration in atmosphere was about 750 ppb (parts per billion) in pre-industrial level, *i.e.*, 1750 AD., which had become 1750 ppb in 2000 AD. Methane is generally produced as a result of incomplete decomposition by a group of bacteria called **methanogens**, under anaerobic conditions. The concentration of this gas is rising at a rate of about one per cent per year. Between the years 1980-90 about 15 per cent of the total warming has been attributed to this gas alone.

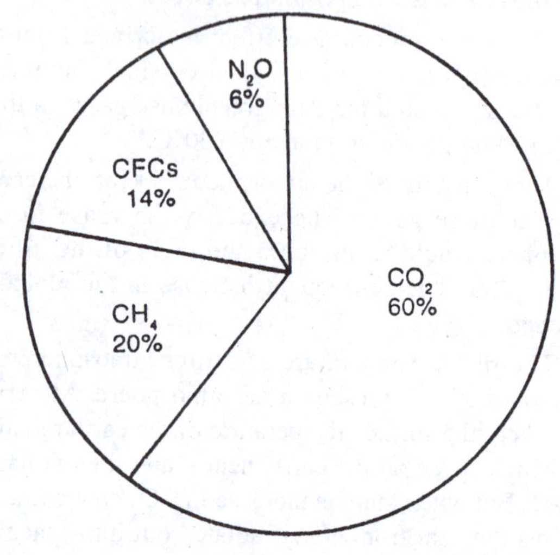

Fig. 30.3. Contribution of percentage of greenhouse gases to global warming.

The major sources of methane are : freshwater wetlands, enteric fermentation in cattle, and flooded rice fields. Biomass burning is also a source of methane. In the atmosphere, methane undergoes oxidation to carbon dioxide and water, both of which tend to emphasize the greenhouse effect.

Nitrous oxide. N_2O is another troublesome greenhouse gas. Other oxides of nitrogen are reacted upon or cleared rapidly, while nitrous oxide undergoes decomposition very slowly, and therefore, it tends to accumulate in the atmosphere. The concentration of nitrous oxide in the atmosphere has increased from about 270 ppb in pre-industrial time (*i.e.*, 1750 AD.) to about 316 ppb (parts per billion) in year 2000 AD.

The main sources of nitrous oxide are agriculture, biomass burning and other industrial processes. N_2O is produced during nylon production,burning of nitrogen-rich fuels, livestock waste, breakdown of nitrogen-rich fertilisers in soil and nitrate contaminated ground water.

Chlorofluorocarbons (CFCs). They represent a group of man-made, colourless, odourless, easily liquifiable chemicals which have more potential for global warming than any other greenhouse molecules. They are very stable compounds which may persist in the atmosphere for periods as long as 80 to 260 years. CFCs are synthetic gaseous compounds of carbon and halogens. Though first introduced only in the fifties of twentieth century, chlorofluorocarbons have rapidly attained the concentration of CFC-11 (chloro-fluorocarbons) and HFC-23 (hydro-fluorocarbons) in the air about 282 ppt (parts per trillion) in 2000 AD. Till 1985, about 15 million tons of these compounds had been released in the atmosphere. Inspite of international efforts to check the use of these chemicals, CFCs are still rising at a rate of about 5 per cent per year.

Major sources of CFCs are : leaking air conditioners, refrigeration units and evoporation of industrial solvents; and production of plastic foams and propellants in aerosol spray cans.

The human activities are mainly responsible for the accumulation of greenhouse gases in the atmosphere. Energy of huge livestock population, use of chlorofluorocarbons, land use modification and industrial production are some of the aspects of human activity which are responsible for accumulation of greenhouse gases.

Table 30.1. Sources of Greenhouse Gases

Gas	*Source of discharge*	*Percentage of global warming*
Carbon dioxide (CO_2)	Burning of fossil fuels; deforestation; change in land use.	60 per cent
Methane (CH_4)	Freshwater wetlands; enteric fermentation in cattle; flooded rice fields and biomass burning.	20 per cent
Chlorofluorocarbons (CFCs)	Leaking air conditioners, refrigeration units and evaporation of industrial solvents; production of plastic foams and propellants in aerosol spray cans.	14 per cent
Nitrous oxide (N_2O)	Agriculture, biomass burning; nylon production, burning of nitrogen-rich fuels, livestock waste, breakdown of nitrogen-rich fertilisers in soil and nitrate contaminated ground water.	6 per cent

Percentage of greenhouse gases. It is estimated that carbon dioxide contributes about 60 per cent of the total global warming. The share of methane is 20 per cent, while that of chlorofluorocarbons is 14 per cent. However, a smaller contribution to global warming is made by nitrous oxide, *i.e.*, 6 per cent.

EFFECTS OF GREENHOUSE GASES

The increasing abundance of greenhouse gases in the atmosphere has the following three main effects :

(*i*) CO_2 fertilisation effect on plants

(*ii*) Global warming and

(*iii*) Depletion of ozone (O_3) layer in the stratosphere.

Carbondioxide (CO_2) Fertilisation Effect on Plants

The data produced in USA have shown that atmospheric carbondioxide concentration has been rapidly rising since 1959 as shown in the graph. If such rising trend continues, by the end of

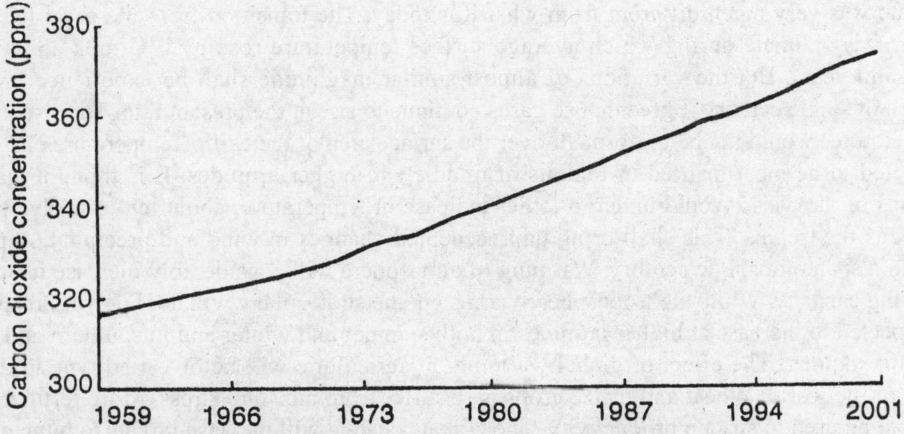

Fig. 30.4. Increase in CO_2 concentration in atmosphere from 1959 to 2001.

twenty first century the atmospheric concentration of CO_2 shall increase to a level between 540 and 970 ppm.

When the CO_2 concentration of the atmosphere is more or less doubled, the growth of many plants, *i.e.*, C_3 plants in particular, under favourable conditions of water, nutrients, light and temperature, could increase by about thirty per cent on average, in the short term of few years or so.

The response of plants to elevated concentrations of CO_2 is called **carbon dioxide fertilisation effect**.

Due to increased carbondioxide concentration, the rate of photosynthesis also increases, and the stomatal conductance decreases due to partial closure of stomata. Hence, the transpiration rate reduces, and water-use efficiency increases.

Such effect allows many species to grow successfully in regions of water scarcity.

Under higher atmospheric carbon dioxide concentrations, plants allocate, a greater proportion of photosynthate to roots. However, greater root production increases development of mycorrhiza and fixation of nitrogen in root nodules, thus, makes possible the plants to grow in soils which are poor in nutrition.

However, in natural conditions the beneficial effects of increased carbon dioxide may not be there because of negative effects of global warming.

Possible Effects of Global Warming

Global records of earth's surface temperatures indicate that a warming of about 0.6° (*i.e.*, 0.3° – 0.7°C) has occurred during the last century alone. Results from recent climatic models suggest that mean global temperatures would rise by 2° – 6°C during the twenty first century, if we assume that carbon dioxide concentration increases to 600 ppm (parts per million). The predicted global warming in near future has the potential to affect the (*i*) weather and climate, (*ii*) sea level change, (*iii*) distribution and phenology of species and (*iv*) food production and the fishery resources in the oceans.

(*i*) **Effects on weather and climate.** As already mentioned in the preceding paragraph, the global mean temperature has increased about 0.6°C in the twentieth century. It is thought, that mean global temperature may increase by 1.4 to 5.8°C by the end of twenty first century, (*i.e.*, 2100 AD).

The projected change in mean surface temperatures may appear insignificant because variatious of this magnitude are experienced in course of seasonal or even daily weather. The world climate was very much different from what it is today. The transition from the great Ice Age to present-day climate during which average surface temperature rose by 5°C took about twelve thousand years. But the variations of almost similar magnitude shall be experienced within a course of single century if greenhouse gases continue to rise at the present rate. The rise in global temperatures would not be uniform all over the surface area of the earth. Temperature changes are expected to be most marked in regions of middle and higher altitudes. It is thought that polar regions of the world would undergo larger increase in temperature, about ten to twelve times as much as the tropics. This shall bring unprecedented changes in wind and precipitation patterns within a span of a single century. Warming of atmosphere will considerably increase its moisture carrying capacity. While the troposphere warms up, the stratosphere will cool down. Precipitation is expected to increase at higher latitudes in both summer and winter and in southern and eastern Asia in summer. The effect of global warming on agriculture will be of varied type in different parts of the world. Wheat and maize crops may suffer from moisture stress. More fertilizers shall have to be used to sustain productivity. The climate change will increase threats to human health, particularly in tropical and sub-tropical regions of the world, due to change in ranges of disease vectors, water-born pathogens, etc. Global warming shall bring with it an entirely new environment in which life though not impossible yet its existence shall be tougher to maintain.

(*ii*) **Sea level change.** Though there are so many uncertainties regarding the consequences of global warming, one of the obvious results of the general heating up of earth's surface shall be a rather rapid rise in mean sea level. Evidences suggest that in Ice Age, about 12,000 years back it was nearly 100 m lower than the present-day level. It is expected that global rise in temperature shall further enhance the rate of already rising sea level in two ways : (*i*) Large deposits of ice present on earth's surface shall melt, which will add more water to the oceans. (*ii*) Rise in temperature shall also cause thermal expansion of the upper layers of the water.

Sea level has been raised by 1 to 2 mm per year during the twentieth century. It is expected that by the end of twenty first century (*i.e.*, 2100 AD.), the global mean sea level can increase upto 0.88 m over the 1990 level. A rise of even half a meter in sea level would largely effect human population, one-third of which lives within 60 km of a coastline.

Thus, many of the world's important cities and coastal areas will come under the threat of flood, and several low-lying islands will be submerged.

The flooding of coastal salt marshes and estuaries will deprive many important birds and fish their breeding grounds and their extinction will be imminent.

Thus, sea-level rise is predicted to have negative impacts on human settlements, such as tourism, fresh-water supplies, fisheries, wetlands, etc.

(*iii*) **Effects on distribution and phenology of species.** As the climatic belts shift away from equator towards poles, vegetation will also shift in the same direction to stay in favourable climatic conditions. Those species, which will not do so shall die. There will be losses of genetic resources on large scale. Only hardy and resistant forms shall come up and survive. For example, with a global temperature rise by 2 to 5°C during the twenty first century, the temperate region vegetation may shift 250-600 km poleward.

As trees are more sensitive to temperature stress, a rapid rise in temperature may cause large scale death of trees and there replacement by scrub vegetation.

As temperature changes will affect wind and precipitation patterns also water could play an important role in altering the biotic communities. It has been suggested that some rise in precipitation, however, shall be balanced by an enhanced evaporation and this could lead to water deficit and moisture stress in many regions of the world. Insects and pests may increase as warmer conditions could be more favourable to their growth and coupled with higher humidity pathogens shall multiply.

(*iv*) **Food production.** Increased temperature will cause eruption of plant diseases and pests. Besides, there will be explosive growth of weeds and increased basal rate of respiration of plants. Thus, all these factors will be responsible for decrease of crop production.

However, small temperature increase may slightly enhance crop productivity in temperate regions, but larger temperature changes will reduce crop productivity there also.

Even a small rise in temperature will have detrimental effect on crop-productivity, both in tropical and sub-tropical regions.

It is expected that rice yield alone, in southeast Asia will go down by 5 per cent for each 1°C rise in temperature. However, in spite of beneficial carbon dioxide fertilisation effect, the overall world crop productivity will decline considerably due to projected global warming, and the food supply of the whole world shall be affected.

As the climatic belts will shift away from equator towards poles, vegetation will shift in the same direction to stay in favourable climatic conditions. Those species which shall be unable to do so, they will die.

There will be losses of genetic resources on large scale. Only hardy and resistant species shall come up and survive. An altogether changed biotic spectrum shall replace the earlier ones and almost all important biomes shall be affected.

Alterations in cropping pattern shall occur and pest resistant varieties more suitably to warmer conditions shall have to be developed. In brief greenhouse warming (global warming) shall bring with if an entirely new environment in which life though not impossible yet its existence shall be tougher to meet with.

Approaches to Deal with Global Warming

Some strategies which could reduce the global warming are as follows :

(*i*) Reduction of the greenhouse gas emissions by the use of limited fossil fuels, by developing alternative renewable sources of energy, such as solar energy, wind energy, etc.

(*ii*) By increasing the vegetation cover, particularly the forests, for photosynthetic utilisation of corbon dioxide.

(*iii*) By using minimum nitrogen fertilisers in agriculture, so that, the N_2O emissions may be reduced.

(*iv*) By developing substitutes for chlorofluorocarbons (CFCs).

Besides, above mentioned strategies, adaptations to address localised impacts of climate change shall be important.

Stratospheric Ozone Depletion

Sun emits radiations at all wavelengths. Of these, radiations important to our planet are : ultraviolet (UV) radiations, light rays, and infrared rays or heat waves. Radiations of visible spectrum and infrared rays carry little energy which does not harm living beings. However, the energy content of ultraviolet radiations is larger than the limits of tolerance of a living cell, and therefore, is harmful or even lethal to a living system.

Stratospheric O_3 layer. High up in the stratosphere, that is, about 20 to 26 km above the sea level, short wavelength ultraviolet radiations are absorbed by molecular oxygen which splits up into its constituent atoms. These atoms combine with molecular oxygen to produce ozone.

$$O_2 \xrightarrow{\text{UV rays}} O + O$$

$$O_2 + O \xrightarrow{\text{UV rays}} O_3$$

Thus, ozone is a result of photochemical reactions in which the starting molecule is oxygen. Along with this reaction another photochemical reaction which causes breakdown of ozone molecules due to absorption of radiations also occur :

$$O_3 \longrightarrow O_2 + O$$

$$O_3 + O \longrightarrow 2O_2$$

The two reactions, *i.e.*, the formation and destruction of ozone molecule normally balances each other and ultimately result in effective absorption of short wavelength ultraviolet radiations in the stratospheric region. Life underneath is thus protected from the biocidal solar radiations.

Most of the ozone present in the atmosphere, almost 90 per cent is concentrated in the stratosphere at an altitude of about 20 to 26 km above the surface of earth. The thickness of the vertical column of stratospheric ozone layer, condensed to standard temperature and pressure, averages 0.29 cm above the equator and may exceed 0.40 cm above the poles at the end of the winter season. Though maximum amount of ozone production occurs over equator, yet its concentration is lowest here, as most of the ozone formed is displaced towards the poles with massive movement of atmospheric air.

Ozone shield. The stratospheric ozone layer acts as the ozone shield, which protects the earth biota from harmful effects of strong UV radiation. Absorption of UV radiation by ozone layer increases more rapidly with its thickness. Thus, maximum amount of UV radiation passing through the atmosphere reaches the earth surface in the tropics, *i.e.*, near the equator, and this amount decreases towards the poles.

The concentration of ozone in the stratosphere changes with seasons. In spring season (February to April), this is highest, while during fall season (July to October), this is lowest.

Ozone hole. During the period 1956-1970, the spring season O_3 layer thickness above Antarctica varied from 280 to 325 Dobson Unit (DU), *i.e.*, 1 DU = 1 ppb (part per billion). The thickness was sharply reduced to 225 DU in 1979 and to 136 DU in 1985. Later the O_3 layer thickness continued to decline to about 94 DU in 1994.

The decline in spring time ozone layer thickness is termed **ozone hole**. The ozone hole was first discovered in 1985 over Antarctica. The presence of ozone hole was also discovered above Arctic in 1990. The global average total column ozone amount for the period 1997-2001 was about 3 per cent below the pre-1980 average values.

Ozone hole is formed by intense UV activity on CFCs. It starts appearing at the end of winter, shows maximum display in spring and disappears in summer. This is because in winter the air is heavy and industrial gases like CFCs get concentrated over an area in Antarctica where the maximum damage of ozone layer has been noticed. In Antarctica CFCs bearing winds in winter become frozen to form ice crystals. During spring time when ice melts the trapped CFCs and ClO are freed at once which in turn release nascent chlorine to attack ozone. Thus, ozone hole becomes large in the spring season.

CFCs, CH_4 and N_2O escape into the stratosphere and cause destruction of O_3 there. Most damaging effect is of CFCs (chlorofluorocarbons) which produce active chlorine (*i.e.*, Cl and ClO radicals) in the presence of UV radiation. These radicals catalytically destroy ozone and convert it into oxygen. CH_4 and N_2O also cause ozone destruction through a series of reactions.

For making above mentioned discoveries related to O_3 destruction, Sherwood Rowland and Mario Molina, along with Paul Crutzen, were awarded Nobal Prize for chemistry in 1995.

Effect of ozone depletion. Ozone layer shields us from biocidal UV radiations. Depletion of stratospheric ozone concentration shall result in an increased penetration of UV–B radiations. A higher loss of stratospheric ozone could cause entry of not only UV–B radiations but also UV–C radiations. This could turn the geological clock back to the era when the ozone umbrella was very thin. Terrestrial life shall be drastically affected so also will be aquatic life which occurs in shallow waters.

The thinning of the O_3 layer results in an increase in the UV–B radiation reaching the earth surface. A 5 per cent loss of ozone results in a 10 per cent increase in UV–B radiation.

In humans the increased UV–B radiation increases the chances of cataract and skin cancer. This also diminishes the functioning of immune system.

Increased levels of UV–B radiation affect photosynthesis, and damage nucleic acids in living organisms. UV–B radiation inhibits photosynthesis in most phytoplankton as it penetrates through the clear open sea waters. Thus, the whole food chain of the organisms would be affected that depend on phytoplankton for thier food.

INTERNATIONAL CONCERN FOR MITIGATING GLOBAL CHANGE

Global warming is the global concern. The long term challenge of stabilising the atmospheric concentrations of greenhouse gases requires that global emissions be significantly lowered than what they are today. In this context, the representatives of many countries assembled from time to time to pass some resolutions, which can be implemented.

Montreal Protocol. In 1987, seventy seven industrially developed countries signed the Montreal Protocol. This conference was held at Montreal. This was a landmark international agreement to protect the stratospheric ozone by agreeing to limit the production and use of ozone-depleting substances. It was decided in the conference to limit the use of CFCs and cut level of CFCs production to 50 per cent by 1999. To-date, more than 175 countries have signed the Montreal Protocol. However, India and China did not sign the Protocol. Developed countries have also agreed to establish a multilateral fund under Montreal Protocol to provide financial support to developing countries for that purpose.

London Protocol. In 1990, London Protocol was signed. In London Protocol, India, China and several other countries demanded more time to reduce CFCs consumption. However, the developed countries agreed to stop CFCs consumption completely by 2000 AD, and for developing nations the time limit was extended to 2010 AD.

Toronto Conference. In 1988, the world conference on the changing atmosphere was held in Toronto (Canada). There it was decided to reduce 20 per cent carbon dioxide emissions by the year 2005.

Earth Summit. In 1992, The First Earth Summit was held in Rio de Janeiro, Brazil. The Earth Summit, was held under the banner of UNCED (*i.e.*, The United Nations Conference on Environment and Development). This summit, established the principles for reducing greenhouse gas emission.

Kyoto Protocol. The Kyoto Protocol, approved by a follow-up conference held in Kyoto, Japan, during December 1997, has specified the commuitments of different countries to mitigate climate change. This protocol has established the rules to reduce overall greenhouse gas emissions to a level at least 5 per cent below the 1990 level by 2008-2012.

At present, on the basis of Montreal Protocol, India has established **ozone cell** under the Ministry of Forests and Environment with the help of UNDP (United Nations Development Programme). By now, India has developed R-22, and R-134 as CFCs substitutes which are presently at trial stage.

31

Microbes in Human Welfare

Introduction

Besides macroscopic organisms, microbes are the major components of biokingdom. Such organisms are only microscopic. Microbes are omnipresent, found in soil, water, air, ice, snow, inside bodies of human beings, animals and plants. Some are found in hot springs (upto 80° – 100°C) and even in geysers (**thermal vents**). Microbes have great diversity – they may be monerans, protozoans, fungi and other minute organisms of plant kingdom. Viruses, viriods and prions also are included in microbes.

Several microbes are useful to human beings in various ways, and a subject called 'Industrial Microbiology' has been developed. Some of the major contributions of the microbes are discussed here.

Industrial Microbiology involves the use of microorganisms to produce organic chemicals, antibiotics and other pharmaceuticals and supplements. Microbes are also involved in insects and pests control, and for the improvement and maintenance of environmental quality. The modern development of sanitation and public health has resulted in the reduction of the incidence of many diseases. The use of mineral water contributed a safer environment for man. The major fields of applied microbiology (*i.e.*, microbes in human welfare) are as follows :

Microbes in Household Products

Food microbiology includes the study of those microbes which provide food due to their high protein value, *e.g.*, production of curd from milk. Microbes such as *Lactobacillus* and others commonly called **lactic acid bacteria (LAB)** grow in milk and convert it to curd. During growth, such bacteria produce acids that coagulate and partially digest the milk proteins. A small amount of curd known as starter is added into fresh boiled milk and kept at suitable temperature, where lactic acid bacteria multiply in millions and milk converts into curd that also improves its nutritional quality by increasing vitamin B_{12}.

The lactic acid bacteria (LAB) also play very important role in checking disease causing microbes in our stomach. The dough, which is used for making foods such as **dosa** and **idli** is also fermented by bacteria. The puffed up appearance of dough is due to the production of CO_2 gas.

In the similar way, the dough which is used for making bread, is fermented by baker's yeast (*Saccharomyces cerevisieae*). Besides bread, yeast fermentation are involved in beer, wines, vinegar, etc. Several traditional drinks and foods are also made by fermentation by yeasts. **Toddy,** a traditional drink of coastal regions, in South India is made by fermenting coconut water. The famous wine of Goa, traditionally known as **fenny** is made by fermenting cashew apples.

Special molds (microbes) are useful in the manufacture of certain foods or ingredients of food. Microbes are also used in production of oriental foods, such as soysauce, bamboo pickles, fish, miso, sonti, etc. Some cheese are mold ripened. Cheese, a valuable food and eaten throughout world is one of the oldest food item in which microbes are used. Different varieties of cheese are known by their characteristic texture, flavour and taste which are developed by different specific microbes. The

'Roquefort and Camembert cheese' are ripened by blue - green molds *Penicillium roquefortii* and *P. camembertii* respectively. The large holes in Swiss cheese are developed due to production of a large amount of CO_2 by a bacterium known as *Propionibacterium sharmanii.*

Microbes in Industrial Products

Enzymes, amino acids, vitamins, antibiotics, organic acids and alcohols are commercially produced by microbes. Production on an industrial scale requires growing microbes in very large vessels called **fermenters.** The main function of a fermenter is to provide a controlled environment for growth of a microorganism, or a defined mixture of microorganisms, to obtain a desired product.

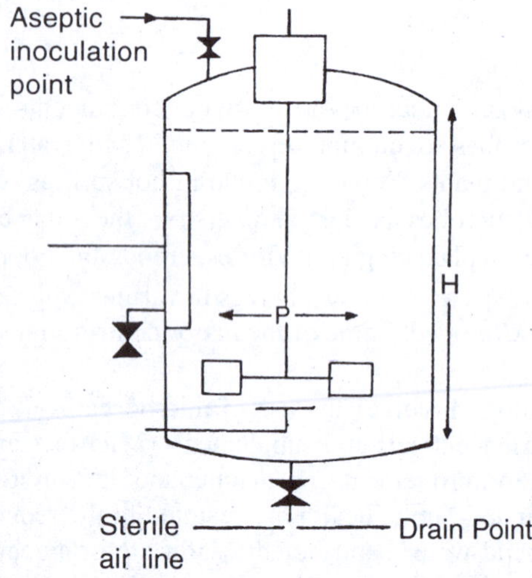

Fig. 31.1. Different parts of fermenter.

Some of the products produced by microbes are used by themselves during their growth and therefore, are called **primary microbial products,** *e.g.*, enzymes, amino-acids and vitamins, while some are not used by the cell for their growth, such as antibiotics, alcohol or organic acids which are known as **secondary microbial products.**

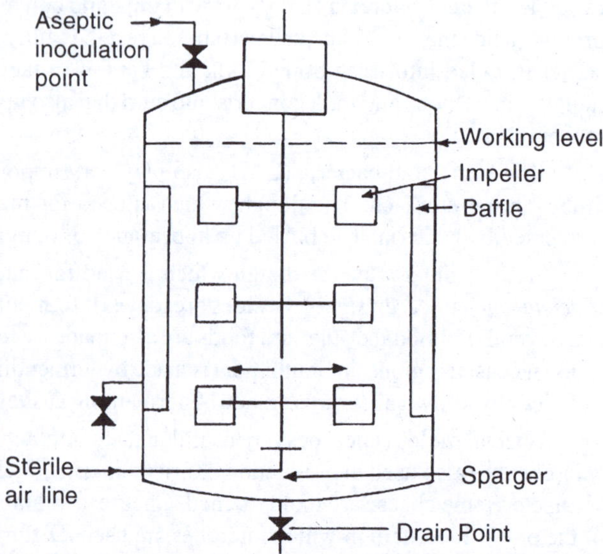

Fig. 31.2. A fermenter with multi-bladed impeller.

Fermented beverages. Microbes (yeasts in particular) have been used from time immemorial for the production of beverages, such as wine, beer, whisky, brandy or rum. For this purpose generally the yeast *Saccharomyces cerevisiae* also known as brewer's yeast is used for fermenting malted cereals and fruit juices to produce ethanol (ethyl alcohol). Depending on the type of raw material used for fermentation and the type of processing, different types of alcoholic beverages are pre-

lipase production.

Lipases have been used to hydrolyse oils for the production of soaps and detergents. They are proved also useful for removing oily stains from the laundry. Microbial lipases can also be used as catalysts for interesterification reactions.

Cellulases. The cellulases are synthesized by a large number of microorganisms including fungi, bacteria, actinomycetes, etc. The extracellular cellulases secreted by *Trichoderma, Penicillium, Fusarium* spp., are among the best examples of cellulases.

Cellulases are being used in clarifying juices and in improvement of beer production. These are also helpful in extraction of essential oils from the plant tissues. Cellulases are also used in bioconversion of lignocellulosic waste into useful products, such as glucose - fructose syrup, single cell protein (SCP) cultivation and as fermentable sugars to alcohol.

Streptokinase enzyme produced by the bacterium *Streptococcus* spp., and modified by genetic engineering is used as a 'clot buster' for removing clots from the blood vessels of patients which may cause heart attack.

Cyclosporin A. This is another bioactive molecule that is used as an immunosuppressive agent in organ transplant patients, is produced by the fungus *Trichoderma polysporum.*

Statins produced by the yeast *Monascus purpureus* have been produced on commercial scale as blood cholesterol lowering agents. It acts competitively inhibiting the enzyme responsible for synthesis of cholesterol.

Microbes in Sewage Treatment

Sewage is the used and wastewater consisting of human excreta, wash waters and industrial and agricultural wastes, such as wastes from live stock, *i.e.*, chicken, cattle, horses, etc., that enter the sewage system. Large quantities of waste water are generated everyday in cities and towns. This municipal wastewater is also called sewage. It contains large amounts of organic matter and microbes many of which are pathogenic. In general sewage contains about 95.5% water and 0.1 to 0.5% organic and inorganic materials. The solid remains in suspended form in water. The celluloses, lignocelluloses,

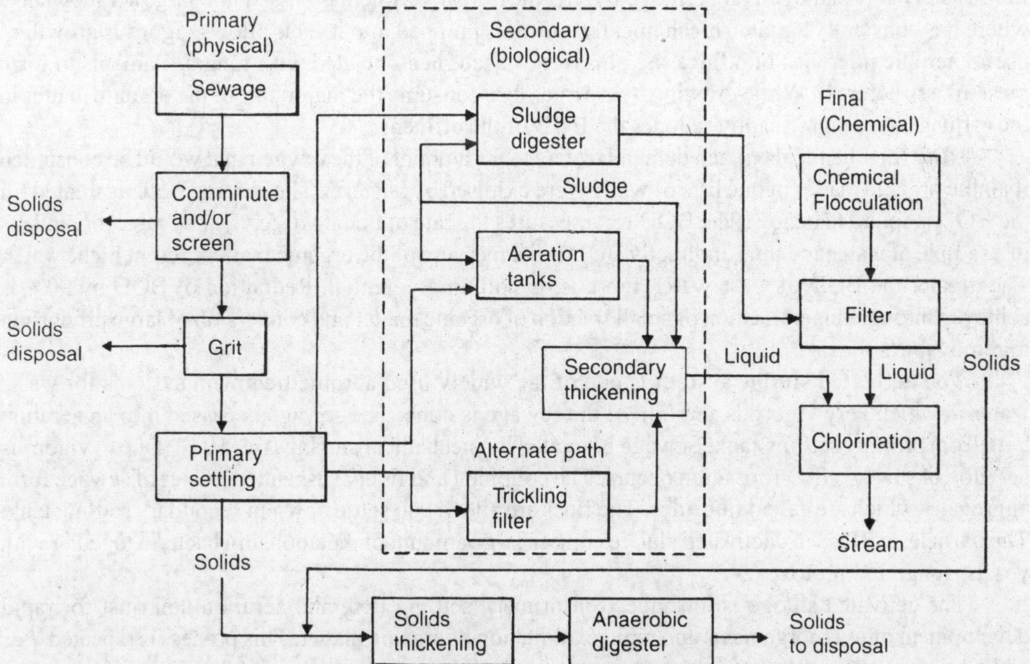

Fig. 31.5. Flow chart of different stages of sewage treatment

proteins and fats are found in colloidal form in water. The other inorganic materials are dissolved and are found in ionic forms. From the industries disposed wastes consist of detergents, antibiotics, paints, biocides, etc. The pulp and paper industries discharge cellulosic and inorganic chemicals. Thus, sewage composition differs with types of industrial effluents discharged into sewage systems. The huge quantity of sewage is disposed off daily. But this cannot be discharged into natural water bodies like rivers and streams directly. Before disposal, hence, sewage is treated in **sewage treatment plants** (STPs) to make it less polluting. Treatment of wastewater is done by the heterotrophic microbes naturally present in the sewage. However, a variety of microorganisms are present in water, for example, bacteria, fungi, protozoa, algae, nematodes, amoebae and viruses of which some are pathogenic. The microorganisms are aerobes, obligate anaerobes and facultative anaerobes present in millions in one ml of sewage. Most of them are intestinal and soil bacteria. The common bacteria are coliforms, streptococci, micrococci, lactobacilli, clostridia, pseudomonads, etc. The other pathogens are those that cause amoebic dysentery, cholera, typhoid fever, bacterial dysentery, polio, hepatitis, etc. These organisms are heterotrophs and thus decompose organic matter of sewage.

The sewage treatment is carried out in *two* main stages. They are as follows :

1. Primary treatment. Primary treatment is the physical removal of 20 - 30% of organic materials present in sewage in particulate form. The particulate material is removed by screening, precipitation of small particulate and settling in basin or tanks where the raw sewage is piped into huge and open tanks. The solid material (sludge) is removed and kept in landfill/composting for anaerobic digestion. The liquid portion is piped into sludge tanks. The accumulated materials in sludge tanks are subjected to aluminium sulphate or the other coagulants so that the suspended particles, organic materials and microorganisms should be trapped as in sedimentation process of water purification. After this, sewage is allowed to go into the primary settling tank where most of the suspended material settles down to form the primary sludge. The effluent from the primary settling tank is taken for secondary treatment.

2. Secondary treatment or Biological treatment. Secondary treatment is also called biological treatment or microbial degradation. By this process 90 - 95% of the **BOD (biochemical oxygen demand)** and many pathogens are removed. Here, the primary effluent is passed into large aeration tanks where it is constantly agitated mechanically and air is pumped into it. This allows vigorous growth of useful aerobic microbes into **flocs** (*i.e.*, masses of bacteria associated with fungal filaments to form mesh like structures). While growing, these microbes consume the major part of the organic matter in the effluent. This significantly reduces the BOD of the effluent.

BOD (biochemical oxygen demand) refers to the amount of the oxygen that would be consumed if all the organic matter in one litre of water were oxidised by bacteria. The sewage water is treated till the BOD is found to be low. The BOD test measures the rate of update of oxygen by microorganisms in a sample of water and thus, indirectly BOD is the measure of the organic matter present in the water. The greater the BOD of wastewater, more is its polluting potential. Reduction of BOD by 90% is achieved through mineralization of small fraction of organic matter and conversion of large proportion to removable solids.

The activated sludge system is one of the widely used aerobic treatment systems for waste water in which very vigorous aeration of the sewage is done. The sewage is passed into an aeration tank from primary settling tank. Sewage is aerated by mechanical stirring (fig. 10.6). Due to vigorous aeration of sewage, floc-formation occurs. The colloided and finely suspended matter of sewage form aggregates which are called floccules. The flocs are allowed to settle down in secondary settling tank. The particles of floc, *i.e.*, activated sludge contain large amount of metabolising bacteria together with yeasts, fungi and protozoa.

The activated sludge is introduced in primary settling tank and aeration tank just for rapid development of microorganisms and rapid exploitation of organic matter. This process is repeated, *i.e.*, addition of settled sludge to fresh sewage, aeration to fresh sewage and so on. This repeating process results in complete flocculation of fresh sewage within a few hours. Activated sludge process reduces the BOD of effluent to 10 - 15% in comparison of raw sewage.

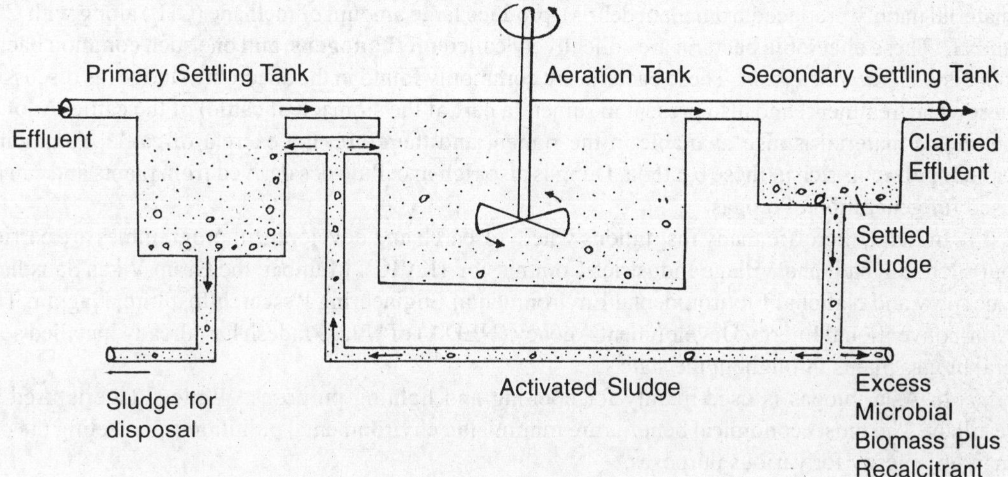

Fig. 31.6. Aerobic activated sludge : secondary sewage treatment system.

The remaining major part of the sludge is pumped into large tanks called anaerobic sludge digestors. Here, other kinds of bacteria, which grow anaerobically, digest the bacteria and the fungi in the sludge. During this digestion, bacteria produce a mixture of gases such as methane, hydrogen sulphide and carbon dioxide. This is biogas and can be used as source of energy as it is inflammable.

The use of activated sludge hastens the efficiency of system. A poor settlement of activated sludge flocs adversely affects the efficiency of sewage treatment plant.

The microorganisms found in activated sludge flocs are the heterotrophs such as Gram - negative rods (*e.g., Escherichia coli, Enterobacter, Pseudomonas, Flavobacterium,* etc.); *Arthrobacter, Corynebacterium, Mycobacterium, Sphaerotilus* are some large filamentous bacteria; and some filamentous fungi, yeast and protozoa. These microbes secrete slime that holds flocs. Thus, the flocs are microbial biomass held together by slime.

The settled sludge should be removed from settling tank time to time otherwise poor settling will result in **bulking of sludge.**

Advantages of using activated sludge process are :

(*a*) significant reduction in BOD and suspended solids, (*b*) reduction in intestinal pathogens, (*c*) requirement of little land, and (*d*) no need of high dilution of final effluent.

The effluent from the secondary treatment plant is generally released into natural water bodies, such as rivers and streams,or it may be subjected to advanced treatment further.

This method has been practiced for more than a century in most of the parts of the world. No other man-made technology has been able to be compared with the microbial treatment of sewage. This is a well known fact that due to increasing urbanization, sewage is being produced in much larger quantities than ever before. However, the number of sewage treatment plants is not sufficient to treat such large quantities. Hence the untreated sewage is discharged directly into rivers that leads to their pollution and increase in water born diseases.

In this direction meagre efforts have been made by the Ministry of Environment and Forests (MOEF). MOEF has initiated Ganga Action Plan and Yamuna Action Plan to save these rivers from pollution, but positive results still awaited.

Microbes in Production of Biogas

Biogas is a mixture of gases (containing predominantly methane) produced by the microbial activity and which may be used as fuel. Many developing countries are encouraging for the installation of biogas plants to meet out the demand of fuel. India is one of the pioneer countries in biogas technology.

Certain bacteria, which grow anaerobically on cellulosic material (*i.e.*, cellulose - containing material mainly produced from plant debris), produce large amount of methane (CH_4) along with CO_2 and H_2. These anaerobic bacteria are collectively called **methanogens,** and one such common bacterium is *Methanobacterium.* These bacteria are commonly found in the anaerobic sludge (as discussed in sewage treatment) and also present in rumen (a part of the stomach of cattle) of the cattle. A lot of cellulosic material is also available in the rumen, and therefore, the excreta of cattle, commonly called '**gobar**' is rich in these bacteria. Dung is also rich in cellulosics derived from plants and can be used for generation of biogas.

In India, there are many institutions where research and development programmes are carried out such as Khadi and Village Industries Commission (KVIC), Mumbai, the Gram Vikas Sansthan, Lucknow and National Environmental Environmental Engineering Research Institure, Nagpur. The Non-conventional Energy Development Agency (NEDA) of Uttar Pradesh has already installed several biogas plants throughout the state.

In Asia, biogas is used mainly for cooking and lighting purposes. While sludge is used as fertilizer. The most economical benefits are minimising environmental pollution and meeting the demand of energy for various purposes.

Generally there are two sources of biomass, *i.e.*, plant and animal for biogas production. The biomass obtained from plants is aquatic or terrestrial in origin, while biomass generated from animals includes cattle dung manure from poultry, goat, sheep and slaughter houses, fisheries waste,etc. However, cattle dung is most potent for biogas production. Besides dung (**gobar**), agricultural residue, apple pomace and deteriorated or dumped wheat grains are also proved to be good source for biogas production.

Production of biogas : The anaerobic digestion is carried out in an airtight cylindrical tank which is called **digestor.** A digester is made up of concrete bricks and cement or steel. It has a side opening (charge pit) into which organic materials for digestion are incorporated. There lies a cylindrical container above the digestor to collect the gas (fig. 31.7). It is noticed that after 50 days sufficient gas is produced in gas tank, which is used for household (cooking, etc.) purposes. Usually, digesters are buried in soil in order to benefit from insulation provided by soil. Commonly, the biogas plants are build in rural areas.

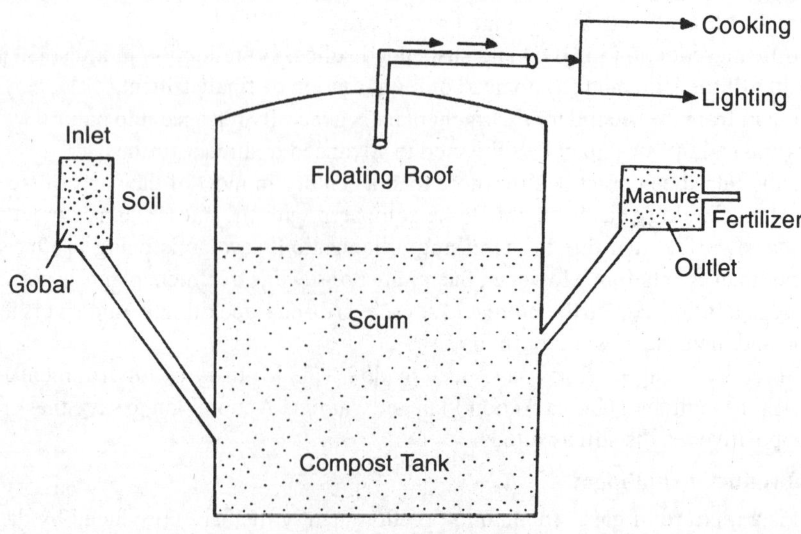

Fig. 31.7. Gobar gas plant

The biogas plant has an outlet which is connected to a pipe to supply biogas for household purposes. The spent slurry is removed through another outlet and may be used as fertilizer.

Microbes as Biocontrol Agents (Biopesticides)

1. Microbial herbicides. The use of endemic or exotic plant pathogens to kill weeds is the effort of phytopathologists. The microbial origin of herbicides is definitely a major contribution and an alternative to chemical weedicides. For example the use of *Cercospora* sp., to control water hyacinth has been established.

2. Bacterial insecticides. There are certain bacteria which are of immense importance for mankind. They are pathogenic to insects, pests and other pathogens and kill these wide range of parasitic organisms.

The bacterium *Bacillus thuringiensis,* now widely known as Bt is the most important, reported to kill a wide range of insects like moths, beetle, mosquitoes, flies, aphids, insects, ants, termites midges, butterflies, even some pathogenic fungi as *Pythium* sp., and *Fusarium* sp., depending upon the host strains of the bacterium. Some of the Bt strains are pathogenic to cockroaches, snails and protozoans. Due to a wide range of host killing, Bt occupies a tremendous significance in agriculture due to exo- and endo - toxin production.

Now more than 50 Bt genes have been isolated, cloned and characterized. Making use of microorganisms and genetic engineering, transgenic microbes and transgenic plants are developed. Many crop plants are colonized by harmless bacteria. Such bacteria are identified and genetically transformed by vectors carrying Bt genes. Certain crop plants are also being protected against insects by genetic mediation, *i.e.*, transgenic plants (resistant to insects) with Bt genes, for example bollworm resistant cotton, stem borer resistant rice, corn borer resistant maize, potato beetle and tuber moth resistant potato, tomato resistant to pinworm, etc.

At I.A.R.I. New Delhi, Bt transgenic cabbage and cauliflower plants have been developed. Apart from crop plants, many forest tree species are also being transformed using Bt genes.

Bt is produced commercially in the form of powder.

3. Entomopathogenic fungi. Many entomopathogenic fungi overcome their hosts only after limited growth and cause death of hosts. Several entomopathogenic fungi attack the insects and kill them. Several examples may be quoted. For example, *Metarhizium* sp., (a fungus) culture filtrate is toxic to *Coleoptera haemocytes in vitro* producing changes in organelles.*Beauveria bassiana* is another fungus species that attacks insects. *Hirsutella thompsonii* (a deuteromycetous fungus) is a potential killer of citrus rust mite. Most of the species of this fungus are pathogenic to invertebrates. *Trichoderma harzianum* and *Trichoderma reesei* are free living fungi, common in soil and root ecosystems. They are effective biocontrol agents of several plant pathogens.

4. Virus insecticides. Several viruses or their products have been commercially exploited in place of chemical pesticides. Viruses of the family Baculoviridae called Baculoviruses are pathogens that attack insects and other arthropods. Baculoviruses are restricted in their host ranges. They do not infect vertebrates, non-arthropod invertebrates, microorganisms or plants. They infect only a few arthropod species. However, the majority of Baculoviruses used as biological control agents are in the genus *Nucleopolyhedrovirus.* They have been shown to have no negative impacts on plants, mammals, birds, fish or even on non-target insects. This is a desirable factor for Integrated Pest Management (IPM) programme, or when an ecologically sensitive area is treated.

Microbes as Biofertilizers

With day-by-day increasing the population, especially in developing countries like India, the stress on agriculture is also increasing continuously. The land area under farming is also decreasing day-by-day due to urbanization. Therefore, the land available for agriculture should be economically utilised and maximum results be obtained.

Most of our agricultural lands are deprived of either one mineral or the other. These minerals are essential for the growth and development of plants. One of the nutrients for any crop is **nitrogen.** This is a major element required by any crop for its development. The nitrogen is provided in the form of chemical fertilizer. Such chemical fertilizers pose health hazards and pollution problem in soil, besides they are quite expensive. Therefore, **biofertilizers** are being recommended in place of chemical fertilizers. Biofertilizers are organisms that enrich the nutrient quality of the soil. The main sources of biofertilizers are bacteria, fungi and cyanobacteria.

Biofertilizers are the formulations of living organisms which are able to fix atmospheric nitrogen in the available form for plants (nitrate form) either by living freely in the soil or associated symbiotically with plants. Although nitrogen fixers are present in the soil, enrichment of soil with effective microbial strains is much beneficial for the crop yields. It has been proved that biofertilizers are cost effective, cheap and renewable source to supplement the chemical fertilizers.

Green manuring. Green manuring is defined as a "farming practice where a leguminous plant which has derived enough benefits from its association with appropriate species of *Rhizobium* is ploughed into the field soil and then a nonlegume is sown and allowed to get benefitted from the already present nitrogen fixer." Some of the cultivated legumes and annual legumes, such as *Crotolaria juncea, Glycine wightii, Sesbania rostrata, Leucaena leucocephala,* etc., contribute nitrogen. Rhizobia, that fix atmospheric nitrogen in the form of nitrates, live in nodules formed on the roots of leguminous plants and nutrients are used by the plants of other nonleguminous crops. Other bacteria, such as *Azospirillum* and *Azotobacter* while free living in soil can fix atmospheric nitrogen in the soil, thus enriching nitrogen content of the soil.

Mycorrhizal biofertilizer. Several fungi are also known to form symbiotic associations with plants, *i.e.*, **mycorrhiza.** Over 90% of vascular plants of world flora form **vesicular - arbascular - mycorrhiza** (VAM). The mycosymbionts are widespread among both cultivated and wild plants. The fungi forming VAM belong to family Endogonaceae of Zygomycotina. The intracellular hyphae either become coiled or differentiated into densely branched arbuscules. Arbuscules function as haustoria and involved in interchange of materials between plant and fungus. Many members of the genus *Glomus* form VAM. The fungal symbionts in these associations absorbs phosphorus from soil and passes it to the

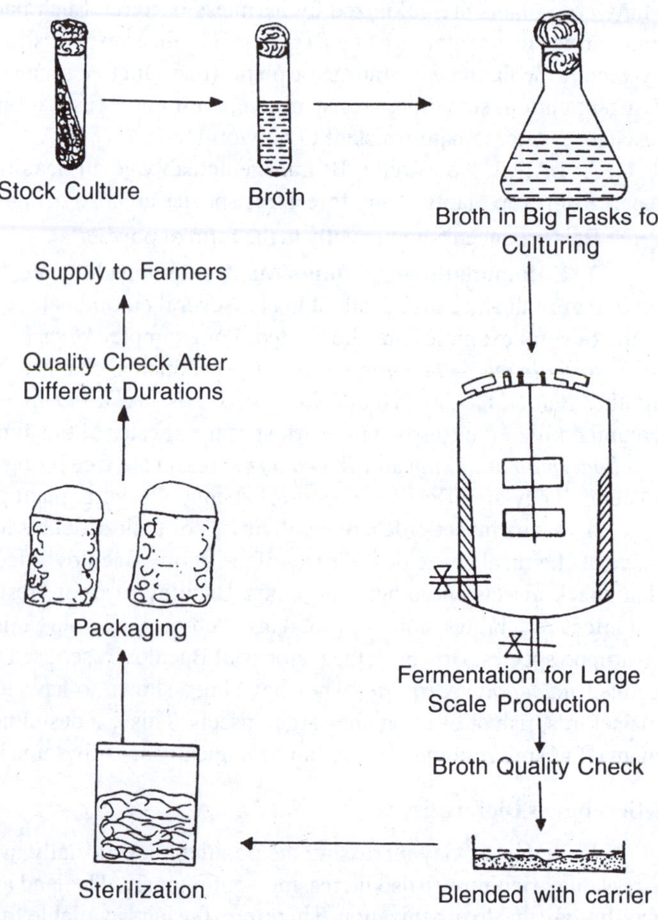

Fig. 31.8. Commerical production of biofertilizer

plant. Plant having such associations show other benefits also, such as resistance to root-borne pathogens, tolerance to salinity and drought, and an overall increase in plant growth and development.

Mycorhizosphere shows increased microbial community that leads to 'mycorhizosphere effect'. The photosynthates flow into soil through roots and mycorrhizae support a diverse community of soil microorganisms, many of which influence plant growth.

Mycorrhizae also increase the absorptive surface of root resulting in increased uptake of water and nutrients from the soil. The ectomycorrhizal fungi translocate phosphorus, nitrogen, calcium and amino acids and increase translocation of Zn, Na and other minerals to the hosts.

Cyanobacteria bio-fertilizers. Biological nitrogen fertilizers play a vital role to solve the problems of soil fertility, soil productivity and environmental quality. *Anabaena azollae,* a cyanobacterium lives in symbiotic association with the free floating water fern *Azolla. Anabaena azollae* can grow photoautotrophically and fixes atmospheric nitrogen. The inoculation of cyanobacteria in rice crop significantly influenced the growth of rice crop by secretion of ammonia in flood water. Besides *Anabaena,* other nitrogen fixing cyanobacteria like *Aulosira, Nostoc, Oscillatoria, Scytonema, Tolypothrix, Calothrix, Hapa-losiphon,* etc., have been held responsible for the spontaneous fertility of the tropi rice fields.

Currently in our contry a number of biofertilizers are available commercially in the market and farmers use these regularly in their fields to replenish soil nutrients and to reduce dependence on chemical fertilizers.

Production of biofertilizer. For production of a good and efficient biofertilizer, first of all an efficient nitrogen fixing strain is required, then its inoculum (*i.e.*, the form in which the strain is to be applied in fields) is produced. Packing, storing and maintenance are other aspects of biofertilizer production. (see fig. 31.8).

32

Some Topics of Interest

1. DOMESTICATION OF PLANTS

Domestication of Plants. Domestication is the process of bringing a species under human management.

The most remarkable fact concerning the food plants in use in the world today, and for the matter the industrial plants as well, is their great antiquity. Most of them were domesticated from wild ancestors long before the beginning of the historical period, and all available records indicate that they were as familiar to the people of the ancient world as they are to us. Comparatively few plants have been developed during the last 2,000 years, although the older ones have been altered and improved in response to the increasing complexity of man's existence. The history of our useful plants and their influence on civilization has always been of interest to the personnel of natural sciences. Many investigations have been carried on in an attempt to determining their age and place of origin, as well as their cultural history. In the history of human civilization, the earliest stages in crop cultivation are believed to be linked with the setting of man in different areas. The earlist sites where societies started growing were around the Nile in Egypt, the Chinese river valley and the northen Indian plains. The ancient civilizations were restricted in area, for they developed only in those regions where the useful plants that were the foundation stones of their existence were native. The presence of valuable plants in all these regions was the most important factor in the successful development of agriculture, although in all these areas climate and soil conditions were very favourable. The climate was equable, with no extremes of heat and cold; the soil was fertile; and there was either ample rainfall or irrigation could be practiced.

Origin and distribution of food plants. It is thought that there was atleast four main centres in which our economic plants originated, and from which they were later dispersed all over the world. These centres were : (1) South-western or Central Asia, the mountainous region from India to Asia Minor and Transcaucasia; (2) the Mediterranean region; (3) South-eastern Asia; and (4) The highlands of tropical America.

According to scientists the man had his origin in Central Asian Plateau, and from which the human race was dispersed. Thus from the very beginning man had at his disposal various food plants, and he must have been dependent on them to a great extent for his existence. For countless ages he was a nomad, wandering from place to place, contend merely to gather the edible fruits, grains, seeds and tubers as he needed them. Later on, he began to make primitive attempts at cultivating these useful plants by sowing seeds in same favourable location. Such attempts at agriculture were of profound improtance for they changed the whole nature of existence. Then he had to give up his nomadic life and remain in one place at least long enough to harvest his crop. In so doing he took the first step toward becoming civilized, for agriculture is the only mode of existence that has enabled man to live together in communities an accumulate the necessitities of life. The establishment of agriculture was of the utmost importance to man and represents the most significant single advance in his development.

In Central Asia the native plants inculded alfalfa, apple, barley, broadbean, buckwheat, cherry, flax, garden peas, garlic, hemp, lentil, mulberry, olive, onion, pomegranate, plum, quince, radish, rye and spinach.

The Mediterranean region was the home of artichoke, asparagus, cabbage, cauliflower, cotton, fig, horseradish, millets, parsnip, parsley and rhubarb. Common to both these areas were the almond, carrot, carob, celery, chestnut, grape, lettuce, mustard, turnip and walnut. Wheat is also a native of some part of his combined area.

In south-eastern Asia the banana, breadfruit, millet, persimmon, orange, rice, soybean, sugar-cane and yam were native; in the America area cacao, American cotton, kidney and lima beans, maize, potato, squash, tobacco, and tomato were indigenous.

The Aryan drifting from the Central Asia region which later invaded this civilization also knew the skills of agriculture–the use of bullocks and plough, and were able to manage their cattle. They later spread themselves numerically and culturally, all over the north Indian plains and further up to the extreme south.

In India, the rate of incerase in population after 1921 started putting an increasing demand on the food supply. The phase between 1921 and 1951 showed the average annual rate of increase in polulation of about 1.4 per cent coupled with a decrease in the average per hectare yield of foodgrains such as wheat and rice. The production of wheat averaged around 798 Kg/hectare in 1901 but only 645 kg./hectare in 1952. Similarly, the figures for rice were 1,050 kg/hectare in 1901but only 800 kg./hectare in 1952. The corresponding current figure can be placed approximately at 1,400 and 1,800 kg./hectare for wheat and rice, respectively. The area under irrigated cultivation has been increased with more intensive cultivation that has marked the "green revoution" in many parts of the country. The important factors responsible for green revolution have been (*i*) development and introduction of high-yielding varieties of crops, (*ii*) extension of the high-yielding varieties over larger areas in the country, (*iii*) multiple cropping, using high inputs of water and fertilizers which provide economical yield of two to three crops in a year from the same field, (*iv*) use of crop protection methods against disease and pests, and (*v*) transfer of the technology of scientific farming from research farms to village farmers.

The recent advancement in agriculture has brought about significant changes in the traditional organizational patterns in villages presented earlier. The village is now getting increasingly linked with the rest of the world through the inputs it requires in terms of materials, knowledge and skills and through the output of the production to other areas. The transport of inputs and outputs, coupled with the storage and distribution of the surplus, has created need for better roads, godowns and repair facilities. The rural scene is changing rapidly. Nationalized banks have moved into the rural areas. The farmer is getting integrated with the external world on an ever-increasing scale. The domestication of plants by man has reached such a stage in India that even in the seventies agriculture dominates the national economy. It provides employment to nearly three-quarters of our population and accounts for nearly half of the national income.

At present, of nearly 360,000 plant species known so far only a few are used as food sources, of which only a dozen provide nearly 90 per cent of world's food supply. New dimensions for exploration could be added to this list. A strategy is yet to be marked out for the arid parts of the world. These account for one third of the earth habitable area. Only about 150 million people inhabit these areas at present. Many more could be supported if they can be transformed into green fields by proper water-management.

The phytoplakton of the sea is many times greater a producer of organic matter on the earth than all the land plants put together. Sir John Murray is of opinion that the sea is as important as the land in the production of food materials. Marine Biology is a new field of study and may lead to the sceintists to a new science "mariculture" parallel to "agriculture". Certainly we have not realized the importance of sea vegetation as yet.

Crop Improvement: Crop improvement means combining desirable characteristics in one plant and then multiplying it. The aim of plant breeding is to select plants with desired characters, cross them and then identify the offspring that combine the attributes of both parents. The modern age of plant breeding began in the early part of the twentieth century after Mendel's work was rediscovered. It is now an established fact, that within a given environment, crop improvement has to be achieved through superior heredity. Thus, plant breeding is an applied branch of botany which deals with the improvement of economically important plants.

2. GERMPLASM COLLECTION AND CONSERVATION

Genetic factors form the basis for plant improvement through breeding. The unique problem of the breeder is to make the plant better adapted to the needs of man by bringing about a favourable combination of genes.

The chromosomes in the somatic cells of the higher plants are composed of sets called **genomes**. Each genome contains a particular number of chromosomes. Although most plants have two genomes, the diploid number (2n), a wide variation in number is possible. Variation in chromosome sets is referred to as **euploidy**—that is true ploidy. For example the bread wheat (*Triticum aestivum*) is hexaploid ($6n = 42$).

The genetic material, especially found within the reproductive cells (germ cells) is known as **germplasm**. The germplasm collections are generally made from an area where agriculture is still in primitive state and where the wild relatives of the crop plants are living in wild area such as Peru, Bolivia and Chile, Middle East and South East Asia, Ethiopia and Mediterranean region.

Thus, **germplasm** is the sum total of all the alleles of the genes present in a crop and its related species.

The germplasm of any crop species consists of the undermentioned materials :

(*i*) Cultivated improved varieties
(*ii*) Improved varieties that are no more in cultivation
(*iii*) **'Desi'** varieties
(*iv*) Lines produced by plant breeders, and
(*v*) Wild species related to the crop species.

In fact, germplasm is the building material, out of which improved varieties are made. The above mentioned materials contain valuable alleles of genes which are indispensable for breeding programme.

Barley, coffee and sorghum are the natives of Ethiopia, whereas the native places of beets, cabbage, lettuce, olives, asparagus and oats are around the mediterranean sea. Lentils, rye, almonds, apricots, apples, pears, pistachios, and pomegranates have been originated from Asia minor and Afghanistan. The native place of wheat is South West Asia. Rice, banana, oranges, black pepper, brinjal, pigeon pea, sugarcane and mango are the natives of South-East Asia. China is original home of tea, soybean, onion, persimmon and several others.

The original place of maize is Mexico and Central America and sunflower came from the U.S.A. Pine Apple and Rubber are natives of Brazil. Tomato and Potato came from the Peruvian Andes. These native places of the food crops are vital geographic regions where the genetic diversity for each species is found.

In early days of history plants were migrated from one place to another from one country to another country through, rivers, migratory birds and people. Early introductions of the plants were made by explorers, adventurers and crusaders. In modern days, the introductions have been made by scientists through international agencies under strict plant quarantine regulations. The major crops that originated in the New World and Old World before the famous journey of Columbus *i.e.*, 1492 are being given in the tabular form.

In modern days, agriculture uses improved varieties that are more or less genetically uniform.

Now, the old varieties have been replaced by more uniform new varieties. The collection of germplasm is most important factor for plant breeding.

Table 32.1. New World and Old World Crops Before 1492.

New World	Old World	
1. Maize	1. Wheat	23. Black pepper
2. Potato	2. Rice	24. Cardamom
3. Tapioca	3. Barley	25. Coriander
4. Kidney bean	4. Oat	26. Apple
5. Lima bean	5. Rye	27. Pear
6. Sunflower	6. Sorghum	28. Peach
7. Peanut	7. Lentil	29. Almond
8. Tomato	8. Pea	30. Fig
9. Squash	9. Soybean	31. Banana
10. Pumpkin	10. Mustard	32. Grapes
11. Red pepper	11. Olive	33. Orange
12. Papaya	12. Cocount	34. Lemon
13. Avocado	13. Cabbage	35. Mango
14. Pine apple	14. Radish	36. Date
15. Custard Apple	15. Carrot	37. Pomegranate
16. Guava	16. Onion	38. Water melon
17. Sapota	17. Garlic	39. Jackfruit
18. Cashew	18. Spinach	40. Coffee
19. Cotton	19. Cucumber	41. Tea
20. Quince	20. Brinjal	42. Cotton
21. Cocoa	21. Beet	43. Jute
22. Cola	22. Sugarcane	44. Poppy

Table 32.2 Centres of World Production and Centres of Origin of Some Important Crops.

Crops	Centres of production	Centres of origin
1. Cacao	Africa	Brazil
2. Coffee	Brazil and Central America	Ethiopia
3. Maize	Mid-West U.S.A.	Tropical America
4. Pineapple	Hawaii	Brazil
5. Potato	Eastern Europe	Peru
6. Wheat	N. Central, N. Ameica	Central Asia
7. Rubber	Malaysia, Indonesia	Brazil
8. Oilpalm	Malaysia	Tropical Africa
9. Groundnut	India	Peru, Brazil

Genetic Conservation

The genetic resources of plants normally have a genetic diversity, *i.e.*, variety of genes which is always changing because of intra-and interspecific breeding in nature. Some of the genes are more desirable than the others. The species having desirable genes are supposed to be significant for the improvement of crops. The **gene pool** is known as the total sum and variety of all the **genes** and their alleles present in a population of species and the **genetic erosion** is the loss of genes from the gene pool due to the factors such as deforestation, urban expansion, shifting cultivation, damage to ecosystem and adoption of certain genetically uniform crop plants.

The plant genetic resources are conserved as follows. This includes four basic ways - (*i*) They are conserved in their natural habitats, such as forests. (*ii*) They are preserved in botanical gardens. (*iii*) They are introduced into the agriculture and horticulture, so that they are cultivated by people. (*iv*) They are preserved in the form of seeds, or other material.

The conservation that maintains the plant diversity in natural reserves is known as *in situ* conservation. It is generally done in National Parks, Wildlife Sanctuaries and Biosphere Reserves. This is meant for the protection of the species that are threatened with loss of extinction and can serve as the source of genetic conservation for the improvement of agricultural crops and domesticated animals. Such achievements can be obtained by the conservation of forests and wildlife in their natural form. *In situ* conservation of wild plants and wild life help to protect species towards extinction, supply materials to restore degraded lands and provide material for genetic improvement of crop plants.

Importance of germ plasm of wild species. The wild plants along with the wild life thrive on their own, without getting any human protection against predators, pests and diseases, whereas the crop plants and the domestic animals get every type of such protection. On the other hand, the agricultural fields are well prepared by ploughing, irrigation and using fertilizers, by farmers. The wild plants normally possess defensive traits such as genes for disease resistance and tolerance to environmental stresses. Such traits have been proved of great advantage for the improvement of agricultural crops and domestication of animals. The examples of such traits are as follows — The common cultivated potato *Solanum tuberosum* has greatly benefitted from its wild allies. For example, from *Solanum demissum* comes resistance to *Phytophthora infestans* the causal organism of late blight of potato. On the other hand, *Solanum acaule* has provided resistance to potato virus-X, and potato leafroll virus; *S. spegazzini* has given resistance to nematodes (*Globodera*) as well as to *Fusarium coeruleum; S. stoloniferum* has given resistance to potato virus-Y.

The wild *Saccharum spontaneum* (**Kans**) has provided resistance to red rot of sugarcane (*Saccharum officinarum*). Much work for the improvement of sugarcane varieties has been done at Sugarcane Breeding Institute, Coimbatore.

The genetic diversity has been maintained since pre-historic times by saving seeds or other propagules for the next planting. In ancient times, the seeds were stored properly in basket and earthen pots appropriately sealed. Rhizomes, bulbs, tubers, etc., were saved for subsequent propagation in cool and dry conditions and the fresh cuttings were planted immediately after harvest.

For the maintenance of plant diversity the samples of wild and cultivated species are collected and properly stored in botanical gardens and gene banks. The botanical gardens are the primary stores for wild plants and have a tremendous role in conservation and education. However, because of limited space, such gardens can accomodate only few plants of a particular species which would represent a fraction of gene pool of a species as a whole. On the other hand, the gardeners prefer to grow more plants of ornamental value, and so the variation of genetic diversity is reduced too much.

Gene Banks

For breeding programme the plant breeders need germplasm collections to develop new, resistant and high yielding varieties. To achieve all that the breeders require access to the entire array of genes available in one species. The practical approach for such achievement is that the collections of germplasm from all corners of world should be stored safely in such gene banks.

A gene bank may be defined as an institution where valuable plant material likely to be lost in the wild or cultivated plants is preserved in a viable condition. The gene banks conserve stock of both seed and vegetative material. In present usage, gene is a functional unit of DNA. The gene bank should not be considered as a plant museum. Here, the stored germplasm preserves the species towards extinction, and it is also utilized by the breeders all over the world to develop improved, resistant and high yielding varieties. Generally the germplasm is stored in the form of seeds. The seed is a living material and can be kept long in a dormant state. When storing of seeds is required for longer periods, it becomes necessary to avoid conditions favourable for enzymatic activity and respiration. This may be done by controlling moisture, temperature and oxygen availability. Generally the seeds remain viable for longest periods when dried and stored at low temperatures. Usually the moisture content of seeds is reduced to 5 per cent or so. If dried further, seeds will die. Temperature may be brought down to – 10°C to – 20°C. The seeds which can face the reduction in moisture and temperature are called **orthodox seeds.** They may be cereals (wheat, rice, maize, etc.) and legumes which can withstand a temperature as low as -196°C (*i.e.,* the temperature of liquid nitrogen). The seeds which are killed by drying and freezing temperature are called **recalcitrant seeds.** The crops with such seeds are —*e.g., tea*, cocoa, rubber, jackfruit, litchi, coconut and several other tropical fruit plants.

In situ conservation is a practical way by which the crops with recalcitrant seeds are conserved. Such seeds can also be treated with fungicides and kept moist with access to oxygen when stored for short periods. However, in modern days, the technique of tissue culture has been developed to store the plants with recalcitrant seeds.

The introduction of the dwarfing genes namely Dee-geo-woo-gen in rice from Taiwan, and Norin—10 from Japan in wheat gave the world new high yielding varieties of rice and wheat. The painstaking research in producing new Mexican wheat varieties by the distinguished plant breeder Norman E. Borlaug ended hunger in Mexico, India and several other parts of Asia. For this great achievement, Borlaug was awarded the Nobel Prize for peace in 1970. The seeds of dwarf Mexican wheats Sonora-64 and Lerma rojo-64 were introduced to India in 1963, and they were integrated to Indian agriculture. The researches of Borlaug, Dr. M.S. Swaminathan and other Indian Scientists have paved the way for green revolution, and India became self-sufficient in food grains for the first time. The green revolution in the developing countries shows the importance of international cooperation in germplasm exchange and breeding programmes. The development of a most widely cultivated rice variety IR 36 by the International Rice Research Institute (IRRI) at Manila, shows the importance of gene banks, and international cooperation in conservation, exchange and breeding of germplasm.

About 350,000 plant species have been described by the botanists, out of these a few hundred have been selected for everyday use. Thirty top crops are enlisted here in table.

Table 32.3. Crops of Today.

Annual World Production in Millions of Metric tons.					
1.	Wheat	450	16.	Tomatoes	45
2.	Corn	400	17.	Sugar beets	35
3.	Rice	395	18.	Rye	30
4.	Potatoes	295	19.	Apples	30
5.	Barley	180	20.	Coconuts	30
6.	Sweet Potatoes	155	21.	Cotton seed oil	30
7.	Cassavas	115	22.	Peanuts	25
8.	Soybeans	105	23.	Yams	25
9.	Grapes	80	24.	Watermelons	20
10.	Oats	65	25.	Cabbages	15
11.	Sorghum	60	26.	Onions	15
12.	Sugarcane	55	27.	Beans	10
13.	Oranges	50	28.	Peas	10
14.	Millets	50	29.	Sunflower seeds	10
15.	Bananas	45	30.	Mangoes	10

Primary sources: USDA, FAO, World Bank and CSTA

3. NEW CROPS

Scientists are in constant search for less known, under-utilized and under-exploited plants to meet the rising demands of the increasing population. Examples of such plants are discussed here.

Guayule

The botanical name of this plant is *Parthenium argentatum.* This is a perennial rubber-bearing shrub. This plant is wild in the Mexican desert and South Western United States. It thrives on poor desert soils and marginal lands where the other plants cannot survive. The rubber obtained from this plant is chemically similar to para rubber. Thus, this wild shrub makes a potential source of natural rubber. The wild plants yield about 12 per cent rubber (dry weight basis), whereas the improved varieties can yield upto 20 per cent.

Subabul

The botanical name of this leguminous plant (of family Mimosaceae) is *Leucaena leucocephala.* This is a fast growing shrub or small tree. This tree is generally used for revegetating deforested tropical lands. The trees are grown as windbreaks, firebreaks, for providing shade and ornamentation. They are generally used as avenue trees in dry and desert lands. The plant has the ability to grow on poor, wornout marginal lands. The cut tree stumps readily regrow and produce healthy multiple coppice shoots. Being a leguminous plant, it fixes nitrogen in the soil and the foliage make a good source of green manure.

The shrub/tree produces highly palatable nutritious forage for cattle, buffaloes and goats. The wood of this tree is a good source of fuel and charcoal. This plant is of much value of afforestation. In India, the plant has been introduced in arid and dry zones with success. Vast areas in the developing countries of Asia have been planted with subabul (*Leucaena leucocephala*).

Winged Bean

The botanical name of this plant is *Psophocarpus tetragonolobus.* This is a leguminous herbaceous plant. It fixes nitrogen in the soil and enriches it. It is vine and can climb a height of 3 metres on some support. It bears long pods with four prominent wings, and thus, known as winged beans.

The seeds, pods, leaves and shoots are rich in protein and are consumed by humans and livestock. The tender pods, leaves, and shoots are used as vegetable. The ripe seeds are roasted and eaten as pea nuts. The unripe tender seeds may be used in soups. The ripe seeds contain 34 per cent protein and 18 per cent oil.

Jojoba

In Mexican deserts, a plant known as **jojoba** is found. Its botanical name is *Simmondsia chinensis.* The seeds of this plant contain about 50 per cent by weight of a liquid wax, which is used in cosmetics. The jojoba plant is dioecious. The female plants start fruiting when they are 3 or 4 years old. They produce maximum yield after 10 years.

This plant is a drought-resistant desert shrub. It can thrive under poor soil and low moisture conditions. Its cultivation in arid regions of this country and other parts of the world help in economic development.

4. UNDER-UTILIZED PLANTS OF INDIA

Vegetable Oils

Usually the vegetable oils are stored up in seeds of the plants. Many vegetable oils are edible and used as cooking media. Some vegetable oils are of much importance, while on the other hand, certain under-utilized plants also yield edible oils, which are of less importance. Besides eating purposes, many vegetable oils are used in several industries. The other potential oil crops of less importance are buffalo gourd (*Cucurbita foetidissima),* bitter colocynth (*Citrullus colocynthis*), neem (*Azadirachta indica*), sal (*Shorea robusta*), mahua (*Madhuca indica*), pilu (*Salvadora persica*), kusum (*Schleichera oleosa*), Kekuna (*Aleurites moluccana*), Bhanjira (*Perilla frutescens*), surpan-laurelwood (*Calophyllum inophyllum*), Karanj (*Pongamia pinnata*) and Znitum (*Olea europea*). From the seeds and other parts of these plants the edible oils which are used in several industries may be obtained.

Medicinal Plants

Thousands of plants are screened every year by drug scientists throughout the world, and thus the search for new medicinal plants continues.

The roots of *Atropa belladona* are used as sedative, stimulant and antispasmodic. The root of colocynth (*Citrullus colocynthis*) is bitter, pungent, cooling, antipyretic, carminative and anthelmintic. The stems of *Ephedra gerardiana* are the source of famous drug known as ephedrine. The leaves of *Adhatoda vasica* make a powerful, expectorant and antispasmodic commonly used in chest diseases. The flowers of *Mesua ferrea* **(Nagkesar)** are astringent and stomachic. The dried flowers are given in vomitting, dysentery, coughs, etc. The ripe dry fruits of *Terminalia bellirica* (**Bahera**) are astringent, bitter, tonic and laxative. The wild herb *Solanum nigrum* (**Makoy**) is a cardiac tonic, alterative, diuretic, sedative, diaphoretic and cathartic. It is used as a decoction in dropsy, enlargement of the liver and jaundice.

Non-alcoholic Beverages

Tea, coffee, cocoa and cola are the main source of non-alcoholic beverage plants. However, some other sources of non-alcoholic beverages are less known and obtained from some tropical under-utilized plants. Such plants of less economic value are - *Ilex paraguariensis* (Mate) grow wild and the beverage is obtained from leaves; *Paullinia cupana* (Guarana), a wild plant, its seeds are used by natives instead of coffee. The leaves of *Caltha edulis* (Khat) are used in Arab Countries to yield **khat**, a principle beverage of natives. *Ilex vomitoria* (Cassine) yields a tea like beverage. The shrub is found in Mexico in wild state. *Paullinia yoco* (Yoco) is an important beverage plant and the beverage is popular among natives of Peru.

Less Known Spices

There is a long list of the plants yielding spices and condiments. However, some plants of this category are used by native and are supposed to be of less economic value. Basil (*Ocimum basilicum*), tonka beans (*Dipteryx odorata*), sweet marjorana (*Majorana hortensis*), savory (*Satureja hortensis*), spearmint (*Mentha spicata*), bay (*Laurus nobilis*), thyme (*Thymus vulgaris*), catnip (*Nepeta cataria*), celary sage (*Salvia sclarea*), hyssop (*Hyssopus officinalis*), chervil (*Anthriscus pulegium*), lovage (*Levisticum officinale*), rue (*Ruta graveolens*), tancy (*Tanacetum vulgare*) and several other plants are less known plants producing spices and condiments.

Table 32.4. Some Important Fodder Trees Of India.

	Name of Tree	Plant parts used as fodder
1.	*Acacia nilotica* (Babul, Kikar)	Leaves and pods are relished by goats and sheep.
2.	*Albizia lebbek* (siris) and *A. procera*	The leaves are eaten by livestock (especially goats and sheep).
3.	*Anogeissus latifolia* (Dhawa)	The foliage is used as fodder for cattle.
	and *A. pendula*	The foliage is relished by cattle and livestock.

4.	*Ficus religiosa* (Peepal) and *F. benghalensis* (Bargad)	The young foliage, and fruits are used to feed live-stock.
5.	*Moringa oleifera* (Sainjana)	The leaves are used as fodder by livestock.
6.	*Morus alba* (Shehtoot)	The leaves are generally eaten by goats, sheep and cattle.
7.	*Quercus semecarpifolia* (Maru)	A tree found in the hills of Himachal and Uttar Pradesh. The twigs and leaves are fed to livestock and cattle.
8.	*Sesbania grandiflora* (Agati)	The leaves are fed to cattle, goats and sheep.
9.	*Zizyphus mauritiana* (Ber)	The leaves make good fodder for cattle, camels, goats and sheep.

5. BIOFERTILIZERS AND BIOLOGICAL PEST CONTROL

Several methods are known for controlling the loss of soil fertility. Here only the tertilizers of biological origin, *i.e.*, *(i)* green manure and (*ii*) biofertilizers have been discussed.

Green Manures

Green manuring is the practice of growing, ploughing under and mixing of green crops with soil to improve soil fertility and productivity. Its effects of soils are similar to those of farmyard manures. It is cheap and best method to increase soil fertility as it can supplement farmyard and other organic manures without involving much cost. Green manures add nitrogen and organic matter to the soil for the improvement of crop yield. Through green manuring mobilization of minerals, reduction of organic nutrient losses due to erosion, leaching and percolation improvement in physical, chemical and biological activities of the soil can be achieved. Green manuring also improves soil aeration and drainage conditions. For green manuring, both leguminous and non-leguminous crops are used. In India leguminous crops such as sannhemp (**sanai**), dhaincha, berseem, clover *Phaseolus,* cowpea, are generally used for green manuring.

Table 32.5. Plants Used as Green Manures in India.

English and Hindi names	Botanical name
1. Sann hemp (**Sanai**)	*Crotalaria juncea*
2. Lentil (**Masur**)	*Lens esculenta*
3. Egyptian clover (**Berseem**)	*Trifolium alexandrium*
4. Sesbania (**Dhaincha**)	*Sesbania aculeata*
5. Cluster bean (**Guar**)	*Cyamopsis tetragonoloba*
6. Cowpea	*Vigna sinensis*
7. Horse-gram	*Macrotyloma uniflorum*

Other Organic Manures

The organic content of the soil which is a good source of plant nutrients contributes most to the fertility of the soil. Organic manures improve soil fertility in the several ways.

Organic manures are of several kinds some of which are discussed below :

(*i*) Farmyard manures. Solid and liquid excreta as dung and urine of all farm animals are termed farmyard manures. They are readymade manures and contain nitrogen, phosphorus and potassium. The farmyard manures of different animals vary greatly in their composition but they are good for all types of soils and all the crops. Farmyard manure when collected in a field in exposed condition for several months shows considerable loss of fertilizing value as upon decomposition a considerable amount of ammonia is lost by volatilization. Therefore it is important to keep manure protected from weather and manure preparation should be carried out in trenches of about a metre depth. When the trenches are filled the top surface is covered with cowdung-earth slurry. In about 2 months the manure is ready for use.

***(ii)* Compost.** Compost manure can be prepared from a variety of refuge materials such as straw, sugarcane refuge, rice hulls, forest litter, weeds, leaves, kitchen wastes. It is prepared in pits usually 6-8 m long, $1\frac{1}{2}$ to 2 m wide and one metre deep. In the pits, 30 cm thick layer of plant residues moistened with cowdung, urine and water is formed and then a second layer of about 30 cm thickness of mixed refuge is spread over it and moistened with slurry. The operation is repeated until the heap rises to a height of 50 cm above ground level. The top is then covered with a thin layer of moist earth. After three months of decomposition, the material is well mixed and again covered. After a couple of months the manure is ready for use.

***(iii)* Sawdust.** Sawdust can be used as bedding material to conserve animal urine or for making compost. It is a low fertilizing material but it is definitely richer than wheat straw in calcium.

***(iv)* Sewage.** In modern system of sanitation, water is used for removal of human excreta and other wastes. Sewage consists of two components :

(*a*) the solid part, called the **sludge** and (*b*) the liquid part, called **effluent** or **sewage water**. Sewage is quite rich in several plant nutrients and can be used for fertilizing the crop by irrigating the soil directly with sewage water but there is a danger for the spread of several human diseases.

Biofertilizers

The available green manures are not sufficient to meet the existing needs of agriculture, and therefore, they are supplemented by the use of **biofertilizers.** The biofertilizers are the organisms which enrich the soil and enhance the fertility. The scientists are busy in evolving the methods for using nitrogen-fixing microorganisms to reduce dependence on chemical fertilizers. The main sources of biofertilizers are bacteria (*e.g., Rhizobium, Azotobacter, Clostridium,* etc.), cyanobacteria (*e.g., Nostoc, Anabaena,* etc.) and Fungi.

The nitrogen fixation is being done in plants by two main sources, *i.e.,* symbiotic nitrogen fixing bacteria and free living bacteria.

1. Symbiotic N_2 fixation. In this system, fixation of molecular nitrogen occurs from a mutualistic association between leguminous or non-leguminous plants and bacterium. The most interesting symbiotic relationship is seen between leguminous plants and genus *Rhizobium.* Infection of bacteria into roots of legumes forms root nodules. It takes nitrogen directly from the air, and converts into nitrogenous components such as amino acid and polypeptides. This conversion is the result of the combined action of plant cells and bacterial cells. Nodule formation occurs in about 90 per cent of leguminous plants. The *Rhizobium* species is always specific to the species of legumes. A single strain of *Rhizobium* can infect only a certain species of legumes.

In addition to legume-Rhizobium symbiosis, certain non-leguminous plants can also fix nitrogen with the symbiosis of bacteria other than Rhizobia. The best example is *Psychotria* plant which in association with the bacterium *Klebsiella* fixes nitrogen in the roots.

Free-living N_2 fixation. Winogradsky (1895) discovered first of all that anaerobic species of *Clostridium pasteurianum* bacteria possess the property of nitrogen fixation without symbiotic aid. Beijerinck (1901) observed aerobic non-symbiotic nitrogen fixing bacteria *Azotobacter* and *Desulfovibrio,* etc., also show non-symbiotic nitrogen fixing phenomenon.

Genus *Azotobacter* is aerobic, non-sporing and motile having peritrichously arranged flagella. It has highest rate of oxygen up-take than any of the living organism. They grow in the soil autotrophically, but require organic substances as the source of energy. These organic substances are decomposed into cellulose starch and proteins, etc. If the carbohydrates such as starch and molasses wastes are added to the soil, nitrogen accumulation will be stimulated through the growth of *Azotobacter* and other non-symbiotic N_2-fixing bacteria. They take N_2 from atmosphere, combine with their composition and release it in organic form as secretion and on death.

Table 32.6 Nitrogen Fixing Bacteria.

Symbiotic		Free Living		
		Aerobic		Anaerobic
Leguminous plants	Non-Leguminous plants	Photosynthetic	Photosynthetic	Heterotrophic
Many genera of family Leguminosae in association with the genus *Rhizobium*	*Psychotria* with *Klebsiella* sp.	*Azotobacter* sp., *Azotomonas, Pseudomonas, Spirillum, Beijerinckia Nocardia* sp.	*Chromatium Chlorobium, Rhodospirillum rubrum, Rhodopseudomonas* sp. *Methano–bacterium.*	*Ceostridium, Aerobacter aerogenus Bacillus, Polymyxa, Desulpho vibrio desulphuricans, Achromobacter*

2. Azolla-Anabaena symbiosis. *Azolla* is a fast-growing water fern (a pteridophyte). A cyanobacterium, known as *Anabaena azollae* is found in the cavities of this fern. The cyanobacterium fixes nitrogen from the air and excretes the nitrogenous compounds into the leaf cavity of *Azolla. A. .pinnata* has been proved to be an excellent biofertilizer for rice.

3. Cyanobacteria. The cyanobacteria (also known as blue-green algae) such as *Anabaena, Nostoc* and *Aulosira* obtain the energy required for nitrogen fixation through photosynthesis. These cyanobacteria help in fixing atmospheric nitrogen in soil. These organisms have been used as biofertilizers at Pantnagar University and Indian Agricultural Research Institute, New Delhi.

4. Mycorrhiza. This is a from of symbiotic association of certain fungi with the roots of higher plants. Generally there are two types of **mycorrhiza-(*i*) ectomycorrhiza** and (*ii*) **endomycorrhiza.**

Ectomycorrhiza. This is a form of symbiotic association in which a well-developed mycelium forms a mantle on the outside of the root. Such mycorrhiza increases the surface of interface between soil and plant roots. The mycorrhiza enhances water and nutrient uptake. Thus increasing plant growth, vigour and yield. Ectomycorrhizae are commonly found on the roots of pine, oak, peach and eucalyptus trees. They absorb and store nitrogen, phosphorus, potassium and calcium in their fungal mantle. They also convert the complex organic molecules into simpler ones.

Endomycorrhiza. This is another form of mycorrhizal association where the fungus lives in the intercellular spaces as well as in the cortical cells. Such symbiotic associations are commonly found in the roots of many herbs such as orchids and some other plants. Such mycorrhizae are produced both by septate and aseptate fungi. In a form of endomycorrhiza, a fungi lives between the cells of the cortex and develops temporary hyphal projections that penetrate the cells of cortex. They are known as **vesicular-arbuscular mycorrhiza** (VAM). Such mycorrhizae are important in the phosphate nutrition of herbs.

Biopesticides

The living organisms employed to destroy the undesirable organisms are called biopesticides. The biopesticides can be divided into two groups- (*a*) bioherbicides and (*b*) bioinsecticides.

(*a*) Bioherbicides. The biological control of weeds involves utilization of insects that would feed selectively on a weed or use of certain microorganisms which all produce diseases in the weeds and eliminate them. When insects are to be used as bioherbicides, their specificity should be carefully tested. One should make sure that in the absence of a specific weed, the insect would not feed on some crop of economic value. In this country and Australia the over growth of *Opuntia* spp., (cacti) was checked by the introduction of the cochineal insect (*i.e., Cactoblastis cactorum*). The fungus *Phytophthora palmivora* has been employed to check the growth of milk-weed vines in citrus orchards.

(*b*) Bioinsecticides control by pathogens and parasites. A number of harmful insects for the crops can by reduced by other insects, diseases, parasites and predators. To control the aphids, this method has been applied by the use of ladybugs. If it is found, that the biopesticide cannot adapt to the new environment where a pest has to be destroyed, it is multiplied by artificial breeding and are released at the appropriate time.

Biological control by sterilization technique. A major pest known as screw worm has been eradicated by raising a population of sterile males (which were developed by irradiation technique), and releasing them among the natural fertile individuals to compete.

Insect hormones. The insect hormones called **pheromones** have been utilized for pest control. The female insects produce pheromones to attract the male partners for mating. The pheromones are useful for sending alarm signals for this behaviour. Controlling the mating behaviour of the insect by the use of pheromones, the population of the concerned insect pest can be checked. Pheromones are secreted by females and detected by the antennae of males. They are species specific. Traps containing pheromones of the gypsy moth are kept in the infested fields. Now the males approach the trap and they stick to the hollow cylinder coated inside with a sticky substance, and the males are not available for reproduction.

In another method, known as 'confusion technique' large amounts of hydrophobic paper with pheromones are dropped in an infested field. Here the males fail to locate the females because of the smell of sex hormone all over and ultimately they die.

The insect hormones known as **Juvenile hormone** and **molting hormone** or **ecdysone** are needed for the development of the young ones to the adult stage. The periodical shedding of insect cuticle is brought about by the hormone ecdysone. Juvenile hormone is needed in the early stage of development to check the early maturation. If this hormone is given artificially at a later stage, the insects turn into giant larvae and immature adults which die rather early. Thus the insect pests meet the early death.

Natural insecticides. The natural insecticides are obtained from some microorganisms and some plants. The insecticides of vegetable origin possess low mammalian toxicity and do not harm the predators in food chain. In ancient times, the Chinese people and Romans used the insecticides obtained from plants and animals. The alkaloids such as nicotine (obtained from *Nicotiana tabacum*), and pyrethrum (obtained from *Chrysanthemum* spp.) make good insecticides. The extracts from neem tree (*Azadirachta indica*) contain **azadirachtin** which keep away the insects. The foliage coated with neem **extract repel** leaf-eating pests.

Integrated Pest Management (IPM)

The integrated pest management seems to be most essential these days. The IPM involves several 'cultural controls' or 'cultural practices' to control the crop pests.

Soil conditioning, crop rotation and sanitation practices may be used to avoid several pest problems.

The crops can be protected by eliminating the pests by starvation.

The crops of economic value may be grown mixed with a specific weed that may be consumed by insect pests instead of the main crop.

In conclusion, it can be said that the use of chemical pesticides should be at minimum level, and instead biological control methods should be practiced. Thus the environment can be saved from agrochemical pollution.

Introduction and Acclimatization

When a variety, line or population of a plant species is taken from one area into a new area is called **plant introduction.**

'Plant introduction and acclimatization' is the easiest and most rapid method of crop improvement where the acclimatization follows the introduction and both the processes go side by side. "The process of introducing new plants into a locality or introducing the plants from their growing place to a new locality with different climates is known as **plant introduction** and their adjustment under this changed climate is termed as **acclimatization**. Both, new crops and new varieties of crops, may be introduced either in the form of seeds or cuttings. In vegetatively propagated crops the cuttings are imported and in sexually propagated crops the seeds are imported. In the latter case, the cross-pollinated crops are generally introduced since in them there is greater frequency of gene recombinations owing to the frequent cross pollinations and some of the recombinations may be more favourably adapted in the new environment. The crops may be introduced into a locality either from outside the country or from different regions within the country. Introduction within the country are very convenient but for introductions from another country there is a definite procedure. The material desired is demanded from the concerned authority of agency of the foreign country through the plant introduction organization of the country. The material is packed properly and sent by sea. Before entering the country, it is inspected at sea port by the plant protection authorities and, if found fit according to quarantine rules, is permitted to enter the country. In the country, it is handed over to the concerned workers or institutions. There it is grown under local conditions and tested for the acclimatization and the presence of desired characters. If found suitable for both, it is either utilized as such after selection or utilized in hybridization for transferring the desired characters into the local varieties.

This method of crop improvement has been proved of great importance, especially in recent years, as a source of resistance material to some crop diseases.

Such introductions have given us many valuable crops, *e.g.,* potato, tomato, cauliflower, grapes, guava, etc. Soybean is also an example of introduction. Introductions have also provided us improved varieties, like Sonora-64 of wheat and Taichung Native 1 of rice. These dwarf, varieies provided starting material for 'green revolution' in india.

Quarantine

Along with new introduction, new weeds, insect pests and diseases could also come into our country, *e.g., Argemone mexicana,* a prickly weed is an introduction.

All introductions should carefully be examined prior to their introduction in new area. The introduction must be free from pests, patogens and weeds. This type of checking and formation of rules and regulation, are called **'quarantine'.** The introduction, free from pathogens, pests and weeds are allowed to enter into a country or new area Quarantine is also applied to animals and humans to reduce risk of entry of pathogen in the country.

33

Forests of India

COUNTRY'S FORESTS

In dimensions India ranks as the seventh largest country in the world with a spread of 3287.3 thousand sq kilometers. While its population is overbearingly large compared to its geographical area and is the second highest with a census figure of 845 million (1991) (now 1210 million), in number of cattle and domestic animals, it stands at the top of the list of countries. Its length and breadth coupled with the varied relief features bestow on it a variety of climates and diverse life forms (both flora and fauna). The climatic conditions experienced over different parts of the Indian land mass range from the arctic to the moist tropical with every known climatic type represented but with an element indigenous to the land. This climatic diversity along with the edaphic elements endow the country with forests of many kinds.

The overall forest area as per the latest estimate of 1991 stands at 640.70 thousand sq km or 19.49% of the country's geographical area, registering an increase of 560 sq km over the last inventory. The reversal of the decreasing trends in forest areas towards an increase, particularly in the tribal hill states of Mizoram and Meghalaya where indiscriminate jhooming is practiced and the dry, desert tracts of Rajasthan and Punjab with little forest, speaks of the determination of the nation to conserve and enlarge its forest area. Yet, the per capita forest areas of the country by states/region varies widely from as low as 0.1 ha for Punjab (Chandigarh .005 ha) to 6.00 ha for Arunachal Pradesh and indicates the need to improve the situation, so as to ensure self sufficiency in forest produce in the different regions and thereby save the natural forests which are unevenly distributed over the country and mainly confined to a few zones.

ENVIRONMENT AND FORESTS

A sound environmental policy must simultaneously focus on the conservation and protection of resources and take due account of the needs of those who depend on the resources for their livelihood. Resource and environmental conservation are implicitly linked with the economic status of the people and perpetuation of poverty decreases the chances of conservation in the long run. The extrication of the people from poverty dictate utilisation of resources for production of goods but not without ensuring sustainability of the resources on which production is based, otherwise a decline in productivity will be experienced sooner or later and magnify poverty and hunger for resources. Therefore, the focus on sustained availability of resources is integral to the overall strategy for development and providing an opportunity for sustainable livelihood to all people. Forest and woodlands have a major role in contributing to the well being of the rural masses who represent about seventy five per cent of the country's population. The economy of our rural people being basically biomass based, forests and forest produce contribute sizeably, either directly or indirectly. In addition there are demands of urban population competing for resource needs from forest lands along with the local population. The forest lands and wastelands must, therefore, be developed to support multiple roles, i.e. ecological, economic, social and cultural. To enhance the effectiveness of management, conservation and sustainable development of forests and ensure adequate production of forests' goods/services and sustainable utilisation thereof, it becomes expedient to build up technical skills and improve the quality of human resources, research capability and support, forestry extension, information dissemination systems and public education. An environmentally sound forest management strategy must be duly supported by a rational and holistic approach to the existing constraints and working of solutions thereto. The securing of multiple roles of forests/forest lands using well deliberated and researched solutions has been repeatedly emphasised in the national/ international forums by many organisations related to the field.

Protection of the atmosphere and reduction in green house gases are major environmental issues of the present. Land and resource use both influence the atmospheric changes and redefining of land use particularly forest/waste lands can appreciably influence green house gas sinks and regulate atmospheric emissions. Judicious use of forests and rehabilitation of wastelands will also help in reducing loss of biodiversity which increase resilience of ecosystems to climatic changes and pollution damage by effluents.

An annual afforestation of 2 million hectares as targeted and the cessation of deforestation have the potential of removing from the atmosphere, of as much as 12 million tons of carbon annually if trees could be successfully established and a growth rate of 15 cu m/ha/annum achieved and maintained.

Reforestation, however, withdraws carbon from the atmosphere for only as long as the forests are gaining mass. After maturity, forests are approximately in balance with respect to carbon, neither accumulating nor releasing it. On the other hand, reforestation could reduce the emission of CO_2 to the atmosphere indefinitely if wood fuels were to replace fossil fuels.

Presently, other than slowing the rate of release of green house gases by changes in fuel, more efficient engines and protecting forests, the atmospheric changes so far brought about can only be reversed by more trees or forests so that the atmospheric carbon can be captured by photosynthesis to be deposited as terrestrial carbon pools. A concerted effort for afforestation could enable us to lock atmospheric carbon in an appreciable measure as biomass over long durations, leading to an effective decrease in green-house gases. Forest plantations could control the global carbon balance providing the most efficient sink for atmospheric carbon. The forests have a crucial role in the global cycling of carbon and forestry may rescue the humanity from the lurking dangers of global warming. Land use is inextricably linked with atmospheric changes and may have far reaching effects on forest, biodiversity and other biological resources, impacting equally on the most vital of all economic activity, agriculture.

DESERTS AND ARIDITY

Desertification affects a sizeable land mass in the states of Rajasthan, Haryana and Gujarat. A permanent vegetal cover could promote and stabilize the hydrological balance in the semi desert and adjoining dry area, thus preventing the march of the deserts and maintaining land productivity. Tree and vegetation also effectively stabilise sand dunes and enhance the carrying capacity of the desert area to support life in these fragile ecosystems. Agrisilvi and Silvi-pastoral systems could effectively be used to contribute in rehabilitating the desert lands.

Afforestation technologies using drought resistant, fast growing species particularly leguminous for providing multiple benefits in the form of green belts or in association with agriculture/pastures are required to be researched and tested towards ameliorating the desert, semi desert tracts. Also needing emergent conservation measures and research attention in the deserts tracts are areas of special ecological significance holding rare biodiversity.

MOUNTAIN ECOSYSTEMS AND WATER RESOURCES

Some of the ecosystems in the country namely the Himalayas, Western and Eastern Ghats, Vindhyas and Satpura mountains, have unique features and resources, some most vital to the nation's ecological security. Sustainable development of the mountain ecosystems which are fragile and already stressed to the limits of their recuperative capacity are of utmost importance for regulating the water in the rivers and streams that originate therefrom. The aquafers in these mountains have from times immemorial supported life in the hills and dales and the plains below, nurturing the people and their civilization. The rapid loss of tree cover from these mountains have rendered them susceptible to accelerated soil erosion, landslides and loss of habitat and genetic diversity. The unprecedented degradation of the mountain ecosystems is probably the foremost threat to the ecological safety and security of the dwellers in the watersheds of the rivers emanating therefrom. The rehabilitation of the mountains and proper management of its resources, particularly the common resources of forests assume special significance in the existing scenario.

Water is acutely in short supply. The ever increasing demands of water for industrial, agricultural uses and in urban dwellings have led to a perceptible shortage of this prime resource which sustains all life forms and economic activity of humans. In future the forests in the watersheds will have to be primarily managed for regulating water and its harvesting. The recurring floods and droughts experienced in our country demand concerted efforts towards expeditious rehabilitation of the watersheds and the verduring with trees of the flood plains of river-systems to increase charging of the aquafers, prevent floods and siltation of reservoirs to regulate and sustain supply of water around the year. Managing mountain forests and their rehabilitation towards the newly defined goals call for earnest efforts in providing answers to many complex social and technical issues before embarking on achieving the objectives.

Mountains are extremely vulnerable to external forces whether induced by humans or natural. Mountains and hills both vertically and horizontally nurture a variety of ecosystems because of the gradients of temperature, precipitation and insolation. Thus, holding a large biodiversity over a limited area. The ecological complexity of the mountain watersheds and the impact of present land use and management practices are yet to be fully comprehended and appreciated for harnessing their potential on a sustainable basis. The rich mineral wealth of the Himalayas which are now being exploited on a commercial scale has added to their ecological degradation. Mineral excavation is a violent activity disregardful of the sanctity of mountains and leads to large scale destruction over areas far and wide. Unless and until concurrent measures are initiated to rehabilitate the mined slopes, the consequential events that may follow could be horrendous. As a result of the indiscriminate mining over the past, not only have these areas been denuded of vegetation and their visage scarred but more gravely the life giving water sources have dried up and the streams in the catchments choked with sediment and boulders. During monsoons they get inundated and overflow the banks destroying the fertile lands alongside the streams and rivers.

An economical and effective technology for the rehabilitation of these mined sites is an urgent need both towards the ecological well being of the local inhabitants and those residing in the lower parts of the catchments.

SHIFTING CULTIVATION

The mountains of the north eastern moist tropical zone which in the not long past was once resplendent with a verdure coat of priceless natural vegetation today lie bereft of its cover due to the relentless assault of shifting cultivation. Extent of land area stressed beyond the recuperative capabilities of nature are ever on the increase. The severity of the onslaught of shifting cultivation if unchecked may inflict irreversible damage to the delicate natural ecobalance of moist tropical forests evolved over the millennium.

The north eastern states of India, Assam, Meghalaya, Nagaland, Manipur, Tripura, Mizoram and Arunachal Pradesh today are the worst affected by shifting cultivation, the other states being Orissa, Madhya Pradesh, Andhra Pradesh and Kerala. Quantum jump in populations of societies mainly tribal practicing shifting cultivation, have forced adoption of ever shorter fallow periods, which on the average presently is 3-5 years compared to 25-30 years half a century ago. The productivity of lands under shifting cultivation are directly dependent on the rest or fallow periods, during which such lands rebuild their store of organic matter, essential for remunerative agricultural yields. In the tropical forest ecosystems nutrients are predominantly stored in its vegetation rather than the soil. The fertility of soils depend on recycled organic detritus falling as litter. Unfortunately shifting cultivation of the present day are approaching nearer permanent cultivations minus the cultural practices and inputs needed in maintaining their productive potential. Consequences of indiscriminate shifting cultivation far outreach the evils of soil loss only, as beside, water run off rates increase substantially in denuded catchment areas due to poorer infiltration, reduced evapo transpiration and absence of physical barriers earlier provided by the destroyed multistoried vegetation, inundating extensive areas as an annual feature. Moreover, deforestation by burning for preparing lands for agricultural persuits is today considered as a major anthropogenic source of green house gas emissions responsible for global warming.

Generally tribals practice shifting cultivation on the hill slopes as soils are poor and prone to landslides rendering permanent cultivation impossible. As an alternative introduction of high value exotic species has been mooted in these alien eco-systems which though well meaning and cannot be contested regarding their likely contribution to the tribal economy, but would require technical expertise with which the tribal may not be at home. Moreover, these species would generally demand heavy external inputs to check insects pests and pathogens, competition from natural vegetation and addition of nutrients. Whereas, on the other hand species that are members of the natural plant community would be much easier to cultivate. A more pragmatic approach would therefore be to encourage culture of indigenous forest crops of high value with a ready market along with horticultural and agricultural crops. Research at developing such agroforestry models as are in harmony with the ecological and cultural ethos, of the tribals can provide tested alternatives to this practice. The ICFRE has initiated a special project for researching technical, social and economic solutions to the problems of rehabilitation of the Himalayas particularly in respect of mined and shifting cultivation sites.

MANGROVES

The littoral boundaries of India are over 7500 km in run and the coasts are sizeably populated. To sustain and develop our coastal and marine areas and their resources are therefore of importance to our national economy. Mangroves the brackish wetland forests along the river deltas and estuary constitute a very important segment of coastal resources as well are ecologically knitted to other valuable marine resources. These forests simultaneously prevent, erosion of the coast line by the sea and provide a natural barrier against hurricanes and coastal storms.

The recent estimate of Forest Survey of India based on satellite imagery of 1987-89 place the mangrove forests of the country to be 4244 sq km in extent. About 85% of the mangrove areas lie in Sunderbans delta of West Bengal and sea coast of Andamans. The Andaman group consists of more than 350 islands with a land area of about 8500 sq km, of which 1190 sq km area are covered with mangrove forests. In these islands are located some of the least disturbed and best preserved mangroves. Andamans represent a large group of rare endemic flora consisting of 79 species of which 66 are in the endangered list. Further, Andamans represent a unique eco-system where mangrove forests are adjacent to *Dipterocarpus* forests growing along the sea shore and creeks. Except for the works of pioneers like Sahni (1959), Mathuda (1957), Thothathri (1960, 62), Balakrishnan (1979) and Jain (1980) these mangroves have not received much attention for their scientific exploration. The increasing pressures on mangroves for timber and firewood has subjected these ecosystems to severe stresses which demand urgent attention towards their protection and conservation.

Studies so far carried on mangrove ecosystems pertain primarily to the fields of ion balance, salt regulation mechanisms, salt dependence of species; physiology, biochemistry and photosynthetic behaviour of different species; plant systematics, floral biology; effect of pollutants on species etc. However, a lot more requires to be done in the field of research to standardise management practices for sustained production from mangroves. In order to achieve the targeted goals research on mangrove systems is being strengthened in the Council's Institutes at Jorhat and Bangalore.

FOREST INDUSTRIES

The National Forest Policy of 1988 envisaging the shortages of forest produce directs that concessions on raw material supplies to forest based industries should cease and industry encouraged to raise raw materials needed by them in direct collaboration with farmers. The growing stock of wood in the country is 4196 million m^3 with a net annual increment of 52 million m^3 or say 1.24 per cent of the total growing stock. However, the average production of wood from the forest is low (0.7 m^3/ha), as also the conversion yields which are 40-45 per cent of round wood. Logs available for veneering and plywood are generally of low girth; wood joinery, plywood and pulp and paper industry is to a great extent dependent on imported raw material. Towards conserving forest resources and self sufficiency in wood emphasis is being laid on encouraging technology based on utilisation of small size wood being increasingly cultivated by farmers and agricultural residues and in the up-grading of wood quality through seasoning and preservation.

Wood processing industry in the country comprises saw milling, wood working including furniture and joinery, sport goods, wood based panels and pulp and paper. This sector has high potential for labour employment. Moreover some wood based industries have an export potential in a variety of items. Export of finished articles of wood and wood products are on the rise. Industries are receiving active support in the promotion of their products in domestic and foreign markets. Quality control assurance provided to markets by the Indian Standards Institute with the assistance of FRI, Dehra Dun has helped in expanding foreign trade. Raw material requirements for wood based industries, are on a steady rise and have to be met by utilization of juvenile timbers grown in the private sector within the country. The ICFRE institutes have contributed valuably in this direction by research on sawing, seasoning, finishing and preservation technology for small size timber grown by farmers which are generally prone to developing defects and refractory to preservative treatment. Research on utilisation of small timbers from *Eucalyptus* and poplars for furniture making, carving and in manufacture of reconstituted wood like, MDF and laminated wood is being vigorously persued and commendable success achieved. As a result Eucalyptus and Poplars have found acceptance for use in manufacture of furniture and eucalyptus investigated for making doors and window. The Haryana and Punjab forest corporations were assisted by the Forest Research Institute, Dehra Dun for popularising the utilization of these species which are being grown and marketed on a large scale. The diverse utilization of wood from the farms will also ensure remunerative returns to the growers and sustain incentives for enlarging wood production.

The shortage of desirable biomass for paper and pulp has prompted the Council's institute to undertake research on substitutes for conventional biomass for pulp making and in the genetic improvement for higher biomass production in species yielding quality pulp. Screening of the genotypes with requisite traits for establishing clonal multiplication gardens has been initiated for *Anthocephalus chinensis, Eucalyptus* spp. and *Casuarina equisetifolia* towards increasing biomass production of useful products initiated towards making the process efficient and environmentally less hazardous.

BIODIVERSITY

India is fortuitously endowed with a wide spectrum of biodiversity in plant genetic resources to be recognised as one of the world's top "12 mega diversity" nations. The geological history of the sub continent bestow on the country a range of landscapes and environmental regimes, that nurture a rainbow of flora, fauna in groups and sprinkles, African, Eurasian and the orient. The vastness of the plant genetic resources available to us may be gauged from the fact that 238 families comprising 16,000 species nearly 16 per cent of flowering plants of the 425 families recognised to be existing in the world, are known to exist in our sub continent. Overall the plant wealth of the country is assessed to be 45,000 species.

India's contribution to the plant species cultivated over the world is also significant. It has contributed some very useful crops such as rice, sugarcane, minor millets, brassicas, rice-bean, asiatic vignas, egg plant, banana, citrus, mango, jackfruit, cardamom, tree cotton, jute, edible dioscoreas, vegetable and seed amaranths, colocaciaa, black pepper, turmeric, ginger, cucurbits and umbellifers, rauvolfia and several herbal drugs, rhododendrons, jasmines, bamboos, ornamental orchids, betelnut etc. Some of the biospheres are of vital importance to the country. For instance, the Silent valley and the adjoining regions have contributed 20 major genes for diseases and pest resistance for improving rice crops. The animal wealth of India is over 68,300 spp of which 60,000 are insects, about 1690 are fish and 370 mammals.

Anthropogenic disturbances in the natural environment have caused destruction of plant and animal habitats to varying degrees results in reversible and irreversible changes in the various ecosystems. Increased developmental activities accelerate these changes consequently nearly 33% of the plant species have become endemic, located mainly in 26 endemic centres in India. The rate of endemism is 33% in reptiles and 62% in amphibians. The red book published recently enlists 814 sensitive species.

Significant losses of biological diversity could affect the future well being of human life. Simplification of ecosystem and extinction of species diminish future resource options and the ability of nature to provide life supporting ecological services. Conservation is an approach that balances peoples short and long term needs for natural resources through management of lands, workers and biota to restore, maintain or enhance basic land health and productivity. Efforts at conservation are only fruitful with a holistic approach to, the eco-systems capacity peculiarities which demands coordination between the organisations engaged in this task. The creation and preservation of biodiversity reserves, their monitoring and evaluation and networking for information to enhance the gains and sharing of information generated by these endeavours is some of the important programme of the ICFRE.

FLORISTIC RESERVES OF INDIA

Biosphere reserves and preservation plots: Multipurpose protected areas known as biosphere reserves are meant to preserve the genetic diversity in the representative ecosystems. As per the recommendation of a Core Advisory Group set up by the Government of India in 1979, 14 potential sites were identified for setting up of Biosphere Reserves in the country and so far seven biosphere reserves have been set up.

Preservation plots are also termed as miniature nature reserves. They are demarcated forest areas set aside for the preservation of the forest in perpetuity permitting only such human interference as necessary for their protection and maintenance.

There are at present 309 preservation plots throughout the country, 187 in natural forests and 122 in plantations covering an area of about 8500 ha. Such permanently protected areas are indispensable to all branches of sciences dealing with natural processes of biological evolution that go in the particular habitat. In addition several sample plots have been established in almost all forest types to study and compare the changes under undisturbed conditions and management systems.

Protected trees: Trees of special botanical interest are protected as long as possible (preferably upto physical age). There are 537 such protected trees existing in various states of the country. A list of such trees has been reported by Ghosh and Kaul (1977), Bhadran (1958) and Champion and Seth (1968). A few of these trees are:

Protected Giant trees of India

Species	Locality	Girth (cm)	Height (m)
Dipterocarpus gracilis	Andamans	724	43.9
Eucalyptus globulus	Kodaikanal	960	36.6
Michelia champaca	Karnataka	2,035	65.5
Pinus roxburghii	Uttar Pradesh	411	65.5
Pterocarpus dalbergioides	Andamans	602	51.5
Shorea robusta	Uttar Pradesh	734	41.4
Tectona grandis	Karnataka	602	25.9

EX SITU CONSERVATION

Various research institutes/universities/states have their Botanical gardens, Arboreta, Pinata, Populata, Zoological parks, Zoos etc. Endangered and also economically and scientifically important species are reared in these places and may be released to their respective habitats when the species may require rehabilitation.

NETWORKING

India is a vast country both in geographical extent and richness in biodiversity. The natural home of biological diversity is the undisturbed countryside which is maintained and managed generally by the state forest departments. Forest department a well knit organisation in the country has helped in establishment of protected areas/preservation areas to approximately 5% of her geographical area representing biodiversity of nearly all kind found in the country. These areas have been constantly monitored over the year by the forest departments and research institutes and findings, observations recorded. In the beginning there was only one research agency i.e., Forest Research Institute, Dehra Dun that was engaged in forestry research and its coordination with the state forest departments and other user agencies. Of late a number of organisations i.e. ICFRE and its institutes, Botanical Survey of India, Zoological Survey of India, Wildlife Institute of India, State Forest Research Institute/departments, agriculture/ traditional universities etc., have been created to contribute to forestry research and forest conservation. These organisations engaged in research and documentation, dissemination of information on plant and animal species existing in these areas must collaborate in documenting their findings to draw meaningful results from the information generated by them. Major work on floristic reserves, insects and micro fauna is being carried out by ICFRE and its institutes whereas Wildlife Institute of India is engaged in monitoring the larger animal species and conducting research on their population dynamics in protected areas/wildlife reserves. Similarly, the Botanical survey, Zoological survey etc. also conduct surveys and research related to a fragment of the various ecosystems. A collective effort only can translate information gathered by different agencies into comprehensive knowledge on the researched ecosystems.

BIOFERTILIZERS

In the present scenario when the environmental pollution has assumed global concern, indiscriminate use of fertilizers and pesticides for improved growth is likely to accentuate the problem. Check on use of hazardous chemicals is the first step to arrest the increasing deterioration of the environment. In this context, use of mycorrhizae as a biofertilizer and a biocontrol agent for root diseases assumes significance being non-obnoxious and non-polluting. This is all the more important in forestry practices where dependency on fertilizers for plant growth and toxic chemicals to control diseases is not only economically prohibitive but also environmentally unsafe.

The Forest Research Institute commenced scientific investigations on mycorrhizae a long time ago and since has accumulated considerable scientific information on their usage in increasing productivity of forest plantations.

Verbuscular Arbuscular Mycorrhizae (VAM) not only helps in making available unsoluble phosphates and potassium to the plant by solubilising them for the uptake of plants but the hyphae of the mycorrhizae also serve as an effective extension of the plant root system providing it access to larger quantity of nutrients and water. Also VAM has been established to enhance the plants resistance to diseases and pests. Thus mycorrhizae quantitatively increase productivity. Though mycorrhizae as ubiquitously present but to be effective a sufficiently large population on the host trees is required. Moreover every species prefers a specific mycorrhizae which may vary with age of the tree and therefore the best host mycorrhizae relations must be identified for inoculation to provide optimum growth. The ICFRE institutes are screening plantations and forests for mycorrhizae associations, culturing VAM spore for multiplication and inoculating nursery seedlings to ensure sufficient mycorrhizae populations of the matched type in plantations for im proving their productivity. Rhizobiurn culture has also been simultaneously commenced and best species strains for different leguminous tree species are being identified. It has been revealed that a combination of VAM and rhizobia provides the best result compared to either of them applied singly.

BIOPESTICIDES

Plants including several tree species are known to possess pesticidal properties. Extract of different parts of plants/ trees like *Azadirachta indica, Anona squamosa, Derris indica, Melia azedarach, Chrysanthemum cinerariaefolium*, and many plants belonging to the genera *Derris* show pesticidal activities against different insects. Research on developing biopesticides from plant products has been initiated by the institutes under the Indian Council of Forestry Research and Education for controlling pests affecting forest and agricultural crops. Various extractives of *Dalbergia stipulacea, Melia azedarach, Hopea parviflora, Mangifera indica* and *Hevea brasiliensis* have been found to have insecticidal and fungicidal properties and are being investigated for practical application as pesticides and fungicides.

BIOLOGICAL REJUVENATION

Soil is a living entity. However, in some soils biological activity either ceases or is severly retarded due to disruption of the soil's environment viz., high salt concentration, mining, denudation of tree cover, and other land rendering them unproductive potential. Introduction of a permanent vegetative cover on these soils have the promise of biologically rejuvenating them back to their original potential unlike other systems of production e.g. agriculture. Indian Council of Forestry Research and Education has taken the task of researching techniques for the biological rejuvenation of 93.6 million hectare of India's wastelands which are presently almost barren. Of these, salt affected soils constitute about 7 million hectares.

These problem soils are formed mainly due to combined effect of climate, topography and hydrology. The problems encountered in their management are the poor physical, chemical and biological conditions. The physical attributes that render them inhospitable for plant growth are low infiltration rate, impeded drainage, poor soil structure, like hardness when dry, water logging during rains and severe moisture stress during dry period. In chemical properties they are poor and imbalanced in nutrient availability, contain excessive water soluble salts and the dominance of sodium in them reduce availability of other essential nutrients to the plants. Occasionally an impervious calcium carbonate pan develops at a depth of 30 cm to 200 cm which acts as a physical barrier for roots and removal of excessive salt concentration.

Trees through their deep and sturdy root system open up the soil and improve soil structure. Further, addition of organic litter from trees helps in amelioration of these problem soils. Studies reveal that *Prosopis juliflora* growing on sodic soils return 74.39 q/ha/yr of organic litter and *Acacia nilotica* 59.17 q/ha/yr adding valuable nutrients to these impoverished soils for rebuilding the biological activity. Besides this, roots of the trees also contribute organic matter/nutrients to the soils and render them hospitable to useful micro and macro flora and fauna essential towards improving and sustaining the production from land.

The Forest Research Institute, Dehra Dun and Arid Forest Research Institute, Jodhpur of the Indian Council of Forestry Research and Education have conducted successful research trials for biologically rejuvenating salinesodic soil where not even a single blade of grass was found growing. Research included establishment of critical limits of a species salt tolerance, identification of stress tolerant species, standardisation of soil amelioration methods and optimal season of planting, application of mycorrhizae and with leguminous trees rhizobium.

Based on the research results, a package of practice was developed for bio-rejuvenation of salt affected soils and the same was demonstrated on farmers field at Kanaksinghpur village, Sultanpur district, in collaboration with the Indian Farmers and Fertillizer Cooperative (IFFCO), Farm Forestry Division, Sultanpur. The areas adopted for biological rejuvenation in the villages of Sultanpur district were initially cultivated for growing paddy with a liberal helping of gypsum. However, over the years the yield continued to decline and the cultivation turned to be unremunerative and therefore abandoned. It is heartening to note that the farmers and the supervisors, who had no experience of afforestation works with the assistance of the councils scientists successfully rehabilitated these problem areas with a survival of about 80%. Subsequently, the IFFCO has formed 17 Primary farm forestry cooperatives in the districts of Sultanpur (8 cooperatives), Rae Bareli (4 cooperatives), Pratapgarh (3 cooperatives), and Allahabad (2 cooperatives) where agriculture lands had been rendered barren due to salanity-sodacity. Of the 3026 ha of saline sodic wasteland included in these co-operatives, a total of 2045 ha was afforested with Neem, Karanj, Arjun, Mahua, Shisham, Siris, Eucalyptus, Subabul, Kikar, Dhaincha, Ber, Guava, Aonla, Bel and Karaunda.

Over the time indigenous grasses and flora has appeared in the rejuvenated areas and grass species *Bracharia mutica, Panicum antidotale, Chloris gayana* and *Cynodon dactylon* provide a mat on these past wastelands. The production of grasses has enabled the villagers to sustain rearing of milch cattle, every house hold now owns atleast one and the milk produced contributes to their income. Similar work on rehabilitating lateritic soils, ravinous and gullied areas has been initiated by the Council's institute at Jabalpur and Jodhpur.

EXO-RESTORATION OF MINED AREAS

Land after mining generally does not remain conducive to tree growth. The problem is much more vexed in the Himalayan highlands where the extensive surface exeavation of sloping land not only damages the ecosystem within the corrals of the mined area, but also sets a chain of ecological disturbances for beyond, down the watersheds. Therefore, the rehabilitation of these areas assumes special significance.

The ecological rehabilitation of mined areas in the Himalaya is being attempted, using plant species of economic value to local populations and as ecologically compatible with such degraded sites. Results have shown that mined areas may be ecologically rehabilitated and biologically rejuvenated for sustained production of goods of economic value to locals (within a short span of 8-9 years).

The main features of the technology are:

(*i*) Selection and planting of suitable grass and shrub species as ecologically compatible with the site, during years 1 and 2.

(*ii*) Planting of native species of trees and shrubs as socio-economically valued by local populations during years 3 and 4.

Rock phosphate mines in Dehra Dun and Mussoorie region have been successfully reclaimed in this manner. The techniques are being replicated in limestone mines of Mussoorie (Uttarakhand),

Rajban (Himachal Pradesh) and magnesite mines of Almora (Uttarakhand). The socio-economic impact of mining and their subsequent restoration in Dehradun-Mussoorie region is being studied.

The restored rock phosphate areas after 9 years are now a dense conglomeration of diverse vegetation providing stability to the site and producing 12 tons/ha of fuel, 22 tons/ha fodder and 775 kg/ha of litter returning 0.24 tonnes/ha of N, P, K, Ca and Mg. Whereas sediment concentration in stream water after site restoration reduced from 240 mg/1 to 120-130 mg/1.

FORESTS AND TRIBALS

The National Forest Policy 1988 directs that forests and forestry should be managed and developed to improve the socio economic conditions of the rural poor and tribal communities. Tribals inhabit nearly 40 per cent of the country's area constituting 51 million people or 8 per cent of the country's population and are by and large dependent on forests for their livelihood.

The rich biodiversity of the forests provide them a variety of produce to subsist and for trade, permitting them economic independence and self sufficiency in the goods/ services required in their daily lives. Forests nurture the tribals with wholesome food, rich in nutrients, vitamins and mineral in form of meat, fruits, vegetables, honey, mushrooms etc., raw materials for household articles, constructional material, grazing for animals and manure for agriculture, medicines and a host of marketable goods the earnings from which enable them meet their meagre needs of clothing and food when scarce. The decline in the area and density of forests has adversely impacted on the economy of tribals. The well being of tribals is perceptibly linked with the conservation of the forest ecosystem to which they integrally belong. The forest biodiversity in such areas therefore must be secured, developed and managed, to augment and sustain a respectable living for the tribal folks in harmony with their cultural ethos. Renewal of the tree vegetation which are traditionally a source of nutritious food for the tribals and development of non wood forest product resources as provide employment to local artisans and skilled, non skilled labour can substantially contribute to local tribal economy. In certain states the incomes from non wood forest produce constitute between 30 to 50 per cent of tribal earnings. Herbs of value as cosmetics and medicine and traditional tribal handicrafts are becoming popular with the affluent section of the society and foreign clients and cultivation, processing and marketing of forest produce of value can make an appreciable impact on improving the living standard of tribals who find it increasingly difficult to eke out a living in the present circumstances.

Forests and wastelands cover over a third of the country's geographical area. Moreover, a majority of the tribals and rural poor living in and around forest are, by and large, dependent on forests for their livelihood. Our rural economy too, in general, is integrally linked with forest resources, rendering them important for the well being of rural folks. Forest development programmes compliment agriculture farming systems and sustain and secure them ecologically. Forests and trees conserve soil and moisture; help cycle nutrients; improve microclimate; reduce desiccating winds; increase water infiltration; provide grazing and fodder for farm animals, fuel and timber for the household, nutritious food; and, a host of goods and services for the sustenance of rural folks.

The contributions of forests and trees in directly and indirectly upgrading the rural economy are immense and immeasurable. Forests and tree plantation programmes provide employment to local people and the accruing goods, in turn, regenerate opportunities for employment in their harvesting, transport, conversion and processing. A host of other employment opportunities and that for gathering food may be added for tribals and others dwelling in the vicinity of forests by way of non wood forest products' cultivation and gathering in association with tree produce, e.g., bamboos, canes, medicinal plants, mushrooms, species yielding fibres, oil seed bearing trees/ shrubs, sericulture, apiculture etc. Many of these raw materials originating from forest can be further processed for value added products on cottage scale.

LAND DEGRADATION AND IMPROVING FOREST PRODUCTIVITY

Land degradation is probably the foremost environmental problem requiring attention on priority. Productivity of large areas is declining with the population increasing unabatedly and making ever more demands on land for food, fibre, fuel, fodder and timber. Wild plant resources have the potential to meet future needs for food and, if land degradation continues unchecked, these would be lost to the humanity for ever. Another natural corollary to land degradation is nutrient depletion and, if more and more fragile ecosystems are cultivated, the problem is likely to aggravate. The goal, therefore, should be to increase food production without compromising soil fertility. Afforestation of marginal and wastelands and development of agroforestry systems for conserving soil and water and increasing nutrient cycling can provide remedy for land rehabilitation, improvement of soil fertility for sustainable agricultural production and maintenance of land productivity. Integrated land development in consonance with land capabilities, with forestry as a component, can provide answers to many problems being faced today in regard to fodder, fuel and timber shortage, floods and soil erosion, siltation of water reservoirs, water scarcity, resources for forest based industries, unemployment etc.

DEMAND AND SUPPLY

The estimated annual demand of industrial timber and firewood in the country is around 27.5 million cu m and 235 million cu m respectively against an estimated production of 12 million cu m and 40 million cu m. The average annual production of wood per ha in India is a paltry 0.5 cu m compared to the world average of 2.1 cu m. India, with plentiful sunshine and being only second to Brazil in annual precipitation though seasonal, can increase its forest productivity manifolds. Use of improved clones and site matched varieties in afforestation programmes by themselves have the capability of raising the productivity five folds, whereas genetic improvement can further add to these gains.

FODDER

The Ministry of Agriculture's Committee on Livestock Food and Fodder, in 1973, estimated the fodder requirement at 339.5 million tonnes of dry fodder and 340.4 million tonnes of green fodder. The National Commission on Agriculture (1976) estimated the green and dry fodder requirement for the year 2000 at 590.1 and 373.0 million tonnes respectively. According to the estimates given in the Handbook of Agriculture (1980), the availability of green and dry fodder is estimated at about 224.08 and 231.05 million tonnes respectively and, in 2000, it is expected to be 575.0 and 356.8 million tonnes respectively. The ICFRE institutes are working on the cultivation of multipurpose tree species which also have good fodder value. In this direction, *Ulmus laevigata and Paulownia fortunei/P. kawakamii* are being investigated vigorously. It is hoped that these species will replace *Eucalyptus* and Poplar in the northern Indian gangetic plains and provide timber, fuel and fodder in abundance.

FUELWOOD

According to an ADE publication of 1984, 'A Perspective on Demand for Energy in India up to 2004-2005', the population of India in 2004-2005 would be 1046 million. The estimated energy requirement for this population would be 312 million tonnes of firewood, 17 million tonnes of kerosene, 94 million tonnes of agro-waste, 207 million tonnes of cattle dung, 176 million cu. meters of biogas and 100 twh of electricity. The Working Group on Energy Policy (1979) estimated the consumption of fuelwood, agriculture waste and animal dung to be 133.1, 41.0 and 73.0 million tonnes respectively, whereas the total production of fuelwood was estimated at 49 million tonnes by the Fuelwood Committee, Planning commission (1982). Therefore, research in increasing the production of fuelwood assumes great significance. A number of steps towards the objectives e.g., high density planting, introduction of fast growing species, genetic improvement for higher biomass production and afforestation of stress sites etc., have been advanced by the Council.

NON-WOOD FOREST PRODUCTS

The ubiquitous wood is better recognised as a forest product contributing to our economy even though non wood produce have a greater role in the daily lives and prosperity of the common man and perhaps also provide raw materials for luxury products the preserve of the rich and affluent. The better known among the non wood produce of socio-economic importance are fodder, bamboos, gums, resin, leaves, oils and fruits. Their contribution to the economy is largely unquantified as more than 60% of such production is locally consumed by the population. The harvesting and utilization of non wood produce is largely handled by the non commercial sector and mainly by local populations for meeting day to day needs. Therefore, it has failed to attract the attention it deserves, relegating the scientific management of these valuable resources to an insignificant place.

SOCIAL FORESTRY

The National Commission on Agriculture (NCA) in 1976 directed that forest resources should also be developed outside the reserve and protected forest areas with the active participation of local communities to bridge the ever increasing gap between demand and supply and commonly termed the programme as Social forestry. It is, thus, essentially a programme for developing forest resources in rural and urban areas for meeting the fuel wood, fodder, timber, environmental and aesthetic needs of the people with the active support of the beneficiaries/ communities.

The ICFRE institutes' contribution in social forestry programmes encompasses research regarding selection of species for various components, improved quality planting material, studies relating to issues involving peoples' participation, increasing participation and constraints thereto, financing, marketing and utilization of juvenile timbers 'produced under these programmes.

AGROFORESTRY

India's economy being largely biomass based, forestry the foster mother of agriculture has as vital a role to play in agriculture. A sizeable part of the energy cycled in the agricultural systems is contributed by trees and forests, for example, fodder and grazing for animals, manure for fields, timber for construction, wood for agricultural implements, and fuel wood etc. The indiscriminate expansion of agriculture during earlier phases of the Five Year Plans, at the expense of forests and forest lands, has eroded the very foundation on which ecologically secure agriculture must be based, i.e., forestry. Over use of lands has resulted in the creation of wastes, development of salinity and alkalinity, soil erosion, depletion of water resources, shortages in fuel, fodder and timber and a host of other problems, rendering agriculture less economical and environmentally unfriendly. Agriculture, which probably could be sustained by ecologically sound land use, is now increasingly becoming energy intensive and dependent on external inputs.

Agroforestry, the land use system that incorporates woody perennials with agricultural and animal husbandry etc., to compliment each other, simultaneously ensures ecological security for food production. This system in the past has been a well recognised land use system in India ensuring a large measure of self sufficiency for the rural masses providing sustained production of every day household needs, food, firewood, timber, fodder, oil seeds, etc. Over the years, the system crumbled under the pressure of raising agricultural production which, though achieved self sufficiency in food, created acute shortage of forest/ tree based biomass required in the daily lives of rural people. Self sufficiency in forest produce can save the natural forests from destruction by external pressures making revival of agroforestry expedient and to be practised in a more scientific manner than in the past.

The ICFRE institutes pursue research in the trials and introduction of high yielding tree species of value in agriculture systems and designing of models for improving land use in the different agro-ecological regions, i.e., on likely tree-crop production systems, their spatial geometries and effects on production of goods, services and ecological impacts thereof, evaluate the information gathered for the weaknesses, constraints and potential of the existing land use systems and conduct economic analysis of existing promising agroforestry systems.

PARTICIPATORY FOREST MANAGEMENT

It is being reiterated that any forest management divorced from the aspirations of the communities directly dependent on them cannot ensure conservation and sustainability of forests. The growing ambitions of local populations have increased their stake in the forest resources adjoining the habitations and their exclusion from the process of management is spelling disaster. Policing of forests has proved ineffective and inadequate for ensuring protection. In management, the present exigencies outweigh the future concerns. To assure the continuity of forests, a high degree of flexibility is demanded in their present management and accommodating the views of the local users of forest resources is called for. A number of state forest departments are developing processes and policies towards cooperative management of forest resources with the participation of the local community. The forest departments of West Bengal, Haryana and Gujarat have developed joint management systems and have received a positive response from the local communities in the regeneration and ecological rehabilitation endeavours of problem areas. Reliance is now being placed on collaborative management for sustaining forest production with local community participation, in place of policing and departmental controlled production. This has helped restore trust and confidence of the locals in the state forest departments as evidenced by the successful community rehabilitation programmes of the degraded Sal areas of Arabari, Pukharia and Bankura in West Bengal. The conservation of forests is directly linked with increasing the productivity of forest lands and a sustained flow of a reasonable annual income therefrom for the community to maintain their interest in the protection of resources. The West Bengal's Arabari and similar models would need an indepth study and a critical appraisal for the soundness of the technical and socio-economic premises on which founded. A thorough examination of these participatory management systems can help in crystallising lessons and determining the general guiding principles for participatory management. Such collaborative management ventures are of utmost importance for ensuring forest conservation and increasing/ diversifying forest production for the welfare of communities/populations, particulary those living in and around forest.

FORESTRY EDUCATION

Forestry education in India has a history of over 100 years old though rich and varied in academic content, it has a bias towards providing a professional training to inservice officers at different levels. Formal forestry education in the strict sense was not available in India till the near past. In pursuance of the recommendations of National Commission on Agriculture (1976), forestry education was introduced in the universities. Presently, 23 universities, including 18 State Agriculture Universities, are imparting forestry education at graduate levels. Of these five offer postgraduate forestry courses. Forestry education in the universities is still developing and a number of inadequacies, viz., lack of trained faculty, the existing teachers being primarily agricultural scientists, comprehend forestry to be akin to agriculture, absence of exposure to practical aspects of forestry, insufficient infrastructural facilities for research and specialisation, research divorced from practical needs of the forestry sector, inadequate library and documentation services etc.

The Forest Research Institute, Dehra Dun possesses a sound infrastructure and the longest tradition, in Asia, of imparting forestry education and training to professionals along with forestry research its primary function. Though, the status of a Deemed University has been granted to FRI only lately, the institute has been the pioneer and the only body which has conducted and organised basic and applied forestry research in the country till recently and, therefore, recognised by many universities for guiding research in forestry and allied fields. The existing documented literature on forestry has primarily been contributed by research efforts of this institute.

FORESTRY EXTENSION

Depending on land ownership, forestry extension needs to be targeted at individuals or the community. The procedure and approach in the two cases would vary widely due to differing objectives. Also, the factors limiting forestry extension are many and complex, making the training and visit system (TNVs) adopted in agriculture as ineffective. Other models have been experimented in integrated projects like Dhauladhar project undertaken by German collaboration in Kangra of Himachal Pradesh and was based on winning the trust and confidence of the targeted community and named Truco.

In order to translate the National Forest Policy of 1988 into action and keeping with its directives in obtaining the willing support and co-operation of the people for the success of the forest conservation programmes, a strong research and education support base in forestry extension should be built at the national level and provided leadership by the ICFRE. In fact, systematic development of forestry extension methods has been postponed for too long inspite of realisation for their urgent need, impeding the transfer of technological skill, developed at the research institutes of ICFRE and elsewhere to the grass roots, where it is expediently needed for increasing success rates of forestry schemes and enhancing productivity as envisaged. The present endeavours of the state forest agencies, though commendable, are based on perceptions of individual members or groups of such organisations and need critical appraisal cum research support to be referred and crystallised into sound strategies for developing extension models and references as general-specific recommendations for initiating and nurturing afforestation and forest rehabilitation programmes that are in harmony with ecological settings and socio-economic aspiration of the ethnic groups sequestered in different zones. The ICFRE has also embarked on the publication of brochures and handouts on tree species of economic importance towards their popularisation among masses and for information of the general public.

34

Forest Products of India

BIOFERTILIZERS

Biofertilizers constitute beneficial microorganisms found in nature which provide nutrients to the plants in available form through natural processes. Among the biofertilizers which are now being using in forestry research in India are *Rhizobium*, *Azotobacter*, *Vesicular-Arbuscular Mycorrhizae* (VAM) and *Frankia*. Rhizobium is a symbiotic nitrogen fixing bacterium which forms nodules on the roots of leguminous plants. Its role in increasing production of agricultural crops is now well established. *Azotobacter* is a free living form that also fixes atmospheric nitrogen besides making available growth promoting substances known to stimulate plant growth. *Frankia* is an actinomycete which is, however, restricted in distribution in forest trees such as *Casuarina* and *Alnus*. This symbiotic actinomycete fixes atmospheric nitrogen and is of great significance particularly in coastal plantations of *Casuarina* where there is a problem of limited amount of nitrogen available to the plants in the soil. VAM which occur commonly in forest trees are of much relevance to forestry. The role of VAM in increasing the absorptive surface of roots through extended hyphal network in the soil and thereby increasing nutrient uptake and plant growth is now well established. Besides, VAM also increases plant resistance to unfavourable conditions such as drought, frost, heavy metal toxicity, and extremes of soil acidity. They decrease transplantation shock and thereby increase the chances of better establishment and subsequent improved growth of seedlings in outplantings. Alleviation of soil compaction, more efficient use of water and nutrients and improvement in nodulation are other benefits which VAM are known to confer to the plants. Further, their role in biocontrol of root diseases and imparting resistance to the plants are important attributes which preclude the use of chemicals to combat diseases. In the present scenario when use of inorganic fertilizers and pesticides is fraught with the danger of aggravating the already grim environmental situation, use of VAM as biofertilizer and biocontrol agent is very much desired because of their being respectful to the environment.

BIOPESTICIDES

Plants including several tree species are known to possess pharmacological and pesticidal properties. For centuries, man has been using different parts of plants for treatment of various ailments all over the world.

Discovery of certain biologically active principles in various parts of plants/tree species such as Neem *(Azadirachta indica)* and several other plant species like *Anona squamosa, Chrysanthemum cineraraefolium, Derris indica* and *Melia azedarach* etc. as well as their potentiality as biopesticides against harmful insect species have of late, drawn world wide attention of the professional entomologists, research organisations and industry. In fact, the Neem tree has been nicknamed as *"Wonder tree for the future"* particularly in the third world countries for reforestation and a whole range of useful products including pest control agents against harmful insects. As a part of research and development activities, different institutes under the Indian Council of Forestry Research and

Education are engaged in isolating biopesticides from plants for controlling harmful forest insect pests. These pesticides are environmentally safe and without adverse side effects of conventional chemical pesticides.

NEEM EXTRACTIVES

Neem leaves and the Neem products, particularly the Neem seeds have been in use in India since time immemorial against insect pest of food grain, books and cloth, etc. In Forestry, the work on these aspects is of recent origin. A broad programme has been drawn for isolation, identification and developing technology for preservation of biologically active ingredients, having gustatory phagodeterrent/ antifeedant properties.

The studies were further extended to include several other plant species possessing pesticidal properties and were tested against some of the important forestry pests.

Methanol extractives of Bhekal *(Prinsepia utilis)* seeds were tested against the Poplar defoliator, *Clostera cupreata* for antifeedant properties. The extractive showed good biological activity at 0.25 per cent dilution against the 3rd instar larvae of C. *cupreata.* Acetone and alcohol extracts of bark and root of *Dalbergia stipulacea* and alcohol extract of the leaves of *Adina cordifolia* were effective as food poison against the moth *Corcyra cephalanica* which infests rice, Wheat, Sorghum, etc. They were also effective as contact poison to larvae of Poplar defoliator, *Colstera cupreata* and caused 70 per cent mortality.

Several other plant species such as *Acorus calamus, Lantana camara, Adhatoda vasica* and *Melia azedarach* were screened for their insecticidal activities against *Ailanthus* webworm *Atteva fabriciella* Swed. (Lepidoptera: Yponomeutidae) in laboratory. Crude extracts in concentrations of 1.0, 2.0 and 5.0 per cent in different solvents were tried against larvae of *Ailanthus* webworm.

WOOD EXTRACTIVES

Studies for evolving measures to control biodeterioration with the help of wood extractives, other plant derivatives, bioactive substances and biological agencies were undertaken. An alcholic bark extract of *Eucalyptus tereticornis* was tested against the weed parthenium. All the leaves of weed, sprayed with this bark extract, wilted and eventually dried up. The drooping and drying of the inflorescence was also noticed. The flower buds failed to blossom. No pollination was noticed. Ultimately, these buds dried up. Dried plants were uprooted and burnt. In the next season however, very few plant of this weed sprouted up, probably because of wind dispersal of seeds to this area. More work on the eradication of parthenium with the help of extractives and replacement of plant species is under progress.

Ethanol soluble phenolic extractives of *Hopea parviflora* were concentrated to powder form and tried at different concentrations for finding out the degree of resistance induced in the wood against fungi and termites and the results were encouraging. It may now be possible to increase the dosage and concentrtation for obtaining maximum resistance in wood.

FOOD FROM FORESTS

Forests are a precious endowment of nature to mankind, bountifully bearing omnifarious resources for sustenance of living beings. The man of pre agricultural era was nurtured exclusively by food and other resources from the wild. Even today a sizeable section of humanity particularly tribals draw on forests for food. Some food from forests remain gourmet delight even for the sophisticated people in advanced societies. Forests are the repository of a variety of nutritious food gifted by innumerable species in the form of edible tubers, roots, shoots, leaves, flowers, fruits, nuts, seeds, mushroom, game meat, honey, etc. Forests are also the treasure house of genes capable of contributing to the improvement and security of cultivated food and horticultural crops.

Energy, protein, vitamins A, B, C, D, folacin, thiamine, riboflavin, niacin and minerals such as iron, zinc, and calcium are nutrients essential for good health. However, their availability is very often inadequate in drylands and backward areas. The primary nutritional deficiencies encountered are Protein-Energy Malnutrition (PEM), anaemia and vitamin A deficiency. Malnutrition has serious implications because PEM as well as other nutrient deficiencies make children susceptible to infection and thus increase incidence of mortality. Vitamin 'A' deficiency in its severe form results in permanent blindness. Poor dietary selection with inclusion of inadequate amount of food stuffs which are rich in vitamin A (β-carotene), iron and calcium lead to nutritional problems. Fruits and vegetables collected from the forest areas are rich in vitamins and minerals.

Forest plants that may be utilized for food are many and diverse. Depending upon the plant, any of its part like fruits, seeds, leaves, flowers rhizomes, roots are eaten by people of different regions.

NUTRITIVE VALUE OF SOME FOOD PLANTS FROM FOREST
PER 100 GMS OF EDIBLE PORTION

LEAVES

1. Moisture, 2. Protein, 3. Fat, 4. Minerals, 5. Fibre, 6. Other Carbohydrates, 7. Calories

SI.No.	Botanical Name	1	2	3	4	5	6	7
1	2	3	4	5	6	7	8	9
1.	*Alternanthera sessiles*	77.4	5.0	0.7	2.5	2.8	11.6	73
2.	*Amaranthus caudatus*	90.0	3.0	0.7	3.3	1.0	2.0	26
3.	A. *spinosus*	85.0	3.0	0.3	3.6	1.1	7.0	43
4.	A. *tricolor*	85.7	4.0	0.5	2.7	1.0	6.3	46
5.	A. *viridis*	81.8	5.2	0.3	2.8	6.1	3.8	38
6.	*Antidesma acidum*	7.2	7.2	4.8	9.5	13.5	57.8	303
7.	*Basella alba*	90.8	2.8	0.4	1.8	–	4.2	32
8.	*Bauhinia purpurea*	78.1	3.6	1.0	2.1	5.5	9.7	62
9.	*Cassia tora*	84.9	5.0	0.8	1.7	2.1	5.5	49
10.	*Celosia argentea*	87.4	1.1	0.8	2.6	3.7	4.4	29
11.	*Chenopodium album*	89.6	3.7	0.4	2.6	0.8	2.9	30
12.	*Cleome viscosa*	80.4	5.6	1.9	3.8	–	8.3	73
13.	*Colocasia esculenta* (Black variety)	78.8	6.8	2.0	2.5	1.8	8.1	77
14.	C. *esculenta* (Green variety)	82.7	3.9	1.5	2.2	2.9	6.8	56
15.	*Commelina benghalensis*	92.2	2.1	0.4	2.0	0.8	2.5	22
16.	*Corchorus olitorius*	83.6	6.0	1.0	1.8	2.8	4.8	52
17.	*Euphorbia hirta*	78.1	4.7	1.7	3.2	–	12.3	83
18.	Hi*biscus cannabinus*	86.4	1.7	1.1	0.9	–	9.9	56
19.	*Malva parviflora*	86.2	4.3	0.6	2.1	1.2	5.6	45
20.	*Moringa oleifera*	75.9	6.7	1.7	2.3	0.9	12.5	92
21.	*Polygonum plebejum*	83.2	3.2	0.7	3.9	2.1	6.9	46
22.	*Portulaca oleracea*	90.5	2.4	0.6	2.3	1.3	2.9	27
23.	*Sauropus androgynous*	73.6	6.8	3.2	3.4	1.4	11.6	103
24.	*Solanum nigrum*	82.1	5.9	1.0	2.1	-	8.9	68
25.	T*ribulus terrestris*	79.1	7.2	0.5	4.6	–	8.6	68
26.	*Trianthema portulacastrum*	91.3	2.0	0.4	2.2	0.9	3.2	24
27.	*Tamarindus indica*	70.5	5.8	2.1	1.5	1.9	18.2	115

(Contd)

Seeds

1	2	3	4	5	6	7	8	9
1.	*Anacardium occidentale*	5.9	21.2	6.9	2.4	1.3	22.3	596
2.	*Areca catechu*	31.3	4.9	4.4	1.0	7.2	47.2	248
3.	*Eurayle ferox*	12.8	9.7	0.1	0.5	–	76.9	347
4.	*Nelumbo nucifera*	10.0	17.2	2.4	3.8	2.6	64.0	346
5.	*Nymphaea nouchali*	10.0	8.3	1.0	0.9	4.2	75.6	345
	Flowers							
1.	*Calligonum polygonoides*	6.2	18.0	–	6.8	–	–	–
2.	*Madhuca latifolia*	18.6	4.4	0.6	2.7	1.7	72.0	311
3.	*Moringa oleifera*	85.9	3.6	0.8	1.3	1.3	7.1	50
	Fruits							
1.	*Aegle marmelos*	61.5	1.8	0.3	1.7	2.9	31.8	137
2.	*Anacardium occidentale*	86.3	0.2	0.1	0.2	0.9	12.3	51
3.	*Annona reticulata*	76.8	1.4	0.2	0.7	5.2	15.7	70
4.	*A. squamosa*	69.3	1.7	0.6	1.1	3.7	23.6	107
5.	*Antidesma ghaesembilla*	72.3	1.9	1.0	1.1	13.1	10.6	59
6.	*Artocarpus helerophyllus*	76.2	1.9	0.1	0.9	1.1	19.8	88
7.	*A. lacucha*	82.1	0.7	0.1	0.8	2.0	13.3	66
8.	*Averrhoa carambola*	91.9	0.7	0.1	0.4	0.8	6.1	28
9.	*Bauhinia vahlii*	25.5	23.7	28.2	3.0	2.3	17.3	418
10.	*Buchanania lanzan*	74.3	2.2	0.8	1.7	1.5	19.5	94
11.	*Capparis zeylanica*	67.0	6.1	3.8	2.0	9.6	11.5	105
12.	*Carissa carandus*	91.0	1.1	2.9	0.6	1.5	2.9	42
13.	*Catunaregwn uligiosa*	81.7	1.0	0.2	0.7	3.9	12.5	56
14.	*Coccinia cordifolia*	93.5	1.2	0.1	0.5	1.6	3.1	18
15.	*Cordia dichotona*	82.5	1.8	1.0	2.2	0.3	12.2	65
16.	*Cydonia oblonga*	85.7	0.3	0.1	0.3	1.7	11.9	50
17.	*Dillenia indica*	82.3	0.8	0.2	0.8	2.5	13.4	59
18.	*Diospyros kaki*	80.0	0.7	0.2	0.3	0.9	17.9	76
19.	*D. malabarica*	69.6	1.4	0.1	0.8	1.5	25.6	113
20.	*D. melanoxylon*	70.6	0.8	0.2	0.8	0.8	26.8	112
21.	*Lirnonia acidissima*	64.2	7.1	3.7	1.9	5.0	18.1	134
22.	*Ficus benghalensis*	74.1	1.7	2.0	1.9	8.5	11.8	72
23.	*F. religiosa*	62.4	2.5	1.7	2.3	9.9	21.2	110
24.	*Flacourtia indica*	67.8	1.7	1.8	1.3	4.7	22.7	114
25.	*F. jangomas*	77.7	0.5	0.1	0.8	1.0	19.9	83
26.	*Fragaria vesca*	87.8	0.7	0.2	0.4	1.1	9.8	44
27.	*Grewia asiatica*	80.8	1.3	0.9	1.1	1.2	14.7	72
28.	*Madhuca latifolia*	74.4	2.2	1.2	1.1	1.5	19.7	98
29.	*Mangifera indica*	86.1	0.6	0.1	0.3	1.1	11.8	51
30.	*Manilkara hexandra*	68.6	0.5	2.4	0.8	–	27.7	134
31.	*Mimusops elengi*	54.7	1.8	1.0	2.3	4.3	35.9	160

(*Contd*)

1	2	3	4	5	6	7	8	9
32.	*Momordica cochinchinensis*	90.4	0.6	0.1	0.9	1.6	6.4	29
33.	*M. dioica*	84.1	3.1	1.0	2.3	3.0	7.7	52
34.	*Moringa oleifera*	86.9	2.5	0.1	2.0	4.8	3.7	26
35.	*Phoenix sylvestris*	59.2	1.2	0.4	1.7	3.7	33.8	144
36.	*Pithecellobium dulce*	79.2	2.7	0.4	0.7	1.0	16.0	78
37.	*Prunus arrneniaca* (fresh)	85.3	1.0	0.3	0.7	1.1	11.6	53
38.	*P. arrneniaca* (dried)	19.4	1.6	0.7	2.8	2.1	73.4	306
39.	*P. persica*	80.4	0.3	0.3	0.4	5.2	13.4	58
40.	*P. salicina*	35.5	0.3	0.3	1.7	2.0	60.4	246
41.	*Schleichera oleosa*	86.2	1.5	0.8	1.0	0.6	9.9	53
42.	*Spondias pinnata*	90.3	0.7	3.0	0.5	1.0	4.5	48
43.	*Syzygium cumini*	83.7	0.7	0.3	0.4	0.9	14.0	62
44.	*S. jambos*	89.1	0.7	0.2	0.3	1.2	9.7	43
45.	*Tamarindus indica*	20.9	3.1	0.1	2.9	5.6	67.4	283
46.	*Vaccinium symplocifolium*	79.5	0.8	0.6	0.3	7.3	11.5	55
47.	*Zizyhus m.auriliana*	81.6	0.8	0.3	0.3	–	17.0	74
48.	*Z. rugosa*	55.3	3.2	1.3	2.0	4.9	33.3	158
	Underground Parts							
1.	*Amorphophallus paeoniifolius*	78.7	1.2	0.1	0.8	0.8	18.4	79
2.	*Colocasia esculenta*	73.1	3.0	0.1	1.7	1.0	21.1	97
3.	*Dioscorea alala*	79.6	1.3	0.1	0.8	0.1	18.1	79
4.	*D. bulbifera*	67.6	4.5	1.8	1.3	1.1	24.7	124
5.	*D. hamillonii*	66.7	1.8	0.2	1.0	1.5	28.8	124
6.	*D. pentaphylla*	79.6	2.9	0.3	0.8	0.9	15.5	43
7.	*Nelumbo nucifera*	85.9	1.7	0.1	0.2	0.8	11.3	53
8.	*Nymphaea nouchali* (white)	62.5	3.1	0.3	1.3	1.1	31.7	142
9.	*N. lolus* (red)	49.1	4.1	0.3	1.6	1.5	43.4	193
	Stem/Shoots							
1.	*Aternanthera sessiles*	77.4	5.0	0.7	2.5	2.8	11.6	73
2.	*Amaranthus caudatus*	90.00	3.0	0.7	3.7	1.0	2.0	26
3.	*A. spinosus*	85.0	3.0	0.3	3.6	1.1	7.0	43
4.	*A. tricolor*	85.7	4.0	0.5	2.7	1.0	6.3	46
5.	*A. viridis*	81.8	5.2	0.3	2.8	6.1	3.8	38
6.	*Basella alba*	92.1	2.0	0.7	1.7	0.6	2.9	26
7.	*Euphorbia hirta*	78.1	4.6	–	–	–	–	–
8.	*Lepidium latifolium*	–	29.6	4.1	–	11.1	31.6	–
9.	*Portulaca oleracea*	92.2	1.3	0.3	–	0.8	3.6	22
10.	*Rhodiola tiberica*	–	11.5	3.9	–	15.6	42.2	–
11.	*Salsola kaki*	–	11.6	3.0	–	–	–	–
12.	*Sauropus androgynous*	73.6	6.8	1.4	3.4	11.6	–	–
13.	*Solanum nigrum*	82.1	5.9	1.0	2.1	–	3.9	–
14.	*Tribulus terrestris*	79.0	7.2	–	–	–	–	–
15.	*Urtica hyperborean*	–	26.7	6.8	–	5.5	34.6	–

Seeds of *Pinus gerardiana* are commonly eaten in hilly areas specially in cold desert of Himachal Pradesh. Conifer Research Centre, Shimla is engaged in the task of carrying out studies for the propagation and conservation of this important species.

Sesbania is a fast growing plant having very high nutritional value. It is rich in Vitamin A. Tropical Forest Research Institute, Jabalpur is conducting experiments on space and water regime, and fertilizer trials (including biofertilizers) for this species.

Number of agroforestry models have been developed on socioeconomic and ecological considerations at ICFRE institutes. Planting of multipurpose tree species yielding nutritive food (such as edible bamboos, sesbania, drumstick) is under demonstration in selected villages.

Mushrooms are eaten both by rich people in the cities and poor forest tribals. FRI, Dehradun has been successful in cultivating Shiitake *(Lentinus edodes),* a lignicolous Japanese mushroom, on Oak logs. The technique to increase its yield is being standardised. Cultivation of mushrooms has also been undertaken at TFRI, Jabalpur.

FODDER FROM FORESTS

India has the largest livestock population in the world, as per 1982 census, it was 415.9 million. This population is about one-sixth of the total livestock population of the world.

LIVESTOCK POPULATION 1951-1982

Category of livestock	Livestock population in (million) different years				% increase in 1982 over 1951
	1951	1961	1972	1982	
Cattle	155.30	175.56	178.36	190.79	22.85
Buffaloes	43.35	51.21	57.34	69.00	59.17
Sheep	38.43	40.02	39.99	48.07	25.08
Goats	47.08	60.86	61.52	94.72	101.19
Horses and ponies	1.51	1.33	0.94	0.93	–99.07
Pigs	4.42	5.18	6.90	9.58	–90.42
Camels	0.63	0.90	1.11	1.03	63.94
Others	1.30	1.15	1.11	1.82	40.00
Total	292.02	336.21	353.34	415.94	42.43

SOURCE: LIVER CENSUS REPORTS

The increase in livestock population was not uniform through the different states of the country. In almost all the states, density of livestock is in excess of the carrying capacity of the land area. Effective steps, therefore, are expedient to reduce the livestock population particularly in ecologically fragile areas.

REQUIREMENT OF FODDER

Presently, the livestock is grossly underfed and the available fodder resources on the average can only provide for one third the maintenance ration prescribed i.e. 4.53 kg of roughage per day plus 2.72 kg of green grass for a body weight of 227 kg. The prevailing situation calls for an urgent need to increase fodder resources. Cultivation of fodder crops on agriculture lands, as a solution to overcome the acute fodder shortage, is impractical due to constraints of availability of land and other inputs for the purpose. The only practicable alternative available is encouraging the propagation and planting of fodder tree species on village, waste and marginal lands under various afforestation programmes.

In many regions of India, the only fodder for livestock available is the straw from cultivated cereals. Available grazing is poor during large part of ear except for a short period during and after the monsoon when the available feed is sufficient to support the existing cattle population. Pastures as they exist in other countries such as United States, Australia and New Zealand are very rare in India, because of high biotic pressure. It is not feasible to set apart portions of the cultivable lands exclusively for fodder production, the dovetailing of arable farming and forestry i.e. agroforestry is yet to be earnestly adopted in the intensely cultivated plains of India. Therefore, selection of hardy species of fodder value for large scale plantations towards fodder production from marginal lands appears the only feasible solution having appeal for general adoption. The fodder resources may be augmented and improved by employing better management techniques and judicious mix of species of nutritious fodder, i.e. trees, shrubs, herbs and grasses and improving the grazing grounds near forests and villages. Also important is the preservation of surplus fodder available during and after the rainy season.

The assessment of fodder demands for such a wide variety of livestock as we have in the country is difficult. The fodder requirements depend on the method of animal rearing and feeding, type of livestock, socio-economic status of the people, climate, etc. Many attempts have been made to assess the fodder livestock requirement in our country. The estimates made vary widely due to different feeding schedules adopted for such calculations. The Committee on Livestock Feed and Fodder, Ministry of Agriculture estimated the fodder required in 1973 as 339.5 million tonnes of dry fodder and 340.4 million tonnes of green fodder as against the availability of 198.8 million tonnes of dry fodder and 227.5 million tonnes of green fodder. The National Commission on Agriculture (1976) estimated the green and dry fodder requirement for the year 2000 as 590.1 and 373.0 million tonnes respectively as against the estimated availability of 575.0 and 356.8 million tonnes respectively. Recent estimates prepared by the Committee on Fodder and Grass is given below.

ESTIMATED FODDER REQUIREMENT FOR LIVESTOCK IN INDIA

Type of fodder	Estimated production (million tonnes)	
	Dry fodder	Green fodder
1985	780	932
1990	832	992
1995	890	1062
2000	949	1136

FODDER PRODUCTION

Reliable estimates on fodder production in India are not available. The assessment of fodder production in our country was attempted by the Committee on Livestock Feed and Fodder and the National Commission on Agriculture (1976). According to the estimates given in Handbook of Agriculture (1980), the availability of green and dry fodder is estimated to be about 224.08 and 231.05 million tonnes respectively. The Committee on Fodder and Grasses estimated present fodder production is given below.

ESTIMATED FODDER PRODUCTION IN INDIA

Type of fodder	Estimated production (million tonnes)	
	Dry fodder	Green fodder
Agriculture crop	–	–
Residues	236	–
Grass	205	–
Green fodder		
(i) Cultivated green fodder		208
(ii) Top feed including sugarcane tops		4
(iii) Weeds		4
(iv) Leaf fodder from trees		24
Total	441	250

Source: Anon., 1987a

GAP BETWEEN DEMAND AND SUPPLY

A comparison of the estimated requirement and production of fodder given in the above tables indicate that the production of fodder in the country falls short of the requirements and the shortage of green fodder is more acute than that of dry fodder. As a result of continuing degradation of the forests and grasslands, dry fodder production in the form of grass is expected to further decrease. The possibility of increasing dry fodder production from agricultural areas is limited. There is also little likelihood of finding additional land for green fodder production. The gap between demand and supply of fodder in the country will thus further widen if effective steps for increasing fodder availability are not taken.

The increasing shortage of fodder for the livestock calls for better utilisation of fodder produced in the country. In many areas, the dry stalks of maize, sorghum and pearl millet are fed to the livestock without chaffing. The practice results in avoidable wastage of fodder. The grasses are normally harvested in a dry condition after seed shedding. The nutritive value of such grasses is very low. Harvesting grasses at the appropriate time and adoption of proper storage methods can help in improving the nutritive value of fodder. The agriculture area presently utilised for green fodder is only about 4 per cent. In view of the acute shortages of green fodder, more area should be brought under green fodder production. Also while forests and grasslands near the habitations are overgrazed, the grass production from some remote localities is not fully utilized. The grass from such areas may be harvested and stored as a fodder bank for meeting fodder demand in times of scarcity. The production from grassland in general is very low and has the potential for considerable increase by fertilizer application and introduction of legumes.

FODDER FROM FORESTS

In India, forest is a major source of fodder for a large livestock population. Fodder is either available directly in the form of grasses, which are grazed by livestock or collected by the livestock owner, or produced by lopping the leaves of certain trees and shrubs. The number of animals that graze. in forests rose from 35 million in 1957-68 to 60 million in 1973-74 and it was estimated to be 90 million in 1987. About 25 per cent cattle in the country are dependent on forests for fodder. The dependence of cattle on forest for fodder has resulted in unlimited and uncontrolled grazing inside forests. This heavy uncontrolled grazing has caused the trampling of young regeneration, disturbance and/ or compacting of soil, transmission of disease to wildlife and overall deterioration of the forest ecosystem. A large number of forest trees yield valuable leaf fodder and they are lopped for this purpose. The leaf fodder of some trees is almost as nutritious as that of leguminous fodder crops. The trees are capable of producing almost as much green fodder per unit area as agricultural fodder. The tree fodder production does not suffer during drought when most annual grasses dry up. The leaf fodder production offers an added advantage of producing fuelwood as a by-product which helps in meeting the energy requirements of the rural population. The trees and annual fodder grasses together utilise the site better. Leguminous fodder trees enrich the site through fixation of nitrogen. This practice also helps ir effective soil and water conservation.

PRODUCTION OF GRASS AND LEAF FODDER

The quality of the grass produced depends on the type of areas, soil fertility, type of grass, rainfall, other climatic conditions and biotic factors which vary significantly along the length and breadth of the country, However, most of the grass lands in the country are in poor condition, degraded, over grazed and its productivity varies from 0.5 to 6.0 tonnes/ha/yr. Production of grasses in forest areas is higher and productivity varies from 1.5 to 3.0 tonnes/ha/yr but only about 50 per cent of grass produced in forest is utilisable due to remoteness of forest area, closure and other factors. Estimated total utilisable dry grass production in the country is about 205 million tonnes, green production in the country is about 205 million tonnes, green fodder cultivated from about 4 per cent agriculture area (20 per cent irrigated and 80 per cent rainfed) is 208 million tonnes, sugar cane tops is 4 million tonnes. The leaf fodder produced from trees constitute an important source of fodder in many parts of the country particularly in hilly and arid areas. However, it is further difficult to estimate the quantity of leaf fodder available from forest due to its dependence on type of forest, density, remoteness, proportion of fodder tree, practice and intensity of harvesting, etc. Based on the fact given in forest working plan the dry deciduous forest near Varanasi produce about 4.9 tonnes/ha/yr of dry leaf fodder and assuming that green fodder is approximately four times of dry fodder and taking 2 per cent of leaf fodder as utilisable, the green leaf fodder production is estimated at 0.4 tonnes/ha/yr i.e., 25 million tonnes from forest areas.

NUTRITIVE VALUE OF LEAF FODDER FROM TREES

The nutritive value of fodder tree species have been worked out by several workers. ICFRE institutes are working on finding out the nutritive value of the fodder species of local importance.

LOPPING EXPERIMENTS

The experiments on lopping intensity and lopping regime suggest that for optimum fodder production and growth of trees, only about one-third of the crown should be lopped. The branches thicker than 8 cm in girth should not be lopped. The trees need rest for 2-3 years before they are fit for being lopped again. Most research evidence suggests that 3-4 year's rest is essential.

SILVIPASTURE TO MAXIMISE FODDER PRODUCTION

The grassland in arid areas of Rajasthan, Gujarat and Haryana consists of several types of grasses, namely, *Urochloa* spp., *Eragrostis* spp., *Dichanthium annulatum, Lasiurus sindicus, Panicum antidotale, Cenchrus setigerus,* C. *ciliaris, Aristida* spp., *Chloris* spp., *Heteropogon contortus, Sehima nervosum, Tetrapogon* spp., *Erenmopogon* spp., etc. The important fodder tree species which occur scattered naturally in these grasslands is *Prosopis cineraria* and the shrub species is *Zizyphus nummularia.* These two species playa vital role in the silvipasture system followed in these areas. These village grazing lands in Rajasthan are called 'oran' or 'bir'. In individual birs and also in community birs, these two species are encouraged and lopped for fodder. *Prosopis cineraria* is a very good fodder tree for arid areas. The tree can be lopped and the firewood and timber can also be used for household purposes. If it is cut at the ground level, it coppices profusely, giving rise to about 6 to 12 coppice shoots. If grazing or browsing is allowed the tree forms a bushy structure. Under severe grazing intensities, the branches spread horizontally, which facilitates grazing for sheep. The shrub of *Zizyphus nummularia* grow in grasslands in arid areas. It regenerates usually through the root suckers. It has been found that 14 per cent density of Z. *nummularia* bush is optimum for obtaining the highest yield of leaf fodder and grass.

IMPROVED WOOD UTILIZATION AND SUBSTITUTION

Until the recent past, wood from traditional species remained primary, raw material for many wood based industries due to abundant natural availability of these species. Now the picture has changed as forests have been depleted beyond proportion, thus putting a question mark on the industrial utilization of these woods. Attention on this was focussed during the past two decades and multipurpose massive afforestation programmes were launched. Now stocks of wood from new fast growing species are available in place of the conventional woods. However, their introduction and utilization in the industrial sector requires a great deal of research on their properties, primary and secondary processing techniques, besides functional tests on their performance in the end uses. A great deal of research has been carried out in the area of improved wood utilization and sincere efforts have been made on -wood susbstitution by undertaking R&D work on reconstituted wood and wood products.

Due to juvenile nature of these plantation woods, several processing problems have cropped up which were however, not prominent in conventional species. Some of these are, distortion in sawing;

variation in density and mechanical properties; problematic behaviour in seasoning, preservative treatments; and wood working and finishing. The pioneering research work done during the past decade covering these areas has yielded many promising results towards improved wood utilization.

WOOD PROPERTIES

As a first step, during the last 5-7 years, wood from about 20 plantation species grown under various forestry programmes has been evaluated for density, strength and shrinkage characteristics. The suitability indices for different industrial and engineering uses with respect to *Tectona grandis* (teak) as 100 of two most important species *Eucalyptus* hybrid and *Populus deltoides* which have been planted very widely in different regions of the country have been worked out.

The data of plantation timbers has formed the basis for recommending mechanical suitability of these woods for various end uses such as construction, joinery, tool handles, packing cases, furniture, transmission line poles, dunnage pallets, defence stores and a number of other industrial uses. Studies have revealed that plantation timbers have immense potential for substituting conventional timbers for many end uses. Examples are: *Casuarina equisetifolia,. Eucalyptus* hybrid, *Leucaena leucocephala* and *Robinia pseudoacacia* (for construction); *Acacia nilotica, Azadirachta indica, E.*hybrid and *Hevea brasiliensis* (for doors and windows), *Acacia tortilis,* hybrid and *Robinia pseudoacacia* (furniture and joinery); *Azadirachta indica,. Casuarina equisetifolia, Syzygium cumini* and *Terminalia arjuna* (mine timber) and *Acacia tortilis,* E. hybrid and *Leucaena leucocephala* (tools handles). These plantation timber species have since been incorporated in the relevant Indian Standard Specifications.

Extensive studies have been made on *Eucalyptus* hybrid with respect to variation in strength from pith to periphery, top to bottom, age of the tree and locality of plantation. It has been found that there is no significant different in the strength of *E. hybrid* from Punjab, Haryana & Uttar Pradesh. The strength has been found to vary with the age and increases significantly upto the age of 14-15 years and thereafter no appreciable change has been observe (The rotation period of E. hybrid for obtaining mechanically matured wood in suitable girth for joinery and furniture may therefore be taken as 15 years. The strength increases from pith to periphery in the heart wood region and the falls in the sap wood portion. Though the sap wood strength is lower than that of immediate outer heartwood, there is no significant difference between the strengths of overall heart and sapwoods. The central portion is uniformly and significantly weaker than the remaining part. It is also prone to develop defects like cracking and collapse and should therefore be boxed out during sawing. The strength is uniform up to 15 meter height and so logging car advantageously be done upto 15 meters provided suitable girth is available. In case of *Populus deltoides* strength is found to increase from pith to periphery A special study on fast grown plantation of *Tectona grandis* introduced in Mizoram in 1962-63 indicated that strength of wood from this plantation is comparable to normal *Tectona grandis* wood grown in the other parts of the country. A significant observation made is that the strength from pith to periphery is uniform contrary to what is observed in case of fast grown plantations of exotics, indicating desirability of taking-up plantations of important indigenous species.

Classification and grading of bamboo on the basis of strength properties for structural utilisation was done. Out of 20 species of bamboo tested so far, 16 species were found suitable for constructional use and were classified in 3 structural groups. Criteria for evaluation of safe working stresses were worked out and safe working stresses of the 16 species were evaluated. Grading rules for structural use of bamboo were also evolved. This classification will promote utilisation of bamboo for constructional purposes.

WOOD SEASONING

Interest in wood seasoning has been growing in the past decades not only for timbers used for manufactured products for export but also or furniture and joinery wood. Economic compulsions to use secondary species i.e., other than teak, for general purposes has rendered adoption of proper seasoning procedures essential. Seasoning of timber in Indian conditions to 12 per cent moisture content not only upgrades the timber quality but cause considerable econnomy in utilization. The studies carried out at F.R.I., Dehra Dun formed the basis for establishment of the indigenous kiln manufacturing and kiln seasoning industries in India and drawing up of specifications on seasoning practices, which are (i) evolving kiln drying schedules for about 150 commercial species, recording air drying behaviour for more than 250 species, developing end coating compositions and correct stacking practices (ii) evolving atleast 6 standard design of steam heated kiln and two different designs of wood based fired furnace kilns (iii) development of baffling profiles, locations and orientation to improve air distribution uniformily in three different kiln designs. A better humidity control base kiln drying schedule has been evolved at FRI, Dehra Dun. It has been show that the bulk of sawn planks (except those containing the weaker core woo from the log) can be seasoned free from surface cracks by properly applying this schedule. Demonstration of these innovative processing methods have enabled *E.* hybrid to be commercially adopted for furniture, doors and windows at various places in the contry with cost economies of 33 to 50 per cent compared to traditional species like *Dalbergia sissoo.*

High costs and energy requirements for seasoning of sawn wood have hindered its adoption in the past. To provide an accelerated yet economically feasible seasoning method for the large number of small wood base, industries in the country, F.R.I. has developed a solar heated kiln. Recent innovation in the existing F.R.I. Solar kiln was made by increasing the solar energy collection by way of increasing absorbing surface area by sheathing the East, West and South walls from inside with black painted G-I sheet. Energy gains and losses from solar kiln were measured. Net energy available daily for evaporation of moisture from the timber was of the order of 20 KWH.

The recently introduced dehumidification drying process has beer evaluated and found to reduce energy requirements by 24-43 per cent compared to standard steam heated kilns. Experimental

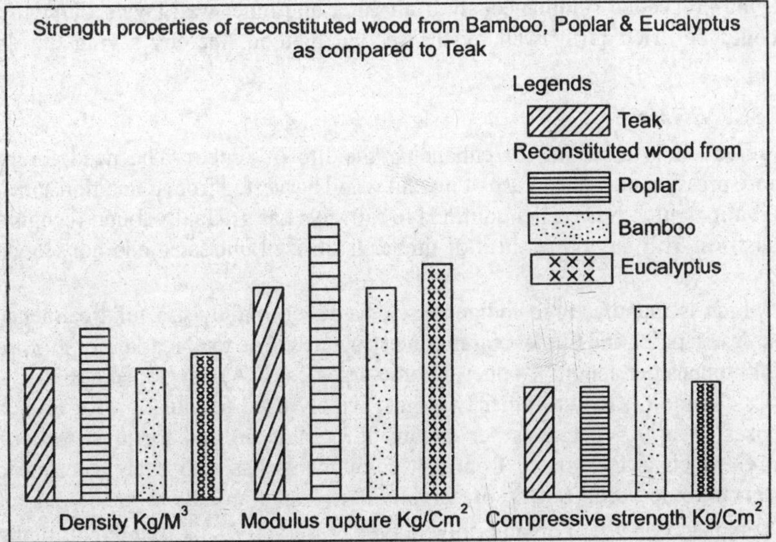

results revealed that only 0.86-0.98 KWH of electric power per kg of water evaporated are required as against 1.37-1.51 KWH per kg in steam heated kilns. This process is getting popular due to energy saving and operating costs and being automatically controlled and pollution free.

The comparative cost in different seasoning methods for representative out put is given below:

Seasoning method	Seasoning capacity (Annual) (m^3)	Investment (Rs.in lakhs)	Seasoning costs (Rs./m^3)	Fuel
Steam kilns	850	7.74	639–974	Steam coal
			429–654	Rice husk
Dehumidification	750	8.36	508–762	Electric power
FRI Solar kiln running fan	315	3.38	236–394	Electricity for
Solar predrying (6–10 days)–cum steam kilns (7–10days).	1200	12.43	380–560	Rice husk.

Severe warpage problem associated with processing of *Populus deltoides* could be solved by adopting the saw dry rip (SDR) process. Kiln drying schedules for *Prosopis juliflora* and *Acacia tortilis* were evolved for degrade free seasoning. The earlier trial and error approach to formulation of kiln drying schedules by applying arbitrarily graded schedule of increasing intensity and observing the degrade obtained was substituted by a new method based on diffusion theory of drying and using experimentally determined values of critical stage moisture content and specific gravity of the new species.

Bamboo is extensively used in split as well as round form for making a variety of items. In both forms, it needs proper seasoning to prevent it from cracking, splitting and discolouration in service. Split bamboo in the form of strips does not offer any major problems in seasoning but in the round form, it is beset with the problem of serious degrade and suffers serious end splitting and cracking at the nodes even in slow air seasoning in mild weather. Kiln seasoning under controlled conditions is also not successful in preventing these defects. Studies have shown that *Dendrocalamus giganteus* in round form can be successfully seasoned free from cracking, splitting, fungal decay or insect attack after giving a treatment with an aqueous solution of PEG-600, boric acid and sodium pentachlorophenate at 45°C to the freshly felled bamboo pieces. In another study, the nodal partitions were punctured, solar heated air was passed through bamboo tube and bamboo we dried in the open sun so that drying proceeded simultaneously from outer and inner wall layers. *Bambusa nutans* in round form could be dried fairly well by the second method, thereby saving the drying time; considerably.

WOOD PRESERVATION

Wood preservation is essential for enhancing the life of timber. The need for preservative treatment is more pressing now due to use of juvenil wood harvested from plantation species. Interest in wood preservatior which was earlier confined to railways has gradually found recognition by few other user industries. Still large quantities of timber is utilized untreated and consequently have a low life span.

Creosote, which is manufactured indigenously, has been mainly use for treatment of railway sleepers, transmission poles and timbers for marine structures. The total market for this preservatives is around 15,000 tonnes per annum. Copper-chrome-arsenic (CCA) developed at F.R.I., Dehra Dun and acid-copper-chrome (ACC) are other two most widely used formulation for timbers used for different purposes. Nearly 500 tonnes per annum of CCA is utilised for preservation. An equal quantity of ACC which is formulated from salt components has also been reported to be used. Recently copper- chrome-boron (CCB) preservative has made an advent in the market.

Earlier, the emphasis on use of organic solvent type preservatives like penta-chlorophenol (PCP) and copper/zinc naphthenates or abitates was low because of the high costs of solvents. Till recently PCP was not manufactured indigenously. With the local availability of the solvents these preservative have made a big entry in the markets. Copper and zinc napththenates manufactured from imported naphthenic acid are also used in large quantities by defence stores departments.

With decrease in supply of durable timbers from natural forests, need for the use of Secondary wood species after preservative treatment was felt, as it could increase the life of non-durable timbers by 3-5 times. Therefore, basic data on natural durability, treatability, treatment schedules, and also on optimizing absorptions of the preservatives for use of treated timber in adverse conditions of use was generated. Specifications on these aspects were also framed for the benefit of wood treatment industries, suppliers and users. Simultaneously efforts were also made to develop effective non leachable preservative compositions so that scope of use of treated timbers could be widened.

Protection of freshly felled wood at the stump site and in the depots is very important. The moment tree is felled, fungal and insect attack begins, besides, physical degradation, due to rapid loss of moisture from the ends and the surface. Anti-splitting compositions based on coal tar and coaltar-cum--preservative coatings have been developed at F.R.I., Dehra Dun which are being used for protection of logs in the country.

It is estimated that nearly 18–20 per cent of the wood raw material is lost due to biological deterioration alone during transit and storage. Trials on storage of bamboo, *Eucalyptus* and *Populus* have demonstrated that prophylactic treatment with sodium pentachlorophenate and its mixture with boric acid-borax can arrest stain, fungal and borer attack. Field trials conducted by F.R.I., Dehra Dun at Bengal Paper Mills, Raniganj, Andhra Pradesh Paper Mills, Rajahmundri and Sirpur Paper Mills, Kogaz Nagar have confirmed the effectiveness of above composition on the protection of bamboo for pulp and paper.

Wood being lignocellulosic material, is also prone to fire hazard. Fire-retardant-cum-antiseptic composition developed at F.R.I., Dehra Dun has been found very effective for the protection of timber, thatch grass, reeds, palm leaves etc., for interior applications. Efforts have also been made to develop non leachable fire retardant composition for exterior use.

Poles of *Casuarina* and *Eucalyptus* hybrid treated with toxic Copper Chrome Arsenic (CCA), Acid Cupric Chromate (ACC) and Copper Chrome Boric (CCB) compositions remain in good condition whereas, untreated ones decay within two years. At present, durability tests on *Acacia tortilis, Prosopis chilensis* and two bamboo species are being conducted besides E. hybrid, which seems to be moderately durable on the basis of field test carried out at Dehra Dun.

Pretreatments like ammoniation, steaming, ponding and treatment with mold fungus *Trichoderma viride* were found to aid permeation of the preservative in the wood of *Eucalyptus* hybrid.

A new preservative system, ammonical copper-arsenite, developed at Forest Research Institute has worked well on E. hybrid and *Prosopis chilensis.* This system has also shown 4-6 mm of preservative penetration on dry solid bamboo *(Dendrocalamus strictus)* with an adequate retention (7-8 kg per cum in 13 days soaking) recommended in Indian standards for exterior use.

Besides preservation, prophylactic treatments are also essential to prevent wood and other lignocellulosic materials from sap stain, and termite attack. In this regard about 26 insecticide/pesticide formulations were evaluated. Chlorinated homologs showed the best anti-sap stain property and defence against termites could be achieved with 0.25 per cent Sodium Pentachloro Phenate (NaPCP) & 2 per cent CCA mixture.

Preservative-treated (polyurethane-coated) panels (27 nos) of marine plywood, exposed on 21.2.1991 at Goa, were examined regularly almost every month and their condition noted. The treated ones also indicated heavy damage by *Martesia striata* within 24 months (Controls were heavily destroyed within 6 months).

CCA treated catamarans made of *Bombax ceiba* (3 nos. launched in 1986) and *Albizia chinensis* (2 nos. launched in 1968) at Visakahapatnam remained under service trial. *B. ceiba* catamarans continued to be in excellent condition after 7 years of service, and *A. chinensis* catamarans remained serviceable after 25 years use. *Albizia falcataria* catamarans treated with CCA under service test at Madras (since 1990) and South Andhra Coast (since 1991) continued to be in good condition and free from marine borer attack.

WOOD WORKING AND FINISHING

Studies on wood working, carving and finishing properties of wood from species are quite important for improving wood utilization. In this regard, working qualities under six major wood working operations, viz., planning, turning, shaping, boring, martizing, sanding, and carving behaviour under conventional chiselling, punching, scooping & fret saw were studied for *Acacia tortilis, Abies pindrow, Mangifera andamanica* and *Neonauclea gageana.* All timbers are comparable to teak *(Tectona grandis)* in working and carving properties except *Abies pindrow* in carving.

Imaginative and judicious shaping of sandal wood roots in various non-geometric but figurative art forms (antiques) indicated that roots of this species are excellent to work with chiselling and sanding inspite of heavy interlocking of grains. Utilization of roots is thus possible and adds new dimension to wood working and waste wood (stumps-roots-drift wood) utilization.

WOOD PACKAGING

The wood packaging research is linked with improved wood utilization in two ways. Firstly, as introduction of new species in the trade and secondly for improving design parameters and effecting economy in wood use for different loads, as wood consumption in this sector is around 25 per cent of the total industrial wood requirement. Method of manufacturing wooden boxes have not changed significantly and considerable quantity of wood is wasted during conventional manufacturing process.

To cope up with this problem, research on packing case design and testing has been carried out recently with special reference to fruits, for which alone, the requirement of wood is 5.6 m.cum per annum. Conventional designs were tested in the first instance and results have indicated that these boxes were quite weak thereby causing fruit wastage to the tune of 30 per cent during transportation and handling.

A series of experiments were conducted by making different permutations and combinations of boxes using conventional as well as plantation grown species having low girth. In a generalised way, it was observed that economy of wood is possible by using standard sizes of improved designs.

COMPOSITE WOOD AND IMPROVED WOOD

A great deal of research work done on various aspects of improved utilization of raw materials of plantation origin has led to substitution of solid wood for several uses resulting in the conservation of valuable forest resources in the country. Some of these are discussed here in brief.

PLYWOOD FROM PLANTATION SPECIES

To overcome the problem of short supply of traditional plywood logs, the properties of plywood made using Urea Formaldehyde (UF) and Phenol Formaldehyde (PF) glue and plantation species like *Ulmus wallichiana, Dracontomelum mangiferum, Azadirachta indica, Alnus nitida, Lagerstroemia parviflora, Quercus* spp., *Millingtonia hortensis, Melia azedarach, Artocarpus lacucha* and *Populus deltoides* were studied and satisfactory plywood has been prepared. Based on the results, these species have been included in the list of timber species for plywood in relevant Indian specifications. At present *Populus deltoides* is being used in large quantities for production of various grades of plywood in the country.

Protection of glued assemblies against the attack of fungi and insects has also received attention. Treatment schedules for plywood from a large number of species have been worked out for the benefit of the industry and users.

COMPRESSED WOOD FROM SOCIAL FORESTRY SPECIES

Research was carried out on Indian timbers, including fast growing social forestry species like *Azadirachta indica, Anogeissus latifolia, Acacia tortilis, Eucalyptus* hybrid, *Melia azedarach, Populus ciliata* and *Populus deltoides,* for finding suitable substitutes for imported woods for the manufacture of shuttles for use in textile and jute industries. It has been established that the timber from these species could be compressed to densities ranging between 1.05-1.35 g/cm^3. Strength properties of these compressed timbers compare favourably with conventional and imported timbers and are suitable for shuttle manufacture.

PARTICLE BOARD FROM BAGASSE

Suitability of bagasse for particle board has been evaluated and results indicated that it was a suitable raw material for the manufacture of medium density particle board. Good quality boards meeting requirements of the specification con be prepared using 8 per cent phenol formaldehyde resin on dry basis of particles. Veneered particle boards from bagasse using 8 percent phenol formaldehyde resin meet the requirement is specification for veneered particle board.

UTILISATION OF SAW DUST FOR PARTICLE BOARD

Satisfactory medium density particle boards meeting the requirements of specification were prepared from pine saw dust using 12 per cent phenol formaldehyde resin. The fraction of saw dust passing through 20 mesh but retained over 80 mesh sieve which constitute the bulk (about 75%) of the saw dust could be used for making particle board. The properties of the board were further improved by mixing about 20 per cent wood particles in the saw dust furnish. Veneered particle boards were also made from saw dust using 1.2 per cent phenol formaldehyde resin and 0.5 per cent wax emulsion as sizing agent.

FIBREBOARD FROM *EUCALYPTUS* HYBRID

Eucalyptus hybrid wood, lops and tops, twigs, thinnings, branches etc, were evaluated for making fibre hard boards. The bark content in lops and tops varying from 8.7 to 14.5 per cent did not have adverse effect and satisfactory boards could be prepared by cooking the material with 2 per cent alkali for 2 hours before defibration. Water absorption of the boards was considerably reduced by adding about 1 per cent wax emulsion as sizing agent.

FIBREBOARD FROM POPLARS

One and two years old *Populus deltoides* containing respectively 30 and 21 per cent bark was found suitable for hardboard manufacture. However, 2 years old plants gave better properties than the 1 year old. It was observed that bark to the extent of 20 per cent helps in improving the physical and mechanical properties of the boards as compared to the boards prepared from wood alone.

Populus ciliata without bark was evaluated for making hard boards. It was found suitable for making fibre hard board.

FIBREBOARD FROM SU-BABUL

Suitability of Su-babul *(Leucaena leucocephala)* with and without bark was evaluated for making fibre hard boards. Average bark content in the bolt of Su-babul (13 to 22 em. in girth) was found to be 10.2 per cent on oven dry weight basis. The results indicated that Su-babul wood containing about 10 per cent bark could be used for making satisfactory fibre hard boards.

FIBRE HARDBOARDS FROM SPENT CHIPS OF KHAIR *ACACIA CATECHU*

Acacia catechu (Khair) is a very important species used for extraction of Katha and Cutch. Extracted chips are used as a fuel and do not find any economic use at present. Suitability of these chips was evaluated for making hardboards. The boards, meeting the requirements of specification for standard hardboard, could be prepared with 3-7 per cent alkali concentration and cooking for 3 hours.

BUILDING BOARDS FROM *LANTANA CAMARA*

Lantana camara, due to its prolific growth and wide adaptability has overrun large areas in India and has developed into an obnoxious weed. Lantana is a source of fire hazard in deciduous forests and causes serious destruction of forest crops. Although Lantana is available in large areas in the country, it has not found any economic use so far. Keeping this in view suitability of *Lantana camara* was evaluated for fibre hardboards and particle board.

For making fibre hardboards, Lantana sticks were cut into pieces, digested with varying amounts of alkali for different cooking periods, defibrated and formed into mat and pressed. Suitable hardboards meeting the requirements of specification were obtained from *Lantana camara.* An attempt was also made to develop hardboards from mixed pulps of Lantana and pine needles. By mixing about 50 per cent lantana pulp with pine needle pulp satisfactory boards were obtained.

For making particle boards, Lantana sticks were converted into particles, screened and dried. Particles were mixed with 10 per cent urea formaldehyde resin on the weight of particles, formed into mat and pressed. Suitable particle boards meeting the requirements were obtained from *Lantana camara.*

MEDICINES AND AROMA FROM FORESTS

India is a veritable emporium of medicinal and aromatic plants. Presently more than 5,000 medicinal plants are known to be utilized in Indian systems of medicine for treating the human sufferings throughout the country. Various systems like Ayurveda, Unani and Homoeopathy have been utilizing plants for their respective preparations and have now assumed great importance owing to side effects of synthetic drugs. Indian flora has innumerable medicinal plants which are collected from the forests. Many of them are also being exported to developed countries and earn handsome foreign exchange.

CULTIVATION OF MEDICINAL PLANTS

A number of important medicinal plants known for their therapeutic efficacy have become rare or nearly extinct owing to over-exploitation, deforestation, biotic interferences and unscientific collection. As such, there is an urgent need to conserve and utilize these species in a scientific manner, so as to ensure a sustained supply of raw material for various pharmaceutical industries. Some of such species are: *Aconitum heterophyllum, Nardostachys jatamansi, Orchis latifolia, Picrorhiza kurroa, Dioscorea deltoidea, Podophyllum hexandrum, Saussurea lappa, Crocus sativus, Rheum emodi, Bunium persicum* etc.

Maintaining ethnobotanical herbarium and museum of plants was the strategy followed for conservation of heritage of threatened plant species. However, this strategy cannot serve the urgent demand of conserving endangered flora and gene pool of valuable plant species. The most useful method would be to preserve and multiply the plants in botanical gardens, nurseries and through large scale cultivation. The latter one could be a strategy for those species which are commercially important and which could be adopted by farmers and landless labourers. However, before the medicinal plants could be adopted by farmers, research cum demonstration centres shall have to be established in different agro-climatic zones. Package of practices shall have to be evolved for recommending to the prospective farmers.

For *ex-situ* conservation of rare and endangered species a germplasm bank has also been created at NWFP nursery at Chakrata. Methods for mass propagation by tissue culture are being developed to provide planting material on commercial scale. Some of important plants which have been raised at this nursery are: *Mentha piperita, M. citrata, M. spicata, Digitalis purpurea, D. lanata, Picrorhiza kurroa, Bunium persicum, Angelica glauca, Podophyllum hexandrum, Pyrethrum* sp., *Atropa acuminata, Skimmia laureola, Salvia lanata, Aconitum heterophyllum, Ephedra gerardiana, Orchis latifolia, Dioscorea deltoidea, Hyoscyamus niger, Colchicum luteum, Rheum emodi, Nardostachys jatamansi, Aconitum napellus, Onosma echioidea, Artemisia brevifolia, Rheum nobile, Jurinea dolomiaca, Saussurea costus, Gentiana kurroo.*

Cultivation techniques for the medicinal plants of commercial importance are also being standardised at FRI, Dehra Dun, TFRI, Jabalpur and other centres. The species which are being experimented at these centres are: *Rauvolfia serpentina* (Sarpagandha), *Withania somnifera* (Ashwagandha), *Mentha arvensis* (Pepermint), *Urgenia indica* (Jangli pyaj), *Allium purpureum* (Jangli lahsoon), *Curcuma caesia* (Kali Haldi), *Acarus calamus* (Bach), *Hibiscus cancellatus* (Kamraj), *Curcuma aromatica* (Jangli Haldi), *Curcuma amada* (Amahaldi), *Cymbopogon martinii* (Roshagrass), *Vetiveria zizanioides* (Khash grass), *Curcuma angustifolia* (Tikhur), *Ancylus pyrethrum* (Akalkara), *Swertia chirayata* (chirayata), *Glycyrrhiza glabra* (Mulethi), *Chlorophytum tuberosum* (Safed musli), *Asparagus racemosus* (Satavar), *Spilanthus oleracea* (Akarkara), *Gloriosa superba* (Kalihari), *Adhatoda vasica* (Adusa), *Hedychium coronarium* (Gulbakaoti), *Costus speciosus* (Keokand), *Zingiber purpureum* (Jangli adarakh), *Dioscorea bulbifera, Pimpinella anisum.*

A naturally occurring rare medicinal plant *Aleetra parasitica,* parasitic on *Vitex negundo* and known to cure leprosy has been identified; the plant has been multiplied in NWFP nursery at FRI, Dehra Dun and TFRI, Jabalpur. Attempts have been made to re-introduce *Catharanthus roseus, Plumbago zeylanica, Withania somnifera, Acorus calamus, Vetiveria zizanioides* and *Chlorophytum tuberosum* in natural forest conditions at TFRI, Jabalpur, and their performance is being observed.

ECONOMIC RETURNS

Cultivation of medicinal plants promises good returns. Cultivation trials of some important medicinal plants carried out at Amarkantak revealed that *Acarus calamus* (Buch) can fetch annual returns of Rs. 20,000–27,000/ha, *Curcuma angustifolia* (Tikhur) Rs. 27,040/ha, and *Rauvolfia serpentina* - (Sarpagandha) Rs. 48,040/ha. Economics of cultivation of other important species like *Withania somnifera, Plantago ovata, Plumbago zeylanica, Curcuma caesia* are also being evaluated.

EXPORT

A survey of export potential of medicinal plants especially the Bombay market has indicated that there is enough potential for the export of medicinal plants. Quantity wise maximum demand is of 'Senna' leaves followed by Isabgol seeds and husk. The seeds of *Cassia tora* are in great demand in Japan and Taiwan. In general, medicinal herbs have great demand in Middle East, West Germany, U.S.A and Japan.

TAXUS BACCATA

Recently "Taxol", a chemical from *Taxus baccata,* known for its anti--cancer properties, has gained attention of scientists all over the world and the leaves/ twigs of the species are in great demand internationally. Medicinal tincture made from the young shoots has long been in use for the treatment of headache, giddiness, feeble and falling pulse, coldness of the extremities, diarrhoea and severe biliousness. The leaves are credited with emmenagogue and antispasmodic properties. They are employed for the treatment of hysteria, epilepsy and nervousness, and as a lithic in calculous complaints. *T. baccata* has been reported to be the source of the drug "ZARNAB", which is very frequently prescribed in the Unani system of medicine.

Taxus baccata is found in the temperate Himalayas at altitudes between 1,800 and 3,000 m and presently enlisted in the rare and endangered category by the Botanical Survey of India in their Red Data Book. Therefore, the development of a strategy for their conservation and sustainable supply to meet the demand of the market is expedient.

NARDOSTACHYS JATAMANSI

It is an erect, perennial plant found in upper regions of the Himalayas. The plant is propagated by cuttings of underground parts and sometimes by seeds. It is valued for its rhizomes. It has been used in the indigenous systems of medicine in India. It is used as sedative for various mental afflictions like hysteria and epilepsy. It is claimed to be an exellent drug for pulpitation and as a good tranquilliser. Jatamansi root in hair oils promote growth of the hair. Jatamansi is also used as a substitute for valerian *(Valeriana officinalis* Linn). It yields up to 1-9 per cent of a pale yellow essential oil with pleasant colour.

On further chemical investigation, a special fraction have been isolated which is known as *NARDOSAN* which possesses considerable sedative activity both by oral and parenteral routes.

Since the *N. jatamansi,* is found in the Alpine Himalayas from Punjab to Sikkim and Bhutan at altitudes of 3000-5000 m, the development of a strategy for its conservation and sustainable supply to meet the demand of the market is expedient.

PICRORHIZA KURROA (KATKI)

Picrorhiza kurroa grows in Kashmir and North Western Himalayas at an altitude of 5000 to 11000'. The important drug which is isolated from the roots of the plant is known as *Gentian* possesses bitter and tonic properties. A number of compounds viz. kutkin, D. mannitol, vanillic acid and others have been isolated and characterised from the roots of *P. kurroa.* Picrolive isolated from *Picrorhiza kurroa* possesses curative and preventive effect against hepatic damage.

AROMA FROM FORESTS

The essential oils and balms (resin + essential oil) occur in almost ant parts of the plants of the *Coniferopsida* and *Taxopsida* amongst Gymnosperms. They are without exception stored in schizogenous excretary cavities and canals. About 1/3 (about 87 families) of Angiosperms are known to contain essential oils. Some important essential oil bearing families are *Labiatae, Rutaceae, Umbelliferae, Geraniaceae, Lauraceae, Rosaceae, Compositae, Zingiberaceae, Myrtaceae* and *Graminae,* etc. Essential oils and their constituents present in plants have been used as one of the characters in the study of variation and of speciation in nature (Biosystematics) or in the plant classification by Taxonomists. Essential oil bearing plants or aromatic plants hold a prestigious position as they are the sources of perfumes, cosmetics, flavouring agents and aromatic chemicals and for their value as antiseptic, deodorants, repellents, and medicines.

SANDAL

Sandal *(Santalum album* Linn) is one of the important tree species of our forests. The tree is the source of heartwood from which the fragrant 'East Indian Sandal Oil' is distilled. It is found in many of the states of the country. The total area under mixed sandal forests is estimated to be about 9600 km^2. Karnataka and Tamil Nadu account for major areas extending over 5000 km^2 and 3600 km^2 respectively. The tree is capable of growing from sea level to 1800 m and with annual rainfall of 300-2540 mm. It grows on a variety of soils, under varying conditions of soil pH, and in places where there is little top soil, but is prevalent mostly on red loam soils. Research on various aspects of sandal as indicated below is being carried out at IWST, Bangalore.

VARIATION

Sandal is a highly variable polymorphic species. Trees vary between themselves significantly for leaf length and leaf width except in plus trees. Three distinct types of sandal have been found on morphological characters: Robust type (Srinivasapura in Kolar Division and other areas of Karnataka); Chickaballapur type (small bluish green leaves); and Thindlu type (narrow sapwood). The colour of the heartwood varies from light yellow to dark brown. The sandalwood oil yield from the light coloured wood is 3 - 6 percent, whereas, from the brown and dark brown wood, it is up to 2.5 per cent. The variation in the colour of the heartwood under similar soil and climatic conditions suggests that these variations are genetically controlled. Herbarium specimens of all morphological variants of sandal and wood samples are being maintained. *In situ* conservation has been undertaken at Janiguda in Orissa State and at Thindlul in Karnataka State. Under *ex situ* conservation programme, three clonal banks have been set up, at Gottipura, Bangalore Division (Karnataka), Karvetinagar, Chittor Division (Andhra Pradesh) and Kurumbapatti, Salem Division (Tamil Nadu) with ramets from 58, 10 and 3 plus trees respectively. In all 79 plus trees have been selected. In the germplasm bank at Gottipura, morphological variants have been planted in 27 plots.

GROWTH AND YIELD

The growth in the form of increment in girth per annum varies from 1 to 5 cm under favourable soil and moisture conditions. 250 well growing trees with an annual increment of one kg per year can give an annual income of Rs. 10,000/ha assuming the cost of sandal wood as Rs. 40/kg.

TREE IMPROVEMENT

The work in progress is to produce sandalwood trees which are capable of yielding more heartwood and sandal oil in a minimum period of time and are resistance to pests and diseases.

SEED STANDS

Four seed stands have been identified: at Anchalpatty, Marayoor (Kerala), Chitteri, Dharmapuri (Tamil Nadu), Rayalpad, Kolar (Karnataka) and Thangli, Chickmagalur (Karnataka). Janiguda is also a source of good seeds.

PROGENY TRIALS

Progenies of selected plus trees and morphologically different phenotypes have been raised. So far three half-sib progeny trial plots containing 19, 16 and 28 plus tree families have been established at Nallal (Karnataka).

CLONAL SEED ORCHARD

Two clonal seed orchards based on a computerised design to maximise inter-breeding have been established at Nallal and Tirupathi (Andhra Pradesh). Clonal material required for establishing such clonal seed orchards were also supplied to Forest Departments of Karnataka and Tamil Nadu.

PHYSIOLOGY OF PARASITISM IN SANDAL

Sandal is an obligatory parasite. Four species namely *Acacia nilotica, Wrightia tinctoria, Melia dubia* and *Casuarina equisetifolia* have been found to serve as good hosts for sandal. Studies on the physico-chemical properties and nutritional levels of soils and the mineral uptake of leaves of sandal in the host trial plots have shown that sandal depends on its host for phosphorous, potassium and magnesium and in absence of a host it is not capable of growing normally.

BIOCHEMICAL AND PHYSIOLOGICAL STUDIES

A study of isoenzyme pattern in respect of peroxidase, malate dehydogenase and esterase in the fully expanded green leaves of healthy sandal plants showing variations in the leaf shape revealed (a) both at vegetative and flowering stages, characteristic differences exist in the pattern of isoenzymes of peroxidase and malate dehydrogenase of different leaf types, (b) there are 3 sub-types in big ovate leaves, and (c) that the sandal plants with normal ovate, wavy and normal ovate and non-wavy leaves are genetically very close.

A study of the content and chemical composition of the sandal oil from the heartwood at different heights of the bole from root to tip showed that there is a decrease of 45 per cent in sandal oil and 35 per cent in santalol while there is an increase of 45 per cent in sandtalenes and 35 per cent in santalyly acetate. The study also revealed that there is a decrease of 20 per cent in sandal oil and 25 per cent in santalol and an increase of 30 per cent in santalene and 60 per cent in santalyly acetate from core to periphery of the heartwood taken from breast height.

Incorporation of polyethylene glycol 200 (PEG) helped in total recovery of sandal oil form sandal concretes. The ethyl acetate or acetone extracts of sandal powder when steam or vacuum distilled in the presence of PEG gave 10-15 per cent more oil as compared to the yield by the steam distillation method.

A new essential oil (HESP) was obtained from the acid hydrolysate of the acetone extract of the exhausted sandal powder.

Sandal seed kernel contains a drying oil (55-60%) rich in santalbic glyceride. Catalytic hydrogenation of the oil gave a semi-drying oil possessing characteristics of oils normally used in soap industry.

EUCALYPTUS

Eucalyptus (Family Myrtaceae) is large genus of evergreen aromatic trees. It was first introduced in India, as early as 1790, at Nandi Hills (Mysore). In 1843, it was successfully introduced at the Nilgiri Hills in South India. Among many species so far tried in India, the Nandi provenance of *Eucalyptus tereticornis* popularly known as *Eucalyptus* hybrid or Mysore Gum has been the most widely used species for raising plantations in denuded and barren areas and also for replanting low-value natural crops. The total area under its plantation is about 4.28 lakh hectares in different states. Its wood finds ready use in making pulp and paper, furniture and as fuel.

Essential oil obtained from the leaves of *Eucalyptus* hybrid has been investigated in detail in the institutes under ICFRE. It consists of *alpha*–pinene (11.11-37.80 %), *beta*-pinene (11.87-31.79 %), *alpha*- terpinene (1.14--40.95%), *beta*-phellandrene (0.60-11.87%), *para*-cymene (7.40-44.12%) terpinyl acetate (2.15-9.68%), and cineole (10.95-50.72%). Being a hybrid, the composition of the oil varies from tree to tree. Statistically significant correlation was found between the refractive index and cineole content of the essential oil. It was observed that almost all the oil samples with refractive index below 1.4700 were rich in cineole and hence can be used as medicinal oil. Another method based on the stomata count per unit area of the leaf measured on a thermacole peal of leaf imprint was also developed. Lesser number of stomata per unit area appears to be the most important character of cineole rich leaf. Gel electrophoretic study of the iso-enzyme (peroxidase) pattern indicated distinct differences between cineole - rich trees and others. *Eucalyptus globulus* ssp *bicostata* leaves gave 6.58% medicinal oil with 73.09% of 1,8- Cineole. From the essential oils of *Eucalyptus camaldulensis, Eucalyptus* hybrids FRI-4, FRI-5, *Eucalyptus deglupta* and *Eucalyptus robusta,* fifteen constituents, namely *alpha*-pinene, Camphene, *beta*-pinene, phellandrene, limonene, 1,8- Cineole, r-terpinene, p-cymene, citronellal, linalool, terpen-I-ene-4-ol, citronellyl acetate, borneol, alpha-terpineol and piperitone were characterised in varying compositions in all the species.

The essential oils from E. *tereticornis,* E. *camaldulensis,* E. hybrid FRI-4 and E. hybrid FRI-5 have been tested for antimicrobial activity against eight bacteria and six fungi. These essential oils have maximum antifungal as well as antibacterial activity when it is undiluted. The activity decreases as the dilution increases.

BURSERA

Bursera penicillata is an exotic from Mexico, introduced in India about 90 years ago at Bangalore. It is popularly known as Indian Lavender. The fruit husk of this plant on steam distillation yields 10 to 14 per cent essential oil rich in linalyl acetate and linalool, which is used in soap making, perfumery and cosmetics. Introduction trials initiated at the campus of TFRI, Jabalpur and Tilda near Raipur have revealed that this species can be successfully grown on degraded soils in Central India. It has shown a consistent growth of about 40-50 cm/ year at both the places. Plants have started bearing flowers and fruits or 3 years age.

GRASS OILS

Some of the grasses are rich in essential oil. Important grass oils are Lemon grass oil, Palmarosa oil, Ginger grass oil, Citronella grass oil and Vetiver oil. Lemon grass oil is obtained from *Cymbopogon flexuosus* (Lemon grass) and is used in perfumery, soap and cosmetic industries. This oil is rich in citral which is the starting material for the synthesis of ionones, methylionone and vitamin - *A*. *Palmarosa* oil is obtained from the grass *Cymbopogon martinii* var. *motia* (Palmarosa grass, Rosha grass) and is used extensively as a base for perfumes and cosmetics, particularly in soaps. *Cymbopogon martinii* var. *sofia* (Ginger grass) is the source of Ginger grass oil which is used in perfume blends for scenting the cheaper varieties of soaps. Citronella oil is obtained from Citronella grass, *(Cymbopogon nardus)* and is extensively used in the preparation of low cost perfumes. Vetiver oil (also known as khus oil). is obtained by the steam distillation of the roots of the grass *Vetiveria zizanioides* (Vetiver Khus) densely tufted perennial grass found through out the plains and lower hills up to an altitude of 1,200 m mainly in Uttar Pradesh, Punjab and Rajasthan. It is also cultivated in parts of Tamil Nadu and Kerala. While the roots of this plant are sweet scented, the leaves are odourless. There are two varieties of the grass, one flowering and the other non-flowering. The North Indian variety is non-flowering, while both the varieties are found in South India. Vetiver oil is widely used in perfumes, cosmetics and soaps. It imparts a pleasing and lasting note to perfume and serves as fixative. The institutes under ICFRE have developed suitable technologies for the distillation of essential oils from these and similar grasses.

The recent survey on occurrence of medicinal and aromatic plants revealed that Vetiver and Palmarosa grasses occur in patches, mostly confined to water logged and slopy areas. Vidharbha region of Maharashtra hold a great promise for undertaking large scale cultivation of these grasses, especially in the forests and hilly regions of Melghat, Mehkar, Buldhana, Chandrapur and Gadchiroli. Based on the survey the potential areas suited for cultivation of these grasses in Maharashtra and Madhya pradesh have been suggested as follows:

Essential oil yielding grass	Existing area	Potential area for extension
1	2	3
A. **Maharashtra**		
1. Palmarosa (Rosha, Tikhadi grass for geraniol)	Dharni, Melghat, Achalpur (for Motia) Chandrapur, Gadchiroli, Tadoba National park.	Ramtek, Akola (Medshi, Washim, Akot)
2. Lemon grass	Ratnagiri, Thane, Palghar, Dapoli, Kolaba, Chandrapur Dahanu, Gadchiroli	Chikhaldara, Dharni, Mohur Achalpur, Kinwat, Nanded
3. Khus	Sawant wadi, Ratnagiri, Amboli, Lonawala, Matheran, Chandrapur, Gadchiroli.	Bhandara, Ycotmal, Wardha, Achalpur, Buldhana, Akola,
B. **Madhya Pradesh**		
1. Palmarosa	Seoni, Chhindwara, Shahadol Bilaspur Satna, Sagar, Chhatarpur.	Jabalpur, Raipur Balaghat, Rewa,
2. Lemon grass	Sarguja	Mandla, Jashpur, Bastar
3. Khus	Amarkantak, Bandhargarh, Saguja (Kudargarh), Balaghat	Jabalpur, Seoni Jashpur

Since these grasses are not commercially cultivated actual/recorded yield per hectare is not available. Even then, it has been predicted that a yield of 200-250 kg/ha essential oil can be obtained from Palmarosa grass in 4 years and 500 to 530 kg/ha essential oil can be obtained from Lemon grass in 6 years. The estimated expenses for the commercial cultivation of Palmarosa grass is about Rs.25,000/ha and that for Lemon grass is Rs.18,000/ha. India exports a limited quantity of these oils.

Trade in essential oil and perfumery chemicals is expanding rapidly. Innovative techniques of isolation of fragrant principle employing super critical fluids which can extract the odouriferous principle with its original fragrence, development of superior strains of aromatic plants rich in active principles and sophistication in pre and post harvest techniques have contributed immensely to the development of production and trade in other countries. Despite the rich wealth of aromatic plants in our country, our share in the international trade is very low. Therefore, the institutes under ICFRE will give emphasis to research on aromatic plants. The following species will be taken up on priority for their cultivation, exploitation, storage, etc., to boost the yield of essential oil production for the use of industries:

(*a*) *Matricaria chamamilla*
(*b*) *Acarus calamus*
(*c*) *Mentha* sp
(*d*) *Rasmarinus afficinalis*
(*e*) *Vetiveria zizanioides*
(*f*) *Glycyrrhiza glabra*

This would be achieved through the following:-

(*i*) Identification of superior strains rich in the perfumery constituents.
(*ii*) Fertilizer trials on the quantity and quality of the oil.
(*iii*) Vegetative propagation of superior strains
(*iv*) Introduction of mycorrhiza to increase biomass and essential oil production.
(*v*) To study the antifungal and antibacterial efficacy of oils.
(*vi*) To develop efficient methods of extraction of oil and isolation of major ingredients.

India imports a few essential oils to meet the requirement of the perfumery industries. Therefore, the institutes under ICFRE will also carry out research on the introduction of exotics at suitable sites to avoid/reduce the import of essential oils.

SOCIAL FORESTRY

The concept of social forestry was evolved in India during the early seventies. The programme expanded in scope and quantum during the Fifth Five Year Plan and then during the Sixth Five Year Plan with higher allocation of funds and generous assistance provided by the international aid agencies, viz., World Bank, CIDA, SIDA, USAID, ODA, etc. The National Forest Policy (1988) also placed major emphasis on social forestry and afforestation programmes.

OBJECTIVES

(*a*) To meet the needs of the society for fuelwood, small timber, bamboo, fodder and other NWFP on sustainable basis.
(*b*) To release cow-dung for use as manure for agriculture by supplying more of fuelwood.
(*c*) To provide gainful employment to the rural population. To ensure efficient soil and water conservation.
(*d*) To ensure efficient soil and water conservation.
(*e*) To raise the income of rural households by adopting appropriate on-farm technologies.
(*f*) To reclaim the degraded lands for productive use.
(*g*) To contribute to general improvement of environment by increasing tree cover and thus improve the quality of life.

Botanical Name	Climatic zone	Annual Rainful (mm)	Soil condi-tion	Form	Height (meter)	Ability to fix atm. N	Shelter belt/ wind break	Soil pro-tec-tion	Orna-ment-al/ shade	Ani-mal fodder	Fuel wood	Saw-logs	Poles/ posts	Other uses
1	2	3	4	5	6	7	8	9	10	11	12	13	14	15
1. *Acacia auriculi formis*	M/D tr	1000-2000	Alkaline, water, logged	T	8-30	X	X	X	S	–	X	X	X	Pulp, Tannin
2. *Acacia mearnsii*	M/D tr	600-925	Sandy clay	Sm.T	6-20	X	X	X	–	–	X	X	X	Pulp, Tannin
3. *Acacia sylvestris*	M/Str.	800-1100	Sandy loam	Sm.T	8-30	X	X	X	S	–	X	X	X	Tannin
4. *Acacia tortilis*	D	300-450	Alkaline	Sm.T	4-15	–	X	X	–	X	X	–	–	Agric, implements
5. *Acacia senegal*	Arid-semi-arid	300-450	Sandy	Sm.T	10-13	X	X	X	–	X	X	–	X	Pharm-aceutical
6. *Acacia nilotica*	W/M/Dtr	300-450	Heavy	Sm.T	10-20	X	–	X	–	X	X	X	X	Gum, Tannin Nector, Gum
7. *Albizia procera*	M/Dtr	1000-1750	Sandy loan	Sm.T	7-15	X	X	X	S	–	S	–	X	Tannin
8. *Alnus nepalensis*	W/M/Str W/M Temp	600-450	Moist Sandy Clay	T	20-30	X	–	X	–	X	X	X	X	Timber
9. *Azadirachta indica*	W/M/dtr M/D Str	350-600	Moist sandy clay loam	T	12-15	–	X	X	S	–	X	X	X	Medicine
10. *Ailanthus excelsa*	W/M/Dtr & Str	350-600	Sandy loam	Sm.T	6-10	–	X	X	S	X	–	–	X	Pulp wood
11. *Anogeissus latifolia*	Mtr & Str	500-600	Dry sandy	T	15-20	–	X	X	–	X	X	X	X	Pulp wood, Tannin, Skin-worm Gum dye
12. *Adhatoda vasica*	M/Dtr M.Temp	500-600	Dry	Sm.T	2.5-6	–	–	X	–	–	X	X	X	Green manure, dye, nector, medicine.
13. *Brossonetia papyrifera*	D.Str	600-925	Moist sandy loam	Sm.T	12-15	–	–	X	S	–	–	–	X	Paper making
14. *Bambusa tulda*	W/Mtr	1000-2000	Alluvial	Culms.	25	–	–	–	–	–	–	–	X	Building furniture, basket, mats.
15. *Bauhinia variegala*	Mtr&	760-1270	Dry hilly	Sm,T	10-15	–	–	X	X	X	X	–	–	Agric ulture. Implements
16. *Casuorina equisetifolia*	W/Mtr	200-300	Clay saline	T	15-25	X	X	X	O	–	X	X	X	Pulp, Tannin
17. *Dodonea viscosa*	M/Dtr	120-1580	Alkaline, Salinealkali	Shrub-Sm.T	1-5	–	–	–	S	X	X	X	–	Fuel, timber fruit
18. *Dendrocalamus strictus*	M/Dtr	1000-2000	Sandy loam, Hilly	Culms	8-20	–	–	–	–	–	–	–	–	Building, Furniture,

1	2	3	4	5	6	7	8	9	10	11	12	13	14	15
19. *Dalbergia sissoo*	W/M/Dtr & M/D Str	500-2000	Sandy, alluvial	T	20-30	X	O/S	X	X	X	X	X	X	basket, mats construction, door, window, pharmaceu- tical
20. *Eucalyplus spp.*	M/Dtr	600-950	Moist sandy loam to alkaline	T	10-20	–	X	–	–	–	X	X	X	Pulp, oil, furniture, construction, pharma- ceutical
21. *Gmelina arborea*	W/Mtr	750-4500	Moist, base rich	T	20-30	–	–	X	S	–	X	X	X	Pulp, Nector
22. *Grevia optiva*	M.Temp	1299-2500	Sandy loam, hilly	T	15-18	–	–	X	–	X	–	X	–	Fibre
23. *Leucaena leucocephala*	Dtr	600-1500	Heavy soil	T	15-20	X	–	X	–	X	X	X	–	Pulp
24. *Morus alba*	W/MTemp & Str	750-1500	Sandy loam	Sm.T	12-25	–	–	X	S	X	X	–	X	Silk-worm
25. *Populus spp.*	Mtr to	800-1100	Sandy loam	T	20	–	X	X	O	–	X	X	X	Match-wood, pulp wood, packing case, plywood.
26. *Prosopis cineraria*	D tr	750-850	Alkaline, arid	Sm.T	5-9	X	X	X	–	X	–	X	X	Agri, implements, Building.
27. *Prosopis juliflora*	D tr	150-750	Sandy, soil	Sm.T	5-10	X	X	X	–	X	–	X	X	Nector, Agri. implements, building
28. *Pithec ellobium dulce*	M/D tr	450-600	Sandy loam calcareous	T	15-20	X	X	X	–	X	X	–	X	Tannin
29. *Quercus species*	W Temp	600-925	Moist sandy clay	Sm.T	15-20	–	–	X	S	X	X	X	X	Nector Gum. Agri. implements
30. *Robinia pseudoacacia*	W Temp	1000-1500	Loamy, hilly	T	15-18	X	X	X	O	X	X	–	X	Tool-handle
31. *Santalum album*	D tr	750-1500	Red Soil loamy	T	15-20	–	–	X	–	–	–	X	X	Pharmace- utical
32. *Syzygium cuminii*	W/M/ Dtr	1500-10000	Sandy	T	15-30	–	X	X	O/S	–	X	–	X	Agri. imp- lements food, nector
33. *Tamarix aphylla*	Dtr	350-500	Alkaline sandy	Sm.T	8-12	–	X	X	S	–	X	–	X	Plough furmiture
34. *Zizyphus mauritiana*	Dtr	300-500	Sandy loam	Sm.T	10-12	–	–	–	–	X	X	–	–	Agri. implements tannin food

atm N-Atmospheric Nitrogen; M tr – Moist tropical; S tr – Sub-tropical; Dtr- Dry-tropical; M temp – Moist temperate; W temp- Wet temperate; T-Tree; Sm-Small; X-Yes; '–' –No; O – Ornamental; S– Shape; mm. – Mili meter.

Components

(i) Strip plantations

(ii) Community woodlots

(iii) Farm forestry

(iv) Rehabilitation of degraded forest lands

Other components are popularisation of wood saving devices and rural development programmes.

SPECIES TRIALS

Species trials for farm forestry were undertaken on farmer's fields growing wheat, paddy etc. The studies at FRl brought out that *Bombax ceiba, Tectona grandis, Toona ciliata, Dalbergia sissoo, Grevillea robusta, Terminalia arjuna* are suitable tree species for planting on bunds under agroforestry conditions in Western UP and Haryana. Similarly *Grewia optiva* and *Bauhinia* species were suitable fodder tree species to combine with.

Restriction on felling and transport of trees: Market studies and social survey's by FRl, Dehradun revealed that legal restrictions on harvesting, transportation and marketing of tree products were found affecting individual's ability to enjoy benefits of tree planting. Various tree species under the group were found with different restrictions. Some species were declared reserved and their ownership vested with government only. For others, there were transport regulations only. Yet another group of species had both harvesting and transport regulations. These restrictions varied from state to state. Authority to grant permission also vested in different authorities in different states. The situation was complicated because of vested interests who prevent dissemination of proper information regarding restrictions and procedures. These affected the harvest benefits of tree planting to discourage farmers. Some departments have reduced restrictions on felling to encourage farmers. *Eucalyptus* and *Populus* spp. have been freed of restrictions in U.P., Haryana and Punjab. In U.P. there are Zonal restrictions on species like Khair. In Gujarat *Eucalyptus, Casuarina* and Babool are free from restrictions. These relaxation help only to limited extent. They affect movement of wood from one zone to other. Also the information about relaxed rules has not been adequately disseminated. Study of laws in various states and bottlenecks in their implementation need to be continuously evaluated with reference to progress of farm forestry programme in the states.

Availability of market infrastructure: No organised market infrastructure for the sale of farm forestry products exists and the farmers are exploited by middlemen. Middlemen thrive on legal restrictions and limited market opportunities through supply contacts. Few developed market centres like Saharanpur, Yamunanagar too are beyond reach of growers due to legal bottlenecks in felling and transport. Tree growers cooperatives were non-existent in entire nothern region. Markets for farm forestry products are not developed to cope up with the expansion of farm forestry. Due to crashed prices of pulpwood and fuel wood in some areas, and the farmers could not make the expected profits.

Capital requirements: Trees require a longer time frame to mature and generate income therefrom. The available financial institutions favour short term loans to support agriculture. Need for special provisions were felt to suit farm-forestry efforts by farmers. Insufficiency of funds had led to lack of after care of plantations which are essential for their survival.

JOINT FOREST MANAGEMENT

Of recent planners, managers and executers view Joint Forest Management (JFM) as the only alternative to preserve and conserve the existing forest resources. A survey was conducted to monitor the JFM in Haryana and the following constraints were observed: resource harvest and sharing of usufructs, village commitments to protection and regeneration of forest resources, dominance of management groups by powerful leaders, inter and intra village conflicts over common resources and legal and tenurial issues.

AGRICULTURAL CROPS ON THE HILL TERRACES. Species trial for various pines viz. *Pinus patula, P. greggii, P. taeda, P. kesiya* and *P. roxburghii* taken up in degraded and wasteland around Mussoorie indicated higher survival for *P. greggii* followed by *P. patula.* Species planted on shoulder bunds of terraces under small farmers production system showed better performance in respect of *Celtis australis* and *Robinia pseudoacacia.* The studies at IFGTB, Coimbatore indicated that *Azadirachta india* and *Gliricidia maculata* are found compatible under agroforestry conditions.

AGROFORESTRY STUDIES

Poplar cuttings of 30 cm length and 1.5 cm diameter gave maximum height and diameter. The cuttings soaked in water for 12 hours and kept in moist gunny bags improved the vigour of plants. FYM applied @ 400 quintals/ha gave the maximum height diameter growth. Data on costs and returns from Poplars (8 years and 12 years rotations) raised by Forest Department on Government lands and on cultivators fields in Central Tarai Forest Division, Haldwani and Rudrapur, indicated that raising of Poplar at 8 year rotation with agriculture crop on farm land is a more profitable proposition with benefit/cost ratio of 3.22 (with 12 percent interest in 8 year rotation) as against Poplar with and without agriculture crop on forest land,(benefit/ cost ratio 2.15 and 1.51 respectively).

FOREST BASED INDUSTRIES

In view of increasing demand of wood for various wood based industries in the country, apart from efforts for increasing production of raw materials, special emphasis is being attached to increasing efficiency of the primary wood processing industry viz., saw milling, wood seasoning, wood working, etc., and the secondary wood processing industry like pulp and paper and wood based panels. Further, restriction on total ban on export of round timbers have been imposed for certain species. Therefore, greater volumes of finished wood products like plywood, veneers, and other processed items are expected to be exported to the countries in place of round timbers being exported so far. Thus, there is ample scope for expansion in this sector.

SAW MILLING

The saw milling industry is very important and needs proper development to meet the present and future demands of timber in the country.

There are nearly 23,220 saw mills in the country employing nearly 1,39,650 workers. The installed capacity of these sawmills is estimated at 27.12 million m^3 per annum. Ninety eight percent of these saw mills are small and have a log-intake capacity of nearly 3,000 m^3 annually, i.e., on an average 10 logs per day, accounting for 82 per cent of the total sawn timber production in the country. These sawmills convert approximately, 13.48 million m^3 of logs annually and cater for different end uses as shown below:

Item	Sawn timber (%)
Construction	28
Box wood (packing)	18
Joinery	27
Furniture	11
Sleepers	8
Others	8

It is estimated that the conversion yield from log form to graded sawn timber is only 40-50 per cent which is considerably low compared to conversion achieved in the developed countries, which is 55 to 65 per cent.

Wood wastes of the conversions are rarely put to any commercial use, nearly 90 per cent of it being utilised as fuel wood. Efforts have been made to work out tables for economic conversion of logs.

WOOD WORKING INDUSTRIES

The demand of products of wood working industries viz., joinery, furniture and fixtures etc., for meeting the housing requirements of the population is increasing. Demand of products like textile mill accessories, sports goods, picture frames, battery separators and other items has increased manifolds. This sector is labour intensive and has tremendous employment potential and therefore should be encouraged. In wood processing industries capital/labour ratio is as low as 696.51 compared to capital/labour ratios in light metal fabrication (850.78), textiles (766.93), paper and paper products (1402.22). Semi-automatic technology which offers employment opportunities to traditionally skilled workers is most suited to this sector. Although there is ample man power available, there is shortage of trained personnel. Recent survey revealed that in this sector semi-skilled and unskilled workers lack formal institutional training and among the skilled only 3.3 percent have formal taining. Survey on the level of automation in furniture industry revealed that 60 per cent wood work is produced manually, 8.5 per cent in semi-automatic units and 31.5 per cent in units using automatic technology.

The products of wood working industries range widely from furniture and cabinet making, textile and jute mill accessories, sports goods, musical instruments, drawing and mathematical instruments etc., to wood carvings and handicrafts. Like other wood processing industries there is shortage of wood raw material necessitating search of suitable alternate species. Forest Research Institute, Dehra Dun is engaged in fulfilling the research needs of the industry in this regard and to a great extent it has succeeded too. Physical and mechanical properties, anatomical features, wood working behaviour etc., of a large number of wood species have been studied and recommended for different end uses. Indian Standards have also been formulated for a variety of items of wood working industry.

WOOD BASED PANELS

In India three main types of wood based panels viz., plywood including blockboards and flushdoors, fibreboards and particleboards are manufactured. Demand of such products has increased considerably in recent times. Wood based panels are now extensively being used in building construction, furniture, cabinet making, boat and ship building etc.. Wood based panel industry in the country is well established and to a great extent it has been possible through consistent research and development efforts made at the Forest Research Institute Dehra Dun.

The use of panel products, wherever possible, as a replacement for solid wood is desirable for several reasons. Primary wood species like Deodar *(Cedrus deodara),* Teak *(Tectona grandis)* etc., are in short supply. In view of the large demands, these have become costlier and beyond the reach of common man. On the other hand, bulk of the secondary non-durable wood like Mango *(Mangifera indica),* Chir *(Pinus roxburghii)* etc., which, are comparatively cheaper and abundantly available. However, these are upgraded and judiciously used in panel products. Thus, it ultimately helps in relieving heavy strain on our meagre forest resources. Forest wastes, logging wastes, wood wastes, and several other lignocellulosic materials can also be used in the manufacture of products like particleboard, fibreboard etc.

Timber and synthetic resin adhesives constitute the two basic raw materials for the manufacture of wood based panel products. Availability of sufficient quantity of wood has been a problem ever since the production of commercial plywood picked up in the country. There are 77 timber species recommended for manufacture of different grades of plywood and 20 species for decorative veneers. Earlier logs of over 120 cm girth were used to be accepted for veneering and now the scarcity of timber logs of even 60 cm girth are accepted. All out efforts are being made to raise plantation of fast growing species suitable for plywood.

The basic difference between the raw materials required for plywood and particleboard is that. former requires better quality of timber, while, for the latter wood wastes, including lops and tops of trees, wood residues from wood processing industry and other fibrous agricultural wastes can also be used. The nonwood fibrous material play an important role in fibreboard manufacture in reducing the cost of the product and in substituting wood which is becoming scarcer day by day.

Urea formaldehyde and phenol formaldehyde are the two important adhesives used in this industry. However, the cost of synthetic resins is high in the country as compared to developed countries. Therefore, there is a need to reduce the cost component of the resin and also that of wood raw material to make products more competitive.

Plywood : Plywood was introduced in India to substitute 2.5 cm thick wood tea-chests in the early twentieth century. Plywood tea-chest being light got wider acceptability. Plywood manufacturing units were set up here. Veneer for plywood, however, used to be imported. Necessity of data on veneer quality and gluing behaviour of Indian timbers was felt and over 100 different species were studied at F.R.I., Dehra Dun of which nearly 60 found use in the plywood manufacture. This has resulted in the vast expansion of the industry. Whereas, the production of plywood in 1947 was mainly confined to tea-chest plywood, in 1987 i.e., after a gap of 40 years the country is producing almost all grades of plywood and allied products valued approximately at Rs. 400 crores.

Particleboard: A large number of raw material including wood wastes and agricultural wastes, have been studied for their suitability of making particleboards using different binders, viz., cashew nut shell liquid, urea formaldehyde, phenol formaldehyde resins, and tannin adhesives. Experiments have been conducted on the suitability of bagasse, saw dust, pine needles and lantana etc., for developing particleboards. Effect of resin content on properties of particleboard and the relationship between resin content and density of wood on the properties of particleboard were also studied. Tempering of phenolic bonded particleboard by post heat and oil tempering improved the static bending properties and reduced the water absorption of the board. The particle board prepared from acetylated particles absorb less water and also show lesser swelling as compared to that prepared from unacetylated particles. Research data on suitability of various raw materials has to a great extent helped the industry in the country in meeting its material requirements and in formulating various Indian Standards on the products.

Fibreboard: A large number of raw materials, including wood wastes and agricultural residues have been tested for their suitability for hardboard and insulation board making. Hardboards have been developed from mixed species and also from grasses, reeds, barks lantana, etc. Recently *Eucalyptus* hybrid, spent chips of Khair, bagasse, Poplar, Su-babul etc., have been tested and found suitable. Effect of bark on properties of hardboards from *Eucalyptus* hybrid and *Populus deltoides* has been studied. Experiments to replace plywood in tea chests by hardboards were carried out. Hardboards from pine needle, and from mixture of pine needles and fir pulps, pine needle and lantana pulps for packing boxes for apples were prepared with satisfactory results.

PULP AND PAPER INDUSTRY

There are about 300 pulp and paper mills in India. The total production of paper and board including newsprint is around 1.8 million tonnes as against the installed capacity of about 3.3 million tonnes. The per capita consumption of paper, paper board and newsprint is about 3.3 kg at present. Growth of population, increase in development work in almost all sectors of industries, consequently increase in professional and technical jobs in the work force, and the growing economic activities in the country are bound to raise the consumption rate of pulp, paper, board and newsprint.

The regionwise locational distribution of paper mills (big and small) in the country is given below:

Locational distribution of paper mills

Region	Total mills %	Large paper mills %	Small paper mills %
Eastern	23.0	41.6	4.3
Western	27.3	14.4	40.2
Northern	22.0	8.3	34.8
Southern	27.7	35.7	20.7

Large paper mills are forest based, while small paper mills are based on agricultural residues (wheat straw, rice straw and bagasse etc.), grasses, waste paper.

The advent of pulping process by William Raitt at F.R.I. (1922-24) led to the establishment of bamboo as main raw material for pulp and paper making in India. It still continues to be the main raw material for pulping in the large mills. Although the annual consumption of bamboo by the pulp and paper industry now stands about 1.0 million tonnes, the demand would increase to 4.0 million tonnes in the next ten years, with the present pattern of raw material consumption. Bamboo forests in our country, except a few pockets, have been more or less fully exploited. A few pockets of bamboo forest not yet fully tapped are in the North-eastern hill states and the Bastar region in Madhya Pradesh: Availability of bamboo in the North-eastern region covering an, area of 125 million hectares could be about 2.5 million tonnes. However, 'the region lacks infrastructural facilities and therefore extraction of bamboo in these regions is expensive.

In view of the limited availability of bamboo, Indian paper industry has been using hardwoods which are locally available from forests or have been planted. Utilisation of hardwoods started in India in the early sixties based on the research work carried out by the F.R.I.. It is envisaged that, for pulp, paper and newsprint manufacture, utilization of bamboo and hardwood would continue to dominate.

Requirement of forest raw material for meeting the future demand for pulp, paper, paper board and newsprint as worked out by the Development Council for Paper, Pulp and Allied Industries is given below. These projections are based on the assumption that 30 per cent of the total requirement would be met from agricultural residues and recycled fibres.

Requirement of forest based raw materials.

Year	Requirement of paper and paper board (million tonne)	Requirement of raw material (million air dry tonne)	Availability (million air dry tonne)	Shortfall (million air dry tonne)
		Paper and paper board		
1986	1.76	3.45	2.85	0.60
1991	2.24	4.40	2.85	1.55
1996	2.80	5.50	2.85	2.65
2000	3.40	6.70	2.85	3.85
		Newsprint		
1986	0.48	0.67	0.36	0.31
1991	0.62	0.87	0.36	0.51
1996	0.80	1.12	0.36	0.76
2000	1.03	1.45	0.36	1.09

2.8 air dry tonne = 1 tonne of paper/paper board

2.0 air dry tonne = 1 tonne of newsprint.

The availability of forest raw material, which is already in short supply, being inadequate to sustain even the present level of production, is expected to become further critical. At present, a higher proportion of bamboo is being utilized in the raw material furnish.

The situation of shortage of forest raw materials led to establishment of a number of small pulp and paper mills based on research mainly carried out at F.R.I. on agricultural residues and other secondary raw material which have now assumed a significant role in meeting the demand of cellulose fibre for the paper industry. The most important of these are: rice straw, wheat straw, bagasse, grasses, jute, kenaf and waste paper.

To meet the estimated requirement of 3.40 million tonnes per annum of paper and boards and 1.03 million tonne per annum newsprint, about 8.15 million tonnes per annum of fibrous raw material would be required by the end of the century. The fast depleting forest cover coupled with an ever increasing price trend will only worsen the scenario.

Increasing Ecological consciousness further restricts the exploitation of forests. Therefore, in order to maintain present tempo of production and keep the prices of end-product in check, adequate emphasis for improving the availablity of raw material and maximizing pulp and paper production from available raw materials has to be given.

Large pulp and paper mills: A major constraint in the growth of paper industry is its dwindling raw material base of forest origin. It has, therefore, been recognised that pulpwood plantation specific to the requirements of the paper industry have to be taken up. However, before this effort yields results, there may be an intervening period when acute scarcity of raw material is likely to be felt. F.R.I. has done extensive work on production of Eucalyptus, Poplar, Su-babul, Casurina and their technical assesment for pulp and paper manufacture. Work done at F.R.I on production of improved high yield pulps as a measure to maximise production has led to development of new processes for making the lignin hydrophilic and consequently increasing the strength/ surface properties of high yield pulps.

A majority of paper mills were established in the country 30 to 50 years ago and predominant practice is to produce pulp through batch digester using kraft process. The industry could not yet reap the advantage of the better technology i.e. continous digesters, mainly because of the following reasons; (a) higher outlays for machinery replacements in the existing units; (b) relatively smaller size of Indian mills renders uneconomical use of continous digester. Due to low thermodynamic efficiency requirement of power in terms of coal and electricity is very high, as a result of which the supply of energy required on sustained basis has always been a problem for number of pulp and paper mills. High cost of inputs, lack of infrastructural support, and high incidence of taxation has been attributed to the poor financial health of the industry.

Rayon grade pulp industry: At present there are five units in the country producing about 2-5 lakh tonnes of rayon grade pulp annually from bamboo and hardwoods by both prehydrolysis kraft and sulphite processes. The country is importing 0.15 million tonnes of dissolving pulp made out of spruce, beech, birch imported from Europe and Canada.

In Asia India is one of the pioneer country in production of Rayon grade pulp. It has given technical know how to Indonesia, Thialand and Phillipines in setting up dissolving pulp plants based on the local available raw materials. Capacity of plants are being increased and pulp industries are also setting up integrated staple fibre and cellulose derivatived plants.

OTHER INDUSTRIES

Other forest based industries that merit mention are Katha and Cutch, Rosin and Terpentine, Bidi, Match box and Sticks" Oxylic acid from tree bark, Sandal wood oil, Eucalyptus oil, Agarbatti, lac, Silk etc. ICFRE institutes are engaged in research for augmenting the raw material production from suitable species and their optimum utilisation.

35

Forest Research Centres of India

FOREST RESEARCH INSTITUTE, DEHRADUN

This Institute caters to forestry research needs of Uttar Pradesh, Punjab and Haryana. The Institute continues to provide leadership in research areas relating to plant systematics, forest mycology and pathology (including mycorrhiza), forest entomology, forest soils and application of bio-technology in forestry.

The Institute is a 'Deemed University' and presently offers three post-graduate courses, viz., Plantation Technology, Pulp and Paper Technology and Wood Science and Technology. It also has a number of research scholars on its roll who are pursuing research work towards the award of doctorate degrees. Research on clonal multiplication of promising phenotypes sign of (phenotypes/ genotypes) genotypes of many important species like Chir pine, Shisham, Eucalyptus, Teak etc., for raising quality planting material are presently being pursued and an appreciable headway has been made.

Juvenile shoot production was stimulated by "pinning down" and "hedging" techniques for use as vegetative propagules. Pinned down seedlings produced a larger number of axillary shoots, of uniform size, as compared to hedged plants, where variation in the size of the shoots was more. Species that responded to the pinning down technique were - *Dalbergia sissoo, Celtis australis, Grewia optiva, Bauhinia variegata* while *Ulmus laevigata. Cassia* responded poorly to this technique.

A low cost vegetative propagation unit was developed which consists of a wooden frame covered with polythene sheets. Different rooting media were used in side this structure and it was found that sand and small pebbles, of less than 2 mm in diameter, gave results at par with vermiculite. Juvenile cuttings of *Ulmus laevigata, Celtis australis, Acer oblongum, Quercus leucotrichophora, Dalbergia sissoo, Michelia champaca, Bauhinia variegata* were successfully rooted in this chamber. Juvenile cuttings of Ul*mus laevigata* and *Acer oblongum* were rooted in open beds with over 85 and 80 percent success respectively.

Seeds of *Prosopis cineraria* belonging to 14 sources/provenances were irradiated with continuous and fractionated doses of gamma rays ranging from 10 to 80 Kr. Stimulation in growth was observed in dose 10 Kr. (Fractionated) and 40 Kr. (continuous). Cytological studies were conducted on root tips of seedlings raised from gamma ray irradiated seeds.

Seed/shoot cuttings/root cuttings from *Dalbergia sissoo* plus trees from Haryana and U.P., were collected for clonal multiplication.

Plantations/Forests of 25 forest tree species of West Bengal were surveyed along with state silviculturists to identify seed stands, plus trees and establish seed orchards. More than 40 seed stands of 23 species and a few plus trees of each species were chosen for seed collection/genetic improvement work.

Methods for mass clonal multiplication by nodal segment culture have been developed for *Eucalyptus* hybrid, *Dalbergia sissoo, Melia azedarach, Gmelina arborea* and *Azadirachta indica.* A method, using cladode segments has been developed which can be very useful for large scale production of superior clonal material of *Casuarina equisetifolia* and similar cladode bearing species. Protocols have been developed for vegetative propagation of *Dendrocalamus hamiltonii* and *Bambusa arundinacea* two important bamboo species by macroproliferation method. The method is simple, cos⁺ effective and easily understandable by the field staff and lay-man.

Detailed descriptions of 40 species of Andamans have been written up and keys for identification and punched card key features prepared. Epidermal studies under light microscope have been completed on two genera viz., *Melocalamus* and *Thamnocalamus.*

Efforts were continued to restore lime stone mined areas. After successful establishment of grass-shrub cover on these sites, planting of *Cupressus* sp. were taken up (2 x 2m spacing). Studies on survival and vegetation dynamics were taken up to ascertain changes in plant population over the last three years. Soil samples were collected and analysed for moisture and nutrient status. Litter production, nutrient cycling and biomass production in reclaimed rock phosphate mines is also being studied.

The effect of distillery effluents on the growth of tree species *Acacia catechu, Dalbergia sissoo* and *Morus alba* which are commonly planted under social forestry schemes is also being investigated.

VAM and rhizobium are being screened at FRI for their associations with seedlings of tree species in nurseries and tested for best VAM rhizobium seedling combinations towards increasing productivity. Culture of VAM and rhizobium has also been initiated for distribution to the State Silviculturists. Techniques have been developed for bulk culturing of *Glomus musseae,* G. *fasciculatum* and G. *aggregatum. Acacia nilotica* was found to respond positively to dual inoculation of G. *fasciculatum* and rhizobium increasing biomass yields up to 174 percent.

Lentinus edodes, an edible Japanese mushroom was successfully cultivated on oak logs at Solan and about 5 Kg mushrooms were harvested during 1992. Application of growth hormones i.e., IAA and IBA and GA3 induced early development of sporophores.

Efforts at enhancing forest productivity are being supplemented by researching protection measures against insect pests and diseases. Studies have been successfully conducted to combat pests of *Holoptelea integrifolia, Bombax ceiba* and coneworm of *Picea smithiana.* Investigations on biological control of insect pests in plantations, particularly of poplars, have indicated the promise of some of their natural enemy complexes as control agents and an appreciable headway has been made in culturing the potential enemy species populations for release in the field to test their efficacy.

Predation studies on *Isyndus heros* Fab. a new predator of *Heteropsylla cubana* revealed that the assassin bug can devour 20 jumping plant lice per day.

Over the years, this institute has done pioneering research work in the field of forest product utilisation. The current emphasis is on utilisation of juvenile timbers of different species being harvested from short rotation plantations. To enhance utility of such timbers, development of suitable sawing methods, seasoning schedules, preservative treatments and research on technology for manufacturing modified woods - laminated wood and medium density fibre board etc., based on such material, have been initiated. Commendable work has been done in the development of sawing techniques and utilization of juvenile timbers of Eucalyptus and Poplar species, as a result of which, these species have now found new markets in the manufacture of furniture, carved panels, door and window frames. Lesser known species are also being tested for various commercial end uses as substitutes for traditional timber utilised by cottage industries in furniture making and wood carving.

The scientists of Cellulose and Paper division are researching on environment friendly bleaching processes, bio-pulping and chemical modification of spent liquor for manufacturing industrial chemicals. Semi-chemical kraft pulps from *Populus deltoides* were produced by optimising pulping conditions and 60 per cent yield obtained. Pulps were. bleached using oxygen and calcium hypochlorite under varying conditions and chemical charges. The pulps obtained were of good strength properties. Chemi-thermomechanical pulping of *Eucalyptus tereticornis* was carried out by pretreatment of chips with NaOH under varying concentration and refining conditions. Modification of these pulps by oxygen and H_2O_2 yielded highly improved pulps with respect to strength/optical properties. To reduce the kappa number and pollution, oxygen was used as bleaching agent in place of chlorine. The pulps were analysed for pentosan, lignin, kappa number and strength properties. Further work on multistage bleaching system is in progress. Milled wood lignin isolated from control and decayed *Eucalyptus globulus* wood were subjected to physico-chemical and functional group analysis. Nitrobenzene oxidation products were characterized by HPLC. Some other research fields relate to developing processes for de-inking of newspaper, recycling of waste paper from offices, computers and old magazines.

The non-wood forest product are of utmost importance for the socio-economic upliftment of forest dwellers. Special efforts are being made to identify and standardize cultivation practices, grading and processing methods for medicinal plants and oil seed species of forest origin. Wood and other wastes from forests, e.g., bark, leaves, fruits, seed etc. is also being investigated for recovery of chemicals, oils, gums etc. of industrial value. The bark of *Anogeissus pendula* and *Acacia catechu* has been found to yield economical quantities of oxalic acid, the technology for which has been perfected on pilot scale trials. Similarly, leaves of some tree species have been found to contain chemicals with high therapeutic value and have been referred to the drug laboratories for testing.

Picrorhiza kurrooa, Nardostachys jatamansi, Orchis latifolia, Aconitum heterophyllum, Aconitum napellus and *Onosma echioides* being endangered species were planted in nurseries for their *ex situ* conservation and studies on their active ingredients.

A study was conducted to assess the resin yield from the healed surfaces of previously tapped wounds (tapped by cup and lip method) by rill method which revealed a possibility of very high yields from such surfaces nearly 5.65 kg/tree/season. Trees tapped earlier by cup and lip method in the Pine forest of U.P., H.P. and J&K could be profitably retapped by rill method of tapping to augment resin production.

TROPICAL FOREST RESEARCH INSTITUTE, JABALPUR

Tropical Forest Research Institute, Jabalpur has the responsibility for meeting forestry research needs of the states of Maharashtra, Madhya Pradesh and Orissa and, in addition, research on non-wood forest products and forest protection at the national level. The major research fields covered by this Institute pertain to silviculture and management of teak, sal, bamboos and other important species of tropical deciduous forests, forest fires, grazing, forest conservation and non wood products in relation to tribal development.

The Institute's area of operation corrals some of the country's most valuable tropical forests having immense biological diversity and a large tribal population. A number of mountain chains namely Vindhyans, Satpuras and Amarkantak adorn the landscape of the area harnessing life giving waters which are channelled by these watersheds into the rivers Narmada, Tapti, Mahanadi and Godavari. The forests of this region, therefore, have a multiple role to perform, regulation of water, production of valuable forest goods and providing socio-economic security to the forest dwellers or tribals.

Among the notable research achievements of the Institute are:

(*a*) Establishment of a herbal garden in the Institute, housing over 500 species of medicinal plants some being rare and threatened. This garden serves as an educational centre for students of indigenous systems of medicine and a live reference collection. It also helps to extend cultivation practices of these herbs among progressive farmers of the region, in addition to being the gene resource for propagation and research.

(*b*) A survey conducted in the National parks at Bandhavgarh (M.P.) and Melghat (M.S.) revealed the occurrence of 230 medicinal plant species in these areas. The phenology and taxonomy of these plants was studied. *Gloriosa superba, Dioscorea daemona, Chlorophytum arundinaceum, Litsea* sp., *Trichosanthes cucumerina, Oroxylum indicum* and a few other threatened species, which have a large market potential, were found to occur in these areas. Based on study, recommendations were made to the Maharashtra forest department to initiate steps for the *in situ* conservation of these medicinal plants especially the threatened species of Melghat forests which have been accepted and being implemented.

(*c*) Research conducted by the Institute on harvesting of tendu leaves; collection methods, storage and drying of non–wood forest products like mahua, aonla, belguda and medicinal plants have helped towards optimal utilization of these products and conservation of these resources.

(*d*) *Bursera penicillata* an important essential oil yielding plant, adapted to degraded sites, has been successfully introduced in the Institute's campus and near Raipur in Madhya Pradesh.

(*e*) The causal fungus for blight disease responsible for heavy mortality in *Bambusa nutans* cultivated by the farmers of the coastal belt of Orissa was identified as *Sarocladium oryzae,* and control measures developed to contain the mortality.

(*f*) Thirteen new fungal diseases were recorded for the first time in the country whereas 75 new fungal diseases were recorded in M.P. The taxonomy of *Cercospora boswelliae,* C. *carissai, Didymella tandonii, Phomopsis indica, Phaeoseptoria paulowniae* have been described for the first time in India. Three seed borers of *Prosopis juliflora, Lagerstroemia parviflora* and *Bauhinia variegata* were also recorded for the first time in the Indian subcontinent.

(*g*) Mass multiplication of VAM and *Rhizobium* was initiated for inoculating seedlings in nurseries and plantations towards increasing productivity in the region.

(*h*) Studies conducted on biological control of some insect pests and fungal diseases have revealed that:

(*i*) The microbe *Bacillus thuringiensis* could effectively control larvae of *Atteva fabriciella*, the defoliator of *Ailanthus excelsa* and *Tectona grandis.*

(*ii*) Extracts of plants *Cuscuta reflexa* and *Vitex negundo* were effective in the control of Fusarium wilt of *Albizia lebbek* and *Dalbergia sissoo.*

(*iii*) Leaf extract of Marigold and culture filtrate of *Aspergilius terreus* and Neem seed oil controlled *Funalia leonina* a fungus causing wood decay.

(*i*) Species trial on skeletal soil of Sambalpur brought out that *Albizia procera, Gmelina arborea, Peltophorum ferrugineum* and *Dalbergia sissoo* were the most suited species for afforesting such lands.

(*j*) Experiment in bhata land near Raipur (M.P.) indicated that *Emblica officinalis, Gmelina arborea* and *Hardwickia binata* perform the best on lateritic soils.

(*k*) Studies on suitability of tree species for planting mine overburdens and their growth performance indicate that the species *Pithecellobium dulce, Acacia auriculiformis, Derris indica* and *Eucalyptus camaldulensis* are the best suited for mined areas.

(*l*) A silvipastoral model with two fodder grasses viz., *Pennisetum pedicellatum and Panicum maximum raised in the plots of Dalbergia sissoo and Daris indica* was developed. The dry mass yields of *P. pedicellatum* and *P. maximum* were recorded to be 15.37 tonnes/ ha and 12.73 tonnes/ ha respectively.

ARID FOREST RESEARCH INSTITUTE, JODHPUR

Located in the heart of Thar desert at Jodhpur, Arid Forest Research Institute was created in the year 1988 to conduct forestry research primarily towards controlling desertification and improving land productivity of arid lands. The institute, in general is to meet the forestry research needs of Rajasthan, Gujarat and Dadar Nagar Haveli. The Institute focuses on research in the fields of traditional land use systems and their improvement, agroforestry, sand dune stabilisation, silvipastures; afforestation of wastelands, such as arid saline lands, the raun of Kachchh, Arawali hills and sandy plains, development of irrigated plantations alongside Indira Gandhi Nahar Pariyojana (IGNP); regeneration and management of natural forests.

The main research studies pursued during the year are as follows:

(*a*) Phenology and silviculture of important trees and shrubs of the arid region. The studies included phenology, seed morphology and biology, germination, nursery practices, vegetative propagation, etc. The tree species covered being *Tecomella undulata*, *Acacia* spp., *Prosopis cineraria* and *Tamarix* spp. and the shrubs *Calligonum polygonoides, Capparis* spp., *Haloxylon salicornicum* and *Atriplex* spp.

(*b*) The efficacy of seven species were studied for the stabilisation of shifting sand dunes trials at the end of 18 months indicated that *Prosopis juliflora, Acacia tortilis* and *A. planifrons* perform better than *Zizyphus nummularia, Tecomella undulata, Prosopis cineraria* and *Faderbida albida* which performed poorly.

(*c*) Studies on moisture management in dry areas were undertaken for determining the most effective method of moisture harvesting. Of the different methods tried inter-row ridges were found to be the most effective for harvesting rain water as was evidenced by the biomass production in the species *Azadirachta indica, Prosopis cineraria* and *Tecomella undulata* associated with these experiments.

(*d*) Of the various soil working techniques followed in plantations in the arid areas, trench cum mound and ring pit method of soil working were found to be superior compared to others as evidenced by the growth response of three species raised namely *Albizia lebbek, Azadirachta indica* and *Prosopis cineraria.*

(*e*) In the arid zone, it is customary to provide some irrigation during the initial years of planting. An experiment was conducted to ascertain the optimum watering regime which indicated that *Prosopis cineraria* and *Tecomella undulata* did not respond to watering while *Azadirachta indica, Albizia lebbek* and *Acacia nilotica, A. planifrons* responded positively to watering and mulching.

(*f*) Along IGNP area, studies were conducted on application methods and frequency of irrigation, observations at the end of 30 months indicate that the method basin flooding with 30 mm water was the best irrigation system.

Nutrient management studies in arid areas indicated that addition of farm yard and phosphorus fertilizer together provides better results than addition of farm yard manure only.

Similarly, application of N and P together helped to provide better growth than addition of only nitrogen.

(*g*) Several species of the exotic *Atriplex* and some local species namely *Salvadora* spp., *Tamarix* spp. were used for the revegetating highly saline lands. Of the species introduced *Atriplex lentiformis* indicated the best promise providing higher survival and biomass production.

(*h*) Provenance research has been initiated in the species *Acacia nilotica, Tecomella undulata* and *Azadirachta indica.* Preliminary observations indicate that *Acacia nilotica* provenance from Gurgaon and Hastinapur perform better in the desert region.

(*i*) Studies on neem seed storage indicated that seeds mixed with ash stored in polybags demonstrated higher retention of viability.

(*j*) Research on micropropagation and macro propagation of some species were initiated and chambers for tissue culture/propagule hardening were erected.

(*k*) Insect and disease survey has highlighted several damaging insect pests and diseases. Biological control measures for controlling babul defoliator *Taragama siva* using dipterous parasite are being investigated.

(*l*) Studies on peoples participation in social forestry programmes were initiated and are in progress, in Jodhpur and Mehsana districts.

(*m*) Research relating to screening and multiplication of VAM fungi associated with important tree species has been taken up recently and being pursued vigorously.

INSTITUTE OF RAIN AND MOIST DECIDUOUS FORESTS RESEARCH, JORHAT

Institute of Rain and Moist Deciduous Forests Research, Jorhat was established in the year 1988 with the objective to strengthen forestry research in the states of Assam, Nagaland, Meghalaya, Manipur, Tripura, Mizoram, Arunachal Pradesh, Sikkim and Andaman and Nicobar Islands. The research fields include basic research in the Rain and Moist Deciduous forests and Mangrove forest ecosystem, forest hydrology, ecology, ethnobotany and shifting cultivation. As the institute is still in its formative stage preliminary work towards achieving the ultimate objectives have been initiated by establishing and developing research capabilities in the various disciplines relating to the ecosystem components of forests of north-eastern states.

Mycorrhizal associations of forest tree species of the moist zone are being investigated for improving nursery stock and productivity of plantation being raised in the eastern zone. Seasonal variation in Vascular Arbuscular Mycorrhiza (VAM) spore populations and root colonization in the tree species *Mesua ferrea, Aquilaria agallocha, Lagerstroemia speciosa, Artocarpus chama, Artocarpus lacucha* were investigated and associated VAM spores identified and multiplied.

Monitoring in the nurseries and field for incidence of pathological diseases in the following species was conducted, *Acacia auriculiformis, Ficus lepidosa, F. rumphii, Emblica officinalis, Bambusa* spp., *Mensonia dipikae, Samanea saman, Mimusops elengi, Albizia procera, Artocarpus lacucha.*

Systematic studies on the leaf spot disease occurring in the seedlings of *Mensonia dipidae* in forest nurseries and blight disease of Bamboo plantation were undertaken. Biological control of pests of Bamboos and other tree species like *Aquilaria agallocha, Gmelina arborea* is being attempted using natural enemy complexes comprising viruses and other insects.

The work on micro and macro propagation of two Bamboo species' namely *Dendrocalamus hamiltonii* and *Bambusa tulda* has been initiated and success achieved in producing propagules *in vitro* using tissues from seedlings. Steps are underway on *in vitro* culture of active rhizome buds and shoots for seedling production and many culture medias are being tested towards initiating differentiation.

Studies relating to nutrient cycling in different forest stands shows that soil nutrients are maximum in natural mixed forest stands and higher in comparison to Bamboo and other forest stands composed of pure *Mesua ferrea* and *Dipterocarpus macrocarpa.* Litter fall studies in natural forest indicated maximum leaf litter fall during December to February. Species like *Mesua ferrea, Castanopsis* spp. produced maximum litter. Apart from these studies detailed taxonomical work on Jorhat district flora and ethnobotany has also been undertaken.

The institute is actively engaged in implementing the UNDP project in 5 villages in Jorhat district. Socio-economical survey of these five villages has been completed.

INSTITUTE OF FOREST GENETICS AND TREE BREEDING, COIMBATORE

The primary objective of the Institute is to conduct, at the national level, research on all aspects of genetic improvement and tree propagation of forest species. It also conducts research in eco-restoration of western ghats, mangrove forests of the west and the east coasts, and conservation of genetic diversity of the tropical evergreen forests of western ghats and is responsible for attending to the overall forestry research needs of the states of Tamil Nadu, Kerala and Pondichery.

The Institute has initiated both short term and long term projects related to forest genetics and tree breeding. The short term projects include selection from natural populations and plantations; provenance testing and establishment of first generation clonal seed orchards of *Tectona grandis, Casuarina equisetifolia, Acacia nilotica, Albizia lebbek, Azadirachta indica, Bambusa* sp., *Tamarindus indica, Eucalyptus grandis, Bambusa bambos (Bambusa arundinacea), Bambusa vulgaris* and *Dendrocalamus* sp. The long term projects relate to their reproductive biology. A botanical garden/ arboretum is maintained by the Institute and has a collection of 350 species of tropical forest plants, some of which are rare, threatened and endemic. Nearly 150 different species of medicinal plants of high value are also maintained in the garden.

The Institute designed a low cost double roof coir mat shade house for rooting cuttings using conventional approaches and for hardening of tissue culture raised seedlings, achieving improved success in vegetative propagation of *Acacia nilotica, Albizia lebbek, Gliricidia sepium;* nodal cuttings of *Eucalyptus;* cladodes of *Casuarina equisetifolia.*

Micropropagation rooting protocols for *Acacia. nilotica* and seedlings of *Dendrocalamus strictus* have been achieved in MS liquid medium supplemented with BAP in stationary cultures. Using combinations of micro and macro propagation protocols, a provenly superior tree of *Eucalyptus tereticornis* was cloned by rejuvenating and subsequent multiplication in the mist chamber.

Research on time of seed collection and maturity indices for the species *Azadirachta indica, Acacia nilotica, Hardwickia binata, Derris indica, Tamarindus indica* and *Albizia lebbek* were undertaken. Two methods of drying of seeds were experimented, viz., drying out in the sun and drying in the shade. Seeds dried out in the sun showed greater variation in moisture content for the seeds of *Azadirachta indica* and *Derris indica.* Though seeds are stored best at moisture level of 8% $\frac{1}{n}$ 11 %, these two species indicated a higher requirement of moisture level, over 20%.

Studies of Teak plantations of Tamil Nadu and Kerala were conducted to estimate biomass productivity and nutrient status of soils at different ages which was correlated with productivity. Regression equations were fitted between easily measurable parameters and biomass.

Multipurpose tree species were screened for their suitability for intercropping with agricultural crops. Initial results have indicated *G. maculata* and *Azadirachta indica* to be the most promising agroforestry species.

Plantations of *Eucalyptus globulus, Pinus patula* and *Acacia mearnsii* growing in the Nilgiris were sampled for estimates on their biomass productivity at different ages and their nutrient status have also been investigated for correlations with productivity.

Research on ecorestoration of degraded lands in the western ghat ecosystem indicated the suitability of multipurpose trees *Acacia planifrons* and *A. nilotica* for rehabilitation and fuel wood production. *Casuarina equisetifolia* plantations were studied for their growth performance, productivity at different ages and nutrient dynamics.

Studies were initiated on evaluation, selection and application of *Mycorrhizae* and root nodule *microsymbionts* in forestry species. Three different soil micro-organisms, i.e., Rhizobium, Azospirillum and *Phosphobacterium* were selected and the inoculum was applied to the roots of the seedlings of *Derris indica, Albizia lebbek, Tamarindus indica, Acacia ferrugenia, Ceiba pentandra, Delonix regia, Cassia siamea* and *Mellisoma hopensis* with promising results.

Studies on pest problems in nurseries and young plantations indicated *Selepa celtis, Dasychira mendosa* and a lepidopteran leaf webber to be causing considerable destruction in nurseries. Search for biocontrol agents has revealed the presence of two species of endoparasitoids viz., *Coccygomimus* sp. and *Apanteles taprobanae* and one species of ectoparasitoid namely *Euplectrus euplexiae* operating as biocontrol agents of *Selepa celtis* causing upto 15% pest mortality. The predators *Stegodyphus* sp. (Araneae) and two species of Reduvids were also found functioning as predators of plantation defoliator insects. The parasitic and predatory potential of many biocontrol agents are being evaluated and suitable candidates reared in the laboratory. Preliminary screening of plants raised from six seed sources from Tamil Nadu, for insect resistance indicated that plants of Vathalagundu origin exhibit maximum resistance closely followed by Mannar Koil origin.

Studies on the profitability of cultivation of different forestry species, marketing position, social and institutional acceptance of developed technology and man-forest interaction are also being conducted by the Institute.

A survey of five villages under UNDP programme was conducted in Coimbatore district of Tamil Nadu for evaluating farmes' needs and choice of tree species. The assessment based on this evaluation will form the basis for planning tree introduction on farm lands.

INSTITUTE OF WOOD SCIENCE AND TECHNOLOGY, BANGALORE

Institute of Wood Science and Technology, Bangalore, caters to the research needs of the states of Karnataka, Andhra Pradesh and Goa. The main research thrusts, in the institute was earlier confined to the disciplines of wood properties and utilization, wood seasoning and preservation, chemistry of forest products and wood biodegradation. From the current year research on tree improvement, silviculture, propagation, production and economics, etc. has been resumed and strengthened.

The Institute maintains three field stations for forestry research in Karnataka state, and two outstation marine centres at Vishakapatnam and Cochin (Kochi).

The Institute is fully equipped for research and possesses a timber seasoning kiln, preservative treatment plant, wood workshop, mist chamber and greenhouse, and research nurseries. Expertise and facilities exist for soil and plant analysis (for both macro-and micro-nutrients), seed testing, propagation (including mist chamber) and for investigations on pests and diseases, identification and testing of timbers and wood products. The Institute is well recognised for its expertise on identification of terrestrial and marine wood-borers, marine fouling organisms, and wood-decaying fungi and bacteria.

Amongst the main achievements of the Institute are, research on anatomy of 40 lesser known Andaman timbers; their properties, uses and the publication thereon; information on properties of 15 timbers of Malaysian species and one, *i.e, Pterocarpus angolensis* from Africa; studies on enzymatic hydrolysis of saw dust bark and wood shavings for production of higher value goods, i.e., animal feed, soil conditioners, and as a source of carbohydrates; studies on *Eucalyptus* bark for manufacture of oxalic acid; investigations on biocidal activities of some plant extractives e.g., red sandars have shown anti-bacterial and hypoglygemic activity in preliminary tests and is being further investigated; evaluation of several natural gums mixtures for their burning and coal defusing properties to substitute 'Jigat' in Agarbatti manufacture; researches on wood seasoning and preservation relating to rubber wood a species prone to warping; improvement in seasoning and preservation treatment schedules for rubber wood and for enhancing resistance to fungi and insects. Simple methods for treating freshly felled poles by sap displacement method were popularised among the people of Hasan district. Studies on relation between moisture content and natural resistance of timbers against powder post beetles and termites under field conditions indicated *Acacia auriculiformis* and *Eucalyptus camaldulensis* to demonstrate high resistance against infection even after 36 months. Use of alcoholic extract of *Eucalyptus tereticornis* bark as a weedicide for the control of parthenium weed and *Hopea parviflora* wood extractives to be anti wood against fungi and termite. The copper-chrome-boron (CCB) a chemical of low toxcity compared to C.C.A. was tested for its efficacy against termite attack of wood. It was found that C.C.B was a better preservative than C.C.A. Seasonality of a recruitment and intensity of attack of wood boring and fouling organisms were monitored at 3 borers in Andhra Pradesh and a number of wood borers encountered and identified. Ecological studies in respect of these borers are being studied. Wood borers from test panels of various species and marine structures including fisheries boats along the Goa coast and in Cochin waters were monitored and examined. Preservative treated timbers were tested at Vishakapatnam harbour for standardising doses of preservative application.

Destruction of mangrove vegetation along the Goan coast by Marine Organisms are under investigation. Fifteen species of marine borers were identified and their ecology, posts specificity and associate borers are being studied. Considerable data has been generated on harmful effects of settlement by fouling organisms like birnacles, oysters and gastropodes on mangrove saplings. The damage to mangrove regeneration was found to be severe and nearly 90 percent.

Different phenotypes of sandal wood from 79 superior trees identified from Karnataka, Tamil Nadu, Andhra Pradesh and Kerala have been collected at the field stations of Gottipura, Nanlal and Yelawala (Mysore). Clonal seed orchards and progeny trials of sandal wood have been scientifically laid for improving the species. Seed of promising clones from the seed orchards has been distributed to the state forest departments and the private agencies for raising plantations. Clonal propagation of trees, bamboos and other priority species using propagules from superior trees has been initiated by the Institute.

The research has been organised on sandal viz., nursery practices, cultural practices, genetics and tree improvement, bio-ecology, control of spike diseases, micro and macro propagation and management. A research laboratory at Vishakapatnam is being established for providing trials for utilization of secondary species for making catamarans and their preservation.

Five villages have been identified in Bangalore and in Chittur district for socio-economic survey. Two Scientists of the institute were awarded the Ron Cockcraft award during the year 1992-93 by International Research Group of Wood Preservation, Sweden.

TEMPERATE RESEARCH CENTRE, SHIMLA

Temperate Research Centre, Shimla is responsible for the forestry research needs of the Hill

States of Himachal Pradesh and Kashmir. Its research mandates relate to regeneration of temperate forests afforestation/rehabilitation of the cold deserts, forest conservation and ecological rehabilitation of the Himalayas.

The major thrust areas of research of this Centre are artificial regeneration of Silver· Fir and Spruce; seed storage; nursery and planting technology; nutrient cycling in coniferous forests of western Himalayas; development of afforestation/rehabilitation techniques for the cold desert utili sing indigenous species like junipers, Hippophae etc.; management of / alpine pastures; and cultivation of some medicinal and aromatic plants of the temperate zone.

CENTRE FOR FOREST PRODUCTIVITY, RANCHI

Centre for Forest Productivity, Ranchi is responsible for catering to the forestry research needs relating to the states of Bihar and West Bengal. The centre has initiated ecological research on the mangroves of Sunderbans and Eastern Himalayan forest ecosystems. Other spheres of centres research activity pertain to the disciplines of development of lac cultivation and extension; plantation forestry; social and agro-forestry; rehabilitation of degraded forests; non–wood forest products; cash crops of forest origin; and socio-economic inputs from forests towards tribal development.

CENTRE FOR SOCIAL FORESTRY AND ENVIRONMENT, ALLAHABAD

Centre for Social Forestry and Environment, Allahabad, caters to the research needs of Gangetic plains of D.P., Bihar and northern Vindhyan tracts of Madhya Pradesh. The main research thrust of this Centre pertains to ecological rehabilitation of gangetic plains and northern Vindhyas, biological rejuvenation of soil of high salt concentration, afforestation of ravines and development of watersheds. The centre is mainly concentrating on various aspects of social forestry.

Glossary of Medical Terms

Abortifacient. An agent that produces abortion.

Acne. A pimple–like eruption of the sebaceous glands of the skin, with accumulation of yellow secretion and black overgrowth of the horny layer of the skin.

After pains. Painful contraction of the womb after child birth.

Alopecia. A disease of the scalp resulting in complete or partial baldness.

Alterative. A drug which corrects disordered process of nutrition and restores the normal function of an organ or of the system.

Amenorrhoea. Abnormal suppression of menses.

Anaemia. A deficiency of blood or of red blood-cells.

Anasarca. Dropsy.

Angina pectoris. A disease of the heart marked by severe constricting pains in the chest.

Anodyne. A drug that relieves pain.

Antacid. A drug which neutralizes the acidity of the gastric juice.

Anthelmintic. A drug that kills intestinal worms.

Antihydrotic. A drug which checks sweating.

Antilithic. A drug which counteracts the development of stone.

Antiperiodic. A drug that cures periodic attacks.

Antiphlegistic. A drug which counteracts inflammation.

Antipyretic. A drug which reduces fever.

Antiscorbutic. A drug which cures scurvy.

Antispasmodic. A drug which counteracts spasmodic disorders.

Aperient. A mild purgative.

Aphrodisiac. A drug which promotes sexual desire.

Aphthae. Minute white ulcers on the tongue and in the mouth.

Apoplexy. Sudden loss of consciousness with some paralysis.

Ardor urine. A burning sensation on urinating.

Aromatic. A drug which is fragrant, spicy and mildly stimulant.

Ascaris. Intestinal parasitic round worms.

Asthma. A chronic disorder of the bronchial tubes.

Astringent. A drug which checks secretion or bleeding.

Atony. Lack of muscular power.

Bechic. A remedy for cough.

Bedsores. Ulceration on any part of body exposed to pressure of bed-ridden patient.

Beriberi. A deficiency disease caused by lack of vitamins especially B_1.

Bronchitis. An inflammation of the air passages.

Bronchorrhoea. Excessive discharge from the bronchial mucous membrane.

Calculus. A hard and solid concretion formed in the body, especially in the urinary organs; it may be sand, gravel or stone.

Cancer. Any malignant growth.

Caries. Decay of teeth.

Carminative. A drug which relieves flatulence.

Cathartic. A drug which induces active movement of the bowels.

Cholagogue. A drug which promotes flow of bile.

Colic. Pain due to spasmodic contraction of the abdomen.

Congestion. An abnormal collection of blood in the blood vessels of any organ or part of the body.

Coujunctivitis. Inflammation of the conjunctiva, the mucous membrane covering the eyeball and lining the eyelids.

Contusion. An injury to the soft parts without breaking the skin.

Cystitis. Inflammation of the bladder.

Dandruff. An inflamed condition of the scalp characterized by the presence of white scales in the hair due to the exfoliation of the horny cells of the scalp.

Demulcent. An agent having a soothing effect on the skin and mucous membranes.

Deobstruent. A drug that removes an obstruction to secretion or excretion by opening the natural passages or pores of the body.

Depilatory. An agent that removes or destroys hair.

Diabetes. A wasting disease of metabolism; abundant sugar is present continuously in the urine.

Diaphoretic. A drug that induces copious perspiration.

Diphtheria. An infectious disease of the throat and the air passage.

Discutient. A drug which disperses or absorbs a tumour or any coagulated fluid in the body.

Diuretic. A drug which increases the secretion and discharge of urine.

Dropsy. A disease marked by an excessive collection of a watery fluid in the tissues or the cavities of the body.

Dysentery. An infectious disease of which the chief symptoms are acute diarrhoea and discharge of mucus and blood.

Dysmenorrhoea. Usually painful and difficult menstruation.

Dyspepsia. Indigestion.

Dysuria. Painful and difficult urination.

Eczema. A skin disease accompanied by swelling, redness and exudation of lymph.

Elephantiasis. A disease of the skin caused by a tiny worm and attended with hypertrophy of the affected parts.

Emetic. A drug which induces vomiting.

Emmenagogue. A drug which promotes menstruation or regulates the menstrual periods.

Emollient. A drug which allays irritation of the skin and alleviates swelling and pain.

Enteritis. Inflammation of the intestines.

Epilepsy. A chronic nervous disorder marked by attacks of unconsciousness or convulsions.

Expectorant. A drug that promotes the removal of catarrhal matter and phlegm from the bronchial tubes.

Febrifuge. An agent used for reducing fever.

Fistula. An abnormal channel which connects one cavity of the body with another, or which opens out from a cavity to the surface of the body.

Flatulence. A disorder in which there is an excessive collection of the gas in the stomach.

Freckles. Coloured spots on the exposed parts of the skin.

Galactagogue. An agent that promotes secretion and flow of milk.

Gleet. A chronic discharge from the urethra.

Glycosuria. A diseased condition of the urine in which sugar is excreted.

Goitre. A chronic enlargement of the thyroid gland.

Gonorrhoea. An infectious venereal disease marked by an inflammatory discharge from the genital organs.

Haemoptysis. Spitting of blood from the lungs or bronchial tubes.

Haemorrhage. Bleeding, especially profuse, from any part of the body.

Heartburn. A burning feeling in the regions of the chest and stomach, generally due to indigestion.

Hemiplegia. Paralysis of one side of the body.

Hepatic. Pertaining to the liver.

Hepatitis. Inflammation of the liver.

Hernia. Rupture; protrusion through its covering of any organ of the body.

Hypnotic. A drug which induces sleep.

Hypochondriasis. A mental disorder in which the patient is tormented by meloncholy views, particularly about its health.

Hysteria. A disease in which a physically healthy patient has lost control over acts and feelings and suffers from imaginary ailments.

Intermittent fever. Fever which is marked by intervals of normal temperature between periods of rise of temperature.

Itch. An infectious skin disease, caused by a mite, without specific lesions and marked by excessive itching.

Jaundice. A diseased condition in which there is a yellowish staining of the tissues and excretions with bile.

Lactagogue. Galactagogue.

Laryngitis. Inflammation of the larynx.

Leprosy. A chronic wasting disease caused by germ; the disease generally results in mutilations and deformities.

Leucoderma. A condition of the skin in which there is loss of pigment wholly or partially.

Lithontriptic. A drug used for removing calculi of stones formed in the urinary system.

Malaria. A recurrent disease marked by bouts of shivering, sudden rise of temperature and general aching of the body.

Menorrhagia. Abnormally excessive menstruation.

Metrorrhagia. Bleeding from the womb.

Migraine. Periodic attack of headache affecting one side of the head.

Narcotic. A drug which induces deep sleep.

Nausea. A feeling that vomiting is about to take place.

Nephritis. Inflammation of the kidney.

Neuralgia. Pain felt above a nerve.

Night blindness. A disease in which the patient is incapable of seeing in the dark.

Ophthalmia. Conjunctivitis.

Orchitis. Inflammation of the testicles.

Otitis. Inflammation of the ear.

Paralysis. A disease in which there is loss of power of voluntary movement in any part of the body.

Pectoral. A drug to cure disorders of the chest.

Pharyngitis. Inflammation of the pharynx.

Phythisis. Consumption; tuberculosis of the lungs.

Piles. An inflamed condition of the veins in the rectal region.

Pneumonia. Inflammation of the lungs.

Prophylactic. An agent that prevents disease.

Psoriasis. A common chronic inflammation of the skin, marked by rounded reddened patches which are covered with dry silvery scales.

Pyrrhoea. A disease marked by purulent discharge from the gums.

Refrigerant. A drug which relieves feverishness or produces a feeling of coolness.

Rheumatism. An indefinite term used for pains in the muscles, joints and certain tissues.

Ringworm. A parasitic skin disease usually marked by red, scaly, circular patches.

Rubefacient. A mild counter-irritant.

Scabies. An itching skin disease caused by a mite.

Scorbutic. Suffering from scurvy.

Scurvy. A deficiency disease due to lack of vitamin C.

Sedative. A drug which reduces excitement, irritation and pain.

Sialagogue. A drug which promotes salivation.

Soporific. A drug that induces sleep.

Stomachic. A drug that strengthens the stomach and promotes its action.

Stomatitis. Inflammation of the mouth.

Styptic. An agent which checks bleeding.

Syphilis. A chronic venereal disease.

Tetanus. An infectious disease, marked by painful contraction in the muscles.

Tonsillitis. Inflammation of the tonsils.

Typhoid fever. An acute infectious disease characterized by ulceration of the intestines, eruption of rose-coloured spots, and a typical course of temperature.

Ulcer. An open sore on the skin.

Urethritis. Inflammation of the urethra.

Vermifuge. A drug which expels intestinal worms.

Vulnerary. A drug which promotes healing of wounds.

Whooping cough. An acute infectious disease of coughing.

Question Bank

LONG ANSWER QUESTIONS

1. Give an account of drugs obtained from plants (15 minutes; 15 marks).

2. Assume that you and a small group of your friends are transported to an uninhabited island in the Indian Ocean to live and that you are permitted to take the seeds or propagative organs of 12 species of plants with you. List them in order of preference (mention the botanical name and family) and give reasons for your selection (10 mts.; 10 mks.).

3. By means of outline sketches only, show the morphology of the economically important of any five of the following plants: saffron, opium, poppy, cocoa, black pepper, citrus, coffee, and clove; give the botanical name of each source and comment on the uses (15 mts.; 15 mks.).

4. State the botanical names, families and morphology of the parts from which the following are obtained: (*a*) cotton, (*b*) jute, (*c*) coir, (*d*) hemp. How would you differentiate one from the other? (15 mts.; 15 mks.).

5. Name five important drug plants of India and write detailed account of any two of them. (15 mts.; 15 mks.).

6. List the chief fatty oil-yielding crops in India. Classify the oils on the basis of their drying properties and mention their uses. What do you mean by the term hydrogenation. (15 mts.; 15 mks.).

7. By simple sketches only, show the morphology of the economically important parts of the following: clove, cashew, coriander and black pepper. Mention their latin names and uses. (15 mts.; 15 mks.).

8. What are essential oils? Give a brief account of the botany, cultivation, extraction and uses of important essential oils produced in India. (15 mts.; 15 mks.).

9. Give the method of extraction, properties and economic uses of the following: (*a*) groundnut oil, (*b*) linseed oil and (*c*) castor oil. (10 mts.; 10 mks.).

10. Name any *four* fibre yielding plants, giving their families. Describe the methods of fibre extraction from them and their commercial uses. (15 mts.; 15 mks.).

11. Give the methods of cultivation of any *one* of the following: (*a*) cotton; (*b*) rubber. (10 mts.; 10 mks.).

12. Explain how the latex is processed to get rubber? (10 mts.; 10 mks.).

13. The whole coconut tree is economically useful. Justify the above statement. (20 mts.; 10 mks.).

SHORT ANSWER QUESTIONS

1. Write what do you know of the source and economic importance of the cork. (3 minutes; 3 marks).

2. Write what do you know of the source and economic importance of coconut. (5 mts.; 5 mks.).

3. List the botanical names of six plants from Cruciferae (Brassicaceae) used as vegetables and mention the morphology of the useful part in each. (5 mts.; 5 mks.).

4. Name the species of cottons exploited for commercial purposes. How would you distinguish between the Old World and New World cottons? (5 mts.; 5 mks.).

5. How is opium obtained and how is it used? (5 mts.; 5 mks.).

6. Write note on any *one* of the following:

(*a*) *Oryza*. (5 mts.; 5 mks.).

(*b*) *Rauwolfia*. (5 mts.; 5 mks.).

(*c*) *Cannabis*. (6 mts.; 6 mks.).

7. Give botanical names of *three* important timber-yielding plants of India. (3 mts.; 3 mks.).

8. Name the sources of any *one* of the following products; mention the plant part from which they are obtained and indicate the main areas of cultivation of these plants in India.

(*a*) Ephedrine. (3 mts.; 3 mks.).

(*b*) Tobacco. (3 mts.; 3 mks.).

(*c*) Clove. (3 mts.; 3 mks.).

(*d*) Saffron. (3 mts.; 3 mks.).

9. Describe the medicinal importance of *Adhatoda*, *Zingiber* and *Penicillium*. (6 mts.; 6 mks.).

10. What is the morphology of the edible part in (*i*) orange, (*ii*) potato, (*iii*) papaya? (3 mts.; 3 mks.).

11. Distinguish between any *one* pair of the following:

(*a*) Ring porous and diffuse porous. (3 mts.; 3 mks.).

(*b*) Sap wood and heart wood. (3 mts.; 8 mks.).

(*c*) Beedi and cigarette. (3 mts.; 3 mks.).

(*d*) Green tea and black tea. (3 mts.; 3 mks.).

(*e*) Lint and fuzz. (3 mts.; 3 mks.).

12. Name a plant each yielding a worldwide commodity which owes its origin to (*i*) epidermis, (*ii*) tuberous stem, (*iii*) periderm, and (*iv*) whole leaf.

13. what plant materials (mention Latin names) are used in manufacture of essential oils. (3 mts.; 3 mks.).

14. Write short note on any *one* of the following:

(*a*) Indigenous raw materials of paper production. (5 mts.; 5 mks.).

(*b*) Wood seasoning. (5 mts.; 5 mks.).

(*c*) Molasses. (5 mts.; 5 mks.).

(*d*) Papain. (5 mts.; 5 mks.).

15. How would you distinguish any *one* of the following pairs:

(*a*) *Curcuma* from *Zingiber*. (3 mts.; 3 mks.).

(*b*) Starch of wheat from that of rice and maize (3 mts.; 3 mks.).

(*c*) Essential oil from fatty oil. (3 mts.; 3 mks.).

(*d*) Fibres of cotton from those of jute. (3 mts.; 3 mks.).

16. Give the botanical names, morphological nature of plant parts yielding economic products, and uses of any *one* of the following:

(*a*) Sugarcane. (3 mts.; 3 mks.).

(*b*) Jute. (3 mts.; 3 mks.).

(*c*) Potato. (3 mts.; 3 mks.).

(*d*) Wheat. (3 mts.; 3 mks.).

17. Write botanical names of any *six* plants in which either the stem or the leaf is of economic importance.

18. Explain giving suitable reasons for the following:

Farmers generally use leguminous plants, with nodulated roots, for rotation of crops. (5 mts.; 5 mks.).

19. Write briefly: Extraction of essential oils from rose and Sandal wood. (3 mts.; 3 mks.).

20. In India cotton textile industry is largely confined to Bombay and other parts of Western India, whereas jute industry is mostly restricted to Calcutta and other parts of West Bengal. Comment. (5 mts.; 5 mks.).

21. Write short note on any one of the following:

(*a*) Bye-products of sugar industry. (5 mts.; 5 mks.).

(*b*) Extraction of essential oils. (5 mts.; 5 mks.).

(*c*) Important Indian spices. (5 mts.; 5 mks.).

22. *Cannabis sativa* hits the front lines in many of the leading magazines of the world. Comment. (5 mts.; 5 mks.).

23. Write a short note on Soybean (5 mts.; 5 mks.).

24. Distinguish between any *one* of the following pairs:

(*a*) Pine wood and teak wood. (3 mts.; 3 mks.).

(*b*) Fennel and Coriander. (3 mts.; 3 mks.).

(*c*) Essential oil and fatty oil. (3 mts.; 3 mks.).

(*d*) Flue curing and sun curing. (3 mts.; 3 mks.).

25. Name the principal states in India in which the following are cultivated:
(*i*) Jute, (*ii*) coconut, (*iii*) tea, (*iv*) mango, and (*v*) black pepper. (3 mts.; 3 mks.).
26. How does milling affect the quality of rice, (3 mts.; 3 mks.).
27. Name drugs of plant origin which could be used in the following:
Malaria, hypertension, rheumatism, heart disorders and dysentery. (3 mts.; 3 mks.).
28. Write what you know of the source and economic importance of Ginger. (3 mts.; 3 mks.).
29. Write briefly: Tobacco and health hazards. (5 mts.; 5 mks.).
30. Write briefly: Harvesting, grading and processing of tea. (5 mts.; 5 mks.).
31. Write briefly: Origin of wheat (5 mts.; 5 mks.).
32. Write briefly: Principal uses of wood. (5 mts.; 5 mks.).
33. Write briefly: Extraction of sugar and by-products of sugar industry. (5 mts.; 5 mks.).
34. What are the principal sources of vegetable oils in India. List the *four* major uses of oils.
35. Give the botanical name of the plant yielding any *one* of the following product and explain its morphology and economic importance:
(*a*) Reserpine. (5 mts.; 5 mks.).
(*b*) Coffee. (5 mts.; 5 mks.).
(*c*) Rose-wood. (5 mts.; 5 mks.).
36. Name a useful plant product that is derived from (*i*) epidermis, (*ii*) whole leaf, (*iii*) mesocarp, (*iv*) periderm, (*v*) tuberous root and (*vi*) heart wood.
37. Mention some algae as source of food to human beings. (3 mts.; 3 mks.).
38. Mention the importance of Gramineae (Poaceae) as a source of food. (3 mts.; 3 mks.).
39. Write notes on: Economic uses of tannins. (6 mts.; 6 mks.).
40. Write any *three* economic uses of Euphorbiaceae members. (3 mts.; 3 mks.).
41. Outline the processing of coffee by wet method. (3 mts.; 3 mks.).
42. What kind of soil will be preferred for growing rubber. (2 mts.; 2 mks.).
43. How will you start to process the Paddy grains soon after harvesting them (Mention only the first two steps.). (2 mts.; 3 mks.).
44. Mention 5 rubber yielding plants and their families, cultivated in India. (2 mts.; 5 mks.).
45. The natural habit of tea is tree but under cultivation it is a shrub. How is it so? (2 mts.; 2 mks.).
46. Mention the other uses of Paddy, other than the grains being used as a staple food. (2 mts.; 2 mks.).
47. Give the technical name of the fruit and mention the morphology of the useful parts and their actual use in the fruits of the following:
(*a*) Mango, (*b*) Cashew and (*c*) Jack fruit. (8 mts.; 12 mks.).
48. State what you know about the morphological nature of the following: (*i*) Apple, (*ii*) Cauliflower, (*iii*) Radish.
49. Give the technical name of the fruit and mention the morphology of the edible portion of the following:
(*i*) Tomato, (*ii*) Rice, (*iii*) Coco, (*iv*) Apple and (*v*) Groundnut. (5 mts.; 10 mks.).
50. Which is the active principle in coffee? (2 mts.; 2 mks.).
51. How is the mesocarp of coconut fruit used in industry? (3 mts.; 2 mks.).
52. Name any *two* economic uses of cotton other than using it in textile industry. (2 mts.; 3 mks.).

SIMPLE QUESTIONS

1. Arrange the following in the order of their food value: tapioca, groundnut, soybean, potato, rice and *Citrus*. (2 mts.; 2 mks.).
2. Name *five* plant products which earn foreign exchange for our country (2 mts.; 2 mks.).

3. Mention the titles and names of authors of four books on economic botany. (2 mts.; 2 mks.).

4. What is the nature of the chief food reserve in (*i*) tapioca, (*ii*) mango, (*iii*) soybean, (*iv*) maize. (2 mts.; 2 mks.).

5. Which of the Indian states are the leading producers for Cane sugar, Sandal wood, Potato, Jute and Teak. (2 mts.; 2 mks.).

6. What are the 'active' chemical principles in the following: Coffee, Ergot, Opium, Rauwolfia and Tobacco? (3 mts.; 3 mks.).

7. What plant materials (mention Latin names) are used in the manufacture of any *one* of the following:

(*a*) Sugar. (2 mts.; 2 mks.).
(*b*) Spice powder for flavouring foods. (2 mts.; 2 mks.).
(*c*) Paper. (2 mts.; 2 mks.).
(*d*) Soap. (2 mts.; 2 mks.)

8. Name the Research Institute along with its location where improvement work in any *one* of the following crop plants is being carried out in India:

(*a*) Jute. (1 mt.; 1 mk.).
(*b*) Wheat. (1 mt.; 1 mk.).
(*c*) Sugarcane. (1 mt.; 1 mk.).
(*d*) Rice. (1 mt.; 1 mk.).
(*e*) Potato. (1 mt.; 1 mk.).

9. Give the botanical name from which the following is obtained:

(*a*) Caffeine. (1 mt.; 1 mk.).
(*b*) Papain. (1 mt.; 1 mk.).
(*c*) Morphine. (1 mt.; 1 mk.).
(*d*) Hashish. (1 mt.; 1 mk.).

10. Which of the Indian states are leading producers of the following:

(*a*) Tea. (1 mt.; 1 mk.).
(*b*) Tobacco. (1 mt.; 1 mk.).
(*c*) Rubber. (1 mt.; 1 mk.).
(*d*) Banana. (1 mt.; 1 mk.).
(*e*) Belladonna. (1 mt.; 1 mk.).
(*f*) Deodar. (1 mt.; 1 mk.).

11. What is the centre of the origin of the following:

(*a*) Coffee. (1 mt.; 1 mk.).
(*b*) Cocoa. (1 mt.; 1 mk.).
(*c*) Clove. (1 mt.; 1 mk.).
(*d*) Cinchona. (1 mt.; 1 mk.).
(*e*) Groundnut. (1 mt.; 1 mk.).
(f) Black pepper (1 mt.; 1 mk.).
(*g*) Eucalyptus. (1 mt.; 1 mk.).
(*h*) Mango. (1 mt.; 1 mk.).
(*i*) Red pepper. (1 mt.; 1 mk.).
(*j*) Soybean. (1 mt.; 1 mk.).

12. Mention one important source (botanical name) for the following:

(*a*) Atropine. (1 mt.; 1 mk.).
(*b*) Cigar. (1 mt.; 1 mk.).
(*c*) Match sticks. (1 mt.; 1 mk.).
(*d*) Tapioca. (1 mt.; 1 mk.).

13. How is hydrogenation of oil done? (2 mts.; 2 mks.).

14. In what ways does paddy differ from rice? (2 mts.; 2 mks.).
15. Where did maize originate? (2 mts.; 2 mks.).
16. What makes a wood heavy? (2 mts.; 2 mks.).
17. Give the botanical name and the morphology of the product used in
(*a*) Cinchona. (1 mt.; 1 mk.).
(*b*) Groundnut. (1 mt.; 1 mk.).
(*c*) Rubber. (1 mt.; 1 mk.).
(*d*) Coir. (1 mt.; 1 mk.).
(*e*) Sorghum. (1 mt.; 1 mk.).
(*f*) Cardamum. (1 mt.; 1 mk.).
(*g*) Castor. (1 mt.; 1 mk.).
18. Describe how *Cichorium intybus* is used as an economically important product. (2 mts.; 2 mks.).
19. Name any *two* fibre-yielding plants. (2 mts.; 2 mks.).
20. Give the botanical name which yields the following: (*a*) Rose wood, (*b*) Satin wood, (*c*) Sandal wood, (*d*) Laurel wood. (3 mts.; 4+ mks.).
21. Name *two* essential oil-yielding plants. (2 mts.; 2 mks.).
22. Describe the morphology of the useful part in: (*a*) *Ganja*. (2 mts.; 2 mks.), (*b*) Taro. (2 mts.; 2 mks.), (*c*) Cardamom (2 mts.; 2 mks.), (*d*) Groundnut (2 mts.; 2 mks.), (*e*) Pepper (2 mts.; 1 mk.), (*f*) Coriander (2 mts.; 1 mk.), (*g*) Cloves. (2 mts.; 1 mk.).
23. Name the plant which yields Quinine. How is it extracted? (5 mts.; 5 mks.).
24. Give the botanical name for the following:
(*a*) Bengal gram, (*b*) Green gram, (*c*) Black gram, (*d*) Red gram. (1 mt.; 4+ mks.).
25. Give the botanical name of the following:
(*a*) The plant produces Aconitum, (*b*) The plant produces cabbage, (*c*) The plant produces Melia oil, (*d*) The plant produces Bengal gram. (2 mts.; 2 mks.).
26. Name a plant for which desuckering is done. (1 mt.; 1 mk.).
27. Mention one cultivated species and one wild species of sugarcane. (1 mt.; 1 mk.).
28. Give the use of bagasse. (1 mt.; 2 mks.).
29. What is meant by retting? (2 mts.; 1 mk.).
30. Which plant produces chicory? (1 mt.; 1 mk.).
31. Give the name of the plant which produces coffee. (1 mt.; 1 mk.).
32. Describe the morphology of the useful parts of cotton. (2 mts.; 2 mks.).
33. Give the names of any *two* major rice research stations, their location in India. (2 mts.; 2 mks.).
34. Describe the morphology of the useful part in groundnut and cotton. (2 mts.; 2 mks.).
35. Arrange in order of highest protein value: (*a*) Soybean, (*b*) Potato, (*c*) Rice, (*d*) Wheat, (*e*) Sugarcane.
36. Give one example each of plant/plant-products used in the manufacture of following: (*a*) Macroni, (*b*) rum, (*c*) morphine, (*d*) haematoxylin. (2 mts.; 2 mks.).

MULTIPLE CHOICE

1. Botanical name of jute is:
(*a*) *Cannabis sativa* (*b*) *Linum usitatissimum*
(*c*) *Corchorus capsularis* (*d*) None of the above
(1 minute; 1 mark)
2. Teak wood which is obtained from *Tectona grandis* belongs to the family of:
(*a*) Labiatae (lamiaceae) (*b*) Verbenaceae
(*c*) Rubiaceae (*d*) Apocynaceae
(1 mt.; 1 mk.).

3. *Digitalis purpurea* belongs to the family of:
(*a*) Papilionanceae (*b*) Acanthaceae
(*c*) Scrophulariaceae (*d*) Solanaceae
(1 mt.; 1 mk.).

4. Sunhemp is obtained from:
(*a*) *Crotolaria juncea* (*b*) *Cannabis sativa*
(*c*) *Musa textiles* (*d*) *Cocos nucifera*
(1 mt.; 1 mk.).

5. Cloves are obtained from *Eugenia aromatica* and the morphology of the product being:
(*a*) product petioles (*b*) dried pedicels
(*c*) dried flower buds (*d*) dried seeds
(1 mts.; 1 mk.).

6. Botanical name of the plant that yields pararubber:
(*a*) *Ficus elastica* (*b*) *Hevea brasiliensis*
(*c*) *Manihot glaziovii* (*d*) *Castilla elastica*
(1 mt.; 1 mk.).

7. Reserpine, a drug is extracted from:
(*a*) *Rauwolfia serpentina* (*b*) *Ferula asafoetida*
(*c*) *Atropa belladona* (*d*) *Digitalis purpurea*
(1 mt.; 1 mk.).

8. Plant parts useful for extraction of opium from *Papaver somniferum* are:
(*a*) young seedlings (*b*) old leaves
(*c*) unripe fruits (*d*) ripened seeds
(1 mt.; 1 mk.).

9. Botanical name of cauliflower is:
(*a*) *Brassica oleracea* var. *botrytis* (*b*) *Brassica obracea* var. *gongylodes*
(*c*) *Brassica obracea* var. *capitata* (*d*) *Brassica obracea* var. *gemmifera*
(1 mt.; 1 mk.).

10. The common name of *Eleusine coracana* is:
(*a*) Ragi (*b*) Barley
(*c*) Wheat (*d*) Oats

11. The cotton fibre from cotton is obtained from:
(*a*) roots (*b*) stems
(*c*) seeds (*d*) leaves

12. The commercial jute fibres are:
(*a*) Phloem fibre (*b*) Xylem fibres
(*c*) Interxylary fibres
(1 mt.; 1 mk.).

13. Cotton fibre is derived from:
(*a*) Phloem fibres (*b*) Epidermal hairs on seed
(*c*) Outgrowth on the stem (*d*) Sclerenchymatous cells
(1 mt.; 1 mk.).

14. Paddy is suitable for cultivation in:
(*a*) Red soils (*b*) Dry soils
(*c*) Black soils (*d*) Irrigated soils
(1 mt.; 1 mk.).

15. The economic product of tobacco plant is:
(*a*) Flowers (*b*) Leaves
(*c*) Roots (*d*) Stems
(1 mt.; 1 mk.).

16. Fibre yielding plant is:
(*a*) *Triticum* (*b*) *Pennisetum*
(*c*) *Gossypium* (*d*) *Rauwolfia*
(1 mt.; 1 mk.).

17. The quality of tobacco depends mainly on:
(*a*) Curing process (*b*) Variety and curing process
(*c*) Climatic conditions in which plant grows. (*d*) Nutrition of the plant in the field
(1 mt.; 1 mk.).

18. Opium is obtained from:
(*a*) dried leaves (*b*) roots
(*c*) latex from unripe capsules (*d*) seeds that are fried
(1 mt.; 1 mk.).

19. Coir is obtained from
(*a*) roots of date palm tree (*b*) leaf bases of areca nut palm
(*c*) mesocarp of coconut (*d*) leaves of *Cocos nucifera*
(1 mt.; 1 mk.).

20. The centre of origin of rice plant is:
(*a*) India (*b*) Indo-Malayan region
(*c*) India and Africa (*d*) Africa
(1 mt.; 1 mk.).

21. Rubber is collected from:
(*a*) Crushing the stem of *Euphorbias*. (*b*) Tapping the stem of *Carica papaya*
(*c*) Tapping the stem of *Hevea brasiliensis* (*d*) Crushing the fruits and collecting the latex of *Achras sapota*
(1 mt.; 1 mk.).

22. Essential oils are those:
(*a*) oils which are essential for human beings. (*b*) oils which are essential to the plants which produce them
(*c*) oils which are used as lubricants (*d*) oils which yield perfumes
(1 mt.; 1 mk.).

23. Asafoetida is obtained by:
(*a*) exudation from the stem (*b*) extraction from the fruits
(*c*) extraction from the root (*d*) extracted from the leaves.
(1 mt.; 1 mk.).

24. Short fibres are known as:
(*a*) lint (*b*) fluff
(*c*) fuzz (*d*) flint (1 mt.; 1 mk.).

25. Long fibres are known as:
(*a*) flint (*b*) lint
(*c*) fluff (*d*) fuzz
(1 mt.; 1 mk.).

26. The bark of *Cinnamomum zeylanicum* is used as spice because of:
(*a*) A sweet flavour (*b*) Aromatic oils secreted
(*c*) Ethereal oils secreted
(1 mt.; 1 mk.).

27. Corms and root tubers of Araceae have a pungent acrid taste because they contain:
(*a*) Aromatic oil (*b*) Oxalate crystals
(*c*) Alkaloid (*d*) Latex
(1 mt.; 1 mk.).

28. The composition of the cotton fibre is:
(*a*) cellulose (*b*) callose
(*c*) chitin (*d*) pectin
(1 mt.; 1 mk.).

29. Tea can be grown:
(*a*) at sea level (*b*) sea shore
(*c*) in elevated areas (*d*) in dry places
(1 mt.; 1 mk.).

30. If tea leaves are kept in hot water for longer period, the liquid becomes bitter because:
(*a*) of the defect in the tea leaves (*b*) the volatile oil in the leaves dissolves out
(*c*) the thein dissolves out (*d*) the tannin dissolves out
(1 mt.; 1 mk.).

31. Tea and coffee can be classified as:
(*a*) distilled beverages (*b*) non-alcoholic beverages
(*c*) fermented beverages (*d*) alcoholic beverages
(1 mt.; 1 mk.).

32. The products of commercial importance, yielded by coconut palm are:
(*a*) latex and oil (*b*) rubber and wood
(*c*) oil and fibre (*d*) fibre and rubber
(*e*) medicines
(1 mt.; 1 mk.).

33. Potatoes are usually propagated by vegetative means because:
(*a*) they do not produce seeds (*b*) by this method it is possible to maintain genetic quality
(*c*) by this method incidence of diseases may be reduced (*d*) potato seeds have long dormancy period
(1 mt.; 1 mk.).

34. The banana plant is:
(*a*) tree (*b*) shrub
(*c*) herb
(1 mt.; 1 mk.).

35. The true stem of banana is:
(*a*) Bulb (*b*) Rhizome
(*c*) Corm
(1 mt.; 1 mk.).

36. The major composition of banana fruit is:
(*a*) starch (*b*) protein
(*c*) glucose (*d*) fat
(1 mt.; 1 mk.).

37. Commercial tea is prepared from:
(*a*) stem (*b*) leaves
(*c*) flowers (*d*) root
(*e*) bark
(1 mt.; 1 mk.).

38. Rubber is obtained from:
(*a*) cell sap (*b*) gum
(*c*) resin (*d*) latex
(1 mt.; 1 mk.).

39. The commercial coffee is prepared from:

(*a*) leaves (*b*) flowers

(*c*) fruits (*d*) root

(*e*) seed

(1 mt.; 1 mk.).

40. Most of the rubber plants belong to the family

(*a*) Euphorbiaceae (*b*) Cannaceae

(*c*) Rubiaceae (*d*) Rutaceae

(1 mt.; 1 mk.).

41. Latex cells occur in:

(*a*) xylem (*b*) cambium

(*c*) bark (*d*) cortex

(1 mt.; 1 mk.).

42. The artificial ripening of banana is done by:

(*a*) Keeping them in a room having temperature below 0^0C.

(*b*) Keeping them at room temperature.

(*c*) Keeping them in a room where the temperature is high.

(1 mt.; 1 mk.).

43. Coconut oil derived from the copra can be classified as:

(*a*) essential oil (*b*) drying oil

(*c*) semi-drying oil (*d*) vegetable fat.

(1 mt.; 1 mk.).

44. What is the source of chewing gum:

(*a*) Sugar from sugar-cane (*b*) Gum arabic from *Acacia* spp.

(*c*) Latex from *Achras sapota* (*d*) Fluid from *Musa* spp.

(1 mt.; 1 mk.).

Bibliography

LIST OF BOOKS

Bailey, L.H. Manual of Cultivated Plants. *The Macmillan Co., New York*, 1949.

Barrett, O.W. The Tropical Crops. *The Macmillan Co., New York*, 1928.

Beaven, E.S. Barley. *Duckworth, London*, 1947.

Bentley, R. and Trimen, H. Medicinal Plants. *J. & A. Churchill, London*, 4 vols, 1880.

Blank, F.C. Handbook of Food and Agriculture. *Reinhold Publishing Corp., New York*, 1955.

Blatter, E. Palms of British India and Ceylon, *Oxford University Press, London*, 1926.

Blow, C.M. & Stokes, Natural Rubber Latex, No. 2—Latex Casting. *Rubber Development Board, London*, 1952.

Bor, N.L. Manual of Indian Forest Bombay, *Oxford University Press, London*, 1953.

Bourdillon, T.F. The Forest Trees of Travancore, *Government of Travancore*, 1908; reprinted, 1937.

Brautlecht, C.A. Starch; Its Sources, Production and Uses. *Reinhold Publishing Corp., New York*, 1953.

Burkill, I.H. A Dictionary of the Economics Products of the Malay Peninsula. *Crown Agents for the Colonies, London*, 2 vols., 1935.

Chopra, R.N. Indigenous Drugs of India; Their Medicinal and Economic Aspects. *The Art Press, Calcutta*, 1933.

Chopra, R.N. Indigenous Plants of India, *Manager of Publications, Delhi*, 1949.

Chopra, R.N. et al. Glossary of Indian Medicinal Plants. *C.S.I.R. New Delhi*, 1956.

Choudhari, B.L. Vegetable Gardening in the Plains. *Industry Publishers, Ltd., Calcutta*, 1947.

Colthrust Ida. Familiar Flowering Trees in India. *Thacker, Spink & Co. Ltd., Calcutta*, 1937.

Cotton in India. A Monograph. 4 vols. *Ind. Coun. Agri. Res., New Delhi.*

Cowan, A.M. & Cowan, J.M. The trees of Northern Bengal. *Govt. of Bengal, Calcutta*, 1929.

Dalziel, J.M. The Useful Plants of West tropical Africa. *Crown Agents for the Colonies, London*, 1937: reprinted 1948.

Dastur, J.F. Medicinal Plants of India and Pakistan. *D.B. Taraporevala Sons & Co. Ltd., Bombay*, 1951.

Dastur, J.F. Useful Plants of India and Pakistan. *D.B. Taraporevala Sons & Co. Ltd., Bombay*, 1951.

Desch, H.E. Manual of Malayan Timbers, Vol. II. *Malaya Publishing House Ltd., Singapore, Malayan Forest Records*, No. 15, 1954.

Dhingra, D.R. Development of Essential Oil Industry in Uttar Pradesh; a summary of the work done under Essential Oil Scheme at *H.B. Technological Institute, Kanpur, revised edn.*, 1958.

Dijkman, M.J. Hevea; Thirty Years of Research in the *Far East University of Miami Press, Florida*; 1951.

Duthie, J.F. & Fuller, J.B. Field and Gardan Crops of the N.W. Provinces and Oudh. *Govt. of N.W. Province & Oudh*, 3 vols., 1882—1893.

Dutt, C.P. & Pugh, B.M. Principles and Practices of Crop Production of India. *Allahabad Agricultural Institute, Allahabad*, 1940.

Dymock, W., Warden C.J. & Hopper, D. Pharmacographia Indica, *Trubner & Co. London*, 1890—99.

Eckey, E.W. Vegetable Fats and Oils. *Reinhold Publishing Corp. New York*, 1954.

Edlin, H.L. British Plants and Their Uses. *B.T. Batsford Ltd., London*, 1951.

Finnemore, H. The Essential Oils. *Ernest Benn Ltd., London*, 1926.

Gamble, J.S. A Manual of Indian Timbers, *Sampson, Low, Marston & Co. Ltd., London*, 1922.

Gildemeister, E. and Fr. Hoffmann. The Volatile Oils. Longmans. *Green & Co. London*, vol. 2, 1916.

Gopalaswamiengar, K.S. Complete Gardening in India. *The Hosali Press, Bangalore, revised edn.* 1951.

Griffiths, C. Oils, Fats and Waxes. *Science Publications (Great Britain) Ltd., London*, 1954.

Guenther, E. The Essential Oils. *D. Van Nostrand Co. Ltd., New York*, 6 vols. 1948—52.

Haarer, A.E. Jute Substitute Fibres. *Wheatland Journals Ltd., Great Britain*, 1952.

Harris, M. Handbook of Textile Fibres. *Harris Research Laboratories, Inc., Washington*, 1954.

Hector, J.M. Introduction to the Botany of Field crops. *Central News Agency, Johannesburg*, 2 vols., 1936.

Henry, T.A. The Plant Alkaloids. *J. & A. Churchill Ltd., London, 4th Edn.*, 1949.

Hill, A.F. Economic Botany; A Textbook of Useful Plants and Plant Products. *McGraw-Hill Book Co., Inc., New York, 2nd Edn.* 1952.

Howard, A.L. A Manual of the Timber of the World: Their Characteristics and Uses. *Macmillan & Co. Ltd. London, 3rd Edn.*, 1948.

Howes, F.N. Nuts: Their Production and Everyday Uses. *Faber & Faber Ltd., London*, 1948.

Howes, F.N. Vegetable Gums & Resins. *The Chronica Botanica Co., Waltham*, 1949.

Howes, F.N. Vegetable Tanning Materials. *Butterworths Scientific Publications, London*, 1953.

Hunter, H. The Barley Crop. *Crosby Lockwood & Son Ltd., London*, 1952.

Hurt, E.F. Sunflower: For Food, Fodder and Fertility. *Faber & Faber Ltd., London*, 1948.

Hutchinson, J. & Melville, R. The Story of Plants and Their Uses to Man. *P.R. Gawthorn, Ltd., London*, 1933.

Jamieson, G.S. Vegetable Fats & Oils. *Reinhold Publishing Corp., New York, 2nd Edn.* 1943.

Jellinek, P. The Practice of Modern Perfumery. *Interscience Publishers, Inc., New York*, 1954.

Jordan L.A. et al. Oils for the Paint Industry. *Paint Research Station, Teddington, Middlesex*, 1951.

Kanny Lall Dey. The Indigenous Drugs of India. *Thacker, Spink & Co., Calcutta, 3rd edn.*; 1896.

Kar, S.N. The Jute Fibre. *Book Society Calcutta*, 1954.

Kaul, S.N. Forest Products of Jammu & Kashmir, *Kashmir Pratap Steam Press, Srinagar*, 1928.

Kierstead, S.P. Natural Dyes. *Bruce Humphries, Inc., Boston*, 1950.

Kirtikar, K.R. Basu. B.D. & an I.C.S. (retd.); *Lalit Mohan Basu Allahabad, 4 vols.*; *2nd. edn.*, 1935.

Koman, M.C. Report on the Investigations of Indigenous Drugs. *Govt. Press Madras* 1918, 1919, 1920.

Krishnamurthi, S. Horticultural and Economic Plants of the Nilgiris. *Coimbatore Co-operative Printing Works Ltd., Coimbatore*, 1953.

Krishnamurthi Naidu G. Commercial Guide to the Forest Economic Products of Mysore. *Govt. Press Bangalore*, 1917.

Krumbeigel, G.H. List of Economic Plants Imported in Lal Bagh Botanic Garden. *Bangalore, Govt. Press, Bangalore*, 1948.

Lock, R.H. Rubber and Rubber Planting. *University Press, Cambridge*, 1913.

Maheshwari, P. & Singh, U. Dictionary of Economic Plants in India I.C.A.R. New Delhi, 1965.

Mauersberger, H.R. Matthews' Textile Fibres: Their Physical, Microscopic and Chemical Properties. *John Wiley & Sons, Inc., New York, 6th edn.*, 1954.

McCann, C. Trees of India. *B.D. Taraporevala Sons & Co., Bombay.*

Modi, J.P. A Text Book of Medical Jurisprudence and Toxicology. *Tripathi Ltd., Bombay*, 1945.

Mudaliar, V.T.S. Common Cultivated Crops of South India. *Anaudha Nilayam Private Ltd., Madras*, 1955.

Nadkarni, K.M. Indian Materia Medica. *Popular Book Depot, Bombay, 2 vols., 3rd Edn*; 1954.

Naidu, S.B. Common Commercial Timbers of the U.P. *Govt. Press, Allahabad*, 1934.

Nelson, A. Medical Botany. E. & S. *Livingstone Ltd., Edinburgh*, 1951.

Pal, B.P. Wheat. Ind. Coun. Agri. Res, New Delhi, 1966.

Panshin, A.J. et al. Forest Products: Their Sources, Production and Utilization. *McGraw-Hill Book Co., Inc., New York*, 1950.

Parry, J.W. The Spice Handbook: Spices, Aromatic Seeds and Herbs. *Chemical Publishing Co., Inc. Brooklyn, N.Y.* 1945.

Pearson, R.S. & Brown, H.P. Commercial Timbers of India. *Central Publication Branch, Calcutta* 2 vols., 1932.

Pharmacognosy of Ayurvedic Drugs, Kerala. *University of Travancore, Trivandrum, Ser. I.* 1951.

Poisonous Plants of India. 2 vols. *Ind. Coun. Agri. Res. New Delhi.*

Radley, J.A. Starch and its Derivatives. *Chapman, London, 2 vols., 3rd Edn.*, 1953.

Record, S.J. & Hess, R.W. Timbers of the New World, *Yale University Press, New Haven*, 1944.

Review of Research on Spices and Cashewnut. *Ind. Coun. Agri. Res.* New Delhi.

Rodger, A. A Handbook of the Forest Products of Burma. *Times of India Press, Bombay*, 1943.

Roy, S.C. Monograph on the Gur Industry of India. *Indian Central Sugarcane Committee, New Delhi*, 1951.

Schery, R.W. Plants for Man. *Prentice-Hall, Inc., New York*, 1952.

Shivnath Rai. Foodstuffs of India. *Corporation of Calcutta*, 1940.

Steinmetz, E.F. Materiea Medica Vegetables, *Holland.* 3 vols., 1954.

Stevens, H.P. Natural Rubber Latex and its Application: No. 1 An Introduction to its Origin, Properties and Manufacture. *The British Rubber Development Board, London,* 1952.

Stevens, H.P. & Stevens, W.H. Rubber Latex. *The British Rubber Development Board, London,* 1948.

Thompson, H.C. Vegetable Crops. *McGraw-Hill Book Co., Inc., N.Y. 4th Edn.,* 1949.

Trease, C.E. A Text Book of Pharmacognosy. *Bailliere, Tindall Co. London, 6th edn.,* 1952.

Trotter, H. Manual of Indian Forest Utilisation. *Oxford University Press, London,* 1940.

Trotter, H. The Common Commercial Timbers of India and their Uses. *Manager, Vasant Press, Dehra Dun, 3rd Edn.,* 1944.

Use of Leguminous Plants. *International Institute of Agriculture, Rome,* 1936.

Van Royen, W. Atlas of World's Resources; Vol. I—The Agricultural Resources of the World. *Prentice Hall, Inc., N.Y.,* 1954.

Wallis, T.E. Text Book of Pharmacognosy. *J. & A. Churchill Ltd., London,* 1946.

Watt, G. The Commercial Products of India. *John Murray, London,* 1908.

Watt, G. A Dictionary of the Economic Products of India. *Supdt., Govt. Printing, Calcutta,* 6 vols., 1889—99.

Wilson, H.K. Grain Crops. *McGraw-Hill Book Co. Inc. N.Y.,* 1948.

The Wealth of India: A Dictionary of Indian Raw Materials and Industrial Products. *C.S.I.R. New Delhi,* Industrial Products.

The Wealth of India: A Dictionary of Indian Raw Materials and Industrial products, *C.S.I.R. New Delhi.* Raw Materials.

Yegna Narayan Aiyer A.K. Field Crops of India with special reference to Mysore. *The Bangalore Printing & Publishing Co., Ltd., Bangalore,* 3rd edn., 1950.

OTHER REFERENCES

Agricultural Situation in India. *Min. of Food and Agr., Govt. of India.* 12 : 1339-1373, 1958.

Aiyer, R.K. Economics of the Coir Industry. *Coir* 1 : 5-8; 39, 1956.

Badhwar, R.L; G.V. Karira, and S. Ramaswami. *Rauvolfia serpentina*: The Wonder Drug of India (Rauvolfa, *sarpagandha*). *Ind. For.* 81 (4) : 258-268, 1955.

Biswas, K. Cultivation of Rauvolfia in West Bengal, *Ind, Jour, Pharm.* 18 (5) : 170-175. 1956.

Cashew and Pepper Bulletin. 1958.

Cotton in India. 1950-51 to 1954-55. *Min. Food and Agr., Govt. of India.* 1956.

Dutta, P.K., I.C. Chopra. and L.D. Kapoor. Cultivation of *Rauvolfia serpentina* in India. *Economic Botany.* 17 (4) : 243-251. 1963.

Ghose. R.L.M., Ghatge, M.B. and Subrahmanyan, V. Rice in India. *Ind. Coun. Agr. Res.* 1956.

Iyer, M.A.S. Millets. *Ind. Coun. Agr. Res. Silver Jubilee Souvenir.* 1954. pp. 21-23.

Jain, A.P. Agricultural Situation in India. *Min. Food and Agr. Govt. of India.* 10 : 510-519. 1955.

Maheshwari, P. and Tandon, S.L. Agriculture and Economic Development in India. *Economic Botany.* 13 (3) : 205-242. 1959.

Mehta, T.R. Pulses *Ind. Coun. Agri. Res, Silver Jubilee Souvenir,* 1954. pp. 27-30.

Menon, N.S. *Rubber Board Bulletin.* 4 : 61-65. 1957.

Narayanan, B.T. Processing for Promotion of Quality in Coffee. *Indian Coffee.* 22 : 24-29; 31. 1958.

Pal, B.P. Advances in plant breeding and genetics in relation to crop improvement in India in the last twenty-five years. *Emp. Jour. Exp. Agri.* 26 : 123-132. 1958.

Patel, M.S. Tobacco Cultivation and Industry. *Commerce.* 95 : 118-120. 1957.

Report on the marketing of sesamum and niger seed in India. *Min. Food and Agr., Govt. of India.* 1951.

Sankaran, R. Expansion of castor cultivation in India. *Ind. Oilseeds Jour.* 5 : 67-71. 1958.

Tea in India 1952. *Min. Food and Agr., Govt. of India.* 1955.

LIST OF PERIODICALS

Acta Botanica Indica, *School of Plant Morphology,* Meerut.

Advisory Circular. *Rubber Research Institute, Ceylon.*

Agriculture and Animal Husbandry in India, *Calcutta.*

Agriculture and Animal Husbandry, *Uttar Pradesh, Lucknow.*

Agricultural Journal of India, *Pusa.*

Agricultural Situation in India, *New Delhi.*
American Journal of Botany, *Lancaster, Pa.*
American Naturalist, *Lancaster, Pa.*
Andhra Agricultural Journal, *Bapatla, Andhra.*
Annual Report of the Indian Central Oilseeds Committee; Oilseeds Series, *Hyderabad.*
Annual Report of the Indian Council of Agricultural Research. *New Delhi.*
Botanical Review. *Lancaster, Pa.*
Bulletin. Agricultural Research Institute, *Pusa. Calcutta.*
Bulletin of Applied Botany and Plant Breeding, *Leningrad.*
Bulletin of the Botanical Society of Bengal. *Calcutta.*
Bulletin. Central Food and Technology Research Institute. *Mysore.*
Bulletin. Department of Agriculture, Assam. *Shillong.*
Bulletin. Department of Agriculture, Bombay. *Bombay.*
Bulletin. Department of Agriculture, Madras, *Madras.*
Bulletin. Indian Central Coconut Committee. *Ernakulam.*
Bulletin. Indian Council of Agricultural Research. *New Delhi.*
Bulletin. Pharmacognosy Laboratory, Ministry of Health, Govt. of India. *New Delhi.*
Bulletin of the Central Research Institute, University of Travancore. *Trivandrum.*
Coconut Bulletin. *Ernakulam.*
Composite Wood. *Dehra Dun.*
Current Science. *Bangalore.*
Economic Botany. *Lancaster, Pa.*
Farmer. *Bombay.*
Fibres. *London.*
Food Science Abstracts. *London.*
Forestry Abstracts. Commonwealth Agricultural Bureaux, *Farnham Royal England.*
Forest Research in India. *Calcutta.*
Horticultural Abstracts. Indian Council of Agricultural Research. *New Delhi.*
Indian Farming. *New Delhi.*
Indian Farming. New Series. *New Delhi.*
Indian Forester. *Dehra Dun.*
Indian Forest Bulletin. *Dehra Dun.*
Indian Forest Bulletin. New Series. *Dehra Dun.*
Indian Forest Leaflets. *Dehra Dun.*
Indian Forest Memoirs. Forest Botany Series. *Calcutta.*
Indian Forest Records. *Dehra Dun.*
Indian Forest Records. New Series. Botany. *Dehra Dun.*
Indian Forest Records. New Series. Utilization. *Dehra Dun.*
Indian Journal of Agricultural Science. *New Delhi.*
Indian Journal of Genetics and Plant Breeding. *New Delhi.*
Indian Journal of Horticulture. *New Delhi.*
Indian Journal of Sugarcane Research and Development. *New Delhi.*
Indian Pulp and Paper. *Calcutta.*
Indian Rubber Board Bulletin. *Kottayam, Kerala.*
Indian Rubber Board. Serial Pamphlet. *Kottayam, Kerala.*
Indian Rubber Bulletin. *Calcutta.*
Indian Rubber Statistics. Ministry of Food and Agriculture, Govt. of India. *New Delhi.*
Indian Sugar. *Kanpur.*
Indian Textile Journal. *Bombay.*
I. & S. Bulletin. Ministry of Industry and Supply, Govt. of India. *New Delhi.*
Journal of the Indian Botanical Society. *Madras.*
Journal of the Indian Institute of Science. *Bangalore.*
Journal of Royal Horticultural Society. *London.*
Journal of the Science of Food and Agriculture. *London.*

Journal of Scientific and Industrial Research. *New Delhi.*
Journal of the Textile Institute. *Manchester.*
Jute Bulletin. *Calcutta.*
Leaflet. Department of Agriculture, Assam. *Shillong.*
Madras Agricultural Journal. *Coimbatore.*
Madras Forest College. Magazine. *Coimbatore.*
Memoirs of the Department of Agriculture in India. Botanical Series. *Pusa.*
Memoirs of the Department of Agriculture, Madras. *Madras.*
Miscellaneous Bulletin. I.C.A.R. *New Delhi.*
Nature. *London.*
Oilseeds Series. Indian Central Oilseeds Committee. *Hyderabad.*
Perfumery and Essential Oil Record. *London.*
Proceedings of the Bihar Academy of Agricultural Science. *Sabour, Bihar.*
Proceedings of National Academy of Sciences, India. *Allahabad.*
Proceedings of the National Institute of Sciences of India. *Calcutta.*
Report. I.C.A.R. *Calcutta.*
Rubber Board Bulletin. *Kottayam, Kerala.*
Rubber India. *Bombay.*
Rubber in India. Ministry of Food and Agriculture, Govt. of India. *New Delhi.*
Science. *New York.*
Science and Culture. *Calcutta.*
Scientific Monograph. Indian (Imperial) Council of Agricultural Research, India. *Calcutta.*
Scientific Reports of the Indian (Imperial) Agricultural Research Institute. *New Delhi.*
Tanner. *Bombay.*
Tea Board India. *Calcutta.*
The Encyclopaedia. Indian Tea Association. Scientific Department, *Tocklai. Assam.*
Tropical Agriculture. *Trinidad.*
Tropical Woods. *New Haven, Conn.*

FARM BULLETINS AND INFORMATION LEAFLETS

Farm Bulletins being issued by the Farm Information Unit, Directorate of Extension, Ministry of Food and Agriculture, New Delhi:

Farm Bulletin No. 1. Rice Cultivation in India.
Farm Bulletin No. 2. Groundnut Cultivation in India.
Farm Bulletin No. 3. Onion and Garlic Cultivation in India.
Farm Bulletin No. 5. Cotton Cultivation in India.
Farm Bulletin No. 6. The Mango in India.
Farm Bulletin No. 8. Potato Cultivation in India.
Farm Bulletin No. 9. Cashewnut Cultivation in India.
Farm Bulletin No. 10. Tobacco Cultivation in India.
Farm Bulletin No. 11. The Chiku in India.
Farm Bulletin No. 12. Grape Culture in India.
Farm Bulletin No. 14. Arecanut Cultivation in India.
Farm Bulletin No. 15. Oranges, Lemons and Limes in India.
Farm Bulletin No. 17. Tapioca Cultivation in India.
Farm Bulletin No. 18. The Ber in India.
Farm Bulletin No. 19. Temperate Fruits in India.
Farm Bulletin No. 21. The Breadfruit in India.
Farm Bulletin No. 22. The Banana in India.
Farm Bulletin No. 24. The Pineapple in India.
Farm Bulletin No. 25. The Papaya in India.
Farm Bulletin No. 26. Guava in India.
Farm Bulletin No. 27. Wheat Cultivation in India.
Farm Bulletin No. 28. The Mangoesteen in India.

Farm Bulletin No. 29. The Phalsa in India.
Farm Bulletin No. 33. The Jackfruit in India.
Farm Bulletin No. 35. Sugarcane Cultivation in Peninsular India.
Farm Bulletin No. 36. Vegetable Cultivation in North India.
Farm Bulletin No. 38. Aonla, Mulbery and Karonda in India.
Farm Bulletin No. 39. The Loquat in India.
Farm Bulletin No. 41. Passion Fruit and Jaman in India.
Farm Bulletin No. 43. Home Preservation of Fruits.
Farm Bulletin No. 44. The Litchi in India.
Farm Bulletin No. 45. Ginger Cultivation in India.
Farm Bulletin No. 48. Bajra Cultivation in India.
Farm Bulletin No. 50. Castor in Cultivation India.
Farm Bulletin No. 51. Jute Cultivation in India.
Farm Bulletin No. 53. Turmeric Cultivation in India.
Farm Bulletin No. 54. Sitaphal and other Annona Fruits.
Farm Bulletin No. 55. Pepper Cultivation in India.

Farm Bulletins and Leaflets being issued by the Farm Information Unit, Directorate of Extension, Ministry of Agriculture and Irrigation, New Delhi:

Farm Bulletin. Betelvine Cultivation in India.
Farm Bulletin. Safflower Cultivation in India.
Farm Bulletin. Lemongrass in India.
Farm Bulletin. Cultivation of Sweetpotato.
Farm Bulletin. Cultivation of Toria.
Farm Bulletin. Coconut Cultivation in India.
Farm Bulletin. Jackfruit.
Farm Bulletin. Cardamom in India.
Farm Bulletin. Fig Cultivation in India.
Farm Bulletin. Cultivation of Mints.
Farm Bulletin. Temperate Fruit in India.
Farm Bulletin. Bhindi Cultivation in India.
Farm Bulletin. Isbagol Cultivation in India.
Farm Bulletin. Sugarcane Cultivation in Northern India.
Farm Bulletin. A Guide to Mushroom Cultivation.
Farm Bulletin. Beans in India.
Farm Bulletin. The Grape in India.
Farm Bulletin. Rauvolfia serpentina.
Farm Bulletin. Cultivation of Cucumber & Kakri.
Information Leaflet. Sesamum.
Information Leaflet. Silk Cotton tree.
Information Leaflet. Nut Meg.
Information Leaflet. Clove.
Information Leaflet. Cinnamon.
Information Leaflet. Raise Sunflower.
Information Leaflet No. 17. Raise better Barley crop.
Information Leaflet No. 18. Getting good Gram yields.
Information Leaflet No. 20. The way to higher Wheat yields.
Information Leaflet No. 30. Rapes and Mustards.

PERIODICALS ON PHARMACOGNOSY

1. Annual Reports of the Calcutta School of Tropical Medicine; 2. Bulletin of the Royal Botanic Gardens, Kew; 3. Chemical Gazette; 4. Indian Forest Records; 5. Indian Forester; 6. Indian Journal of Medical and Physical Science; 7. Indian Journal of Medical Research; 8. Indian Medical Gazette; 9. Indian Medical Journal; 10. Indian Medical Record; 11. Journal of Ayurveda; 12. Journal of the Bombay Natural History Society; 13. Journal of Indian Chemical Society; 14. Journal and Proceedings of the Asiatic Society of Bengal; 15. Kew Bulletin; 16. Lancet; 17. Pharmaceutical Journal.

Index

1. A fruit shop at Kullu, India.
Transparency : Courtesy - Sanjeeva Pandey, I.F.S.

2. A spice shop at Kullu, India.
Transparency : Courtesy - Sanjeeva Pandey, I.F.S.

3. Toddy palms at Rajpipla, India.
Transparency : Courtesy - Sanjeeva Pandey, I.F.S.

4. The cashewnuts in Kerala State, India.

5. The Indian tea, Nilgiri hills, India.

6. The rubber plantation, Kottayam, India.

7. Nursery of Sarpagandha (*Rauvolfia serpentina*) plants at F.R.I. Dehradun.

8. Solar seasoning kiln at F.R.I. Dehradun.

9. Composite wood laboratory at F.R.I. Dehradun.

10. The bamboo plantation at F.R.I. Dehradun.

11. Tobacco cultivation at Anand, Gujarat, India.
Transparency : Courtesy - Sanjeeva Pandey, I.F.S.